Lecture Notes in Computer Science 6571

Commenced Publication in 1973
Founding and Former Series Editors:
Gerhard Goos, Juris Hartmanis, and Jan van Leeuwen

Lecture Notes in Computer Science 6571

Dario Catalano Nelly Fazio
Rosario Gennaro Antonio Nicolosi (Eds.)

Public Key Cryptography – PKC 2011

14th International Conference
on Practice and Theory in Public Key Cryptography
Taormina, Italy, March 6-9, 2011
Proceedings

Springer

Volume Editors

Dario Catalano
Università di Catania, Italy
E-mail: catalano@dmi.unict.it

Nelly Fazio
City University of New York, NY, USA
E-mail: fazio@cs.ccny.cuny.edu

Rosario Gennaro
IBM T.J. Watson Research Center, Hawthorne, NY, USA
E-mail: rosario@us.ibm.com

Antonio Nicolosi
Stevens Institute of Technology, Hoboken, NJ, USA
E-mail: nicolosi@cs.stevens.edu

ISSN 0302-9743
e-ISSN 1611-3349
ISBN 978-3-642-19378-1
e-ISBN 978-3-642-19379-8
DOI 10.1007/978-3-642-19379-8
Springer Heidelberg Dordrecht London New York

Library of Congress Control Number: 2011921312

CR Subject Classification (1998): E.3, K.6.5, C.2, D.4.6, K.4.4, E.4

LNCS Sublibrary: SL 4 – Security and Cryptology

Typesetting: Camera-ready by author, data conversion by Scientific Publishing Services, Chennai, India

Printed on acid-free paper

Springer is part of Springer Science+Business Media (www.springer.com)

Preface

The 14th International Conference on Practice and Theory in Public Key Cryptography (PKC 2011) was held March 6–9, 2011 in Taormina, Italy. PKC 2011 was sponsored by the International Association for Cryptologic Research (IACR).

The conference received 103 submissions and each submission was assigned to at least three committee members. Submissions co-authored by members of the Program Committee were assigned to at least five committee members. Due to the large number of high-quality submissions, the review process was challenging and we are deeply grateful to the 30 committee members and the 96 external reviewers for their outstanding work. After extensive discussions, the Program Committee selected 28 submissions for presentation during the conference: these are the articles that are included in this volume. The review process was run using Shai Halevi's software, and we are indebted to him for his help in setting it up and running it.

The program also included one invited talk: "New Developments in Leakage-Resilient Cryptography" given by Vinod Vaikuntanathan, whom we thank for accepting our invitation and for contributing to the success of the conference.

Our thanks go also to Springer for publishing the proceedings in the *Lecture Notes in Computer Science* series.

March 2011

Rosario Gennaro
Nelly Fazio
Antonio Nicolosi
Dario Catalano

PKC 2011

The 14th IACR International Conference on
Practice and Theory of Public Key
Cryptography

Taormina, Italy, March 6–9, 2011

Program Chair

Rosario Gennaro	IBM Research, USA

General Chairs

Nelly Fazio	City University of New York, USA
Antonio Nicolosi	Stevens Institute of Technology, USA

Local Arrangements Chair

Dario Catalano	Università di Catania, Italy

Program Committee

Masayuki Abe	NTT Labs, Japan
John R. Black	University of Colorado, USA
Alexandra Boldyreva	Georgia Institute of Technology, USA
Colin Boyd	Queensland University of Technology, Australia
Emmanuel Bresson	EADS, France
Melissa Chase	Microsoft Research, USA
Paolo D'Arco	Università di Salerno, Italy
Alexander W. Dent	Royal Holloway University of London, UK
Stefan Dziembowski	Università di Roma "La Sapienza", Italy
Dario Fiore	École Normale Supérieure, France
Marc Fischlin	Technische Universität Darmstadt, Germany
Carmit Hazay	Aarhus Universitet, Denmark
Martin Hirt	ETH Zurich, Switzerland
Stanislaw Jarecki	University of California at Irvine, USA
Eike Kiltz	Ruhr-Universität Bochum, Germany
Kaoru Kurosawa	Ibaraki University, Japan
Yehuda Lindell	Bar-Ilan University, Israel
Sebastià Martin	Universitat Politecnica de Catalunya, Spain
Alexander May	Ruhr-Universität Bochum, Germany

Jesper Buus Nielsen	Aarhus Universitet, Denmark
Bryan Parno	Microsoft Research, USA
Mario Di Raimondo	Università di Catania, Italy
Mike Rosulek	University of Montana, USA
Guy Rothblum	IAS Princeton, USA
Kazue Sako	NEC, Japan
Berry Schoenmakers	T.U. Eindhoven, The Netherlands
Thomas Shrimpton	Portland State University, USA
Nigel Smart	Bristol University, UK
Edlyn Teske-Wilson	University of Waterloo, Canada
Muthu Venkitasubramaniam	New York University, USA

PKC Steering Committee

Ronald Cramer	CWI and Universiteit Leiden, The Netherlands
Yvo Desmedt	University College London, UK
Hideki Imai	Chuo University and AIST, Japan
David Naccache	École Normale Supérieure, France
Tatsuaki Okamoto	NTT Labs, Japan
David Pointcheval	École Normale Supérieure, France
Moti Yung (Secretary)	Google Inc. and Columbia University, USA
Yuliang Zeng (Chair)	University of North Carolina at Charlotte, USA

External Reviewers

Michel Abdalla
Divesh Aggarwal
Joel Alwen
Nuttapong Attrapadung
Kfir Barhum
Aurélie Bauer
Rikke Bendlin
Dan J. Bernstein
Olivier Billet
Christina Brzuska
David Cash
Dario Catalano
Céline Chevalier
Sandro Coretti
Emiliano De Cristofaro
Özgür Dagdelen
Ivan Damgård
Alfredo Rial Duran
Oriol Farràs
Jean-Charles Faugere
Ignacio Fernández-Rúa
Anna Lisa Ferrara
Eiichiro Fujisaki
Jun Furukawa
Steven Galbraith
Sanjam Garg
Juan Gonzalez
Choudary Gorantla
Ignacio Gracia
Jens Groth
Shai Halevi
Goichiro Hanaoka
Kristiyan Haralambiev
Swee-Huay Heng
Mathias Herrmann
S.J.A. de Hoogh
Toshiyuki Isshiki
Tibor Jager
Antoine Joux
Bhavana Kanukurthi
Hugo Krawczyk
Tanja Lange
Peter van Liesdonk
Richard Lindner
Georg Lippold
Patrick Longa
Adriana Lopez-Alt
Christoph Lucas
Hemanta Maji
Mark Manulis
Barbara Masucci
Sarah Meiklejohn
Sigurd Torkel Meldgaard
Petros Mol
Paz Morillo
Ryo Nojima
Adam O'Neill

Miyako Ohkubo
Tatsuaki Okamoto
Maria Cristina Onete
Claudio Orlandi
Carles Padró
Dan Page
Anat Paskin
Valerio Pastro
Ludovic Perret
Le Trieu Phong
Krzysztof Pietrzak
Angel Perez del Pozo
Pavel Raykov
Markus Rückert
Carla Ràfols
Yusuke Sakai
Dominique Schröder
Sven Schage
Hovav Shacham
Francesco Sica
Joseph Silverman
Claudio Soriente
Björn Tackmann
Keisuke Tanaka
Stefano Tessaro
Enrico Thomae
Tomas Toft
Joana Treger
Daniele Venturi
Damien Vergnaud
Jorge L. Villar
Ivan Visconti
Benne de Weger
Christopher Wolf
Qianhong Wu
Shota Yamada
Go Yamamoto
Bo-Yin Yang
Angela Zottarel

Table of Contents

Protocols

Chosen-Ciphertext Security

Invited Talk

Encryption

Signatures II

Zero-Knowledge

Cryptanalysis

Linearly Homomorphic Signatures over Binary Fields and New Tools for Lattice-Based Signatures

Dan Boneh* and David Mandell Freeman**

Stanford University
{dabo,dfreeman}@cs.stanford.edu

Abstract. We propose a linearly homomorphic signature scheme that authenticates vector subspaces of a given ambient space. Our system has several novel properties not found in previous proposals:

- It is the first such scheme that authenticates vectors defined over *binary fields*; previous proposals could only authenticate vectors with large or growing coefficients.
- It is the first such scheme based on the problem of *finding short vectors in integer lattices*, and thus enjoys the worst-case security guarantees common to lattice-based cryptosystems.

Our scheme can be used to authenticate linear transformations of signed data, such as those arising when computing mean and Fourier transform or in networks that use network coding. Our construction gives an example of a cryptographic primitive — homomorphic signatures over $\mathbb{F}_2$ — that can be built using lattice methods, but cannot currently be built using bilinear maps or other traditional algebraic methods based on factoring or discrete log type problems.

Security of our scheme (in the random oracle model) is based on a new hard problem on lattices, called k-SIS, that reduces to standard average-case and worst-case lattice problems. Our formulation of the k-SIS problem adds to the "toolbox" of lattice-based cryptography and may be useful in constructing other lattice-based cryptosystems.

As a second application of the new k-SIS tool, we construct an ordinary signature scheme and prove it k-time unforgeable in the standard model assuming the hardness of the k-SIS problem. Our construction can be viewed as "removing the random oracle" from the signatures of Gentry, Peikert, and Vaikuntanathan at the expense of only allowing a small number of signatures.

Keywords: Lattice-based cryptography, homomorphic signatures.

1 Introduction

A *linearly homomorphic signature scheme* signs n-dimensional vectors $\mathbf{v}_1, \ldots, \mathbf{v}_k$ defined over some finite field $\mathbb{F}_p$ and outputs one signature per vector. The linear homomorphic property is that given these k signatures, anyone can produce a signature on

* Supported by NSF.
** Supported by an NSF Mathematical Sciences Postdoctoral Research Fellowship.

D. Catalano et al. (Eds.): PKC 2011, LNCS 6571, pp. 1–16, 2011.

any vector $\mathbf{v}$ in the $\mathbb{F}_p$-linear span of $\mathbf{v}_1, \ldots, \mathbf{v}_k$. The signature is secure if it is difficult to produce a signature on any vector $\mathbf{v} \in \mathbb{F}_p^n$ outside the linear span of $\mathbf{v}_1, \ldots, \mathbf{v}_k$. We give precise definitions in Section 2.

The original motivation for linearly homomorphic signatures comes from the *network coding* routing mechanism [12, 11, 25, 4, 13]. In a computer network that uses network coding, a message sender signs a number of "augmented" message vectors and transmits the resulting vector-signature pairs through the network to a recipient. Each router along the way receives a number of signed vectors and creates a random linear combination $\mathbf{v}$ of the vectors it receives. The router uses the homomorphic property to derive a signature on $\mathbf{v}$ and forwards $\mathbf{v}$ and its signature to the next router, which then does the same with the signed vectors it receives. The ultimate recipient obtains several random linear combinations of the original message vectors, discards all vectors that are not properly signed, and recovers the original message by solving a full-rank linear system over $\mathbb{F}_p$. Security of the signature scheme ensures that the recipient obtains the originally transmitted message vectors. In implementations there is a desire to use network coding with addition over $\mathbb{F}_2$, so that computations on messages are simple XORs and decoding amounts to solving a linear system over $\mathbb{F}_2$.

Beyond network coding, linearly homomorphic signatures enable linear computations on authenticated data. For example, consider a server that stores signed data samples $s_1, \ldots, s_n$ in $\mathbb{F}_p$. The signature on sample s_i is actually a signature on the vector $(s_i|\mathbf{e}_i) \in \mathbb{F}_p^{n+1}$, where $\mathbf{e}_i$ the ith unit vector in $\mathbb{F}_p^n$. The server stores (i, s_i) and a signature on $(s_i|\mathbf{e}_i)$. (The vector $\mathbf{e}_i$ need not be stored with the data and can be reconstructed from i when needed.) Using the homomorphic property, the server can compute a signature σ on the sum $(\sum_{i=1}^n s_i, 1, \ldots, 1)$. If σ reveals no other information about the original samples, then the server can publish the sum $\sum_{i=1}^n s_i$ and the signature σ on the sum while maintaining privacy of the original data. The "augmentation" $(1, \ldots, 1)$ proves that the published message really is the claimed sum of the original samples.[1] More generally, the server can publish an authenticated inner product of the samples $\mathbf{s} := (s_1, \ldots, s_n)$ with any known vector $\mathbf{c} \in \mathbb{F}_p^n$ without leaking additional information about the samples. This is needed, for example, to publish an authenticated Fourier coefficient from the Fourier transform of $\mathbf{s}$. It can also be used to compute an authenticated least squares fit for a set of signed data.

Previous results on linearly homomorphic signatures make use of groups in which the discrete logarithm problem is hard [18, 11, 25, 4] or the RSA assumption holds [13]. In the former case, signatures are linearly homomorphic over $\mathbb{F}_p$ for some large p, while in the latter case, signatures are homomorphic over the integers (with some bound on the size of the coefficients allowed in linear combinations). In particular, no previous scheme can support linear operations over a small field such as $\mathbb{F}_2$. This appears to be an inherent limitation of discrete log-type systems, since the discrete log problem is not hard in $\mathbb{F}_2$. A similar limitation prevents an RSA-based system over $\mathbb{F}_2$.

More distantly related to our work is the notion of "redactable" signatures [24, 16, 15, 3, 22, 21, 10, 8, 7]. These schemes have the property that given a signature on a message, anyone can derive a signature on subsets of the message. Our focus here is

[1] Strictly speaking, in order to prevent mix-and-match attacks between different data sets one needs to link the n samples with a random tag that uniquely identifies the data set.

quite different — we look at linear operations on tuples of authenticated vectors rather than a subset operation on a single message.

Our Contributions

- **Homomorphic signatures over $\mathbb{F}_2$:** We construct the first unforgeable linearly homomorphic signature scheme that authenticates vectors with coordinates in $\mathbb{F}_2$. Our construction gives an example of a cryptographic primitive that can be built using lattice methods, but cannot currently be built using bilinear maps or other traditional algebraic methods based on factoring or discrete log type problems. Our scheme can be easily modified to authenticate vectors with coefficients in other small fields, including prime fields and extension fields such as $\mathbb{F}_{2^d}$. In addition, our scheme is *private*, in the sense that a derived signature on a vector $\mathbf{v}$ leaks no information about the original signed vectors beyond what is revealed by $\mathbf{v}$.
- **A simple k-time signature without random oracles:** We describe a stateless signature scheme and prove it secure in the standard model when used to sign at most k messages, for small values of k. The public key of our scheme is significantly smaller than that of any other stateless lattice-based signature scheme that can sign multiple large messages and is secure in the standard model. Our construction can be viewed as "removing the random oracle" from the signature scheme of Gentry, Peikert, and Vaikuntanathan [14], but only for signing k messages.
- **New tools for lattice-based signatures:** Unforgeability of both of our schemes is based on a new hard problem on lattices, which we call the *k-Small Integer Solutions* (k-SIS) problem. We show that k-SIS reduces to the standard *Small Integer Solution* (SIS) problem, which is known to be as hard as standard worst-case lattice problems [20].

  Unforgeability of our k-time signature scheme depends on bounds for the length of vectors sampled from discrete Gaussian distributions. We prove both upper and lower bounds that are essentially as tight as possible. Our upper bound improves on a result of Micciancio and Regev [20, Lemma 4.4], and our lower bound is (to our knowledge) the first such bound in the literature.

  Privacy of our linearly homomorphic scheme depends on a new result on discrete Gaussian distributions, namely, that the distribution of a sum of samples from a discrete Gaussian is statistically close to a discrete Gaussian distribution that depends *only on the sum* and not on the individual samples. While the analogous result for *continuous* Gaussians is well known, this is (to our knowledge) the first such result for discrete Gaussians.

Overview of the Homomorphic Signature Scheme. Our construction builds on the signature scheme of Gentry, Peikert, and Vaikuntanathan [14], in which signatures are short vectors σ in lattices defined modulo some large integer q. The key idea in our construction is to use short vectors σ in (cosets of) lattices defined modulo $2q$, which allows us to encode different information modulo 2 and modulo q: $\sigma \bmod 2$ encodes information about the vector being signed, while $\sigma \bmod q$ encodes a solution to a hard problem, ensuring that an adversary cannot forge the signature.

The fact that σ is a short *integer* vector ensures that the two parts cannot be attacked independently. Specifically, applying the Chinese remainder theorem to two vectors σ_2 and σ_q that are correct mod 2 and mod q, respectively, does not produce a short integer vector. This property appears to be unique to lattice- based cryptography: if we attempted a similar construction in discrete log groups of order $2q$, we would easily be able to attack the order 2 and order q parts independently.

Concretely, our construction works as follows. Let q be an odd prime. To sign a vector subspace $V = \text{span}(\mathbf{v}_1, \ldots, \mathbf{v}_k)$ of $\mathbb{F}_2^n$, we define a matrix $\mathbf{A}_V \in \mathbb{Z}_{2q}^{m \times n}$ and then sign each basis vector $\mathbf{v}_i$. A signature on $\mathbf{v}_i \in \mathbb{F}_2^n$ is a low-norm vector $\sigma_i \in \mathbb{Z}^m$ such that $\mathbf{A}_V \cdot \sigma_i = q \cdot \mathbf{v}_i \mod 2q$. A signature $\sigma \in \mathbb{Z}^m$ on a vector $\mathbf{y} \in \mathbb{F}_2^n$ is valid if σ has small norm and $\mathbf{A}_V \cdot \sigma = q \cdot \mathbf{y} \bmod 2q$.

Producing such a signature requires knowing a short basis for the integer lattice defined by the kernel of $\mathbf{A}_V$; to obtain such a basis we combine the trapdoor generation algorithm of Alwen and Peikert [2] with the basis delegation mechanism of Cash, Hofheinz, Kiltz, and Peikert [9].

The homomorphic property of our scheme is now immediate: if we are given arbitrary vector-signature pairs $(\mathbf{v}_j, \sigma_j) \in \mathbb{F}_2^n \times \mathbb{Z}^m$ for $j = 1, \ldots, \ell$, we can create a signature on $\mathbf{v} = \mathbf{v}_1 + \cdots + \mathbf{v}_\ell \in \mathbb{F}_2^n$ by computing $\sigma = \sigma_1 + \cdots + \sigma_\ell \in \mathbb{Z}^m$. Since the σ_j are all valid signatures on the $\mathbf{v}_j$, we see that $\mathbf{A}_V \cdot \sigma = q \cdot \mathbf{v} \bmod 2q$ and σ has low norm (if ℓ is sufficiently small), so σ is a valid signature on $\mathbf{v}$.

Security and the k-SIS Problem. To prove unforgeability, we need to show that given signatures on basis vectors of V, it is impossible to generate a signature on a vector outside of V. To do so we define the k-SIS problem, which, roughly speaking, is as follows:

> Given a matrix $\mathbf{A} \in \mathbb{Z}_q^{n \times m}$ and k short vectors $\mathbf{e}_1, \ldots, \mathbf{e}_k \in \mathbb{Z}^m$ satisfying $\mathbf{A} \cdot \mathbf{e}_i = 0 \bmod q$, find a short vector $\mathbf{e} \in \mathbb{Z}^m$ satisfying $\mathbf{A} \cdot \mathbf{e} = 0 \bmod q$, such that $\mathbf{e}$ is not in $\mathbb{Q}$-span$(\{\mathbf{e}_1, \ldots, \mathbf{e}_k\})$.

When $k = 0$ this is the standard SIS problem [20].

In Section 5 we show that an adversary that breaks the homomorphic signature scheme (defined mod $2q$) in the random oracle model can be used to solve the k-SIS problem (defined mod q). In Section 4 we show that the k-SIS problem is as hard as the SIS problem. Our reduction degrades exponentially in k, which forces us to use a constant-size k if we want our linearly homomorphic scheme to be provably secure based on worst-case lattice problems. It is an important open problem to give either a tighter reduction to SIS or a direct reduction from k-SIS to worst-case lattice problems.

For some applications of linearly homomorphic signatures it is desirable that the derived signatures be private; that is, a derived signature on a vector $\mathbf{v}$ in span$(\mathbf{v}_1, \ldots, \mathbf{v}_k)$ should not leak information about the original vectors $\mathbf{v}_1, \ldots, \mathbf{v}_k$ beyond what is revealed by $\mathbf{v}$. For our construction, to prove privacy it suffices to show that the distribution obtained by summing independent discrete Gaussians depends only on the coset of the sum and the linear combination being computed. We prove this statement in Section 4.

Overview of the k-Time Signature Scheme. Our goal is to use the same mechanism as in the homomorphic signature scheme to construct an ordinary signature scheme.

Since homomorphic signatures are not existentially unforgeable, we must find a way to remove the homomorphic property. To do this, we impose the requirement that the length of a valid signature $\sigma \in \mathbb{Z}^m$ be very close to the expected length of the vector produced by the signing algorithm. We then show that any linear combination of valid signatures will be too long to satisfy this tight bound, so the homomorphic property is of no use to an adversary.

As with the homomorphic scheme, the security of our k-time signature scheme follows from hardness of the k-SIS problem. We prove security (in the standard model) against a static attacker, who submits all of his message queries before receiving the public key. By a standard transformation using *chameleon hashes* [17], this implies the existence of a scheme secure against an adaptive attacker, also in the standard model. Our security proof requires tight bounds on the length of a vector sampled from a discrete Gaussian distribution. We use new upper and lower bounds that are essentially as tight as possible.

Lyubashevsky and Micciancio [19] give a lattice-based one-time signature scheme secure in the standard model and show how it can be converted to sign k messages. For small values of k, our construction gives a more efficient stateless k-time signature than is produced by this conversion.

Outline of the Paper. Section 2 gives a formal definition and security model for linearly homomorphic signatures. In Section 3 we review facts about lattices. Section 4 describes the new tools we use in our constructions, including the k-SIS problem and our reduction of k-SIS to SIS. In Section 5 we present our homomorphic scheme, and in Section 6 we present our k-time signature scheme. Finally, in Section 7 we describe extensions of our scheme and open questions. Because of space considerations, proofs have been omitted and can be found in the full version of this paper [6].

2 Linearly Homomorphic Signatures

We define linearly homomorphic signatures over any principal ideal domain R. These signatures authenticate tuples (a.k.a. vectors) of elements of R. This definition encompasses the homomorphic signatures over finite fields defined by Boneh et al. [4] as well as the signatures over $\mathbb{Z}$ and $\mathbb{Z}_N$ defined by Gennaro et al. [13]. While we describe the system in terms of a fixed ring R, it may be that R is determined by the Setup algorithm, as in the case where the size of R depends on the system's security parameter.

To prevent "mix-and-match" attacks, each set of vectors signed is given a unique identifier id, which serves to tie together all vectors that belong to the same file or data set. Our security model requires that this identifier be unpredictable; in our scheme it is chosen at random by the signer.

Definition 1 (adapted from [4]). Let R be a principal ideal domain. A *linearly homomorphic signature scheme over* R is a tuple of probabilistic, polynomial-time algorithms (Setup, Sign, $\mathsf{Combine}$, Verify) with the following functionality:

– $\mathsf{Setup}(n, \mathsf{params})$. On input a security parameter n (in unary) and additional public parameters params that include the dimension N of the ambient space and the dimension k of subspaces to be signed, this algorithm outputs a public key pk and a secret key sk.

- $\mathsf{Sign}(\mathsf{sk}, \mathsf{id}, \mathbf{v})$. On input a secret key sk, an identifier $\mathsf{id} \in \{0,1\}^n$, and a vector $\mathbf{v} \in R^N$, this algorithm outputs a signature σ.
- $\mathsf{Combine}(\mathsf{pk}, \mathsf{id}, \{(\alpha_i, \sigma_i)\}_{i=1}^{\ell})$. On input a public key pk, an identifier id, and a set of tuples $\{(\alpha_i, \sigma_i)\}_{i=1}^{\ell}$ with $\alpha_i \in R$, this algorithm outputs a signature σ. (This σ is intended to be a signature on $\sum_{i=1}^{\ell} \alpha_i \mathbf{v}_i$.)
- $\mathsf{Verify}(\mathsf{pk}, \mathsf{id}, \mathbf{y}, \sigma)$. On input a public key pk, an identifier $\mathsf{id} \in \{0,1\}^n$, a vector $\mathbf{y} \in R^N$, and a signature σ, this algorithm outputs either 0 (reject) or 1 (accept).

We require that for each $(\mathsf{pk}, \mathsf{sk})$ output by $\mathsf{Setup}(n, \mathsf{params})$, we have:

1. For all id and $\mathbf{y} \in R^N$, if $\sigma \leftarrow \mathsf{Sign}(\mathsf{sk}, \mathsf{id}, \mathbf{y})$ then $\mathsf{Verify}(\mathsf{pk}, \mathsf{id}, \mathbf{y}, \sigma) = 1$.
2. For all $\mathsf{id} \in \{0,1\}^n$ and all sets of triples $\{(\alpha_i, \sigma_i, \mathbf{v}_i)\}_{i=1}^{\ell}$, if it holds that $\mathsf{Verify}(\mathsf{pk}, \mathsf{id}, \mathbf{v}_i, \sigma_i) = 1$ for all i, then

$$\mathsf{Verify}\left(\mathsf{pk},\ \mathsf{id},\ \textstyle\sum_i \alpha_i \mathbf{v}_i,\ \mathsf{Combine}\left(\mathsf{pk},\ \mathsf{id},\ \{(\alpha_i, \sigma_i)\}_{i=1}^{\ell}\right)\right) = 1.$$

In our lattice-based linearly homomorphic signature scheme, we cannot combine arbitrarily many valid signatures and still guarantee successful verification. We capture this property by saying that the scheme is *L-limited* if correctness property (2) holds for all $\ell \leq L$ whenever the σ_i are output by the Sign algorithm.

Unforgeability. The security model for linearly homomorphic signatures allows an adversary to make adaptive signature queries on files of his choosing, with the signer randomly choosing the identifier id for each file queried. The winning condition captures the fact that there are two distinct types of forgeries: a vector-signature pair $(\mathbf{y}^*, \sigma^*)$ that verifies for some file *not* queried to the signer (a *type 1 forgery*), or a pair $(\mathbf{y}^*, \sigma^*)$ that verifies for some file that *was* queried to the signer, but for which $\mathbf{y}^*$ is not a linear combination of the vectors queried (a *type 2 forgery*).

Definition 2 (adapted from [4]). A homomorphic signature scheme $\mathcal{S} = (\mathsf{Setup}, \mathsf{Sign}, \mathsf{Combine}, \mathsf{Verify})$ over R is *unforgeable* if the advantage of any probabilistic, polynomial-time adversary $\mathcal{A}$ in the following security game is negligible in the security parameter n:

Setup: The challenger runs $\mathsf{Setup}(n, \mathsf{params})$ to obtain $(\mathsf{pk}, \mathsf{sk})$, and gives pk to $\mathcal{A}$.

Queries: Proceeding adaptively, $\mathcal{A}$ specifies a sequence of k-dimensional subspaces $V_i \subset R^N$, represented as a k-tuples of basis vectors $\mathbf{v}_{i1}, \ldots, \mathbf{v}_{ik}$. For each i, the challenger chooses id_i uniformly from $\{0,1\}^n$ and gives to $\mathcal{A}$ the identifier id_i and the j signatures $\sigma_{ij} \leftarrow \mathsf{Sign}(\mathsf{sk}, \mathsf{id}_i, \mathbf{v}_{ij})$ for $j = 1, \ldots, k$.

Output: $\mathcal{A}$ outputs $\mathsf{id}^* \in \{0,1\}^n$, a *non-zero* vector $\mathbf{y}^* \in R^N$, and a signature σ^*.

The adversary *wins* if $\mathsf{Verify}(\mathsf{pk}, \mathsf{id}^*, \mathbf{y}^*, \sigma^*) = 1$, and either (1) $\mathsf{id}^* \neq \mathsf{id}_i$ for all i (a *type 1 forgery*), or (2) $\mathsf{id}^* = \mathsf{id}_i$ for some i but $\mathbf{y}^* \notin V_i$ (a *type 2 forgery*). The *advantage* HomSig-Adv$[\mathcal{A}, \mathcal{S}]$ of $\mathcal{A}$ is defined to be the probability that $\mathcal{A}$ wins the game.

Privacy. Given signatures on vectors $\mathbf{v}_1, \ldots, \mathbf{v}_k$ in R^N, it is desirable that derived signatures on a vector $\mathbf{v}$ in span$(\mathbf{v}_1, \ldots, \mathbf{v}_k)$ not leak any information about $\mathbf{v}_1, \ldots, \mathbf{v}_k$ beyond what is revealed by $\mathbf{v}$. We are not trying to hide the fact the derivation took place or the function that was used to compute $\mathbf{v}$, merely the inputs to the function.

More precisely, we define privacy for linearly homomorphic signatures using a variation of a definition from [8]. The definition captures the idea that given signatures on a number of derived vectors in one of two different vector spaces, the attacker cannot tell which space the derived signatures came from. This indistinguishability holds even if the secret key is leaked. We call signatures with this property *weakly context hiding*. The reason for "weak" is that we are not hiding the fact that derivation took place or the computed function and we assume the original signatures are not public. Ahn et al. [1] define a stronger notion of privacy, called *strong context hiding*, that requires that derived signatures be distributed as independent fresh signatures on the same message; this requirement ensures privacy even if the original signatures are exposed.

Definition 3. A homomorphic signature scheme $\mathcal{S} = (\mathsf{Setup}, \mathsf{Sign}, \mathsf{Combine}, \mathsf{Verify})$ over R is *weakly context hiding* if the advantage of any probabilistic, polynomial-time adversary $\mathcal{A}$ in the following security game is negligible in the security parameter n:

Setup: The challenger runs $\mathsf{Setup}(n, \mathsf{params})$ to obtain $(\mathsf{pk}, \mathsf{sk})$ and gives pk and sk to $\mathcal{A}$.

Challenge: $\mathcal{A}$ outputs $(V_0, V_1, f_1, \ldots, f_s)$ where V_0 and V_1 are linear spaces over R^N represented as k-tuples of vectors $(\mathbf{v}_1^{(b)}, \ldots, \mathbf{v}_k^{(b)})$ for $b = 0, 1$. The functions $f_1, \ldots, f_s$ are R-linear functions[2] on $(R^N)^k$ satisfying

$$f_i(\mathbf{v}_1^{(0)}, \ldots, \mathbf{v}_k^{(0)}) = f_i(\mathbf{v}_1^{(1)}, \ldots, \mathbf{v}_k^{(1)}) \qquad \text{for all } i = 1, \ldots, s.$$

In response, the challenger generates a random bit $b \in \{0, 1\}$ and a random tag $\tau \in \{0, 1\}^n$ and signs the vector space V_B using the tag τ. Next, for $i = 1, \ldots, s$ the challenger uses $\mathsf{Combine}$ to derive signatures σ_i on $f_i(\mathbf{v}_1^{(b)}, \ldots, \mathbf{v}_k^{(b)})$ and sends $\sigma_1, \ldots, \sigma_s$ to $\mathcal{A}$. The functions $f_1, \ldots, f_s$ can be output adaptively after V_0, V_1 are output.

Output: $\mathcal{A}$ outputs a bit b'.

The adversary $\mathcal{A}$ wins the game if $b = b'$. The *advantage* of $\mathcal{A}$ is the probability that $\mathcal{A}$ wins the game.

Winning the context hiding game game means that the attacker was able to determine whether the challenge signatures were derived from signatures on V_0 or from signatures on V_1. We note that for discrete log-based linearly homomorphic signatures such as those of [4], weak context hiding follows from the uniqueness of the signature.

3 Background on Lattices

In this section we describe the lattices we will be using and their properties. Precise statements of these results can be found in the full version of this paper [6].

[2] If the scheme is L-limited, we require the f_i to have at most L nonzero coefficients.

Notation. For any integer $q \geq 2$, we let $\mathbb{Z}_q$ denote the ring of integers modulo q. When q is prime, $\mathbb{Z}_q$ is a field and is sometimes denoted $\mathbb{F}_q$. We let $\mathbb{Z}_q^{n \times m}$ denote the set of $n \times m$ matrices with entries in $\mathbb{Z}_q$. We denote matrices by capital boldface letters and vectors by lowercase boldface letters. We say a function $f : \mathbb{Z} \to \mathbb{R}^+$ is *negligible* if it is $O(n^{-c})$ for all $c > 0$, and we use $\text{negl}(n)$ to denote a negligible function of n. The function $\lg x$ is the base 2 logarithm of x.

Lattices. An m-dimensional lattice Λ is a full-rank discrete subgroup of $\mathbb{R}^m$. We will be interested in *integer lattices* Λ, i.e., those whose points have coordinates in $\mathbb{Z}^m$. For any integer $q \geq 2$ and any $\mathbf{A} \in \mathbb{Z}_q^{n \times m}$, we define

$$\Lambda_q^{\perp}(\mathbf{A}) := \{\mathbf{e} \in \mathbb{Z}^m : \mathbf{A} \cdot \mathbf{e} = \mathbf{0} \bmod q\}, \; \Lambda_q^{\mathbf{u}}(\mathbf{A}) := \{\mathbf{e} \in \mathbb{Z}^m : \mathbf{A} \cdot \mathbf{e} = \mathbf{u} \bmod q\}.$$

Alwen and Peikert [2, Theorem 3.2] describe an algorithm $\mathsf{TrapGen}$ that outputs an (almost) uniformly random matrix $\mathbf{A} \in \mathbb{Z}_q^{n \times m}$ along with a "short" basis for $\Lambda_q^{\perp}(\mathbf{A})$. Cash et al. [9, Lemma 3.2] give an algorithm $\mathsf{ExtBasis}$ that "delegates" a basis: given two matrices $\mathbf{A}, \mathbf{A}'$ in $\mathbb{Z}_q^{n \times m}$ and a short basis for $\Lambda_q^{\perp}(\mathbf{A})$, the algorithm computes a short basis for $\Lambda_q^{\perp}(\mathbf{A} \| \mathbf{A}')$. The length of a basis $\mathbf{T}$ is measured by the *Gram-Schmidt norm* and denoted by $\|\widetilde{\mathbf{T}}\|$.

For any real $\epsilon > 0$, the *smoothing parameter* $\eta_\epsilon(\Lambda)$ is defined to be the smallest positive s such that $\rho_{1/s}(\Lambda^* \setminus \{0\}) \leq \epsilon$ [20].

Gaussian Distributions. Let L be a subset of $\mathbb{Z}^m$. For any vector $\mathbf{c} \in \mathbb{R}^m$ and any positive parameter $\sigma \in \mathbb{R}_{>0}$, let $\rho_{\sigma,\mathbf{c}}(\mathbf{x}) := \exp\left(-\pi \|\mathbf{x} - \mathbf{c}\|^2 / \sigma^2\right)$ be a Gaussian function on $\mathbb{R}^m$ with center $\mathbf{c}$ and parameter σ. Let $D_{\sigma,\mathbf{c}}$ be the continuous Gaussian distribution over $\mathbb{R}^m$ with center $\mathbf{c}$ and parameter σ, with $D_{\sigma,\mathbf{c}}(\mathbf{x}) = \rho_{\sigma,\mathbf{c}}(\mathbf{x})/\sigma^n$. Let $\rho_{\sigma,\mathbf{c}}(L) := \sum_{\mathbf{x} \in L} \rho_{\sigma,\mathbf{c}}(\mathbf{x})$ be the discrete integral of $\rho_{\sigma,\mathbf{c}}$ over L. Finally, let $\mathcal{D}_{L,\sigma,c}$ be the discrete Gaussian distribution over L with center $\mathbf{c}$ and parameter σ. In particular, for all $\mathbf{y} \in L$, we have $\mathcal{D}_{L,\sigma,\mathbf{c}}(\mathbf{y}) = \frac{\rho_{\sigma,\mathbf{c}}(\mathbf{y})}{\rho_{\sigma,\mathbf{c}}(L)}$. For notational convenience, $\rho_{\sigma,\mathbf{0}}$ and $\mathcal{D}_{L,\sigma,\mathbf{0}}$ are abbreviated as ρ_σ and $\mathcal{D}_{L,\sigma}$, respectively.

Gentry, Peikert, and Vaikuntanathan [14, Theorems 4.1 and 5.9] describe algorithms $\mathsf{SampleGaussian}$ and $\mathsf{SamplePre}$ that output vectors sampled from distributions statistically close to $\mathcal{D}_{\Lambda_q^{\perp}(\mathbf{A}),\sigma,\mathbf{c}}$ and $\mathcal{D}_{\Lambda_q^{\mathbf{u}}(\mathbf{A}),\sigma}$, respectively.

Hardness Assumption. The security of our signature schemes is based on the problem of finding short vectors in $\Lambda_q^{\perp}(\mathbf{A})$ for random $\mathbf{A}$. This is known as the *Small Integer Solution* (SIS) problem, and is defined as follows.

Definition 4. An instance of the $\mathsf{SIS}_{q,m,\beta}$ problem is a matrix $\mathbf{A} \in \mathbb{Z}_q^{n \times m}$. A solution to the problem is a nonzero vector $\mathbf{v} \in \mathbb{Z}^m$ such that $\|\mathbf{v}\| \leq \beta$ and $\mathbf{A} \cdot \mathbf{v} = \mathbf{0} \bmod q$ (i.e., $\mathbf{v} \in \Lambda_q^{\perp}(\mathbf{A})$). If $\mathcal{B}$ is an algorithm that takes as input a matrix $\mathbf{A} \in \mathbb{Z}_q^{n \times m}$, we define $\mathsf{SIS}\text{-}\mathrm{Adv}[\mathcal{B}, (q, m, \beta)]$ to be the probability that $\mathcal{B}$ outputs a solution to a uniformly random $\mathsf{SIS}_{q,m,\beta}$ problem instance $\mathbf{A}$.

Micciancio and Regev [20] and Gentry et al. [14] show that the (average case) SIS problem for $\beta = \text{poly}(n)$ is hard assuming worst-case hardness of certain standard approximation problems on lattices.

4 New Tools

4.1 A "One-More" SIS Problem

The security of most lattice-based signature schemes depends on the adversary's inability to find a short vector in $\Lambda_q^\perp(\mathbf{A})$ for some public matrix $\mathbf{A}$. However, for our linearly homomorphic signatures this criterion is insufficient. Roughly speaking, an adversary in our scheme will be given several short vectors $\mathbf{e}_1, \ldots, \mathbf{e}_k \in \Lambda_q^\perp(\mathbf{A})$ and must produce a short vector in $\Lambda_q^\perp(\mathbf{A})$ that is not in the span of the $\mathbf{e}_i$. This is a "one-more" variant of the standard SIS problem, analogous to the "one-more discrete logarithm" problem in group-based cryptography (see e.g., [23]). We will see in Section 4.3 that for certain choices of parameters the problem is equivalent to finding *any* short vector in $\Lambda_q^\perp(\mathbf{A})$ distinct from $\{\pm\mathbf{e}_i\}$, making the "one-more" analogy even more appropriate. We now formally define the problem.

Definition 5. For any integer $k \geq 0$, an instance of the k-$\mathsf{SIS}_{q,m,\beta,\sigma}$ problem is a matrix $\mathbf{A} \in \mathbb{Z}_q^{n \times m}$ and a set of k vectors $\mathbf{e}_1, \ldots, \mathbf{e}_k \in \Lambda_q^\perp(\mathbf{A})$. A solution to the problem is a nonzero vector $\mathbf{v} \in \mathbb{Z}^m$ such that

1. $\|\mathbf{v}\| \leq \beta$,
2. $\mathbf{A} \cdot \mathbf{v} = \mathbf{0} \bmod q$ (i.e., $\mathbf{v} \in \Lambda_q^\perp(\mathbf{A})$), and
3. $\mathbf{v} \notin \mathbb{Q}\text{-span}(\{\mathbf{e}_1, \ldots, \mathbf{e}_k\})$.

If $\mathcal{B}$ is an algorithm that takes as input a matrix $\mathbf{A} \in \mathbb{Z}_q^{n \times m}$ and vectors $\mathbf{e}_i \in \mathbb{Z}^m$ for $i = 1, \ldots, k$, we define k-SIS-$\mathrm{Adv}[\mathcal{B}, (q, m, \beta, \sigma)]$ to be the probability that $\mathcal{B}$ outputs a solution to a k-$\mathsf{SIS}_{q,m,\beta,\sigma}$ problem instance $(\mathbf{A}, \mathbf{e}_1, \ldots, \mathbf{e}_k)$ over uniformly random $\mathbf{A}$ in $\mathbb{Z}_q^{n \times m}$ and $\mathbf{e}_i$ drawn from the distribution $\mathcal{D}_{\Lambda_q^\perp(\mathbf{A}),\sigma}$.

The main result of this section is to show that an adversary $\mathcal{A}$ that solves the k-SIS problem in dimension m can be used to solve the SIS problem in dimension $m - k$. More precisely, we have the following:

Theorem 6. *Let q be a prime, and let m, β, σ, and k, be polynomial functions of a security parameter n. Suppose that $m \geq 2n \lg q$, $m/k > n$, $\sigma > \omega(\sqrt{\log m})$, $t > \omega(\sqrt{\log n})$, and $q > \sigma \cdot \omega(\sqrt{\log m})$.*

Let $\beta' = \beta \cdot (k^{3/2} + 1)k!(t\sigma)^k$. Let $\mathcal{A}$ be a polynomial-time adversary for the k-$\mathsf{SIS}_{q,m,\beta,\sigma}$ problem. Then there exists a polynomial-time algorithm $\mathcal{B}$ that solves $\mathsf{SIS}_{q,m-k,\beta'}$, such that

$$\mathsf{SIS}\text{-}\mathrm{Adv}[\mathcal{B}, (q, m-k, \beta')] \geq k\text{-}\mathsf{SIS}\text{-}\mathrm{Adv}[\mathcal{A}, (q, m, \beta, \sigma)] - \mathrm{negl}(n).$$

Since the SIS problem is only assumed to be hard for parameters $\beta = \mathrm{poly}(n)$, the fact that the above reduction degrades exponentially in k means that k must be chosen to be small enough so that β' is still polynomial in n. In our application the parameter σ is $\omega(\sqrt{n})$, which means that k must be chosen to be $O(1)$. In this case, if we take $t = O(\log \sigma)$ and $\beta' = \beta \cdot O(\sigma^k \log^k \sigma)$, then Theorem 6 shows that if the $\mathsf{SIS}_{q,m-k,\beta'}$ problem is hard, then the k-$\mathsf{SIS}_{q,m,\beta,\sigma}$ problem is also hard.

The idea of the proof of Theorem 6 is as follows: given an SIS challenge $\mathbf{A}' \in \mathbb{Z}_q^{n\times(m-k)}$, we can choose k random vectors $\mathbf{e}_i$ from a Gaussian distribution over $\mathbb{Z}^m$ and append k columns to $\mathbf{A}'$ to create a matrix $\mathbf{A} \in \mathbb{Z}_q^{n\times m}$ such that the $\mathbf{e}_i$ are in $\Lambda_q^\perp(\mathbf{A})$. If the k-SIS adversary $\mathcal{A}$ outputs a short vector $\mathbf{e}^* \in \Lambda_q^\perp(\mathbf{A})$ that is $\mathbb{Q}$-linearly independent of the $\{\mathbf{e}_i\}$, then we can use Gaussian elimination over $\mathbb{Z}$ (via Cramer's rule) to compute a short vector $\mathbf{v} \in \Lambda_q^\perp(\mathbf{A})$ with zeroes in the last k entries. Reading off the first $m-k$ entries of $\mathbf{v}$ gives us a short vector in $\Lambda_q^\perp(\mathbf{A}')$. Details can be found in the full version of this paper [6].

4.2 Tight Bounds on the Length of Gaussian Samples

Signatures in our schemes will be "short" vectors sampled from Gaussian distributions over cosets of a particular lattice. To quantify what "short" means, we must demonstrate an upper bound on the length of a vector sampled from a Gaussian.

Micciancio and Regev [20, Lemma 4.4] show that if σ is larger than the smoothing parameter of the n-dimensional lattice Λ, then with overwhelming probability the length of a vector sampled from $\mathcal{D}_{\Lambda,\sigma,\mathbf{c}}$ is at most $\sigma\sqrt{n}$. They also show that the *expected* length of such a sample is at most $\sigma\sqrt{n/2\pi} + \mathrm{negl}(n)$. Our result below "bridges the gap" between the upper bound and the expected length. Furthermore, we show an equally strong *lower* bound on the length of the Gaussian sample.

Proposition 7. *Let $\Lambda \subset \mathbb{R}^n$ be a lattice. Suppose $\sigma \geq \eta_\epsilon(\Lambda)$ for some negligible ϵ. Let $\mathbf{c} \in \mathbb{R}^m$ be any vector. Then for any constant $\alpha > 0$ we have*

$$\Pr\left[(1-\alpha)\sigma\sqrt{\tfrac{n}{2\pi}} \leq \|\mathbf{x}-\mathbf{c}\| \leq (1+\alpha)\sigma\sqrt{\tfrac{n}{2\pi}} : x \xleftarrow{\mathrm{R}} \mathcal{D}_{\Lambda,\sigma,\mathbf{c}}\right] \geq 1 - \mathrm{negl}(n).$$

4.3 Removing Linear Independence from the k-SIS Problem

We now show that for small values of k and tight length bounds, we can relax the linear independence condition in the statement of the k-SIS problem. Specifically, if we can find *any* nonzero vector $\mathbf{e}^*$ of the required length not equal to $\pm\mathbf{e}_i$ for any of the k vectors $\mathbf{e}_i$ in the problem statement, then with overwhelming probability $\mathbf{e}^*$ is not in the linear span of the $\mathbf{e}_i$.

Proposition 8. *Suppose $m \geq 2n\lg q$ and $k \cdot \omega(\sqrt{\log n}) < \min(\sigma, m^{1/4})$. Let $(\mathbf{A}, \mathbf{e}_1, \ldots, \mathbf{e}_k)$ be a k-SIS challenge with $\mathbf{A}$ chosen uniformly at random from $\mathbb{Z}_q^{n\times m}$ and $\mathbf{e}_i$ sampled from $\mathcal{D}_{\Lambda_q^\perp(\mathbf{A}),\sigma}$. Then with overwhelming probability, the only nonzero vectors of length at most $1.1\cdot\sigma\sqrt{m/2\pi}$ in $\mathbb{Q}$-span$(\mathbf{e}_1, \ldots, \mathbf{e}_k)$ are the vectors $\pm\mathbf{e}_i$ for $i = 1, \ldots, k$.*

4.4 Linear Combinations of Discrete Gaussians

The privacy property of our linearly homomorphic scheme will follow from the fact that the distribution obtained by summing independent discrete Gaussian samples is itself a discrete Gaussian distribution that depends only on the coset of the sum and the linear combination being computed.

Theorem 9. *Let $\Lambda \subseteq \mathbb{Z}^m$ be a lattice and $\sigma \in \mathbb{R}$. For $i = 1, \ldots, k$ let $\mathbf{t}_i \in \mathbb{Z}^m$ and let X_i be mutually independent random variables sampled from $\mathcal{D}_{\Lambda+\mathbf{t}_i,\sigma}$. Let $\mathbf{c} = (c_1, \ldots, c_k) \in \mathbb{Z}^k$, and define*

$$g := \gcd(c_1, \ldots, c_k), \qquad \mathbf{t} := \sum_{i=1}^k c_i \mathbf{t}_i.$$

Suppose that $\sigma > \|\mathbf{c}\| \cdot \eta_\epsilon(\Lambda)$ for some negligible ϵ. Then $Z = \sum_{i=1}^k c_i X_i$ is statistically close to $\mathcal{D}_{g\Lambda+\mathbf{t},\|\mathbf{c}\|\sigma}$.

In the full version of this paper [6] we generalize this theorem to $\bar{Z} = A \cdot \bar{X}$, where $\bar{X} = (X_1, \ldots, X_k)$ and A is an $s \times k$ matrix.

5 A Linearly Homomorphic Signature Scheme over $\mathbb{F}_2$

We now describe our linearly homomorphic signature scheme over $\mathbb{F}_2$. Our construction is inspired by the signature scheme of Gentry, Peikert, and Vaikuntanathan [14]. In the GPV scheme, signatures are short vectors in $\Lambda_q^{\mathbf{u}}(\mathbf{A})$, where $\mathbf{u}$ is the hash of the message to be signed. The key idea in our construction of homomorphic signatures is to work simultaneously modulo 2 and modulo an odd prime q. Specifically, a signature on a vector $\mathbf{v} \in \mathbb{F}_2^n$ is a short vector $\mathbf{e} \in \mathbb{Z}^m$ such that $\mathbf{e}$ is in both $\Lambda_q^\perp(\mathbf{A})$ and $\Lambda_2^{\mathbf{v}}(\mathbf{A})$. The mod 2 part ties the signature to the message, while the mod q part ensures that the signature cannot be forged. By the Chinese remainder theorem, such a vector $\mathbf{e}$ is in the lattice $\Lambda_{2q}^{q\cdot\mathbf{v}}(\mathbf{A})$.

In order to be able to sign multiple files, the matrix $\mathbf{A}$ must be different for every file, yet still have a trapdoor that allows us to generate signatures using the SamplePre algorithm. To achieve this, we divide $\mathbf{A}$ into two parts. The left half is a public matrix generated by the TrapGen algorithm; the right half depends on the identifier of the file being signed. Given the secret basis output by TrapGen, we can use the ExtBasis algorithm to compute a short basis for $\Lambda_{2q}^\perp(\mathbf{A})$.

Our scheme is as follows:

Setup(n, params). Given a security parameter n and parameters params $= (N, k, L, m, q, \sigma)$, where $N = n$ is the dimension of vectors to be signed, $k < n$ is the dimension of subspaces to be signed, $L \geq 1$ is the maximum number of linear combinations that can be authenticated, $m(n, L) > n$ is an integer, $q(n, L)$ is an odd prime, and $\sigma(n, L)$ is a real number, do the following:

1. Run TrapGen($n, m, 2q$) to generate a matrix $\mathbf{A} \in \mathbb{Z}_{2q}^{n\times m}$ and a basis $\mathbf{T}$ of $\Lambda_{2q}^\perp(\mathbf{A})$ such that $\|\widetilde{\mathbf{T}}\| \leq 30\sqrt{n \lg 2q}$.
2. Let $H : \{0,1\}^* \to \mathbb{Z}_{2q}^{n\times m}$ be a hash function, viewed as a random oracle.
3. Output the public key pk $\leftarrow (\mathbf{A}, H)$, and the private key sk $\leftarrow (\mathbf{A}, H, \mathbf{T})$.

Sign(sk, id, $\mathbf{v}$). Given secret key sk $= (\mathbf{A}, H, \mathbf{T})$, identifier id $\in \{0,1\}^n$, and a vector $\mathbf{v} \in \mathbb{F}_2^n$, do the following:

1. Set $\mathbf{B} \leftarrow \mathbf{A} \| H(\mathsf{id}) \in \mathbb{Z}_{2q}^{n\times 2m}$.
2. Let $\mathbf{S} \leftarrow$ ExtBasis($\mathbf{T}, \mathbf{B}$) be a basis for $\Lambda_{2q}^\perp(\mathbf{B})$ with $\|\widetilde{\mathbf{S}}\| = \|\widetilde{\mathbf{T}}\|$.
3. Output $\mathbf{e} \leftarrow$ SamplePre($\mathbf{B}, \mathbf{S}, \sigma, q \cdot \mathbf{v}$).

$\mathsf{Combine}(\mathsf{pk}, \mathsf{id}, \{(\alpha_i, \mathbf{e}_i)\}_{i=1}^{\ell})$. Given a public key pk, an identifier id, and pairs $\{(\alpha_i, \mathbf{e}_i)\}_{i=1}^{\ell}$ with $\alpha_i \in \mathbb{F}_2 = \{0, 1\}$, output $\mathbf{e} \leftarrow \sum_{i=1}^{\ell} \alpha_i \mathbf{e}_i \in \mathbb{Z}^{2m}$.

$\mathsf{Verify}(\mathsf{pk}, \mathsf{id}, \mathbf{y}, \mathbf{e})$. Given a public key $\mathsf{pk} = (\mathbf{A}, H)$, an identifier id, a signature $\mathbf{e} \in \mathbb{Z}^{2m}$, and a vector $\mathbf{y} \in \mathbb{F}_2^n$, do the following:

1. Set $\mathbf{B} \leftarrow \mathbf{A} \| H(\mathsf{id}) \in \mathbb{Z}_{2q}^{n \times 2m}$.
2. If (a) $\|\mathbf{e}\| \leq L \cdot \sigma\sqrt{2m}$ and (b) $\mathbf{B} \cdot \mathbf{e} = q \cdot \mathbf{y} \bmod 2q$, output 1. Otherwise output 0.

Proposition 10. *Suppose* $\sigma \geq 30\sqrt{n \lg 2q} \cdot \omega(\sqrt{\log n})$. *Then the scheme described above is an L-limited linearly homomorphic signature scheme over* $\mathbb{F}_2$.

Unforgeability. We prove unforgeability of our linearly homomorphic signature scheme over $\mathbb{F}_2$ in the random oracle model. Given an adversary that breaks the signature scheme over $\mathbb{Z}_{2q}$, we construct an adversary that simulates the signature scheme and the hash function H and solves the k-SIS problem over $\mathbb{Z}_q$. By Theorem 6, this adversary can in turn be used to solve the SIS problem over $\mathbb{Z}_q$.

Our simulation begins by guessing which of the adversary's signature and hash queries will correspond to the file identifier id^* associated with the adversary's forgery and outputting a public key $\mathbf{A}$ derived from the k-SIS challenge matrix. For queries *not* associated with id^*, the simulator "swaps the roles" of the public key and hash function as follows: we use TrapGen to program the random oracle with a matrix $H(\mathsf{id})$ for which we know a short basis, and we use ExtBasis to compute a short basis for $\mathbf{A} \| H(\mathsf{id})$. We can then compute the signatures as in the real system.

For the query id^*, we construct $H(\mathsf{id}^*)$ so that the k-SIS challenge vectors are valid signatures for the vectors queried by the adversary. We construct the mod q part of $H(\mathsf{id}^*)$ using the fact that valid signatures are elements of $\Lambda_q^{\perp}(\mathbf{A} \| H(\mathsf{id}^*))$, and we construct the mod 2 part of $H(\mathsf{id}^*)$ using the fact that the k-SIS challenge vectors are statistically close to random mod 2.

With this setup, a forged signature is exactly a solution to the k-SIS problem mod q. We now give the theorem; the proof appears in the full paper [6].

Theorem 11. *Let* $\mathcal{N}$ *be the linearly homomorphic signature scheme over* $\mathbb{F}_2$ *described above. Suppose that* $m = \lceil 6n \lg 2q \rceil$ *and* $\sigma = 30\sqrt{n \lg 2q} \log n$. *Let* $\beta = L \cdot \sigma\sqrt{2m}$. *Then* $\mathcal{N}$ *is unforgeable in the random oracle model assuming that* $k\text{-}\mathsf{SIS}_{q,2m,\beta,\sigma}$ *is infeasible.*

Since the SIS problem is only assumed to be hard for $\beta' = \mathrm{poly}(n)$, by Theorem 6 our choice of σ forces k to be $O(1)$ to ensure security based on SIS.

Privacy. In our linearly homomorphic signature scheme, one derives a signature on a linear combination $\mathbf{v}$ of messages by taking a linear combination of the signatures on the original messages $\mathbf{v}_1, \ldots, \mathbf{v}_k$. Hence, the derived signature on $\mathbf{v}$ is a linear combination of short vectors in cosets of some lattice Λ. To show that this derived signature does not leak information about the original signatures, we use Theorem 9 to show that a linear combination of k signatures generated by our signing algorithm is itself a short vector sampled from a distribution that depends only on the function computed and the message $\mathbf{v}$ output by the function. In particular, the derived signature does not depend on the original vectors $\mathbf{v}_1, \ldots, \mathbf{v}_k$ (up to negligible statistical distance). It follows that the derived signature does not leak any information about $\mathbf{v}_1, \ldots, \mathbf{v}_k$ beyond what is revealed by $\mathbf{v}$. We note that the length of σ reveals information about the computed linear function, but not about the original messages.

The above argument implies that a single derived signature is private; for multiple signatures we use a similar argument with an appropriate generalization of Theorem 9. The proof of our privacy theorem can be found in the full paper [6].

Theorem 12. *Let $\mathcal{N}$ be the linearly homomorphic signature scheme over $\mathbb{F}_2$ described in Section 5. Suppose that k is constant, $m = \lceil 6n \lg 2q \rceil$ and $\sigma = 30\sqrt{n \lg 2q} \log n$. Then $\mathcal{N}$ is weakly context hiding.*

6 k-Time GPV Signatures without Random Oracles

In this section we give a second application of the k-SIS mechanism described in Section 4, namely, a stateless variant of the the signature scheme of Gentry, Peikert, and Vaikuntanathan [14] that is k-time unforgeable in the standard model. The notion of k-time security means that a signing key can only be used to sign k messages. In particular, a forger is allowed at most k signing queries.

The main idea is to construct signatures as in our homomorphic scheme of Section 5, but remove the homomorphic property by setting the bound on the length of a valid signature to be very close to the expected length of the signature. Since we expect a small number of such vectors to form a set that is nearly orthogonal, any linear combination of signatures will produce a vector that is too long to be accepted as a valid signature.

We prove our signature scheme *weakly unforgeable*; i.e., unforgeable under a static chosen-message attack, in which the adversary must submit all signature queries before seeing the public key. A standard transformation using *chameleon hashes* [17] produces a scheme that is unforgeable under the usual notion of adaptive chosen-message attack.

We now describe our weakly unforgeable signature scheme, which is essentially a GPV signature in which hashing is replaced with the Chinese remaindering of the message (viewed as a vector in $\mathbb{F}_2^n$) with the zero vector in $\mathbb{Z}_q^n$.

$\mathsf{Setup}(n, \mathsf{params})$. Given a security parameter n that is also the bit length of messages to be signed, do the following:

1. Choose an odd prime q. Set $m \leftarrow \lceil 6n \lg 2q \rceil$. Set $\sigma \leftarrow 30\sqrt{n \lg 2q} \log n$.
2. Run $\mathsf{TrapGen}(n, m, 2q)$ to generate a matrix $\mathbf{A} \in \mathbb{Z}_{2q}^{n \times m}$ and a basis $\mathbf{T}$ of $\Lambda_{2q}^{\perp}(\mathbf{A})$ such that $\|\widetilde{\mathbf{T}}\| \leq \sigma / \log n$.
3. Output the public key $\mathsf{pk} \leftarrow (\mathbf{A}, \sigma)$ and the private key $\mathsf{sk} \leftarrow (\mathbf{A}, \sigma, \mathbf{T})$.

$\mathsf{Sign}(\mathsf{sk}, \mathbf{v})$. Given secret key $\mathsf{sk} = (\mathbf{A}, \sigma, \mathbf{T})$, and a message $\mathbf{v}$ (interpreted as a vector in $\mathbb{F}_2^n$), output $\mathbf{e} \leftarrow \mathsf{SamplePre}(\mathbf{A}, \mathbf{T}, \sigma, q \cdot \mathbf{v})$.

$\mathsf{Verify}(\mathsf{pk}, \mathbf{e}, \mathbf{v})$. Given a public key $\mathsf{pk} = (\mathbf{A}, \sigma)$, a signature $\mathbf{e} \in \mathbb{Z}^m$, and a message $\mathbf{v} \in \mathbb{F}_2^n$, output 1 if (a) $0 < \|\mathbf{e}\| \leq 1.1 \cdot \sigma\sqrt{m/2\pi}$ and (b) $\mathbf{A} \cdot \mathbf{e} = q \cdot \mathbf{v} \bmod 2q$. Otherwise output 0.

Correctness of our scheme follows from Proposition 7 (with $\alpha = 0.1$), using [14, Lemma 5.3] to bound the smoothing parameter. In our fully unforgeable scheme, the vector $\mathbf{v}$ used in signing is not the message but rather $H(M, r)$, where M is the message, r is random, and H is a chameleon hash function. The signature includes the randomness r in addition to the vector $\mathbf{e}$. For a discussion of lattice-based chameleon hash functions, see [9, §2.2].

An adversary attacking our k-time signature scheme requests k signatures $\mathbf{e}_i$ on messages of his choice, receives a public key and the signatures, and then outputs a message $\mathbf{v}^*$ and a signature $\mathbf{e}^*$. The adversary wins the game if $\mathsf{Verify}(\mathsf{pk}, \mathbf{e}^*, \mathbf{v}^*) = 1$ and $\mathbf{v}^*$ is not equal to any of the messages queried.

As was the case for our homomorphic scheme of Section 5, a valid forgery is a short vector $\mathbf{e}^*$ in $\Lambda_q^\perp(\mathbf{A})$. The key idea in our security proof is that the length bound on a valid $\mathbf{e}^*$ is so tight that by Proposition 8, the only nonzero integer vectors of comparable length in the $\mathbb{Q}$-span of the requested signatures $\mathbf{e}_i$ are the vectors $\pm\mathbf{e}_i$. (Since $-\mathbf{e}_i$ authenticates the same message as $\mathbf{e}_i$, the signature $-\mathbf{e}_i$ is not a valid forgery.) Thus $\mathbf{e}^*$ is outside the linear span of the $\mathbf{e}_i$, and we can use it to solve the k-SIS instance in which the $\mathbf{e}_i$ are the challenge vectors.

Our security theorem is as follows; the proof appears in the full paper [6].

Theorem 13. *Let S be the signature scheme described above. Suppose k is constant and $\beta = 1.1 \cdot \sigma\sqrt{m/2\pi}$. Then S is a weakly unforgeable k-time signature scheme assuming that k-$\mathsf{SIS}_{q,m,\beta,\sigma}$ is infeasible.*

7 Further Directions

Extending the Linearly Homomorphic System. While our linearly homomorphic scheme in Section 5 authenticates vectors with coordinates in $\mathbb{F}_2$, the same construction works for any field $\mathbb{F}_p$ where p is a small prime. We simply set $m = \lceil 6n \lg pq \rceil$ and $\sigma = 30\sqrt{n \lg pq} \log n$, and sign a vector $\mathbf{v} \in \mathbb{F}_p^n$ using the lattice $\Lambda_{pq}^{q \cdot \mathbf{v}}(\mathbf{A})$. If p is odd and we identify $\mathbb{F}_p$ with $\{-(p-1)/2, \dots, (p-1)/2\}$, then the output of $\mathsf{Combine}$ on ℓ vectors can be up to $\ell(p-1)$ times as long as the largest input vector. An argument as in Proposition 10 shows that the resulting system is $L/(p-1)$-limited.

More interestingly, our system can also be used to authenticate vector spaces defined over non-prime fields. Suppose for concreteness that our vectors live in $(\mathbb{F}_{2^d})^n$. If we fix a basis for $\mathbb{F}_{2^d}$ over $\mathbb{F}_2$, then when computing signatures we may view the vectors as elements of $(\mathbb{F}_2)^{nd}$ and compute signatures in exactly the same manner as above. The difference comes when computing linear combinations over $\mathbb{F}_{2^d}$: in our representation multiplying an element $x \in \mathbb{F}_{2^d}$ by an element $\alpha \in \mathbb{F}_{2^d}$ consists of multiplying the corresponding vector $\mathbf{x} \in \mathbb{F}_2^d$ by a matrix $M_\alpha \in \mathbb{F}_2^{d \times d}$. To compute this action on the signature vector $\mathbf{e} \in \mathbb{Z}^m$, we lift M_α to an integer matrix with entries in $\{0, 1\}$ and group the elements of $\mathbf{e}$ into d-tuples corresponding to the underlying elements of $\mathbb{F}_{2^d}$. Multiplying each d-tuple by M_α now has the effect of multiplying the underlying elements of $\mathbb{F}_{2^d}$ by α. We see that this action increases the length of $\mathbf{e}$ by a factor of at most d, so combining ℓ vectors gives an output that is up to ℓd times as long as the largest input vector. By the same argument as above, the system over $\mathbb{F}_{2^d}$ is L/d-limited.

Open Problem. An important open problem inspired by our construction is to find a tight reduction of k-SIS to worst-case lattice problems, either by improving on the reduction to SIS given by Theorem 6 or by a direct argument. An improved reduction would support the use of the k-SIS problem in developing cryptosystems for other applications and would also allow us to implement our systems with smaller parameters.

An Alternative Construction. In recent work [5], we have developed a different construction of linearly homomorphic signatures over small fields. Unforgeability of the new scheme (in the random oracle model) reduces directly to the SIS problem, without going through the intermediate k-SIS reduction. As a result, the new scheme can authenticate linear combinations of polynomially many vectors, whereas the one in this paper requires the number of vectors combined to be constant. Signatures in the two schemes are of comparable length.

Acknowledgments. We thank Chris Peikert for helpful discussions, and in particular for providing the idea of using continuous Gaussians to prove Proposition 7. We thank Susan Hohenberger, Hugo Krawczyk, Daniele Micciancio, and Brent Waters for useful conversations. We also thank the anonymous reviewers for helpful suggestions.

References

1. Ahn, J.H., Boneh, D., Camenisch, J., Hohenberger, S., Shelat, A., Waters, B.: Computing on authenticated data (2010) (manuscript)
2. Alwen, J., Peikert, C.: Generating shorter bases for hard random lattices. In: STACS, pp. 75–86 (2009) `http://www.cc.gatech.edu/~cpeikert/pubs/shorter.pdf`
3. Ateniese, G., Chou, D.H., de Medeiros, B., Tsudik, G.: Sanitizable signatures. In: di Vimercati, S.d.C., Syverson, P.F., Gollmann, D. (eds.) ESORICS 2005. LNCS, vol. 3679, pp. 159–177. Springer, Heidelberg (2005)
4. Boneh, D., Freeman, D., Katz, J., Waters, B.: Signing a linear subspace: Signature schemes for network coding. In: Jarecki, S., Tsudik, G. (eds.) PKC 2009. LNCS, vol. 5443, pp. 68–87. Springer, Heidelberg (2009)
5. Boneh, D., Freeman, D.M.: Homomorphic signatures for polynomial functions (2010) (manuscript)
6. Boneh, D., Freeman, D.M.: Homomorphic signatures over binary fields and new tools for lattice-based signatures. Cryptology eprint report 2010/453 (2010), `http://eprint.iacr.org/2010/453`, Full version of this paper
7. Brzuska, C., Busch, H., Dagdelen, Ö., Fischlin, M., Franz, M., Katzenbeisser, S., Manulis, M., Onete, C., Peter, A., Poettering, B., Schröder, D.: Redactable signatures for tree-structured data: Definitions and constructions. In: Zhou, J., Yung, M. (eds.) ACNS 2010. LNCS, vol. 6123, pp. 87–104. Springer, Heidelberg (2010)
8. Brzuska, C., Fischlin, M., Freudenreich, T., Lehmann, A., Page, M., Schelbert, J., Schröder, D., Volk, F.: Security of sanitizable signatures revisited. In: Jarecki, S., Tsudik, G. (eds.) PKC 2009. LNCS, vol. 5443, pp. 317–336. Springer, Heidelberg (2009)
9. Cash, D., Hofheinz, D., Kiltz, E., Peikert, C.: Bonsai trees, or how to delegate a lattice basis. In: Gilbert, H. (ed.) EUROCRYPT 2010. LNCS, vol. 6110, pp. 523–552. Springer, Heidelberg (2010)
10. Chang, E.-C., Lim, C.L., Xu, J.: Short redactable signatures using random trees. In: Fischlin, M. (ed.) CT-RSA 2009. LNCS, vol. 5473, pp. 133–147. Springer, Heidelberg (2009)

11. Charles, D., Jain, K., Lauter, K.: Signatures for network coding. International Journal of Information and Coding Theory 1, 3–14 (2009)
12. Fragouli, C., Soljanin, E.: Network coding fundamentals. Found. Trends Netw. 2, 1–133 (2007)
13. Gennaro, R., Katz, J., Krawczyk, H., Rabin, T.: Secure network coding over the integers. In: Nguyen, P.Q., Pointcheval, D. (eds.) PKC 2010. LNCS, vol. 6056, pp. 142–160. Springer, Heidelberg (2010)
14. Gentry, C., Peikert, C., Vaikuntanathan, V.: Trapdoors for hard lattices and new cryptographic constructions. In: Ladner, R.E., Dwork, C. (eds.) STOC, pp. 197–206. ACM, New York (2008)
15. Haber, S., Hatano, Y., Honda, Y., Horne, W., Miyazaki, K., Sander, T., Tezoku, S., Yao, D.: Efficient signature schemes supporting redaction, pseudonymization, and data deidentification. In: ASIACCS 2008, pp. 353–362 (2008)
16. Johnson, R., Molnar, D., Song, D., Wagner, D.: Homomorphic signature schemes. In: Preneel, B. (ed.) CT-RSA 2002. LNCS, vol. 2271, pp. 244–262. Springer, Heidelberg (2002)
17. Krawczyk, H., Rabin, T.: Chameleon signatures. In: Network and Distributed System Security Symposium (NDSS) (2000)
18. Krohn, M., Freedman, M., Mazières, D.: On-the-fly verification of rateless erasure codes for efficient content distribution. In: Proc. of IEEE Symposium on Security and Privacy, pp. 226–240 (2004)
19. Lyubashevsky, V., Micciancio, D.: Asymptotically efficient lattice-based digital signatures. In: Canetti, R. (ed.) TCC 2008. LNCS, vol. 4948, pp. 37–54. Springer, Heidelberg (2008)
20. Micciancio, D., Regev, O.: Worst-case to average-case reductions based on Gaussian measures. In: 45th Annual IEEE Symposium on Foundations of Computer Science — FOCS 2004, pp. 372–381 (2004)
21. Miyazaki, K., Hanaoka, G., Imai, H.: Digitally signed document sanitizing scheme based on bilinear maps. In: ACM Symposium on Information, Computer and Communications Security — ASIACCS 2006, pp. 343–354 (2006)
22. Miyazaki, K., Iwamura, M., Matsumoto, T., Sasaki, R., Yoshiura, H., Tezuka, S., Imai, H.: Digitally signed document sanitizing scheme with disclosure condition control. IEICE Transactions on Fundamentals E88-A, 239–246 (2005)
23. Paillier, P., Vergnaud, D.: Discrete-log-based signatures may not be equivalent to discrete log. In: Roy, B. (ed.) ASIACRYPT 2005. LNCS, vol. 3788, pp. 1–20. Springer, Heidelberg (2005)
24. Steinfeld, R., Bull, L., Zheng, Y.: Content extraction signatures. In: Kim, K.-c. (ed.) ICISC 2001. LNCS, vol. 2288, pp. 285–304. Springer, Heidelberg (2002)
25. Zhao, F., Kalker, T., Médard, M., Han, K.: Signatures for content distribution with network coding. In: Proc. Intl. Symp. Info. Theory (ISIT) (2007)

Homomorphic Network Coding Signatures in the Standard Model

Nuttapong Attrapadung[1] and Benoît Libert[2,⋆]

[1] Research Center for Information Security, AIST, Japan
n.attrapadung@aist.go.jp
[2] Université catholique de Louvain, Crypto Group, Belgium
benoit.libert@uclouvain.be

Abstract. Network coding is known to provide improved resilience to packet loss and increased throughput. Unlike traditional routing techniques, it allows network nodes to perform transformations on packets they receive before transmitting them. For this reason, packets cannot be authenticated using ordinary digital signatures, which makes it difficult to hedge against pollution attacks, where malicious nodes inject bogus packets in the network. To address this problem, recent works introduced signature schemes allowing to sign linear subspaces (namely, verification can be made w.r.t. any vector of that subspace) and which are well-suited to the network coding scenario. Currently known network coding signatures in the standard model are not homomorphic in that the signer is forced to sign all vectors of a given subspace at once. This paper describes the first homomorphic network coding signatures in the standard model: the security proof does not use random oracles and, at the same time, the scheme allows signing individual vectors *on-the-fly* and has constant per-packet overhead in terms of signature size. The construction is based on the dual encryption technique introduced by Waters (Crypto'09) to prove the security of hierarchical identity-based encryption schemes.

Keywords: Network coding, homomorphic signatures, provable security, standard model.

1 Introduction

Network coding [1,18] is an attractive paradigm that offers an interesting alternative to traditional routing mechanisms. Instead of merely storing and forwarding packets in transit, intermediate nodes are allowed to modify them: typically, at each node, outgoing packets contain vectors that are calculated as linear combinations of vectors conveyed by incoming packets. In *random linear network coding*, packets are combined using coefficients which each node chooses at random, independently of its neighbors. Still, receiving nodes are able to recover

⋆ This author acknowledges the Belgian National Fund for Scientific Research (F.R.S.-F.N.R.S.) for his "chargé de recherches" fellowship and the BCRYPT Interuniversity Attraction Pole.

D. Catalano et al. (Eds.): PKC 2011, LNCS 6571, pp. 17–34, 2011.

the original file from any set of, say $m > 1$, valid packets containing linearly independent vectors and without *a priori* knowing the coefficients chosen by intermediate nodes on the road. This technique has been shown (see [10] for instance) to provide many advantages such as an improved resilience to random packet loss or a substantially increased throughput in certain topologies.

Unfortunately, network coding is highly sensitive to *pollution attacks*, where malicious nodes inject invalid packets (*i.e.*, nodes outside the linear span of the received packets) in the network in order to prevent target nodes from recovering the original file. Since network nodes perform linear transformations over all their incoming packets, even a single faulty packet is likely to contaminate the entire network and eventually hinder the decoding process. To address this concern, intermediate good nodes need a method to verify the validity of incoming packets and sieve out bad ones. Obviously, the problem cannot be resolved by ordinary digital signatures since transmitted packets are modified by the network and cannot be merely signed by the source. For this reason, cryptographic approaches rely on techniques allowing to authenticate packets using homomorphic hash functions [16,13,25] or homomorphic signatures [9,7,12]. These primitives are designed in such a way that a signature (resp. a hash value) of a vector $\vec{v}$ can be obtained from signatures (resp. hash values) of several vectors that $\vec{v}$ is a linear combination of.

In contrast to information-theoretic approaches (like [14,15]) that defend against network faults by introducing redundancies in packets, cryptographic techniques do not place restrictions on the adversary's behavior (e.g. by limiting his ability to eavesdrop the network or the fraction of nodes he can corrupt): as long as the receiver obtains sufficiently many correct packets, he can always recover the file regardless of the number of faults. On the other hand, these techniques typically require computational assumptions and sometimes appeal to idealizations such as the random oracle model [4]. This paper aims at making another step towards eliminating the latter.

Related Work. Homomorphic signatures were first suggested by Johnson, Molnar, Song and Wagner [20]. Their definition was adapted to the network coding scenario by Boneh, Freeman, Katz and Waters [7] who designed an efficient homomorphic NCS scheme in the random oracle model using bilinear maps. At the expense of losing the homomorphic property, they also showed how to build a network coding signature in the standard model. In [7], signature sizes were proved asymptotically optimal since a signature on any subspace necessarily grows with the dimension of that subspace. Recently, Gennaro, Katz, Krawczyk and Rabin gave a homomorphic signature [12] based on the RSA assumption (in the random oracle model) and showed how to work with small coefficients over the integers (instead of finite fields) in networks of bounded size. At the same time, Agrawal, Boneh, Boyen and Freeman [3] considered the situation of network nodes mixing packets from multiple distinct sources and described a multi-source network coding signature (without the homomorphic property) in the standard model.

In the secret-key setting, Agrawal and Boneh [2] considered how to improve upon the speed of network coding public key signatures and designed message authentication codes with homomorphic properties. Assuming that a bounded number of verifiers may collude, they also showed how intermediate nodes can verify the integrity of network-coded data. More recently, Li *et al.* [19] gave a MAC-based approach supporting in-network verification and resisting an arbitrary number of collusions.

OUR CONTRIBUTION. To the best of our knowledge, in the public key setting, known *homomorphic* network coding signatures [7,12] all rely on random oracles in their security proof. Indeed, existing NCS schemes in the standard model (*i.e.*, the second scheme of [7] and the multi-source system in [3]) can only be used to sign all the base vectors of a subspace at once. This requires the source to be aware of the entire file before sending the first packet.

This paper describes the first homomorphic NCS scheme with a security proof outside the random oracle methodology. Our construction is based on the dual encryption paradigm, introduced by Waters [24] and developed in [17], the purpose of which was initially to build fully secure (hierarchical) identity-based encryption [22,6] schemes. We pinpoint an intuitive connection between NCS schemes and the spatial encryption primitive of Boneh and Hamburg [8], where the receiver's ability to decrypt is made contingent on his knowledge of a private key for a subspace containing the vector assigned to the ciphertext. We explain that such a scheme can be turned into a (not necessarily homomorphic) NCS scheme when the file identifier can be suitably tied up to the signed subspace. The homomorphic property is then achieved by carefully re-using the signer's random coins across all vectors of the same linear subspace: by deriving these coins from the file identifier using a pseudorandom function, the signer can start transmitting packets before the file to be sent is completely known.

In order to prove security in the sense of the definition of Boneh *et al.* [7], we use groups of composite order and apply the technique of Lewko and Waters [17] in the context of signatures. One difficulty to deal with is that, unlike previous homomorphic NCS schemes [7,12], the system uses a randomized signing algorithm and signatures on distinct vectors must be generated using partially identical randomness in order to be linearly combinable. We thus have to take special precautions to prevent malicious nodes from re-randomizing signatures and wrongly accuse the signer of flooding the network with signatures that cannot be combined.

Since we work in groups of composite order N, vector coordinates and network coefficients must be chosen in a ring $\mathbb{Z}_N$ instead of a prime field as in [7]. Nevertheless, the scheme has counterparts in prime order groups. While Freeman's framework [11] does not seem to apply (given that it does not apply to the Lewko-Waters techniques [17], as mentioned in [11]) to generically transform the scheme into an instantiation in prime-order groups, the system can be adapted in asymmetric pairing-friendly groups of prime order in the same way as the Lewko-Waters IBE [17]. It is also translatable in groups with symmetric pairings using the techniques of [24]. In the paper, we chose to describe it in composite order groups for simplicity.

ORGANIZATION. In the following, we first review the notion of network coding signatures in section 2. Our homomorphic scheme and its proof are detailed in sections 3.1 and 3.2, respectively.

2 Background and Definitions

2.1 Network Coding

This section briefly recalls the idea of linear network coding. Consider a network with one source node and a subset of nodes called "target nodes". The purpose is to have the source transmit a file to all target nodes, where a file is represented as a matrix containing the m row vectors $\vec{v}_1, \ldots, \vec{v}_m \in \mathbb{Z}_N^k$ over a ring or a field $\mathbb{Z}_N$. The source node first creates m augmented vectors $\vec{w}_1, \ldots, \vec{w}_m \in \mathbb{Z}_N^n$, with $n = k + m$, by setting

$$\begin{pmatrix} —\vec{w}_1— \\ —\vec{w}_2— \\ \vdots \\ —\vec{w}_m— \end{pmatrix} = \left(\begin{array}{c|cccc} —\vec{v}_1— & 1 & 0 & \cdots & 0 \\ —\vec{v}_2— & 0 & 1 & \cdots & 0 \\ \vdots & & & \ddots & \\ —\vec{v}_m— & 0 & 0 & \cdots & 1 \end{array}\right). \tag{1}$$

The source then sends these augmented vectors to its neighbor nodes.

We notice that the span of row vectors of the above matrix will generate a vector subspace $V \subset \mathbb{Z}_N^n$ of dimension m with the basis $\vec{w}_1, \ldots, \vec{w}_m$. As defined in [7], when the basis is in the above form (in the right-hand side of Equation (1)), it is called a *properly augmented basis.*

Each honest intermediate node in the network processes the incoming packets as follows. Upon receiving vectors $\vec{y}_1, \ldots, \vec{y}_\ell \in \mathbb{Z}_N^n$ on its ℓ incoming edges, it computes a new vector for each outgoing edge as a linear combination of the vectors it received. Namely, at the jth outgoing edge, the vector $\vec{z}_j \in \mathbb{Z}_N^n$ will have the form $\vec{z}_j = \sum_{i=1}^{\ell} \alpha_{i,j} \vec{y}_i$, for some (typically random) coefficients $(\alpha_{1,j}, \ldots, \alpha_{\ell,j}) \in \mathbb{Z}_N^\ell$.

A target node will recover the file using a set of vectors from its incoming edges. This can be done if they consist of m vectors $\{\vec{y}_i = (\vec{x}_i || \vec{u}_i)\}_{i=1}^m$ where $\vec{u}_1, \ldots, \vec{u}_m$ are linearly independent (here, $\vec{x}_i \in \mathbb{Z}_N^k, \vec{u}_i \in \mathbb{Z}_N^m$). The original file is then recovered as

$$\begin{pmatrix} —\vec{v}_1— \\ —\vec{v}_2— \\ \vdots \\ —\vec{v}_m— \end{pmatrix} = \begin{pmatrix} —\vec{u}_1— \\ —\vec{u}_2— \\ \vdots \\ —\vec{u}_m— \end{pmatrix}^{-1} \begin{pmatrix} —\vec{x}_1— \\ —\vec{x}_2— \\ \vdots \\ —\vec{x}_m— \end{pmatrix},$$

which is computable thanks to to the linear independence of $\vec{u}_1, \ldots, \vec{u}_m$.

2.2 Definitions

We first recall the definition of network coding signatures from [7].

Definition 1. *A network coding signature (NCS) scheme consists of a triple of efficient algorithms* $\Sigma = (\mathsf{Keygen}, \mathsf{Sign}, \mathsf{Verify})$ *with the following specifications.*

Keygen(λ, n)**:** *is a probabilistic algorithm that takes as input a security parameter* $\lambda \in \mathbb{N}$ *and an integer* $n \in \mathsf{poly}(\lambda)$ *denoting the length of vectors to be signed. It outputs a positive integer* $N \in \mathbb{N}$*, a public key* pk*, the corresponding private key* sk *and the description of an efficiently samplable file identifier space* $\mathcal{I}$*.*

Sign$(\mathsf{sk}, \mathsf{id}, V)$**:** *is a (possibly probabilistic) algorithm that takes as input a private key* sk*, a file identifier* $\mathsf{id} \in \mathcal{I}$ *and a vector subspace* V *(described as a set of linearly independent vectors* $\vec{v}_1, \ldots, \vec{v}_m \in \mathbb{Z}_N^n$*) of dimension* $m < n$*. It outputs a signature* σ*.*

Verify$(\mathsf{pk}, \mathsf{id}, \vec{y}, \sigma)$**:** *is a deterministic algorithm that takes as input a public key* pk*, a file identifier* $\mathsf{id} \in \mathcal{I}$*, a vector* $\vec{y}$ *and a signature* σ*. It outputs* 1 *or* 0*.*

Correctness requires that, for all $\lambda \in \mathbb{N}$, all integers $n \in \mathsf{poly}(\lambda)$ and all triples $(\mathsf{pk}, \mathsf{sk}, \mathcal{I}) \leftarrow \mathsf{Keygen}(\lambda, n)$, it holds that for all $\mathsf{id} \in \mathcal{I}$ and all vector subspace $V \subset \mathbb{Z}_N^n$, if $\sigma = \mathsf{Sign}(\mathsf{sk}, \mathsf{id}, V)$, then $\mathsf{Verify}(\mathsf{pk}, \mathsf{id}, \vec{y}, \sigma) = 1$ for all $\vec{y} \in V$.

In what follows, we define homomorphic network coding signature schemes. Unlike previous homomorphic schemes [7,12], the construction in this paper uses a probabilistic signing algorithm. To make it possible to publicly combine signatures on distinct vectors from the same file, the signer has to re-use part of his random coins to sign all vectors of the subspace. As long as these signatures are generated using the appropriate coins, network nodes can always combine them. However, attention must be paid to the fact that anyone can attempt to re-randomize signatures so as to prevent them from being combinable later on and disrupt the system. For this reason, network nodes have to make sure that valid signatures of vectors from the same file were produced using compatible randomness before combining them. We thus slightly modify the specification of homomorphic NCS schemes [7] and add a compatibility-checking algorithm that allows testing whether signatures are indeed combinable.

Definition 2. *A homomorphic network coding signature scheme is a tuple of efficient algorithms* $\Sigma = (\mathsf{Keygen}, \mathsf{Sign}, \mathsf{CompatibilityCheck}, \mathsf{Combine}, \mathsf{Verify})$

Keygen(λ, n)**:** *is a probabilistic algorithm that takes as input a security parameter* $\lambda \in \mathbb{N}$ *and an integer* $n \in \mathsf{poly}(\lambda)$ *denoting the length of vectors to be signed. It outputs a key pair* $(\mathsf{pk}, \mathsf{sk})$ *and the description of a file identifier space* $\mathcal{I}$*.*

Sign$(\mathsf{sk}, \mathsf{id}, \vec{v})$**:** *is a possibly randomized algorithm that takes in a private key* sk*, a file identifier* $\mathsf{id} \in \mathcal{I}$ *and a vector* $\vec{v}$*. It outputs a signature* σ*.*

CompatibilityCheck$(\mathsf{pk}, \mathsf{id}, \{\sigma_i\}_{i=1}^{\ell})$**:** *takes as input a public key* pk*, a file identifier* id *and a set of* ℓ *signatures* $\{\sigma_i\}_{i=1}^{\ell}$*. It outputs* 1 *if these signatures are deemed compatible for combination and* 0 *otherwise.*

Combine$(\mathsf{pk}, \mathsf{id}, \{(\beta_i, \sigma_i)\}_{i=1}^{\ell})$: *is a (possibly randomized) algorithm that takes as input a public key* pk, *a file identifier* id *and* ℓ *tuples* (β_i, σ_i), *each one of which consists of a weight* β_i *and a signature* σ_i. *Intuitively, the output is a signature* σ *on the vector* $\vec{y} = \sum_{i=1}^{\ell} \beta_i \vec{v}_i$, *where* σ_i *is a signature on* $\vec{v}_i$.

Verify$(\mathsf{pk}, \mathsf{id}, \vec{y}, \sigma)$: *is a deterministic algorithm that takes as input a public key* pk, *a file identifier* $\mathsf{id} \in \mathcal{I}$, *a signature* σ *and a vector* $\vec{y}$. *It outputs* 0 *or* 1.

Correctness is formulated by mandating that, for all security parameters $\lambda \in \mathbb{N}$, all integers $n \in \mathsf{poly}(\lambda)$ and all triples $(\mathsf{pk}, \mathsf{sk}, \mathcal{I}) \leftarrow \mathsf{Keygen}(\lambda, n)$, the following holds.

1. For all $\mathsf{id} \in \mathcal{I}$ and all n-vectors $\vec{y}$, if $\sigma = \mathsf{Sign}(\mathsf{sk}, \mathsf{id}, \vec{y})$, then we necessarily have $\mathsf{Verify}(\mathsf{pk}, \mathsf{id}, \vec{y}, \sigma) = 1$.
2. For all $\mathsf{id} \in \mathcal{I}$, any $\ell > 0$ and any set of vectors $\{\vec{v}_i\}_{i=1}^{\ell}$, if $\sigma_i = \mathsf{Sign}(\mathsf{sk}, \mathsf{id}, \vec{v}_i)$ for $i = 1$ to ℓ, then $\mathsf{CompatibilityCheck}(\mathsf{pk}, \mathsf{id}, \{\sigma_i\}_{i=1}^{\ell}) = 1$.
3. For all $\mathsf{id} \in \mathcal{I}$, any $\ell > 0$ and any set of triples $\{(\beta_i, \sigma_i, \vec{v}_i)\}_{i=1}^{\ell}$, if the following two conditions are satisfied
   a. $\mathsf{Verify}(\mathsf{pk}, \mathsf{id}, \vec{v}_i, \sigma_i) = 1$ for each $i \in \{1, \ldots, \ell\}$,
   b. $\mathsf{CompatibilityCheck}(\mathsf{pk}, \mathsf{id}, \{\sigma_i\}_{i=1}^{\ell}) = 1$,

   then it must hold that $\mathsf{Verify}\big(\mathsf{pk}, \mathsf{id}, \mathsf{Combine}(\mathsf{pk}, \mathsf{id}, \{(\beta_i, \sigma_i)\}_{i=1}^{\ell}), \sigma\big) = 1$.

In the following, we say that signatures $\{\sigma_i\}_{i=1}^{\ell}$ are *compatible* if they correspond to the same $\mathsf{id} \in \mathcal{I}$ and if it holds that $\mathsf{CompatibilityCheck}(\mathsf{pk}, \mathsf{id}, \{\sigma_i\}_{i=1}^{\ell}) = 1$.

When $\{\sigma_i\}_{i=1}^{\ell}$ is a set of compatible signatures, we say that σ is compatible with $\{\sigma_i\}_{i=1}^{\ell}$ if $\{\sigma\} \cup \{\sigma_i\}_{i=1}^{\ell}$ forms a set of compatible signatures. In particular, when a signature $\tilde{\sigma}$ of a subspace V consists of signatures $(\sigma_1, \ldots, \sigma_m)$ on independent vectors $\vec{v}_1, \ldots, \vec{v}_m \in V$, we say that σ is compatible with $\tilde{\sigma}$ if it is compatible with all $\{\sigma_i\}_{i=1}^{m}$.

Conversion. We recall how a homomorphic network coding signature allows signing vector subspaces, as noted in [7]. Let scheme $\Sigma_2 = (\mathsf{Keygen}_2, \mathsf{Sign}_2, \mathsf{CompatibilityCheck}_2, \mathsf{Combine}_2, \mathsf{Verify}_2)$ be a homomorphic NCS scheme. An ordinary network coding signature $\Sigma_1 = (\mathsf{Keygen}_1, \mathsf{Sign}_1, \mathsf{Verify}_1)$ can be obtained as follows.

$\mathsf{Keygen}_1(\lambda, n) = \mathsf{Keygen}_2(\lambda, n)$

$\mathsf{Sign}_1(\mathsf{sk}, \mathsf{id}, V) = (\sigma_1, \ldots, \sigma_m)$, where $\sigma_i = \mathsf{Sign}_2(\mathsf{sk}, \mathsf{id}, \vec{v}_i)$ for $i = 1$ to m and $\vec{v}_1, \ldots, \vec{v}_m$ is a properly augmented basis of $V \subseteq \mathbb{Z}_N^n$.

$\mathsf{Verify}_1(\mathsf{pk}, \mathsf{id}, \vec{y}, \sigma) =$ outputs 1 if and only if

$$\begin{cases} \mathsf{CompatibilityCheck}_2(\mathsf{pk}, \mathsf{id}, \{\sigma_i\}_{i=1}^{m}) = 1 \\ \mathsf{Verify}_2(\mathsf{pk}, \mathsf{id}, \vec{y}, \mathsf{Combine}_2(\mathsf{pk}, \mathsf{id}, \{(y_{n-m+i}, \sigma_i)\}_{i=1}^{m})) = 1. \end{cases}$$

Security. The security definition hereafter slightly generalizes the one of [7]. It requires that it be infeasible to publicly destroy the "combinability" of valid signatures without rendering them invalid when they are considered individually.

Our goal is to guarantee that, if valid signatures of several vectors from the same file have incompatible randomness, the signer is necessarily deviating from the specification of the scheme. When such a bogus or misbehaving signer is detected, honest network nodes may simply stop processing their packets.

Definition 3. *A network coding signature scheme $\Sigma = (\mathsf{Keygen}, \mathsf{Sign}, \mathsf{Verify})$ is secure if no probabilistic polynomial time (PPT) adversary has non-negligible advantage (as a function of the security parameter $\lambda \in \mathbb{N}$) in the following game:*

1. *The adversary $\mathcal{A}$ chooses an integer $n \in \mathbb{N}$ and sends it to the challenger who runs $\mathsf{Keygen}(\lambda, n)$ and obtains $(\mathsf{pk}, \mathsf{sk})$ before sending pk to $\mathcal{A}$.*
2. *On polynomially-many occasions, $\mathcal{A}$ chooses a linear subspace $V_i \subset \mathbb{Z}_N^n$ of dimension $m_i < n$. The challenger replies by choosing a file identifier $\mathsf{id}_i \in \mathcal{I}$ from the identifier space $\mathcal{I}$ and returns id_i and $\sigma_i = \mathsf{Sign}(\mathsf{sk}, \mathsf{id}_i, V_i)$ to $\mathcal{A}$.*
3. *$\mathcal{A}$ outputs an identifier $\mathsf{id}^\star$, a signature $\sigma^\star$ and a vector $\vec{y} \in \mathbb{Z}_N^n$. The adversary $\mathcal{A}$ is deemed successful if $\mathsf{Verify}(\mathsf{pk}, \mathsf{id}^\star, \vec{y}^\star, \sigma^\star) = 1$ and either of the following holds:*
   - *(Class i): $\mathsf{id}^\star \neq \mathsf{id}_i$ for any i and $\vec{y}^\star \neq \vec{0}$.*
   - *(Class ii): $\mathsf{id}^\star = \mathsf{id}_i$ for some $i \in \{1, \ldots, q\}$ and the signature $\sigma^\star$ is not compatible with σ_i.*
   - *(Class iii): $\mathsf{id}^\star = \mathsf{id}_i$ for some $i \in \{1, \ldots, q\}$ and $\vec{y}^\star \notin V_i$.*

$\mathcal{A}$'s advantage is defined as his probability of victory taken over all coin tosses.

As in [7], a homomorphic NCS scheme Σ' is said to be secure if the network coding signature constructed via the conversion presented above is secure.

2.3 Complexity Assumptions

We consider groups $(\mathbb{G}, \mathbb{G}_T)$ of composite order $N = p_1p_2p_3$ for which a bilinear map $e : \mathbb{G} \times \mathbb{G} \to \mathbb{G}_T$ is computable. In the following, for each $i \in \{1, 2, 3\}$, we denote by $\mathbb{G}_{p_i}$ the subgroup of order p_i. Also, for all distinct $i, j \in \{1, 2, 3\}$, we call $\mathbb{G}_{p_ip_j}$ the subgroup of order p_ip_j.

An important property of composite order groups is that pairing two elements of order p_i and p_j, with $i \neq j$, always gives the identity element $1_{\mathbb{G}_T}$.

In this setting, we rely on the following assumptions introduced in [17].

Assumption 1. Given $g \stackrel{R}{\leftarrow} \mathbb{G}_{p_1}, X_3 \stackrel{R}{\leftarrow} \mathbb{G}_{p_3}$, and T, it is infeasible to efficiently decide if $T \in_R \mathbb{G}_{p_1p_2}$ or $T \in_R \mathbb{G}_{p_1}$.

Assumption 2. Let $g, X_1 \stackrel{R}{\leftarrow} \mathbb{G}_{p_1}, X_2, Y_2 \stackrel{R}{\leftarrow} \mathbb{G}_{p_2}, Y_3, Z_3 \stackrel{R}{\leftarrow} \mathbb{G}_{p_3}$. Given a tuple (g, X_1X_2, Z_3, Y_2Y_3) and T, it is hard to decide if $T \in_R \mathbb{G}$ or $T \in_R \mathbb{G}_{p_1p_3}$.

Assumption 3. Let $g \stackrel{R}{\leftarrow} \mathbb{G}_{p_1}$, $X_2, Y_2, Z_2 \stackrel{R}{\leftarrow} \mathbb{G}_{p_2}$, $X_3 \stackrel{R}{\leftarrow} \mathbb{G}_{p_3}$ and $\alpha, s \stackrel{R}{\leftarrow} \mathbb{Z}_N$. Given $(g, g^\alpha X_2, X_3, g^s Y_2, Z_2)$, it is infeasible to compute $e(g, g)^{\alpha s}$.

We note that, while Lewko and Waters rely on the decisional variant of Assumption 3 (according to which $e(g, g)^{\alpha s}$ is indistinguishable from a random element of $\mathbb{G}_T$), its computational counterpart suffices here.

3 Homomorphic NCS Scheme in the Standard Model

Intuitively, the construction is based on an observation that network coding signatures can be seen as an implication of the spatial encryption primitive introduced by Boneh and Hamburg [8] in the same way as identity-based encryption implies digital signatures (according to an observation by Naor reported in [6]). In spatial encryption, private keys are associated with affine subspaces while ciphertexts correspond to vectors. Decryption is possible when the ciphertext's vector lies in the subspace of the key. By applying Naor's transformation to the spatial encryption scheme of [8], one readily obtains a sort of selectively secure network coding signature, modulo some twist to bind the file identifier to the subspace which is being signed. By itself, this transformation does not provide the homomorphic property that we are after. To obtain it, we need to start from a specific variant of the NCS scheme derived from the spatial encryption system of [8] and carefully re-use the same randomness to separately sign vectors of the same subspace. Full security (as opposed to selective security [5]) is obtained using the Lewko-Waters techniques to build (hierarchical) identity-based encryption schemes [17].

More precisely, the public key comprises the description of bilinear groups $(\mathbb{G}, \mathbb{G}_T)$ of order $N = p_1p_2p_3$, a number of $\mathbb{G}_{p_1}$ elements $(g, u, \{h_i\}_{i=0}^n)$ as well as $e(g, g)^\alpha$ for some $\alpha \stackrel{R}{\leftarrow} \mathbb{Z}_N$. The first two components of each signature form a selectively-secure Boneh-Boyen signature [5] $(\sigma_1, \sigma_2) = (g^\alpha \cdot (u \cdot h_0^{\mathsf{id}})^r, g^r)$ on the file identifier id. As implicitly showed in [17], this signature can be proved fully secure if g, u and h_0 live in the subgroup of order p_1 and if σ_1, σ_2 are multiplied by a random element of $\mathbb{G}_{p_3}$. This signature (σ_1, σ_2) is then augmented with an element $\sigma_3 = (\prod_{j=1}^n h_j^{v_j})^r$, where $(v_1, \ldots, v_n) \in \mathbb{Z}_N^n$ is the vector to be signed. If all the vectors of $\mathrm{span}(\vec{v}_1, \ldots, \vec{v}_m)$ were signed altogether (by introducing one σ_3 per base vector), signatures would have nearly the same shape as private keys in the spatial encryption scheme of [8]: the only difference is the introduction of a file identifier in σ_1. Fortunately, base vectors can be signed separately as long as they are signed using the same exponent r. In this case, anyone can publicly compute a signature on any linear combination of $\vec{v}_1, \ldots, \vec{v}_m$.

To save the signer from maintaining a state and remember which random exponents were used to sign the vectors of all subspaces, $r \in \mathbb{Z}_N$ can be derived by applying a pseudorandom function to the file identifier id so as to be re-computable later on. We emphasize that the use of a PRF is not meant to de-randomize the scheme in an attempt to obtain unique signatures. The goal is simply to render the signer stateless.

To achieve security in the sense of definition 3, we need to keep signatures from being publicly re-randomizable in their $\mathbb{G}_{p_1}$ components. A simple solution is to compute (σ_1, σ_2) as a signature on a hash value of both id and $e(g, g)^r$, which prevents from altering the underlying r without invalidating the signature. Although this simple trick would not work with Waters signatures [23] (because their security proof would cease to go through), it is compatible with the dual encryption technique [24,17] which is used to prove security. In addition, anyone can detect if vectors of the same file are signed using different values of r and only the signer can be blamed in such a situation.

3.1 Construction

Keygen(λ, n)**:** given a security parameter $\lambda \in \mathbb{N}$ and an integer $n \in \mathsf{poly}(\lambda)$, choose bilinear groups $(\mathbb{G}, \mathbb{G}_T)$ of order $N = p_1 p_2 p_3$, where $p_i > 2^\lambda$ for each $i \in \{1,2,3\}$. Choose $\alpha \xleftarrow{R} \mathbb{Z}_N$, $g \xleftarrow{R} \mathbb{G}_{p_1}$, $X_{p_3} \xleftarrow{R} \mathbb{G}_{p_3}$, $b, a_i \xleftarrow{R} \mathbb{Z}_N$ for $i = 0$ to n. Then, select a collision-resistant hash function $H : \{0,1\}^* \to \mathbb{Z}_N$, an identifier space $\mathcal{I}$ and pick a random seed $\kappa \xleftarrow{R} \{0,1\}^\tau$ for a pseudorandom function $\Psi : \{0,1\}^\tau \times \mathcal{I} \to \mathbb{Z}_N$, where $\tau \in \mathsf{poly}(\lambda)$. The private key is $\mathsf{sk} := \big(g^\alpha, \kappa\big)$ while the public key is

$$\mathsf{pk} := \Big((\mathbb{G}, \mathbb{G}_T),\ g,\ e(g,g)^\alpha,\ u = g^b,\ \{h_i = g^{a_i}\}_{i=0,\ldots,n},\ X_{p_3},\ H\Big).$$

Sign$(\mathsf{sk}, \mathsf{id}, \vec{v})$**:** on input of a vector $\vec{v} = (v_1, \ldots, v_n) \in \mathbb{Z}_N^n$, a file identifier $\mathsf{id} \in \mathcal{I}$ and the private key $\mathsf{sk} = \big(g^\alpha, \kappa\big)$, conduct the following steps. First, compute a pseudorandom scalar $r = \Psi(\kappa, \mathsf{id}) \in \mathbb{Z}_N$. Then, compute $\mathsf{id}' = H(\mathsf{id}, e(g,g)^r) \in \mathbb{Z}_N$, choose $R_3, R_3', R_3'' \xleftarrow{R} \mathbb{G}_{p_3}$ and compute a signature $\sigma = (\sigma_1, \sigma_2, \sigma_3)$ as

$$\sigma_1 = g^\alpha \cdot (u \cdot h_0^{\mathsf{id}'})^r \cdot R_3, \qquad \sigma_2 = g^r \cdot R_3', \qquad \sigma_3 = \big(h_1^{v_1} \cdots h_n^{v_n}\big)^r \cdot R_3''$$

CompatibilityCheck$(\mathsf{pk}, \mathsf{id}, \{\sigma_i\}_{i=1}^{\ell})$**:** parses σ_i as $(\sigma_{i,1}, \sigma_{i,2}, \sigma_{i,3}) \in \mathbb{G}^3$ for $i = 1$ to ℓ. The algorithm will return 1 if and only if all $\sigma_{i,2}$ have the same $\mathbb{G}_{p_1}$ component: for $i = 2$ to ℓ, it checks if $e(\sigma_{1,2}/\sigma_{i,2}, g) = 1_{\mathbb{G}_T}$ and returns 0 otherwise. If all checks succeed, it returns 1.

Combine$(\mathsf{pk}, \mathsf{id}, \{(\beta_i, \sigma_i)\}_{i=1}^{\ell})$**:** given pk, a file identifier id and ℓ tuples (β_i, σ_i), parse σ_i as $\sigma_i = (\sigma_{i,1}, \sigma_{i,2}, \sigma_{i,3})$ for $i = 1$ to ℓ. Set $\sigma_1 = \sigma_{1,1} \cdot R_3$, $\sigma_2 = \sigma_{1,2} \cdot R_3'$ for randomly chosen $R_3, R_3' \xleftarrow{R} \mathbb{G}_{p_3}$. Then, compute $\sigma_3 = \prod_{i=1}^{\ell} \sigma_{i,3}^{\beta_i} \cdot R_3''$, with $R_3'' \xleftarrow{R} \mathbb{G}_{p_3}$, and output $(\sigma_1, \sigma_2, \sigma_3)$.

Verify$(\mathsf{pk}, \mathsf{id}, \vec{y}, \sigma)$**:** given a public key $\mathsf{pk} = \big(g, e(g,g)^\alpha, u, \{h_i\}_{i=0}^n, X_{p_3}\big)$, a signature $\sigma = (\sigma_1, \sigma_2, \sigma_3)$ and a vector $\vec{y} = (y_1, \ldots, y_n) \in (\mathbb{Z}_N)^n$, compute $\mathsf{id}' = H(\mathsf{id}, e(\sigma_2, g))$ and return 1 if and only if

$$e(\sigma_1, g) = e(g,g)^\alpha \cdot e(u \cdot h_0^{\mathsf{id}'}, \sigma_2) \quad \text{and} \quad e(\sigma_3, g) = e(\sigma_2, h_1^{y_1} \cdots h_n^{y_n}). \qquad (2)$$

Verifying the correctness of the scheme is straightforward since pairing an element of $\mathbb{G}_{p_1}$ with an element of $\mathbb{G}_{p_3}$ always gives the identity element in $\mathbb{G}_T$.

Efficiency. Signatures only consist of 3 group elements. Without optimizations, verifying individual signatures entails to compute four pairings. However, when multiple signatures must be checked before being combined, a constant number of pairing evaluations suffices when randomized batch verification techniques are used.

Indeed, when network nodes process ℓ signatures $\{(\sigma_{i,1}, \sigma_{i,2}, \sigma_{i,3})\}_{i=1}^{\ell}$ from the same file identified by $\mathsf{id} \in \mathcal{I}$, they can first check that all $\{\sigma_{i,2}\}_{i=1}^{\ell}$ have the same $\mathbb{G}_{p_1}$ component by testing if $e\big(\prod_{i=2}^{\ell}(\sigma_{1,2}/\sigma_{i,2})^{\omega_i}, g\big) = 1_{\mathbb{G}_T}$ for randomly chosen $\omega_2, \ldots, \omega_\ell \stackrel{R}{\leftarrow} \mathbb{Z}_N$. If this test is satisfied, $\sigma_{1,2}, \ldots, \sigma_{\ell,2}$ all correspond to the same r, with overwhelming probability, and the same $\sigma_{1,2}$ can be used to verify equations (2) for $i = 1$ to ℓ. Namely, if $\sigma_i = (\sigma_{i,1}, \sigma_{i,2}, \sigma_{i,3})$ pertains to $\vec{v}_i = (v_{i,1}, \ldots, v_{i,n})$, signatures are all valid if $e(\sigma_{i,1}, g) = e(g, g)^{\alpha} \cdot e(u \cdot h_0^{\mathsf{id}'}, \sigma_{1,2})$ and $e(\sigma_{i,3}, g) = e(\sigma_{1,2}, \prod_{k=1}^{n} h_k^{v_{i,k}})$ for $i = 1$ to ℓ. Then, if the network node picks randomizers $\delta_i, \delta_i' \stackrel{R}{\leftarrow} \mathbb{Z}_N$, for $i = 1$ to ℓ, all signatures can be batch-verified by testing if

$$e(g, \prod_{i=1}^{\ell} \sigma_{i,1}^{\delta_i} \cdot \prod_{j=1}^{\ell} \sigma_{j,3}^{\delta_j'}) = e(g, g)^{\alpha \cdot \sum_{i=1}^{\ell} \delta_i} \cdot e\big(\sigma_{1,2}, (u \cdot h_0^{\mathsf{id}'})^{\sum_{i=1}^{\ell} \delta_i} \cdot \prod_{k=1}^{n} h_k^{\sum_{j=1}^{\ell} \delta_j' v_{jk}}\big).$$

When verification fails, recent techniques [21] can be adapted to determine which signatures are bad and which packets should be filtered.

As in earlier standard model constructions [7,3], the public key size is linear in the dimension n of vectors. We leave it as an interesting open problem to avoid this dependency without resorting to random oracles.

Converted Scheme. From the homomorphic network coding signature, one can obtain an ordinary network coding signature via the generic conversion given by Boneh *et al.* [7] (and recalled in section 2.2). Applying this conversion to our scheme results in the signature of the form $\{(\sigma_{i,1}, \sigma_{i,2}, \sigma_{i,3})\}_{i=1}^{m}$. This scheme is redundant and we can reuse the first two elements for all i. Indeed to sign a subspace V where $\vec{v}_1, \ldots, \vec{v}_m$ is the properly augmented basis, the signing algorithm outputs $\sigma = (\sigma_1, \sigma_2, \{\sigma_{3,i}\}_{i=1}^{m})$ where

$$\sigma_1 = g^{\alpha} \cdot (g^{b + a_0 \mathsf{id}'})^r \cdot R_3, \qquad \sigma_2 = g^r \cdot R_3', \qquad \sigma_{3,i} = (g^{\langle \vec{a}, \vec{v}_i \rangle})^r \cdot R_{3,i}'',$$

where $R_3, R_3', R_{3,i}'' \in_R \mathbb{G}_{p_3}$ and we denote $\vec{a} = (a_1, \ldots, a_n)$. In the next section, we will prove the security of this scheme instead of the scheme converted with the generic conversion.

3.2 Security Proof

We first give a simple lemma describing the general form of signatures that are accepted by the verification of the proposed NCS scheme (with redundancy cut as mentioned above).

Lemma 1. *For any identifier-vector-signature tuple* $(\mathsf{id}, \vec{y}, \sigma = (\sigma_1, \sigma_2, \{\sigma_{3,i}\}_{i=1}^{m}))$, *if it holds that* $\mathsf{Verify}(\mathsf{pk}, \mathsf{id}, \vec{y}, \sigma) = 1$, *then we have*

$$\sigma_1 = g^{\alpha} \cdot (g^{b + a_0 \mathsf{id}'})^r \cdot Z_1, \qquad \sigma_2 = g^r \cdot Z_2, \qquad \sigma_{3,i} = (g^{\langle \vec{a}, \vec{v}_i \rangle})^r \cdot Z_{3,i}, \qquad (3)$$

where $\mathsf{id}' = H(\mathsf{id}, e(\sigma_2, g))$*, for some* $r \in \mathbb{Z}_N$*,* $Z_1, Z_2, Z_{3,i} \in \mathbb{G}_{p_2p_3}$ *and some vectors* $\vec{v}_1, \dots, \vec{v}_m \in \mathbb{Z}_N^n$ *such that*

$$\vec{a}(U(\vec{y}^R)^\top - \vec{y}^\top) = 0, \text{ where } U = \begin{pmatrix} | & & | \\ \vec{v}_1^\top & \cdots & \vec{v}_m^\top \\ | & & | \end{pmatrix} \tag{4}$$

where we write $\vec{y} = \vec{y}^L || \vec{y}^R$ *with* $\vec{y}^L \in \mathbb{Z}_N^{n-m}, \vec{y}^R \in \mathbb{Z}_N^m$*.*

Proof. Let an id-vector-signature tuple $(\mathsf{id}, \vec{y}, \sigma = (\sigma_1, \sigma_2, \{\sigma_{3,i}\}_{i=1}^m))$ be a valid tuple, that is $\mathsf{Verify}(\mathsf{pk}, \mathsf{id}, \vec{y}, \sigma) = 1$. We will prove that σ will have the form of equation (3). First, since the tuple is accepted, we have

$$e(\sigma_1, g) = e(g, g)^\alpha \cdot e(g^b \cdot (g^{a_0})^{\mathsf{id}'}, \sigma_2) \tag{5}$$

$$e(\prod_{i=1}^m \sigma_{3,i}^{y_{n-m+i}}, g) = e(\sigma_2, g^{\langle \vec{a}, \vec{y} \rangle}), \tag{6}$$

where $\mathsf{id}' = H(\mathsf{id}, e(\sigma_2, g))$. Since $\sigma_2 \in \mathbb{G}$, we can write $\sigma_2 = g^r Z_2$ for some $r \in \mathbb{Z}_N$ and $Z_2 \in \mathbb{G}_{p_2p_3}$. Equation (5) then implies $\sigma_1 = g^\alpha \cdot (g^{b+a_0\mathsf{id}'})^r \cdot Z_1$ for some $Z_1 \in \mathbb{G}_{p_2p_3}$, as claimed. Similarly, we have $\sigma_{3,i} = (g^{\beta_i})^r \cdot Z_{3,i}$ for some $\beta_i \in \mathbb{Z}_N$. It remains to prove the property of β_i. Equation (6) implies that $\sum_{i=1} \beta_i y_{n-m+i} = \langle \vec{a}, \vec{y} \rangle$. If we write $\beta_i = \langle \vec{a}, \vec{v}_i \rangle$ for some $\vec{v}_i \in \mathbb{Z}_N^n$, then the equation (4) is obtained. This concludes the proof. □

Theorem 1. *The scheme is a secure homomorphic network coding signature if* Ψ *is a secure pseudorandom function, if* H *is a collision-resistant hash function and if Assumption 1, Assumption 2 and Assumption 3 all hold.*

Proof. The proof follows the dual system methodology used in [24,17]. From Lemma 1, any valid identifier-vector-signature triple $(\mathsf{id}, \vec{y}, \sigma)$ will have the following generic form:

$$\sigma_1 = g^\alpha \cdot (g^{b+a_0\mathsf{id}'})^r \cdot g_2^{w_1} \cdot R_1, \qquad \sigma_2 = g^r \cdot g_2^{w_2} \cdot R_2, \tag{7}$$

$$\sigma_{3,i} = (g^{\langle \vec{a}, \vec{v}_i \rangle})^r \cdot g_2^{w_{3,i}} \cdot R_{3,i}, \tag{8}$$

where $\mathsf{id}' = H(\mathsf{id}, e(\sigma_2, g))$, for some $r \in \mathbb{Z}_N$, $w_1, w_2, w_{3,i} \in \mathbb{Z}_N$, some group elements $R_1, R_2, R_{3,i} \in \mathbb{G}_{p_3}$ and vectors $\vec{v}_1, \dots, \vec{v}_m \in \mathbb{Z}_N^n$ which satisfied Eq. (4).

We will distinguish two types of signatures as follows.

- Type A: $(w_1, w_2, w_{3,1}, \dots, w_{3,n}) = (0, 0, 0, \dots, 0) \bmod p_2$.
- Type B: $(w_1, w_2, w_{3,1}, \dots, w_{3,n}) \neq (0, 0, 0, \dots, 0) \bmod p_2$.

We will call Type A forgery (resp. Type B forgery) a fake signature of Type A (resp. Type B) which is produced by the forger in the game of definition 3.

The proof considers a sequence of $q + 3$ games. It starts with the real attack game Game_{real} followed by $\mathrm{Game}_1, \mathrm{Game}_2, \mathrm{Game}_3, \mathrm{Game}_{4.1}, \dots, \mathrm{Game}_{4.q}$. In the following we let $V^{(j)}$ be the j-th query where $j \in \{1, \dots, q\}$ and let $(\sigma_1^{(j)}, \sigma_2^{(j)}, \{\sigma_{3,i}^{(j)}\}_{i=1}^m)$ be the answer to the query.

Game$_1$: REPLACING r WITH RANDOM. This game is identical to as Game$_{real}$ with the difference that the challenger generates all signatures using truly random exponents $r \stackrel{R}{\leftarrow} \mathbb{Z}_N$ (and care is taken to use the same r to sign all vectors of the same subspace) instead of pseudorandom values. Clearly, the security of the PRF implies that Game$_1$ is computationally indistinguishable from Game$_{real}$.

Game$_2$: ELIMINATING COLLISION. It is as Game$_1$ but the game will abort if

- Adversary $\mathcal{A}$ outputs a class-(i) forgery (*i.e.*, $\mathsf{id}^\star \neq \mathsf{id}_j$ for any j and $\vec{y}^\star \neq \vec{0}$) or a class-(ii) forgery (*i.e.*, $\mathsf{id}^\star = \mathsf{id}_j$ for some $j \in \{1, \ldots, q\}$ and $e(\sigma_2^\star, g) \neq e(\sigma_2^{(j)}, g)$) but for which $\mathsf{id}'^\star = \mathsf{id}'_j$. In other words, the collision of H occurs as $H(\mathsf{id}^\star, e(\sigma_2^\star, g)) = H(\mathsf{id}_j, e(\sigma_2^{(j)}, g))$, for some index $j \in \{1, \ldots, q\}$.

It is straightforward to show that under the collision-resistance of H the difference between Game$_1$ and Game$_2$ is negligible.

Game$_3$: RESTRICTION MODULO p_2. It is as Game$_2$ but the game will further abort if either of the following event occurs.

- Adversary $\mathcal{A}$ outputs a class-(i) forgery (*i.e.*, $\mathsf{id}^\star \neq \mathsf{id}_j$ for any j and $\vec{y}^\star \neq \vec{0}$) or a class-(ii) forgery (*i.e.*, for which $\mathsf{id}^\star = \mathsf{id}_j$ for some j and $e(\sigma_2^\star, g) \neq e(\sigma_{2,j}, g)$) but $\mathsf{id}'^\star = \mathsf{id}'_j \bmod p_2$ (even if $\mathsf{id}'^\star \neq \mathsf{id}'_j$) for some index $j \in \{1, \ldots, q\}$.
- Adversary $\mathcal{A}$ outputs a class-(iii) forgery (*i.e.*, $\mathsf{id}^\star = \mathsf{id}_j$ for some j and $\vec{y}^\star \notin V_j$) but for which $\vec{y}^\star \bmod p_2 \in V_j \bmod p_2$. Here, we denote by $V \bmod p_2$ the subspace V reduced in $\mathbb{Z}_{p_2}^n$. More precisely, for any subspace $V = \mathrm{span}(\vec{v}_1, \ldots, \vec{v}_m) \subset \mathbb{Z}_N^n$, the notation $V \bmod p_2$ denotes $\mathrm{span}(\vec{v}_1 \bmod p_2, \ldots, \vec{v}_m \bmod p_2) \subset \mathbb{Z}_{p_2}^n$.

Lemma 2 shows that, under Assumption 1 and Assumption 2, the difference between Game$_2$ and Game$_3$ is negligible. Then, Lemma 3 shows that, if $\mathcal{A}$ can output a Type B forgery in Game$_3$, Assumption 1 is false.

Game$_{4.0}$: SIMPLIFICATION. This is a reformulation of Game$_3$ for ease of reading. The game will accept only the following forgery. (Otherwise, it will abort).

- Adversary $\mathcal{A}$ outputs a forgery with $\mathsf{id}'^\star \neq \mathsf{id}'_j \bmod p_2$ for any j and $\vec{y}^\star \neq \vec{0}$.
- Adversary $\mathcal{A}$ outputs a forgery for which $\mathsf{id}^\star = \mathsf{id}_j$, for some $j \in \{1, \ldots, q\}$, and $\vec{y}^\star \bmod p_2 \notin V_j \bmod p_2$.

We note that in this game, as in Game$_{real}$, $\mathcal{A}$ is only given Type A signatures.

Game$_{4.k}$ ($1 \leq k \leq q$): HYBRID TYPES. It is as Game$_0$ but the adversary obtains Type B signatures at the first k signing queries whereas the challenger answers the remaining $q - k$ signing queries by returning Type A signatures. Lemma 4 shows that, if the adversary has noticeably higher probability to output a Type A forgery in Game$_{4.(k+1)}$ than in Game$_{4.k}$, there must be a breach in Assumption 2.

Game$_{4.q}$: ALL TYPE B. The forger $\mathcal{A}$ only obtains Type B signatures and it becomes easy to prove that any Type A forgery allows breaking Assumption 3, as shown by Lemma 5.

Denote negl as a negligible function in λ. Let $\mathsf{W}_i, \mathsf{W}_i^{\mathsf{A}}, \mathsf{W}_i^{\mathsf{B}}$ be the probability that the adversary successfully outputs a forgery in game i of either type, type A, and type B respectively. We then have that $|\mathsf{W}_{real} - \mathsf{W}_3| \leq \mathsf{negl}$ guaranteed by the security of PRF, collision-resistance hash, and Lemma 2. Also $\mathsf{W}_3^{\mathsf{B}} \leq \mathsf{negl}$, $|\mathsf{W}_{4,0}^{\mathsf{A}} - \mathsf{W}_{4,q}^{\mathsf{A}}| \leq \mathsf{negl}$, and $\mathsf{W}_{4,q}^{\mathsf{A}} \leq \mathsf{negl}$ are implied by Lemma 3, 4, and 5, respectively. Combining the above, we obtain

$$\mathsf{W}_{real} \leq |\mathsf{W}_{real} - \mathsf{W}_3| + \mathsf{W}_3^{\mathsf{B}} + |\mathsf{W}_{4,0}^{\mathsf{A}} - \mathsf{W}_{4,q}^{\mathsf{A}}| + \mathsf{W}_{4,q}^{\mathsf{A}} \leq \mathsf{negl},$$

where we recall that $\mathsf{W}_3 = \mathsf{W}_{4,0}$ and see that $\mathsf{W}_3 = \mathsf{W}_3^{\mathsf{A}} + \mathsf{W}_3^{\mathsf{B}}$. This concludes the proof. □

Lemma 2. *Any significant difference between the adversary's behaviors in Game$_2$ and Game$_3$ contradicts either Assumption 1 or Assumption 2.*

Proof. The two games are identical unless the adversary $\mathcal{A}$ outputs a forgery involving a pair $(\mathsf{id}'^{\star}, \vec{y}^{\star})$ such that either: (1) $\mathsf{id}'^{\star} = \mathsf{id}'_j \bmod p_2$ whereas we have $\mathsf{id}'^{\star} \neq \mathsf{id}'_j \bmod N$ for some $j \in \{1, \ldots, q\}$; (2) there exists $j \in \{1, \ldots, q\}$ such that $\mathsf{id}'^{\star} = \mathsf{id}'_j \bmod N$ but $\det(M) = 0 \bmod p_2$ and $\det(M) \neq 0 \bmod N$, where $M \in \mathbb{Z}_N^{n \times n}$ is the matrix

$$M = \begin{pmatrix} R_{n\times(n-m_j-1)} & \vec{v}_{j,1}^{\top} & \cdots & \vec{v}_{j,m_j}^{\top} & \vec{y}^{\star\top} \end{pmatrix},$$

with $m_j = \dim(V_j) < n$, such that $R_{n\times(n-m_j-1)}$ is a $n \times (n - m_j - 1)$ matrix whose columns are orthogonal to $\mathrm{span}(\vec{v}_{j,1}, \ldots, \vec{v}_{j,m_j}, \vec{y}^{\star})$ (such a matrix can be obtained via the Gram-Schmidt process). The matrix has the desired properties since $\vec{y}^{\star} \bmod p_2 \in V \bmod p_2$ although $\vec{y}^{\star} \notin V$. The simulator $\mathcal{B}$ can extract a non-trivial factor of N by computing $\gcd(\mathsf{id}'^{\star} - \mathsf{id}'_j, N)$ in case (1) or $\gcd(\det(M), N)$ in case (2). As shown in [17][Lemma 5], this allows breaking either Assumption 1 or Assumption 2 depending on which factor is extracted. □

Lemma 3. *Under Assumption*[1] *1, no PPT adversary can output a Type B forgery in Game$_3$.*

Proof. We show that, if the adversary outputs a Type B forgery in Game$_3$, there is an algorithm $\mathcal{B}$ that, given (g, X_3, T), decides if $T \in_R \mathbb{G}_{p_1}$ or $T \in_R \mathbb{G}_{p_1 p_2}$.

The distinguisher $\mathcal{B}$ sets up the public key pk as $e(g,g)^{\alpha}$, $X_{p_3} = X_3$, $u = g^b$, $h_i = g^{a_i}$ for $i = 0$ to n. Denote $\vec{a} = (a_1, \ldots, a_n)$. It answers all private key queries according to the specification of the signing algorithm since it knows the private key.

[1] We note that the lemma holds under a weaker assumption which is the hardness of finding an element of order p_2 or $p_2 p_3$ given (g, X_3).

At the end, $\mathcal{A}$ outputs a file identifier $\mathsf{id}^\star$, a Type-B signature $(\sigma_1^\star, \sigma_2^\star, \{\sigma_{3,i}^\star\}_{i=1}^m)$ and a vector $\vec{y}^\star$. The algorithm $\mathcal{B}$ then computes

$$\eta_1 = \frac{\sigma_1^\star}{g^\alpha \cdot {\sigma_2^\star}^{b+a_0\mathsf{id}'^\star}}, \qquad \eta_2 = \frac{\prod_{i=1}^m {\sigma_{3,i}^\star}^{y_{n-m+i}}}{{\sigma_2^\star}^{\langle \vec{a}, \vec{y}^\star\rangle}}.$$

The $\mathbb{G}_{p_1}$ components of these terms are necessarily canceled out due to equations (3)-(4). Recall that a Type-B signature is in the generic form (7) with $(w_1, w_2, w_{3,1}, \ldots, w_{3,n}) \neq (0, 0, 0, \ldots, 0) \bmod p_2$. For this reason, the $\mathbb{G}_{p_2}$ components in η_1, η_2 will be $g_2^{w_1 - w_2(b+a_0\mathsf{id}'^\star)}$ and $g_2^{\sum_{i=1}^m w_{3,i}y_{n-m+i} - w_2\langle \vec{a}, \vec{y}^\star\rangle}$, respectively. Hence, as long as $b, a_0, \vec{a} \bmod p_2$ are information theoretically hidden to the adversary, there must be an element of $\mathbb{G}_{p_2p_3}$ with non-trivial $\mathbb{G}_{p_2}$ component among η_1, η_2. But this is true since $b, a_0, \vec{a} \bmod p_2$ is uncorrelated to $b, a_0, \vec{a} \bmod p_1$, which is the only information available from the public key. Therefore, our distinguisher $\mathcal{B}$ can conclude that $T \in \mathbb{G}_{p_1p_2}$ if and only if either $e(T, \eta_1) \neq 1_{\mathbb{G}_T}$ or $e(T, \eta_2) \neq 1_{\mathbb{G}_T}$. □

Lemma 4. *The adversary outputs a Type A forgery with negligibly different probabilities in Game$_{4.k}$ and Game$_{4.(k+1)}$ if Assumption 2 holds.*

Proof. Let us assume that a forger $\mathcal{A}$ has significantly better probability of outputting a Type A forgery in Game$_{4.(k+1)}$ than in Game$_{4.k}$. We outline a distinguisher $\mathcal{B}$ that breaks Assumption 2 with non-negligible advantage.

Algorithm $\mathcal{B}$ takes as input $(g, X_1X_2, Z_3, Y_2Y_3, T)$ and uses its interaction with $\mathcal{A}$ to decide if $T \in \mathbb{G}$ or $T \in \mathbb{G}_{p_1p_3}$. Recall that $\mathcal{A}$ must obtain Type B signatures at her first k signing queries and Type A signatures at the last $q-k-1$ queries. We will simulate the interaction so that the kth-query will be a Type A signature (hence Game$_{4.k}$) if $T \in \mathbb{G}_{p_1p_3}$ and a Type B signature (hence Game$_{4.(k+1)}$) if $T \in \mathbb{G}$. We then show that the distinguisher $\mathcal{B}$ can indeed distinguish whether $\mathcal{A}$'s forgery will be of Type A or Type B with overwhelming probability.

To this end, $\mathcal{B}$ prepares the public key pk by choosing $\alpha, b \stackrel{R}{\leftarrow} \mathbb{Z}_N$, $a_i \stackrel{R}{\leftarrow} \mathbb{Z}_N$, for $i = 0$ to n, and setting $u = g^b$, $h_i = g^{a_i}$ for $i = 0$ to n. The public key $\mathsf{pk} = \{g, e(g,g)^\alpha, u, h_0, h_1, \ldots, h_n, Z_3\}$ is given to $\mathcal{A}$. Then, $\mathcal{B}$ answers $\mathcal{A}$'s queries depending on the index $j \in \{1, \ldots, q\}$ of the query.

[**Case** $j < k$]. To sign the jth vector space $V^{(j)} = \mathrm{span}(\vec{v}_1^{(j)}, \ldots, \vec{v}_m^{(j)})$ chosen by $\mathcal{A}$, $\mathcal{B}$ first chooses a random file identifier $\mathsf{id}_j \stackrel{R}{\leftarrow} \mathcal{I}$ and a random exponent $r \stackrel{R}{\leftarrow} \mathbb{Z}_N$. It then chooses $w_1, w_2, \{w_{3,i}\}_{i=1}^m \stackrel{R}{\leftarrow} \mathbb{Z}_N$, $Z_3, Z_3', \{Z_{3,i}''\}_{i=1}^m \stackrel{R}{\leftarrow} \mathbb{G}_{p_3}$. Let $\mathsf{id}_j' = H(\mathsf{id}_j, e(g,g)^r)$. It finally computes a Type-B signature $(\sigma_1, \sigma_2, \{\sigma_{3,i}\}_{i=1}^m)$ on $V^{(j)}$ as

$$\sigma_1 = g^\alpha \cdot (u \cdot h_0^{\mathsf{id}_j'})^r \cdot (Y_2Y_3)^{w_1} \cdot Z_3, \qquad \sigma_2 = g^r \cdot (Y_2Y_3)^{w_2} \cdot Z_3',$$
$$\sigma_{3,i} = (g^{\langle \vec{a}, \vec{v}_i^{(j)}\rangle})^r \cdot (Y_2Y_3)^{w_{3,i}} \cdot Z_{3,i}''.$$

[**Case** $j > k$]. In this case, $\mathcal{A}$ simply computes a Type A signature using the private key g^α as specified by the signing algorithm (except that, as in Game$_1$, r is chosen at random in $\mathbb{Z}_N$ rather than as a pseudorandom value).

[**Case** $j = k$]. To answer the kth private key query $V^{(j)} = \text{span}(\vec{v}_1^{(j)}, \ldots, \vec{v}_m^{(j)})$, $\mathcal{B}$ first picks a random file identifier $\mathsf{id}_j \xleftarrow{R} \mathcal{I}$. It then chooses $w_1, w_2, \{w_{3,i}\}_{i=1}^m \xleftarrow{R} \mathbb{Z}_N$, $Z_3, Z_3', \{Z_{3,i}''\}_{i=1}^m \xleftarrow{R} \mathbb{G}_{p_3}$. It uses its input T to compute a hash value $\mathsf{id}_j' = H(\mathsf{id}_j, e(T, g))$. It finally computes the signature $(\sigma_1, \sigma_2, \{\sigma_{3,i}\}_{i=1}^m)$ on the subspace $V^{(j)}$ as

$$\sigma_1 = g^{\alpha} \cdot T^{b + a_0 \mathsf{id}_j'} \cdot Z_3, \qquad \sigma_2 = T \cdot Z_3', \qquad \sigma_{3,i} = T^{\langle \vec{a}, \vec{v}_i^{(j)} \rangle} \cdot Z_{3,i}''.$$

It is easy to observe that, in the situation where $T \in_R \mathbb{G}$, if we let g_2^x be the $\mathbb{G}_{p_2}$ component of T for some $x \in \mathbb{Z}_{p_2}^*$, we obtain a Type B signature where $w_1 = x(b + a_0 \mathsf{id}_j') \bmod p_2$, $w_2 = x \bmod p_2$, and $w_{3,i} = x(\langle \vec{a}, \vec{v}^{(j)} \rangle) \bmod p_2$ for $i = 1$ to m. In contrast, if $T \in_R \mathbb{G}_{p_1 p_3}$, the above forms a Type A signature.

At the end of the game, $\mathcal{A}$ outputs a forgery $\sigma^\star = (\sigma_1^\star, \sigma_2^\star, \{\sigma_{3,i}^\star\}_{i=1}^m)$ and a vector $\vec{y}^\star$ and a file identifier $\mathsf{id}^\star$ such that the property stated in the Game_0 holds. That is either

- a forgery with $\mathsf{id}'^\star \neq \mathsf{id}_j' \bmod p_2$ for any j and $\vec{y}^\star \neq \vec{0}$.
- a forgery with $\mathsf{id}^\star = \mathsf{id}_j$ for some j and $\vec{y}^\star \bmod p_2 \notin V^{(j)} \bmod p_2$.

At this stage, $\mathcal{B}$ halts and checks whether the forgery is of Type A or B. If the forgery is of Type A, it returns 0 (meaning that $T \in_R \mathbb{G}_{p_1 p_3}$). If the forgery is believed to be of Type B, $\mathcal{B}$ rather bets on $T \in_R \mathbb{G}_{p_1 p_2 p_3}$ and outputs 1.

In order to decide which kind of forgery $\mathcal{A}$ comes up with, $\mathcal{B}$ uses the input value $X_1 X_2$ as follows. The algorithm $\mathcal{B}$ computes $\mathsf{id}'^\star = H(\mathsf{id}^\star, e(\sigma_2^\star, g))$ and

$$\eta_1 = \frac{\sigma_1^\star}{g^{\alpha} \cdot {\sigma_2^\star}^{b + a_0 \mathsf{id}'^\star}}, \qquad \eta_2 = \frac{\prod_{i=1}^m {\sigma_{3,i}^\star}^{y_{n-m+i}}}{{\sigma_2^\star}^{\langle \vec{a}, \vec{y}^\star \rangle}}.$$

The $\mathbb{G}_{p_1}$ component of each term is canceled out due to equations (3)-(4). If $e(X_1 X_2, \eta_1) = 1$ and $e(X_1 X_2, \eta_2) = 1$, then the algorithm $\mathcal{B}$ deduces that $\sigma^\star$ is of Type A. Otherwise, it is seen as a Type B signature. To see why this test works with overwhelming probability, we note that, since $\sigma^\star$ properly verifies, it must be of the form of equation (7) with $(w_1^\star, w_2^\star, w_{3,1}^\star, \ldots, w_{3,m}^\star)$ so that we have

$$e(X_1 X_2, \eta_1) = e(X_2, g_2)^{w_1^\star - w_2^\star (b + a_0 \mathsf{id}'^\star)},$$
$$e(X_1 X_2, \eta_2) = e(X_2, g_2)^{\sum_{i=1}^m w_{3,i}^\star y_{n-m+i} - w_2^\star (\langle \vec{a}, \vec{y}^\star \rangle)}.$$

If $\sigma^\star$ is of Type B, it can only be interpreted as a Type A signature if and only if

$$w_1^\star - w_2^\star (b + a_0 \mathsf{id}'^\star) = 0 \bmod p_2, \text{ and} \tag{9}$$

$$\sum_{i=1}^m w_{3,i}^\star y_{n-m+i} - w_2^\star (\langle \vec{a}, \vec{y}^\star \rangle) = 0 \bmod p_2. \tag{10}$$

We show that this occurs with negligible probability as follows.

- If the forgery is of the first class, that is $\mathsf{id}^\star \neq \mathsf{id}_j$ for any $j \in \{1,\ldots,q\}$, then $b + a_0\mathsf{id}'^\star \bmod p_2$ is independent of $\mathcal{A}$'s view which consists only of $b + a_0\mathsf{id}_k \bmod p_2$. Therefore equation (9) occurs with negligible probability.
- If the forgery is of the second class, that is $\vec{y}^\star \bmod p_2 \notin V^{(j)} \bmod p_2$ for any j, then $\langle \vec{a}, \vec{y}^\star \rangle \bmod p_2$ is independently of $\mathcal{A}$'s view. Indeed, let us consider what $\mathcal{A}$ knows in the information theoretic sense about the values $(a_1,\ldots,a_n)$ taken modulo p_2. It amounts to the right-hand side of the following system of linear equations:

$$\begin{pmatrix} -\vec{v}_1^{(k)}- \\ \vdots \\ -\vec{v}_m^{(k)}- \end{pmatrix} \begin{pmatrix} a_1 \\ \vdots \\ a_n \end{pmatrix} = \begin{pmatrix} t_1 \\ \vdots \\ t_m \end{pmatrix} \bmod p_2,$$

where we let $g_2^{t_i}$ be the $\mathbb{G}_{p_2}$ component of $\sigma_{3,i}^{(k)}$. Since $\vec{y}^\star \bmod p_2 \notin V^{(k)} \bmod p_2$, $\vec{y}^\star$ is not in the row space of the above matrix. Therefore, $\langle \vec{a}, \vec{y}^\star \rangle \bmod p_2$ is independently of $\mathcal{A}$'s view. □

Lemma 5. *Any PPT algorithm $\mathcal{A}$ outputting a Type A forgery in Game$_{4.q}$ allows breaking Assumption 3.*

Proof. We outline an algorithm $\mathcal{B}$ that takes as input $(g, g^\alpha X_2, X_3, g^s Y_2, Z_2)$ and aims at computing $T = e(g,g)^{\alpha s}$ using its interaction with $\mathcal{A}$. To this end, $\mathcal{B}$ generates the public key $\mathsf{pk} = (g, e(g,g)^\alpha, u, \{h_i\}_{i=0,\ldots,n}, X_{p_3})$ by choosing $b, a_0, \ldots, a_n \overset{R}{\leftarrow} \mathbb{Z}_N$ and setting $X_{p_3} = X_3$, $e(g,g)^\alpha = e(g^\alpha X_2, g)$ as well as $u = g^b$ and $h_i = g^{a_i}$ for $i = 0$ to n.

When the forger $\mathcal{A}$ makes a private key query $V^{(j)} = \mathrm{span}(\vec{v}_1^{(j)}, \ldots, \vec{v}_n^{(j)})$, $\mathcal{B}$ chooses $\mathsf{id} \overset{R}{\leftarrow} \mathcal{I}$, $r \overset{R}{\leftarrow} \mathbb{Z}_N$, $w_1, w_2 \overset{R}{\leftarrow} \mathbb{Z}_N$, $R_3, R_3', R_3'' \overset{R}{\leftarrow} \mathbb{G}_{p_3}$, $w_{3,i} \overset{R}{\leftarrow} \mathbb{Z}_N$, $R_{3,i} \overset{R}{\leftarrow} \mathbb{G}_{p_3}$, for $i = 1$ to n. It defines $\mathsf{id}' = H(\mathsf{id}, e(g,g)^r) \in \mathbb{Z}_N$ and computes

$$\sigma_1 = (g^\alpha X_2) \cdot (u \cdot h_0^{\mathsf{id}'})^r \cdot Z_2^{w_1} \cdot R_3, \qquad \sigma_2 = g^r \cdot Z_2^{w_2} \cdot R_3',$$
$$\sigma_{3,i} = (g^{\langle a, \vec{v}_i^{(j)} \rangle})^r \cdot Z_2^{w_{3,i}} \cdot R_{3,i}$$

which has the distribution of a Type B signature.

At the end of the game, $\mathcal{A}$ outputs a valid tuple of a file identifier $\mathsf{id}^\star$, a signature $\sigma = (\sigma_1^\star, \sigma_2^\star, \{\sigma_{3,i}^\star\}_{i=1}^m)$ of Type-A and a vector $\vec{y}^\star$. Algorithm $\mathcal{B}$ then computes

$$T = e\Big(g^s Y_2, \frac{\sigma_1^\star}{\sigma_2^{\star\, b + a_0 \mathsf{id}'^\star}}\Big) = e\Big(g^s Y_2, \frac{g^\alpha \cdot (g^{b + a_0 \mathsf{id}'^\star})^r \cdot Z_1}{(g^r Z_2)^{b + a_0 \mathsf{id}'^\star}}\Big) = e(g,g)^{\alpha s}.$$

where the second equation is due to lemma 1 ($Z_1, Z_2 \in \mathbb{G}_{p_2 p_3}$). Since σ is of Type A signature, therefore $\sigma_1^\star, \sigma_2^\star$ has no $\mathbb{G}_{p_2}$ component. Hence, the component Y_2 is canceled out in the pairing computation. This yields $T = e(g,g)^{s\alpha}$. To conclude, in Game$_{4.q}$, $\mathcal{A}$'s advantage is thus negligible if Assumption 3 holds. □

References

1. Ahlswede, R., Cai, N., Li, S., Yeung, R.: Network Information Flow. IEEE Trans. on Information Theory 46, 1204–1216 (2000)
2. Agrawal, S., Boneh, D.: Homomorphic MACs: MAC-Based Integrity for Network Coding. In: Abdalla, M., Pointcheval, D., Fouque, P.-A., Vergnaud, D. (eds.) ACNS 2009. LNCS, vol. 5536, pp. 292–305. Springer, Heidelberg (2009)
3. Agrawal, S., Boneh, D., Boyen, X., Freeman, D.: Preventing Pollution Attacks in Multi-source Network Coding. In: Nguyen, P.Q., Pointcheval, D. (eds.) PKC 2010. LNCS, vol. 6056, pp. 161–176. Springer, Heidelberg (2010)
4. Bellare, M., Rogaway, P.: Random oracles are practical: A paradigm for designing efficient protocols. In: ACM CCS 1993, pp. 62–73 (1993)
5. Boneh, D., Boyen, X.: Efficient Selective-ID Secure Identity-Based Encryption Without Random Oracles. In: Cachin, C., Camenisch, J.L. (eds.) EUROCRYPT 2004. LNCS, vol. 3027, pp. 223–238. Springer, Heidelberg (2004)
6. Boneh, D., Franklin, M.: Identity-Based Encryption from the Weil Pairing. Journal of Computing 32(3), 586–615 (2003); Boneh, D., Franklin, M.: Identity-Based Encryption from the Weil Pairing. In: Kilian, J. (ed.) CRYPTO 2001. LNCS, vol. 2139, pp. 213–229. Springer, Heidelberg (2001)
7. Boneh, D., Freeman, D., Katz, J., Waters, B.: Signing a Linear Subspace: Signature Schemes for Network Coding. In: Jarecki, S., Tsudik, G. (eds.) PKC 2009. LNCS, vol. 5443, pp. 68–87. Springer, Heidelberg (2009)
8. Boneh, D., Hamburg, M.: Generalized Identity Based and Broadcast Encryption Schemes. In: Pieprzyk, J. (ed.) ASIACRYPT 2008. LNCS, vol. 5350, pp. 455–470. Springer, Heidelberg (2008)
9. Charles, D., Jain, K., Lauter, K.: Signatures for Network Coding. In: 40th Annual Conference on Information Sciences and Systems (CISS 2006) (2006)
10. Fragouli, C., Soljanin, E.: Network Coding Fundamentals. Now Publishers Inc., Hanover (2007)
11. Freeman, D.: Converting Pairing-Based Cryptosystems from Composite-Order Groups to Prime-Order Groups. In: Gilbert, H. (ed.) EUROCRYPT 2010. LNCS, vol. 6110, pp. 44–61. Springer, Heidelberg (2010)
12. Gennaro, R., Katz, J., Krawczyk, H., Rabin, T.: Secure Network Coding over the Integers. In: Nguyen, P.Q., Pointcheval, D. (eds.) PKC 2010. LNCS, vol. 6056, pp. 142–160. Springer, Heidelberg (2010)
13. Gkantsidis, C., Rodriguez, P.: Network Coding for Large Scale Content Distribution. In: IEEE INFOCOM (2005)
14. Ho, T., Leong, B., Koetter, R., Médard, M., Effros, M., Karger, D.: Byzantine Modification Detection in Multicast Networks using Randomized Network Coding. In: International Symposium on Information Theory (ISIT), pp. 144–152 (2004)
15. Jaggi, S., Langberg, M., Katti, S., Ho, T., Katabi, D., Médard, M., Effros, M.: Resilient Network Coding in the Presence of Byzantine Adversaries. IEEE Trans. on Information Theory 54, 2596–2603 (2008)
16. Krohn, M., Freedman, M., Mazieres, D.: On-the-fly Verification of Rateless Erasure Codes for Efficient Content Distribution. In: IEEE Symposium on Security and Privacy, pp. 226–240 (2004)
17. Lewko, A., Waters, B.: New Techniques for Dual System Encryption and Fully Secure HIBE with Short Ciphertexts. In: Micciancio, D. (ed.) TCC 2010. LNCS, vol. 5978, pp. 455–479. Springer, Heidelberg (2010)

18. Li, S.-Y.-R., Yeung, R.-W., Cai, N.: Linear Network Coding. IEEE Trans. on Information Theory 49, 371–381 (2003)
19. Li, Y., Yao, H., Chen, M., Jaggi, S., Rosen, A.: RIPPLE Authentication for Network Coding. In: IEEE INFOCOM 2010 (2010)
20. Johnson, R., Molnar, D., Song, D., Wagner, D.: Homomorphic Signature Schemes. In: Preneel, B. (ed.) CT-RSA 2002. LNCS, vol. 2271, pp. 244–262. Springer, Heidelberg (2002)
21. Matt, B.: Identification of Multiple Invalid Signatures in Pairing-Based Batched Signatures. In: Jarecki, S., Tsudik, G. (eds.) PKC 2009. LNCS, vol. 5443, pp. 337–356. Springer, Heidelberg (2009)
22. Shamir, A.: Identity-Based Cryptosystems and Signature Schemes. In: Blakely, G.R., Chaum, D. (eds.) CRYPTO 1984. LNCS, vol. 196, pp. 47–53. Springer, Heidelberg (1985)
23. Waters, B.: Efficient Identity-Based Encryption Without Random Oracles. In: Cramer, R. (ed.) EUROCRYPT 2005. LNCS, vol. 3494, pp. 114–127. Springer, Heidelberg (2005)
24. Waters, B.: Dual System Encryption: Realizing Fully Secure IBE and HIBE under Simple Assumptions. In: Halevi, S. (ed.) CRYPTO 2009. LNCS, vol. 5677, pp. 619–636. Springer, Heidelberg (2009)
25. Zhao, F., Kalker, T., Médard, M., Han, K.: Signatures for Content Distribution with Network Coding. In: International Symposium on Information Theory (ISIT) (2007)

Efficient Attribute-Based Signatures for Non-monotone Predicates in the Standard Model

Tatsuaki Okamoto[1] and Katsuyuki Takashima[2]

[1] NTT, 3-9-11 Midori-cho, Musashino-shi, Tokyo, 180-8585, Japan
okamoto.tatsuaki@lab.ntt.co.jp
[2] Mitsubishi Electric, 5-1-1, Ofuna, Kamakura, Kanagawa, 247-8501, Japan
Takashima.Katsuyuki@aj.MitsubishiElectric.co.jp

Abstract. This paper presents a *fully* secure (*adaptive*-predicate unforgeable and private) attribute-based signature (ABS) scheme in the *standard* model. The security of the proposed ABS scheme is proven under standard assumptions, the decisional linear (DLIN) assumption and the existence of collision resistant (CR) hash functions. The admissible predicates of the proposed ABS scheme are more general than those of the existing ABS schemes, i.e., the proposed ABS scheme is the first to support general *non-monotone* predicates, which can be expressed using *NOT* gates as well as AND, OR, and Threshold gates, while the existing ABS schemes only support *monotone* predicates. The proposed ABS scheme is efficient and practical. Its efficiency is comparable to (several times worse than) that of the most efficient (almost optimally efficient) ABS scheme the security for which is proven in the generic group model.

1 Introduction

1.1 Background

The concept of digital signatures was introduced in the seminal paper by Diffie and Hellman in 1976. In this concept, a pair comprising a secret signing key, sk, and public verification key, pk, is generated for a signer, and signature σ of message m generated using sk is verified by the corresponding pk. Hence, the signer of (m, σ) using sk is identified through pk. Although it is one of the requirements of signatures, there is no flexibility or privacy in the relationship between signers and claims attested by signatures due to the tight relation between sk and pk.

Recently, versatile and privacy-enhanced variants of digital signatures have been studied, where the relation between a signing key and verification key is more flexible or sophisticated. In this class of signatures, the signing key and verification key are parameterized by *attribute* $\boldsymbol{x}$ and *predicate* $\boldsymbol{v}$, respectively, and signed message (m, σ) generated by the signing key with parameter $\boldsymbol{x}$, $\mathsf{sk}_{\boldsymbol{x}}$, is correctly verified by public-key pk and parameter $\boldsymbol{v}$, $(\mathsf{pk}, \boldsymbol{v})$, iff predicate $\boldsymbol{v}$ accepts attribute $\boldsymbol{x}$, i.e., $\boldsymbol{v}(\boldsymbol{x})$ holds. The privacy of signers in this class of signatures requires that a signature (for predicate $\boldsymbol{v}$) generated by $\mathsf{sk}_{\boldsymbol{x}}$ (where $\boldsymbol{v}(\boldsymbol{x})$ holds) release no information regarding attribute $\boldsymbol{x}$ except that $\boldsymbol{v}(\boldsymbol{x})$ holds.

D. Catalano et al. (Eds.): PKC 2011, LNCS 6571, pp. 35–52, 2011.

When predicate $\boldsymbol{v}$ is the equality with parameter v (i.e., $\boldsymbol{v}(x)$ holds iff $x = v$), the class of signatures for this predicate is *identity-based signatures* (IBS) [25]. Here note that there is no room for privacy in IBS, since predicate $\boldsymbol{v}$ uniquely identifies attribute x of the signer's secret key, sk_x, such that $x = v$.

Group signatures [9] are also in this class of signatures with another type of predicate $\boldsymbol{v}$, where $\boldsymbol{v}(\boldsymbol{x})$ holds iff predicate parameter v is the group identity (or pk_v is a public key identifying group v) and attribute x is a member identity of group v (or sk_x is a secret key of member x of group v). Due to the privacy requirement, signatures generated using sk_x release no information regarding member identity x except that x is a member of group v (Note that the concept of group signatures traditionally requires the *privacy-revocation* property as well as the above-mentioned privacy).

Recently, this class of signatures with more sophisticated predicates, *attribute-based signatures* (ABS), has been extensively studied [11–13, 16–19, 24, 27], where $\boldsymbol{x}$ for signing key $\mathsf{sk}_{\boldsymbol{x}}$ is a tuple of attributes $(x_1, \ldots, x_i)$, and $\boldsymbol{v}$ for verification is a threshold or access structure predicate. The widest class of predicates in the existing ABS schemes are monotone access structures [18, 19], where predicate $\boldsymbol{v}$ is specified by a monotone span program (MSP), (M, ρ), along with a tuple of attributes $(v_1, \ldots, v_j)$, and $\boldsymbol{v}(\boldsymbol{x})$ holds iff MSP (M, ρ) accepts the truth-value vector of $(\mathsf{T}(x_{i_1} = v_1), \ldots, \mathsf{T}(x_{i_j} = v_j))$. Here, $\mathsf{T}(\psi) := 1$ if ψ is true, and $\mathsf{T}(\psi) := 0$ if ψ is false (For example, $\mathsf{T}(x = v) := 1$ if $x = v$, and $\mathsf{T}(x = v) := 0$ if $x \neq v$). In general, such a predicate can be expressed using AND, OR, and Threshold gates.

An example of such monotone predicate $\boldsymbol{v}$ for ABS is (Institute = Univ. A) AND (TH2((Department = Biology), (Gender = Female), (Age = 50's)) OR (Position = Professor)), where TH2 means the threshold gate with threshold value 2. Attribute $\boldsymbol{x}_A$ of Alice is ((Institute := Univ. A), (Department := Biology), (Position := Postdoc), (Age := 30), (Gender := Female))), and attribute $\boldsymbol{x}_B$ of Bob is ((Institute := Univ. A), (Department := Mathematics), (Position := Professor), (Age := 45) (Gender := Male))). Although their attributes, $\boldsymbol{x}_A$ and $\boldsymbol{x}_B$, are quite different, it is clear that $\boldsymbol{v}(\boldsymbol{x}_A)$ and $\boldsymbol{v}(\boldsymbol{x}_B)$ hold, and that there are many other attributes that satisfy $\boldsymbol{v}$. Hence Alice and Bob can generate a signature on this predicate, and due to the privacy requirement of ABS, a signature for $\boldsymbol{v}$ releases no information regarding the attribute or identity of the signer, i.e., Alice or Bob (or other), except that the attribute of the signer satisfies $\boldsymbol{v}$.

There are many applications of ABS such as attribute-based messaging (ABM), attribute-based authentication, trust-negotiation and leaking secrets (see [18, 19] for more details).

The security conditions for ABS are given hereafter (see Section 3.2 for the formal definitions).

Unforgeability: A valid signature should be produced only by a *single* signer whose attribute $\boldsymbol{x}$ satisfies the claimed predicate $\boldsymbol{v}$, not by a collusion of users who pooled their attributes together. More formally, no poly-time adversary can produce a valid signature for a pair comprising predicate and message

$(\boldsymbol{v}, m)$, even if the adversary *adaptively* chooses $(\boldsymbol{v}, m)$ after executing secret-key and signing oracle attacks, provided that $\boldsymbol{x}$ where $\boldsymbol{v}(\boldsymbol{x})$ holds is not queried to the secret-key oracle and $(\boldsymbol{v}, m)$ is not queried to the signing oracle (We simply call this unforgeability "*adaptive*-predicate unforgeability" or more simply "unforgeability").

We can also define a *weaker* class of unforgeability, '*selective*-predicate unforgeability,' where an adversary should choose predicate $\boldsymbol{v}$ for the forgery signature before executing secret-key and signing oracle attacks.

Privacy: A signature for predicate $\boldsymbol{v}$ generated using secret key $\mathsf{sk}_{\boldsymbol{x}}$ releases no information regarding attribute $\boldsymbol{x}$ except that $\boldsymbol{v}(\boldsymbol{x})$ holds.

More formally, for any pair of attributes $(\boldsymbol{x}_1, \boldsymbol{x}_2)$, predicate $\boldsymbol{v}$ and message m, for which $\boldsymbol{v}(\boldsymbol{x}_1)$ and $\boldsymbol{v}(\boldsymbol{x}_2)$ hold simultaneously, the distributions of two valid signatures $\sigma(m, \boldsymbol{v}, \mathsf{sk}_{\boldsymbol{x}_1})$ and $\sigma(m, \boldsymbol{v}, \mathsf{sk}_{\boldsymbol{x}_2})$ are equivalent, where $\sigma(m, \boldsymbol{v}, \mathsf{sk}_{\boldsymbol{x}})$ is a correctly generated signature for $(m, \boldsymbol{v})$ using correct secret key $\mathsf{sk}_{\boldsymbol{x}}$ with attribute $\boldsymbol{x}$ (We simply call this condition "*privacy*").

Full Security: We say that an ABS scheme is *fully* secure if it satisfies *adaptive*-predicate unforgeability and *privacy.*

Maji, Prabhakaran, and Rosulek [18, 19] presented ABS schemes for the widest class of predicates among the existing ABS schemes, monotone access structure predicates, which cover threshold predicates as special cases. The scheme shown in [18] is an almost optimally efficient ABS scheme, but the security was only proven in the generic group model. The scheme shown in [19] is the only existing ABS scheme for which (full) security was proven in the standard model. It is, however, much less efficient and more complicated than the scheme in [18] since it employs the Groth-Sahai NIZK protocols [10] as building blocks.

Li, Au, Susilo, Xie and Ren [16], Li and Kim [17], and Shahandashti and Safavi-Naini [24] presented ABS schemes that are proven to be secure in the standard model. However, the proven security is not the full security, but a weaker level of security with *selective*-predicate unforgeability. Moreover, the admissible predicates in [17] are limited to conjunction or (n, n)-threshold predicates, and those of [16, 24] are limited to (k, n)-threshold predicates.

Guo and Zeng [11] and Yang, Cao and Dong [27] presented ABS schemes for threshold predicates, but their security definitions do not include the *privacy* condition of ABS.

Khader [12, 13] presented ABS schemes for monotone access structure predicates. These schemes, however, do not satisfy the *privacy* condition of ABS, since they only conceal the identity of the signer. They also reveal the attributes that the signer used to generate the signature. In addition, the security is proven in a non-standard model, the random oracle model.

Based on this background, there are two major problems in the existing ABS schemes.

1. No ABS scheme for *non-monotone* predicates, which can be expressed using NOT gates as well as AND, OR and Threshold gates, has been proposed (even in a weaker security notion or a non-standard model).

2. The only fully secure ABS scheme in the *standard* model [19] is much less efficient than the (almost optimally efficient) ABS scheme in the generic group model [18].

Non-monotone predicates should be used in many ABS applications. For example, annual review reports in the Mathematics Department of University A are submitted by reviewers, and these reports are anonymously signed by the reviewers through ABS with some predicates. The predicates may be selected freely by them (signers) except that it should be in the following form: NOT((Institute = Univ. A) AND (Department = Mathematics)) AND $(\cdots)$.

1.2 Our Results

This paper addresses these problems simultaneously.

- This paper proposes the first fully secure (i.e., adaptive-predicate unforgeable and perfectly private) ABS scheme for a wide class of predicates, *non-monotone* access structures, where $\boldsymbol{x}$ for signing key $\mathsf{sk}_{\boldsymbol{x}}$ is a tuple of attributes $(x_1, \ldots, x_i)$, non-monotone predicate $\boldsymbol{v}$ is specified by a *span program* (SP) (M, ρ) along with a tuple of attributes $(v_1, \ldots, v_j)$, and $\boldsymbol{v}(\boldsymbol{x})$ holds iff SP (M, ρ) accepts the truth-value vector of $(\mathsf{T}(x_{i_1} = v_1), \ldots, \mathsf{T}(x_{i_j} = v_j))$. Our scheme can be generalized using non-monotone access structures combined with *inner-product relations* (see Definition 5 and the remark). More precisely, attribute $\boldsymbol{x}$ for signing key $\mathsf{sk}_{\boldsymbol{x}}$ is a tuple of attribute vectors (e.g., $(\vec{x}_1, \ldots, \vec{x}_i) \in \mathbb{F}_q^{n_1+\cdots+n_i}$), and predicate $\boldsymbol{v}$ for verification is a non-monotone access structure or span program (SP) (M, ρ) along with a tuple of attribute vectors (e.g., $(\vec{v}_1, \ldots, \vec{v}_j) \in \mathbb{F}_q^{n_1+\cdots+n_j}$), where the component-wise inner-product relations for attribute vectors (e.g., $\{\vec{x}_{i_\iota} \cdot \vec{v}_\iota = 0$ or not $\}_{\iota \in \{1,\ldots,j\}}$) are input to SP (M, ρ). Namely, $\boldsymbol{v}(\boldsymbol{x})$ holds iff the truth-value vector of $(\mathsf{T}(\vec{x}_{i_1} \cdot \vec{v}_1 = 0), \ldots, \mathsf{T}(\vec{x}_{i_j} \cdot \vec{v}_j = 0))$ is accepted by SP (M, ρ).

  Remark: In our scheme (Section 4), attribute $\boldsymbol{x}$ is expressed by the form $\Gamma := \{(t, x_t) \mid t \in T \subseteq \{1, \ldots, d\}\}$ in place of just an attribute tuple $(x_1, \ldots, x_i)$, where t identifies a sub-universe or category of attributes, and x_t is an attribute in sub-universe t (examples of (t, x_t) are (Name, Alice) and (Age, 38)). Predicate $\boldsymbol{v}$ is expressed by $\mathbb{S} := (M, \rho)$, where ρ is abused as ρ (defined by SP) combined with $\{(t_i, v_i) \mid i = 1, \ldots, \ell\}$ (see Definitions 4 and 5 for the difference regarding ρ in SP and $\mathbb{S}$).
- The proposed ABS scheme is proven to be fully secure under standard assumptions, the *decisional linear (DLIN)* assumption (over prime order pairing groups) and the existence of *collision resistant (CR) hash* functions, in the *standard* model.
- In contrast to the ABS scheme in [19] that employs the Groth-Sahai NIZK protocols, our ABS scheme is more directly constructed without using any general subprotocols like NIZK. Our construction is based on the dual pairing vector spaces (DPVS) proposed by Okamoto and Takashima [14, 20–22], which can be realized from *any type of (e.g., symmetric or asymmetric)*

prime order bilinear pairing groups. See Section 2.1 for the concept and actual construction of DPVS.

- To prove the security (especially the unforgeability), this paper employs the techniques for fully secure functional encryption (FE) [14, 22], which elaborately combine the dual system encryption methodology proposed by Waters [26] and DPVS.

  Note that although the techniques for the FE schemes in [14, 22] can be employed for ABS, it is still a challenging task to construct a fully secure ABS scheme, since the security requirements of ABS and FE differ in some important points, for example, the privacy condition is required in ABS but there is no counterpart notion in FE. This paper develops several novel techniques for our ABS scheme. See Section 4.1 for more details.
- The efficiency of the proposed ABS scheme is comparable to that of the most efficient ABS scheme in the generic group model [18], and better than that of the only existing fully secure ABS scheme in the standard model [19]. See Section 4.4 for a comparison.
- This paper also presents an extension, multi-authority (MA) setting, of the proposed ABS scheme in Section 5. One of the merits of our MA-ABS scheme is that it is seamlessly extended from the original (single-authority (SA)) setting, in which the signing and verification algorithms of the MA-ABS scheme are essentially the same as those of the original ABS (SA-ABS) scheme.

  In MA-ABS, each authority called an attribute authority is responsible for a category of attributes, and a user obtains a part of secret key for each attribute from an attribute authority responsible for the category of the attribute. We follow the model of MA-ABS introduced in [18, 19], where a central trustee in addition to attribute authorities is required but no interaction among attribute authorities (and the trustee) is necessary, and different attribute authorities may not trust each other, nor even be aware of each other.

  We prove that the proposed MA-ABS scheme is fully secure (in the sense of the MA-ABS model of [18, 19]) under the DLIN assumption and CR hash functions in the standard model (see the full version of this paper for the proof). Our MA-ABS scheme is almost as efficient as the original SA-ABS scheme.

1.3 Related Works

- **Ring and mesh signatures:** Ring and mesh signatures [4, 23] are related to ABS.

  In the ring signatures, the claimed predicate on a signature of message m is that m is endorsed by one of the users identified by the list of public keys $(\mathsf{pk}_1, \mathsf{pk}_2, \ldots)$, or the predicate is a disjunction of a list of public keys. A valid ring signature can be generated by one of the listed users.

  The mesh signatures are an extension of ring signatures, where the predicate is an access structure on a list of pairs comprising a message and public key (m_i, pk_i), and a valid mesh signature can be generated by a person who has enough standard signatures σ_i on m_i, each valid under pk_i, to satisfy the given access structure.

A crucial difference between mesh signatures and ABS is the security against the collusion of users. In mesh signatures, several users can collude by pooling their signatures together and create signatures that none of them could produce individually. That is, such collusion is considered to be legitimate in mesh signatures. In contrast, the security against collusion attacks is one of the basic requirements in ABS and MA-ABS, as described in Section 1.1 and Section 5.

- **Anonymous credentials (ACs):** Another related concept is ACs [2, 3, 5–8]. The notion of ACs also provides a functionality for users to demonstrate anonymously possession of attributes, but the goals of ACs and ABS differ in several points.
  As mentioned in [19], ACs and ABS aim at different goals: ACs target very strong anonymity even in the registration phase, whereas under less demanding anonymity requirements in the registration phase, ABS aims to achieve more expressive functionalities, more efficient constructions and new applications. In addition, ABS is a signature scheme and a simpler primitive compared with ACs.

1.4 Notations

When A is a random variable or distribution, $y \xleftarrow{\mathsf{R}} A$ denotes that y is randomly selected from A according to its distribution. When A is a set, $y \xleftarrow{\mathsf{U}} A$ denotes that y is uniformly selected from A. $y := z$ denotes that y is set, defined or substituted by z. When a is a fixed value, $A(x) \to a$ (e.g., $A(x) \to 1$) denotes the event that machine (algorithm) A outputs a on input x. A function $f : \mathbb{N} \to \mathbb{R}$ is *negligible* in λ, if for every constant $c > 0$, there exists an integer n such that $f(\lambda) < \lambda^{-c}$ for all $\lambda > n$.

We denote the finite field of order q by $\mathbb{F}_q$, and $\mathbb{F}_q \setminus \{0\}$ by $\mathbb{F}_q^{\times}$. A vector symbol denotes a vector representation over $\mathbb{F}_q$, e.g., $\vec{x}$ denotes $(x_1, \ldots, x_n) \in \mathbb{F}_q^n$. For two vectors $\vec{x} = (x_1, \ldots, x_n)$ and $\vec{v} = (v_1, \ldots, v_n)$, $\vec{x} \cdot \vec{v}$ denotes the inner-product $\sum_{i=1}^n x_i v_i$. The vector $\vec{0}$ is abused as the zero vector in $\mathbb{F}_q^n$ for any n. X^{T} denotes the transpose of matrix X. A bold face letter denotes an element of vector space $\mathbb{V}$, e.g., $\boldsymbol{x} \in \mathbb{V}$. When $\boldsymbol{b}_i \in \mathbb{V}$ $(i = 1, \ldots, n)$, $\mathsf{span}\langle \boldsymbol{b}_1, \ldots, \boldsymbol{b}_n \rangle \subseteq \mathbb{V}$ (resp. $\mathsf{span}\langle \vec{x}_1, \ldots, \vec{x}_n \rangle$) denotes the subspace generated by $\boldsymbol{b}_1, \ldots, \boldsymbol{b}_n$ (resp. $\vec{x}_1, \ldots, \vec{x}_n$). For bases $\mathbb{B} := (\boldsymbol{b}_1, \ldots, \boldsymbol{b}_N)$ and $\mathbb{B}^* := (\boldsymbol{b}_1^*, \ldots, \boldsymbol{b}_N^*)$, $(x_1, \ldots, x_N)_{\mathbb{B}} := \sum_{i=1}^N x_i \boldsymbol{b}_i$ and $(y_1, \ldots, y_N)_{\mathbb{B}^*} := \sum_{i=1}^N y_i \boldsymbol{b}_i^*$.

2 Preliminaries

2.1 Dual Pairing Vector Spaces by Direct Product of Symmetric Pairing Groups

Definition 1. *"Symmetric bilinear pairing groups"* $(q, \mathbb{G}, \mathbb{G}_T, G, e)$ *are a tuple of a prime* q, *cyclic additive group* $\mathbb{G}$ *and multiplicative group* $\mathbb{G}_T$ *of order* q, $G \neq 0 \in \mathbb{G}$, *and a polynomial-time computable nondegenerate bilinear pairing* $e : \mathbb{G} \times \mathbb{G} \to \mathbb{G}_T$ *i.e.,* $e(sG, tG) = e(G, G)^{st}$ *and* $e(G, G) \neq 1$.

Let $\mathcal{G}_{\mathsf{bpg}}$ be an algorithm that takes input 1^λ and outputs a description of bilinear pairing groups $(q, \mathbb{G}, \mathbb{G}_T, G, e)$ with security parameter λ.

In this paper, we concentrate on the symmetric version of dual pairing vector spaces [14, 20–22] constructed by using symmetric bilinear pairing groups given in Definition 1.

Definition 2. *"Dual pairing vector spaces (DPVS)" $(q, \mathbb{V}, \mathbb{G}_T, \mathbb{A}, e)$ by a direct product of symmetric pairing groups $(q, \mathbb{G}, \mathbb{G}_T, G, e)$ are a tuple of prime q, N-dimensional vector space $\mathbb{V} := \overbrace{\mathbb{G} \times \cdots \times \mathbb{G}}^{N}$ over $\mathbb{F}_q$, cyclic group $\mathbb{G}_T$ of order q, canonical basis $\mathbb{A} := (\boldsymbol{a}_1, \ldots, \boldsymbol{a}_N)$ of $\mathbb{V}$, where $\boldsymbol{a}_i := (\overbrace{0, \ldots, 0}^{i-1}, G, \overbrace{0, \ldots, 0}^{N-i})$, and pairing $e : \mathbb{V} \times \mathbb{V} \to \mathbb{G}_T$.*

The pairing is defined by $e(\boldsymbol{x}, \boldsymbol{y}) := \prod_{i=1}^{N} e(G_i, H_i) \in \mathbb{G}_T$ where $\boldsymbol{x} := (G_1, \ldots, G_N) \in \mathbb{V}$ and $\boldsymbol{y} := (H_1, \ldots, H_N) \in \mathbb{V}$. This is nondegenerate bilinear i.e., $e(s\boldsymbol{x}, t\boldsymbol{y}) = e(\boldsymbol{x}, \boldsymbol{y})^{st}$ and if $e(\boldsymbol{x}, \boldsymbol{y}) = 1$ for all $\boldsymbol{y} \in \mathbb{V}$, then $\boldsymbol{x} = \boldsymbol{0}$. For all i and j, $e(\boldsymbol{a}_i, \boldsymbol{a}_j) = e(G, G)^{\delta_{i,j}}$ where $\delta_{i,j} = 1$ if $i = j$, and 0 otherwise, and $e(G, G) \neq 1 \in \mathbb{G}_T$.

DPVS also has linear transformations $\phi_{i,j}$ on $\mathbb{V}$ s.t. $\phi_{i,j}(\boldsymbol{a}_j) = \boldsymbol{a}_i$ and $\phi_{i,j}(\boldsymbol{a}_k) = \boldsymbol{0}$ if $k \neq j$, which can be easily achieved by $\phi_{i,j}(\boldsymbol{x}) := (\overbrace{0, \ldots, 0}^{i-1}, G_j, \overbrace{0, \ldots, 0}^{N-i})$ where $\boldsymbol{x} := (G_1, \ldots, G_N)$. We call $\phi_{i,j}$ "canonical maps".

DPVS generation algorithm $\mathcal{G}_{\mathsf{dpvs}}$ takes input 1^λ ($\lambda \in \mathbb{N}$) and $N \in \mathbb{N}$, and outputs a description of $\mathsf{param}_{\mathbb{V}} := (q, \mathbb{V}, \mathbb{G}_T, \mathbb{A}, e)$ with security parameter λ and N-dimensional $\mathbb{V}$. It can be constructed by using $\mathcal{G}_{\mathsf{bpg}}$.

The asymmetric version of DPVS, $(q, \mathbb{V}, \mathbb{V}^*, \mathbb{G}_T, \mathbb{A}, \mathbb{A}^*, e)$, is given in the full version of [22]. The above symmetric version is obtained by identifying $\mathbb{V} = \mathbb{V}^*$ and $\mathbb{A} = \mathbb{A}^*$ in the asymmetric version. (For another construction of DPVS using higher genus Jacobians, see [20].)

2.2 Decisional Linear (DLIN) Assumption

Definition 3 (DLIN Assumption). *The DLIN problem is to guess $\beta \in \{0, 1\}$, given $(\mathsf{param}_{\mathbb{G}}, G, \xi G, \kappa G, \delta\xi G, \sigma\kappa G, Y_\beta) \xleftarrow{\mathsf{R}} \mathcal{G}_\beta^{\mathsf{DLIN}}(1^\lambda)$, where*

$$\begin{aligned}
\mathcal{G}_\beta^{\mathsf{DLIN}}(1^\lambda) : &\ \mathsf{param}_{\mathbb{G}} := (q, \mathbb{G}, \mathbb{G}_T, G, e) \xleftarrow{\mathsf{R}} \mathcal{G}_{\mathsf{bpg}}(1^\lambda), \\
&\ \kappa, \delta, \xi, \sigma \xleftarrow{\mathsf{U}} \mathbb{F}_q, \quad Y_0 := (\delta + \sigma)G, \quad Y_1 \xleftarrow{\mathsf{U}} \mathbb{G}, \\
&\ \text{return} \ \ (\mathsf{param}_{\mathbb{G}}, G, \xi G, \kappa G, \delta\xi G, \sigma\kappa G, Y_\beta),
\end{aligned}$$

for $\beta \xleftarrow{\mathsf{U}} \{0, 1\}$. For a probabilistic machine $\mathcal{E}$, we define the advantage of $\mathcal{E}$ for the DLIN problem as: $\mathsf{Adv}_{\mathcal{E}}^{\mathsf{DLIN}}(\lambda) := \left| \mathsf{Pr}\left[\mathcal{E}(1^\lambda, \varrho) \to 1 \,\middle|\, \varrho \xleftarrow{\mathsf{R}} \mathcal{G}_0^{\mathsf{DLIN}}(1^\lambda)\right] - \mathsf{Pr}\left[\mathcal{E}(1^\lambda, \varrho) \to 1 \,\middle|\, \varrho \xleftarrow{\mathsf{R}} \mathcal{G}_1^{\mathsf{DLIN}}(1^\lambda)\right] \right|$. The DLIN assumption is: For any probabilistic polynomial-time adversary $\mathcal{E}$, the advantage $\mathsf{Adv}_{\mathcal{E}}^{\mathsf{DLIN}}(\lambda)$ is negligible in λ.

2.3 Collision Resistant (CR) Hash Functions

Let $\lambda \in \mathbb{N}$ be a security parameter. A collision resistant (CR) hash function family, H, associated with $\mathcal{G}_{\mathsf{bpg}}$ and a polynomial, $poly(\cdot)$, specifies two items:

- A family of key spaces indexed by λ. Each such key space is a probability space on bit strings denoted by KH_λ. There must exist a probabilistic polynomial-time algorithm whose output distribution on input 1^λ is equal to KH_λ.
- A family of hash functions indexed by λ, $\mathsf{hk} \xleftarrow{\mathsf{R}} \mathsf{KH}_\lambda$ and $\mathsf{D} := \{0,1\}^{poly(\lambda)}$. Each such hash function $\mathsf{H}_{\mathsf{hk}}^{\lambda,\mathsf{D}}$ maps an element of D to an element of $\mathbb{F}_q^\times$ with q that is the first element of output $\mathsf{param}_{\mathbb{G}}$ of $\mathcal{G}_{\mathsf{bpg}}(1^\lambda)$. There must exist a deterministic polynomial-time algorithm that on input 1^λ, hk and $\varrho \in \mathsf{D}$, outputs $\mathsf{H}_{\mathsf{hk}}^{\lambda,\mathsf{D}}(\varrho)$.

Let $\mathcal{E}$ be a probabilistic polynomial-time machine. For all λ, we define $\mathsf{Adv}_{\mathcal{E}}^{\mathsf{H},\mathsf{CR}}(\lambda) := \Pr[(\varrho_1, \varrho_2) \in \mathsf{D}^2 \wedge \varrho_1 \neq \varrho_2 \wedge \mathsf{H}_{\mathsf{hk}}^{\lambda,\mathsf{D}}(\varrho_1) = \mathsf{H}_{\mathsf{hk}}^{\lambda,\mathsf{D}}(\varrho_2)]$, where $\mathsf{D} := \{0,1\}^{poly(\lambda)}, \mathsf{hk} \xleftarrow{\mathsf{R}} \mathsf{KH}_\lambda$, and $(\varrho_1, \varrho_2) \xleftarrow{\mathsf{R}} \mathcal{E}(1^\lambda, \mathsf{hk}, \mathsf{D})$. H is a collision resistant (CR) hash function family if for any probabilistic polynomial-time adversary $\mathcal{E}$, $\mathsf{Adv}_{\mathcal{E}}^{\mathsf{H},\mathsf{CR}}(\lambda)$ is negligible in λ.

3 ABS for Non-monotone Predicates

3.1 Span Programs and Non-monotone Access Structures

Definition 4 (Span Programs [1]). *Let $\{p_1, \ldots, p_n\}$ be a set of variables. A span program over $\mathbb{F}_q$ is a labeled matrix, $\hat{M} := (M, \rho)$, where M is a $(\ell \times r)$ matrix over $\mathbb{F}_q$ and ρ is a labeling of the rows of M by literals from $\{p_1, \ldots, p_n, \neg p_1, \ldots, \neg p_n\}$ (every row is labeled by one literal), i.e., $\rho : \{1, \ldots, \ell\} \to \{p_1, \ldots, p_n, \neg p_1, \ldots, \neg p_n\}$.*

A span program accepts or rejects an input by the following criterion. For every input sequence $\delta \in \{0,1\}^n$ define submatrix M_δ of M consisting of those rows whose labels are set to 1 by the input δ, i.e., either rows labeled by some p_i such that $\delta_i = 1$ or rows labeled by some by some $\neg p_i$ such that $\delta_i = 0$. (i.e., $\gamma : \{1, \ldots, \ell\} \to \{0,1\}$ is defined by $\gamma(j) = 1$ if $[\rho(j) = p_i] \wedge [\delta_i = 1]$ or $[\rho(j) = \neg p_i] \wedge [\delta_i = 0]$, and $\gamma(j) = 0$ otherwise. $M_\delta := (M_j)_{\gamma(j)=1}$, where M_j is the j-th row of M.)

Span program $\hat{M}$ accepts δ if and only if $\vec{1} \in \mathsf{span}\langle M_\delta \rangle$, i.e., some linear combination of the rows of M_δ gives the all one vector, $\vec{1}$. (The row vector has the value 1 in each coordinate.) A span program computes boolean function f if it accepts exactly those inputs δ where $f(\delta) = 1$.

A span program is called monotone if the labels of the rows are only the positive literals $\{p_1, \ldots, p_n\}$. Monotone span programs compute monotone functions. (So, a span program in general is "non"-monotone.)

We assume that access structure matrix M (of type $\ell \times r$) satisfies the condition: $M_i \neq \vec{0}$ for $i = 1, \ldots, \ell$.

We now introduce a non-monotone access structure with evaluating map γ by using the inner-product of attribute vectors in a general form. Although we will show the notion, security definition and security proof of the proposed ABS scheme in this general form, we will describe the proposed ABS scheme in a simpler form in Section 4.2. We will show this simpler form of Definition 5 in the remark.

Definition 5 (Inner-Products of Attribute Vectors and Access Structures). *$\mathcal{U}_t$ ($t = 1, \ldots, d$ and $\mathcal{U}_t \subset \{0,1\}^*$) is a sub-universe, a set of attributes, each of which is expressed by a pair of sub-universe id and n_t-dimensional vector, i.e., $(t, \vec{v})$, where $t \in \{1, \ldots, d\}$ and $\vec{v} \in \mathbb{F}_q^{n_t} \setminus \{\vec{0}\}$.*

We now define such an attribute to be a variable, p, of span program $\hat{M} := (M, \rho)$ i.e., $p := (t, \vec{v})$. Access structure $\mathbb{S}$ is span program $\hat{M} := (M, \rho)$ along with variables $p := (t, \vec{v}), p' := (t', \vec{v}'), \ldots$, i.e., $\mathbb{S} := (M, \rho)$ such that $\rho : \{1, \ldots, \ell\} \to \{(t, \vec{v}), (t', \vec{v}'), \ldots, \neg(t, \vec{v}), \neg(t', \vec{v}'), \ldots\}$.

Let Γ be a set of attributes, i.e., $\Gamma := \{(t, \vec{x}_t) \mid \vec{x}_t \in \mathbb{F}_q^{n_t} \setminus \{\vec{0}\}, 1 \le t \le d\}$.

When Γ is given to access structure $\mathbb{S}$, map $\gamma : \{1, \ldots, \ell\} \to \{0, 1\}$ for span program $\hat{M} := (M, \rho)$ is defined as follows: For $i = 1, \ldots, \ell$, set $\gamma(i) = 1$ if $[\rho(i) = (t, \vec{v}_i)] \wedge [(t, \vec{x}_t) \in \Gamma] \wedge [\vec{v}_i \cdot \vec{x}_t = 0]$ or $[\rho(i) = \neg(t, \vec{v}_i)] \wedge [(t, \vec{x}_t) \in \Gamma] \wedge [\vec{v}_i \cdot \vec{x}_t \neq 0]$. Set $\gamma(i) = 0$ otherwise.

Access structure $\mathbb{S} := (M, \rho)$ accepts Γ iff $\vec{1} \in \mathsf{span}\langle (M_i)_{\gamma(i)=1} \rangle$.

Remark: A simpler form of the inner-product relations in the above-mentioned access structures is a special case when $n_t = 2$ for all $t \in \{1, \ldots, d\}$, and $\vec{x} := (1, x)$ and $\vec{v} := (v, -1)$. Hence, $(t, \vec{x}_t) := (t, (1, x_t))$ and $(t, \vec{v}_i) := (t, (v_i, -1))$, but we often denote them shortly by (t, x_t) and (t, v_i). Then, $\mathbb{S} := (M, \rho)$ such that $\rho : \{1, \ldots, \ell\} \to \{(t, v), (t', v'), \ldots \neg(t, v), \neg(t', v'), \ldots\}$ $(v, v', \ldots \in \mathbb{F}_q)$, and $\Gamma := \{(t, x_t) \mid x_t \in \mathbb{F}_q, 1 \le t \le d\}$.

When Γ is given to access structure $\mathbb{S}$, map $\gamma : \{1, \ldots, \ell\} \to \{0, 1\}$ for span program $\hat{M} := (M, \rho)$ is defined as follows: For $i = 1, \ldots, \ell$, set $\gamma(i) = 1$ if $[\rho(i) = (t, v_i)] \wedge [(t, x_t) \in \Gamma] \wedge [v_i = x_t]$ or $[\rho(i) = \neg(t, v_i)] \wedge [(t, x_t) \in \Gamma] \wedge [v_i \neq x_t]$. Set $\gamma(i) = 0$ otherwise.

We now construct a secret-sharing scheme for a (non-monotone) access structure (span program).

Definition 6. *A secret-sharing scheme for access structure $\mathbb{S} := (M, \rho)$ is:*

1. *Let M be an $\ell \times r$ matrix, and column vector $\vec{f}^{\mathrm{T}} := (f_1, \ldots, f_r)^{\mathrm{T}} \xleftarrow{\mathsf{U}} \mathbb{F}_q^r$. Then, $s_0 := \vec{1} \cdot \vec{f}^{\mathrm{T}} = \sum_{k=1}^{r} f_k$ is the secret to be shared, and $\vec{s}^{\mathrm{T}} := (s_1, \ldots, s_\ell)^{\mathrm{T}} := M \cdot \vec{f}^{\mathrm{T}}$ is the vector of ℓ shares of secret s_0 and share s_i belongs to $\rho(i)$.*
2. *If access structure $\mathbb{S} := (M, \rho)$ accepts Γ, i.e., $\vec{1} \in \mathsf{span}\langle (M_i)_{\gamma(i)=1} \rangle$ with $\gamma : \{1, \ldots, \ell\} \to \{0, 1\}$, then there exist constants $\{\alpha_i \in \mathbb{F}_q \mid i \in I\}$ such that $I \subseteq \{i \in \{1, \ldots, \ell\} \mid \gamma(i) = 1\}$ and $\sum_{i \in I} \alpha_i s_i = s_0$. Furthermore, these constants $\{\alpha_i\}$ can be computed in time polynomial in the size of matrix M.*

3.2 Definitions and Security of ABS

Definition 7 (Attribute-Based Signatures : ABS). *An attribute-based signature scheme consists of four algorithms.*

Setup *This is a randomized algorithm that takes as input security parameter and format* $\vec{n} := (d; n_1, \ldots, n_d)$ *of attributes. It outputs public parameters* pk *and master key* sk.

KeyGen *This is a randomized algorithm that takes as input a set of attributes,* $\Gamma := \{(t, \vec{x}_t) | \vec{x}_t \in \mathbb{F}_q^{n_t} \setminus \{\vec{0}\}, 1 \le t \le d\}$, pk *and* sk. *It outputs signature generation key* sk_Γ.

Sig *This is a randomized algorithm that takes as input message* m, *access structure* $\mathbb{S} := (M, \rho)$, *signature generation key* sk_Γ, *and public parameters* pk *such that* $\mathbb{S}$ *accepts* Γ. *It outputs signature* σ.

Ver *This takes as input message* m, *access structure* $\mathbb{S}$, *signature* σ *and public parameters* pk. *It outputs boolean value* accept $:= 1$ *or* reject $:= 0$.

An ABS scheme should have the following correctness property: for all $(\mathsf{sk}, \mathsf{pk}) \xleftarrow{\mathsf{R}} \mathsf{Setup}(1^\lambda, \vec{n})$, all messages m, all attribute sets Γ, all signing keys $\mathsf{sk}_\Gamma \xleftarrow{\mathsf{R}} \mathsf{KeyGen}(\mathsf{pk}, \mathsf{sk}, \Gamma)$, all access structures $\mathbb{S}$ such that $\mathbb{S}$ accepts Γ, and all signatures $\sigma \xleftarrow{\mathsf{R}} \mathsf{Sig}(\mathsf{pk}, \mathsf{sk}_\Gamma, m, \mathbb{S})$, it holds that $\mathsf{Ver}(\mathsf{pk}, m, \mathbb{S}, \sigma) = 1$ with probability 1.

Definition 8 (Perfect Privacy). *An ABS scheme is perfectly private, if, for all* $(\mathsf{sk}, \mathsf{pk}) \xleftarrow{\mathsf{R}} \mathsf{Setup}(1^\lambda, \vec{n})$, *all messages* m, *all attribute sets* Γ_1 *and* Γ_2, *all signing keys* $\mathsf{sk}_{\Gamma_1} \xleftarrow{\mathsf{R}} \mathsf{KeyGen}(\mathsf{pk}, \mathsf{sk}, \Gamma_1)$ *and* $\mathsf{sk}_{\Gamma_2} \xleftarrow{\mathsf{R}} \mathsf{KeyGen}(\mathsf{pk}, \mathsf{sk}, \Gamma_2)$, *all access structures* $\mathbb{S}$ *such that* $\mathbb{S}$ *accepts* Γ_1 *and* $\mathbb{S}$ *accepts* Γ_2, *distributions* $\mathsf{Sig}(\mathsf{pk}, \mathsf{sk}_{\Gamma_1}, m, \mathbb{S})$ *and* $\mathsf{Sig}(\mathsf{pk}, \mathsf{sk}_{\Gamma_2}, m, \mathbb{S})$ *are equal.*

For an ABS scheme with prefect privacy, we define algorithm $\mathsf{AltSig}(\mathsf{pk}, \mathsf{sk}, m, \mathbb{S})$ with $\mathbb{S}$ and master key sk instead of Γ and sk_Γ: First, generate $\mathsf{sk}_\Gamma \xleftarrow{\mathsf{R}} \mathsf{KeyGen}(\mathsf{pk}, \mathsf{sk}, \Gamma)$ for arbitrary Γ which satisfies $\mathbb{S}$, then $\sigma \xleftarrow{\mathsf{R}} \mathsf{Sig}(\mathsf{pk}, \mathsf{sk}_\Gamma, m, \mathbb{S})$. return σ.

Since the correct distribution on signatures can be perfectly simulated without taking any private information as input, signatures must not leak any such private information of the signer.

Definition 9 (Unforgeability). *For an adversary,* $\mathcal{A}$, *we define* $\mathsf{Adv}_{\mathcal{A}}^{\mathsf{ABS,UF}}(\lambda)$ *to be the success probability in the following experiment for any security parameter* λ. *An ABS scheme is existentially unforgeable if the success probability of any polynomial-time adversary is negligible:*

1. *Run* $(\mathsf{sk}, \mathsf{pk}) \xleftarrow{\mathsf{R}} \mathsf{Setup}(1^\lambda, \vec{n})$ *and give* pk *to the adversary.*
2. *The adversary is given access to oracles* $\mathsf{KeyGen}(\mathsf{pk}, \mathsf{sk}, \cdot)$ *and* $\mathsf{AltSig}(\mathsf{pk}, \mathsf{sk}, \cdot, \cdot)$.
3. *At the end, the adversary outputs* $(m', \mathbb{S}', \sigma')$.

We say the adversary succeeds if $(m', \mathbb{S}')$ *was never queried to the* AltSig *oracle,* $\mathbb{S}'$ *does not accept any* Γ *queried to the* KeyGen *oracle, and* $\mathsf{Ver}(\mathsf{pk}, m', \mathbb{S}', \sigma') = 1$.

4 Proposed ABS Scheme

4.1 Construction Ideas

Here, we will show some basic ideas to construct the proposed ABS scheme. Our ABS scheme is constructed on a ciphertext policy (CP) functional encryption (FE) scheme [22], which is adaptively payload-hiding against chosen-plaintext attacks. The description of the CP-FE scheme is given in the full version of [22].

Roughly speaking, a secret signing key, sk_Γ, with attribute set Γ and a verification text, $\vec{\boldsymbol{c}}$, with access structure $\mathbb{S}$ (for signature verification) in our ABS scheme correspond to a secret decryption key, sk_Γ, with Γ and a ciphertext, $\vec{\boldsymbol{c}}$, with $\mathbb{S}$ in the CP-FE scheme, respectively. No counterpart of a signature, $\vec{\boldsymbol{s}}^*$, in the ABS exists in the CP-FE, and the privacy property for signature $\vec{\boldsymbol{s}}^*$ is also specific in ABS. Signature $\vec{\boldsymbol{s}}^*$ in ABS may be interpreted to be a decryption key specialized to decrypt a ciphertext with access structure $\mathbb{S}$, that is delegated from secret key sk_Γ.

The algorithms of the proposed ABS scheme can be described in the light of such correspondence to the CP-FE scheme:

Setup. Almost the same as that in the CP-FE scheme except that $\{\widehat{\mathbb{B}}^*_t\}_{t=1,\ldots,d+1}$ are revealed as a *public* parameter in our ABS, while they are *secret* in the CP-FE scheme. They are published in our ABS for the signature generation procedure Sig to meet the *privacy* of signers (for randomization). This implies an important gap between CP-FE and ABS.

KeyGen. Almost the same as that in the CP-FE scheme except that a (7 dimensional) space with basis $\mathbb{B}^*_{d+1}$ is additionally introduced in our ABS and two elements $\boldsymbol{k}^*_{d+1,1}$ and $\boldsymbol{k}^*_{d+1,2}$ in this space are included in a secret signing key in order to embed the hash value, $\mathsf{H}^{\lambda,\mathsf{D}}_{\mathsf{hk}}(m \,||\, \mathbb{S})$, of message m and access structure $\mathbb{S}$ in signature $\vec{\boldsymbol{s}}^*$.

Sig. Specific in ABS. To meet the privacy condition for $\vec{\boldsymbol{s}}^*$, a novel technique is employed to randomly generate a signature from sk_Γ and $\{\widehat{\mathbb{B}}^*_t\}_{t=1,\ldots,d+1}$.

Ver. Signature $\vec{\boldsymbol{s}}^*$ in the ABS is an endorsement to message m by a signer with attributes accepted by access structure $\mathbb{S}$. The signature verification in our ABS checks whether signature (or specific decryption key) $\vec{\boldsymbol{s}}^*$ works as a decryption key to decrypt a verification text (or a ciphertext) associated with $\mathbb{S}$ and $\mathsf{H}^{\lambda,\mathsf{D}}_{\mathsf{hk}}(m \,||\, \mathbb{S})$.

Security proofs. Roughly speaking, the *adaptive*-predicate unforgeability of the ABS under the KeyGen oracle attacks can be guaranteed by the *adaptive* payload-hiding property of the CP-FE, since a forged signature implies a decryption key specified for the challenge ciphertext to break the payload-hiding. Note that there are many subtleties in the proof of unforgeability for the ABS, e.g., the unforgeability should be ensured in the ABS even when publishing $\{\widehat{\mathbb{B}}^*_t\}_{t=1,\ldots,d+1}$ for the privacy requirement, while they are secret in the CP-FE. We develop a novel technique to resolve the difficulty. See the full version of this paper for more details.

4.2 Construction

For simplicity, here, we describe our ABS scheme for a specific parameter $\vec{n} := (d; 2, \ldots, 2)$ (see the remark of Definition 5). A general form of our ABS scheme is given in the full version.

We define function $\widetilde{\rho} : \{1, \ldots, \ell\} \to \{1, \ldots, d\}$ by $\widetilde{\rho}(i) := t$ if $\rho(i) = (t, v)$ or $\rho(i) = \neg(t, v)$, where ρ is given in access structure $\mathbb{S} := (M, \rho)$. In the proposed scheme, we assume that $\widetilde{\rho}$ is injective for $\mathbb{S} := (M, \rho)$. The full version of this paper shows how to relax the restriction.

$\mathsf{Setup}(1^\lambda,\ \vec{n} := (d; 2, \ldots, 2)):\ \mathsf{param}_{\mathbb{G}} := (q, \mathbb{G}, \mathbb{G}_T, G, e) \xleftarrow{\mathsf{R}} \mathcal{G}_{\mathsf{bpg}}(1^\lambda)$,

$\mathsf{hk} \xleftarrow{\mathsf{R}} \mathsf{KH}_\lambda,\ \psi \xleftarrow{\mathsf{U}} \mathbb{F}_q^\times,\ N_0 := 4,\ N_t := 7$ for $t = 1, \ldots, d+1$,

for $t = 0, \ldots, d+1$, $\mathsf{param}_{\mathbb{V}_t} := (q, \mathbb{V}_t, \mathbb{G}_T, \mathbb{A}_t, e) := \mathcal{G}_{\mathsf{dpvs}}(1^\lambda, N_t, \mathsf{param}_{\mathbb{G}})$,

$X_t := (\chi_{t,i,j})_{i,j} \xleftarrow{\mathsf{U}} GL(N_t, \mathbb{F}_q),\ (\vartheta_{t,i,j})_{i,j} := \psi \cdot (X_t^{-1})^{\mathrm{T}}$,

$\boldsymbol{b}_{t,i} := (\chi_{t,i,1}, \ldots, \chi_{t,i,N_t})_{\mathbb{A}_t},\ \mathbb{B}_t := (\boldsymbol{b}_{t,1}, \ldots, \boldsymbol{b}_{t,N_t})$,

$\boldsymbol{b}^*_{t,i} := (\vartheta_{t,i,1}, \ldots, \vartheta_{t,i,N_t})_{\mathbb{A}_t},\ \mathbb{B}^*_t := (\boldsymbol{b}^*_{t,1}, \ldots, \boldsymbol{b}^*_{t,N_t})$,

$g_T := e(G, G)^\psi,\ \ \mathsf{param}_{\vec{n}} := (\{\mathsf{param}_{\mathbb{V}_t}\}_{t=0,\ldots,d+1},\ g_T)$,

$\widehat{\mathbb{B}}_0 := (\boldsymbol{b}_{0,1}, \boldsymbol{b}_{0,4}),\ \ \widehat{\mathbb{B}}_t := (\boldsymbol{b}_{t,1}, \boldsymbol{b}_{t,2}, \boldsymbol{b}_{t,7})$ for $t = 1, \ldots, d+1$,

$\widehat{\mathbb{B}}^*_t := (\boldsymbol{b}^*_{t,1}, \boldsymbol{b}^*_{t,2}, \boldsymbol{b}^*_{t,5}, \boldsymbol{b}^*_{t,6})$ for $t = 1, \ldots, d+1$,

$\mathsf{sk} := \boldsymbol{b}^*_{0,1},\ \ \mathsf{pk} := (1^\lambda, \mathsf{hk}, \mathsf{param}_{\vec{n}}, \{\widehat{\mathbb{B}}_t\}_{t=0,\ldots,d+1}, \{\widehat{\mathbb{B}}^*_t\}_{t=1,\ldots,d+1}, \boldsymbol{b}^*_{0,3})$.

return sk, pk.

$\mathsf{KeyGen}(\mathsf{pk},\ \mathsf{sk},\ \ \Gamma := \{(t, x_t) \mid 1 \le t \le d\}):$

$\delta \xleftarrow{\mathsf{U}} \mathbb{F}_q^\times,\ \ \varphi_0, \varphi_{t,\iota}, \varphi_{d+1,1,\iota}, \varphi_{d+1,2,\iota} \xleftarrow{\mathsf{U}} \mathbb{F}_q$ for $t = 1, \ldots, d;\ \iota = 1, 2$;

$\boldsymbol{k}^*_0 := (\ \delta,\ 0,\ \varphi_0,\ 0\ )_{\mathbb{B}^*_0}$,

$\boldsymbol{k}^*_t := (\delta(\ 1,\ x_t\ ),\ \ 0,\ 0,\ \ \varphi_{t,1},\ \varphi_{t,2},\ \ 0)_{\mathbb{B}^*_t}$ for $(t, x_t) \in \Gamma$,

$\boldsymbol{k}^*_{d+1,1} := (\ \delta(\ 1,\ 0\ ),\ \ 0,\ 0,\ \ \varphi_{d+1,1,1},\ \varphi_{d+1,1,2},\ \ 0\ )_{\mathbb{B}^*_{d+1}}$,

$\boldsymbol{k}^*_{d+1,2} := (\ \delta(\ 0,\ 1\ ),\ \ 0,\ 0,\ \ \varphi_{d+1,2,1},\ \varphi_{d+1,2,2},\ \ 0\ )_{\mathbb{B}^*_{d+1}}$,

$T := \{0, (d+1, 1), (d+1, 2)\} \cup \{t\ \mid\ 1 \le t \le d,\ (t, x_t) \in \Gamma\}$,

return $\mathsf{sk}_\Gamma := (\Gamma, \{\boldsymbol{k}^*_t\}_{t \in T})$.

$\mathsf{Sig}(\mathsf{pk},\ \mathsf{sk}_\Gamma,\ m,\ \mathbb{S} := (M, \rho)):$ If $\mathbb{S} := (M, \rho)$ accepts $\Gamma := \{(t, x_t)\}$, then

compute I and $\{\alpha_i\}_{i \in I}$ such that $\sum_{i \in I} \alpha_i M_i = \vec{1}$, and

$I \subseteq \{i \in \{1, \ldots, \ell\} \mid\ [\rho(i) = (t, v_i)\ \wedge\ (t, x_t) \in \Gamma\ \wedge\ v_i = x_t]$

$\vee\ \ [\rho(i) = \neg(t, v_i)\ \wedge\ (t, x_t) \in \Gamma\ \wedge\ v_i \ne x_t]\ \}$,

$\xi \xleftarrow{\mathsf{U}} \mathbb{F}_q^\times,\ \ (\beta_i) \xleftarrow{\mathsf{U}} \{(\beta_1, \ldots, \beta_\ell)\ \mid\ \sum_{i=1}^{\ell} \beta_i M_i = \vec{0}\}$,

$\boldsymbol{s}^*_0 := \xi \boldsymbol{k}^*_0 + \boldsymbol{r}^*_0$, where $\boldsymbol{r}^*_0 \xleftarrow{\mathsf{U}} \mathsf{span}\langle \boldsymbol{b}^*_{0,3} \rangle$,

$\boldsymbol{s}^*_i := \gamma_i \cdot \xi \boldsymbol{k}^*_t + \sum_{\iota=1}^{2} y_{i,\iota} \cdot \boldsymbol{b}^*_{t,\iota} + \boldsymbol{r}^*_i$ for $1 \le i \le \ell$, where $\boldsymbol{r}_i^* \xleftarrow{\mathsf{U}} \mathsf{span}\langle \boldsymbol{b}_{t,5}^*, \boldsymbol{b}_{t,6}^* \rangle$, and $\gamma_i, \vec{y}_i := (y_{i,1}, y_{i,2})$ are defined as

$$\text{if } i \in I \ \wedge\ \rho(i) = (t, v_i), \quad \gamma_i := \alpha_i, \quad \vec{y}_i := \beta_i(1, v_i),$$

$$\text{if } i \in I \ \wedge\ \rho(i) = \neg(t, v_i), \quad \gamma_i := \frac{\alpha_i}{v_i - x_t}, \quad \vec{y}_i := \frac{\beta_i}{v_i - y_i}(1, y_i), \quad \text{where } y_i \xleftarrow{\mathsf{U}} \mathbb{F}_q \backslash \{v_i\},$$

$$\text{if } i \notin I \ \wedge\ \rho(i) = (t, v_i), \quad \gamma_i := 0, \quad \vec{y}_i := \beta_i(1, v_i),$$

$$\text{if } i \notin I \ \wedge\ \rho(i) = \neg(t, v_i), \quad \gamma_i := 0, \quad \vec{y}_i := \frac{\beta_i}{v_i - y_i}(1, y_i), \quad \text{where } y_i \xleftarrow{\mathsf{U}} \mathbb{F}_q \backslash \{v_i\},$$

$$\boldsymbol{s}_{\ell+1}^* := \xi(\boldsymbol{k}_{d+1,1}^* + \mathsf{H}_{\mathsf{hk}}^{\lambda,\mathsf{D}}(m \,||\, \mathbb{S}) \cdot \boldsymbol{k}_{d+1,2}^*) + \boldsymbol{r}_{\ell+1}^*, \quad \text{where } \boldsymbol{r}_{\ell+1}^* \xleftarrow{\mathsf{U}} \mathsf{span}\langle \boldsymbol{b}_{d+1,5}^*, \boldsymbol{b}_{d+1,6}^* \rangle,$$

return $\vec{\boldsymbol{s}}^* := (\boldsymbol{s}_0^*, \ldots, \boldsymbol{s}_{\ell+1}^*)$.

$\mathsf{Ver}(\mathsf{pk},\ m,\ \mathbb{S} := (M, \rho), \vec{\boldsymbol{s}}^*):\quad \vec{f} \xleftarrow{\mathsf{U}} \mathbb{F}_q^r,\ \vec{s}^{\mathrm{T}} := (s_1, \ldots, s_\ell)^{\mathrm{T}} := M \cdot \vec{f}^{\mathrm{T}},$

$s_0 := \vec{1} \cdot \vec{f}^{\mathrm{T}},\ \eta_0, \eta_{\ell+1}, \theta_{\ell+1}, s_{\ell+1} \xleftarrow{\mathsf{U}} \mathbb{F}_q,$

$\boldsymbol{c}_0 := (\ -s_0 - s_{\ell+1},\ 0,\ 0,\ \eta_0\ )_{\mathbb{B}_0},$

for $1 \le i \le \ell$,

if $\rho(i) = (t, v_i)$, return 0 if $\boldsymbol{s}_i^* \notin \mathbb{V}_t$, else

$$\boldsymbol{c}_i := (\ s_i + \theta_i v_i,\ -\theta_i,\ \ 0,\ 0,\ \ 0,\ 0,\ \ \eta_i\ )_{\mathbb{B}_t}, \quad \text{where } \theta_i, \eta_i \xleftarrow{\mathsf{U}} \mathbb{F}_q,$$

if $\rho(i) = \neg(t, v_i)$, return 0 if $\boldsymbol{s}_i^* \notin \mathbb{V}_t$, else

$$\boldsymbol{c}_i := (\ s_i(\ v_i,\ -1\ ),\ \ 0,\ 0,\ \ 0,\ 0,\ \ \eta_i\ )_{\mathbb{B}_t}, \quad \text{where } \eta_i \xleftarrow{\mathsf{U}} \mathbb{F}_q,$$

$\boldsymbol{c}_{\ell+1} := (\ s_{\ell+1} - \theta_{\ell+1} \cdot \mathsf{H}_{\mathsf{hk}}^{\lambda,\mathsf{D}}(m \,||\, \mathbb{S}),\ \theta_{\ell+1},\ \ 0,\ 0,\ \ 0,\ 0,\ \ \eta_{\ell+1}\ )_{\mathbb{B}_{d+1}},$

return 0 if $e(\boldsymbol{b}_{0,1}, \boldsymbol{s}_0^*) = 1$,

return 1 if $\prod_{i=0}^{\ell+1} e(\boldsymbol{c}_i, \boldsymbol{s}_i^*) = 1$, return 0 otherwise.

[Correctness] $\prod_{i=0}^{\ell+1} e(\boldsymbol{c}_i, \boldsymbol{s}_i^*)$

$$= e(\boldsymbol{c}_0, \boldsymbol{k}_0^*)^{\xi} \cdot \prod_{i \in I} e(\boldsymbol{c}_i, \boldsymbol{k}_t^*)^{\gamma_i \xi} \cdot \prod_{i=1}^{\ell} \prod_{\iota=1}^{2} e(\boldsymbol{c}_i, \boldsymbol{b}_{t,\iota}^*)^{y_{i,\iota}} \cdot e(\boldsymbol{c}_{\ell+1}, \boldsymbol{s}_{\ell+1}^*)$$
$$= g_T^{\xi\delta(-s_0 - s_{\ell+1})} \cdot \prod_{i \in I} g_T^{\xi\delta\alpha_i s_i} \cdot \prod_{i=1}^{\ell} g_T^{\beta_i s_i} \cdot g_T^{\xi\delta s_{\ell+1}}$$
$$= g_T^{\xi\delta(-s_0 - s_{\ell+1})} \cdot g_T^{\xi\delta s_0} \cdot g_T^{\xi\delta s_{\ell+1}} = 1.$$

4.3 Security

Theorem 1. *The proposed ABS scheme is perfectly private.*

Theorem 2. *The proposed ABS scheme is unforgeable (adaptive-predicate unforgeable) under the DLIN assumption and the existence of collision resistant hash functions.*

For any adversary $\mathcal{A}$, there exist probabilistic machines $\mathcal{E}_1, \mathcal{E}_{2,h}^+, \mathcal{E}_{2,h+1}$ $(h = 0, \ldots, \nu_1 - 1)$, $\mathcal{E}_{3,h}, \mathcal{E}_{4,h}$ $(h = 1, \ldots, \nu_2)$, whose running times are essentially the same as that of $\mathcal{A}$, such that for any security parameter λ,

$$\mathsf{Adv}_{\mathcal{A}}^{\mathsf{ABS,UF}}(\lambda) \le \mathsf{Adv}_{\mathcal{E}_1}^{\mathsf{DLIN}}(\lambda) + \sum_{h=0}^{\nu_1-1}\left(\mathsf{Adv}_{\mathcal{E}_{2,h}^{+}}^{\mathsf{DLIN}}(\lambda) + \mathsf{Adv}_{\mathcal{E}_{2,h+1}}^{\mathsf{DLIN}}(\lambda)\right)$$
$$+ \sum_{h=1}^{\nu_2}\left(\mathsf{Adv}_{\mathcal{E}_{3,h}}^{\mathsf{DLIN}}(\lambda) + \mathsf{Adv}_{\mathcal{E}_{4,h}}^{\mathsf{H,CR}}(\lambda)\right) + \epsilon,$$

where ν_1 *is the maximum number of* $\mathcal{A}$*'s* KeyGen *queries,* ν_2 *is the maximum number of* $\mathcal{A}$*'s* AltSig *queries, and* $\epsilon := ((2d+16)\nu_1 + 18\nu_2 + 2d + 18)/q$.

The proofs of Theorems 1 and 2 are given in the full version of this paper.

4.4 Performance

In this section, we compare the efficiency and security of the proposed ABS scheme with the existing ABS schemes in the standard model (two typical instantiations) [19] as well as the ABS scheme in the generic group model [18] (as a benchmark). Since all of these schemes can be implemented over a *prime order* pairing group, the size of a group element can be around the size of $\mathbb{F}_q$ (e.g., 256 bits). In Table 1, ℓ and r represent the size of the underlying access structure matrix M for a predicate, i.e., $M \in \mathbb{F}_q^{\ell \times r}$. For example, some predicate with 4 AND and 5 OR gates as well as 10 variables may be expressed by a 10×5 matrix, and a predicate with 49 AND and 50 OR gates as well as 100 variables may be expressed by a 100×50 matrix (see the appendix of [15]). λ is the security parameter (e.g., 128).

Table 1. Comparison with the Existing ABS Schemes

	MPR08 [18]	MPR10 [19] (Boneh-Boyen based)	MPR10 [19] (Waters based)	Proposed
Signature size (# of group elts)	$\ell + r + 2$	$51\ell + 2r + 18\lambda\ell$	$36\ell + 2r + 9\lambda + 12$	$7\ell + 11$
Model	generic group model	standard model	standard model	standard model
Security	full	full	full	full
Assumptions	CR hash	q-SDH and DLIN	DLIN	DLIN and CR hash
Predicates	monotone	monotone	monotone	non-monotone
Sig. size example 1 ($\ell = 10, r = 5$, $\lambda = 128$)	17	23560	1534	81
Sig. size example 2 ($\ell = 100, r = 50$, $\lambda = 128$)	152	282400	4864	711

5 Multi-Authority ABS (MA-ABS)

5.1 Definitions and Security of MA-ABS

We follow the model and security definitions of MA-ABS in [18, 19].

Definition 10 (Multi-Authority ABS : MA-ABS). *A multi-authority ABS scheme consists of the following algorithms/protocols.*

TSetup. *This is a randomized algorithm. The signature trustee runs algorithm* $\mathsf{TSetup}(1^\lambda)$ *which outputs trustee public key* tpk *and trustee secret key* tsk.

UserReg. *This is a randomized algorithm. When a user with user id* uid *registers with the signature trustee, the trustee runs* $\mathsf{UserReg}(\mathsf{tpk}, \mathsf{tsk}, \mathsf{uid})$ *which outputs public user-token* $\mathsf{token}_{\mathsf{uid}}$. *The trustee gives* $\mathsf{token}_{\mathsf{uid}}$ *to the user.*

ASetup. *This is a randomized algorithm. Attribute authority* t $(1 \le t \le d)$ *who wishes to issue attributes runs* $\mathsf{ASetup}(\mathsf{tpk})$ *which outputs attribute-authority public key* apk_t *and attribute-authority secret key* ask_t. *The attribute authority,* t*, publishes* apk_t *and stores* ask_t.

AttrGen. *This is a randomized algorithm. When attribute authority* t *issues user* uid *a secret key associated with attribute* x_t*, first it obtains (from the user) her user-token* $\mathsf{token}_{\mathsf{uid}}$*, and runs token verification algorithm* $\mathsf{TokenVerify}(\mathsf{tpk}, \mathsf{uid}, \mathsf{token}_{\mathsf{uid}})$. *If the token is verified, then it runs* $\mathsf{AttrGen}(\mathsf{tpk}, t, \mathsf{ask}_t, \mathsf{token}_{\mathsf{uid}}, x_t)$ *that outputs attribute secret key* usk_t. *The attribute authority gives* usk_t *to the user.*

Sig. *This is a randomized algorithm. A user signs message* m *with claim-predicate (access structure)* $\mathbb{S} := (M, \rho)$*, only if there is a set of attributes* Γ *such that* $\mathbb{S}$ *accepts* Γ*, the user has obtained a set of keys* $\{\mathsf{usk}_t \mid (t, x_t) \in \Gamma\}$ *from the attribute authorities. Then signature* σ *can be generated using* $\mathsf{Sig}(\mathsf{tpk}, \mathsf{token}_{\mathsf{uid}}, \{\mathsf{apk}_t, \mathsf{usk}_t \mid (t, x_t) \in \Gamma\}, m, \mathbb{S})$*, where* $\mathsf{usk}_t \xleftarrow{\mathsf{R}} \mathsf{AttrGen}(\mathsf{tpk}, t, \mathsf{ask}_t, \mathsf{token}_{\mathsf{uid}}, x_t)$.

Ver. *To verify signature* σ *on message* m *with claim-predicate (access structure)* $\mathbb{S}$*, a user runs* $\mathsf{Ver}(\mathsf{tpk}, \{\mathsf{apk}_t\}, m, \mathbb{S}, \sigma)$ *which outputs boolean value* $\mathsf{accept} := 1$ *or* $\mathsf{reject} := 0$.

The definition of perfect privacy for the multi-authority (MA) ABS is essentially the same as that of the single-authority (SA) ABS (Definition 8). The major difference of the unforgeability of MA-ABS and SA-ABS is that adversary $\mathcal{A}$ can corrupt an arbitrary subset of attribute authorities provided that adversary $\mathcal{A}$ cannot make a trivial forgery attack. These definitions are given in the full version of this paper.

5.2 Construction

The key idea of our construction of MA-ABS scheme is to share $G_{\mathsf{uid}} := \delta G_1$ as well as G_0 and G_1 among attribute authorities to generate $\delta \boldsymbol{b}^*_{t,i}$ by each authority t. Hence, G_0 and G_1 are included in tpk and $G_{\mathsf{uid}} := \delta G_1$ is shared with attribute authorities through the user's token $\mathsf{token}_{\mathsf{uid}}$.

For matrix $X := (\chi_{i,j})_{i,j=1,\ldots,N} \in \mathbb{F}_q^{N\times N}$ and element $\boldsymbol{v}$ in N-dimensional $\mathbb{V}$, $X(\boldsymbol{v})$ denotes $\sum_{i=1,j=1}^{N,N} \chi_{i,j}\phi_{i,j}(\boldsymbol{v})$ using canonical maps $\{\phi_{i,j}\}$ (Definition 2). Similarly, for matrix $(\vartheta_{i,j}) := (X^{-1})^{\mathrm{T}}$, $(X^{-1})^{\mathrm{T}}(\boldsymbol{v}) := \sum_{i=1,j=1}^{N,N} \vartheta_{i,j}\phi_{i,j}(\boldsymbol{v})$. It holds that $e(X(\boldsymbol{x}), (X^{-1})^{\mathrm{T}}(\boldsymbol{y})) = e(\boldsymbol{x}, \boldsymbol{y})$ for any $\boldsymbol{x}, \boldsymbol{y} \in \mathbb{V}$.

Moreover, $(\mathsf{G}_{\mathsf{SIG}}, \mathsf{S}, \mathsf{V})$ is a (conventional) unforgeable signature scheme.

$\mathsf{TSetup}(1^\lambda)$: $\mathsf{param}_{\mathbb{G}} := (q, \mathbb{G}, \mathbb{G}_T, G, e) \xleftarrow{\mathsf{R}} \mathcal{G}_{\mathsf{bpg}}(1^\lambda)$,

$\mathsf{hk} \xleftarrow{\mathsf{R}} \mathsf{KH}_\lambda$, $(\mathsf{verk}, \mathsf{sigk}) \xleftarrow{\mathsf{R}} \mathsf{G}_{\mathsf{SIG}}(1^\lambda)$ $N_0 := 4$, $N_{d+1} := 7$, $\kappa, \xi \xleftarrow{\mathsf{U}} \mathbb{F}_q^\times$,

for $t = 0, d+1$, $\mathsf{param}_{\mathbb{V}_t} := (q, \mathbb{V}_t, \mathbb{G}_T, \mathbb{A}_t, e) := \mathcal{G}_{\mathsf{dpvs}}(1^\lambda, N_t, \mathsf{param}_{\mathbb{G}})$,

$X_t := (\chi_{t,i,j})_{i,j} \xleftarrow{\mathsf{U}} GL(N_t, \mathbb{F}_q)$, $(\vartheta_{t,i,j})_{i,j} := (X_t^{-1})^{\mathrm{T}}$,

$\boldsymbol{b}_{t,i} := \kappa(\chi_{t,i,1}, \ldots, \chi_{t,i,N_t})_{\mathbb{A}_t}$, $\mathbb{B}_t := (\boldsymbol{b}_{t,1}, \ldots, \boldsymbol{b}_{t,N_t})$,

$\boldsymbol{b}^*_{t,i} := \xi(\vartheta_{t,i,1}, \ldots, \vartheta_{t,i,N_t})_{\mathbb{A}_t}$, $\mathbb{B}^*_t := (\boldsymbol{b}^*_{t,1}, \ldots, \boldsymbol{b}^*_{t,N_t})$,

$G_0 := \kappa G$, $G_1 := \xi G$, $g_T := e(G,G)^{\kappa\xi}$,

$\widehat{\mathbb{B}}_0 := (\boldsymbol{b}_{0,1}, \boldsymbol{b}_{0,4})$, $\widehat{\mathbb{B}}_{d+1} := (\boldsymbol{b}_{d+1,1}, \boldsymbol{b}_{d+1,2}, \boldsymbol{b}_{d+1,7})$,

$\widehat{\mathbb{B}}^*_{d+1} := (\boldsymbol{b}^*_{d+1,1}, \boldsymbol{b}^*_{d+1,2}, \boldsymbol{b}^*_{d+1,5}, \boldsymbol{b}^*_{d+1,6})$,

$\mathsf{tsk} := (\boldsymbol{b}^*_{0,1}, \mathsf{sigk})$,

$\mathsf{tpk} := (1^\lambda, \mathsf{hk}, \{\mathsf{param}_{\mathbb{V}_t}, \widehat{\mathbb{B}}_t\}_{t=0,d+1}, \boldsymbol{b}^*_{0,3}, \widehat{\mathbb{B}}^*_{d+1}, g_T, G_0, G_1, \mathsf{verk})$,

return $(\mathsf{tsk}, \mathsf{tpk})$.

$\mathsf{UserReg}(\mathsf{tpk}, \mathsf{tsk}, \mathsf{uid})$: $\delta \xleftarrow{\mathsf{U}} \mathbb{F}_q^\times$, $\varphi_0, \varphi_{d+1,1,\iota}, \varphi_{d+1,2,\iota} \xleftarrow{\mathsf{U}} \mathbb{F}_q$, $G_{\mathsf{uid}} := \delta G_1$,

$\boldsymbol{k}^*_0 := (\ \delta,\ 0,\ \varphi_0,\ 0\ )_{\mathbb{B}^*_0}$,

$\boldsymbol{k}^*_{d+1,1} := (\ \delta(1,\ 0),\ \ 0,\ 0,\ \ \varphi_{d+1,1,1},\ \varphi_{d+1,1,2},\ \ 0\ )_{\mathbb{B}^*_{d+1}}$,

$\boldsymbol{k}^*_{d+1,2} := (\ \delta(0,\ 1),\ \ 0,\ 0,\ \ \varphi_{d+1,2,1},\ \varphi_{d+1,2,2},\ \ 0\ )_{\mathbb{B}^*_{d+1}}$,

$\mathsf{usk}_0 := (\boldsymbol{k}^*_0, \boldsymbol{k}^*_{d+1,1}, \boldsymbol{k}^*_{d+1,2})$, $\sigma_{\mathsf{uid}} := \mathsf{S}(\mathsf{sigk}, (\mathsf{uid}, G_{\mathsf{uid}}))$,

return $\mathsf{token}_{\mathsf{uid}} := (\mathsf{uid}, G_{\mathsf{uid}}, \sigma_{\mathsf{uid}}, \mathsf{usk}_0)$.

$\mathsf{ASetup}(\mathsf{tpk})$: $\boldsymbol{u}_{j,i} := (0^{i-1},\ G_j,\ 0^{7-i})$ for $j=0,1; i=1,..,7$, $X_t \xleftarrow{\mathsf{U}} GL(7, \mathbb{F}_q)$,

$\mathbb{B}_t := (\boldsymbol{b}_{t,i})_{i=1,\ldots,7} := (X_t(\boldsymbol{u}_{0,1}), \ldots, X_t(\boldsymbol{u}_{0,7}))$,

$\mathbb{B}^*_t := (\boldsymbol{b}^*_{t,i})_{i=1,\ldots,7} := ((X_t^{-1})^{\mathrm{T}}(\boldsymbol{u}_{1,1}), \ldots, (X_t^{-1})^{\mathrm{T}}(\boldsymbol{u}_{1,7}))$,

$\widehat{\mathbb{B}}_t := (\boldsymbol{b}_{t,1}, \boldsymbol{b}_{t,2}, \boldsymbol{b}_{t,7})$, $\widehat{\mathbb{B}}^*_t := (\boldsymbol{b}^*_{t,1}, \boldsymbol{b}^*_{t,2}, \boldsymbol{b}^*_{t,5}, \boldsymbol{b}^*_{t,6})$,

return $(\mathsf{ask}_t := X_t,\ \mathsf{apk}_t := (\widehat{\mathbb{B}}_t, \widehat{\mathbb{B}}^*_t))$.

$\mathsf{TokenVerify}(\mathsf{tpk}, \mathsf{uid}, \mathsf{token}_{\mathsf{uid}})$ holds iff $\mathsf{V}(\mathsf{verk}, (\mathsf{uid}, G_{\mathsf{uid}}), \sigma_{\mathsf{uid}}) = 1$.

$\mathsf{AttrGen}(\mathsf{tpk}, t, \mathsf{ask}_t, \mathsf{token}_{\mathsf{uid}}, x_t \in \mathbb{F}_q)$: $\varphi_{t,1}, \varphi_{t,2} \xleftarrow{\mathsf{U}} \mathbb{F}_q$,

$\boldsymbol{k}^*_t := (X_t^{-1})^{\mathrm{T}}((\ G_{\mathsf{uid}},\ x_t G_{\mathsf{uid}},\ \ 0,\ 0,\ \ \varphi_{t,1}G_1,\ \varphi_{t,2}G_1,\ \ 0\ ))$,

that is, $\boldsymbol{k}^*_t = (\ \delta,\ \delta x_t,\ \ 0,\ 0,\ \ \varphi_{t,1},\ \varphi_{t,2},\ \ 0\ )_{\mathbb{B}^*_t}$,

return $\mathsf{usk}_t := \boldsymbol{k}^*_t$.

$\mathsf{Sig}(\mathsf{tpk},\ \mathsf{token}_{\mathsf{uid}}, \{\mathsf{apk}_t, \mathsf{usk}_t \xleftarrow{\mathsf{R}} \mathsf{AttrGen}(\mathsf{tpk}, t, \mathsf{ask}_t, \mathsf{token}_{\mathsf{uid}}, x_t) \mid (t, x_t) \in \Gamma\}$, $m, \mathbb{S} := (M, \rho))$ and $\mathsf{Ver}(\mathsf{tpk},\ \{\mathsf{apk}_t\}_{t=1,\dots,d},\ m,\ \mathbb{S} := (M, \rho), \overrightarrow{s}^*)$ are essentially the same as those in Section 4.2.

5.3 Security

Theorem 3. *The proposed MA-ABS scheme is perfectly private.*

Theorem 4. *The proposed MA-ABS scheme is unforgeable (adaptive-predicate unforgeable) under the DLIN assumption and the existence of collision resistant hash functions.*

For any adversary $\mathcal{A}$, there exist probabilistic machines $\mathcal{E}_1, \mathcal{E}_{2,h}^{+}, \mathcal{E}_{2,h+1}$ $(h = 0, \dots, \nu_1 - 1)$, $\mathcal{E}_{3,h}, \mathcal{E}_{4,h}$ $(h = 1, \dots, \nu_2)$, whose running times are essentially the same as that of $\mathcal{A}$, such that for any security parameter λ,

$$\mathsf{Adv}_{\mathcal{A}}^{\mathsf{MA\text{-}ABS,UF}}(\lambda) \leq \mathsf{Adv}_{\mathcal{E}_1}^{\mathsf{DLIN}}(\lambda) + \sum_{h=0}^{\nu_1-1} \left(\mathsf{Adv}_{\mathcal{E}_{2,h}^{+}}^{\mathsf{DLIN}}(\lambda) + \mathsf{Adv}_{\mathcal{E}_{2,h+1}}^{\mathsf{DLIN}}(\lambda)\right)$$
$$+ \sum_{h=1}^{\nu_2} \left(\mathsf{Adv}_{\mathcal{E}_{3,h}}^{\mathsf{DLIN}}(\lambda) + \mathsf{Adv}_{\mathcal{E}_{4,h}}^{\mathsf{H,CR}}(\lambda)\right) + \epsilon,$$

where ν_1 is the maximum number of $\mathcal{A}$'s UserReg *queries, ν_2 is the maximum number of $\mathcal{A}$'s* AltSig *queries, and $\epsilon := ((2d + 16)\nu_1 + 18\nu_2 + 2d + 18)/q$.*

The proofs of Theorems 3 and 4 are given in the full version of this paper.

References

1. Beimel, A.: Secure schemes for secret sharing and key distribution. PhD Thesis, Israel Institute of Technology, Technion, Haifa, Israel (1996)
2. Belenkiy, M., Camenisch, J., Chase, M., Kohlweiss, M., Lysyanskaya, A., Shacham, H.: Randomizable proofs and delegatable anonymous credentials. In: Halevi, S. (ed.) CRYPTO 2009. LNCS, vol. 5677, pp. 108–125. Springer, Heidelberg (2009)
3. Belenkiy, M., Chase, M., Kohlweiss, M., Lysyanskaya, A.: P-signatures and noninteractive anonymous credentials. In: Canetti, R. (ed.) TCC 2008. LNCS, vol. 4948, pp. 356–374. Springer, Heidelberg (2008)
4. Boyen, X.: Mesh signatures. In: Naor, M. (ed.) EUROCRYPT 2007. LNCS, vol. 4515, pp. 210–227. Springer, Heidelberg (2007)
5. Camenisch, J., Groß, T.: Efficient attributes for anonymous credentials. In: CCS 2008, pp. 345–356. ACM, New York (2008)
6. Camenisch, J., Lysyanskaya, A.: An efficient system for non-transferable anonymous credentials with optional anonymity revocation. In: Pfitzmann, B. (ed.) EUROCRYPT 2001. LNCS, vol. 2045, pp. 93–118. Springer, Heidelberg (2001)
7. Camenisch, J., Lysyanskaya, A.: Signature schemes and anonymous credentials from bilinear maps. In: Franklin, M.K. (ed.) CRYPTO 2004. LNCS, vol. 3152, pp. 56–72. Springer, Heidelberg (2004)
8. Chaum, D.: Security without identification: Transaction systems to make big brother obsolete. CACM 28(10), 1030–1044 (1985)
9. Chaum, D., van Heyst, E.: Group signatures. In: Davies, D.W. (ed.) EUROCRYPT 1991. LNCS, vol. 547, pp. 257–265. Springer, Heidelberg (1991)

10. Groth, J., Sahai, A.: Efficient non-interactive proof systems for bilinear groups. In: Smart, N.P. (ed.) EUROCRYPT 2008. LNCS, vol. 4965, pp. 415–432. Springer, Heidelberg (2008)
11. Guo, S., Zeng, Y.: Attribute-based signature scheme. In: ISA 2008, pp. 509–511. IEEE, Los Alamitos (2008)
12. Khader, D.: Attribute based group signatures, ePrint, IACR, `http://eprint.iacr.org/2007/159`
13. Khader, D.: Attribute based group signature with revocation. ePrint, IACR, `http://eprint.iacr.org/2007/241`
14. Lewko, A., Okamoto, T., Sahai, A., Takashima, K., Waters, B.: Fully Secure Functional Encryption: Attribute-Based Encryption and (Hierarchical) Inner Product Encryption. In: Gilbert, H. (ed.) EUROCRYPT 2010. LNCS, vol. 6110, pp. 62–91. Springer, Heidelberg (2010)
15. Lewko, A.B., Waters, B.: Decentralizing attribute-based encryption. ePrint, IACR, `http://eprint.iacr.org/2010/351`
16. Li, J., Au, M.H., Susilo, W., Xie, D., Ren, K.: Attribute-based signature and its application. In: ASIACCS 2010, pp. 60–69. ACM, New York (2010)
17. Li, J., Kim, K.: Attribute-based ring signatures. ePrint, IACR, `http://eprint.iacr.org/2008/394`
18. Maji, H., Prabhakaran, M., Rosulek, M.: Attribute-based signatures: Achieving attribute-privacy and collusion-resistance. ePrint, IACR, `http://eprint.iacr.org/2008/328`
19. Maji, H., Prabhakaran, M., Rosulek, M.: Attribute-based signatures. To appear in CT-RSA 2011, `http://eprint.iacr.org/2010/595`
20. Okamoto, T., Takashima, K.: Homomorphic encryption and signatures from vector decomposition. In: Galbraith, S.D., Paterson, K.G. (eds.) Pairing 2008. LNCS, vol. 5209, pp. 57–74. Springer, Heidelberg (2008)
21. Okamoto, T., Takashima, K.: Hierarchical predicate encryption for inner-products. In: Matsui, M. (ed.) ASIACRYPT 2009. LNCS, vol. 5912, pp. 214–231. Springer, Heidelberg (2009)
22. Okamoto, T., Takashima, K.: Fully secure functional encryption with general relations from the decisional linear assumption. In: Rabin, T. (ed.) CRYPTO 2010. LNCS, vol. 6223, pp. 191–208. Springer, Heidelberg (2010), `http://eprint.iacr.org/2010/563`
23. Rivest, R.L., Shamir, A., Tauman, Y.: How to leak a secret. In: Boyd, C. (ed.) ASIACRYPT 2001. LNCS, vol. 2248, pp. 552–565. Springer, Heidelberg (2001)
24. Shahandashti, S.F., Safavi-Naini, R.: Threshold attribute-based signatures and their application to anonymous credential systems. In: Preneel, B. (ed.) AFRICACRYPT 2009. LNCS, vol. 5580, pp. 198–216. Springer, Heidelberg (2009)
25. Shamir, A.: Identity-based cryptosystems and signature schemes. In: Blakely, G.R., Chaum, D. (eds.) CRYPTO 1984. LNCS, vol. 196, pp. 47–53. Springer, Heidelberg (1985)
26. Waters, B.: Dual system encryption: Realizing fully secure IBE and HIBE under simple assumptions. In: Halevi, S. (ed.) CRYPTO 2009. LNCS, vol. 5677, pp. 619–636. Springer, Heidelberg (2009)
27. Yang, P., Cao, Z., Dong, X.: Fuzzy identity based signature. ePrint, IACR, `http://eprint.iacr.org/2008/002`

Ciphertext-Policy Attribute-Based Encryption: An Expressive, Efficient, and Provably Secure Realization

Brent Waters*

University of Texas at Austin
bwaters@cs.utexas.edu

Abstract. We present a new methodology for realizing Ciphertext-Policy Attribute Encryption (CP-ABE) under concrete and noninteractive cryptographic assumptions in the standard model. Our solutions allow any encryptor to specify access control in terms of any access formula over the attributes in the system. In our most efficient system, ciphertext size, encryption, and decryption time scales linearly with the complexity of the access formula. The only previous work to achieve these parameters was limited to a proof in the generic group model.

We present three constructions within our framework. Our first system is proven selectively secure under a assumption that we call the decisional Parallel Bilinear Diffie-Hellman Exponent (PBDHE) assumption which can be viewed as a generalization of the BDHE assumption. Our next two constructions provide performance tradeoffs to achieve provable security respectively under the (weaker) decisional Bilinear-Diffie-Hellman Exponent and decisional Bilinear Diffie-Hellman assumptions.

1 Introduction

Public-Key encryption is a powerful mechanism for protecting the confidentiality of stored and transmitted information. Traditionally, encryption is viewed as a method for a user to share data to a targeted user or device. While this is useful for applications where the data provider knows specifically which user he wants to share with, in many applications the provider will want to share data according to some policy based on the receiving user's credentials.

Sahai and Waters [35] presented a new vision for encryption where the data provider can express how he wants to share data in the encryption algorithm itself. The data provider will provide a predicate $f(\cdot)$ describing how he wants to share the data and a user will be ascribed a secret key associated with their credentials X; the user with credentials X can decrypt a ciphertext encrypted

* Supported by NSF CNS-0716199, CNS-0915361, and CNS-0952692, Air Force Office of Scientific Research (AFO SR) under the MURI award for "Collaborative policies and assured information sharing" (Project PRESIDIO), Department of Homeland Security Grant 2006-CS-001-000001-02 (subaward 641), a Google Faculty Research award, and the Alfred P. Sloan Foundation.

D. Catalano et al. (Eds.): PKC 2011, LNCS 6571, pp. 53–70, 2011.

with predicate f if $f(X) = 1$. Sahai and Waters [35] presented a particular formulation of this problem that they called Attribute-Based Encryption (ABE), in which a user's credentials is represented by a set of string called "attributes" and the predicate is represented by a formula over these attributes. Several techniques used by SW were inspired by prior work on Identity-Based Encryption [36, 13, 23, 18, 10]. One drawback of the Sahai-Waters approach is that their initial construction was limited to handling formulas consisting of one threshold gate.

In subsequent work, Goyal, Pandey, Sahai, and Waters [27] further clarified the concept of Attribute-Based Encryption. In particular, they proposed two complementary forms of ABE. In the first, Key-Policy ABE, attributes are used to annotate the ciphertexts and formulas over these attributes are ascribed to users' secret keys. The second type, Ciphertext-Policy ABE, is complementary in that attributes are used to describe the user's credentials and the formulas over these credentials are attached to the ciphertext by the encrypting party. In addition, Goyal et al. [27] provided a construction for Key-Policy ABE that was very expressive in that it allowed the policies (attached to keys) to be expressed by *any* monotonic formula over encrypted data. The system was proved selectively secure under the Bilinear Diffie-Hellman assumption. However, they left creating expressive Ciphertext Policy ABE schemes as an open problem.

The first work to explicitly address the problem of Ciphertext-Policy Attribute-Based Encryption was by Bethencourt, Sahai, and Waters [7]. They described an efficient system that was expressive in that it allowed an encryptor to express an access predicate f in terms of any monotonic formula over attributes. Their system achieved analogous expressiveness and efficiency to the Goyal et al. construction, but in the Ciphertext-Policy ABE setting. While the BSW construction is very expressive, the proof model used was less than ideal — the authors only showed the scheme secure in the generic group model, an artificial model which assumes the attacker needs to access an oracle in order to perform any group operations[1].

Recently, ABE has been applied in building a variety of secure systems [34, 40, 9, 8]. These systems motivate the need for ABE constructions that are both foundationally sound and practical.

Ciphertext Policy ABE in the Standard Model. The lack of satisfaction with generic group model proofs has motivated the problem of finding an expressive CP-ABE system under a more solid model. There have been multiple approaches in this direction.

First, we can view the Sahai-Waters[35] construction most "naturally" as Key-Policy ABE for a threshold gate. In their work, Sahai and Waters describe how to realize Ciphertext-Policy ABE for threshold gates by "grafting" so called "dummy attributes" over their basic system. Essentially, they transformed a KP-ABE system into a CP-ABE one with the expressiveness of a single threshold

[1] Alternatively, we could derive a concrete, but interactive and complicated assumption directly from the scheme itself and argue that the scheme is secure under this assumption. However, this view is also not very satisfactory.

gate[2]. Cheung and Newport[22] provide a direct construction for constructing a policy with a single AND gate under the Bilinear Diffie-Hellman assumption. Their approach has the drawbacks that it only allows a fixed number of system attributes and is limited to an AND gate (does not enable thresholds). In retrospect these two limitations actually make it less expressive than the SW transformation, although this wasn't necessarily immediately apparent.

Most recently, Goyal, Jain, Pandey, and Sahai [26] generalized the transformational approach to show how to transform a KP-ABE system into a CP-ABE one using what they call a "universal access tree". In particular, they provided a mapping onto a "universal" (or complete) access tree of up to depth d formulas consisting of threshold gates of input size m, where m and d are chosen by the setup algorithm. They applied a similar "dummy attribute" approach.

In order to accommodate a general access formula of size n, their scheme first translates this into a balanced formula. Under standard techniques a formula of size n can be "balanced" such that any formula (tree) of size n can be covered by a complete tree of size approximately $O(n^{3.42})$. Their work was the first feasibility result for expressive CP-ABE under a non-interactive assumption. Unfortunately, the parameters of ciphertext and private key sizes add encryption and decryption complexity blow up (in the worst case) by an $n^{3.42}$ factor limiting its usefulness in practice. For instance, in a system with U attributes defined and n nodes the ciphertext overhead will be approximately a factor of $U \cdot n^{2.42}$ greater than that of the BSW system. To give a concrete example, for the modest parameters of universe size $U = 100$ attributes and a formula of 20 nodes the blowup factor relative to BSW is approximately 140,000.

Our Contribution. We present a new methodology for realizing Ciphertext-Policy ABE systems from a general set of access structures in the *standard model* under concrete and non-interactive assumptions. Both the ciphertext overhead and encryption time scale with $O(n)$ where n is the size of the formula. In addition, decryption time scales with the number of nodes.

Our first system allows an encryption algorithm to specify an access formula in terms of any access formula. In fact our techniques are slightly more general. We express access control by a Linear Secret Sharing Scheme (LSSS) matrix M over the attributes in the system. Previously used structures such as formulas (equivalently tree structures) can be expressed succinctly [6] in terms of a LSSS. *We do not lose any efficiency by using the more general LSSS representation as opposed to the previously used tree access structure descriptions.* Thus, we achieve the same performance and functionality as the Bethencourt, Sahai, and Waters construction, but under the standard model.

In addition, we provide two other constructions that tradeoff some performance parameters for provable security under the respective weaker assumptions of decisional-Bilinear Diffie-Hellman Exponent (d-BDHE) and decisional-Bilinear Diffie-Hellman assumptions. In Table 1 we summarize the comparisons between our schemes and the GJPS and BSW CP-ABE systems in terms of ciphertext and

[2] The Sahai-Waters construction was given prior to the Key-Policy and Ciphertext-Policy distinction; our interpretation is a retrospective one.

key sizes and encryption and decryption times. Taken all together our first scheme realizes the same efficiency parameters as the BSW encryption scheme, but under a concrete security assumption. At the same time, our d-BDH construction is proved under the same assumption as the GJPS system and achieves significantly better performance.

Our Techniques. Our techniques provide a framework for *directly* realizing provably secure CP-ABE systems. In our systems, the ciphertext distributes shares of a secret encryption exponent s across different attributes according to the access control LSSS matrix M.

A user's private key is associated with a set S of attributes and he will be able to decrypt a ciphertext iff his attributes "satisfy" the access matrix associated with the ciphertext. As in previous ABE systems, the primary challenge is to prevent users from realizing collusion attacks. Our main tool to prevent this is to randomize each key with an freshly chosen exponent t. During decryption, each share will be multiplied by a factor t in the exponent. Intuitively, this factor should "bind" the components of one user's key together so that they cannot be combined with another user's key components. During decryption, the different shares (in the exponent) that the algorithm combines are multiplied by a factor of t. Ultimately, these randomized shares are only useful to that one particular key.

Our construction's structures and high level intuition for security is similar to the BSW construction. The main novelty in our paper is provide a method for proving security of such a construction. The primary challenge one comes across is (in the selective model) how to create a reduction that embeds a complex access structure in a short number of parameters. All prior ABE schemes follow a "partitioning" strategy for proving security where the reduction algorithm sets up the public parameters such that it knows all the private keys that it needs to give out, yet it cannot give out private keys that can trivially decrypt the challenge ciphertext. In prior KP-ABE schemes the challenge ciphertext was associated with a set S^* of attributes. This structure could fairly easily be embedded in a reduction as the public parameter for each attribute was simply treated differently depending whether or not it was in S^*. In CP-ABE, the situation is much more complicated as ciphertexts are associated with a potentially large access structure M^* that includes attributes multiple times. In general, the size of M^* is much larger than the size of the public parameters[3]. Consequently, there is not a simple "on or off" method of programming this into the parameters. Arguably, it is this challenge that lead the BSW paper to apply the generic group heuristic and GJPS paper to translate the problem back to KP-ABE.

In this paper, we create a method for directly embedding any LSSS structure M^* into the public parameters in our reduction. In the proofs of our system a simulator can "program" the LSSS matrix M^* of the challenge ciphertext (in the selective model of security). Consider a LSSS matrix M^* of size $l^* \times n^*$. For each row i of M^* the simulator needs to program in ℓ pieces of information

[3] Here we roughly mean size to be number of rows in the LSSS system or nodes in an access tree.

$(M^*_{i,1}, \ldots, M^*_{i,\ell})$ into the parameters related to the attribute assigned to that row. In our most efficient system we program in M^* using the d-Parallel BDHE assumption; however, in Section 5 we show variations of our construction that are provably secure using similar ideas, but under weaker assumptions.

Our methodology of creating a system and proof that directly addresses CP-ABE stands in contrast to the approach of GJPS which essentially maps CP-ABE requirements onto a KP-ABE scheme.

Table 1. Comparison of CP-ABE systems in terms of ciphertext size, private key size, encryption and decryption times and assumptions. We let n be the size of an access formula , A be the number of attributes in a user's key, and T be (minimum needed) number of nodes satisfied of a formula by a user's attributes, and U be the number of attributes defined in the system. For our d-BDHE construction of the system defines a parameter $k_{\max}$, which is the maximum number of times a single attribute will appear in a particular formula. In the GJPS construction and our d-BDH one of Section 5 the systems define $n_{\max}$ as a bound on the size any formula. The ciphertext and private key sizes are given in terms of the number of group elements, encryption time in terms of number of exponentiations, and decryption in terms of number of pairing operations.

System	Ciphetext Size	Private Key Size	Enc. Time	Assumption
BSW[7]	$\mathcal{O}(n)$	$\mathcal{O}(A)$	$\mathcal{O}(n)$	Generic Group
GJPS[26]	$\mathcal{O}(U \cdot n_{\max}^{3.42})$	$\mathcal{O}(A \cdot n_{\max}^{3.42})$	$\mathcal{O}(U \cdot n_{\max}^{3.42})$	d-BDH
Section 3	$\mathcal{O}(n)$	$\mathcal{O}(A)$	$\mathcal{O}(n)$	d-Parallel BDHE
Full version [42]	$\mathcal{O}(n)$	$\mathcal{O}(k_{\max} \cdot A)$	$\mathcal{O}(n)$	d-BDHE
Section 5	$\mathcal{O}(n^2)$	$\mathcal{O}(k_{\max} \cdot A + n_{\max})$	$\mathcal{O}(n^2)$	d-BDH

1.1 Related Work

Some of the roots of ABE can be traced back to Identity-Based Encryption [36, 13, 23, 18, 10, 41, 24, 14] (IBE). One can view IBE as a very special case of ABE.

Different authors [38, 32, 4, 17, 3, 5] have considered similar problems without considering collusion resistance. In these works a data provider specifies an access formula such that a group of users can decrypt if the *union* of their credentials satisfies the formula. By only requiring the union of the credentials one does not worry about collusion attacks. In these schemes a setup authority simply assigns a separate public key to each credential and gives the corresponding secret key to each user that possesses the credential. Encryption is done by splitting secrets and then encrypting each share to the appropriate public key. Some of these schemes were inspired by earlier work [21, 20].

Since the introduction of Attribute-Based Encryption by Sahai and Waters [35], there have been several papers [27, 7, 19, 33, 26] that have proposed different varieties of ABE. Most of them have been for monotonic access structures over attributes; one exception is the work of Ostrovsky, Sahai, and Waters [33] that showed how to realize negation by integrating revocation schemes into the GPSW ABE cryptosystem.

Most work on ABE is focused on complex access controls for hiding an encrypted payload of data. A related line of work called predicate encryption or searching on encrypted data attempts to evaluate predicates over the encrypted data itself [39, 12, 1, 16, 15, 37, 29]. These systems have the advantages of hiding the associated access structures themselves and thus providing a level of "anonymity". The concept of predicate encryption is more general than the one we consider. However, the predicate encryption systems realized thus far tend to be much less expressive than access control systems that leave the access structures in the clear.

Other examples of encryption systems with more "structure" added include Hierarchical Identity-Based Encryption [28, 25] and Wildcard IBE [2].

Finally, Lewko et. al. [31] recently leveraged the encoding technique from our work to build an ABE system that achieves adaptive (non-selective) security. The system of Lewko et. al. is based in composite order groups, which results in some loss of practical efficiency compared to our most efficient system. In addition, our BDH system is based off of more standard assumptions than those used in Lewko et al.

2 Background

We first give formal definitions for access structures and relevant background on Linear Secret Sharing Schemes (LSSS). Then we give the security definitions of ciphertext policy attribute based encryption (CP-ABE). Finally, we give background information on bilinear maps.

2.1 Access Structures

Definition 1 (Access Structure [6]). *Let* $\{P_1, P_2, \ldots, P_n\}$ *be a set of parties. A collection* $\mathbb{A} \subseteq 2^{\{P_1, P_2, \ldots, P_n\}}$ *is monotone if* $\forall B, C$: if $B \in \mathbb{A}$ *and* $B \subseteq C$ *then* $C \in \mathbb{A}$. *An* access structure *(respectively, monotone access structure) is a collection (respectively, monotone collection)* $\mathbb{A}$ *of non-empty subsets of* $\{P_1, P_2, \ldots, P_n\}$, *i.e.,* $\mathbb{A} \subseteq 2^{\{P_1, P_2, \ldots, P_n\}} \backslash \{\emptyset\}$. *The sets in* $\mathbb{A}$ *are called the* authorized sets, *and the sets not in* $\mathbb{A}$ *are called the* unauthorized sets.

In our context, the role of the parties is taken by the attributes. Thus, the access structure $\mathbb{A}$ will contain the authorized sets of attributes. We restrict our attention to monotone access structures. However, it is also possible to (inefficiently) realize general access structures using our techniques by having the not of an attribute as a separate attribute altogether. Thus, the number of attributes in the system will be doubled. From now on, unless stated otherwise, by an access structure we mean a monotone access structure.

2.2 Linear Secret Sharing Schemes

We will make essential use of linear secret-sharing schemes. We adapt our definitions from those given in [6]:

Definition 2 (Linear Secret-Sharing Schemes (LSSS)). *A secret-sharing scheme Π over a set of parties $\mathcal{P}$ is called* linear *(over $\mathbb{Z}_p$) if*

1. *The shares for each party form a vector over $\mathbb{Z}_p$.*
2. *There exists a matrix an M with ℓ rows and n columns called the share-generating matrix for Π. For all $i = 1, \ldots, \ell$, the i'th row of M we let the function ρ defined the party labeling row i as $\rho(i)$. When we consider the column vector $v = (s, r_2, \ldots, r_n)$, where $s \in \mathbb{Z}_p$ is the secret to be shared, and $r_2, \ldots, r_n \in \mathbb{Z}_p$ are randomly chosen, then Mv is the vector of ℓ shares of the secret s according to Π. The share $(Mv)_i$ belongs to party $\rho(i)$.*

It is shown in [6] that every linear secret sharing-scheme according to the above definition also enjoys the *linear reconstruction* property, defined as follows: Suppose that Π is an LSSS for the access structure $\mathbb{A}$. Let $S \in \mathbb{A}$ be any authorized set, and let $I \subset \{1, 2, \ldots, \ell\}$ be defined as $I = \{i : \rho(i) \in S\}$. Then, there exist constants $\{\omega_i \in \mathbb{Z}_p\}_{i \in I}$ such that, if $\{\lambda_i\}$ are valid shares of any secret s according to Π, then $\sum_{i \in I} \omega_i \lambda_i = s$.

Furthermore, it is shown in [6] that these constants $\{\omega_i\}$ can be found in time polynomial in the size of the share-generating matrix M.

Note on Convention. We note that we use the convention that vector $(1, 0, 0, \ldots, 0)$ is the "target" vector for any linear secret sharing scheme. For any satisfying set of rows I in M, we will have that the target vector is in the span of I.

For any unauthorized set of rows I the target vector is not in the span of the rows of the set I. Moreover, there will exist a vector w such that $w \cdot (1, 0, 0 \ldots, 0) = -1$ and $w \cdot M_i = 0$ for all $i \in I$.

Using Access Trees. Prior works on ABE (e.g., [27]) typically described access formulas in terms of binary trees. Using standard techniques [6] one can convert any monotonic boolean formula into an LSSS representation. An access tree of ℓ nodes will result in an LSSS matrix of ℓ rows. We refer the reader to the appendix of [30] for a discussion of how to perform this conversion.

2.3 Ciphertext-Policy ABE

A ciphertext-policy attribute based encryption scheme consists of four algorithms: Setup, Encrypt, KeyGen, and Decrypt.

Setup(λ, U). The setup algorithm takes security parameter and attribute universe description as input. It outputs the public parameters PK and a master key MK.

*Encrypt(*PK*, $M, \mathbb{A}$).* The encryption algorithm takes as input the public parameters PK, a message M, and an access structure $\mathbb{A}$ over the universe of attributes. The algorithm will encrypt M and produce a ciphertext CT such that only a user that possesses a set of attributes that satisfies the access structure will be able to decrypt the message. We will assume that the ciphertext implicitly contains $\mathbb{A}$.

*Key Generation(*MK*, S).* The key generation algorithm takes as input the master key MK and a set of attributes S that describe the key. It outputs a private key SK.

*Decrypt(*PK, CT, SK*).* The decryption algorithm takes as input the public parameters PK, a ciphertext CT, which contains an access policy $\mathbb{A}$, and a private key SK, which is a private key for a set S of attributes. If the set S of attributes satisfies the access structure $\mathbb{A}$ then the algorithm will decrypt the ciphertext and return a message M.

We now describe a security model for ciphertext-policy ABE schemes. Like identity-based encryption schemes [36, 13, 23] the security model allows the adversary to query for any private keys that cannot be used to decrypt the challenge ciphertext. In CP-ABE the ciphertexts are identified with access structures and the private keys with attributes. It follows that in our security definition the adversary will choose to be challenged on an encryption to an access structure $\mathbb{A}^*$ and can ask for any private key S such that S does not satisfy $\mathbb{A}^*$. We now give the formal security game.

Security Model for CP-ABE.

Setup. The challenger runs the Setup algorithm and gives the public parameters, PK to the adversary.

Phase 1. The adversary makes repeated private keys corresponding to sets of attributes $S_1, \ldots, S_{q_1}$.

Challenge. The adversary submits two equal length messages M_0 and M_1. In addition the adversary gives a challenge access structure $\mathbb{A}^*$ such that none of the sets $S_1, \ldots, S_{q_1}$ from Phase 1 satisfy the access structure. The challenger flips a random coin b, and encrypts M_b under $\mathbb{A}^*$. The ciphertext CT^* is given to the adversary.

Phase 2. Phase 1 is repeated with the restriction that none of sets of attributes $S_{q_1+1}, \ldots, S_q$ satisfy the access structure corresponding to the challenge.

Guess. The adversary outputs a guess b' of b.

The advantage of an adversary $\mathcal{A}$ in this game is defined as $\Pr[b' = b] - \frac{1}{2}$. We note that the model can easily be extended to handle chosen-ciphertext attacks by allowing for decryption queries in Phase 1 and Phase 2.

Definition 3. *A ciphertext-policy attribute-based encryption scheme is secure if all polynomial time adversaries have at most a negligible advantage in the above game.*

We say that a system is *selectively* secure if we add an Init stage before setup where the adversary commits to the challenge access structure $\mathbb{A}^*$. All of our constructions will be proved secure in the selective security model.

2.4 Bilinear Maps

We present a few facts related to groups with efficiently computable bilinear maps and then give our number theoretic assumptions.

Let $\mathbb{G}$ and $\mathbb{G}_T$ be two multiplicative cyclic groups of prime order p. Let g be a generator of $\mathbb{G}$ and e be a bilinear map, $e : \mathbb{G} \times \mathbb{G} \to \mathbb{G}_T$. The bilinear map e has the following properties:

1. Bilinearity: for all $u, v \in \mathbb{G}$ and $a, b \in \mathbb{Z}_p$, we have $e(u^a, v^b) = e(u, v)^{ab}$.
2. Non-degeneracy: $e(g, g) \neq 1$.

We say that $\mathbb{G}$ is a bilinear group if the group operation in $\mathbb{G}$ and the bilinear map $e : \mathbb{G} \times \mathbb{G} \to \mathbb{G}_T$ are both efficiently computable. Notice that the map e is symmetric since $e(g^a, g^b) = e(g, g)^{ab} = e(g^b, g^a)$.

Decisional Parallel Bilinear Diffie-Hellman Exponent Assumption. We define the decisional q-parallel Bilinear Diffie-Hellman Exponent problem as follows. Choose a group $\mathbb{G}$ of prime order p according to the security parameter. Let $a, s, b_1, \ldots, b_q \in \mathbb{Z}_p$ be chosen at random and g be a generator of $\mathbb{G}$. If an adversary is given $\boldsymbol{y}$=

$$g, g^s, g^a, \ldots, g^{(a^q)}, \; , g^{(a^{q+2})}, \ldots, g^{(a^{2q})}$$
$$\forall_{1 \leq j \leq q} \; g^{s \cdot b_j}, \; g^{a/b_j}, \ldots, g^{(a^q/b_j)}, , g^{(a^{q+2}/b_j)}, \ldots, g^{(a^{2q}/b_j)}$$
$$\forall_{1 \leq j,k \leq q, k \neq j} \; g^{a \cdot s \cdot b_k / b_j}, \ldots, g^{(a^q \cdot s \cdot b_k / b_j)}$$

it must remain hard to distinguish $e(g, g)^{a^{q+1}s} \in \mathbb{G}_T$ from a random element in $\mathbb{G}_T$.

An algorithm $\mathcal{B}$ that outputs $z \in \{0, 1\}$ has advantage ϵ in solving decisional q-parallel BDHE in $\mathbb{G}$ if

$$\left| \Pr\left[\mathcal{B}(\boldsymbol{y}, T = e(g, g)^{a^{q+1}s}) = 0\right] - \Pr\left[\mathcal{B}(\boldsymbol{y}, T = R) = 0\right] \right| \geq \epsilon$$

Definition 1. *We say that the (decision) q parallel-BDHE assumption holds if no polytime algorithm has a non-negligible advantage in solving the decisional q-parallel BDHE problem.*

We give a proof that the assumption generically holds in the full version of our paper [42].

3 Our Most Efficient Construction

We now give our main construction that both realizes expressive functionality and is efficient and is provably secure under a concrete, non-interactive assumption.

In our construction the encryption algorithm will take as input a LSSS access matrix M and distribute a random exponent $s \in \mathbb{Z}_p$ according to M. Private keys are randomized to avoid collusion attack.

Setup(U). The setup algorithm takes as input the number of attributes in the system. It then chooses a group $\mathbb{G}$ of prime order p, a generator g and U random group elements $h_1, \ldots, h_U \in \mathbb{G}$ that are associated with the U attributes in the system. In addition, it chooses random exponents $\alpha, a \in \mathbb{Z}_p$.

The public key is published as

$$\text{PK} = \quad g, \; e(g, g)^\alpha, \; g^a, \; h_1, \ldots, h_U.$$

The authority sets MSK $= g^\alpha$ as the master secret key.

*Encrypt(*PK, $(M, \rho), \mathcal{M}$ *).* The encryption algorithm takes as input the public parameters PK and a message $\mathcal{M}$ to encrypt. In addition, it takes as input an LSSS access structure (M, ρ). The function ρ associates rows of M to attributes.

Let M be an $\ell \times n$ matrix. The algorithm first chooses a random vector $\boldsymbol{v} = (s, y_2, ..., y_n) \in \mathbb{Z}_p^n$. These values will be used to share the encryption exponent s. For $i = 1$ to ℓ, it calculates $\lambda_i = \boldsymbol{v} \cdot M_i$, where M_i is the vector corresponding to the ith row of M. In addition, the algorithm chooses random $r_1, \ldots, r_\ell \in \mathbb{Z}_p$.

The ciphertext is published as CT =

$$C = \mathcal{M}e(g,g)^{\alpha s},\ C' = g^s$$
$$(C_1 = g^{a\lambda_1} h_{\rho(1)}^{-r_1},\ D_1 = g^{r_1}), \ldots, (C_\ell = g^{a\lambda_\ell} h_{\rho(\ell)}^{-r_n},\ D_\ell = g^{r_\ell})$$

along with a description of (M, ρ).

*KeyGen(*MSK, S*).* The key generation algorithm takes as input the master secret key and a set S of attributes. The algorithm first chooses a random $t \in \mathbb{Z}_p$. It creates the private key as

$$K = g^{\alpha} g^{at} \quad L = g^t \quad \forall x \in S\ K_x = h_x^t.$$

*Decrypt(*CT,SK*).* The decryption algorithm takes as input a ciphertext CT for access structure (M, ρ) and a private key for a set S. Suppose that S satisfies the access structure and let $I \subset \{1, 2, \ldots, \ell\}$ be defined as $I = \{i : \rho(i) \in S\}$. Then, let $\{\omega_i \in \mathbb{Z}_p\}_{i \in I}$ be a set of constants such that if $\{\lambda_i\}$ are valid shares of any secret s according to M, then $\sum_{i \in I} \omega_i \lambda_i = s$. (Note there could potentially be different ways of choosing the ω_i values to satisfy this.)

The decryption algorithm first computes

$$e(C', K) / \left(\prod_{i \in I} (e(C_i, L) e(D_i, K_{\rho(i)}))^{\omega_i}\right) =$$
$$e(g,g)^{\alpha s} e(g,g)^{ast} / \left(\prod_{i \in I} e(g,g)^{ta\lambda_i \omega_i}\right) = e(g,g)^{\alpha s}$$

The decryption algorithm can then divide out this value from C and obtain the message $\mathcal{M}$.

3.1 Proof

An important step in proving our system secure will be for the reduction to "program" the challenge ciphertext into the public parameters. One significant obstacle that we will encounter is that an attribute may be associated with multiple rows in the challenge access matrix (i.e. the ρ function is not injective). This is analogous to an attribute appearing in multiple leafs in an access tree.

For example, suppose that in our reduction we want to program our parameters such that for h_x based on the i-th row of M^* if $\rho^*(i) = x$. However, if there exist $i \neq j$ such that $x = \rho(i) = \rho(j)$ then there is an issue since we must program *both* row i and row j in the simulation. Intuitively, there is a potential conflict in how to program the parameters.

In this reduction we resolve this by using different terms from the parallel BDHE assumption to program multiple rows of M^* into one group element corresponding to an attribute. The extra terms provided allow us to do so without ambiguity[4]. In Section 5 we show a tradeoff where our reduction can program the information using just the decisional Bilinear Diffie-Hellman assumption, but at some loss of efficiency.

We prove the following theorem.

Theorem 1. *Suppose the decisional q-parallel BDHE assumption holds. Then no polytime adversary can selectively break our system with a challenge matrix of size* $\ell^* \times n^*$*, where* $\ell^*, n^* \leq q$*.*

Suppose we have an adversary $\mathcal{A}$ with non-negligible advantage $\epsilon = \mathsf{Adv}_{\mathcal{A}}$ in the selective security game against our construction. Moreover, suppose it chooses a challenge matrix M^* where both dimensions are at most q. We show how to build a simulator, $\mathcal{B}$, that plays the decisional q-parallel BDHE problem.

Init. The simulator takes in a q-parallel BDHE challenge $\boldsymbol{y}, T$. The adversary gives the algorithm the challenge access structure (M^*, ρ^*), where M^* has n^* columns.

Setup. The simulator chooses random $\alpha' \in \mathbb{Z}_p$ and implicitly sets $\alpha = \alpha' + a^{q+1}$ by letting $e(g,g)^{\alpha} = e(g^a, g^{a^q})e(g,g)^{\alpha'}$.

We describe how the simulator "programs" the group elements $h_1, \ldots, h_U$. For each x for $1 \leq x \leq U$ begin by choosing a random value z_x. Let X denote the set of indices i, such that $\rho^*(i) = x$. The simulator programs h_x as:

$$h_x = g^{z_x} \prod_{i \in X} g^{a M^*_{i,1}/b_i} \cdot g^{a^2 M^*_{i,2}/b_i} \cdots g^{a^{n^*} M^*_{i,n^*}/b_i}.$$

Note that if $X = \emptyset$ then we have $h_x = g^{z_x}$. Also note that the parameters are distributed randomly due to the g^{z_x} value.

Phase I. In this phase the simulator answers private key queries. Suppose the simulator is given a private key query for a set S where S does not satisfy M^*.

The simulator first chooses a random $r \in \mathbb{Z}_p$. Then it finds a vector $\boldsymbol{w} = (w_1, \ldots, w_{n^*}) \in {\mathbb{Z}_p}^{n^*}$ such that $w_1 = -1$ and for all i where $\rho^*(i) \in S$ we have that $\boldsymbol{w} \cdot M^*_i = 0$. By the definition of a LSSS such a vector must exist. *Note that if such a vector did not exist then the vector* $(1, 0, 0, \ldots, 0)$ *would be in the span of* S. *See the discussion in Section 2.*

The simulator begins by implicitly defining t as

$$r + w_1 a^q + w_2 a^{q-1} + \cdots + w_{n^*} a^{q-n^*+1}.$$

It performs this by setting $L = g^r \prod_{i=1,\ldots,n^*} (g^{a^{q+1-i}})^{w_i} = g^t$.

[4] We note that certain assumptions have been leveraged to "program" a large amount of information into single group elements in other contexts. Gentry's reduction [24] embeds a degree q polynomial into a single group element.

We observe that by our definition of t, we have that g^{at} contains a term of $g^{-a^{q+1}}$, which will cancel out with the unknown term in g^{α} when creating K. The simulator can compute K as:

$$K = g^{\alpha'} g^{ar} \prod_{i=2,\ldots,n^*} (g^{a^{q+2-i}})^{w_i}.$$

Now we must calculate $K_x \ \forall x \in S$. First, we consider $x \in S$ for which there is no i such that $\rho^*(i) = x$. For those we can simply let $K_x = L^{z_x}$.

The more difficult task is to create key components K_x for attributes $x \in S$, where x is used in the access structure. For these keys we must make sure that there are no terms of the form g^{a^{q+1}/b_i} that we can't simulate. However, we have that $M_i^* \cdot \boldsymbol{w} = 0$; therefore, all of these terms cancel.

Again, let X be the set of all i such that $\rho^*(i) = x$. The simulator creates K_x in this case as follows.

$$K_x = L^{z_x} \prod_{i \in X} \prod_{j=1,\ldots,n^*} \left(g^{(a^j/b_i)r} \prod_{\substack{k=1,\ldots,n^* \\ k \neq j}} (g^{a^{q+1+j-k}/b_i})^{w_k} \right)^{M^*_{i,j}}$$

Challenge. Finally, we build the challenge ciphertext. The adversary gives two messages $\mathcal{M}_0, \mathcal{M}_1$ to the simulator. The simulator flips a coin β. It creates $C = \mathcal{M}_\beta T \cdot e(g^s, g^{\alpha'})$ and $C' = g^s$.

The tricky part is to simulate the C_i values since this contains terms that we must cancel out. However, the simulator can choose the secret splitting, such that these cancel out. Intuitively, the simulator will choose random $y'_2, \ldots, y'_{n^*}$ and the share the secret using the vector

$$\boldsymbol{v} = (s, sa + y'_2, sa^2 + y'_3, \ldots, sa^{n-1} + y'_{n^*}) \in \mathbb{Z}_p{}^{n^*}.$$

In addition, it chooses random values $r'_1, \ldots, r'_\ell$.

For $i = 1, \ldots, n^*$, we define R_i as the set of all $k \neq i$ such that $\rho^*(i) = \rho^*(k)$. In other words, the set of all other row indices that have the same attribute as row i. The challenge ciphertext components are then generated as

$$D_i = g^{-r'_i} g^{-sb_i}$$

$$C_i = h_{\rho^*(i)}^{r'_i} \left(\prod_{j=2,\ldots,n^*} (g^a)^{M^*_{i,j} y'_j} \right) (g^{b_i \cdot s})^{-z_{\rho^*(i)}} \cdot \left(\prod_{k \in R_i} \prod_{j=1,\ldots,n^*} (g^{a^j \cdot s \cdot (b_i/b_k)})^{M^*_{k,j}} \right)$$

Phase II. Same as phase I.

Guess. The adversary will eventually output a guess β' of β. The simulator then outputs 0 to guess that $T = e(g,g)^{a^{q+1}s}$ if $\beta = \beta'$; otherwise, it outputs 1 to indicate that it believes T is a random group element in $\mathbb{G}_T$.

When T is a tuple the simulator $\mathcal{B}$ gives a perfect simulation so we have that

$$\Pr\left[\mathcal{B}\left(\boldsymbol{y}, T = e(g,g)^{a^{q+1}s}\right) = 0\right] = \frac{1}{2} + \mathsf{Adv}_{\mathcal{A}}.$$

When T is a random group element the message $\mathcal{M}_\beta$ is completely hidden from the adversary and we have $\Pr[\mathcal{B}(\boldsymbol{y},T = R) = 0] = \frac{1}{2}$. Therefore, $\mathcal{B}$ can play the decisional q-parallel BDHE game with non-negligible advantage.

4 Constructions from Weaker Assumptions

Our first construction provided a very efficient system, but under a strong (but still non-interactive) assumption. To bridge this gap we introduce two additional constructions that provide a tradeoff of performance versus strength of assumptions. We effectively explore a spectrum between system efficiency and strength of assumption. The final construction is proven secure under the simple decisional-BDH assumption.

Overview. The primary obstacle in achieving security from weaker assumptions is that we must be able to reflect the challenge access structure M^* in the parameters during the reduction. We create two different constructions using the same framework.

In our full version [42] we give a construction provably secure under the existing d-BDHE assumption introduced by Boneh, Boyen and Goh [11]. To accommodate a weaker assumption we introduce a parameter $k_{\max}$ which is the maximum number of times any one attribute can appear in an access formula. A private key in this system will be a factor of $k_{\max}$ larger than our main construction.

Next, in Section 5 we give a construction provably secure under the much more standard decisional Bilinear Diffie-Hellman assumption. To realize security under this assumption our system must additionally introduce a parameter $n_{\max}$, where performance parameters will be a factor of $n_{\max}$ larger than our most efficient construction.

5 Bilinear Diffie-Hellman Construction

While our unrestricted construction realizes a potentially ideal type of efficiency, we would like to also show that secure CP-ABE systems can be realized from static assumptions. Here we show how to realize our framework under the decisional Bilinear Diffie Hellman d-(BDH) assumption.

The primary challenge with realizing a construction provably secure under BDH is we need a way for a reduction to embed the challenge matrix M^* in the parameters. Since the BDH assumption gives the reduction less components to embed this, there is no obvious path for reducing the previous constructions to d-BDH. We surmount this obstacle by expanding our ciphertexts and public parameter space. By doing this we enable our reduction to embed the challenge matrix.

Our construction is parametrized by a integer $n_{\max}$ that specifies the maximum number of columns in a ciphertext policy. The public parameters, keys and ciphertext size will all grow linearly in this parameter[5].

[5] One could achieve smaller ciphertexts by creating multiple systems with different $n_{\max}$ values and use the one that fit the actual policy most tightly.

Like our first construction we restrict $\rho()$ to be an injective function, but can alleviate this restriction by applying a similar transformation to allow an attribute to appear $k_{\max}$ times for some specified $k_{\max}$. Our construction follows.

Setup($U, n_{\max}$). The setup algorithm takes as input, U, the number of attributes in the system U and $n_{\max}$ the maximum number of columns in an LSSS matrix (or number of nodes in an access formula). It then creates a group $\mathbb{G}$ of prime order p and a generator g and chooses random elements $(h_{1,1}, \ldots, h_{1,U}), \ldots, (h_{n_{\max},1}, \ldots, h_{n_{\max},U})$ In addition, it chooses random exponents $\alpha, a \in \mathbb{Z}_p$.

The public key is published as

$$\text{PK} = \quad g,\ e(g,g)^{\alpha},\ g^{a},$$

$$(h_{1,1}, \ldots, h_{1,U}), \ldots, (h_{n_{\max},1}, \ldots, h_{n_{\max},U})$$

The authority sets MSK $= g^{\alpha}$ as the master secret key.

Encrypt(PK, $(M, \rho), \mathcal{M}$). The encryption algorithm takes as input the public parameters PK and a message $\mathcal{M}$ to encrypt. In addition, it takes as input an LSSS access structure (M, ρ). The function ρ associates rows of M to attributes. In this construction we limit ρ to be an injective function, that is an attribute is associated with at most one row of M.

Let M be an $\ell \times n_{\max}$ matrix. (If one needs to create a policy for $n < n_{\max}$, then one can simply "pad out" the rightmost $n_{\max} - n$ columns with all zeros.) The algorithm first chooses a random vector $\boldsymbol{v} = (s, y_2, ..., y_{n_{\max}}) \in \mathbb{Z}_p^n$. These values will be used to share the encryption exponent s.

The ciphertext is published as

$$\text{CT} = C = \mathcal{M}e(g,g)^{\alpha s},\ C' = g^{s},\ \forall_{\substack{i=1,\ldots,\ell \\ j=1,\ldots,n_{\max}}}\ C_{i,j} = g^{aM_{i,j}v_j} h_{j,\rho(i)}^{-s}$$

along with a description of M, ρ.

KeyGen(MSK, S). The key generation algorithm takes as input the master secret key and a set S of attributes. The algorithm first chooses a random $t_1, \ldots, t_{n_{\max}} \in \mathbb{Z}_p$. It creates the private key as

$$K = g^{\alpha} g^{at_1} \quad L_1 = g^{t_1}, \ldots, L_n = g^{t_{n_{\max}}}$$

$$\forall x \in S \quad K_x = \prod_{j=1,\ldots,n_{\max}} h_{j,x}^{t_j}.$$

Decrypt(CT,SK). The decryption algorithm takes as input a ciphertext CT for access structure (M, ρ) and a private key for a set S. Suppose that S satisfies the access structure and let $I \subset \{1, 2, \ldots, \ell\}$ be defined as $I = \{i : \rho(i) \in S\}$. Then, let $\{\omega_i \in \mathbb{Z}_p\}_{i \in I}$ be a set of constants such that, if $\{\lambda_i\}$ are valid shares of any secret s according to M, then $\sum_{i \in I} \omega_i \lambda_i = s$. (Note there could potentially be different ways of choosing the ω_i values to satisfy this.)

The decryption algorithm first computes

$$e(C',K)/\left(\prod_{j=1,\ldots,n_{\max}} e(L_j,\prod_{i\in I} C_{i,j}^{\omega_i})\right)\prod_{i\in I} e(K_{\rho(i)}^{\omega_i},C')$$

$$\begin{aligned}
&= e(C',K)/\Big(\prod_{j=1,\ldots,n_{\max}} e(g^{t_j},g^{\sum_{i\in I} aM_{i,j}v_j\omega_i})\cdot \\
&\quad e(g^{t_j},\prod_{i\in I} h_{j,\rho(i)}^{-s\omega_i})\Big)\prod_{i\in I} e(K_{\rho(i)}^{\omega_i},g^s) \\
&= e(C',K)/\prod_{j=1,\ldots,n_{\max}} e(g^{t_j},g^{\sum_{i\in I} aM_{i,j}v_j\omega_i}) \\
&= e(C',K)/e(g^{t_1},g^{\sum_{i\in I} aM_{i,1}v_1\omega_i}) \\
&= e(g^s,g^\alpha g^{at_1})/e(g,g)^{at_1 s} \\
&= e(g,g)^{\alpha s}
\end{aligned}$$

The decryptor can then divide out this value from C and obtain the message $\mathcal{M}$.

5.1 Proof

We prove the following theorem.

Theorem 2. *Suppose the decisional BDH assumption holds. Then no polytime adversary can selectively break our system.*

Due to space limitations we defer the proof of the system to our full version [42].

6 Large Universe of Attributes

One aspect of our main construction is that it defines the set of attributes to be used in the parameters. One useful feature is to be able to dynamically use any string as an attribute. In our full version [42] we show how in the random oracle we can realize any number of attributes with constant size parameters by simply hashing the attribute string. Also in our full version provide a large universe construction in the standard model.

Acknowledgements

We thank Matt Green, Kazuki Yoneyama and anonymous reviewers for useful comments.

References

[1] Abdalla, M., Bellare, M., Catalano, D., Kiltz, E., Kohno, T., Lange, T., Malone-Lee, J., Neven, G., Paillier, P., Shi, H.: Searchable encryption revisited: Consistency properties, relation to anonymous IBE, and extensions. In: Shoup, V. (ed.) CRYPTO 2005. LNCS, vol. 3621, pp. 205–222. Springer, Heidelberg (2005)

[2] Abdalla, M., Catalano, D., Dent, A.W., Malone-Lee, J., Neven, G., Smart, N.P.: Identity-based encryption gone wild. In: Bugliesi, M., Preneel, B., Sassone, V., Wegener, I. (eds.) ICALP 2006, Part II. LNCS, vol. 4052, pp. 300–311. Springer, Heidelberg (2006)
[3] Al-Riyami, S.S., Malone-Lee, J., Smart, N.P.: Escrow-free encryption supporting cryptographic workflow. Int. J. Inf. Sec. 5(4), 217–229 (2006)
[4] Bagga, W., Molva, R., Crosta, S.: Policy-based encryption schemes from bilinear pairings. In: ASIACCS, p. 368 (2006)
[5] Barbosa, M., Farshim, P.: Secure cryptographic workflow in the standard model. In: Barua, R., Lange, T. (eds.) INDOCRYPT 2006. LNCS, vol. 4329, pp. 379–393. Springer, Heidelberg (2006)
[6] Beimel, A.: Secure Schemes for Secret Sharing and Key Distribution. PhD thesis, Israel Institute of Technology, Technion, Haifa, Israel (1996)
[7] Bethencourt, J., Sahai, A., Waters, B.: Ciphertext-policy attribute-based encryption. In: IEEE Symposium on Security and Privacy, pp. 321–334 (2007)
[8] Baden, R., Bender, A., Spring, N., Bhattacharjee, B., Starin, D.: Persona: An online social network with user defined privacy. In: ACM SIGCOMM (2009)
[9] Bobba, R., Fatemieh, O., Khan, F., Gunter, A.K.C.A., Khurana, H., Prabhakaran, M.: Attribute-based messaging: Access control and confidentiality (2009) (manuscript)
[10] Boneh, D., Boyen, X.: Efficient selective-ID secure identity-based encryption without random oracles. In: Cachin, C., Camenisch, J.L. (eds.) EUROCRYPT 2004. LNCS, vol. 3027, pp. 223–238. Springer, Heidelberg (2004)
[11] Boneh, D., Boyen, X., Goh, E.-J.: Hierarchical identity based encryption with constant size ciphertext. In: Cramer, R. (ed.) EUROCRYPT 2005. LNCS, vol. 3494, pp. 440–456. Springer, Heidelberg (2005)
[12] Boneh, D., Di Crescenzo, G., Ostrovsky, R., Persiano, G.: Public key encryption with keyword search. In: Cachin, C., Camenisch, J.L. (eds.) EUROCRYPT 2004. LNCS, vol. 3027, pp. 506–522. Springer, Heidelberg (2004)
[13] Boneh, D., Franklin, M.K.: Identity-based encryption from the weil pairing. In: Kilian, J. (ed.) CRYPTO 2001. LNCS, vol. 2139, pp. 213–229. Springer, Heidelberg (2001)
[14] Boneh, D., Gentry, C., Hamburg, M.: Space-efficient identity based encryption without pairings. In: FOCS, pp. 647–657 (2007)
[15] Boneh, D., Waters, B.: Conjunctive, subset, and range queries on encrypted data. In: Vadhan, S.P. (ed.) TCC 2007. LNCS, vol. 4392, pp. 535–554. Springer, Heidelberg (2007)
[16] Boyen, X., Waters, B.: Anonymous hierarchical identity-based encryption (Without random oracles). In: Dwork, C. (ed.) CRYPTO 2006. LNCS, vol. 4117, pp. 290–307. Springer, Heidelberg (2006)
[17] Bradshaw, R.W., Holt, J.E., Seamons, K.E.: Concealing complex policies with hidden credentials. In: ACM Conference on Computer and Communications Security, pp. 146–157 (2004)
[18] Canetti, R., Halevi, S., Katz, J.: A forward-secure public-key encryption scheme. In: Biham, E. (ed.) EUROCRYPT 2003. LNCS, vol. 2656, pp. 255–271. Springer, Heidelberg (2003)
[19] Chase, M.: Multi-authority attribute based encryption. In: Vadhan, S.P. (ed.) TCC 2007. LNCS, vol. 4392, pp. 515–534. Springer, Heidelberg (2007)

[20] Chen, L., Harrison, K., Moss, A., Soldera, D., Smart, N.P.: Certification of Public Keys within an Identity Based System. In: Chan, A.H., Gligor, V.D. (eds.) ISC 2002. LNCS, vol. 2433, pp. 322–333. Springer, Heidelberg (2002)
[21] Chen, L., Harrison, K., Soldera, D., Smart, N.P.: Applications of Multiple Trust Authorities in Pairing Based Cryptosystems. In: Davida, G.I., Frankel, Y., Rees, O. (eds.) InfraSec 2002. LNCS, vol. 2437, pp. 260–275. Springer, Heidelberg (2002)
[22] Cheung, L., Newport, C.C.: Provably secure ciphertext policy abe. In: ACM Conference on Computer and Communications Security, pp. 456–465 (2007)
[23] Cocks, C.: An identity based encryption scheme based on quadratic residues. In: IMA Int. Conf., pp. 360–363 (2001)
[24] Gentry, C.: Practical identity-based encryption without random oracles. In: Vaudenay, S. (ed.) EUROCRYPT 2006. LNCS, vol. 4004, pp. 445–464. Springer, Heidelberg (2006)
[25] Gentry, C., Silverberg, A.: Hierarchical ID-based cryptography. In: Zheng, Y. (ed.) ASIACRYPT 2002. LNCS, vol. 2501, pp. 548–566. Springer, Heidelberg (2002)
[26] Goyal, V., Jain, A., Pandey, O., Sahai, A.: Bounded ciphertext policy attribute based encryption. In: Aceto, L., Damgård, I., Goldberg, L.A., Halldórsson, M.M., Ingólfsdóttir, A., Walukiewicz, I. (eds.) ICALP 2008, Part II. LNCS, vol. 5126, pp. 579–591. Springer, Heidelberg (2008)
[27] Goyal, V., Pandey, O., Sahai, A., Waters, B.: Attribute-based encryption for fine-grained access control of encrypted data. In: ACM Conference on Computer and Communications Security, pp. 89–98 (2006)
[28] Horwitz, J., Lynn, B.: Toward hierarchical identity-based encryption. In: Knudsen, L.R. (ed.) EUROCRYPT 2002. LNCS, vol. 2332, pp. 466–481. Springer, Heidelberg (2002)
[29] Katz, J., Sahai, A., Waters, B.: Predicate encryption supporting disjunctions, polynomial equations, and inner products. In: Smart, N.P. (ed.) EUROCRYPT 2008. LNCS, vol. 4965, pp. 146–162. Springer, Heidelberg (2008)
[30] Lewko, A., Waters, B.: Decentralizing attribute-based encryption. Cryptology ePrint Archive, Report 2010/351 (2010), `http://eprint.iacr.org/`
[31] Lewko, A., Okamoto, T., Sahai, A., Takashima, K., Waters, B.: Fully Secure Functional Encryption: Attribute-Based Encryption and (Hierarchical) Inner Product Encryption. In: Gilbert, H. (ed.) EUROCRYPT 2010. LNCS, vol. 6110, pp. 62–91. Springer, Heidelberg (2010)
[32] Miklau, G., Suciu, D.: Controlling access to published data using cryptography. In: VLDB, pp. 898–909 (2003)
[33] Ostrovsky, R., Sahai, A., Waters, B.: Attribute-based encryption with non-monotonic access structures. In: ACM Conference on Computer and Communications Security, pp. 195–203 (2007)
[34] Pirretti, M., Traynor, P., McDaniel, P., Waters, B.: Secure attribute-based systems. In: ACM Conference on Computer and Communications Security, pp. 99–112 (2006)
[35] Sahai, A., Waters, B.: Fuzzy identity-based encryption. In: Cramer, R. (ed.) EUROCRYPT 2005. LNCS, vol. 3494, pp. 457–473. Springer, Heidelberg (2005)
[36] Shamir, A.: Identity-based cryptosystems and signature schemes. In: Blakely, G.R., Chaum, D. (eds.) CRYPTO 1984. LNCS, vol. 196, pp. 47–53. Springer, Heidelberg (1985)
[37] Shi, E., Bethencourt, J., Chan, H.T.-H., Song, D.X., Perrig, A.: Multi-dimensional range query over encrypted data. In: IEEE Symposium on Security and Privacy, pp. 350–364 (2007)

[38] Smart, N.P.: Access Control Using Pairing Based Cryptography. In: Joye, M. (ed.) CT-RSA 2003. LNCS, vol. 2612, pp. 111–121. Springer, Heidelberg (2003)
[39] Song, D.X., Wagner, D., Perrig, A.: Practical techniques for searches on encrypted data. In: IEEE Symposium on Security and Privacy, pp. 44–55 (2000)
[40] Traynor, P., Butler, K.R.B., Enck, W., McDaniel, P.: Realizing massive-scale conditional access systems through attribute-based cryptosystems. In: NDSS (2008)
[41] Waters, B.: Efficient identity-based encryption without random oracles. In: Cramer, R. (ed.) EUROCRYPT 2005. LNCS, vol. 3494, pp. 114–127. Springer, Heidelberg (2005)
[42] Waters, B.: Ciphertext-policy attribute-based encryption: An expressive, efficient, and provably secure realization. Cryptology ePrint Archive, Report 2008/290 (2008), http://eprint.iacr.org/

Generic Constructions for Chosen-Ciphertext Secure Attribute Based Encryption

Shota Yamada[1], Nuttapong Attrapadung[2], Goichiro Hanaoka[2], and Noboru Kunihiro[1]

[1] The University of Tokyo
{yamada@it.,kunihiro@}k.u-tokyo.ac.jp
[2] National Institute of Advanced Industrial Science and Technology (AIST)
{n.attrapadung,hanaoka-goichiro}@aist.go.jp

Abstract. In this paper we propose generic conversions for transforming a chosen-plaintext (CPA) secure attribute-based encryption (ABE) to a chosen-ciphertext (CCA) secure ABE. The only known generic conversion, to the best of our knowledge, was presented by Goyal et al. in ACM-CCS 2006, which itself subsumes the well-known IBE-to-PKE conversion by Canetti, Halevi, and Katz proposed in Eurocrypt 2004. The method by Goyal et al. has some restrictions that it assumes the delegatability of the original ABE and can deal only with the key-policy type of ABE with large attribute universe. In contrast, our methodology is applicable also to those ABE schemes without known delegatability. Furthermore, it works for both key-policy or ciphertext-policy flavors of ABE and can deal with both small and large universe scheme. More precisely, our method assumes only either delegatability or a newly introduced property called verifiability of ABE. We then exhaustively check the verifiability of existing ABE schemes and found that most of them satisfy such a property, hence CCA-secure versions of these schemes can be obtained automatically.

1 Introduction

Background. Attribute-based encryption (ABE) is a generalized cryptographic primitive from normal public key encryption (PKE) that provides an access control mechanism over encrypted data using access policies and ascribed attributes among private keys and ciphertexts. ABE was introduced first by Sahai and Waters [30]. In an ABE system, a user in the system possesses a key associated with an access policy, stating what kind of ciphertext that she can decrypt. On the other hand, a ciphertext is associated with a set of attributes. The decryption can then be done if the policy associated to the key is satisfied by the attribute set associated to the ciphertext. This setting of ABE is called key-policy ABE (KP-ABE) since a key is associated with a policy. Its dual notion in which the role of policy and attribute set is swapped is called ciphertext-policy ABE (CP-ABE). In this setting, a policy will be associated to a ciphertext while an attribute set will be associated to a key.

D. Catalano et al. (Eds.): PKC 2011, LNCS 6571, pp. 71–89, 2011.

Most of the proposed ABE schemes [21,5,29] in the literature focused on the aspect of extending the expressiveness of policies so as to achieve fine-grained access control (see below). Some other schemes focused on extending to the multi-authority setting [13,14,3,25], while the most recent achievements in this research area were the schemes that attain adaptive security [26,28].

In this paper we focus on another important issue namely the aspect of attaining security against chosen-ciphertext attack (CCA) for ABE in the standard model. The first CCA-secure KP-ABE has already appeared in the seminal paper for the first expressive KP-ABE scheme by Goyal et al. [21]. Their CCA-secure scheme extends the methodology of converting any identity-based encryption (IBE) scheme to CCA-secure PKE scheme Canetti, Halevi, and Katz [12]. Such a technique relies on delegatability, which is the property that allows using a key of one policy $\mathbb{A}$ to construct a key for another policy $\mathbb{A}'$ that is more restricted than $\mathbb{A}$. Their construction is generic: it is a conversion that transforms any CPA-secure KP-ABE with delegation to a CCA-secure one. For the case of CCA-secure CP-ABE, Bethencourt, Sahai, and Waters [5] mentioned that a similar methodology can be used but they omitted to describe the details. We note that [5] also uses another method for their CCA-secure CP-ABE namely the Fujisaki-Okamoto conversion [18] but this can be proven only in the random oracle model. Some specific CCA-secure construction for CP-ABE with only AND-gates was proposed in [15].

To the best of our knowledge, the only generic and standard-model construction for CCA-secure ABE available in the literature is the aforementioned one by Goyal et al. [21]. The scheme works for the key-policy flavor and can deal only with the scheme for a large attribute universe. It works roughly as follows. To encrypt to an attribute set S, Bob first generates a signing and verification key pair (sk, vk) of a one-time signature scheme. Bob then constructs a ciphertext for $S \cup \{vk\}$ of the original scheme, where vk is treated as a dummy attribute, and signs this with sk. Upon receiving, Alice checks the validity of signature and then delegates her key from policy $\mathbb{A}$ to policy $\mathbb{A}$ AND vk. Alice then uses the latter key to decrypt the ciphertext.

OUR CONTRIBUTIONS. We propose eight generic conversions that transform CPA-secure ABE to CCA-secure ABE. The eights conversions comprise all the combinations by three categorization: (1) whether we consider CP-ABE or KP-ABE, (2) the original scheme deals with a small or large universe of attributes, and (3) the conversion uses which methodology out of two that we propose. One methodology is based on delegatability of ABE, while the other is based on a new property called verifiability of ABE. The former methodology is a reminiscent of the method by Goyal et al. [21] as described above. On the other hand, the latter can be considered as its "dual". Roughly speaking, while the delegatability-based method utilizes the AND functionality of ABE, the new verifiability-based method uses the OR functionality.

Before moving further, we point out an apparent strength of our thesis: one methodology has an advantage over the other in the sense that it requires only either one of the two additional properties. The new verifiability-based one does not require delegatability of ABE. It thus applies to those ABE without known delegation, such as the linear secret sharing based KP-ABE of Goyal et al. [21] and non-monotonic KP-ABE of Ostrovsky et al. [29] for instances.

Another advantage is that our methodology is generic: it converts the underlying CPA-secure ABE in the black-box manner. Readers who are familiar with ABE may argue that CCA-secure version of any ABE can be rather easy to construct since, to the best of our knowledge, all the available schemes so far were based on bilinear pairing and with this tool there are some well-known, but specific, techniques such as [12,9] (in the context of IBE) to attain CCA security. However, using such specific techniques requires researchers to construct and prove the security individually each time a new ABE is proposed, which is not quite convenient. Besides, we believe that some new ABE which is not based on pairing will be proposed in the future.

OUR APPROACH. The new verifiability-based method works roughly as follows. We briefly describe here for the case of KP-ABE (with a large universe). The scheme is indeed similar to the aforementioned method, where Bob encrypts to $S \cup \{vk\}$, while Alice holds a key for a policy $\mathbb{A}$, with the only difference in decryption algorithms. Instead of delegating the key, Alice will use the verifiability to check a kind of well-formed-ness of ciphertext before decrypting. Such a verifiability allows to check whether a ciphertext will decrypt to the same result when using either a key for policy $\mathbb{A}$ or a key for the singleton policy $\{\{vk\}\}$. The latter key will be used only in the proof. The ability to use either key to decrypt can be considered intuitively as an (implicit) OR functionality. For the case of CP-ABE, the utilization of OR will become more clear: there, we explicitly use a policy of the form $\mathbb{A}$ OR vk.

The use of OR and some form of verifiability to attain CCA security can be traced back to the classic Naor-Yung two-key paradigm [27] in the context of CCA-secure PKE. However, their scheme poses a strong requirement: the existence of non-interactive zero knowledge proofs, and thus, enhanced trapdoor permutations. In contrast, our newly defined verifiability for ABE is indeed quite a weak requirement. Regarding this, we show the gap between our verifiability and the commonly defined public verifiability. Furthermore, for pairing-based schemes, the verifiability comes for almost free in many schemes.

Finally, we note that the described methods assume that the original ABE can deal with large universe (super-polynomial size). This is since we have to treat vk as dummy attributes. In our methodology, we further propose how to deal with small universe schemes by introducing a twist similar to the well-known technique by Dwork, Dolev, and Naor [17] (in the context of PKE).

RELATED WORKS ON ABE. ABE was first introduced by Sahai and Waters [30] in the context of a generalization of IBE called Fuzzy IBE, which is an ABE that allows only single threshold access structures. The first KP-ABE scheme that allows any monotone access structures was proposed by Goyal et al. [21]. The first such CP-ABE scheme which allows the same expressiveness was proposed by Bethencourt, Sahai, and Waters [5], albeit the security of their scheme was only proved in the generic bilinear group model. Ostrovsky, Sahai, and

Waters [29] then subsequently extended both schemes to handle also any non-monotone structures. Towards constructing a CP-ABE in the standard model, Cheung and Newport [15] proposed a CP-ABE scheme that allows only AND gate, while Goyal et al. [20] proposed a CP-ABE scheme which allows only a-priori bounded size of gates (bounded CP-ABE). Waters [31] then proposed the first fully expressive CP-ABE in the standard model. Herranz et al. [23] proposed the first constant-size ciphertext scheme for CP-ABE allowing threshold gates. Recently, Attrapadung and Libert [1] proposed the first fully expressive KP-ABE with constant-size ciphertexts. All of these works were limited to deal with selective adversaries [11,6] until only two recent works by Lewko et al. [26] and Takashima and Okamoto [28], where they obtained adaptively secure ABE schemes. The aforementioned ABE schemes deal only with single authority, which is the setting that we focus here as well. Some extensions to multi-authority schemes were considered in [13,14,3,25]. It is also worth mentioning that dual-policy ABE, which is a combination of the mentioned two flavors of ABE, was proposed in [2].

Organization of the Paper. In the rest of this paper, we first give syntax and security notion of FE in Sec. 2, give the definition of verifiability and delegatability of FE in Sec. 3, show our general construction in Sec. 4, prove the security of our constructions for the case of CP-ABE in Sec. 5, 6, show instantiations of our generic construction in Sec. 7, show that our definition of verifiability is strictly weaker notion than usual public verifiability in Sec. 8.

2 Definitions

We capture notions of CP-ABE and KP-ABE by providing a unified definition and security notion for functional encryption[1] here and then instantiating to both primitives in the next subsection.

2.1 Syntax and Security Definition for Functional Encryption

Syntax. Let $R:\ \Sigma_k \times \Sigma_e \to \{0,1\}$ be a boolean function where Σ_k and Σ_e denote "key attribute" and "ciphertext attribute" spaces. A functional encryption (FE) scheme for R consists of the following algorithms: **Setup**, **KeyGen**, **Encrypt**, **Decrypt**.

Setup$(\lambda, \mathrm{des}) \to (PK, MSK)$**:** The setup algorithm takes as input a security parameter λ and a scheme description des and outputs a public key PK and a master secret key MSK.

KeyGen$(MSK, PK, X) \to SK_X$**:** The key generation algorithm takes in the master secret key MSK, the public key PK, and a key attribute $X \in \Sigma_k$. It outputs a private key SK_X.

[1] Our definition of FE is not the fully generalized one, as recently defined in [10]. It can be considered as the class of predicate encryption with public index in [10].

Encrypt$(PK, M, Y) \rightarrow CT$**:** The encryption algorithm takes as input a public key PK, the message M, and a ciphertext attribute $Y \in \Sigma_e$. It will output a ciphertext CT. We assume that Y is implicitly included in CT.

Decrypt$(PK, CT, SK_X) \rightarrow M$ or $\perp$**:** The decryption algorithm takes in the public parameters PK, a ciphertext CT, and a private key SK_X. It outputs the message M or $\perp$ which represents that the ciphertext is not in a valid form.

We require the standard correctness of decryption: that is, for all λ, all (PK, MSK) output by **Setup**(λ, des), all $X \in \Sigma_k$, all SK_X output by **KeyGen**(MSK, PK, X), and $Y \in \Sigma_e$,

- If $R(X,Y) = 1$, then **Decrypt**$(PK,$ **Encrypt**$(PK, M, Y)), SK_X) = M$.
- If $R(X,Y) = 0$, then **Decrypt**$(PK,$ **Encrypt**$(PK, M, Y)), SK_X) = \perp$.

SECURITY NOTION. We now give the definition of indistinguishability under chosen ciphertext attack (CCA-security) for FE scheme Π. This is described by a game between a challenger and attacker $\mathcal{A}$. The game proceeds as follows:

Setup. The challenger runs the setup algorithm and gives PK to $\mathcal{A}$.

Phase1. $\mathcal{A}$ may adaptively make queries of the following types:

– ***Key-extraction query.*** $\mathcal{A}$ submits X to the challenger. If the challenger already extracted a private key SK_X for X, then returns it. Otherwise the challenger runs $SK_X \leftarrow$ **KeyGen**(MSK, PK, X) and returns it.

– ***Decryption query.*** $\mathcal{A}$ submits (CT, X) to the challenger and ask for the decryption result of ciphertext CT under private key for X. If the challenger has not previously extracted a private key SK_X for X, then extract it by $SK_X \leftarrow$ **KeyGen**(MSK, PK, X). Then, the challenger returns the output of **Decrypt**(PK, CT, SK_X) to $\mathcal{A}$.

Challenge. $\mathcal{A}$ declares two equal length messages M_0 and M_1 and target ciphertext attribute $Y^* \in \Sigma_e$. Y^* cannot satisfy $R(X, Y^*) = 1$ for any attribute sets X such that $\mathcal{A}$ already queried private key for X. Then the challenger flips a random coin $\beta \in \{0,1\}$, runs **Encrypt**$(PK, M_\beta, Y^*) \rightarrow CT^*$ and gives challenge ciphertext CT^* to $\mathcal{A}$.

Phase2. $\mathcal{A}$ may adaptively make queries as the same as in **Phase1** with following added restriction. $\mathcal{A}$ cannot query to extract a private key SK_X for X such that $R(X, Y^*) = 1$. $\mathcal{A}$ cannot submit (CT, X) such that $R(X, Y^*) = 1$ and $CT = CT^*$.

Guess. $\mathcal{A}$ outputs a guess β' for β.

We say that $\mathcal{A}$ succeeds if $\beta' = \beta$ and denote the probability of this event by $\Pr^{FE}_{\mathcal{A},\Pi}$. The advantage of an attacker $\mathcal{A}$ is defined as $Adv^{FE}_{\mathcal{A},\Pi} = \Pr^{FE}_{\mathcal{A},\Pi} - \frac{1}{2}$.

Definition 1. *We say that an FE scheme Π is $(\tau, \epsilon, q_D, q_E)$* CCA-secure *if for all τ-time algorithms $\mathcal{A}$ who make a total of q_D decryption queries and a total of q_E key-extraction queries, we have that $Adv^{FE}_{\mathcal{A},\Pi} < \epsilon$. We say that an FE scheme Π is* CCA-secure *if for all polynomial τ, q_D, q_E and for all nonnegligible ϵ, Π is $(\tau, \epsilon, q_D, q_E)$ CCA-secure.*

Definition 2. *We say that an FE scheme Π is* (τ, ϵ, q_E) CPA-secure *if Π is* $(\tau, \epsilon, 0, q_E)$ CCA-secure. *We say that an FE scheme Π is CPA-secure if for all polynomial τ, q_E, for all nonnegligible ϵ, Π is (τ, ϵ, q_E) CPA-secure.*

We say that the FE scheme is selectively CCA/CPA-secure if we add an Initial stage **Init** before the setup where the adversary submits the ciphertext attribute $Y^* \in \Sigma_e$.

2.2 Definitions for Attribute-Based Encryption

Definition 3 (Access Structures). *Consider a set of parties $\mathcal{P} = \{P_1, P_2, \ldots, P_n\}$. A collection $\mathbb{A} \subseteq 2^{\mathcal{P}}$ is said monotone if for all B, C we have that if $B \in \mathbb{A}$ and $B \subseteq C$ then $C \in \mathbb{A}$. An access structure (resp., monotonic access structure) is a collection (resp., monotone collection) $\mathbb{A} \subseteq 2^{\mathcal{P}} \setminus \{\emptyset\}$. The sets in $\mathbb{A}$ are called the authorized sets, and the sets not in $\mathbb{A}$ are called the unauthorized sets.*

Definition 4 (KP-ABE). *Let U be an attribute space. A key-policy attribute-based encryption (KP-ABE) for a collection $\mathcal{A}$ of access structures over U is a functional encryption for $R^{\mathsf{KP}} : \mathcal{A} \times 2^U \to \{0,1\}$ defined by $R^{\mathsf{KP}}(\mathbb{A}, \omega) \mapsto 1$ iff $\omega \in \mathbb{A}$. Furthermore, the description* des *consists of the attribute universe U, $\Sigma_k = \mathcal{A}$, and $\Sigma_e = 2^U$.*

Definition 5 (CP-ABE). *A ciphertext-policy attribute-based encryption (CP-ABE) is the dual variant of KP-ABE. More precisely, if we let U be the attribute space, a CP-ABE for a collection $\mathcal{A}$ of access structures over U is a functional encryption for $R^{\mathsf{CP}} : 2^U \times \mathcal{A} \to \{0,1\}$ defined by $R^{\mathsf{CP}}(\omega, \mathbb{A}) \mapsto 1$ iff $\omega \in \mathbb{A}$. Furthermore, the description* des *consists of the attribute universe U, $\Sigma_k = 2^U$, and $\Sigma_e = \mathcal{A}$.*

Some Terminologies. We define some terminologies and properties related to access structures here. Any monotonic (resp., non-monotonic) access structure $\mathbb{A}$ can be represented by a corresponding boolean formula (resp., with negation), which we denote by $\psi(\mathbb{A})$, over variables in U. This is naturally defined in the sense that $S \in \mathbb{A}$ holds iff the evaluation of $\psi(\mathbb{A})$ with the assignment that sets all variables in S to 1 and other variables outside S to 0 yields the value 1.

Consider the case where $\mathbb{A}$ is a monotonic access structure over U. If we denote a minimal representation of $\mathbb{A}$ by $\mathsf{min}(\mathbb{A}) = \{S \in \mathbb{A} \mid \text{there exists no } B \in \mathbb{A} \text{ such that } B \subset S\}$. Then, it is straightforward to see that $\psi(\mathbb{A}) = \vee_{S' \in \mathsf{min}(\mathbb{A})} (\wedge_{P \in S'} P)$.

Next we consider the case where $\mathbb{A}$ that is a non-monotonic access structure over U. We proceed similarly to Ostrovsky et al. [29]. For each $P \in U$ we define another primed attribute P'. Let $\bar{U} = \{P' \mid P \in U\}$. As in [29], we define a monotonic access structure $\tilde{\mathbb{A}}$ over $U \cup \bar{U}$ in such a way that $S \in \mathbb{A} \Leftrightarrow S \cup \{P' \in \bar{U} \mid P \in U \setminus S\} \in \tilde{\mathbb{A}}$. Then, it is not hard to see that $\psi(\mathbb{A})$ can be written as $\psi(\tilde{\mathbb{A}})$ with each primed variable P' being replaced by the negation of P.

For simplicity, we will use the access structure $\mathbb{A}$ and its corresponding boolean formula $\psi(\mathbb{A})$ interchangeably when specifying a policy.

3 Two Properties: Verifiability and Delegatability

In our constructions, we need FE (CP/KP-ABE) to have either verifiability or delegatability. In this section we define both properties. While the former is a new one defined in this paper, the latter one was already defined in [21,5,7] for the KP-ABE, CP-ABE and general FE cases respectively. We note that the notion of delegation for FE subsumes that of hierarchical IBE [22,6]. We also note that similar notion to the verifiability "committing" is defined in the IBE setting in [19].

VERIFIABILITY. Intuitively, we say that an FE scheme has verifiability if it is possible to verify whether a ciphertext will be recovered into the same decryption result under two different decryption keys with two specific attributes.

Definition 6. *An FE scheme* $\Pi = (\textbf{Setup}, \textbf{KeyGen}, \textbf{Encrypt}, \textbf{Decrypt})$, *is said to have verifiability if there also exists a polynomial time algorithm* **Verify** *that takes as inputs* PK, CT, X, X' *and outputs* 0 *or* 1 *or* $\perp$ *according to the following properties. Let* Y *be obtained from parsing* CT. *We require that first if* $R(X,Y) = 0$ *or* $R(X',Y) = 0$, *then* **Verify** *outputs* $\perp$.

Second if $R(X,Y) = R(X',Y) = 1$ *then if we let* $\textbf{Setup}(\lambda, des) \rightarrow (PK, MSK)$ *then we have the following.*

(Soundness). *For all* CT *in the ciphertext space (which might be invalid),*

$$Pr\left[\begin{array}{c|l} \textbf{Decrypt}(PK, CT, SK_X) = & \textbf{Verify}(PK, CT, X, X') = 1, \\ \quad \textbf{Decrypt}(PK, CT, SK_{X'}) & SK_X \leftarrow \textbf{KeyGen}(PK, MSK, X), \\ & SK_{X'} \leftarrow \textbf{KeyGen}(PK, MSK, X') \end{array}\right] = 1.$$

(Completeness). *For all* M *in the message space,*

$$Pr\left[\textbf{Verify}(PK, CT, X, X') = 1 \mid CT \leftarrow \textbf{Encrypt}(PK, M, Y)\right] = 1.$$

Note that our definition of verifiability is weaker notion than usual public verifiability. Namely, in our definition, validity of the ciphertext is not needed to be checked. See Sec. 8 for the gap between the (standard) public verifiability and our verifiability. Moreover, for our conversion, the above definition of verifiability is sufficient but not necessary. See Appendix B for a weaker (but complicated) variant of our verifiability.

DELEGATABILITY. Intuitively, delegatability is the capability to use a key for some key attribute X to derive another key for key attribute X' which is possible if X' is inferior than X when considering a well-defined partial order relation $\succeq$ over the key attribute domain Σ_k. More precisely, one can derive $SK_{X'}$ from SK_X if $X \succeq X'$. In the both of CP-ABE and KP-ABE cases, we define the partial order relation as $X \succeq X'$ iff $X \supseteq X'$. The formal definition is as follows.

Definition 7. *An FE scheme* $\Pi = (\textbf{Setup}, \textbf{KeyGen}, \textbf{Encrypt}, \textbf{Decrypt})$, *is said to have delegatability if there also exists a polynomial time algorithm* **Delegate** *such that the output of* $\textbf{Delegate}(PK, SK_X, X, X')$ *and of* $\textbf{KeyGen}(MSK, PK, X')$ *have the same probability distribution for all* $X, X' \in \Sigma_k$ *such that* $X' \preceq X$ *and for all* SK_X *output by* $\textbf{KeyGen}(MSK, PK, X)$.

Indeed, this definition is stronger than needed to apply our conversion for the KP-ABE case. In such a case, it suffices to require only that the output of **Delegate**$(PK, SK_{\mathbb{A}}, \psi(\mathbb{A}), \psi(\mathbb{A}) \wedge P)$ and of **KeyGen**$(MSK, PK, \psi(\mathbb{A}) \wedge P)$ have the same probability distribution for all access structure $\mathbb{A}$, an attribute P, and all $SK_{\psi(\mathbb{A})}$ output by **KeyGen**$(MSK, PK, \psi(\mathbb{A}))$.

4 General Constructions of CCA-Secure ABE

In this section, we show eight conversions to convert a CPA-secure FE scheme Π to a CCA-secure FE scheme Π'. The eight conversions consist of the combinations of whether the original FE scheme Π is CP-ABE or KP-ABE, Π has verifiability or delegatability, and Π deals with a small or large universe. To describe all conversions in a concise way, we write them all in one construction template below. Each conversion then differs in only the definitions of specific variables in Π' namely X' for key attribute, Y' for ciphertext attribute, W for dummy attribute universe, and a procedure called **Subroutine** used in decryption algorithms. We define W below, while the rest are given in Table 1.

Attribute Universes. ABE can be categorized by the size of the attribute universe that such a scheme can deal with: whether it is of polynomial or super-polynomial size. These are called small and large universe scheme respectively. In our conversions, the converted scheme Π' will be able to deal with the same type as that of its original scheme Π. Suppose that we construct a scheme Π' to work with a universe U, we will utilize a set W of *dummy attributes*, which is disjointed from U. The original scheme Π is then required to deal with universe $U \cup W$. A set of dummy attributes will then be associated to a verification key vk of a one-time signature scheme used in the conversion (see Appendix C.2). We assume that for all vk, $vk \in \{0,1\}^{\ell}$. The set W is defined as follows.

- If Π is a small universe scheme, we set $W = \{P_{1,0}, P_{1,1}, P_{2,0}, P_{2,1}, \ldots, P_{\ell,0}, P_{\ell,1}\}$. We set a dummy attribute set $S_{vk} \subset W$ by setting $S_{vk} = \{P_{1,vk_1}, P_{2,vk_2}, \ldots, P_{\ell,vk_\ell}\}$, where we denote by vk_j the j-th bit of vk.
- If Π is a large universe scheme, we set $W = \{0,1\}^{\ell}$. We set a dummy attribute set $S_{vk} \subset W$ by simply letting $S_{vk} = \{vk\}$.

Construction Template. Given a CPA-secure FE scheme Π = (**Setup**, **KeyGen**, **Encrypt**, **Decrypt**) with verifiability *or* delegatability, we construct another FE scheme Π' = (**Setup**$'$, **KeyGen**$'$, **Encrypt**$'$, **Decrypt**$'$) which is CCA-secure as follows. Let $\Sigma = (\mathcal{G}, \mathcal{S}, \mathcal{V})$ be a one-time signature scheme.

Setup$'(\lambda, U)$. It outputs **Setup**$(\lambda, U \cup W) \to (PK, MSK)$.

KeyGen$'(MSK, PK, X)$. It outputs **KeyGen**$(MSK, PK, X') \to SK_{X'}$. Hence $SK'_X = SK_{X'}$.

Encrypt$'(PK, M, Y)$. It first creates a one-time signature key pair by running $\mathcal{G}(\lambda) \to (vk, sk)$. It then runs **Encrypt**$(PK, M, Y') \to CT$ and $\mathcal{S}(sk, CT) \to \sigma$. It outputs $CT' = (vk, CT, \sigma)$.

Table 1. How to setup X', Y' and **Subroutine** in each case

Conversion CP-ABE1 CPA CP-ABE w/ verifiability $\Rightarrow$ CCA CP-ABE	Conversion KP-ABE1 CPA KP-ABE w/ verifiability $\Rightarrow$ CCA KP-ABE
Attribute set $X' = X$ **Policy** $Y' = Y \vee (\wedge_{P \in S_{vk}} P)$ **Subroutine** If **Verify**$(PK, CT, X, S_{vk}) = 0$ or $\perp$ Return $\perp$. Else Return **Decrypt**$(PK, CT, SK_{X'})$.	**Policy** $X' = X$ **Attribute set** $Y' = Y \cup S_{vk}$ **Subroutine** If **Verify**$(PK, CT, X, \wedge_{P \in S_{vk}} P) = 0$ or $\perp$ Return $\perp$. Else Return **Decrypt**$(PK, CT, SK_{X'})$.
Conversion CP-ABE2 CPA CP-ABE w/ delegation $\Rightarrow$ CCA CP-ABE	Conversion KP-ABE2 CPA KP-ABE w/ delegation $\Rightarrow$ CCA KP-ABE
Attribute set $X' = X \cup W$ **Policy** $Y' = Y \wedge (\wedge_{P \in S_{vk}} P)$ **Subroutine** Run **Delegate**$(PK, SK'_X, X \cup W, X \cup S_{vk})$ $\rightarrow SK_{X \cup S_{vk}}$. Return **Decrypt**$(PK, CT, SK_{X \cup S_{vk}})$.	**Policy** $X' = X$ **Attribute set** $Y' = Y \cup S_{vk}$ **Subroutine** Run **Delegate**$(PK, SK'_X, X, X \wedge (\wedge_{P \in S_{vk}} P))$ $\rightarrow SK_{X \wedge (\wedge_{P \in S_{vk}} P)}$. Return **Decrypt**$(PK, CT, SK_{X \wedge (\wedge_{P \in S_{vk}} P)})$.

Decrypt$'(PK, CT', SK'_X)$. It parses the ciphertext CT' as (vk, CT, σ). If $\mathcal{V}(vk, CT, \sigma) = 0$, then it outputs $\perp$. Otherwise, it runs a subroutine **Subroutine** and outputs its returned value.

Theorem 1. *Let Π be $(\tau, \epsilon_{ABE}, q)$ CPA-secure CP/KP-ABE scheme with verifiability/delegatability, and Σ be a (τ, ϵ_{OTS}) secure one-time signature scheme, then Π' constructed as above is $(\tau - o(\tau), \epsilon_{ABE} + \epsilon_{OTS}, q_D, q_E)$ CCA-secure CP/KP-ABE scheme where $q \geq q_D + q_E$.*

The theorem can be proven from Lemma 1, 2.

Correctness. We prove the correctness of all the conversions as follows.

- In the case of CP-ABE1 and CP-ABE2, assume that the attribute set X satisfies the policy Y (that is $R^{\mathsf{CP}}(X, Y) = 1$). In CP-ABE1, **Verify** outputs 1 since S_{vk} trivially satisfies $\wedge_{P \in S_{vk}} P$ therefore both X and S_{vk} satisfy $Y \vee (\wedge_{P \in S_{vk}} P)$. The correctness then follows from that of the original ABE. In CP-ABE2, since $X \cup W \supseteq X \cup S_{vk}$, **Delegate** outputs secret key for $X \cup S_{vk}$ correctly and it can be easily seen that $X \cup S_{vk}$ satisfies $Y \wedge (\wedge_{P \in S_{vk}} P)$. The correctness follows similarly.
- In the case of KP-ABE1 and KP-ABE2, assume that the attribute set Y satisfies the policy X (that is $R^{\mathsf{KP}}(X, Y) = 1$). In KP-ABE1, **Verify** outputs 1

since $\wedge_{P \in S_{vk}} P$ is trivially satisfied by S_{vk} therefore both X and $\wedge_{P \in S_{vk}} P$ is satisfied by $Y \cup S_{vk}$. The correctness then follows from that of the original ABE. In KP-ABE2, since $X \succeq X \wedge (\wedge_{P \in S_{vk}} P)$, **Delegate** outputs secret key for $X \wedge (\wedge_{P \in S_{vk}} P)$ correctly and it can be easily seen that $X \wedge (\wedge_{P \in S_{vk}} P)$ is satisfied by $Y \cup S_{vk}$. The correctness follows similarly.

Remark. We propose another two variants which are conversions for CP-ABE and KP-ABE based on verifiability in Appendix A. We also note that the conversion KP-ABE2 for the large universe case is exactly the one in [21]. We include it here to cover the big picture of the whole framework.

Efficiency Consideration. We first consider the expansion of attribute sets. This only occurs in CP-ABE2, where we define a key for set $X' = X \cup W$. A problem may occur for the large universe case, since W is of super-polynomial size the key size may also expand enormously depending on the underlying ABE. If such a problem occurs, we use W as defined in the small universe case.

Next we consider the expansion of policies. In all of available constructions of ABE in the literature, an access structure is represented by either of two methods namely an access tree ([21,5]) or a linear-secret sharing scheme (LSSS) matrix ([21,29,31,26,28]). The efficiency, in particular, key sizes and ciphertext sizes, of these respective ABE schemes tend to depend on the size of access trees or LSSS matrices used in such schemes. (See the definition of LSSS in Appendix C.1). Our conversions particularly use policies of the form $\psi(\mathbb{A}) \vee (\wedge_{P \in S_{vk}} P)$ and $\psi(\mathbb{A}) \wedge (\wedge_{P \in S_{vk}} P)$. Therefore, we have to check whether $\psi(\mathbb{A})$ when augmented to each of both forms still can be represented efficiently or not. To this end, the efficiency is guaranteed from the following two observations.

Proposition 1. *Let access structures $\mathbb{A}$ and $\mathbb{B}$ be expressed by access trees using the method in [21] with h_a, h_b nodes and ℓ_a, ℓ_b leaves respectively. Then access structure corresponding to $\psi(\mathbb{A}) \wedge \psi(\mathbb{B})$ and $\psi(\mathbb{A}) \vee \psi(\mathbb{B})$ can be expressed by access trees both with $h_a + h_b + 1$ nodes and $\ell_a + \ell_b$ leaves.*

Proposition 2. *Let access structures $\mathbb{A}$ and $\mathbb{B}$ be expressed by an $\ell_a \times m_a$ and an $\ell_b \times m_b$ LSSS matrix by using the LSSS of [4] respectively. Then the LSSS matrix corresponding to $\psi(\mathbb{A}) \wedge \psi(\mathbb{B})$ and $\psi(\mathbb{A}) \vee \psi(\mathbb{B})$ can be expressed by an $(\ell_a + \ell_b) \times (m_a + m_b)$ and an $(\ell_a + \ell_b - 1) \times (m_a + m_b)$ LSSS matrix respectively.*

To conclude, since $|S_{vk}| = \ell = poly(\lambda)$ in the large-universe construction and $|S_{vk}| = 1$ in the small-universe construction, our conversions can be efficiently implementable.

Selective security. We remark that our conversion can be also applied to selectively (CPA-)secure ABE schemes, and in such cases, resulting CCA-secure schemes are only selectively (CCA-)secure as well.

5 Security of Our Constructions from Verifiability

Security of our constructions from verifiability, i.e. CP-ABE1 and KP-ABE1 is addressed as follows:

Lemma 1. *Let Π be $(\tau, \epsilon_{ABE}, q)$ CPA-secure CP/KP-ABE scheme with verifiability and Σ be a (τ, ϵ_{OTS}) secure one-time signature scheme, then Π' constructed as in Sec. 4 (CP/KP-ABE1) is $(\tau - o(\tau), \epsilon_{ABE} + \epsilon_{OTS}, q_D, q_E)$ CCA-secure CP/KP-ABE scheme where $q \geq q_D + q_E$.*

In the rest of this section, we prove Lemma 1 for the case of CP-ABE. The lemma can also be proven similar way in the case of KP-ABE.

Proof of Lemma 1 for the case of CP-ABE. Assume we are given an adversary $\mathcal{A}$ which breaks CCA-security of the scheme Π' (CP-ABE1) with running time τ, advantage ϵ, q extraction queries, and, q_D decryption queries. We use $\mathcal{A}$ to construct another adversary $\mathcal{B}$ which breaks CPA-security of the scheme Π. Define adversary $\mathcal{B}$ as follows:

Setup. The challenger runs $\mathbf{Setup}(\lambda, U \cup W) \rightarrow (PK, MSK)$. Then $\mathcal{B}$ is given PK and gives it to $\mathcal{A}$. $\mathcal{B}$ also runs $\mathcal{G}(\lambda) \rightarrow (vk^*, sk^*)$.

Phase1. $\mathcal{A}$ may adaptively make queries of the following types:

– ***Key-extraction query.*** When $\mathcal{A}$ submits S, then $\mathcal{B}$ submits same S to challenger. $\mathcal{B}$ is given private key SK_S for S and gives it to $\mathcal{A}$.

– ***Decryption query.*** When $\mathcal{A}$ submits (CT', S) such that $CT' = (vk, CT, \sigma)$, $\mathcal{B}$ respond to $\mathcal{A}$ as follows. First, $\mathcal{B}$ checks whether $\mathcal{V}(vk, CT, \sigma) = 1$ holds. If it does not hold, then $\mathcal{B}$ returns $\perp$. If it holds and $vk^* = vk$, then $\mathcal{B}$ aborts. Otherwise, $\mathcal{B}$ checks whether $\mathbf{Verify}(PK, CT, S_{vk}, S) = 1$. If it does not hold, then $\mathcal{B}$ returns $\perp$. Otherwise $\mathcal{B}$ submits S_{vk} to the challenger and is given $SK_{S_{vk}}$. Then $\mathcal{B}$ returns output of $\mathbf{Decrypt}(PK, CT, SK_{S_{vk}})$ to $\mathcal{A}$.

Challenge. $\mathcal{A}$ declares two equal length messages M_0 and M_1 and an access structure $\mathbb{A}^*$. Then $\mathcal{B}$ declares the same messages M_0, M_1 and $\mathbb{A}^{*\prime}$ for the challenger, where $\mathbb{A}^{*\prime}$ is an access structure such that $\psi(\mathbb{A}^{*\prime}) = \psi(\mathbb{A}^*) \vee (\wedge_{P \in S_{vk^*}} P)$. The challenger flips a random coin $\beta \in \{0, 1\}$, runs $\mathbf{Encrypt}(PK, M_\beta, \psi(\mathbb{A}^{*\prime})) \rightarrow CT^*$ and gives CT^* to $\mathcal{B}$. Then $\mathcal{B}$ runs $\mathcal{S}(sk^*, CT^*) \rightarrow \sigma^*$, and gives $CT^{*\prime} = (vk^*, CT^*, \sigma^*)$ to $\mathcal{A}$ as challenge ciphertext.

Phase2. $\mathcal{B}$ responds to $\mathcal{A}$'s query as the same as in **Phase1**.

Guess. Finally, $\mathcal{A}$ outputs a guess β' for β. Then $\mathcal{B}$ outputs β' as its guess.

Let **Win** denote the event that $\mathcal{A}$ correctly guess β, **Abort** denote the event that $\mathcal{B}$ aborts. If **Abort** does not occur, from the verifiability of the scheme, $\mathcal{B}$'s simulation is perfect. So, $\mathcal{B}$'s advantage for guessing β is estimated as $Pr[\mathcal{B}$ correctly guesses $\beta] - \frac{1}{2} = Pr[\mathbf{Win}|\overline{\mathbf{Abort}}]Pr[\overline{\mathbf{Abort}}] - \frac{1}{2} \geq Pr[\mathbf{Win}] - Pr[\mathbf{Abort}] - \frac{1}{2} \geq \epsilon - Pr[\mathbf{Abort}]$. Since $Pr[\mathbf{Abort}] \leq \epsilon_{OTS}$ holds due to unforgeability of the one-time-signature, the proof is completed. □

6 Security of Our Construction from Delegatability

Security of our constructions from delegatability, i.e. CP-ABE2 and KP-ABE2 is addressed as follows. In this section, we prove Lemma 2 for the case of CP-ABE. The lemma can also be proven by similar way in the case of KP-ABE.

Lemma 2. *Let Π be a $(\tau, \epsilon_{ABE}, q)$ CPA-secure CP/KP-ABE scheme with delegatability and Σ be (τ, ϵ_{OTS}) secure one-time signature, then Π' constructed as in section4 (CP/KP-ABE2) is $(\tau - o(\tau), \epsilon_{ABE} + \epsilon_{OTS}, q_D, q_E)$ CCA-secure CP/KP-ABE scheme where $q \geq q_D + q_E$.*

Proof of Lemma 2 for the case of CP-ABE. Assume we are given an adversary $\mathcal{A}$ which breaks CCA-security of the scheme Π' (CP-ABE2) with running time τ, advantage ϵ, q_E key-extraction queries, and, q_D decryption queries. We use $\mathcal{A}$ to construct another adversary $\mathcal{B}$ which breaks CPA-security of the scheme Π. Define adversary $\mathcal{B}$ as follows:

Setup. The challenger runs **Setup**$(\lambda, U \cup W) \to (PK, MSK)$. Then $\mathcal{B}$ is given PK and gives it to $\mathcal{A}$. $\mathcal{B}$ also runs $\mathcal{G}(\lambda) \to (vk^*, sk^*)$.

Phase1. $\mathcal{A}$ may adaptively make queries of the following types:

– ***Key-extraction query.*** When $\mathcal{A}$ submits S, then $\mathcal{B}$ submits $S \cup W$ to the challenger. $\mathcal{B}$ is given private key $SK_{S \cup W}$ for $S \cup W$ and gives it to $\mathcal{A}$.

– ***Decryption query.*** When $\mathcal{A}$ submits (CT', S) such that $CT' = (vk, CT, \sigma)$, $\mathcal{B}$ respond to $\mathcal{A}$ as follows. First, $\mathcal{B}$ checks whether $\mathcal{V}(vk, CT, \sigma) = 1$ holds. If it does not hold, then $\mathcal{B}$ returns $\perp$. If it holds and $vk^* = vk$, then $\mathcal{B}$ aborts. Otherwise $\mathcal{B}$ submits $S \cup S_{vk}$ to the challenger and is given $SK_{S \cup S_{vk}}$. Then $\mathcal{B}$ rerandomize it by $SK_{S \cup S_{vk}} \leftarrow$ **Delegate**$(PK, SK_{S \cup S_{vk}}, S \cup S_{vk}, S \cup S_{vk})$ and returns output of **Decrypt**$(PK, CT, SK_{S \cup S_{vk}})$ to $\mathcal{A}$.

Challenge. $\mathcal{A}$ declares two equal length messages M_0, M_1 and $\mathbb{A}^*$. Then $\mathcal{B}$ declares the same messages M_0, M_1, and $\mathbb{A}^{*\prime}$ for the challenger, where $\mathbb{A}^{*\prime}$ is an access structure such that $\psi(\mathbb{A}^{*\prime}) = \psi(\mathbb{A}^*) \wedge (\wedge_{P \in S_{vk^*}} P)$. The challenger flips a random coin $\beta \in \{0, 1\}$, runs **Encrypt**$(PK, M_\beta, \psi(\mathbb{A}^{*\prime})) \to CT^*$ and gives CT^* to $\mathcal{B}$. Then $\mathcal{B}$ runs $\mathcal{S}(sk^*, CT^*) \to \sigma^*$ and gives $CT^{*\prime} = (vk^*, CT^*, \sigma^*)$ to $\mathcal{A}$ as challenge ciphertext.

Phase2. $\mathcal{B}$ responds to $\mathcal{A}$'s query as the same as in **Phase1**.

Guess. Finally, $\mathcal{A}$ outputs a guess β' for β. Then $\mathcal{B}$ outputs β' as its guess.

Similar analysis to the previous section shows that $Pr[\mathcal{B}$ correctly guess $\beta] - \frac{1}{2} \geq \epsilon - \epsilon_{OTS}$. □

7 Applications to Existing Schemes

7.1 The Case of ABE by Lewko et al.

In this section, we show some applications of our conversions to the recent CPA-secure CP-ABE by Lewko et al. [26] to achieve CCA-secure schemes. We observe first that neither delegation was presented nor verifiability is available in their ABE. However, we show here that only a slight modification will allow both properties. For self-containment, we briefly describe their scheme here.

Description for CP-ABE of [26]. The scheme works in a bilinear group of composite order $N = p_1 p_2 p_3$. Denote $\mathbb{G}_{p_j}$ the subgroup of order p_j of $\mathbb{G}$. The master key is $MSK = (\alpha \in \mathbb{Z}_N, X_3 \in \mathbb{G}_{p_3})$, while the public key is of the form

$PK = N, g, g^a, e(g,g)^\alpha, \{T_i = g^{s_i}\}_{i \in U}$ where $g \in \mathbb{G}_{p_1}, a, s_i \in \mathbb{Z}_N$. Note that the scheme works with a small universe U. A secret key for set $S \subset U$ is of the form $SK_S = (S,\ K = g^\alpha g^{at} R_0,\ L = g^t R'_0,\ \{K_i = T_i^t R_i\}_{i \in S})$ for random $R_0, R'_0, R_i \in \mathbb{G}_{p_3}, t \in \mathbb{Z}_N$. Denote $B(\mathbb{A}) = \cup_{S \in \mathbb{A}} S$. A ciphertext for policy $\mathbb{A}$ is of the form $CT = (C = Me(g,g)^{s\alpha},\ C' = g^s,\ \{C_x = g^{aA_x \cdot v} T_{\rho(x)}^{-r_x}, D_x = g^{r_x}\}_{x \in B(\mathbb{A})})$ for some random $s, r_x \in \mathbb{Z}_N$ and where $A_x \cdot v$ is the random share for x of the secret s in the LSSS scheme representing the policy $\mathbb{A}$. Decryption can be done if $S \in \mathbb{A}$ by recovering $e(C', K)/\prod_{\rho(x) \in S} (e(C_x, L) e(D_x, K_{\rho(x)}))^{\omega_x} = e(g,g)^{\alpha s}$ where $\{\omega_x\}$ is the reconstruction coefficient of the LSSS scheme.

Slight Modification. The above scheme seems not to have neither delegatability nor verifiability. This is mainly due to the fact that one cannot check whether a ciphertext consists of only elements in $\mathbb{G}_{p_1 p_2}$ or not. Thus, for achieving delegatability and verifiability, we modify the above ABE scheme by simply including also the generator $X_3 \in \mathbb{G}_{p_3}$ in PK. We argue that this modified scheme is still CPA-secure. This can be easily seen since all the three underlying hard problems in [26] that the scheme is based on contains a generator of $\mathbb{G}_{p_3}$ as an input. In the following, we show verifiability and delegatability of the resulting scheme.

Delegatability. We define **Delegate** of the modified scheme as follows.

Delegate$(PK, SK_S, S'(\subseteq S))$ It chooses random $u \in \mathbb{Z}_N$ and random elements $R_0, R'_0, R_i \in \mathbb{G}_{p_3}$. It computes the key for S' as $SK_{S'} = (S',\ K' = K g^{au} R_0,\ L' = L g^u R'_0,\ \{K'_i = K_i T_i^u R_i\}_{i \in S'})$.

It is straightforward to see that the output of **Delegate**$(PK, SK_S, S'(\subseteq S))$ and that of **KeyGen**(MSK, PK, S') have the same probability distribution.

Verifiability. We define **Verify** of the modified scheme as follows.

Verify(PK, CT, S, S') It parses $PK = (N, g, g^a, e(g,g)^\alpha, \{T_i\}_{i \in \mathcal{U}}, X_3)$ and $CT = (\mathbb{A}, C, C', \{C_x, D_x\}_{x \in B(\mathbb{A})})$ then outputs V as

$$V = \begin{cases} \perp \text{ if } S \notin \mathbb{A} \text{ or } S' \notin \mathbb{A}. \\ 1 \text{ if } \prod\limits_{\rho(x) \in S} (e(C_x, g) e(D_x, T_{\rho(x)}))^{w_{x,S}} \\ \quad = \prod\limits_{\rho(x) \in S'} (e(C_x, g) e(D_x, T_{\rho(x)}))^{w_{x,S'}} = e(g^a, C'), & (1) \\ \quad \text{and } e(C', X_3) = 1,\ e(C_x, X_3) = e(D_x, X_3) = 1 \text{ for all } x \in B(\mathbb{A}). & (2) \\ 0 \text{ otherwise.} \end{cases}$$

Here $\omega_{x,S}$ and $\omega_{x,S'}$ are reconstruction coefficients in the LSSS. Hence we have $\sum_{\rho(x) \in S} \omega_{x,S} A_x = \sum_{\rho(x) \in S'} \omega_{x,S'} A_x = (1, 0, \ldots, 0)$. We now prove that **Verify** algorithm defined as above satisfies soundness and completeness properties.

Proving Soundness. Consider $S, S' \in \mathbb{A}$. Assume that **Verify**$(PK, CT, S, S') = 1$ and that $SK_S, SK_{S'}$ are correctly generated. We will prove that **Decrypt** (PK, CT, SK_S) = **Decrypt**$(PK, CT, SK_{S'})$ holds. To this end, we parse

$SK = (S, K, L, \{K_i\}_{i \in S})$, and see that **Decrypt**(PK, CT, SK_S) outputs the following.

$$\begin{aligned}
&\left(C \cdot \prod\nolimits_{\rho(x) \in S} (e(C_x, L)e(D_x, K_{\rho_x}))^{\omega_x}\right) / e(C', K) && \text{by def} \\
&= \frac{\left(C \cdot \prod_{\rho(x) \in S} (e(C_x, g^t \cdot R'_0)e(D_x, T^t_{\rho(x)} R_{\rho(x)}))^{\omega_x}\right)}{e(C', g^\alpha \cdot g^{at} \cdot R_0)} && \text{by def of } SK_S \\
&= \left(C \cdot \prod\nolimits_{\rho(x) \in S} (e(C_x, g)e(D_x, T_{\rho(x)}))^{t\omega_x}\right) / e(C', g^\alpha \cdot g^{at}) && \text{by (2)} \\
&= C \cdot e(g, C')^{at} / (e(C', g)^\alpha e(g, C')^{at}) = C / e(g, C')^\alpha. && \text{by (1)}
\end{aligned}$$

Now since S is arbitrary, the same result holds for S', which concludes the proof.

Proving Completeness. Assume that a ciphertext CT is correctly generated. We will prove that **Verify**$(PK, CT, S, S') = 1$. A correctly generated ciphertext is the form of $CT = (C = Me(g, g)^{s\alpha},\ C' = g^s,\ \{C_x = g^{aA_x \cdot v} T^{-r_x}_{\rho(x)}, D_x = g^{r_x}\}_{x \in B(\mathbb{A})})$. Since all elements are in $\mathbb{G}_{p_1 p_2}$, (2) holds. Equation (1) also holds by straightforward calculation.

Resulting CCA-secure Schemes. We now compare the two CCA-secure CP-ABE constructions converted from the above (slightly modified) CP-ABE of [26] by using the CP-ABE1 (required verifiability) and CP-ABE2 (required delegatability). As for the public key length, ciphertext length, and encryption cost, it seems that former is as efficient as latter. (Ciphertext length and encryption cost depend on the underlying LSSS matrix.) Secret key length of former is shorter than that of latter. As for the decryption cost, latter is more efficient than former since the **Verify** algorithm contains many pairing computation as opposed to the **Delegate** algorithm.

Table 2. ABE with delegatability or verifiability. In the table, "Deleg." and "Verif." denote delegatability and verifiability respectively.

Schemes		KP/CP	Universe	Deleg.	Verif.	Security	Assumption
Goyal et al.	[21, Sect.4]	KP	small	U	✓	selective	DBDH
Goyal et al.	[21, Sect.5]	KP	large	✓	✓	selective	DBDH
Goyal et al.	[21, Sect.A]	KP	small	U	✓	selective	DBDH
Ostrovsky et al.	[29, Sect.3]	KP	large	U	✓	selective	DBDH
Bethencourt et al.	[5]	CP	large	✓	✓	selective	Generic group
Goyal et al.	[20]	CP	small	U	✓	selective	DBDH
Waters	[31, Sect.3]	CP	small	✓	✓	selective	DPBDHE
Lewko et al.	[24, Sect.6]	KP	large	U	✓	selective	q-MEBDH
Attrapadung et al.	[1]	KP	large	U	✓	selective	DBDHE
Lewko et al.	[26, Sect.2]	CP	small	U	U	full	3 assumptions
Section 7.1 (modified from [26])		CP	small	✓	✓	full	3 assumptions
Lewko et al.	[26, Sect.A]	KP	small	U	U	full	3 assumptions
Okamoto et al.	[28]	KP	large	U	U	full	DLIN
Slightly modified	[28]	KP	large	✓	✓	full	DLIN

We remark that KP-ABE scheme in [26] also could be modified to have verifiability by similar technique to the case of CP-ABE.

7.2 Summary for Applications to Existing Schemes

In Table 2, we give an overview of existing ABE schemes and their properties, and from this table, one can see that many of these schemes satisfy verifiability and/or delegatability. We remark that similarly to [26], Okamoto and Takashima's scheme [28] can be also modified to have both delegatability and verifiability. See the full version of our paper for details. In the table, ✓ denotes there is verify or delegate algorithm that satisfies our definition, "U" denotes there is unknown such algorithm.

8 Remark on Verifiability

Our definition of verifiability is considered weaker than that of the standard public verifiability where roughly speaking, we say that an encryption scheme satisfies *public verifiability* if any third party (who does not have any secret) can always verify whether a given ciphertext is one of possible outputs of the encryption algorithm or not. To see this, we show an FE scheme which has verifiability, but does not have public verifiability. We construct such FE scheme $\Pi' = (\mathbf{Setup}', \mathbf{KeyGen}', \mathbf{Encrypt}', \mathbf{Decrypt}', \mathbf{Verify}')$ from FE scheme $\Pi = (\mathbf{Setup}, \mathbf{KeyGen}, \mathbf{Encrypt}, \mathbf{Decrypt}, \mathbf{Verify})$ with verifiability, an one-way function $f : \{0,1\}^n \rightarrow \{0,1\}^{n'}$, and a hardcore function $h : \{0,1\}^n \rightarrow \{0,1\}$ for f. Here, n and n' are polynomials of λ. $\mathbf{Setup}'$ and $\mathbf{KeyGen}'$ are the same as **Setup** and **KeyGen** respectively. $\mathbf{Encrypt}'$ is slightly different from **Encrypt**. $\mathbf{Encrypt}'(PK, M, Y)$ first runs $\mathbf{Encrypt}(PK, M, Y) \rightarrow CT$ and picks a random $x \leftarrow \{0,1\}^n$ independently. Then it compute $(f(x), h(x)) \in \{0,1\}^{n'+1}$ and returns final ciphertext $CT' = (CT, f(x), h(x))$. $\mathbf{Decrypt}'(PK, CT', SK_X)$ first parses CT' as (CT, y, b) and returns $\mathbf{Decrypt}(PK, CT, SK_X)$ where $y \in \{0,1\}^{n'}$, $b \in \{0,1\}$. $\mathbf{Verify}'(PK, CT', X, X')$ first parses CT' as (CT, y, b) as the same as above, then returns $\mathbf{Verify}(PK, CT, X, X')$.

It is clear that Π' has verifiability since **Verify** algorithm defined as above works correctly. However, Π' does not have public verifiability since for verifying validity of a ciphertext in the sense of public verifiability, one has to correctly guess the hardcore bit $h(x)$ from only $f(x)$.

References

1. Attrapadung, N., Libert, B.: Expressive Key-Policy Attribute-Based Encryption with Constant-Size Ciphertexts. In: Catalano, D., et al. (eds.) PKC 2011. LNCS, vol. 6571. Springer, Heidelberg (2011)
2. Attrapadung, N., Imai, H.: Dual-Policy Attribute Based Encryption. In: Abdalla, M., Pointcheval, D., Fouque, P.-A., Vergnaud, D. (eds.) ACNS 2009. LNCS, vol. 5536, pp. 168–185. Springer, Heidelberg (2009)

3. Attrapadung, N., Imai, H.: Conjunctive Broadcast and Attribute-Based Encryption. In: Shacham, H., Waters, B. (eds.) Pairing 2009. LNCS, vol. 5671, pp. 248–265. Springer, Heidelberg (2009)
4. Benaloh, J.C., Leichter, J.: Generalized Secret Sharing and Monotone Functions. In: Goldwasser, S. (ed.) CRYPTO 1988. LNCS, vol. 403, pp. 27–35. Springer, Heidelberg (1990)
5. Bethencourt, J., Sahai, A., Waters, B.: Ciphertext-Policy Attribute-Based Encryption. In: IEEE Symposium on Security and Privacy (S&P), pp. 321–334 (2007)
6. Boneh, D., Boyen, X.: Efficient Selective-ID Secure Identity-Based Encryption Without Random Oracles. In: Cachin, C., Camenisch, J.L. (eds.) EUROCRYPT 2004. LNCS, vol. 3027, pp. 223–238. Springer, Heidelberg (2004)
7. Boneh, D., Hamburg, M.: Generalized Identity Based and Broadcast Encryption Schemes. In: Pieprzyk, J. (ed.) ASIACRYPT 2008. LNCS, vol. 5350, pp. 455–470. Springer, Heidelberg (2008)
8. Boneh, D., Katz, J.: Improved efficiency for CCA-secure cryptosystems built using identity-based encryption. In: Menezes, A. (ed.) CT-RSA 2005. LNCS, vol. 3376, pp. 87–103. Springer, Heidelberg (2005)
9. Boyen, X., Mei, Q., Waters, B.: Direct Chosen Ciphertext Security from Identity-Based Techniques. In: ACM CCS 2005, pp. 320–329 (2005)
10. Boneh, D., Sahai, A., Waters, B.: Functional Encryption: Definitions and Challenges. In: TCC 2011. LNCS, vol. 6597. Springer, Heidelberg (to appear)
11. Canetti, R., Halevi, S., Katz, J.: A Forward-Secure Public-Key Encryption Scheme. In: Biham, E. (ed.) EUROCRYPT 2003. LNCS, vol. 2656, pp. 254–271. Springer, Heidelberg (2003)
12. Canetti, R., Halevi, S., Katz, J.: Chosen-Ciphertext Security from Identity-Based Encryption. In: Cachin, C., Camenisch, J.L. (eds.) EUROCRYPT 2004. LNCS, vol. 3027, pp. 207–222. Springer, Heidelberg (2004)
13. Chase, M.: Multi-authority Attribute Based Encryption. In: Vadhan, S.P. (ed.) TCC 2007. LNCS, vol. 4392, pp. 515–534. Springer, Heidelberg (2007)
14. Chase, M., Chow, S.: Improving privacy and security in multi-authority attribute-based encryption. In: ACM CCS 2009, pp. 121–130 (2009)
15. Cheung, L., Newport, C.: Provably secure ciphertext policy ABE. In: ACM CCS 2007, pp. 456–465 (2007)
16. Damgård, I., Thorbek, R.: Linear integer secret sharing and distributed exponentiation. In: Yung, M., Dodis, Y., Kiayias, A., Malkin, T.G. (eds.) PKC 2006. LNCS, vol. 3958, pp. 75–90. Springer, Heidelberg (2006)
17. Dolev, D., Dwork, C., Naor, M.: Non-Malleable Cryptography (Extended Abstract). In: STOC 1991, pp. 542–552 (1991)
18. Fujisaki, E., Okamoto, T.: Secure Integration of Asymmetric and Symmetric Encryption Schemes. In: Wiener, M. (ed.) CRYPTO 1999. LNCS, vol. 1666, pp. 537–554. Springer, Heidelberg (1999)
19. Green, M., Hohenberger, S.: Blind Identity-Based Encryption and Simulatable Oblivious Transfer. In: Kurosawa, K. (ed.) ASIACRYPT 2007. LNCS, vol. 4833, pp. 265–282. Springer, Heidelberg (2007)
20. Goyal, V., Jain, A., Pandey, O., Sahai, A.: Bounded Ciphertext Policy Attribute Based Encryption. In: Aceto, L., Damgård, I., Goldberg, L.A., Halldórsson, M.M., Ingólfsdóttir, A., Walukiewicz, I. (eds.) ICALP 2008, Part II. LNCS, vol. 5126, pp. 579–591. Springer, Heidelberg (2008)
21. Goyal, V., Pandey, O., Sahai, A., Waters, B.: Attribute-based encryption for fine-grained access control of encrypted data. In: ACM CCS 2006, pp. 89–98 (2006)

22. Gentry, C., Silverberg, A.: Hierarchical ID-Based Cryptography. In: Zheng, Y. (ed.) ASIACRYPT 2002. LNCS, vol. 2501, pp. 548–566. Springer, Heidelberg (2002)
23. Herranz, J., Laguillaumie, F., Ràfols, C.: Constant Size Ciphertexts in Threshold Attribute-Based Encryption. In: Nguyen, P.Q., Pointcheval, D. (eds.) PKC 2010. LNCS, vol. 6056, pp. 19–34. Springer, Heidelberg (2010)
24. Lewko, A., Sahai, A., Waters, B.: Revocation Systems with Very Small Private Keys. In: IEEE Symposium on Security and Privacy (S&P), pp. 273–285 (2010)
25. Lewko, A., Waters, B.: Decentralizing Attribute-Based Encryption. Cryptology ePrint Archive: Report 2010/351 (2010)
26. Lewko, A., Okamoto, T., Sahai, A., Takashima, K., Waters, B.: Fully Secure Functional Encryption: Attribute-Based Encryption and (Hierarchical) Inner Product Encryption. In: Gilbert, H. (ed.) EUROCRYPT 2010. LNCS, vol. 6110, pp. 62–91. Springer, Heidelberg (2010)
27. Naor, M., Yung, M.: Public-key Cryptosystems Provably Secure against Chosen Ciphertext Attacks. In: STOC 1990, pp. 427–437 (1990)
28. Okamoto, T., Takashima, K.: Fully secure functional encryption with general relations from the decisional linear assumption. In: Rabin, T. (ed.) CRYPTO 2010. LNCS, vol. 6223, pp. 191–208. Springer, Heidelberg (2010)
29. Ostrovsky, R., Sahai, A., Waters, B.: Attribute-based encryption with non-monotonic access structures. In: ACM CCS 2007, pp. 195–203 (2007)
30. Sahai, A., Waters, B.: Fuzzy Identity-Based Encryption. In: Cramer, R. (ed.). EUROCRYPT 2005. LNCS, vol. 3494, pp. 457–473. Springer, Heidelberg (2005)
31. Waters, B.: Ciphertext-Policy Attribute-Based Encryption: An Expressive, Efficient, and Provably Secure Realization. In: Catalano, D., et al. (eds.) PKC 2011. LNCS, vol. 6571, pp. 53–70. Springer, Heibelberg (2011)

A Variants of CP-ABE2 and KP-ABE2

In Table 3, we show CP-ABE3 and KP-ABE3 which are variants of CP-ABE2 and KP-ABE2, respectively. These schemes are very similar, differences are that CP-ABE3 and KP-ABE3 are constructed from verifiability whereas CP-ABE2 and KP-ABE2 are constructed from delegatability.

Table 3. How to setup X', Y' and **Subroutine** in CP/KP-ABE3

Conversion CP-ABE3 CPA CP-ABE w/ verifiability ⇒ CCA CP-ABE	Conversion KP-ABE3 CPA KP-ABE w/ verifiability ⇒ CCA KP-ABE
Attribute set $X' = X \cup W$ **Policy** $Y' = Y \wedge (\wedge_{P \in S_{vk}} P)$ **Subroutine** If **Verify**$(PK, CT, X \cup W, X \cup S_{vk}) = 0$ or $\perp$ Return $\perp$. Else Return **Decrypt**$(PK, CT, SK_{X'})$.	**Policy** $X' = X$ **Attribute set** $Y' = Y \cup S_{vk}$ **Subroutine** If **Verify**$(PK, CT, X, X \wedge_{P \in S_{vk}} P) = 0$ or $\perp$ Return $\perp$. Else Return **Decrypt**$(PK, CT, SK_{X'})$.

B Weaker Verifiability

Our conversion still works even if the underlying FE scheme does not satisfy our verifiability but a weaker notion than it. This weaker variant of our verifiability is defined as follows.

Definition 8. *A FE scheme Π is said to have weaker verifiability if there exists a polynomial time algorithm* **Verify** *that takes as We require that if $R(X,Y)=0$ or $R(X',Y)=0$, then* **Verify** *outputs $\perp$. Here, Y is obtained from parsing CT.*

1. *If SK'_X is output of* **KeyGen**(PK, MSK, X), *then* $Pr[$**Decrypt**$(PK, CT, SK_X) =$ **Decrypt**$(PK, CT, SK_{X'})|$
   $SK_X \leftarrow$ **KeyGen**(PK, MSK, X), $SK_{X'}$**KeyGen**(PK, MSK, X'),
   Verify$(PK, CT, X, X', SK'_X) = 1] = 1$ *always holds.*
2. *If $R(X,Y) = R(X',Y) = 1$, then it holds that for all correctly generated PK and for all CT (which might be invalid),*
   $Pr[$**Verify**$(PK, CT, X, X', SK_X) =$ **Verify**$(PK, CT, X', X, SK_{X'})|$
   $SK_X \leftarrow$ **KeyGen**(PK, MSK, X), $SK_{X'} \leftarrow$ **KeyGen**$(PK, MSK, X')] = 1$.
3. *If SK_X is output of* **KeyGen**(PK, MSK, X) *and $R(X,Y) = R(X',Y) = 1$, then It holds that* $Pr[$**Verify**$(PK, CT, X, X', SK_X) = 1|$
   $SK_X \leftarrow$ **KeyGen**(PK, MSK, X), $CT \leftarrow$ **Encrypt**$(PK, M, Y)] = 1$

C Some Omitted Definitions and Descriptions

C.1 Linear Secret Sharing Schemes

Definition 9 (Linear Secret Sharing Scheme). *Let $\mathcal{P}$ be a set of parties. Let M be a $\ell \times k$ matrix. Let $\pi : \{1, \ldots, \ell\} \to \mathcal{P}$ be a function that maps a row to a party for labeling. A secret sharing scheme Π for access structure $\mathbb{A}$ over a set of parties $\mathcal{P}$ is a linear secret-sharing scheme (LSSS) in $\mathbb{Z}_p$ and is represented by (M, π) if it consists of two efficient algorithms:*

Share$_{(M,\pi)}$: *The algorithm takes as input $s \in \mathbb{Z}_p$ which is to be shared. It chooses $a_2, \ldots, a_k \in \mathbb{Z}_p$ and let $\boldsymbol{a} = (s, a_2, \ldots, a_k)^\top$. It outputs $M \cdot \boldsymbol{a}$ as the vector of ℓ shares. The share $\lambda_i := \langle \boldsymbol{M_i}, \boldsymbol{a} \rangle$ belongs to party $\pi(i)$, where $\boldsymbol{M_i}^\top$ denotes the ith row of M.*

Recon$_{(M,\pi)}$: *The algorithm takes as input an access set $S \in \mathbb{A}$. Let $I = \{i | \pi(i) \in S\}$. It outputs a set of constants $\{(i, \mu_i)\}_{i \in I}$ which has a linear reconstruction property: $\sum_{i \in I} \mu_i \cdot \lambda_i = s$.*

C.2 One-Time-Signature

A one-time-signature scheme consists of the following three algorithms, $\mathcal{G}$, $\mathcal{S}$, and $\mathcal{V}$. The key generation algorithm $\mathcal{G}(\lambda)$ takes as input the security parameter λ, and outputs a verification key vk and a signing key sk. The sign algorithm

$\mathcal{S}(sk, m)$ takes as input sk and a message m, and outputs a signature σ. The verify algorithm $\mathcal{V}(vk, m, \sigma)$ takes as input vk, m, and σ, and outputs a bit $b \in \{0, 1\}$. We require that for all honestly generated sk, all m in the message space, and all σ, output by $\mathcal{S}(sk, m)$, we have $\mathcal{V}(vk, m, \sigma) = 1$. Next, we define strong unforgery of a (one-time) signature scheme Σ against chosen message attacks. Security is defined using the following game between an attacker $\mathcal{A}$ and a challenger. Both the challenger and attacker are given λ as input. First, the challenger runs $\mathcal{G}(\lambda)$ to obtain vk and sk. It gives $\mathcal{A}$ vk. Next, $\mathcal{A}$ may issue at most one signing query m^*. The challenger responds with $\sigma^* = \mathcal{S}(sk, m^*)$. Finally, $\mathcal{A}$ outputs (m, σ). We say that $\mathcal{A}$ succeeds to forge if $\mathcal{A}$ outputs (m, σ) such that $(m, \sigma) \neq (m^*, \sigma^*)$ and $\mathcal{V}(vk, m, \sigma) = 1$, and denote the probability of this event by $Adv^{OTS}_{\mathcal{A},\Sigma}$.

Definition 10. *We say that a one-time-signature scheme Σ is (τ, ϵ)-secure if for all τ-time algorithms $\mathcal{A}$ we have that $Adv^{OTS}_{\mathcal{A},\Sigma} < \epsilon$.*

Expressive Key-Policy Attribute-Based Encryption with Constant-Size Ciphertexts

Nuttapong Attrapadung[1], Benoît Libert[2,⋆], and Elie de Panafieu[3]

[1] Research Center for Information Security, AIST, Japan
n.attrapadung@aist.go.jp
[2] Université catholique de Louvain, ICTEAM – Crypto Group, Belgium
benoit.libert@uclouvain.be
[3] Ecole Normale Supérieure, Cachan, France

Abstract. Attribute-based encryption (ABE), as introduced by Sahai and Waters, allows for fine-grained access control on encrypted data. In its key-policy flavor, the primitive enables senders to encrypt messages under a set of attributes and private keys are associated with access structures that specify which ciphertexts the key holder will be allowed to decrypt. In most ABE systems, the ciphertext size grows linearly with the number of ciphertext attributes and the only known exceptions only support restricted forms of threshold access policies.

This paper proposes the first key-policy attribute-based encryption (KP-ABE) schemes allowing for *non-monotonic* access structures (*i.e.*, that may contain negated attributes) and with constant ciphertext size. Towards achieving this goal, we first show that a certain class of identity-based broadcast encryption schemes generically yields monotonic KP-ABE systems in the selective set model. We then describe a new efficient identity-based revocation mechanism that, when combined with a particular instantiation of our general monotonic construction, gives rise to the first truly expressive KP-ABE realization with constant-size ciphertexts. The downside of these new constructions is that private keys have quadratic size in the number of attributes. On the other hand, they reduce the number of pairing evaluations to a constant, which appears to be a unique feature among expressive KP-ABE schemes.

Keywords: Attribute-based encryption, expressivity, efficiency.

1 Introduction

It frequently happens that sensitive data must be archived by storage servers in such a way that only specific parties are allowed to read the content. In these situations, enforcing the access control using ordinary public key encryption schemes is not very convenient as such primitives severely decrease the flexibility of users to share their data.

⋆ This author acknowledges the Belgian National Fund for Scientific Research (F.R.S.-F.N.R.S.) for his "chargé de recherches" fellowship and the BCRYPT Interuniversity Attraction Pole.

D. Catalano et al. (Eds.): PKC 2011, LNCS 6571, pp. 90–108, 2011.

To address these concerns, Sahai and Waters [29] introduced attribute-based encryption (ABE), which refines identity-based encryption [30, 8] by associating ciphertexts and private keys with sets of descriptive attributes. Decryption is then possible when there is a sufficient overlap between the two sets. These results were extended by Goyal, Pandey, Sahai and Waters [20] into richer kinds of attribute-based encryption, where decryption is permitted when the attribute set satisfies a more complex boolean formula specified by an access structure. This paper describes truly expressive ABE systems featuring compact ciphertexts, regardless of the number of underlying attributes.

RELATED WORK. Attribute-based encryption comes in two flavors. In key-policy ABE schemes (KP-ABE), attribute sets are used to annotate ciphertexts and private keys are associated with access structures that specify which ciphertexts the user will be entitled to decrypt. Ciphertext-policy ABE (CP-ABE) proceeds in the dual way, by assigning attribute sets to private keys and letting senders specify an access policy that receivers' attribute sets should comply with.

The ciphertext-policy scenario was first studied in [5, 18]. The construction of [18] only handles AND gates while the first expressive construction [5] was only analyzed in the generic group model. Goyal, Jain, Pandey and Sahai [21] gave a construction in the standard model but its large parameters and key sizes make it impractical for reasonably expressive policies. Efficient and expressive realizations in the standard model were subsequently put forth by Waters [32] and one of them was recently extended by Lewko *et al.* [25], and subsequently by Okamoto and Takashima [31], into schemes providing adaptive security whereas all prior works on ABE were limited to deal with selective adversaries [13, 14, 6] – who have to make up their mind about their target before having seen public parameters – in their security analysis.

In both CP-ABE and KP-ABE schemes, expressivity requires to go beyond what monotonic access structures can express. Ostrovsky, Sahai and Waters [28] considered access structures that may contain negative attributes without blowing up the size of shares or ciphertexts. Their initial construction was recently improved by Lewko, Sahai and Waters [24] who used techniques from revocation systems (which can be seen as negative analogues of identity-based broadcast encryption) to design the most efficient non-monotonic KP-ABE to date.

OUR CONTRIBUTIONS. So far, the research community has mostly focused on the design of expressive schemes – where access structures can implement as complex boolean formulas as possible – without trying to minimize the size of ciphertexts. Indeed, most schemes [20, 28, 32, 25, 24] feature linear-size ciphertexts in the maximal number of attributes that ciphertexts can be annotated with. In the ciphertext-policy setting, Emura *et al.* suggested a scheme with short ciphertexts [19] but policies are restricted to a single AND gate. More recently, Herranz *et al.* [22] described a scheme with threshold access policies and constant-size[1] ciphertexts. Yet, their scheme is still not as expressive as one

[1] By "constant", we mean that the size only depends on the security parameter λ (the number of transmitted bits is typically $O(\lambda)$) and not on the number of ciphertext attributes.

could hope for. In particular, it seems difficult to extend it to support general linear-secret-sharing-realizable (or LSSS-realizable for short) access structures.

In the context of key-policy attribute-based encryption, this paper aims at devising schemes with constant-size ciphertexts[2] (regardless of the number of ciphertext attributes) allowing for as expressive policies as possible. To this end, we first show that a certain class of identity-based broadcast encryption (IBBE) schemes readily yields KP-ABE schemes with monotonic (though LSSS-realizable) access structures via a generic transformation. The latter preserves the ciphertext size and guarantees the resulting scheme to be selectively secure (as defined in [13, 6]) as long as the underlying IBBE system is itself selectively secure. At the expense of quadratic-size private keys (which comprise $O(t \cdot n)$ elements, where n is the maximal number of ciphertext attributes and t is the maximal number of leaf attributes in access trees), this transformation directly provides us with monotonic KP-ABE schemes with $O(1)$-size ciphertexts.

In a second step, we use a particular output of the aforementioned transformation to design a scheme supporting non-monotonic access structures without sacrificing the efficiency. In the resulting construction, the ciphertext overhead reduces to three group elements, no matter how many attributes ciphertexts are associated with. As in the monotonic case, private keys are inflated by a factor of n in comparison with [28, 24]. Nevertheless, these new schemes remain attractive for applications where bandwidth is the primary concern. In mobile Internet connections for instance, users are charged depending on the amount of transmitted messages; while in contrast, the storage is becoming much cheaper nowadays even for a large amount, as evidently in many smart phones.

As an intermediate step towards the new non-monotonic ABE, we design a new identity-based revocation (IBR) mechanism (as defined by Lewko, Sahai and Waters [24]) with $O(1)$-size ciphertexts and a similar structure to that of the monotonic KP-ABE schemes provided by our general construction. This was necessary since prior IBR systems with short ciphertexts [4] were not directly amenable to fulfill these requirements. We believe this new IBR realization to be of independent interest since it performs noticeably better than previous schemes featuring short ciphertexts [4] and still relies a natural (though "q-type") intractability assumption.

The security of our schemes is proved against selective adversaries (that are not allowed to choose their target attribute set adaptively) under a non-interactive assumption. We leave it as an open problem to obtain KP-ABE schemes with compact ciphertexts that can be proven secure against adaptive adversaries (as in the work of Lewko *et al.* [25]).

Other Related Work. The aforementioned realizations all assume ABE schemes with a single authority and we focus on this context as well. Extensions to the multi-authority scenario were investigated in [15, 16] for a conjunctive

[2] As in the literature on broadcast encryption (see, e.g., [9]) where the list of receivers is not included in the ciphertext, we do not count the description of ciphertext attributes as being part of the ciphertext. Indeed, many ciphertexts may have to be encrypted under the same attribute set.

setting and in [3] for a disjunctive setting. Besides the two usual flavors of ABE, another recently considered kind of ABE schemes [2], called dual-policy ABE, mixes features from both KP-ABE and CP-ABE systems.

ORGANIZATION. In the following, we first review various primitives in section 2. Section 3 describes our general construction of monotonic KP-ABE. The new revocation scheme is depicted in section 4. Section 5 finally presents the non-monotonic ABE realization with compact ciphertexts.

2 Background and Definitions

NOTATION. We will treat a vector as a column vector, unless stated otherwise. Namely, for any vector $\boldsymbol{\alpha} = (\alpha_1, \ldots, \alpha_n)^\top \in \mathbb{Z}_p^n$, $g^{\boldsymbol{\alpha}}$ stands for the vector of group elements $(g^{\alpha_1}, \ldots, g^{\alpha_n})^\top \in \mathbb{G}^n$. For $\boldsymbol{a}, \boldsymbol{z} \in \mathbb{Z}_p^n$, we denote their inner product as $\langle \boldsymbol{a}, \boldsymbol{z} \rangle = \boldsymbol{a}^\top \boldsymbol{z} = \sum_{i=1}^n a_i z_i$. Given $g^{\boldsymbol{a}}$ and $\boldsymbol{z}$, $(g^{\boldsymbol{a}})^{\boldsymbol{z}} := g^{\langle \boldsymbol{a}, \boldsymbol{z} \rangle}$ is computable without knowing $\boldsymbol{a}$. We denote by I_n the identity matrix of size n. For a set U, we define $2^U = \{S \mid S \subseteq U\}$ and $\binom{U}{<k} = \{S \mid S \subseteq U, |S| < k\}$ for $k \leq |U|$.

2.1 Syntax and Security Definition for Functional Encryption

We capture notions of KP-ABE, IBBE, IBR by providing a unified definition and security notion for functional encryption[3] here and then instantiating to these primitives in the next subsections.

SYNTAX. Let $R : \Sigma_k \times \Sigma_e \rightarrow \{0, 1\}$ be a boolean function where Σ_k and Σ_e denote "key index" and "ciphertext index" spaces. A functional encryption (FE) scheme for the relation R consists of algorithms: Setup, KeyGen, $\mathsf{Encrypt}$, $\mathsf{Decrypt}$.

$\mathsf{Setup}(\lambda, \mathrm{des}) \rightarrow (\mathsf{mpk}, \mathsf{msk})$: The setup algorithm takes as input a security parameter λ and a scheme description des and outputs a master public key mpk and a master secret key msk.

$\mathsf{KeyGen}(\mathsf{msk}, X) \rightarrow \mathsf{sk}_X$: The key generation algorithm takes in the master secret key msk and a key index $X \in \Sigma_k$. It outputs a private key sk_X.

$\mathsf{Encrypt}(\mathsf{mpk}, \mathsf{M}, Y) \rightarrow C$: This algorithm takes as input a public key mpk, the message M, and a ciphertext index $Y \in \Sigma_e$. It outputs a ciphertext C.

$\mathsf{Decrypt}(\mathsf{mpk}, \mathsf{sk}_X, X, C, Y) \rightarrow \mathsf{M}$ or $\perp$: The decryption algorithm takes in the public parameters mpk, a private key sk_X for the key index X and a ciphertext C for the ciphertext index Y. It outputs the message M or a symbol $\perp$ indicating that the ciphertext is not in a valid form.

Correctness mandates that, for all λ, all $(\mathsf{mpk}, \mathsf{msk})$ produced by $\mathsf{Setup}(\lambda, des)$, all $X \in \Sigma_k$, all keys sk_X returned by $\mathsf{KeyGen}(\mathsf{msk}, X)$ and all $Y \in \Sigma_e$,

- If $R(X, Y) = 1$, then $\mathsf{Decrypt}(\mathsf{mpk}, \mathsf{Encrypt}(\mathsf{mpk}, \mathsf{M}, Y)), \mathsf{sk}_X) = \mathsf{M}$.
- If $R(X, Y) = 0$, then $\mathsf{Decrypt}(\mathsf{mpk}, \mathsf{Encrypt}(\mathsf{mpk}, \mathsf{M}, Y)), \mathsf{sk}_X) = \perp$.

[3] The term "functional encryption" was defined in slightly different manners in [25,4, 31] before recently fully formalized in [11]. Our definition of FE here and throughout the paper refers to the class of predicate encryption with public index of [11].

SECURITY NOTION. We now give the standard security definition for FE schemes.

Definition 1. *A FE scheme for relation R is fully secure if no probabilistic polynomial time (PPT) adversary $\mathcal{A}$ has non-negligible advantage in this game:*

Setup. *The challenger runs* $(\mathsf{mpk}, \mathsf{msk}) \leftarrow \mathsf{Setup}(\lambda, \mathrm{des})$ *and gives* mpk *to* $\mathcal{A}$.

Phase 1. *On polynomially-many occasions, $\mathcal{A}$ chooses a key index X and gets* $\mathsf{sk}_X = \mathsf{Keygen}(\mathsf{msk}, X)$. *Such queries can be adaptive in that each one may depend on the information gathered so far.*

Challenge. *$\mathcal{A}$ chooses messages $\mathsf{M}_0, \mathsf{M}_1$ and a ciphertext index $Y^\star$ such that $R(X, Y^\star) = 0$ for all key indexes X that have been queried at step 2. Then, the challenger flips a fair binary coin $d \in \{0, 1\}$, generates a ciphertext* $C^\star = \mathsf{Encrypt}(\mathsf{mpk}, \mathsf{M}_d, Y^\star)$, *and hands it to the adversary.*

Phase 2. *$\mathcal{A}$ is allowed to make more key generation queries for any key index X such that $R(X, Y^\star) = 0$.*

Guess. *$\mathcal{A}$ outputs a bit $d' \in \{0, 1\}$ and wins if $d' = d$.*

The advantage of the adversary $\mathcal{A}$ is measured by $\mathbf{Adv}(\lambda) := |\Pr[d' = d] - \frac{1}{2}|$.

A weaker notion called selective security [13, 6] can be defined as in the above game with the exception that the adversary $\mathcal{A}$ has to choose the challenge ciphertext index $Y^\star$ before the setup phase but private key queries $X_1, \ldots, X_q$ can still be adaptive. A dual notion called co-selective security [4], in contrast, requires $\mathcal{A}$ to declare q key queries for key indexes $X_1, \ldots, X_q$ before the setup phase, but $\mathcal{A}$ can adaptively choose the target challenge ciphertext index $Y^\star$.

2.2 Key-Policy Attribute-Based Encryption

Before describing the definition of KP-ABE, we first recall the definitions of access structures and linear secret sharing schemes, as defined in [20].

Definition 2 (Access Structures). *Consider a set of parties $\mathcal{P} = \{P_1, P_2, \ldots, P_n\}$. A collection $\mathbb{A} \subseteq 2^{\mathcal{P}}$ is said to be monotone if, for all B, C, if $B \in \mathbb{A}$ and $B \subseteq C$, then $C \in \mathbb{A}$. An access structure (resp., monotonic access structure) is a collection (resp., monotone collection) $\mathbb{A} \subseteq 2^{\mathcal{P}} \setminus \{\emptyset\}$. The sets in $\mathbb{A}$ are called the authorized sets, and the sets not in $\mathbb{A}$ are called the unauthorized sets.*

Definition 3 (Linear Secret Sharing Scheme). *Let $\mathcal{P}$ be a set of parties. Let L be a $\ell \times k$ matrix. Let $\pi : \{1, \ldots, \ell\} \to \mathcal{P}$ be a function that maps a row to a party for labeling. A secret sharing scheme Π for access structure $\mathbb{A}$ over a set of parties $\mathcal{P}$ is a linear secret-sharing scheme (LSSS) in $\mathbb{Z}_p$ and is represented by (L, π) if it consists of two efficient algorithms:*

$\mathsf{Share}_{(L,\pi)}$: *takes as input $s \in \mathbb{Z}_p$ which is to be shared. It chooses $\beta_2, \ldots, \beta_k \stackrel{R}{\leftarrow} \mathbb{Z}_p$ and let $\boldsymbol{\beta} = (s, \beta_2, \ldots, \beta_k)^\top$. It outputs $L \cdot \boldsymbol{\beta}$ as the vector of ℓ shares. The share $\lambda_i := \langle \boldsymbol{L_i}, \boldsymbol{\beta} \rangle$ belongs to party $\pi(i)$, where $\boldsymbol{L_i}^\top$ is the ith row of L.*

$\mathsf{Recon}_{(L,\pi)}$: *takes as input an access set $S \in \mathbb{A}$. Let $I = \{i |\ \pi(i) \in S\}$. It outputs a set of constants $\{(i, \mu_i)\}_{i \in I}$ such that $\sum_{i \in I} \mu_i \cdot \lambda_i = s$.*

In a key-policy attribute-based encryption scheme, ciphertexts are associated with a set of attributes ω and private keys correspond to access structures $\mathbb{A}$. Decryption is possible when the attribute set ω is authorized in the access structure $\mathbb{A}$ (*i.e.*, $\omega \in \mathbb{A}$). We formally define it as an instance of FE as follows.

Definition 4 (KP-ABE). *Let U be an attribute space. Let $n \in \mathbb{N}$ be a bound on the number of attributes per ciphertext. A key-policy attribute-based encryption (KP-ABE) for a collection $\mathcal{AS}$ of access structures over U is a functional encryption for $R^{\mathsf{KP}} : \mathcal{AS} \times \binom{U}{<n} \to \{0,1\}$ defined by $R^{\mathsf{KP}}(\mathbb{A}, \omega) = 1$ iff $\omega \in \mathbb{A}$ (for $\omega \subseteq U$ such that $|\omega| < n$, and $\mathbb{A} \in \mathcal{AS}$). Furthermore, the description* des *consists of the attribute universe U, $\Sigma_k^{\mathsf{KP}} = \mathcal{AS}$, and $\Sigma_e^{\mathsf{KP}} = \binom{U}{<n}$.*

Definition 4 conforms with the original definition of KP-ABE, as in [20, 28, 24, 25, 11]. There is another variant of KP-ABE recently used in [31], that we call KP-ABE with labeling. We re-formalize it in appendix A, for the purpose of comparison in Table 2. We remark that normal KP-ABE implies KP-ABE with labeling.

We note that chosen-ciphertext secure versions of our proposed KP-ABE schemes in this paper can be obtained from recent generic results of [33].

2.3 Identity-Based Broadcast Encryption and Revocation Scheme

An ID-based broadcast encryption, as formalized in [1], allows a sender to encrypt a message to a set of identities, say $S = \{\mathsf{ID}_1, \ldots, \mathsf{ID}_q\}$, where $q < n$ for some a-priori fixed bound $n \in \mathbb{N}$, so that a user who possesses a key for $\mathsf{ID} \in S$ can decrypt. In contrast, an ID-based revocation scheme [24] allows a sender to specify a revoked set S so that only a user with $\mathsf{ID} \notin S$ can decrypt.

Definition 5. *Let $\mathcal{I}$ be an identity space. An ID-based broadcast encryption scheme (IBBE) with the maximal bound n for the number of receivers per ciphertext is a functional encryption for $R^{\mathsf{IBBE}} : \mathcal{I} \times \binom{\mathcal{I}}{<n} \to \{0,1\}$ defined by $R^{\mathsf{IBBE}}(\mathsf{ID}, S) = 1$ iff $\mathsf{ID} \in S$.*

Definition 6. *Let $\mathcal{I}$ be an identity space. An ID-based revocation (IBR) with the maximal bound n for the number of revoked users per ciphertext is a functional encryption for $R^{\mathsf{IBR}} : \mathcal{I} \times \binom{\mathcal{I}}{<n} \to \{0,1\}$ defined by $R^{\mathsf{IBR}}(\mathsf{ID}, S) = 1$ iff $\mathsf{ID} \notin S$.*

Remark 1. Although selective and co-selective security are incomparable in general, we remark that, in IBR schemes, co-selective security implies selective security. To see why, we first recall that selective security for IBR requires the adversary $\mathcal{A}$ to declare the target revoked set $S^\star$ before seeing the public key mpk. Here, phase 1 can be simplified by letting the challenger hand over all the private keys for identities in $S^\star$ at once (along with mpk). On the other hand, co-selective IBR security requires $\mathcal{A}$ to declare the set $\tilde{S}$ of identities that will be queried for private key generation before seeing mpk whereas the target revocation set $S^\star$ does not have to be fully determined before the challenge phase. At the same time as mpk, the challenger then reveals all keys for identities in $\tilde{S}$ at once. Later, the adversary can choose any $S^\star \subseteq \tilde{S}$ in the challenge phase. Selective security corresponds to the special case where $S^\star = \tilde{S}$.

2.4 Complexity Assumptions

We use groups $(\mathbb{G}, \mathbb{G}_T)$ of prime order p with an efficiently computable mapping $e : \mathbb{G} \times \mathbb{G} \to \mathbb{G}_T$ s.t. $e(g^a, h^b) = e(g,h)^{ab}$ for any $(g,h) \in \mathbb{G} \times \mathbb{G}$, $a, b \in \mathbb{Z}$ and $e(g,h) \neq 1_{\mathbb{G}_T}$ whenever $g, h \neq 1_{\mathbb{G}}$. We rely on the DBDHE assumption introduced in [9]. This assumption is shown to hold in the generic group model [7]. In addition, it is non-interactive and falsifiable [26].

Definition 7. *In* $(\mathbb{G}, \mathbb{G}_T)$*, the* q**-Decision Bilinear Diffie-Hellman Exponent** *(q-DBDHE) problem is, given* $(g, g^{\gamma}, g^{(\gamma^2)}, \ldots, g^{(\gamma^q)}, g^{(\gamma^{q+2})}, \ldots, g^{(\gamma^{2q})}, h, T)$ *where* $\gamma \stackrel{R}{\leftarrow} \mathbb{Z}_p$*,* $g, h \stackrel{R}{\leftarrow} \mathbb{G}$ *and* $T \in_R \mathbb{G}_T$*, to decide if* $T = e(g,h)^{(\gamma^{q+1})}$ *or if* T *is a random element of* $\mathbb{G}_T$*.*

3 Monotonic KP-ABE with Short Ciphertexts

Our first goal is to construct monotonic KP-ABE with short ciphertexts. We do so by showing a general transformation that automatically turns any IBBE scheme fitting a certain template into a KP-ABE in the selective security model.

The construction is somewhat similar to the one described by Boyen [12], which transforms IBE in the exponent-inversion framework to ABE. The approach of [12] took advantage of certain linearity properties in a family of IBE schemes. Our approach also exploits some linearity properties, albeit instead of IBE, we use IBBE as the underlying primitive. In contrast to [12], our transformation preserves the ciphertext size, hence using IBBE with short ciphertexts will yield KP-ABE with the same ciphertext size.

3.1 Linear ID-Based Broadcast Encryption Template

We define a template that IBBE schemes should comply with in order to give rise to (selectively secure) KP-ABE schemes. We call this a linear IBBE template. Let $(\mathbb{G}, \mathbb{G}_T)$ be underlying bilinear groups of order p. A linear IBBE scheme is determined by parameter $n_1, n_2 \in \mathbb{N}$, a family $\mathcal{F}$ of vectors of functions, and a function $\mathcal{D}$, of which the latter two are specified by

$$\mathcal{F} \subset \{(f_1, f_2, F) \mid f_1 : \mathbb{Z}_p^* \to \mathbb{G},\ f_2 : \mathbb{Z}_p^* \to \mathbb{G}^{n_1},\ F : (\mathbb{Z}_p^*)^{\leq n-1} \to \mathbb{G}^{\leq n_2}\},$$
$$\mathcal{D} : \mathbb{G}^{n_1+2} \times \mathcal{I} \times \mathbb{G}^{\leq n_2+1} \times \binom{\mathcal{I}}{<n} \to \mathbb{G}_T,$$

with requirements specified below. A linear IBBE scheme works as follows.

▶ $\mathsf{Setup}(\lambda, n)$: Given a security parameter $\lambda \in \mathbb{N}$ and a bound $n \in \mathbb{N}$ on the number of identities per ciphertext, the algorithm selects bilinear groups $(\mathbb{G}, \mathbb{G}_T)$ of prime order p and a generators $g \stackrel{R}{\leftarrow} \mathbb{G}$. It computes $e(g,g)^{\alpha}$ for a random $\alpha \stackrel{R}{\leftarrow} \mathbb{Z}_p^*$ and chooses functions $(f_1, f_2, F) \stackrel{R}{\leftarrow} \mathcal{F}$. The master secret key consists of $\mathsf{msk} := g^{\alpha}$ while the public key is $\mathsf{mpk} := \big(g,\ e(g,g)^{\alpha},\ f_1,\ f_2,\ F,\ n,\ n_1,\ n_2\big)$.
▶ $\mathsf{Keygen}(\mathsf{msk}, \mathsf{ID})$: It picks $r \stackrel{R}{\leftarrow} \mathbb{Z}_p^*$ and computes

$$\mathsf{sk}_{\mathsf{ID}} = (d_1, d_2, d_3) = \big(g^{\alpha} \cdot f_1(\mathsf{ID})^r,\ g^r,\ f_2(\mathsf{ID})^r\big) \in \mathbb{G}^{n_1+2}.$$

► $\mathsf{Encrypt}(\mathsf{mpk}, \mathsf{M}, S)$: It parses S as $S = \{\mathsf{ID}_1, \ldots, \mathsf{ID}_q\}$, where $q < n$. To encrypt $\mathsf{M} \in \mathbb{G}_T$, it chooses a random exponent $s \xleftarrow{R} \mathbb{Z}_p^*$ and computes the ciphertext as

$$C = (C_0, C_1, C_2) = \Big(\mathsf{M} \cdot e(g,g)^{\alpha \cdot s},\ g^s,\ F(\mathsf{ID}_1, \ldots, \mathsf{ID}_q)^s\Big).$$

► $\mathsf{Decrypt}(\mathsf{mpk}, \mathsf{sk}_{\mathsf{ID}}, \mathsf{ID}, C, S)$: It parses $\mathsf{sk}_{\mathsf{ID}} = (d_1, d_2, d_3)$ and $C = (C_0, C_1, C_2)$ then runs

$$\mathcal{D}\big((d_1, d_2, d_3), \mathsf{ID}, (C_1, C_2), S\big) \rightarrow e(g,g)^{\alpha \cdot s},$$

and obtains $\mathsf{M} = C_0 / e(g,g)^{\alpha \cdot s}$. We are now ready to state the requirements: for all $(f_1, f_2, F) \in \mathcal{F}$, the following two properties must hold.

1. **Correctness.** For all $\alpha, r, s \in \mathbb{Z}_p^*$, $\mathsf{ID} \in \mathcal{I}$, $S = \{\mathsf{ID}_1, \ldots, \mathsf{ID}_q\} \in \binom{\mathcal{I}}{<n}$ and $\mathsf{ID} \in S$, we have

$$\mathcal{D}\Big(\big(g^\alpha f_1(\mathsf{ID})^r, g^r, f_2(\mathsf{ID})^r\big), \mathsf{ID}, (g^s, F(\mathsf{ID}_1, \ldots, \mathsf{ID}_q)^s), S\Big) = e(g,g)^{\alpha \cdot s}.$$

2. **Linearity.** For all $\gamma \in \mathbb{Z}_p^*$, $\mathsf{ID} \in \mathcal{I}$, $S \in \binom{\mathcal{I}}{<n}$, $\mathsf{ID} \in S$, $(d_1, d_2, d_3) \in \mathbb{G}^{n_1+2}$, and $(C_1, C_2) \in \mathbb{G}^{\le n_2+1}$, we have

$$\mathcal{D}\Big((d_1, d_2, d_3)^\gamma, \mathsf{ID}, (C_1, C_2), S\Big) = \mathcal{D}\Big((d_1, d_2, d_3), \mathsf{ID}, (C_1, C_2), S\Big)^\gamma.$$

3.2 Generic Conversion from Linear IBBE to KP-ABE

Let $\Pi_{\mathsf{IBBE}} = (\mathsf{Setup}', \mathsf{Keygen}', \mathsf{Encrypt}', \mathsf{Decrypt}')$ be a linear IBBE system. We construct a KP-ABE scheme from Π_{IBBE} as follows.

► $\mathsf{Setup}(\lambda, n)$: It simply outputs $\mathsf{Setup}'(\lambda, n) \rightarrow (\mathsf{msk}, \mathsf{mpk})$.
► $\mathsf{Keygen}(\mathsf{msk}, (L, \pi))$: The algorithm computes a private key for an access structure that is associated with LSSS scheme (L, π) as follows. Let L be $\ell \times k$ matrix. First, it generates shares of 1 with the LSSS (L, π). Namely, it chooses a vector $\boldsymbol{\beta} = (\beta_1, \beta_2, \ldots, \beta_k)^\top \xleftarrow{R} (\mathbb{Z}_p)^k$ subject to the constraint $\beta_1 = 1$. Then for each $i = 1$ to ℓ, it calculates $\lambda_i = \langle \boldsymbol{L_i}, \boldsymbol{\beta} \rangle$, picks $r' \xleftarrow{R} \mathbb{Z}_p$ and sets D_i as follows.

$$\mathsf{Keygen}'(\mathsf{msk}, \pi(i)) \rightarrow (d_{i,1},\ d_{i,2},\ d_{i,3}),$$
$$D_i = \big(d_{i,1}^{\lambda_i} \cdot f_1(\pi(i))^{r'},\ d_{i,2}^{\lambda_i} \cdot g^{r'},\ d_{i,3}^{\lambda_i} \cdot f_2(\pi(i))^{r'}\big).$$

It then outputs the private key as $\mathsf{sk}_{(L,\pi)} = \{D_i\}_{i=1,\ldots,\ell}$.
► $\mathsf{Encrypt}(\mathsf{mpk}, \mathsf{M}, \omega)$: It simply outputs $\mathsf{Encrypt}'(\mathsf{mpk}, \mathsf{M}, \omega) \rightarrow (C_0, C_1, C_2)$.
► $\mathsf{Decrypt}(\mathsf{mpk}, \mathsf{sk}_{(L,\pi)}, (L, \pi), C, \omega)$: Assume first that the policy (L, π) is satisfied by the attribute set ω, so that decryption is possible. Let $I = \{i |\ \pi(i) \in \omega\}$. It calculates the reconstruction constants $\{(i, \mu_i)\}_{i \in I} = \mathsf{Recon}_{(L,\pi)}(\omega)$. It parses C as (C_0, C_1, C_2) and $\mathsf{sk}_{(L,\pi)}$ as $\{D_i\}_{i=1,\ldots,\ell}$ where $D_i = (d'_{i,1}, d'_{i,2}, d'_{i,3})$. For each $i \in I$, it computes

$$\mathcal{D}\big((d'_{i,1}, d'_{i,2}, d'_{i,3}), \mathsf{ID}, (C_1, C_2), S\big) \rightarrow e(g,g)^{\alpha \cdot s \cdot \lambda_i}, \tag{1}$$

which we prove correctness below. It computes $e(g,g)^{\alpha \cdot s} = \prod_{i \in I} \big(e(g,g)^{\alpha \cdot s \cdot \lambda_i}\big)^{\mu_i}$ and finally obtains $\mathsf{M} = C_0 / e(g,g)^{\alpha \cdot s}$, where we recall that $\sum_{i \in I} \lambda_i \mu_i = 1$.

CORRECTNESS. We now verify that equation (1) is correct. First from a property of keys in linear IBBE, we have that $(d_{i,1}, d_{i,2}, d_{i,3})$ will be in the form $\big(g^{\alpha} \cdot f_1(\pi(i))^{r_i}, g^{r_i}, f_2(\pi(i))^{r_i}\big)$ for some $r_i \in_R \mathbb{Z}_p$. Therefore, we have

$$D_i = \big(g^{\alpha\lambda_i} \cdot f_1(\pi(i))^{\tilde{r}_i\lambda_i}, g^{\tilde{r}_i\lambda_i}, f_2(\pi(i))^{\tilde{r}_i\lambda_i}\big) = \big(d_1^{\lambda_i}, d_2^{\lambda_i}, d_3^{\lambda_i}\big),$$

with $\tilde{r}_i = r_i + r'/\lambda_i$ and $(d_1, d_2, d_3) = \mathsf{sk}_{\pi(i)}$ with randomness $\tilde{r}_i$. Hence,

$$\begin{aligned} \mathcal{D}\big((d'_{i,1}, d'_{i,2}, d'_{i,3}), \mathsf{ID}, (C_1, C_2), S\big) &= \mathcal{D}\big((d_1, d_2, d_3), \mathsf{ID}, (C_1, C_2), S\big)^{\lambda_i} \\ &= \big(e(g,g)^{\alpha \cdot s}\big)^{\lambda_i}, \end{aligned}$$

where each equality holds from linearity and correctness of $\mathcal{D}$ respectively.

The construction only guarantees selective security for the resulting KP-ABE. It does not extend to the adaptive scenario because the proof relies on the fact that the reduction knows the forbidden attribute set from the beginning.

Theorem 1. *If the underlying IBBE scheme is selectively secure, then the resulting KP-ABE system is also selectively secure.* (The proof is given in the full version of the paper).

INSTANTIATION EXAMPLE. The large-universe construction of KP-ABE in [20] falls into our framework here. Its underlying IBBE system can be seen as a particular instance of the linear IBBE template with $n_2 = n$, $f_2(\mathsf{ID}) = \emptyset$, $F(\mathsf{ID}_1, \ldots, \mathsf{ID}_q) = (f_1(\mathsf{ID}_1), \ldots, f_1(\mathsf{ID}_q))$, and the form of f_1 can be straightforwardly deduced from [20]. Since the size of an output from F is linear, ciphertexts in the KP-ABE of [20] are also of linear size.

3.3 IBBE Instantiation with Short Ciphertexts

This subsection presents an IBBE scheme with short ciphertexts and shows how to apply the KP-ABE conversion. This specific IBBE can be seen as an instance of the functional encryption (FE) for zero inner-product proposed in [4, Sect.4.1], which itself is implied by spatial encryption of [10]. A FE system for zero inner-product is defined by a relation $R^{\mathsf{ZIP}} : \mathbb{Z}_p \times \mathbb{Z}_p \to \{0,1\}$ where $R^{\mathsf{ZIP}}(\boldsymbol{X}, \boldsymbol{Y}) = 1$ iff $\langle \boldsymbol{X}, \boldsymbol{Y} \rangle = 0$. The technique of deriving an IBBE scheme from a FE scheme for zero inner-product can be traced to [23]. A private key for an identity ID is defined by setting $\boldsymbol{X} = (x_1, \ldots, x_n)^{\top}$, with $x_i = \mathsf{ID}^{i-1}$. To encrypt to a set $S = \{\mathsf{ID}_1, \ldots, \mathsf{ID}_q\}$, one defines $\boldsymbol{Y} = (y_1, \ldots, y_n)^{\top}$ as a coefficient vector from

$$P_S[Z] = \sum_{i=1}^{q+1} y_i Z^{i-1} = \prod_{\mathsf{ID}_j \in S} (Z - \mathsf{ID}_j), \tag{2}$$

where, if $q + 1 < n$, the coordinates $y_{q+2}, \ldots, y_n$ are set to 0. By doing so, we note that $P_S[\mathsf{ID}] = \langle \boldsymbol{X}, \boldsymbol{Y} \rangle$ evaluates to 0 iff $\mathsf{ID} \in S$. We now describe the IBBE instantiated from the FE system of [4]. Its selective security is an immediate consequence of [4], where it is proved under the DBDHE assumption.

► $\mathsf{Setup}(\lambda, n)$: It chooses bilinear groups $(\mathbb{G}, \mathbb{G}_T)$ of prime order $p > 2^\lambda$ with $g \overset{R}{\leftarrow} \mathbb{G}$. It randomly chooses $\alpha, \alpha_0 \overset{R}{\leftarrow} \mathbb{Z}_p$, $\boldsymbol{\alpha} = (\alpha_1, \ldots, \alpha_n)^\top \overset{R}{\leftarrow} \mathbb{Z}_p^n$. It then sets $\boldsymbol{H} = (h_1, \ldots, h_n)^\top = g^{\boldsymbol{\alpha}}$. The master secret key is $\mathsf{msk} = \alpha$, and the public key is $\mathsf{mpk} = \big(g,\ e(g,g)^\alpha,\ h_0 = g^{\alpha_0},\ \boldsymbol{H} = g^{\boldsymbol{\alpha}}\big)$.

► $\mathsf{Keygen}(\mathsf{msk}, \mathsf{ID})$: The algorithm first defines a vector $\boldsymbol{X} = (x_1, \ldots, x_n)^\top$ such that $x_i = \mathsf{ID}^{i-1}$ for $i = 1$ to n. It chooses $r \overset{R}{\leftarrow} \mathbb{Z}_p$ and outputs the private key as $\mathsf{sk}_{\mathsf{ID}} = (D_1, D_2, K_2, \ldots, K_n)$ where

$$D_1 = g^\alpha \cdot h_0^r, \qquad D_2 = g^r, \qquad \Big\{K_i = \big(h_1^{-\frac{x_i}{x_1}} \cdot h_i\big)^r\Big\}_{i=2,\ldots,n}.$$

► $\mathsf{Encrypt}(\mathsf{mpk}, \mathsf{M}, S)$: To encrypt M to the receiver set S (where $|S| < n$), the algorithm defines $\boldsymbol{Y} = (y_1, \ldots, y_n)^\top$ as the coefficient vector of $P_S[Z]$ from equation (2). It then picks $s \overset{R}{\leftarrow} \mathbb{Z}_p$ and computes the ciphertext as

$$C = (C_0, C_1, C_2) = \Big(\mathsf{M} \cdot e(g,g)^{\alpha s},\ g^s,\ \big(h_0 \cdot h_1^{y_1} \cdots h_n^{y_n}\big)^s\Big).$$

► $\mathsf{Decrypt}(\mathsf{mpk}, \mathsf{sk}_{\mathsf{ID}}, \mathsf{ID}, C, S)$: It defines the vector $\boldsymbol{Y} = (y_1, \ldots, y_n)^\top$ from the polynomial $P_S[Z]$ as usual. It then computes

$$e(g,g)^{\alpha \cdot s} = \frac{e(C_1, D_1 \cdot K_2^{y_2} \cdots K_n^{y_n})}{e(C_2, D_2)}, \tag{3}$$

and recovers $\mathsf{M} = C_0 / e(g,g)^{\alpha \cdot s}$.

CORRECTNESS. If $\langle \boldsymbol{X}, \boldsymbol{Y} \rangle = 0$, then decryption recovers M since

$$D_1 \cdot \prod_{i=2}^{n} K_i^{y_i} = g^\alpha \cdot \Big(h_0 \cdot h_1^{-\frac{1}{x_1}(\langle \boldsymbol{X}, \boldsymbol{Y}\rangle - x_1 y_1)} \prod_{i=2}^{n} h_i^{y_i}\Big)^r = g^\alpha \cdot \Big(h_0 \cdot \prod_{i=1}^{n} h_i^{y_i}\Big)^r,$$

so that $e(C_1, D_1 \cdot \prod_{i=1}^n K_i^{y_i}) = e(g,g)^{\alpha s} \cdot e(h_0 \cdot \prod_{i=1}^n h_i^{y_i}, g^{rs})$ equals the product $e(g,g)^{\alpha s} \cdot e(C_2, D_2)$.

APPLYING THE KP-ABE CONVERSION. The above IBBE can be considered as a linear IBBE system with $n_1 = n - 1$, $n_2 = 1$ and the family $\mathcal{F}$ is defined by taking all functions of the following forms ranging over $h_0, h_1, \ldots, h_n \in \mathbb{G}$:

$$f_1(\mathsf{ID}) = h_0,\ f_2(\mathsf{ID}) = \big(h_1^{-\mathsf{ID}} h_2, \ldots, h_1^{-\mathsf{ID}^{n-1}} h_n\big),\ F(\mathsf{ID}_1, \ldots, \mathsf{ID}_q) = h_0 \prod_{i=1}^{q+1} h_i^{y_i},$$

where the vector $\boldsymbol{Y} = (y_1, \ldots, y_n)^\top$ is defined from the polynomial $P_S[Z]$ in equation (2) as usual. In addition, the function $\mathcal{D}$ is the computation in equation (3), which can be shown to have linearity as required.

The resulting KP-ABE has constant-size ciphertexts. This comes with the expense of longer private keys of size $O(t \cdot n)$, where t is the number of attributes in the access structure. It is also worth mentioning that we can obtain another IBBE with short ciphertexts from the spatial encryption scheme of [10] since it also falls into our framework and thus produces another KP-ABE scheme.

Our goal in this paper is to construct KP-ABE with non-monotonic structures. We will combine the monotonic KP-ABE system in this subsection with new ID-based revocation in the next section.

4 Revocation Scheme with Very Short Ciphertexts

This section describes a new ID-based revocation system which is tailored to the needs of our application. Analogously to the case of IBBE, an IBR scheme can be instantiated from a FE system for *non-zero* inner-product relations. Two such existing IBR schemes [4, Sect. 5.1 and 5.2] already provide constant-size ciphertexts. When it comes to construct a non-monotonic KP-ABE however, these schemes seem hardly compatible with the monotonic KP-ABE of section 3.3 as they rely on different assumptions. We thus describe a new IBR scheme for this purpose. Its structure is similar to that of revocation schemes given in [4] but it provides a better efficiency and relies on the DBDHE assumption.

▶ $\mathsf{Setup}(\lambda, n)$: It chooses bilinear groups $(\mathbb{G}, \mathbb{G}_T)$ of prime order $p > 2^\lambda$ and a generator $g \stackrel{R}{\leftarrow} \mathbb{G}$. It randomly picks $\alpha \stackrel{R}{\leftarrow} \mathbb{Z}_p$, $\boldsymbol{\alpha} = (\alpha_1, \ldots, \alpha_n)^\top \stackrel{R}{\leftarrow} \mathbb{Z}_p^n$ and sets $\boldsymbol{H} = (h_1, \ldots, h_n)^\top = g^{\boldsymbol{\alpha}}$. The master secret key is $\mathsf{msk} = \alpha$, while the public key is $\mathsf{mpk} = \big(g,\ e(g,g)^\alpha,\ \boldsymbol{H} = g^{\boldsymbol{\alpha}}\big)$.
▶ $\mathsf{Keygen}(\mathsf{msk}, \mathsf{ID})$: The algorithm first defines a vector $\boldsymbol{X} = (x_1, \ldots, x_n)^\top$ such that $x_i = \mathsf{ID}^{i-1}$ for $i = 1$ to n. It chooses $r \stackrel{R}{\leftarrow} \mathbb{Z}_p$ and outputs the private key as $\mathsf{sk}_{\mathsf{ID}} = (D_1, D_2, K_2, \ldots, K_n)$ where

$$D_1 = g^\alpha \cdot h_1^r, \qquad D_2 = g^r, \qquad \Big\{K_i = \big(h_1^{-\frac{x_i}{x_1}} \cdot h_i\big)^r\Big\}_{i=2,\ldots,n}.$$

Indeed, we can also write $K_{\boldsymbol{X}} = (K_2, \ldots, K_n) = g^{r \cdot M_{\boldsymbol{X}}^\top \boldsymbol{\alpha}}$, where the matrix $M_{\boldsymbol{X}} \in (\mathbb{Z}_p)^{n \times (n-1)}$ is defined by $M_{\boldsymbol{X}} = \begin{pmatrix} -\frac{x_2}{x_1} & -\frac{x_3}{x_1} & \cdots & -\frac{x_n}{x_1} \\ & I_{n-1} & & \end{pmatrix}$.
▶ $\mathsf{Encrypt}(\mathsf{mpk}, \mathsf{M}, S)$: To encrypt M with the revoked set S (where $|S| < n$), the algorithm defines $\boldsymbol{Y} = (y_1, \ldots, y_n)^\top$ as the coefficient vector of $P_S[Z]$ from equation (2). It then picks $s \stackrel{R}{\leftarrow} \mathbb{Z}_p$ and computes the ciphertext as

$$C = (C_0, C_1, C_2) = \Big(\mathsf{M} \cdot e(g,g)^{\alpha \cdot s},\ g^s,\ \big(h_1^{y_1} \cdots h_n^{y_n}\big)^s\Big).$$

▶ $\mathsf{Decrypt}(\mathsf{mpk}, \mathsf{sk}_{\mathsf{ID}}, \mathsf{ID}, C, S)$: It defines $\boldsymbol{X}$ from ID and $\boldsymbol{Y}$ from S as usual. It then successively computes elements $K = \prod_{i=2}^n K_i^{y_i} = \big(h_1^{-\langle \boldsymbol{X}, \boldsymbol{Y}\rangle / x_1} \cdot h_1^{y_1} \cdots h_n^{y_n}\big)^r$, $\tau = \Big(\frac{e(K, C_1)}{e(C_2, D_2)}\Big)^{-\frac{x_1}{\langle \boldsymbol{X}, \boldsymbol{Y}\rangle}} = e(g, h_1)^{rs}$, and then obtains $\mathsf{M} = C_0 \cdot e(C_1, D_1)^{-1} \cdot \tau$.

Correctness. We first observe that

$$K = \big(h_1^{-(\langle \boldsymbol{X}, \boldsymbol{Y}\rangle - x_1 y_1)/x_1} \prod_{i=2}^n h_i^{y_i}\big)^r = \big(h_1^{-\langle \boldsymbol{X}, \boldsymbol{Y}\rangle / x_1} \prod_{i=1}^n h_i^{y_i}\big)^r$$

so that whenever $\langle \boldsymbol{X}, \boldsymbol{Y}\rangle \neq 0$ (*i.e.*, $\mathsf{ID} \notin S$), the following computation can be done.

$$\tau = \Big(\frac{e(K, C_1)}{e(C_2, D_2)}\Big)^{-\frac{x_1}{\langle \boldsymbol{X}, \boldsymbol{Y}\rangle}} = \Big(\frac{e(h_1^{-\langle \boldsymbol{X}, \boldsymbol{Y}\rangle / x_1} \prod_{i=1}^n h_i^{y_i}, g^{rs})}{e(\prod_{i=1}^n h_i^{y_i}, g^{rs})}\Big)^{-\frac{x_1}{\langle \boldsymbol{X}, \boldsymbol{Y}\rangle}} = e(g, h_1)^{rs}.$$

Finally, we have $e(C_1, D_1)\cdot\tau^{-1} = e(g,g)^{\alpha\cdot s}\cdot e(g^s, h_1^r)\cdot e(g, h_1)^{-rs} = e(g,g)^{\alpha\cdot s}$. We note that the decryption algorithm can be optimized by computing the plaintext as $\mathsf{M} = C_0 \cdot e\big(C_2, D_2^{x_1/\langle \boldsymbol{X},\boldsymbol{Y}\rangle}\big)\cdot e\big(C_1, D_1^{-1}\cdot K^{-x_1/\langle \boldsymbol{X},\boldsymbol{Y}\rangle}\big)$.

At a high level, it shares the same structure (including the form of the public key and the ciphertext) as the IBBE in section 3.3 and relies on the same assumption. Intuitively, these similarities make it possible to assemble both constructions in the design of a non-monotonic ABE system in section 5.

We now prove the co-selective security of the scheme. It is also worth recalling that co-selective security for IBR also implies selective security.

Theorem 2. *The above ID-based revocation scheme with the maximal bound n for the number of revoked users (i.e., $|S| < n$) is co-selectively secure if the n-DBDHE assumption holds in $(\mathbb{G}, \mathbb{G}_T)$.*

Proof. We show an algorithm $\mathcal{B}$ that receives $(g, h, z_1, \ldots, z_n, z_{n+2}, \ldots, z_{2n}, T)$ in $\mathbb{G}^{2n+1}\times\mathbb{G}_T$, where $z_i = g^{(\gamma^i)}$, and decides if $T = e(g,h)^{(\gamma^{n+1})}$ using the co-selective adversary $\mathcal{A}$.

At the outset of the game, the adversary $\mathcal{A}$ declares the set $\tilde{S} = \{\mathsf{ID}_1, \ldots, \mathsf{ID}_q\}$, where $q \le n-1$, of identities for which she wishes to obtain private keys. Let $\boldsymbol{X}_1, \ldots, \boldsymbol{X}_q$ the corresponding vectors. That is, $\boldsymbol{X}_k = (1, \mathsf{ID}_k, \mathsf{ID}_k^2, \ldots, \mathsf{ID}_k^{n-1})$. To prepare the public key, $\mathcal{B}$ chooses $\delta_0 \stackrel{R}{\leftarrow} \mathbb{Z}_p$ and computes $e(g,g)^\alpha = e(z_1, z_n)^{\delta_0}$, which implicitly defines $\alpha = \gamma^{(n+1)}\cdot\delta_0$. Elements $\boldsymbol{H} = (h_1, \ldots, h_n)^\top$ are then defined as follows. For each $k\in[1,q]$, $\mathcal{B}$ considers the vector $\boldsymbol{X}_k = (x_{k,1}, \ldots, x_{k,n})^\top$ and selects $\boldsymbol{b}_k \in \mathbb{Z}_p^n$ such that

$$\boldsymbol{b}_k^\top \cdot M_{\boldsymbol{X}_k} = \boldsymbol{b}_k^\top \cdot \begin{pmatrix} -\frac{x_{k,2}}{x_{k,1}} & -\frac{x_{k,3}}{x_{k,1}} & \cdots & -\frac{x_{k,n}}{x_{k,1}} \\ & I_{n-1} & & \end{pmatrix} = \mathbf{0}. \qquad (4)$$

The simplest candidate consists of the vector $\boldsymbol{b}_k = (1, \frac{x_{k,2}}{x_{k,1}}, \frac{x_{k,3}}{x_{k,1}}, \ldots, \frac{x_{k,n}}{x_{k,1}})^\top$. Then, $\mathcal{B}$ considers the $n\times n$ matrix $B = \big(\boldsymbol{b}_1|\ldots|\boldsymbol{b}_q|\mathbf{0}|\ldots|\mathbf{0}\big)$ whose k^{th} column consists of $\boldsymbol{b}_k$, for $k=1$ to q, and where the $n-q$ remaining columns are $\mathbf{0}$. It defines $\boldsymbol{a} = (a_1, \ldots, a_n)^\top \in (\mathbb{Z}_p)^n$ such that $a_i = \gamma^{n+1-i}$ by setting $g^{\boldsymbol{a}} = (z_n, \ldots, z_1)^\top$. Then, it implicitly sets $\boldsymbol{\alpha} = B\cdot\boldsymbol{a} + \boldsymbol{\delta}$ by randomly choosing $\boldsymbol{\delta} \stackrel{R}{\leftarrow} \mathbb{Z}_p^n$ and defining $\boldsymbol{H} = g^{B\cdot\boldsymbol{a}}\cdot g^{\boldsymbol{\delta}}$, which is uniformly distributed as required.

Due to (4), the matrix B is defined in such a way that, for each $k\in[1,q]$, the kth column of $M_{\boldsymbol{X}_k}^\top \cdot B \in (\mathbb{Z}_p)^{(n-1)\times n}$ is $\mathbf{0}$, so that $M_{\boldsymbol{X}_k}^\top\cdot B\cdot\boldsymbol{a}$ does not contain $a_k = \gamma^{n+1-k}$. Then, a private key for the identity ID_k (and thus the vector $\boldsymbol{X}_k$) can be obtained by implicity defining $\tilde{r}_k = r_k - \delta_0\gamma^k$ for a random $r_k \stackrel{R}{\leftarrow} \mathbb{Z}_p$. Indeed, with the above choice of B, the first coordinate of $\boldsymbol{\alpha} = \boldsymbol{\delta} + \sum_{j=1}^q a_j\boldsymbol{b}_j$ equals $\alpha_1 = \delta_1 + \sum_{j=1}^q a_j = \delta_1 + \sum_{j=1}^q \gamma^{(n+1-j)}$, so that $\mathcal{B}$ is able to compute

$$D_1 = g^\alpha \cdot h_1^{\tilde{r}_k} = g^{(\gamma^{n+1})\delta_0}\cdot h_1^{r_k}\cdot\Big(g^{\delta_1}\cdot\prod_{j=1}^q z_{n+1-j}\Big)^{-\delta_0\gamma_k}$$

$$= h_1^{r_k}\cdot\Big(z_k^{\delta_1}\cdot\prod_{j=1, j\neq k}^q z_{n+1-j+k}\Big)^{-\delta_0}$$

and $D_2 = g^{r_k} \cdot z_{n+1-k}^{-\delta_0}$. As for the delegation component $K_{\boldsymbol{X}_k} = g^{\tilde{r}_k M_{\boldsymbol{X}_k}^\top \boldsymbol{\alpha}}$, $\mathcal{B}$ is also able to compute it from available values since $M_{\boldsymbol{X}_k}^\top \boldsymbol{\alpha} = M_{\boldsymbol{X}_k}^\top \cdot B \cdot \boldsymbol{a} + M_{\boldsymbol{X}_k}^\top \cdot \boldsymbol{\delta}$ is independent of $a_k = \gamma^{n+1-k}$ (recall that the kth column of $M_{\boldsymbol{X}_k}^\top \cdot B$ is $\mathbf{0}$) and no term γ^{n+1} appears in the exponent in $K_{\boldsymbol{X}_k}$.

In the challenge phase, $\mathcal{B}$ chooses $\mathsf{M}_0, \mathsf{M}_1 \in \mathbb{G}_T$ and a revocation set S corresponding to a vector $\boldsymbol{Y} = (y_1, \ldots, y_n)^\top$ that must satisfy $\langle \boldsymbol{X}_k, \boldsymbol{Y} \rangle = 0$ for $k = 1$ to q. This amounts to say that $\boldsymbol{Y} = M_{\boldsymbol{X}_k} \cdot \boldsymbol{w}$, where $\boldsymbol{w} = (y_2, \ldots, y_n)^\top$, for each $k \in [1, q]$. We claim that $\boldsymbol{Y}^\top \cdot B \cdot \boldsymbol{a} = 0$. Indeed,

$$\boldsymbol{Y}^\top \cdot B \cdot \boldsymbol{a} = \boldsymbol{Y}^\top \cdot \Big(\sum_{k=1}^{q} a_k \cdot \boldsymbol{b}_k \Big) = \sum_{k=1}^{q} a_k \cdot \boldsymbol{Y}^\top \cdot \boldsymbol{b}_k = \sum_{k=1}^{q} a_k \cdot \boldsymbol{w}^\top \cdot M_{\boldsymbol{X}_k}^\top \cdot \boldsymbol{b}_k$$

and $M_{\boldsymbol{X}_k}^\top \cdot \boldsymbol{b}_k = \mathbf{0}$ for each $k \in [1, q]$. Therefore, it comes that $\langle \boldsymbol{Y}, \boldsymbol{\alpha} \rangle = \langle \boldsymbol{Y}, \boldsymbol{\delta} \rangle$, so that $\mathcal{B}$ can generate a challenge ciphertext (C_0, C_1, C_2) as

$$C_0 = \mathsf{M}_d \cdot T^{\delta_0}, \qquad C_1 = h, \qquad C_2 = h^{\langle \boldsymbol{Y}, \boldsymbol{\delta} \rangle},$$

for a random bit $d \stackrel{R}{\leftarrow} \{0,1\}$. If $T = e(g,h)^{(\gamma^{n+1})}$, $C = (C_0, C_1, C_2)$ forms a valid encryption of M_d. If T is random, $\mathcal{A}$'s advantage is clearly zero. □

Efficiency Comparisons. We believe this IBR scheme to be of interest in its own right. If we compare it with the scheme of [4, Sect.5.2] (called AL2 here), which also features short ciphertexts, it relies on a stronger assumption (since no "q-type" assumption is needed in [4] or in LSW2 [24]) but provides significantly shorter ciphertexts (as the ciphertext overhead is decreased by more than 75%)[4] and requires fewer pairing evaluations to decrypt (only 2 instead of 9). Another IBR scheme (dubbed AL1 in the table) with a better efficiency than AL2 was described in [4, Sect.5.1]. Still, the new scheme is slightly more efficient and relies on a weaker assumption since q-DBDHE is weaker and appears more natural than the q-type assumption (MEBDH) used in [24, 4].

Table 1. Performances of revocation systems

Schemes	Ciphertext overhead	Private key size	Decryption cost		Assumption
	$\lvert\mathbb{G}\rvert$	$\lvert\mathbb{G}\rvert$	pair.	exp.	
LSW1 [24]	$(2\bar{n}+1)$	3	3	$O(\bar{n})$	n-MEBDH
LSW2 [24]	$(2\bar{n}+7)$	7	9	$O(\bar{n})$	DLIN, DBDH
AL1 [4]	3	$(n+2)$	3	$O(n)$	n-MEBDH
AL2 [4]	9	$(n+2)$	9	$O(n)$	DLIN, DBDH
This work	2	$(n+2)$	2	$O(n)$	n-DBDHE

† $\bar{n}$ = # of revoked users = $|S|$; n = the maximal bound for $\bar{n}$. (*i.e.*, $|S| < n$).
‡ pair.,exp. shows # of pairing and exponentiation computation.

[4] We compare by simple element counting. In a stricter sense, one may want to also consider the compensation due to the attack on q-type assumptions by Cheon [17].

In comparison with the schemes of Lewko, Sahai and Waters, the disadvantage lies in that a bound on the number of revocations must be chosen when the system is set up. A comparative efficiency of known IBR schemes is given in the table hereafter.

5 Non-monotonic KP-ABE with Short Ciphertexts

Ostrovsky, Sahai and Waters [28] suggested a technique to move from monotonic to non-monotonic access structures without incurring an immoderate private key size. They assume a family $\{\Pi_{\mathbb{A}}\}_{\mathbb{A}\in\mathcal{AS}}$ of linear secret-sharing schemes for a set of monotone access structures $\mathbb{A}$. For each such access structure $\mathbb{A} \in \mathcal{AS}$, the set $\mathcal{P}$ of underlying parties is defined in such a way that parties' names can be normal (like x) or primed (like x'). Prime attributes are conceptually seen as the negation of unprimed attributes. In addition, it is required that, if $x \in \mathcal{P}$, then $x' \in \mathcal{P}$ and vice versa.

A family $\mathcal{AS}$ of non-monotone access structures can be defined as follows. For each access structure $\mathbb{A} \in \mathcal{AS}$ over a set of parties $\mathcal{P}$, one defines a possibly non-monotonic access structure $NM(\mathbb{A})$ over the set $\tilde{\mathcal{P}}$ of all unprimed parties in $\mathcal{P}$. An operator $N(.)$ is then defined as follows. For every set $\tilde{S} \subset \tilde{\mathcal{P}}$, one imposes $\tilde{S} \subset N(\tilde{S})$. Also, for each $x \in \tilde{\mathcal{P}}$ such that $x \notin \tilde{S}$, $x' \in N(\tilde{S})$. Finally, $NM(\mathbb{A})$ is defined by saying that $\tilde{S}$ is authorized in $NM(\mathbb{A})$ if and only if $N(\tilde{S})$ is authorized in $\mathbb{A}$ (so that $NM(\mathbb{A})$ has only unprimed parties in its access sets). For each access set $X \in NM(\mathbb{A})$, there is a set in $\mathbb{A}$ containing the elements in X and primed elements for each party not in X.

In [28], the above technique was combined with the Naor-Pinkas revocation method [27] to cope with non-monotonic access structures. Lewko, Sahai and Waters provided improvements using a revocation system with short keys [24] instead of [27]. In the following, we apply the same technique to our revocation mechanism and combine it with the monotonic KP-ABE derived from the IBBE scheme of section 3.3 in order to handle non-negated attributes.

► $\mathsf{Setup}(\lambda, n)$: Given a security parameter $\lambda \in \mathbb{N}$ and a bound $n \in \mathbb{N}$ of the number of attributes per ciphertext, it chooses bilinear groups $(\mathbb{G}, \mathbb{G}_T)$ of prime order $p > 2^\lambda$ and $g \stackrel{R}{\leftarrow} \mathbb{G}$. It defines $\boldsymbol{H} = (h_1, \ldots, h_n)^\top$ and $\boldsymbol{U} = (u_0, \ldots, u_n)^\top$ such that $h_i = g^{\alpha_i}$, $u_j = g^{\beta_j}$ for each $i \in \{1, \ldots, n\}$ and $j \in \{0, \ldots, n\}$ where $\boldsymbol{\alpha} = (\alpha_1, \ldots, \alpha_n)^\top \stackrel{R}{\leftarrow} \mathbb{Z}_p^n$ and $\boldsymbol{\beta} = (\beta_0, \beta_1, \ldots, \beta_n)^\top \stackrel{R}{\leftarrow} \mathbb{Z}_p^{n+1}$. It then picks $\alpha \stackrel{R}{\leftarrow} \mathbb{Z}_p^*$ and computes $e(g,g)^\alpha$. The master secret key is $\mathsf{msk} = \alpha$ and the master public key is $\mathsf{mpk} = \big(g,\ e(g,g)^\alpha,\ \boldsymbol{H} = g^{\boldsymbol{\alpha}},\ \boldsymbol{U} = g^{\boldsymbol{\beta}}\big)$.

► $\mathsf{Keygen}(\mathsf{msk}, \tilde{\mathbb{A}})$: Given a non-monotonic access structure $\tilde{\mathbb{A}}$ such that we have $\tilde{\mathbb{A}} = NM(\mathbb{A})$ for some monotonic access structure $\mathbb{A}$ over a set $\mathcal{P}$ of attributes and associated with a linear secret sharing scheme Π, the algorithm applies Π to obtain shares $\{\lambda_i\}$ of the master secret key α. The party corresponding to share λ_i is denoted by $\breve{x}_i \in \mathcal{P}$, where x_i is the underlying attribute, and can be primed (*i.e.*, negated) or unprimed (non-negated). For each i, the algorithm chooses $r_i \stackrel{R}{\leftarrow} \mathbb{Z}_p$, defines $\boldsymbol{\rho}_i = (\rho_{i,1}, \ldots, \rho_{i,n})^\top = (1, x_i, x_i^2, \ldots, x_i^{n-1})^\top$. That is $\rho_{i,j} = x_i^{j-1}$. It then does as follows.

- For each i such that $\breve{x}_i$ is unprimed (*i.e.*, non-negated), the algorithm computes a tuple $D_i = (D_{i,1}^{(1)}, D_{i,2}^{(2)}, K_{\boldsymbol{\rho}_i,i}^{(3)}) \in \mathbb{G}^{n+1}$ where the first two elements are $(D_{i,1}^{(1)}, D_{i,2}^{(1)}) = \big(g^{\lambda_i} \cdot u_0^{r_i},\ g^{r_i}\big)$ and

$$K_{\boldsymbol{\rho}_i,i}^{(1)} = (K_{i,2}^{(1)}, \ldots, K_{i,n}^{(1)}) = \Big(\big(u_1^{-\frac{\rho_{i,2}}{\rho_{i,1}}} \cdot u_2\big)^{r_i}, \ldots,\ \big(u_1^{-\frac{\rho_{i,n}}{\rho_{i,1}}} \cdot u_n\big)^{r_i}\Big) = g^{r_i \cdot M_{\boldsymbol{\rho}_i}^{\top} \boldsymbol{\beta}},$$

where $M_{\boldsymbol{\rho}_i} \in (\mathbb{Z}_p)^{n \times (n-1)}$ is the matrix $M_{\boldsymbol{\rho}_i} = \begin{pmatrix} -\frac{\rho_{i,2}}{\rho_{i,1}} & -\frac{\rho_{i,3}}{\rho_{i,1}} & \cdots & -\frac{\rho_{i,n}}{\rho_{i,1}} \\ & I_{n-1} & & \end{pmatrix}$.
- For each i such that $\breve{x}_i$ is primed (*i.e.*, negated), one computes a tuple $D_i = (D_{i,1}^{(2)}, D_{i,2}^{(2)}, K_{\boldsymbol{\rho}_i,i}^{(2)}) \in \mathbb{G}^{n+1}$ where $(D_{i,1}^{(2)}, D_{i,2}^{(2)}) = \big(g^{\lambda_i} \cdot h_1^{r_i},\ g^{r_i}\big)$ and

$$K_{\boldsymbol{\rho}_i,i}^{(2)} = (K_{i,2}^{(2)}, \ldots, K_{i,n}^{(2)}) = \Big(\big(h_1^{-\frac{\rho_{i,2}}{\rho_{i,1}}} \cdot h_2\big)^{r_i}, \ldots,\ \big(h_1^{-\frac{\rho_{i,n}}{\rho_{i,1}}} \cdot h_n\big)^{r_i}\Big) = g^{r_i \cdot M_{\boldsymbol{\rho}_i}^{\top} \boldsymbol{\alpha}}.$$

The private key is $\mathsf{sk}_{\tilde{\mathbb{A}}} = \{D_i\}_{\breve{x}_i \in \mathcal{P}} \in \mathbb{G}^{\ell \times (n+1)}$.

► $\mathsf{Encrypt}(\mathsf{mpk}, \mathsf{M}, \omega)$: To encrypt M for a set ω (with $|\omega| < n$), it first defines $\boldsymbol{Y} = (y_1, \ldots, y_n)^{\top}$ as the vector whose first $q+1$ coordinates are the coefficients of the polynomial $P_\omega[Z] = \sum_{i=1}^{q+1} y_i Z^{i-1} = \prod_{j \in \omega}(Z - j)$. If $q+1 < n$, set $y_j = 0$ for $q+2 \leq j \leq n$. Then it picks $s \xleftarrow{R} \mathbb{Z}_p$ and computes

$$C = (C_0, C_1, C_2, C_3) = \Big(\mathsf{M} \cdot e(g,g)^{\alpha \cdot s},\ g^s,\ \big(u_0 \cdot \prod_{i=1}^{n} u_i^{y_i}\big)^s,\ \big(\prod_{i=1}^{n} h_i^{y_i}\big)^s\Big).$$

► $\mathsf{Decrypt}(\mathsf{mpk}, \mathsf{sk}_{\tilde{\mathbb{A}}}, \tilde{\mathbb{A}}, C, \omega)$: It parses C as (C_0, C_1, C_2, C_3) and the private key $\mathsf{sk}_{\tilde{\mathbb{A}}}$ as $\mathsf{sk}_{\tilde{\mathbb{A}}} = \{D_i\}_{\breve{x}_i \in \mathcal{P}} \in \mathbb{G}^{\ell \times (n+1)}$. The algorithm outputs $\perp$ if $\omega \notin \tilde{\mathbb{A}}$. Otherwise, since $\tilde{\mathbb{A}} = NM(\mathbb{A})$ for some access structure $\mathbb{A}$ associated with a linear secret sharing scheme Π, we have $\omega' = N(\omega) \in \mathbb{A}$ and we let $I = \{i : \breve{x}_i \in \omega'\}$. Since ω' is authorized in $\mathbb{A}$, the receiver can efficiently compute coefficients $\{\mu_i\}_{i \in I}$ such that $\sum_{i \in I} \mu_i \lambda_i = \alpha$ (although the shares are not known to the receiver). Let $\boldsymbol{Y} = (y_1, \ldots, y_n)^{\top}$ be the vector containing the coefficients of the polynomial $P_\omega[Z] = \prod_{j \in \omega}(Z - j) = \sum_{i=1}^{q+1} y_i Z^{i-1}$.

- For every positive attribute $\breve{x}_i \in \omega'$ (for which $x_i \in \omega$), the decryption procedure computes $\tilde{D}_{i,1}^{(1)} = D_{i,1}^{(1)} \cdot \prod_{j=2}^{n} K_{i,j}^{(1)\,y_j} = g^{\alpha} \cdot \big(u_0 \cdot u_1^{y_1} \cdots u_n^{y_n}\big)^{r_i}$, and then $e(g,g)^{\lambda_i s} = e(C_1, \tilde{D}_{i,1}^{(1)}) / e(C_2, D_{i,2}^{(1)})$.
- For each negated attribute $\breve{x}_i \in \omega'$ (for which $x_i \notin \omega$), the receiver sets $\boldsymbol{\rho}_i = (1, x_i, \ldots, x_i^{n-1})^{\top}$ and successively computes

$$K_i^{(2)} = \prod_{j=2}^{n} K_{i,j}^{(2)\,y_j} = \big(h_1^{-\langle \boldsymbol{\rho}_i, \boldsymbol{Y} \rangle / x_1} \cdot h_1^{y_1} \cdots h_n^{y_n}\big)^{r_i},$$

$$\tau_i = \Big(\frac{e(K_i^{(2)}, C_1)}{e(C_3, D_{i,2}^{(2)})}\Big)^{-\frac{\rho_{i,1}}{\langle \boldsymbol{\rho}_i, \boldsymbol{Y} \rangle}} = e(g, h_1)^{r_i s}$$

and then $e(g,g)^{\lambda_i s} = e(C_1, D_{i,1}^{(2)})^{-1} \cdot \tau_i^{-1}$.

Finally, decryption computes $\mathsf{M} = C_0 \cdot \prod_{i \in I} e(g,g)^{-\mu_i \lambda_i s}$.

If we split I into $I_0 \cup I_1$, where I_0 and I_1 correspond to unprimed and primed attributes, respectively, decryption can more efficiently compute

$$e(g,g)^{\alpha \cdot s} = e\Big(C_1, \prod_{i \in I_0} \tilde{D}_{i,1}^{(1)^{\mu_i}} \cdot \prod_{i \in I_1} \big(D_{i,1}^{(2)} \cdot K_i^{(2)\,\frac{\mu_i \cdot \rho_{i,1}}{\langle \rho_i, \mathbf{Y} \rangle}}\big)\Big) \cdot e\Big(C_2, \prod_{i \in I_0} D_{i,2}^{(1)^{\mu_i}}\Big) \cdot e\Big(C_3, \prod_{i \in I_1} D_{i,2}^{(2)\,\frac{\mu_i \cdot \rho_{i,1}}{\langle \rho_i, \mathbf{Y} \rangle}}\Big),$$

so that only three pairing evaluations are necessary.

Theorem 3. *The above KP-ABE system with the maximal bound n for the number of attributes per ciphertext (i.e., $|\omega| < n$) is selectively secure if the n-DBDHE assumption holds.* (The proof is given in the full version of the paper).

6 Comparisons

Table 2 compares efficiency among available expressive KP-ABE schemes that support non-monotonic access structures. Comparisons are made in terms of ciphertext overhead, private key size as well as in the number of pairing evaluations and exponentiations (in $\mathbb{G}$ and $\mathbb{G}_T$) upon decryption.

We remark that the functionality of KP-ABE in [31] is slightly different from the original one [20]. For self-containment, we re-formalize it in appendix A, where we also briefly propose a modification of KP-ABE [31] so as to have the same functionality as the original ABE. We also include this modified scheme in Table 2. Note that [31] has a unique feature of being adaptively secure.

Table 2. Efficiency of non-monotonic KP-ABE schemes

Schemes	Ciphertext overhead $\lvert\mathbb{G}\rvert$	Private key size $\lvert\mathbb{G}\rvert$	Decryption cost pair.	Decryption cost exp.	Assumption
OSW [28]	$O(\bar{n})$	$O(t \cdot \log n)$	$O(t)$	$O(t \cdot \bar{n})$	DBDH
LSW [24]	$O(\bar{n})$	$O(t)$	$O(t)$	$O(t \cdot \bar{n})$	n-MEBDH
OT [31]	$O(\bar{n} \cdot \varphi)$	$O(t \cdot \varphi)$	$O(t \cdot \varphi)$	$O(t)$	DLIN
OT^{modified}	$O(\bar{n} \cdot n)$	$O(t \cdot n)$	$O(t \cdot n)$	$O(t)$	DLIN
This work	3	$O(t \cdot n)$	3	$O(t)$	n-DBDHE

† $\bar{n} = |\text{attribute set}| = |\omega|$ for a ciphertext; n =the maximal bound for $\bar{n}$ (*i.e.*, $|\omega| < n$); t = # of attributes in an access structure for a key; φ =maximum size for repetition of attribute label per key (only for the KP-ABE with labeling, formalized in appendix A).

‡ pair., exp. shows # of pairing and exponentiation computation (in $\mathbb{G}$ or $\mathbb{G}_T$), respectively.

7 Concluding Remarks

This paper presented the first results for expressive KP-ABE schemes with constant-size ciphertexts. In the future, it will be interesting to see if shorter private keys can be obtained without affecting the expressivity or the size of ciphertexts and to construct adaptively secure such schemes. Another challenging problem is to achieve similar results in the expressive ciphertext-policy setting.

References

1. Abdalla, M., Kiltz, E., Neven, G.: Generalized Key Delegation for Hierarchical Identity-Based Encryption. In: Biskup, J., López, J. (eds.) ESORICS 2007. LNCS, vol. 4734, pp. 139–154. Springer, Heidelberg (2007)
2. Attrapadung, N., Imai, H.: Dual-Policy Attribute Based Encryption. In: Abdalla, M., Pointcheval, D., Fouque, P.-A., Vergnaud, D. (eds.) ACNS 2009. LNCS, vol. 5536, pp. 168–185. Springer, Heidelberg (2009)
3. Attrapadung, N., Imai, H.: Conjunctive Broadcast and Attribute-Based Encryption. In: Shacham, H., Waters, B. (eds.) Pairing 2009. LNCS, vol. 5671, pp. 248–265. Springer, Heidelberg (2009)
4. Attrapadung, N., Libert, B.: Functional Encryption for Inner Product: Achieving Constant-Size Ciphertexts with Adaptive Security or Support for Negation. In: Nguyen, P.Q., Pointcheval, D. (eds.) PKC 2010. LNCS, vol. 6056, pp. 384–402. Springer, Heidelberg (2010)
5. Bethencourt, J., Sahai, A., Waters, B.: Ciphertext-Policy Attribute-Based Encryption. IEEE Symposium on Security and Privacy (S&P), pp. 321–334 (2007)
6. Boneh, D., Boyen, X.: Efficient Selective-ID Secure Identity-Based Encryption Without Random Oracles. In: Cachin, C., Camenisch, J.L. (eds.) EUROCRYPT 2004. LNCS, vol. 3027, pp. 223–238. Springer, Heidelberg (2004)
7. Boneh, D., Boyen, X., Goh, E.-J.: Hierarchical Identity Based Encryption with Constant Size Ciphertext. In: Cramer, R. (ed.) EUROCRYPT 2005. LNCS, vol. 3494, pp. 440–456. Springer, Heidelberg (2005)
8. Boneh, D., Franklin, M.: Identity-Based Encryption from the Weil Pairing. Journal of Computing 32(3), 586–615 (2003); Boneh, D., Franklin, M.: Identity-Based Encryption from the Weil Pairing. In: Kilian, J. (ed.) CRYPTO 2001. LNCS, vol. 2139, pp. 213–229. Springer, Heidelberg (2001)
9. Boneh, D., Gentry, C., Waters, B.: Collusion Resistant Broadcast Encryption with Short Ciphertexts and Private Keys. In: Shoup, V. (ed.) CRYPTO 2005. LNCS, vol. 3621, pp. 258–275. Springer, Heidelberg (2005)
10. Boneh, D., Hamburg, M.: Generalized Identity Based and Broadcast Encryption Schemes. In: Pieprzyk, J. (ed.) ASIACRYPT 2008. LNCS, vol. 5350, pp. 455–470. Springer, Heidelberg (2008)
11. Boneh, D., Sahai, A., Waters, B.: Functional Encryption: Definitions and Challenges. In: TCC 2011 (2011) (to appear)
12. Boyen, X.: General *Ad Hoc* Encryption from Exponent Inversion IBE. In: Naor, M. (ed.) EUROCRYPT 2007. LNCS, vol. 4515, pp. 394–411. Springer, Heidelberg (2007)
13. Canetti, R., Halevi, S., Katz, J.: A Forward-Secure Public-Key Encryption Scheme. In: Biham, E. (ed.) EUROCRYPT 2003. LNCS, vol. 2656, pp. 254–271. Springer, Heidelberg (2003)
14. Canetti, R., Halevi, S., Katz, J.: Chosen-Ciphertext Security from Identity-Based Encryption. In: Cachin, C., Camenisch, J.L. (eds.) EUROCRYPT 2004. LNCS, vol. 3027, pp. 207–222. Springer, Heidelberg (2004)
15. Chase, M.: Multi-authority Attribute Based Encryption. In: Vadhan, S.P. (ed.) TCC 2007. LNCS, vol. 4392, pp. 515–534. Springer, Heidelberg (2007)
16. Chase, M., Chow, S.: Improving privacy and security in multi-authority attribute-based encryption. In: ACM-CCS 2009, pp. 121–130 (2009)
17. Cheon, J.-H.: Security Analysis of the Strong Diffie-Hellman Problem. In: Vaudenay, S. (ed.) EUROCRYPT 2006. LNCS, vol. 4004, pp. 1–11. Springer, Heidelberg (2006)

18. Cheung, L., Newport, C.: Provably secure ciphertext policy ABE. In: ACM-CCS 2007, pp. 456–465 (2007)
19. Emura, K., Miyaji, A., Nomura, A., Omote, K., Soshi, M.: A Ciphertext-Policy Attribute-Based Encryption Scheme with Constant Ciphertext Length. In: Bao, F., Li, H., Wang, G. (eds.) ISPEC 2009. LNCS, vol. 5451, pp. 13–23. Springer, Heidelberg (2009)
20. Goyal, V., Pandey, O., Sahai, A., Waters, B.: Attribute-based encryption for fine-grained access control of encrypted data. In: ACM CCS 2006, pp. 89–98 (2006)
21. Goyal, V., Jain, A., Pandey, O., Sahai, A.: Bounded Ciphertext Policy Attribute Based Encryption. In: Aceto, L., Damgård, I., Goldberg, L.A., Halldórsson, M.M., Ingólfsdóttir, A., Walukiewicz, I. (eds.) ICALP 2008, Part II. LNCS, vol. 5126, pp. 579–591. Springer, Heidelberg (2008)
22. Herranz, J., Laguillaumie, F., Ràfols, C.: Constant Size Ciphertexts in Threshold Attribute-Based Encryption. In: Nguyen, P.Q., Pointcheval, D. (eds.) PKC 2010. LNCS, vol. 6056, pp. 19–34. Springer, Heidelberg (2010)
23. Katz, J., Sahai, A., Waters, B.: Predicate Encryption Supporting Disjunctions, Polynomial Equations, and Inner Products. In: Smart, N.P. (ed.) EUROCRYPT 2008. LNCS, vol. 4965, pp. 146–162. Springer, Heidelberg (2008)
24. Lewko, A., Sahai, A., Waters, B.: Revocation Systems with Very Small Private Keys. In: IEEE Symposium on Security and Privacy (S&P) (2010)
25. Lewko, A., Okamoto, T., Sahai, A., Takashima, K., Waters, B.: Fully Secure Functional Encryption: Attribute-Based Encryption and (Hierarchical) Inner Product Encryption. In: Gilbert, H. (ed.) EUROCRYPT 2010. LNCS, vol. 6110, pp. 62–91. Springer, Heidelberg (2010)
26. Naor, M.: On Cryptographic Assumptions and Challenges. In: Boneh, D. (ed.) CRYPTO 2003. LNCS, vol. 2729, pp. 96–109. Springer, Heidelberg (2003)
27. Naor, M., Pinkas, B.: Efficient Trace and Revoke Schemes. In: Frankel, Y. (ed.) FC 2000. LNCS, vol. 1962, pp. 1–20. Springer, Heidelberg (2001)
28. Ostrovsky, R., Sahai, A., Waters, B.: Attribute-based encryption with non-monotonic access structures. In: ACM CCS 2007, pp. 195–203 (2007)
29. Sahai, A., Waters, B.: Fuzzy Identity-Based Encryption. In: Cramer, R. (ed.) EUROCRYPT 2005. LNCS, vol. 3494, pp. 457–473. Springer, Heidelberg (2005)
30. Shamir, A.: Identity-Based Cryptosystems and Signature Schemes. In: Blakely, G.R., Chaum, D. (eds.) CRYPTO 1984. LNCS, vol. 196, pp. 47–53. Springer, Heidelberg (1985)
31. Okamoto, T., Takashima, K.: Fully secure functional encryption with general relations from the decisional linear assumption. In: Rabin, T. (ed.) CRYPTO 2010. LNCS, vol. 6223, pp. 191–208. Springer, Heidelberg (2010)
32. Waters, B.: Ciphertext-Policy Attribute-Based Encryption: An Expressive, Efficient, and Provably Secure Realization. In: Catalano, D., et al. (eds.) PKC 2011. LNCS, vol. 6571, pp. 90–107. Springer, Heidelberg (2011)
33. Yamada, S., Attrapadung, N., Hanaoka, G., Kunihiro, N.: Generic Constructions for Chosen-Ciphertext Secure Attribute Based Encryption. In: Catalano, D., et al. (eds.) PKC 2011. LNCS, vol. 6571, pp. 90–107. Springer, Heidelberg (2011)

A Variant: KP-ABE with Labeling

We re-formalize the KP-ABE definition of [31] in our context as follows. Intuitively, the difference from normal KP-ABE is that an attribute is required to be labeled with a number $j \in [1, n]$ and that each attribute in the set associated to a ciphertext is required to be labeled uniquely, namely 1 to n. The scheme of [31] further restricts the maximum repetition allowed for labels in one policy, which we denote by φ in Table 2.

Definition 8 (KP-ABE with labeling). *Let U be an attribute space and let an integer $n \in \mathbb{N}$. Define $U' = \{(j, u) \mid j \in [1, n], u \in U\}$. Define the ciphertext index domain as $\Sigma_e^{\mathsf{KP}'} = \{\{(1, u_1), \ldots, (n, u_n)\} \mid u_1, \ldots, u_n \in U\}$. A KP-ABE with labeling for a collection $\mathcal{AS}'$ of access structures over U' is a functional encryption for $R^{\mathsf{KP}'} : \mathcal{AS}' \times \Sigma_e^{\mathsf{KP}'} \to \{0, 1\}$ defined by $R^{\mathsf{KP}'}(\mathbb{A}, \omega) = 1$ iff $\omega \in \mathbb{A}$ (for $\omega \in \Sigma_e^{\mathsf{KP}'}, \mathbb{A} \in \mathcal{AS}'$).*

We observe that KP-ABE with large universe $U = \{0, 1\}^*$, *e.g.*, [20, 28] and ours, implies KP-ABE with labeling. This is since $U' \subset U$, $\Sigma_e^{\mathsf{KP}'} \subset \Sigma_e^{\mathsf{KP}}$, $\Sigma_k^{\mathsf{KP}'} \subset \Sigma_k^{\mathsf{KP}}$, and $R^{\mathsf{KP}'} \Leftrightarrow R^{\mathsf{KP}}$ holds and the implication comes from the embedding lemma [10, 4]. To the best of our knowledge, the converse is yet known to hold.

We now briefly propose a KP-ABE that conforms with the normal definition by modifying [31]. We construct by instantiating the general KP-FE scheme of [31] with $d = 1$, and with the inner product relation being instantiated to IBBE, similarly as we did in section 3.3, and setting the bound $\varphi = n$.

Faster and Lower Memory Scalar Multiplication on Supersingular Curves in Characteristic Three

Roberto Avanzi[1,⋆] and Clemens Heuberger[2,⋆⋆]

[1] Faculty of Mathematics and Horst Görtz Institute for IT Security
Ruhr-University Bochum, Germany
roberto.avanzi@ruhr-uni-bochum.de

[2] Institut für Mathematik B, Technische Universität Graz, Austria
clemens.heuberger@tugraz.at

Abstract. We describe new algorithms for performing scalar multiplication on supersingular elliptic curves in characteristic three. These curves can be used in pairing-based cryptography. Since in pairing-based protocols besides pairing computations also scalar multiplications are required, and the performance of the latter is not negligible, improving it is clearly important as well. The techniques presented here bring noticeable speed ups (up to 30% for methods using a variable amount of memory and up to 46.7% for methods with a small, fixed memory usage), while at the same time bringing substantial memory reductions – factors like 3 to 8 are not uncommon.

The starting point for our methods is a structure theorem for unit groups of residue classes of a quadratic order associated to the Frobenius endomorphism of the considered curves. This allows us to define new digit sets whose elements are products of powers of certain generators of said groups. There are of course several choices for these generators: we chose generators associated to endomorphisms for which we could find efficient explicit formulae in a suitable coordinate system. A multiple-base-like scalar multiplication algorithm making use of these digits and these formulae brings the claimed speed up.

Keywords: Supersingular elliptic curves, pairing-friendly elliptic curves, scalar multiplication, Frobenius expansion, explicit formulae.

1 Introduction

The following elliptic curves over fields of characteristic three

$$\mathcal{E}_{3,\mu} : Y^2 = X^3 - X - \mu \qquad \text{with } \mu \in \{\pm 1\} \tag{1}$$

⋆ Supported by the European Commission through the IST Programme under contract ICT-2007-216676 ECRYPT II.

⋆⋆ Supported by the Austrian Science Foundation FWF, project S9606, that is part of the Austrian National Research Network "Analytic Combinatorics and Probabilistic Number Theory."

D. Catalano et al. (Eds.): PKC 2011, LNCS 6571, pp. 109–127, 2011.

are supersingular with embedding degree 6. They thus offer less security per bit than ordinary curves for DL-based applications. However, in 1998 Koblitz [20] studied their arithmetic and found their performance to be competitive with other public-key cryptosystems even after the security parameters were adjusted accordingly. Recently, pairing-based cryptography has revived the interest in these curves (together with all other types of pairing-friendly curves [16]). Most of the current research is devoted to the optimization of the field arithmetic and the pairing operation (just a few examples are [1,9,7,8]).

The performance gap between pairing operations and scalar multiplication has steadily decreased in the last years. In some cases the two operations have comparable performance: In [1] variable base point scalar multiplication takes between a half and a third of the time for a Tate pairing; in [9] a η_T pairing over $\mathbb{F}_{3^{97}}$ is computed in 678 field multiplications, whereas a scalar multiplication requires, with current state-of-the art methods, at least 230 multiplications using normal bases and 296 using polynomial bases (cf. Tables 1 and 2 on page 125 under the columns labeled "BMX", that correspond to the current state of the art [10]); one of the fastest implementations of characteristic three arithmetic and of pairings in general, due to Mitsunari [22], requires 0.181 μsec, resp. 0.149 μsec for a field multiplication, resp. an η_T pairing (single threaded) over $\mathbb{F}_{3^{97}}$ – with our operation counts for [10] we extrapolate a timing of about 53.5 μsec for a scalar multiplication; over the field $\mathbb{F}_{3^{193}}$ the same implementation takes 0.624 μsec, resp. 975 μsec for the two operations – similarly we extrapolate 314.2 μsec for a scalar multiplication in this case.

For other characteristics, in [12] extensive experiments are reported not only for trace-zero varieties but also for elliptic curves, and in many cases a η_T pairing evaluation is even faster than a scalar multiplication on the same curve.

Now, most pairing-based protocols – for instance Direct Anonymous Attestation [11] – also require scalar multiplications, the latter being often performed in computationally restricted environments, such as TPM modules. Since in these cases the pairings are usually computed on faster architectures, the question of the performance of scalar multiplication becomes more, not less, severe. Thus it is an important question whether one can either speed up this operation and/or reduce its memory requirements. This is the problem studied in this paper: *We show how to perform scalar multiplication on supersingular Koblitz curves in characteristic three with extremely reduced memory requirements with respect to the state of the art while attaining better performance.*

Usually, scalar multiplication on supersingular Koblitz curves in characteristic three is done by a base three expansion [18] or by an expansion to the base of τ [20,10], where τ is a complex number associated to the *Frobenius* endomorphism of the curve. Double base methods have been suggested in this context, for instance in [2], but only with the rational bases 2 and 3.

For the characteristic two case, it has been suggested to use the so-called *Verschiebung* endomorphism besides τ, as in [3,4]. This is the endomorphism associated to the algebraic conjugate $\bar{\tau}$ of τ. In characteristic two τ and $\bar{\tau}$ can be used as a double bases, but, as we shall see in the next section, this cannot extended to the case considered here.

Our goal to reduce memory requirements and speeding-up scalar multiplication on the curves (1)*, is attained by finding alternative endomorphisms that can serve as additional bases to be used beside* τ*.*

This is essentially done in Section 2 by Theorem 1, that describes the structure of the group of units in the ring of residue classes of $\mathbb{Z}[\tau]$ modulo powers of the prime ideal generated by τ; the generators of this group are chosen in such a way that they can be implemented efficiently, as explained in Section 4. Their usage in practice is described in Sections 3 and 5. The techniques presented here bring noticeable speed ups – from a few percent to a near halving of computational time – while often bringing substantial memory reductions at the same time: reducing the memory requirement to just a third or even one eighth of the memory usage of the current state of the art is not uncommon. Detailed comparisons can be found in Section 5 and concluding remarks are in Section 6.

2 Digit Sets and the Structure of the Unit Group

We begin by recalling some facts on the Frobenius endomorphism τ and the associated ring $\mathbb{Z}[\tau]$.

For cryptographic applications one works in the group $\mathcal{E}_{3,\mu}(\mathbb{F}_{3^m})$ of the $\mathbb{F}_{3^m}$-rational points of the curve $\mathcal{E}_{3,\mu}$, where m is an integer not divisible by either 2 or 3, and it is usually assumed that $\mathcal{E}_{3,\mu}(\mathbb{F}_{3^m})$ contains a unique large subgroup G of prime order p. Cryptographic computations then take place in G.

The Frobenius endomorphism

$$\tau \,:\, \mathcal{E}_{3,\mu}(\mathbb{F}_{3^m}) \to \mathcal{E}_{3,\mu}(\mathbb{F}_{3^m}) \ , \quad (x,y) \mapsto (x^3,y^3)$$

is such that $\tau^2(P) - 3\,\mu\,\tau(P) + 3 \cdot P = 0$ for all points P on the curve; in other words, the relation

$$\tau^2 - 3\,\mu\,\tau + 3 = 0 \tag{2}$$

holds in the endomorphism ring of the curve. Prompted by this, we can identify τ with a imaginary quadratic solution of the last equation

$$\frac{3\mu + \sqrt{-3}}{2} \ , \tag{3}$$

which we also call τ. This induces an isomorphism (of rings with unit) between the ring $\mathbb{Z}[\tau]$ and the endomorphism ring of $\mathcal{E}_{3,\mu}$, which also maps 1 to the identity map on the curve and any rational integer n to the multiplication-by-n isogeny. This endomorphism allows for a fast scalar multiplication based on expansions of scalars to the base of τ.

We recall that $\mathbb{Z}[\tau]$ is a factorial ring (it is a quadratic order of class number 1). Also, since τ is prime, an element of $\mathbb{Z}[\tau]$ is coprime to τ if and only if τ does not divide it. Let

$$\zeta := \frac{1 - \mu\sqrt{-3}}{2} \ , \tag{4}$$

such that ζ is a primitive sixth root of unity. The complex conjugate of τ will be denoted by $\bar{\tau}$, and $\bar{\tau} = \zeta\tau$ holds. These numbers act as follows as endomorphisms on the curve:

$$\zeta : (x, y) \mapsto (x + \mu, -y) ,$$
$$\bar{\tau} : (x, y) \mapsto (x^3 + \mu, -y^3) .$$

Since $3 = \tau\bar{\tau} = \zeta\tau^2$, tripling is an efficient operation as well:

$$3 : (x, y) \mapsto (x^9 + \mu, -y^9) .$$

We see that τ and $\bar{\tau}$ are conjugated in the sense that their ratio is a unit, hence they are essentially the same base and they cannot be used in a double-base or in a double-loop scalar multiplication methods such as those from [2,4].

Now we consider the construction of useful digit sets using the algebraic numbers we have just defined.

Similarly to Solinas [24], one can take one representative from each residue class modulo τ^w which is relatively prime to τ (a *reduced residue system* modulo τ) together with the zero to form a digit set $\mathcal{D}$ for expansions of integers $z \in \mathbb{Z}[\tau]$:

$$z = \sum_{i=0}^{\ell} d_i \tau^i . \tag{5}$$

If, for a given scalar z, such an expansion exists, we can use it to design a scalar multiplication method on $\mathcal{E}_{3,\mu}(\mathbb{F}_{3^m})$: First, we precompute and store all elements of the form $d \cdot P$ for $P \in \mathcal{E}_{3,\mu}(\mathbb{F}_{3^m})$, and then we can compute $z \cdot P$ by a Horner scheme:

$$\tau\bigl(\tau\bigl(\cdots\tau\bigl(\tau\bigl(\tau(d_\ell P) + d_{\ell-1}P\bigr) + d_{\ell-2}P\bigr) + \cdots + d_2 P\bigr) + d_1 P\bigr) + d_0 P .$$

An important desirable property for expansions (5) is that each block of w consecutive digits $d_i, d_{i+1}, \ldots, d_{i+w-1}$ contains at most one non-zero. Such an expansion is called a $\mathcal{D}$-w-NAF and a digit set $\mathcal{D}$ is called a w-Non-Adjacent-Digit-Set, or w-NADS, if every integer $z \in \mathbb{Z}[\tau]$ admits a $\mathcal{D}$-w-NAF. In analogy to the characteristic two case (cf. Solinas [24]), we can choose the elements of the digit set to be of minimal norm in their residue classes modulo τ^w. However, this only guarantees the existence of expansions for $w \geqslant 2$. Smart [23] works in a more general setting with, essentially, $w = 1$ and smallest rational integer digits (and works in general odd characteristic): in characteristic three, in order to guarantee termination, he needs an expansion with digits $\{0, \pm 1, \pm 2\}$ where *at most one digit* is exceptionally allowed to take value 2 or -2. In fact, Blake, Kumar Murty and Xu [10], observed that $\{0, \pm 1\}$ is not a 1-NADS: Indeed, if we try to expand ζ we obtain the infinite expansion

$$\zeta = -1 - \mu\tau - \tau^2 - \mu\tau^3 - \tau^4 - \mu\tau^5 - \ldots .$$

In what follows we shall therefore assume $w \geqslant 2$. Blake, Kumar Murty and Xu proved that a digit set of minimal norm representatives is a w-NADS. One of Koblitz' results can be formulated as saying that $\{0\} \cup \langle\zeta\rangle$ is a 2-NADS.

A big difference with respect to the characteristic two case is that in a given residue class modulo τ^w an element of minimal norm is not necessarily unique (in [5] it is fully explained when this happens, but we do not need this here). However, one is not forced to take an element from *each* of the $6 \times 3^{w-2}$ residue classes coprime to τ and store all the corresponding precomputations: Blake, Kumar Murty and Xu [10, Section 4.2] use a *signed* expansion to reduce the memory requirement by a factor of 2, i.e., to 3^{w-1} points.

But, there are other ways of constructing digit sets, and their structure can be used to design alternative scalar multiplication schemes. To achieve this, we first prove a structure theorem for the unit group of $\mathbb{Z}[\tau]$ modulo τ^w, for each natural number w. In particular we shall prove that this group is the direct product of (up to) three cyclic subgroups. Clearly, taking representatives of each class in this group will yield a reduced residue system modulo τ^w. Since the digit sets we consider in this paper consist of zero and of a reduced residue system modulo τ^w, the structure theorem will allow us to construct digit sets whose elements are products of powers of three fixed elements.

The decomposition of $(\mathbb{Z}[\tau]/\tau^w\mathbb{Z}[\tau])^\times$ can also be derived, with some effort, from Nakagoshi's much more general results for generic unit groups of residue classes of orders of number fields modulo powers of arbitrary prime ideals [21] or from Halter-Koch's classification [17] for quadratic orders. However, the proof we give here is direct and much simpler, the expressions for the generators are explicit, and the generators also enjoy the property that they correspond to endomorphisms of $\mathcal{E}_{3,\mu}$ that lend themselves to efficient evaluation.

Theorem 1. *We have*

$$\left(\frac{\mathbb{Z}[\tau]}{\tau\mathbb{Z}[\tau]}\right)^\times = \langle -1 \rangle \simeq \frac{\mathbb{Z}}{2\mathbb{Z}},$$

$$\left(\frac{\mathbb{Z}[\tau]}{\tau^w\mathbb{Z}[\tau]}\right)^\times = \langle \zeta \rangle \times \left\langle 1+\mu\tau^3 \right\rangle \times \langle -2 \rangle \simeq \frac{\mathbb{Z}}{6\mathbb{Z}} \times \frac{\mathbb{Z}}{3^{\lfloor w/2 \rfloor - 1}\mathbb{Z}} \times \frac{\mathbb{Z}}{3^{\lceil w/2 \rceil - 1}\mathbb{Z}}, \quad w \geqslant 2.$$

Here $\langle \alpha \rangle$ denotes the group generated by α in $(\mathbb{Z}[\tau]/\tau^w\mathbb{Z}[\tau])^\times$.

Remark 1. For $w = 2$, the subgroups generated by $1 + \mu\tau^3$ and -2 in the unit group $(\mathbb{Z}[\tau]/\tau^w\mathbb{Z}[\tau])^\times$ are degenerated, i.e., they are the trivial group with one element, and for $w = 3$ only $\left\langle 1+\mu\tau^3 \right\rangle$ is trivial. For $w \geqslant 4$ all three factor subgroups are not trivial.

Once the structure of $(\mathbb{Z}[\tau]/\tau^w\mathbb{Z}[\tau])^\times$ is given, a digit set for integer expansions can be built in such a way that we take one element from each residue class modulo τ^w that is not divisible by τ. It suffices to take each $d = d_1 d_2 d_3$ with $d_1 \in \langle \zeta \rangle$, $d_2 \in \left\langle 1+\mu\tau^3 \right\rangle$, and $d_3 \in \langle -2 \rangle$. The resulting digit set is invariant by multiplication by ζ, i.e., under rotation of the complex plane by $\pi/3$. The structure of this digit set will be exploited in the scalar multiplication algorithms that will be introduced in Sections 3 and 5.

In order to prove Theorem 1 we need first to compute the orders of the asserted generators modulo τ^w.

Lemma 1. *Let $k \geqslant 2$, $a \in \mathbb{Z}[\tau]$ with $\tau \nmid a$ and $w \geqslant k-1$. Then*

$$\operatorname{ord}_{\tau^w}(1 + a\tau^k) = 3^{\lceil (w-k)/2 \rceil} . \tag{6}$$

In particular, we have

(a) $\operatorname{ord}_{\tau^w}(1+\mu\tau^3) = 3^{\lfloor w/2 \rfloor - 1}$ *for* $w \geqslant 2$,
(b) $\operatorname{ord}_{\tau^w}(-2) = 3^{\lceil w/2 \rceil - 1}$ *for* $w \geqslant 1$,
(c) $\operatorname{ord}_{\tau^w}(1+\mu\tau) = 3^{\lceil w/2 \rceil - 1}$ *for* $w \geqslant 3$,
(d) $\operatorname{ord}_{\tau^w}(\zeta) = 6$ *for* $w \geqslant 2$,

where $\operatorname{ord}_{\tau^w}(\alpha)$ *denotes the order of* α *in* $(\mathbb{Z}[\tau]/\tau^w\mathbb{Z}[\tau])^\times$, *i.e., the least positive exponent* r *such that* $\alpha^r \equiv 1 \pmod{\tau^w}$.

Proof (Lemma 1). We first prove

$$(1 + a\tau^k)^{3^\ell} \equiv 1 + a\zeta^\ell \tau^{k+2\ell} \pmod{\tau^{2k+2\ell}} . \tag{7}$$

by induction on $\ell \geqslant 0$. For $\ell = 0$, (7) holds trivially. Assume that

$$(1 + a\tau^k)^{3^\ell} = 1 + a\zeta^\ell \tau^{k+2\ell} + b\tau^{2k+2\ell}$$

for some $b \in \mathbb{Z}[\tau]$. Then we have

$$\begin{aligned}(1 + a\tau^k)^{3^{\ell+1}} &= (1 + a\zeta^\ell \tau^{k+2\ell} + b\tau^{2k+2\ell})^3 \equiv 1 + 3a\zeta^\ell\tau^{k+2\ell} \\ &\equiv 1 + a\zeta^{\ell+1}\tau^{k+2\ell+2} \pmod{\tau^{2+2k+2\ell}} ,\end{aligned}$$

as $3 = \zeta\tau^2$, which concludes the proof of (7).

We set $\ell = \left\lceil \frac{w-k}{2} \right\rceil$, which results in $k + 2(\ell-1) < w \leqslant k + 2\ell$. As $\tau \nmid a\zeta^{\ell-1}$ (note that ζ is a unit in $\mathbb{Z}[\tau]$), this leads to

$$(1 + a\tau^k)^{3^\ell} \equiv 1 \pmod{\tau^w} , \qquad (1 + a\tau^k)^{3^{\ell-1}} \not\equiv 1 \pmod{\tau^w} .$$

Thus $\operatorname{ord}_{\tau^w}(1 + a\tau^k)$ divides 3^ℓ, but does not divide $3^{\ell-1}$. We conclude that $\operatorname{ord}_{\tau^w}(1 + a\tau^k) = 3^{\lceil (w-k)/2 \rceil}$, as requested.

The assertion for the special case $1 + \mu\tau^3$ follows immediately from (6) by noting that $\left\lceil \frac{w-1}{2} \right\rceil = \left\lfloor \frac{w}{2} \right\rfloor$.

In order to determine the order of -2, we note that $-2 = 1 - 3 = 1 - \zeta\tau^2$, from which the result follows immediately.

Next, we have $(1+\mu\tau)^3 = 1 + (4 - 2\mu\tau)\tau^4$. Thus $\operatorname{ord}_{\tau^w}((1+\mu\tau)^3) = 3^{\lceil w/2 \rceil - 2}$ for $w \geqslant 3$, which implies $\operatorname{ord}_{\tau^w}(1+\mu\tau) = 3^{\lceil w/2 \rceil - 1}$ for $w \geqslant 3$.

Finally, as ζ is a primitive sixth root of unity, we have $\zeta^6 \equiv 1 \pmod{\tau^w}$. As $\zeta^3 = -1$ and $\tau \nmid 2$, we have $\zeta^3 \not\equiv 1 \pmod{\tau^w}$. Finally, $\zeta^2 - 1 = -\mu\tau$ is divisible by τ, but not τ^2, whence $\zeta^2 \not\equiv 1 \pmod{\tau^w}$ for $w \geqslant 2$. Thus $\operatorname{ord}_{\tau^w}(\zeta) = 6$ for $w \geqslant 2$. Of course, this is also a consequence of $\mathcal{D}_2$ being a 2-NADS. □

We are now able to prove the theorem.

Proof (Theorem 1). We prove the assertion by induction on w. We set $\alpha_1 = \zeta$, $\alpha_2 = 1 + \mu\tau^3$ and $\alpha_3 = -2$.

For $w = 1$, there is nothing to show. As $\langle\zeta\rangle$ is known to be a reduced residue system modulo τ^2 and $\langle\alpha_2\rangle$ and $\langle\alpha_3\rangle$ are both the trivial group, we are done for $w = 2$.

Assume that the result holds for some $w \geqslant 2$. We first prove that α_1, α_2, α_3 are independent modulo τ^{w+1}. So we assume that

$$\alpha_1^{a_1}\alpha_2^{a_2}\alpha_3^{a_3} \equiv 1 \pmod{\tau^{w+1}} \tag{8}$$

for some a_j with $0 \leqslant a_j < \mathrm{ord}_{\tau^{w+1}}(\alpha_j)$ for $j \in \{1,2,3\}$. Reducing the relation modulo τ^w, i.e.,

$$\alpha_1^{a_1}\alpha_2^{a_2}\alpha_3^{a_3} \equiv 1 \pmod{\tau^{w}} ,$$

yields $\mathrm{ord}_{\tau^w}(\alpha_j) \mid a_j$ for all $j \in \{1,2,3\}$.

As $\mathrm{ord}_{\tau^{w+1}}(\alpha_1) = \mathrm{ord}_{\tau^w}(\alpha_1) = 6$ by Lemma 1, we immediately get $a_1 = 0$.

By Lemma 1 we also have

$$\begin{aligned}\mathrm{ord}_{\tau^{w+1}}(\alpha_j) &= \mathrm{ord}_{\tau^w}(\alpha_j)\\ \mathrm{ord}_{\tau^{w+1}}(\alpha_k) &= 3 \cdot \mathrm{ord}_{\tau^w}(\alpha_k)\end{aligned}$$

where $\{j,k\} = \{2,3\}$, the appropriate permutation depending on the parity of w. More precisely, it is $(j,k) = (2,3)$ for even w and $(j,k) = (3,2)$ for odd w.

Thus we also have $a_j = 0$ and (8) reduces to $\alpha_k^{a_k} \equiv 1 \pmod{\tau^{w+1}}$, which immediately implies $a_k = 0$, too. This concludes the proof of the independence of α_1, α_2, α_3 modulo τ^{w+1}.

As

$$|\langle\alpha_1\rangle \times \langle\alpha_2\rangle \times \langle\alpha_3\rangle| = 6 \cdot 3^{\lfloor (w+1)/2\rfloor - 1} \cdot 3^{\lceil (w+1)/2\rceil - 1} = 2 \cdot 3^w = \left|\left(\frac{\mathbb{Z}[\tau]}{\tau^w\mathbb{Z}[\tau]}\right)^{\times}\right| ,$$

the generated group has the right cardinality, so α_1, α_2, α_3 do generate the unit group. □

Remark 2. (i) For $w \geqslant 3$, the generator -2 can be replaced by $1 + \mu\tau$. This results from

$$(1 + \mu\tau) = \zeta^4(1 + \mu\tau^3)(-2)^{-1} \tag{9}$$

and $\mathrm{ord}_{\tau^w}(1 + \mu\tau) = \mathrm{ord}_{\tau^w}(-2)$ for $w \geqslant 3$.

(ii) For even values of w only,

$$(\mathbb{Z}[\tau]/\tau^w\mathbb{Z}[\tau])^{\times} = \langle\zeta\rangle \times \langle 1 + \mu\tau\rangle \times \langle -2\rangle ,$$

but not for odd w. This also follows from (9) and from the fact that $\mathrm{ord}_{\tau^w}(1 + \mu\tau) = \mathrm{ord}_{\tau^w}(-2)$. For odd w these elements generate a subgroup of index 3 of the unit group.

(iii) Further choices of generators are possible, beside -2, $1 + \mu\tau$ and $1 + \mu\tau^3$. These elements have been chosen for two reasons: (a) their relatively small norm leads to an overall well bounded norm of the digit set elements, and (b) we were able to find efficient explicit for them, cf. Section 4.

3 Scalar Multiplication Using a Factored Digit Set

Our next goal is to use the digit sets implied by the decomposition of the unit group of Theorem 1 in a precomputationless scalar multiplication algorithm similar to the one presented in [3] for Koblitz curves in characteristic two. In that case the unit group had a much simpler structure than in the present context, that will require a deeper study. From Theorem 1 we know that for $w \geqslant 2$ a decomposition $(\mathbb{Z}[\tau]/\tau^w\mathbb{Z}[\tau])^\times = \langle\zeta\rangle\times\langle\phi\rangle\times\langle\psi\rangle \simeq \mathbb{Z}/6\mathbb{Z} \times \mathbb{Z}/3^a\mathbb{Z} \times \mathbb{Z}/3^b\mathbb{Z}$ exists, where ϕ and ψ are suitable elements of $\mathbb{Z}[\tau]$ identified with the corresponding elements of the endomorphism ring of $\mathcal{E}_{3,\mu}$. Clearly, $a+b = w-2$. Assume further that we can write a scalar z in the form

$$z = \sum_{i=0}^{m} \varepsilon_i \left(\phi^{f_i}\psi^{g_i}\right)\tau^i \tag{10}$$

where $0 \leqslant f_i < 3^a$, $0 \leqslant g_i < 3^b$, and $\varepsilon_i = 0$ or $\varepsilon_i = \zeta^\ell$ with $0 \leqslant \ell < 6$. This is a τ-adic expansion where the digit set is *factored* as the product of three subgroups $\langle\zeta\rangle$, $\langle\phi\rangle$ and $\langle\psi\rangle$.

We can thus perform a scalar multiplication by means of scalar multiplication Algorithm 1 on the next page, whose correctness is an easy fact, as it simply consists of three nested Horner schemes, the two outer ones looping on the exponents of ϕ and ψ, the inner one on the exponents of τ. Note that for $w = 2$ we must have $a = b = 0$ and for $w = 3$ one of a, b must be zero (cf. Remark 1), in which cases the algorithm simplifies because there will be less nested loops.

By an easy generalization of Koblitz' arguments (cf. the end of the proof of Theorem 1 in [20]) it can be proved that the expected density of a w-NAF expansion is $2/(2w+1)$. Note that these arguments do not apply exclusively to expansions using minimal norm digits.

The expansion (10) can also be viewed as a *triple base* representation of the scalar z. Double base representations have been already considered for supersingular Koblitz curves, see for instance [2], where the possibility of using the bases 2 and 3 is mentioned but not analyzed, and in particular a rotational symmetry of the digit set, such as the one that we exploit in our algorithms, is not present.

From Theorem 1 and Remark 2 we know that we can take ζ, -2 and, depending on the parity of w, either $1+\mu\tau$ or $1+\mu\tau^3$ to generate a reduced residue set modulo τ^w, and thus a digit set $\mathcal{D}_w$. In order to guarantee that any τ-adic expansion terminates, we follow the same approach as in [3,4], which consists in reducing the value of the parameter w if the norm of the input becomes too small. As in [3,4] it is easy to verify that this has a minimal and asymptotically negligible impact on the weight of the expansion.

In the next section we consider the efficient implementation of the endomorphisms associated to the ring elements -2, $1+\mu\tau$ and $1+\mu\tau^3$. Since the innermost loops of Algorithm 1 are performed more often, it is a good idea to place the most expensive operation in the outermost loop and the computationally cheapest one in the innermost loop – in other words ψ should be less expensive than ϕ. In Section 5 we shall reconsider Algorithm 1 and some variants, and estimate the costs of scalar multiplication.

Algorithm 1. Low-memory τ-adic Scalar Multiplication on Koblitz Curves

INPUT: $P = (x, y) \in \mathcal{E}_{3,\mu}(\mathbb{F}_{3^m})$, scalar z represented as in Equation (10)
OUTPUT: zP

```
Q ← 0
for j = 3^a − 1 to 0 do
    Q ← φQ                                        [skip first time]
    R ← 0
    for k = 3^b − 1 to 0 do
        R ← ψR                                    [skip first time]
        S ← 0
        for i = m − 1 to 0 do
            S ← τS                                [skip first time]
            if (ε_i ≠ 0 and f_i = j and g_i = k) then
                let ε_i = ζ^ℓ with 0 ⩽ ℓ ⩽ 5
                switch ℓ
                    case 0: S ← S + (x, y)
                    case 1: S ← S + (x + μ, −y)
                    case 2: S ← S + (x − μ, y)
                    case 3: S ← S + (x, −y)
                    case 4: S ← S + (x + μ, y)
                    case 5: S ← S + (x − μ, −y)
        R ← R + S
    Q ← Q + R
return Q
```

4 Group Operations on the Curve

In this section we show how to evaluate $1 + \mu\tau$ efficiently when the curve $\mathcal{E}_{3,\mu}$ is represented using different coordinate systems. We could not find an optimized evaluation of $1 + \mu\tau^3$, that is, just adding P and $\mu\tau^3(P)$ together seems to be the most efficient way of evaluating this endomorphism.

The costs of the various operations on the curve in different coordinate systems are compared § 4.5, in order to choose the best coordinate system for our scalar multiplication algorithms.

4.1 Explicit Formulae for $(1 + \mu\tau)$ in Affine Coordinates

In this subsection as well as in the following three we discuss how to explicitly compute the image of a point P on $\mathcal{E}_{3,\mu}(\mathbb{F}_{3^m}) \setminus \mathcal{E}_{3,\mu}(\mathbb{F}_3)$ under the endomorphism $(1 + \mu\tau)P$ using different coordinate systems. We begin with affine coordinates.

For three points $P_i := (x_i, y_i)$, $i = 1, 2, 3$ on the curve, in the case that $P_1 \neq \pm P_2$, the expression $P_1 + P_2 = P_3$ holds with

$$x_3 = \left(\frac{y_2 - y_1}{x_2 - x_1}\right)^2 - (x_1 + x_2) \quad \text{and} \quad y_3 = y_1 + y_2 - \left(\frac{y_2 - y_1}{x_2 - x_1}\right)^3 . \tag{11}$$

If we take $P_2 = \mu\tau P_1$, then $(x_2, y_2) = (x_1^3, \mu y_1^3)$. For simplicity let us now omit the index 1 in what follows. We obtain

$$x_3 = \left(\frac{\mu y^3 - y}{x^3 - x}\right)^2 - (x + x^3) \quad \text{and} \quad y_3 = y + \mu y^3 - \left(\frac{\mu y^3 - y}{x^3 - x}\right)^3 .$$

Making use of the facts that $x^3 - x = y^2 + \mu$ and $\mu^2 = 1$ and that we are over a field of characteristic three, after a few manipulations we obtain compact expressions for x_3 and y_3,

$$x_3 = x + \mu - \frac{x^3 - x - \mu}{(x^3 - x)^2} = x + \mu - \frac{y^2}{(x^3 - x)^2} \quad \text{and} \quad y_3 = y - \frac{y^3}{(x^3 - x)^3} , \tag{12}$$

that will be the starting point to obtain explicit formulae in different coordinate systems.

4.2 Projective Coordinates

This is a standard coordinate system for elliptic curves, where a finite point (x, y) is represented as $[X{:}Y{:}Z]$ with $x = X/Z$ and $y = Y/Z$. For the purpose of computing on supersingular elliptic curves over fields of characteristic three it has been first used by Koblitz [20]. Koblitz' formulae have been improved by Barreto, Kim, Lynn, and Scott [6].

To obtain an explicit formula consider expressions (12), first replace x, y with X/Z and Y/Z in the two formulae for x_3 and in the formula for y_3. Upon simplification, two rational expressions are obtained. Making their denominators equal and taking this value as the Z-coordinate Z_3 finally yields:

$$\begin{cases} X_3 = ((X + \mu Z)(X^3 - XZ^2)^2 - Y^2 Z^5)(X^3 - XZ^2) , \\ Y_3 = Y(X^3 - XZ^2)^3 - Y^3 Z^7 , \\ Z_3 = Z(X^3 - XZ^2)^3 . \end{cases}$$

This gives us the following operation sequence, where the symbols M, C, I shall denote a multiplication, a cubing and an inversion in $\mathbb{F}_{3^m}$. (We do not distinguish between multiplication and squaring.)

Computing $(1 + \mu\tau)$ in projective coordinates Input: $[X{:}Y{:}Z]$ – Output: $[X_3{:}Y_3{:}Z_3] = (1 + \mu\tau)[X{:}Y{:}Z]$		
Operation	Cost	Remark
$A \leftarrow X^3 - XZ^2$	2 M + 1 C	save Z^2
$B \leftarrow A^3$	1 C	—
$Z_3 \leftarrow ZB$	1 M	—
$Y_3 \leftarrow YB - (YZ^2)^3 Z$	3 M + 1 C	save YZ^2
$X_3 \leftarrow XB + \mu Z_3 - (YZ^2)^2 ZA$	4 M	—
Total cost:	10 M + 3 C	

4.3 Jacobian Coordinates

These coordinates have been introduced in the context of curves in characteristic three by Harrison, Page and Smart in [18], where they are called projective, but they are long known, see for instance [13] and, for cryptographic applications [14]. In order to distinguish them from those described in § 4.2 and in accordance with the rest of the literature on elliptic curves we instead call them *Jacobian*. In Jacobian coordinates the affine point (x, y) is represented as $\langle X{:}Y{:}Z\rangle = (x, y)$, where $x = X/Z^2$ and $y = Y/Z^3$.

As in the projective case our starting point are the addition formulae (11) specialized for the case where $P_2 = \mu\tau P_1$ and simplified, i.e., (12). Putting $x = X/Z^2$ and $y = Y/Z^3$ in the formulae for x_3 and in the formula for y_3 and proceeding as above we obtain

$$\begin{cases} X_3 = ((X + \mu Z^2)(X^3 - XZ^4)^2 - Y^2Z^8)Z(X^3 - XZ^4) \ , \\ Y_3 = Y(X^3 - XZ^4)^3 - Y^3Z^{12} \ , \\ Z_3 = Z^3(X^3 - XZ^4)^3 \ . \end{cases}$$

We obtain the following operation sequence:

Computing $(1 + \mu\tau)$ in Jacobian coordinates Input: $\langle X{:}Y{:}Z\rangle$ – Output: $\langle X_3{:}Y_3{:}Z_3\rangle = (1 + \mu\tau)\langle X{:}Y{:}Z\rangle$		
Operation	Cost	Remark
$A \leftarrow X^3 - XZ^4$	2 M + 2 C	save Z^4
$B \leftarrow A^3$	1 C	—
$Z_3 \leftarrow (ZA)^3$	1 M + 1 C	save ZA
$Y_3 \leftarrow YB - (YZ^4)^3$	2 M + 1 C	save YZ^4
$X_3 \leftarrow [(X + \mu Z^2)A^2 - (YZ^4)^2](ZA)$	5 M	—
Total cost:	10 M + 5 C	

4.4 Modified Jacobian Coordinates

With these coordinates, introduced by Kim and Negre [19], an affine point (x, y) on $\mathcal{E}_{3,\mu}$ is represented by the quadruple $\langle X{:}Y{:}Z{:}T\rangle$, where $x = X/Z^2$ and $y = Y/Z^3$ as before, and $T = Z^2$. The explicit formula in this case is obtained by modifying the formula in § 4.3.

Computing $(1 + \mu\tau)$ in modified Jacobian coordinates Input: $\langle X{:}Y{:}Z{:}T\rangle$ – Output: $\langle X_3{:}Y_3{:}Z_3{:}T_3\rangle = (1 + \mu\tau)\langle X{:}Y{:}Z{:}T\rangle$		
Operation	Cost	Remark
$A \leftarrow X^3 - XT^2$	2 M + 1 C	save T^2
$B \leftarrow A^3$	1 C	—
$Z_3 \leftarrow (ZA)^3$	1 M + 1 C	save ZA
$Y_3 \leftarrow YB - (YT^2)^3$	2 M + 1 C	save YT^2
$X_3 \leftarrow [(X + \mu T)A^2 - (YT^2)^2](ZA)$	4 M	—
$T_3 \leftarrow Z_3^2$	1 M	—
Total cost:	10 M + 4 C	

4.5 Costs of Operations in Different Systems

We now tabulate the costs of several operations on an elliptic curve $\mathcal{E}_{3,\mu}$.

Coordinates → / ↓ Operation	Affine	Projective	Jacobian	Modified Jacobian
ADD	1 I + 3 M	14 M + 1 C	12 M + 4 C	11 M + 4 C
mADD	NA	9 M + 2 C	8 M + 3 C	7 M + 3 C
DBL (also −2)	1 I + 2 M	11 M + 1 C	7 M + 2 C	6 M + 4 C
TPL	4 C	6 C	1 M + 6 C	8 C
τ	2 C	3 C	3 C	4 C
$1+\mu\tau$	1 I + 2 M + 3 C	10 M + 3 C	10 M + 5 C	10 M + 4 C
$1+\mu\tau^3$	1 I + 3 M + 6 C	14 M + 10 C	12 M + 13 C	11 M + 16 C

ADD, DBL, and TPL denote addition of two different points, doubling and tripling of a point, respectively. The prefix m is used to denote a *mixed* addition, i.e., addition of a point given in affine coordinates to a point in a non-affine coordinate system (in other words, $Z_2 = 1$), with a result in the same coordinate system of the second point. We did not find gains with *repeated additions*, i.e. when a given point is added to several inputs, except with standard Jacobian coordinates, where one M can be saved in the ADD. In Jacobian and modified Jacobian coordinates we save a cubing for the generic addition and nothing for the mixed addition. As symbols, τ, $1+\mu\tau$, and $1+\mu\tau^3$ denote the application of the corresponding endomorphisms.

Remark 3. (i) The costs for the projective operations have been inferred from the formulae in [20] (with some obvious simplifications) and [6].

(ii) For mADD in Jacobian coordinates we started with the formula for ADD in [18], which we report here for the reader's convenience:

$$\lambda_1 \leftarrow X_1 Z_2^2 \ , \quad \lambda_2 \leftarrow X_2 Z_1^2 \ , \quad \lambda_3 \leftarrow \lambda_1 - \lambda_2 \ , \quad \lambda_4 \leftarrow Y_1 Z_2^3 \ ,$$
$$\lambda_5 \leftarrow Y_2 Z_1^3 \ , \quad \lambda_6 \leftarrow \lambda_4 - \lambda_5 \ , \quad \lambda_7 \leftarrow \lambda_1 + \lambda_2 \ , \quad \lambda_8 \leftarrow \lambda_4 + \lambda_5 \ ,$$
$$Z_3 \leftarrow Z_1 Z_2 \lambda_3 \ , \quad X_3 \leftarrow \lambda_6^2 - \lambda_7 \lambda_3^2 \ , \quad Y_3 \leftarrow \lambda_8 \lambda_3^3 - \lambda_6^3 \ .$$

Putting $Z_2 = 1$ we save 1 M + 1 S in first step, then 1 M + 1 C in the fourth step. Note that $\lambda_1 = X_1$ and $\lambda_4 = Y_1$, but no further savings come from this. However, in the computation of Z_3 one last M is saved. This brings the total cost to 8 M + 3 C.

(iii) The formulae and relative costs for ADD, DBL, and TPL in modified Jacobian coordinates have been taken from [19].

Modified Jacobian coordinates are then the fastest system, as long as a field inversion is slow. In fact, according to [18] and [1], a field inversion costs more than ten field multiplications already for relatively small fields. Therefore, we shall use the modified Jacobian coordinate system in the sequel.

5 Further Improvements to Scalar Multiplication

Let us have another look at Algorithm 1: it will be the starting point for a few additional scalar multiplication methods.

Trading Frobenius Operations for Basis Conversions: If cubings are not *extremely inexpensive* or even essentially for free (such as with normal bases) they can easily become the dominant operation in Algorithm 1. In fact, there are $2 \cdot 3^w\, m$ of them.

Therefore one can envision the alternative approach of converting the coordinates of S to a normal basis representation before applying a power of τ (that in normal basis representation has nearly no computational cost) and then converting back to polynomial basis representation before adding $\zeta^\ell P$ to it. By means of this, $2 \cdot 3^w m$ cubings are replaced by two basis conversion per coordinate of S for each non-zero digit, i.e., about $8 \cdot \frac{2}{2w+1}\, m$ basis conversions (assuming modified Jacobian coordinates). However, there is a more efficient approach based on the same idea: Instead of applying powers of τ to S and adding P, we convert instead the base point P to normal basis representation and apply τ^i before converting it back, applying ζ^ℓ and adding the resulting point to S. This is Algorithm 2. Only 2 basis conversions at the beginning plus 2 more for each non-zero digit in the expansion are necessary, i.e., about $2 + 2 \cdot \frac{2}{2w+1}\, m$. This is similar to the method used in [4] for elliptic Koblitz curves in characteristic two. This approach can be advantageous, but only for relatively large values of m and w, since a basis conversion can be quite expensive: Whereas for characteristic two one such operation takes about the same time as one polynomial basis multiplication [15], in characteristic three the cost can be between two and three polynomial basis multiplications.

A Different Kind of Tradeoff: The biggest advantages of Algorithms 1 and 2 lie in their minimal memory requirements, but if cubings or basis conversions are not completely negligible, performance will not be their biggest strength. However, one can exploit the structure of the unit group in a more subtle way, to store only about $O(3^{w/2})$ precomputed points instead of $O(3^w)$ to perform a width-w windowed scalar multiplication.

In the notation of Section 3 and in particular of Equation (10), this idea consists in (i) precomputing and storing all the points $\phi^j(P)$ for $0 \leqslant j < 3^a$, and then (ii) using a double Horner scheme on double base representation $z = \sum_{i=0}^{m} \left(\varepsilon_i \phi^{f_i}\right) \psi^{g_i} \tau^i$ with bases τ and ψ, and digits $\varepsilon_i \phi^{f_i}$, in place of the triple Horner scheme of Algorithm 1. It is clear now how to write down the methods: Algorithms 3 and 4 on page 124 are the "square root sized digit set" variants of Algorithms 1 and 2, respectively.

Comparisons: We now compare the performance and memory consumption of these algorithms to other methods presented in the scientific literature. The results are summarized in Tables 1 and 2 on page 125. We now describe our approach to the comparisons:

(i) We consider here the simple τ-adic scalar multiplication from Koblitz [20], corresponding to a τ-adic 2-NAF with digit set $\{0\} \cup \langle\zeta\rangle$, the windowed

Algorithm 2. Low-memory τ-adic Scalar Multiplication on Koblitz Curves with basis conversions

INPUT: $P = (x, y) \in \mathcal{E}_{3,\mu}(\mathbb{F}_{3^m})$, scalar z represented as in Equation (10)
OUTPUT: zP

$Q \leftarrow 0$
$\widehat{P} \leftarrow \text{normal\_basis}(P)$
for $j = 3^a - 1$ **to** 0 **do**
  $Q \leftarrow \phi Q$ [skip first time]
  $R \leftarrow 0$
  for $k = 3^b - 1$ **to** 0 **do**
    $R \leftarrow \psi R$ [skip first time]
    $S \leftarrow 0$
    for $i = 0$ **to** $m - 1$ **do**
      if ($\varepsilon_i \neq 0$ **and** $f_i = j$ **and** $g_i = k$) **then**
        $S \leftarrow S + \varepsilon_i \, \text{polynomial\_basis}(\tau^i \widehat{P})$
    $R \leftarrow R + S$
  $Q \leftarrow Q + R$
return Q

method from [10] for $w \geqslant 3$, and our four algorithms. Note that we extend the method from [10] also to $w = 5$ (since, for large m, $w = 3$ or $w = 4$ are no longer optimal). For completeness we also report the operation counts for Smart's method [23] specialized to characteristic three.

(ii) Seven different field sizes $\mathbb{F}_{3^m}$ with $m = 97, 163, 193, 239, 509, 773$ and 1223, and two representations of the fields – normal basis and polynomial basis – are considered. The first five are fields already considered in the literature, the last two have been chosen to see how the methods scale with the field size, but are not necessarily tied to particular applications.

(iii) All the computational costs are expressed in field multiplications. The cost of a field inversion is taken to be equal to 15, 17, 20, 30, 40, 60 and 80 multiplications, respectively for the seven chosen values of m, and a cubing is equal to 0.15, 0.10, 0.09, 0.07, 0.045, 0.037 and 0.033 multiplications, respectively. *Whereas using normal bases a cubing is essentially for free, the cost cannot be ignored when using a polynomial basis, because of the cost of the reduction of a polynomial of degree up to* $3(m-1)$ *modulo the defining polynomial of the field extension.* These values are approximate distillates of the values found in [18,1] and of our own implementation experiments, and checked against Mitsunari's code [22].

(iv) For each scalar multiplication algorithm parametrized by a "window width" w, the cost corresponding to the optimal value of w is given.

(v) For all our algorithms the generators are chosen following the considerations in Remark 2: *For even* w, we take $\{\phi, \psi\} = \{-2, 1 + \mu\tau\}$ and from Lemma 1 we get $a = b = 3^{w/2-1}$. *For odd* w we take $\{\phi, \psi\} = \{-2, 1+\mu\tau^3\}$;

from Lemma 1 we know that $\mathrm{ord}_{\tau^w}(-2) = 3^{(w-1)/2}$, $\mathrm{ord}_{\tau^w}(1+\mu\tau^3) = 3^{(w-3)/2}$. If $\psi = -2$, then $a = (w-1)/2$, $b = (w-3)/2$, otherwise (i.e., if $\phi = -2$) $a = (w-3)/2$, $b = (w-1)/2$.
When different choices of the generators affect the performance, as in Algorithms 3 and 4, we make further case distinctions in the comparisons.

(vi) Memory consumption is given as the number of registers that are required for storing input-dependent points: The method from [10] needs to store the precomputed points other than the base point P itself, and uses an extra variable in the Horner scheme; Algorithm 1 needs to store Q, R and S and Algorithm 2 also storage $\widehat{P}$ and a copy of $\tau^i \widehat{P}$ in polynomial basis (cf. in Step 11); Algorithm 3 needs storage for the $3^a - 1$ points $\phi^j P$ with $j > 0$, as well as R, S; with respect to Algorithm 3, Algorithm 4 needs one register for P in normal basis as well, and one for the point converted in Step 7.
We do not consider the memory conversion matrices (that only apply to Algorithms 2 and 4) since they can be stored statically.

(vii) Algorithms 2 and 4 are not relevant for the normal basis comparison.

A comparison to expansions to the base of three, such as those in [18], seems due. A tripling requires twice as many cubings as a Frobenius operation. Since the density of a simple base-three expansion is 1/2 – higher than the 2/5 of Koblitz' espansion – the method is slower than Koblitz' τ-adic method. Similarly, their nonary method requiring 7 precomputations is slower than the method of Blake, Kumar, and Xu already for $w = 3$, with comparable memory requirements.

Double base chains with bases $(2, 3)$ such as those presented in [2] make sense when the doubling and tripling operation have both non-trivial costs. While computing the operation chain for a given scalar z one observes that it may end it with: (a) a doubling if $2 \mid z$; (b) a tripling if $3 \mid z$; (c) a doubling and an addition if $2 \nmid z$; or (d) a tripling and an addition/subtraction if $3 \nmid z$. The rest of the chain is the one associated to integers $z/2$, $z/3$, $(z \pm 1)/2$ and $(z \pm 1)/3$ respectively. Now, tripling in our case is always very efficient, but not doubling. Hence, options (c) and (d) are almost always more convenient than (a) also by virtue of of the faster reduction of the intermediate results. Therefore, double base chains almost always degenerate to base-three expansions, which we have just considered.

6 Conclusions and Final Remarks

It is clear from Tables 1 and 2 on page 125 that the new methods provide a substantial improvement w.r.t. the state of the art.

1. In the case of fields represented with a polynomial basis, we see that speedups are attained already for small curves. If $w = 97$, for instance For instance, the method from [10] is already beaten by Algorithm 2 with a much lower memory usage. For $m = 509$, we obtain similar or slightly better performance using Algorithms 3 and 4, but the memory reduction goes from a factor

Algorithm 3. Square-root memory usage τ-adic Scalar Multiplication on Koblitz Curves

INPUT: $P = (x, y) \in \mathcal{E}_{3,\mu}(\mathbb{F}_{3^m})$, scalar z represented as in Equation (10)
OUTPUT: zP

```
for j = 0 to 3^a − 1 do Precompute and store φ^j P
R ← 0
for k = 3^b − 1 to 0 do
    R ← ψR, S ← 0
    for i = m − 1 to 0 do
        S ← τS                                          [skip first time]
        if (ε_i ≠ 0 and g_i = k) then
            S ← S + ε_i(φ^{f_j} P)
                  [Use idea from Algo. 1, Steps 11–18 with (x, y) = φ^{f_j} P from table]
    R ← R + S
return R
```

Algorithm 4. Square-root memory usage Scalar Multiplication on Koblitz Curves with basis conversions

INPUT: $P = (x, y) \in \mathcal{E}_{3,\mu}(\mathbb{F}_{3^m})$, scalar z represented as in Equation (10)
OUTPUT: zP

```
for j = 0 to 3^a − 1 do Precompute and store normal_basis(φ^j P)
R ← 0
for k = 3^b − 1 to 0 do
    R ← ψR, S ← 0
    for i = 0 to m − 1 do
        if (ε_i ≠ 0 and g_i = k) then
            S ← S + ε_i polynomial_basis(τ^i(φ^{f_j} P))
    R ← R + S
return R
```

$2.25 = 27/12$ to $6.75 = 27/4$. For even larger fields, such as $m = 1223$, the method from [10] with $w = 5$ uses 81 registers and has similar performance to Algorithm 3, but the latter uses only 10 memory registers, which therefore is about one eighth than the previous state of the art. Speed improvements are often up to 7% for variable memory usage methods (but with reduced memory usage) to 24% for methods with fixed memory usage (i.e., then comparing our first two algorithms to Koblitz' algorithm).

We also note that whereas Algorithm 4 needs static storage for basis conversion matrices, these are not needed in Algorithm 3, that is usually just a bit slower and still faster than previous methods.

Table 1. Cost – expressed in field multiplications – and random access memory usage – expressed as the number of precomputed and intermediate points – of scalar multiplication on curves over fields represented in polynomial basis. In each entry the computational cost is above, memory usage and, if applicable, the value of w between parentheses, are below.

m	Previous methods			New methods					
	Smart [23]	Koblitz [20]	BMX [10] extended	Algo. 1	Algo. 2 $\psi=-2$	Algorithm 3 $\psi=-2$	Algorithm 3 $\phi=-2$	Algorithm 4 $\psi=-2$	Algorithm 4 $\phi=-2$
97	535.6	339.2	296.1	392.4	331.7	278.1	385.0	278.1	278.1
	1	1	9 (3)	3 (3)	5 (3)	4 (3)	2 (3)	6 (4)	6 (4)
163	854.4	533.5	436.5	547.5	494.2	418.5	493.3	408.8	432.0
	1	1	9 (3)	3 (3)	5 (4)	4 (3)	4 (4)	6 (4)	6 (5)
193	1001.9	624.1	503.6	621.3	567.9	483.6	553.5	468.7	492.1
	1	1	9 (3)	3 (3)	5 (4)	4 (3)	4 (4)	6 (5)	6 (5)
239	1211.8	748.7	595.5	705.3	679.2	572.5	616.9	566.0	566.0
	1	1	27 (3)	3 (3)	5 (4)	4 (3)	4 (4)	12 (6)	12 (6)
509	2509.0	1537.0	1035.4	1325.0	1300.5	1032.5	1115.0	1024.1	1024.1
	1	1	27 (4)	3 (3)	5 (5)	10 (5)	4 (4)	12 (6)	12 (6)
773	3775.2	2305.9	1528.4	1926.4	1830.4	1439.8	1597.8	1419.6	1472.9
	1	1	81 (5)	3 (3)	5 (5)	10 (5)	4 (4)	12 (6)	12 (7)
1223	5923.8	3608.0	2177.4	2930.4	2733.8	2113.5	2400.3	2111.6	2111.7
	1	1	81 (5)	3 (3)	5 (5)	10 (5)	4 (4)	30 (8)	30 (8)

Table 2. Computational costs and memory consumption as in Table 1 but when using a normal basis

m	Previous methods			New methods		
	Smart [23]	Koblitz [20]	BMX [10] extended	Algo. 1	Algorithm 3 $\psi=-2$	Algorithm 3 $\phi=-2$
97	449.2	264.6	230.0	207.1	183.9	183.9
	1	1	9 (3)	3 (3)	4 (4)	4 (4)
163	757.2	449.4	362.0	339.0	286.6	286.6
	1	1	9 (3)	3 (3)	4 (4)	4 (4)
193	897.2	533.4	410.2	388.2	333.2	316.6
	1	1	27 (4)	3 (3)	4 (4)	4 (4)
239	1111.8	662.2	487.7	466.5	399.2	375.2
	1	1	27 (4)	3 (4)	10 (5)	4 (5)
509	2371.8	1418.2	947.8	885.8	701.2	701.2
	1	1	27 (4)	3 (4)	10 (6)	10 (6)
773	3603.8	2157.4	1395.8	1246.8	985.5	985.5
	1	1	81 (5)	3 (5)	10 (6)	10 (6)
1223	5703.8	3417.4	2008.5	1819.6	1470.0	1410.0
	1	1	81 (5)	3 (5)	10 (6)	10 (7)

2. With a normal basis the improvements are even more impressive, going from 20% when $m = 97$ to 26% for $m = 509$ and then reaching nearly 30% for $m = 1223$, in all cases with vastly reduced memory usage.
3. If we compare methods with fixed memory consumption, we see that Algorithm 2 consistently outperforms the method of Koblitz, the speed up ranging from a few percent to 26.4% for $m = 509$ and even 46.7% for $m = 1223$ in the normal basis case (the price to pay being the usage of three intermediate registers in place of just one).

The techniques introduced in this paper therefore bring substantial speedups to scalar multiplication on supersingular Koblitz curves in characteristic three, at the same time reducing the memory footprint – by a factor roughly up to 8 in the examples we explicitly computed.

For extremely restricted environments, with no additional memory for code, the simple simple τ-adic method by Koblitz may of course still be preferable.

References

1. Ahmadi, O., Hankerson, D., Menezes, A.: Software Implementation of Arithmetic in $\mathbb{F}_{3^m}$. In: Carlet, C., Sunar, B. (eds.) WAIFI 2007. LNCS, vol. 4547, pp. 85–102. Springer, Heidelberg (2007)
2. Avanzi, R.M., Dimitrov, V.S., Doche, C., Sica, F.: Extending scalar multiplication using double bases. In: Lai, X., Chen, K. (eds.) ASIACRYPT 2006. LNCS, vol. 4284, pp. 130–144. Springer, Heidelberg (2006)
3. Avanzi, R.M., Heuberger, C., Prodinger, H.: On redundant τ-adic expansions and non-adjacent digit sets. In: Biham, E., Youssef, A.M. (eds.) SAC 2006. LNCS, vol. 4356, pp. 285–301. Springer, Heidelberg (2007)
4. Avanzi, R.M., Heuberger, C., Prodinger, H.: Redundant τ-adic Expansions I: Non-Adjacent Digit Sets and their Applications to Scalar Multiplication, Design, Codes and Cryptography (2010) (to appear)
5. Avanzi, R.M., Heuberger, C., Prodinger, H.: Arithmetic of Koblitz Curves in Characteristic Three (2010) (preprint)
6. Barreto, P.S.L.M., Kim, H.Y., Lynn, B., Scott, M.: Efficient Algorithms for Pairing-Based Cryptosystems. In: Yung, M. (ed.) CRYPTO 2002. LNCS, vol. 2442, pp. 354–368. Springer, Heidelberg (2002)
7. Beuchat, J.-L., Brisebarre, N., Detrey, J., Okamoto, E., Rodríguez-Henríquez, F.: A Comparison between Hardware Accelerators for the Modified Tate Pairing over F_{2^m} and F_{3^m}. In: Galbraith, S.D., Paterson, K.G. (eds.) Pairing 2008. LNCS, vol. 5209, pp. 297–315. Springer, Heidelberg (2008)
8. Beuchat, J.-L., López-Trejo, E., Martínez-Ramos, L., Mitsunari, S., Rodríguez-Henríquez, F.: Multi-core implementation of the tate pairing over supersingular elliptic curves. In: Garay, J.A., Miyaji, A., Otsuka, A. (eds.) CANS 2009. LNCS, vol. 5888, pp. 413–432. Springer, Heidelberg (2009)
9. Beuchat, J.-L., Shirase, M., Takagi, T., Okamoto, E.: An Algorithm for the η_T Pairing Calculation in Characteristic Three and its Hardware Implementation. In: ARITH 2007, pp. 97–104. IEEE Computer Society, Los Alamitos (2007)
10. Blake, I.F., Murty, V.K., Xu, G.: Efficient algorithms for Koblitz curves over fields of characteristic three. J. Discrete Algorithms 3(1), 113–124 (2005)

11. Brickell, E., Chen, L., Li, J.: A new direct anonymous attestation scheme from bilinear maps. In: Lipp, P., Sadeghi, A.-R., Koch, K.-M. (eds.) Trust 2008. LNCS, vol. 4968, pp. 166–178. Springer, Heidelberg (2008)
12. Cesena, E.: Trace Zero Varieties in Pairing-based Cryptography. Ph.D. Thesis, Università degli Studi Roma TRE (2010)
13. Chudnovsky, D.V., Chudnovsky, G.V.: Sequences of numbers generated by addition in formal groups and new primality and factorization tests. Advances in Applied Math. 7, 385–434 (1986)
14. Cohen, H., Miyaji, A., Ono, T.: Efficient elliptic curve exponentiation using mixed coordinates. In: Ohta, K., Pei, D. (eds.) ASIACRYPT 1998. LNCS, vol. 1514, pp. 51–65. Springer, Heidelberg (1998)
15. Coron, J.-S., M'Raïhi, D., Tymen, C.: Fast generation of pairs $(k,[k]P)$ for koblitz elliptic curves. In: Vaudenay, S., Youssef, A.M. (eds.) SAC 2001. LNCS, vol. 2259, pp. 151–164. Springer, Heidelberg (2001)
16. Freeman, D., Scott, M., Teske, E.: A Taxonomy of Pairing-Friendly Elliptic Curves. J. Cryptology 23(2), 224–280 (2010)
17. Halter-Koch, F.: Einseinheitengruppen und prime Restklassengruppen in quadratischen Zahlkörpern. Journal of Number Theory 4, 10–17 (1972)
18. Harrison, K., Page, D., Smart, N.: Software Implementation of Finite Fields of Characteristic Three, for Use in Pairing Based Cryptosystems. LMS Journal of Computation and Mathematics 5, 181–193 (2002)
19. Kim, K.-H., Nègre, C.: Point multiplication on supersingular elliptic curves defined over fields of characteristic 2 and 3. In: SECRYPT 2008. INSTICC Press, pp. 373–376 (2008)
20. Koblitz, N.: An elliptic curve implementation of the finite field digital signature algorithm. In: Krawczyk, H. (ed.) CRYPTO 1998. LNCS, vol. 1462, pp. 327–337. Springer, Heidelberg (1998)
21. Nakagoshi, N.: The structure of the multiplicative group of residue classes modulo $\mathfrak{p}^{N+1}$. Nagoya Mathematical Journal 73, 41–60 (1979)
22. Mitsunari, S.: A fast implementation of η_T pairing in characteristic three on intel processor. Cryptology ePrint Archive, report 2009/032 (2009)
23. Smart, N.: Elliptic Curve Cryptosystems over Small Fields of Odd Characteristic. J. Cryptology 12, 141–151 (1999)
24. Solinas, J.A.: Efficient arithmetic on Koblitz curves. Design, Codes and Cryptography 19, 195–249 (2000)

On the Correct Use of the Negation Map in the Pollard rho Method

Daniel J. Bernstein[1], Tanja Lange[2], and Peter Schwabe[2]

[1] Department of Computer Science
University of Illinois at Chicago, Chicago, IL 60607–7045, USA
djb@cr.yp.to
[2] Department of Mathematics and Computer Science
Technische Universiteit Eindhoven, P.O. Box 513, 5600 MB Eindhoven, Netherlands
tanja@hyperelliptic.org, peter@cryptojedi.org

Abstract. Bos, Kaihara, Kleinjung, Lenstra, and Montgomery recently showed that ECDLPs on the 112-bit secp112r1 curve can be solved in an expected time of 65 years on a PlayStation 3. This paper shows how to solve the same ECDLPs at almost twice the speed on the same hardware. The improvement comes primarily from a new variant of Pollard's rho method that fully exploits the negation map without branching, and secondarily from improved techniques for modular arithmetic.

Keywords: Elliptic curves, discrete-logarithm problem, negation map, branchless algorithms, SIMD.

1 Introduction

In July 2009, Bos, Kaihara, Kleinjung, Lenstra, and Montgomery announced breaking a discrete-logarithm problem on an elliptic curve over a 112-bit prime field using a cluster of 214 PlayStation 3 (PS3) game consoles. The initial announcement was [3]; more details on the same computation were published in [5], [4], and [6]. The overall attack costs were estimated to be about 60 PS3 years. This computation was, and still is, the largest ECDLP for which a successful solution has been publicly announced.

Our main result is that discrete-logarithm problems on the same curve (or any other curve of the form $y^2 = x^3 - 3x + b$ over the same field) can be solved at almost twice the speed on exactly the same hardware. We performed extensive computational experiments to verify our scalability and performance claims; details of these experiments appear in Section 6.

This work was supported by the National Science Foundation under grant ITR–0716498, by the European Commission under Contract ICT-2007-216499 CACE, and by the European Commission under Contract ICT-2007-216646 ECRYPT II. Computer time was provided by the Jülich and Barcelona supercomputer centers. Permanent ID of this document: dde51a91feeb8d746756566ac14323d1. Date: 2010.12.22. See http://cr.yp.to/papers.html#negation for the full version of this paper.

D. Catalano et al. (Eds.): PKC 2011, LNCS 6571, pp. 128–146, 2011.

This result combines several different optimizations. A large part of our work consists of faster algorithms for arithmetic modulo the prime $(2^{128} - 3)/76439$ that defines this field; these algorithms use a different approach from the previous papers, and we describe this approach in detail. We also introduce several smaller refinements in rho computations, often co-designing our choice of iteration function with our lower-level optimizations. However, the largest part of our improvement, gaining a factor of almost $\sqrt{2}$, comes from careful use of the negation map.

The conventional wisdom is that the negation map has been known for ten years and trivially gains a factor of $\sqrt{2}$. However, [3] said "We did not use the common negation map since it requires branching and results in code that runs slower in a SIMD environment." SIMD (single instruction, multiple data) is a critical feature of modern CPU designs, including the Cell processor used in the PS3.

Similarly, [5] said that the benefit of negation was outweighed by "the conditional branches required to check for fruitless cycles." The paper [6] observed that most of the negating rho algorithms stated in the literature were nonfunctional (i.e., had negligible chance of succeeding in the claimed amount of time); considered a huge array of 126 different combinations of negation options; and concluded that the best option did save time *for non-SIMD architectures*, but with a speedup far below $\sqrt{2}$. The paper continued to dispute the possibility of a negation speedup for SIMD architectures:

> If the Pollard rho method is parallelized in SIMD fashion, it is a challenge to achieve any speedup at all. ... Dealing with cycles entails administrative overhead and branching, which cause a non-negligible slowdown when running multiple walks in SIMD-parallel fashion. ... [This] is a major obstacle to the negation map in SIMD environments.

This paper resolves this dispute by explaining how to use the negation map without branching and without significant overhead handling cycles. We demonstrate a speedup very close to $\sqrt{2}$ on the PS3. We comment that the speedup on non-SIMD architectures would be even closer to $\sqrt{2}$.

2 Review of Pollard's rho Method

This section gives an overview of Pollard's rho method. In this paper we will use this method to compute discrete logarithms on elliptic curves but the first subsections apply to any finite cyclic group $G = \langle P \rangle$ and $Q \in G$. Computing the discrete logarithm of Q to the base P means computing an integer k such that $Q = kP$. The integer k is unique modulo ℓ where ℓ is the order of P; we assume for simplicity that ℓ is an odd prime.

The generic rho method. Pollard's rho method [17] is a low-memory algorithm that finds a discrete logarithm by finding a collision in the map

$$(a, b) \mapsto aP + bQ \qquad \text{where} \qquad a, b \in \mathbb{Z}.$$

Finding a collision usually reveals the discrete logarithm k of Q to the base P: if $aP + bQ = a'P + b'Q$ and $b \not\equiv b' \pmod{\ell}$ then $k \equiv (a - a')/(b' - b) \pmod{\ell}$.

A generic way to find this collision is to iterate this function. Define maps a and b from $\langle P \rangle$ to $\mathbb{Z}$ and compute $W_{i+1} = f(W_i) = a(W_i)P + b(W_i)Q$, starting from some initial combination $W_0 = a_0P + b_0Q$. If any W_i and W_j collide then also $W_{i+1} = W_{j+1}$, $W_{i+2} = W_{j+2}$, etc. This means that the sequence enters a cycle. This can be detected efficiently using, e.g., Floyd's cycle-finding method.

If the functions $a(R)$ and $b(R)$ are random modulo ℓ then the iterations perform a random walk in $\langle P \rangle$. The random walk can be modeled as drawing objects with repetition from an urn containing ℓ elements. A collision corresponds to drawing the same element twice. A standard birthday-paradox calculation implies that a collision occurs after approximately $\sqrt{\pi\ell/2}$ iterations on average.

Use of group automorphisms. If the functions a and b are chosen such that $f(W_i) = f(-W_i)$ then the walk is actually defined on equivalence classes under $\pm$. There are only $\lceil \ell/2 \rceil$ different classes. This reduces the average number of iterations by a factor of almost exactly $\sqrt{2}$.

More generally, Pollard's rho method can be combined with any easily computed group automorphism σ of small order. One chooses a and b so that $f(W_i) = f(\sigma(W_i)) = f(\sigma^2(W_i)) = \cdots$. The walk is then defined on equivalence classes under the automorphisms, reducing the average number of iterations accordingly. However, for most elliptic curves the only easily computed group automorphism of small order is the negation map.

The parallel rho method. To spread the computations in Pollard's rho method across multiple computers one replaces the long walk by a collection of short walks, as proposed by van Oorschot and Wiener in [16]. Some fixed subset of $\langle P \rangle$ is declared to be the *set of distinguished points*. Whenever a walk reaches a distinguished point it reports this to a central server and the server stores the distinguished point along with the values for a and b. If two walks reach the same distinguished point the server notices a collision.

This parallelization requires the server to receive, store, and sort all distinguished points. Tradeoffs are possible. If distinguished points are chosen to be rare then a small number of very long walks will be performed, reducing the number of distinguished points sent to the server but increasing the delay before a collision is recognized. If distinguished points are frequent then many shorter walks will be performed.

Additive walks. The generic rho method described above requires two scalar multiplications for each iteration. One can merge the two scalar multiplications into a 2-scalar multiplication, and further merge the 2-scalar multiplications across several parallel iterations, to reduce the number of group additions required for each iteration; but it is much simpler to use an *additive walk*, requiring only one addition for each iteration.

An additive walk is defined as $W_{i+1} = f(W_i) = W_i + R_{h(W_i)}$. Here h maps from $\langle P \rangle$ to $\{0, 1, \ldots, r-1\}$, and $R_0, R_1, \ldots, R_{r-1}$ are precomputed as known combinations of P and Q: say $R_j = c_jP + d_jQ$ for each $j \in \{0, 1, \ldots, r-1\}$.

One then also knows that $W_i = a_iP + b_iQ$, where a and b are defined recursively as follows: $a_{i+1} = a_i + c_{h(W_i)}$ and $b_{i+1} = b_i + d_{h(W_i)}$.

Additive walks have disadvantages. The walks are noticeably nonrandom and need more iterations than the generic rho method to find a collision. This effect disappears as r grows, but if r is large then the precomputed table $R_0, \ldots, R_{r-1}$ does not fit into fast memory. Additive walks also have trouble with automorphisms; see the discussion of fruitless cycles below.

Pollard's original proposal was to use $r = 3$. Instead of 3 different precomputed points, Pollard mixed 2 points with a doubling $W_{i+1} = 2W_i$; note that Pollard was computing discrete logarithms in multiplicative groups, where a doubling (i.e., a squaring) is faster than a general addition (i.e., a multiplication). Experiments by Teske in [21] showed that larger values of r, such as $r = 20$, are much closer to random walks. A general heuristic due to Brent and Pollard implies that nonrandomness slows down this type of walk by a factor $\sqrt{1 - 1/r}$; for further discussion of such heuristics see, e.g., [1].

Eliminating coefficients. In all of the discrete-logarithm computations described above, coefficients a_i, b_i are stored and used to compute the final discrete logarithm. In the parallel rho method these coefficients are communicated along with each distinguished point sent from a client to the server. Computing these coefficients in additive walks requires each client to implement arithmetic modulo ℓ or at least to allocate space for counters to keep track of how often each R_j is used.

An alternative approach, introduced recently as part of the ECC2K-130 attack [1], eliminates the coefficients a_i and b_i. Clients compute W_i without keeping track of a_i and b_i. When a client encounters a distinguished point W_i, it reports the point (or a hash of the point) along with a seed identifying the start of the walk, and then starts a new walk. This saves code in the clients, storage in the clients, communication between the clients and the server, and storage on the server. When the server encounters a collision, it recomputes the two walks involved in the collision, starting from the same seeds but now computing a_i and b_i. This is done only rarely, ideally just once.

Fruitless cycles. Fast negation on elliptic curves reduces the number of iterations in the generic rho method by a factor $\sqrt{2}$, as discussed above. However, a non-negating additive walk is faster than a negating generic walk: the extra speed of each iteration in the non-negating additive walk outweighs the smaller number of iterations in the negating generic walk.

One might think that a negating additive walk combines the advantages of a small number of iterations and a very fast iteration function. It is easy to make a canonical choice $|W_i|$ between W_i and $-W_i$; to define $h(W_i)$ as a function of $|W_i|$, so that $h(W_i) = h(-W_i)$; and to define $f(W_i) = |W_i| + R_{h(W_i)}$. Then f is a walk on equivalence classes under $\pm$, and computing f takes only one addition.

The problem is that a negating additive walk does not behave randomly: it quickly enters short fruitless cycles. For example, if $|W_{i+1}| = -W_{i+1}$ and $h(W_{i+1}) = h(W_i)$ then $W_{i+2} = f(W_{i+1}) = -W_{i+1} + R_{h(W_i)} = -(|W_i| + R_{h(W_i)}) + R_{h(W_i)} = -|W_i|$ so $|W_{i+2}| = |W_i|$. One expects this to occur with probability $1/(2r)$ at each step, and if it does occur then the walk enters a 2-cycle. Subsequent iterations will not proceed to a distinguished point; subsequent computations will be wasted.

It is also possible to have fruitless cycles of larger lengths, although these are less frequent. Heuristics are given in [9]. Choosing a larger r reduces the chance of entering fruitless cycles, but for large-scale computations the cycles will occur and will occur frequently.

There are many different attempts in the literature to work around this problem. The first proposals were introduced independently by Harley and Singer in [12], Wiener and Zuccherato in [23], and Gallant, Lambert, and Vanstone in [10]; further analyses appeared in [9] and much more recently in [6]. A detailed review of these proposals appears in the full version of this paper. Some of the proposals are successful in eliminating fruitless cycles, but all of these proposals involve frequent conditional operations and, as stated in [6], perform poorly in a SIMD environment.

3 How to Use Negation in Pollard's rho Method

This section presents an efficient branchless negating rho algorithm to compute elliptic-curve discrete logarithms. For simplicity we restrict to curves of the form $y^2 = x^3 - 3x + b$ over large prime fields $\mathbb{F}_p$.

This algorithm uses, on average, $(1 + o(1))\sqrt{\pi\ell/4}$ iterations for a group of prime order ℓ, assuming standard heuristics; here $o(1)$ means something that converges to 0 as $\ell \to \infty$. Each iteration uses 5 multiplications mod p, 1 squaring mod p, and an asymptotically negligible amount of extra work.

We emphasize that we use a *branchless* sequence of iterations, always performing the same operations in the same order. This is of theoretical interest: any bounded-time algorithm can be made branchless by standard conversions, but these conversions usually lose efficiency. This is also of practical interest: the algorithm is well suited for modern SIMD CPUs such as the Cell CPU in the PS3, as discussed in subsequent sections of the paper.

We also emphasize that the number of iterations is $(1 + o(1))\sqrt{\pi\ell/4}$, not $(1 + o(1))\sqrt{\pi\ell/2}$. We use the fast elliptic-curve negation map to save a factor of $\sqrt{2}$; we do this without branching, and we do it with asymptotically zero compromise in iteration speed.

Eliminating fruitless cycles. We begin with the simplest type of negating additive walk stated in the literature. The walk starts at the point $W_0 = |b_0Q|$ where b_0 is chosen randomly, and then computes $W_1, W_2, \ldots$ by the rule $W_{i+1} = |W_i + R_{h(W_i)}|$. Here $|(x, y)|$ means (x, y) for $y \in \{0, 2, 4, \ldots, p - 1\}$ or $(x, -y)$ for $y \in \{1, 3, 5, \ldots, p - 2\}$; the hash function h maps points to elements of $\{0, 1, 2, \ldots, r - 1\}$; and the points $R_0, R_1, \ldots, R_{r-1}$ are known multiples of P.

We modify this walk by *occasionally* checking for fruitless cycles of length 2. Specifically, for a sparse pattern of indices i discussed below, we change the definition of W_i as follows. After computing W_{i-1}, we check whether $W_{i-1} = W_{i-3}$.

In the common case that $W_{i-1} \neq W_{i-3}$, we define $W_i = W_{i-1}$. In the unusual case that $W_{i-1} = W_{i-3}$, we define $W_i = |2\min\{W_{i-1}, W_{i-2}\}|$, where min means lexicographic minimum and 2 means doubling.

We further modify this walk by occasionally, with even lower frequency, checking for fruitless cycles of length 4. Specifically, for an even more sparse pattern of indices i discussed below, we redefine W_i as W_{i-1} if $W_{i-1} \neq W_{i-5}$, and we redefine W_i as $|2\min\{W_{i-1}, W_{i-2}, W_{i-3}, W_{i-4}\}|$ if $W_{i-1} = W_{i-5}$.

We continue with analogous modifications for fruitless cycles of lengths 6, 8, etc., up to the smallest even length that exceeds $(\log \ell)/(\log r)$.

Eliminating branches. The sequence of iterations described above might seem to include branches: a branch to replace y by $-y$ if y is odd, for example, and a branch to conditionally compute $W_i = |2\min\{W_{i-1}, W_{i-2}\}|$. However, one can easily simulate all of these branches by a straight-line program with negligible loss of efficiency, as described in the following paragraphs.

First, $|(x, y)|$ is the same as $(x, (1-2\epsilon)y)$ where $\epsilon = y \bmod 2$. The implicit reduction modulo p here is not an asymptotic bottleneck: it takes linear time (even without branching), while all known multiplication algorithms take superlinear time. We prefer to make the implicit reduction explicit, computing $|(x, y)|$ as $(x, y + \epsilon(p - 2y))$; the addition and subtraction take linear time.

Second, we amortize min computations such as $\min\{W_{i-1}, W_{i-2}, W_{i-3}, W_{i-4}\}$ across all relevant iterations: after computing W_{i-3} we initialize a running minimum $W_{\min}$ as $\min\{W_{i-4}, W_{i-3}\}$, then replace it with $\min\{W_{\min}, W_{i-2}\}$ after computing W_{i-2}, then replace it with $\min\{W_{\min}, W_{i-1}\}$ after computing W_{i-1}. These computations are performed for only a small fraction of all indices i, so the loss of efficiency is negligible. See below for a more detailed cost analysis.

Third, we compute doublings such as $D_i = 2\min\{W_{i-1}, W_{i-2}, W_{i-3}, W_{i-4}\} = 2W_{\min}$ for all of the selected indices, whether or not the doublings will actually be used. We then compute W_i without branches by selecting between W_{i-1} and $|D_i|$, the same way that $|(x, y)|$ selects between (x, y) and $(x, -y)$. The selection bit is the output of a branch-free comparison between W_{i-1} and W_{i-5}, or in general between W_{i-1} and W_{i-1-c} for detecting fruitless cycles of length c.

Note that each of these selections and comparisons takes linear time per iteration, and is therefore asymptotically negligible compared to a multiplication.

Eliminating inversions. The bottleneck in each iteration is now exactly one elliptic-curve operation: usually an elliptic-curve addition, but an elliptic-curve doubling for occasional iterations.

The standard formulas for elliptic-curve addition in affine coordinates are as follows. Let $P = (x_1, y_1), R = (x_2, y_2)$ with $x_1 \neq x_2$. Then $P + R = (x_3, y_3)$ where $\lambda = (y_1 - y_2)/(x_1 - x_2)$, $x_3 = \lambda^2 - x_1 - x_2$, and $y_3 = \lambda(x_1 - x_3) - y_1$. These formulas use 1 inversion, 2 multiplications, 1 squaring, and 6 subtractions. The formulas for a doubling, where $R = P$, are very similar; only the computation of $\lambda = 3(x_1^2 - 1)/(2y_1)$ is different. We ignore, without further comment, the extraordinarily unlikely event that $R = -P$.

Inversions are the most expensive operations in finite fields. Standard practice in rho computations is to perform m independent walks in parallel, and to use

Montgomery's trick [15], which computes a batch of m inversions using just 1 inversion and $3(m-1)$ multiplications.

We choose a branchless inversion method, specifically computing the $(p-2)$nd power using $O(\lg p)$ multiplications. We then choose m to grow asymptotically more quickly than $\lg p$: in other words, $\lg p \in o(m)$. A batch of m elliptic-curve additions then costs $5m-3$ multiplications, m squarings, $6m$ subtractions, and 1 inversion. The subtractions and inversion are negligible, so each elliptic-curve addition costs 5 multiplications and 1 squaring.

A batch of m elliptic-curve doublings is slightly more expensive (costing m extra squarings), but occurs for only a small fraction of iterations, as discussed below. Each iteration therefore uses 5 multiplications and 1 squaring.

Analysis and optimization. Fruitless cycles of length 2 appear with probability approximately $1/(2r)$. These cycles persist after they appear, wasting subsequent iterations (in the sense that new points and new collision opportunities do not occur), until we check for them. If we check every w iterations then we expect a cycle to appear with probability approximately $w/(2r)$, and for it to waste approximately $w/2$ iterations on average if it does appear.

This does not mean that w should be chosen as small as possible. If a cycle has *not* appeared then checking for it wastes an iteration. The overall loss is approximately $1+w^2/(4r)$ iterations out of w. To minimize the quotient $1/w+w/(4r)$ we take $w \approx 2\sqrt{r}$.

More generally, fruitless cycles of small length $2c$ appear with probability approximately proportional to $1/r^c$, so the optimal checking frequency is approximately proportional to $1/r^{c/2}$. The loss here rapidly disappears as c increases.

To summarize, fruitless cycles slow down this algorithm by a factor $1+\Theta(1/\sqrt{r})$. This negation overhead $\Theta(1/\sqrt{r})$ is on a larger scale than the overhead $\Theta(1/r)$ from the nonrandomness of r-adding walks, but both overheads become asymptotically negligible if r is chosen so that $r \to \infty$ as $p \to \infty$.

As an illustration of these optimizations, our PS3 software takes $r = 2048$, checks for 2-cycles every 48 iterations, and checks for larger cycles much less frequently. To simplify the software we unify the checks for 4-cycles and 6-cycles into a check for 12-cycles every 49152 iterations. If we had instead taken $r = 512$ then we would have checked for 2-cycles every 24 iterations. In general, the $\Theta(1/\sqrt{r})$ asymptotic means that the negation overhead approximately doubles when the table size is reduced by a factor of 4.

Storage reduction. The storage overhead for detecting and escaping a fruitless cycle consists of storing $W_{\min}$ and W_{i-1-c}. For the latter it is enough to store one of the coordinates.

We further reduce storage by avoiding having all iterations check for cycles at the same time. For example, with a batch of 224 iterations running in parallel, we have just 14 iterations checking for 2-cycles and consuming extra space. All iterations perform 2 addition steps, and then these 14 iterations perform a masked doubling while the remaining 210 iterations perform another addition. We then rotate the batch so that the next 14 iterations check for 2-cycles.

4 Low-Cost Arithmetic in $\mathbb{Z}/(2^{128} - 3)$

Elements of the prime field $\mathbb{F}_p$, where $p = (2^{128} - 3)/76439$, can be represented redundantly as elements of the ring $\mathbb{Z}/(2^{128} - 3)$. Instead of reducing sums and products modulo p one reduces them modulo $2^{128} - 3$. This representation requires about 15% more space per field element, but the sparsity of $2^{128} - 3$ makes reductions much faster.

The prime p was chosen with this small sparse multiple precisely to allow this speedup for cryptographic operations on the secp112r1 curve; see [7, page 3, bottom, and page 6, bottom]. The same type of redundant representation can of course also be used by the cryptanalyst attacking the ECDLP on secp112r1, as in [3], [5], [4], and this paper. The critical problem is then to perform fast arithmetic modulo $2^{128} - 3$.

This section explains how to efficiently decompose multiplications and squarings modulo $2^{128} - 3$ into operations on 16-bit integers and 32-bit integers. Here a *b-bit integer* is an integer in the interval $[-2^{b-1}, 2^{b-1} - 1]$. The next section applies this decomposition to the Cell, obtaining faster arithmetic than [3] et al.

Model of computation. This section uses a simplified model of computation that counts $16 \times 16 \to 32$ multiplications and certain other operations. Specifically, algorithms in this section are branchless sequences of the following operations:

- multiplication: $a, b \mapsto ab$ where a, b are 16-bit integers and ab is a 32-bit integer;
- multiply-add: $a, b, c \mapsto ab + c$ where a, b are 16-bit integers and $c, ab + c$ are 32-bit integers;
- 16-bit addition: $a, b \mapsto a + b$ where $a, b, a + b$ are 16-bit integers;
- 32-bit addition: $a, b \mapsto a + b$ where $a, b, a + b$ are 32-bit integers;
- 32-bit subtraction: $a, b \mapsto a - b$ where $a, b, a - b$ are 32-bit integers;
- 32-bit right shift by 12 bits: $a \mapsto \lfloor a/2^{12} \rfloor$ where a is a 32-bit integer;
- 32-bit right shift by 13 bits: $a \mapsto \lfloor a/2^{13} \rfloor$ where a is a 32-bit integer;
- 32-bit mask clearing 12 bits: $a \mapsto 2^{12} \lfloor a/2^{12} \rfloor$ where a is a 32-bit integer; and
- 32-bit mask clearing 13 bits: $a \mapsto 2^{13} \lfloor a/2^{13} \rfloor$ where a is a 32-bit integer.

We assign cost 1 to each of these operations, except that we assign cost 0.5 to the 16-bit addition operation. The next section explains how this cost model is related to PS3 speed.

Note that these operations are not defined for all pairs of inputs. For example, 32-bit addition is not permitted to add 2^{30} to 2^{30}, because 2^{31} is too large to be a 32-bit integer. One could of course define an extended 32-bit addition operation that handles all cases, working modulo 2^{32} to handle overflows, but there are no overflows in the algorithms in this section. One could also define shifts (and masks) for distances other than 12 and 13, but the only distances used in this section are 12 and 13.

Representing integers modulo $2^{128} - 3$. We represent an element f of the ring $\mathbb{Z}/(2^{128} - 3)$ as a sequence of 10 coefficients $(f_0, \ldots, f_9)$ such that $f = \sum_{0 \le i \le 9} f_i 2^{\lceil i \cdot 12.8 \rceil}$; i.e.,

$$f = f_0 + f_1 2^{13} + f_2 2^{26} + f_3 2^{39} + f_4 2^{52} + f_5 2^{64} + f_6 2^{77} + f_7 2^{90} + f_8 2^{103} + f_9 2^{116}.$$

Note that the exponents $0, 13, 26, 39, 52, 64, 77, 90, 103, 116$ are not exactly evenly spaced. Our non-integer radix $2^{12.8}$ follows the use of radix $2^{25.5}$ by Bernstein in [2], and follows ideas from Costigan and Schwabe in [8] on making best use of SIMD instructions.

We call a coefficient f_i *reduced* if $|f_i| \le 1.01 \cdot 2^{12}$. We call $(f_0, \ldots, f_9)$ *reduced* if all coefficients are reduced.

Polynomial multiplication and polynomial reduction. It is easy to check that if $f = \sum_{0 \le i \le 9} f_i 2^{\lceil i \cdot 12.8 \rceil}$ and $g = \sum_{0 \le i \le 9} g_i 2^{\lceil i \cdot 12.8 \rceil}$ then the product fg equals $\sum_{0 \le i \le 9} r_i 2^{\lceil i \cdot 12.8 \rceil}$ in $\mathbb{Z}/(2^{128} - 3)$ where

$$\begin{aligned}
r_0 &= f_0g_0 + 3(2f_9g_1 + 2f_8g_2 + 2f_7g_3 + 2f_6g_4 + f_5g_5 + 2f_4g_6 + 2f_3g_7 + 2f_2g_8 + 2f_1g_9),\\
r_1 &= f_1g_0 + f_0g_1 + 3(2f_9g_2 + 2f_8g_3 + 2f_7g_4 + f_6g_5 + f_5g_6 + 2f_4g_7 + 2f_3g_8 + 2f_2g_9),\\
r_2 &= f_2g_0 + f_1g_1 + f_0g_2 + 3(2f_9g_3 + 2f_8g_4 + f_7g_5 + f_6g_6 + f_5g_7 + 2f_4g_8 + 2f_3g_9),\\
r_3 &= f_3g_0 + f_2g_1 + f_1g_2 + f_0g_3 + 3(2f_9g_4 + f_8g_5 + f_7g_6 + f_6g_7 + f_5g_8 + 2f_4g_9),\\
r_4 &= f_4g_0 + f_3g_1 + f_2g_2 + f_1g_3 + f_0g_4 + 3(f_9g_5 + f_8g_6 + f_7g_7 + f_6g_8 + f_5g_9),\\
r_5 &= f_5g_0 + 2f_4g_1 + 2f_3g_2 + 2f_2g_3 + 2f_1g_4 + f_0g_5 + 3(2f_9g_6 + 2f_8g_7 + 2f_7g_8 + 2f_6g_9),\\
r_6 &= f_6g_0 + f_5g_1 + 2f_4g_2 + 2f_3g_3 + 2f_2g_4 + f_1g_5 + f_0g_6 + 3(2f_9g_7 + 2f_8g_8 + 2f_7g_9),\\
r_7 &= f_7g_0 + f_6g_1 + f_5g_2 + 2f_4g_3 + 2f_3g_4 + f_2g_5 + f_1g_6 + f_0g_7 + 3(2f_9g_8 + 2f_8g_9),\\
r_8 &= f_8g_0 + f_7g_1 + f_6g_2 + f_5g_3 + 2f_4g_4 + f_3g_5 + f_2g_6 + f_1g_7 + f_0g_8 + 3(2f_9g_9),\\
r_9 &= f_9g_0 + f_8g_1 + f_7g_2 + f_6g_3 + f_5g_4 + f_4g_5 + f_3g_6 + f_2g_7 + f_1g_8 + f_0g_9.
\end{aligned}$$

The factors of 2 arise from the uneven exponent spacing mentioned above: for example, the product of $f_1 2^{13}$ and $g_4 2^{52}$ is $2f_1g_4 2^{64}$, contributing $2f_1g_4$ to $r_5 2^{64}$. The factors of 3 arise from reducing 2^{128} in $\mathbb{Z}/(2^{128} - 3)$.

If $(f_0, \ldots, f_9)$ and $(g_0, \ldots, g_9)$ are reduced then a sum of any 10 products of the form f_ig_j can be computed with cost 10: specifically, 1 multiplication followed by 9 multiply-add operations in any convenient order. The sums $r_0, r_1, \ldots, r_9$ are slightly more expensive because of the extra factors of 2 and 3. We precompute $3g_1, 3g_2, \ldots, 3g_9$ and $2f_1, 2f_2, 2f_3, 2f_4, 2f_6, 2f_7, 2f_8, 2f_9$ (skipping $2f_5$); recall that a 16-bit addition costs only 0.5, so these 26 additions cost only 13. Each r_i is then a sum of 10 products of known 16-bit quantities, costing 100, for a total cost of 113.

Each r_i, and each intermediate result in this multiplication algorithm, is bounded in absolute value by $10 \cdot 3 \cdot 2 \cdot (1.01 \cdot 2^{12})^2 < 0.96 \cdot 2^{30}$. The same algorithm also works if $(f_0, \ldots, f_9)$ is a sum or difference of two reduced vectors while $(g_0, \ldots, g_9)$ is reduced: then each result is bounded in absolute value by $0.96 \cdot 2^{31}$, still safely below 2^{31}.

Coefficient reduction. The product coefficients $r_0, r_1, \ldots, r_9$ constructed above are usually not reduced. Some extra work is required to compute a reduced product suitable for use as input to subsequent multiplications.

For $i \in \{0,1,2,3,5,6,7,8\}$ we reduce r_i by *carrying* from r_i to r_{i+1}. This means changing (r_i, r_{i+1}) into $(r_i - 2^{13}c, r_{i+1} + c)$ where $c = \lfloor (r_i + 2^{12})/2^{13} \rfloor$. This leaves the sum $r_i 2^{\lceil i \cdot 12.8 \rceil} + r_{i+1} 2^{\lceil (i+1) \cdot 12.8 \rceil}$ unaffected, and increases the maximum possible r_{i+1} only slightly, while guaranteeing that the new r_i is between -2^{12} and 2^{12}. This costs 5: an addition, a right shift, a mask, another addition, and a subtraction.

We similarly reduce r_4 by carrying from r_4 to r_5. This has a slightly different definition, to accommodate the uneven spacing of exponents: it means changing (r_4, r_5) into $(r_4 - 2^{12}c, r_5 + c)$ where $c = \lfloor (r_4 + 2^{11})/2^{12} \rfloor$. This guarantees that the new r_4 is between -2^{11} and 2^{11}.

The most expensive step is to reduce r_9 by carrying from r_9 to r_0. This means changing (r_9, r_0) into $(r_9 - 2^{12}c, r_0 + 3c)$ where $c = \lfloor (r_9 + 2^{11})/2^{12} \rfloor$. This leaves $r_0 + r_9 2^{116}$ unaffected modulo $2^{128} - 3$, while guaranteeing that the new r_9 is between -2^{11} and 2^{11}. This costs 7 rather than 5: the computation of $3c$ uses two extra additions.

We use the carry chain $r_0 \to r_1 \to r_2 \to r_3 \to r_4 \to r_5 \to r_6 \to r_7 \to r_8 \to r_9 \to r_0 \to r_1 \to r_2$: we first carry from r_0 to r_1, then from r_1 to r_2, etc., then from r_9 to r_0, then again from r_0 to r_1, then again from r_1 to r_2. Tracing the bounds on each r_i shows that the final r_2 is reduced, and therefore that $(r_0, r_1, \ldots, r_9)$ is reduced.

This carry chain costs 62. The complete multiplication algorithm, taking reduced representations of f and g as input and producing a reduced representation of fg as output, costs 175.

Squaring. Squaring in $\mathbb{Z}/(2^{128} - 3)$ is very similar to multiplication except that several intermediate results are reused: for example, $f_1 g_0 + f_0 g_1$ becomes just $2f_0 f_1$. We begin with a cost-10 computation of $2f_0, \ldots, 2f_9$; $3f_5, \ldots, 3f_9$; $6f_5, \ldots, 6f_9$. We then obtain $r_0, \ldots, r_9$ straightforwardly at cost 55, and apply the same carry chain as for multiplication. The complete squaring algorithm costs 127.

5 Fast Iterations on the PlayStation 3

This section analyzes and optimizes the performance of rho iterations modulo $p = (2^{128} - 3)/76439$ on the PS3. This optimization makes some changes to the iteration function; we take advantage of the flexibility of co-designing our iteration function with our arithmetic algorithms.

The Cell SPEs. The CPU in the PS3 is the Cell Broadband Engine. The main computational power of the Cell is in 8 Synergistic Processor Elements (SPEs). These SPEs are arranged around a central 64-bit PowerPC core. The PS3 makes only 6 of these SPEs available for computations.

We report the performance that we achieve with 6 SPEs, leaving the central PowerPC core mostly idle. This does not make full use of the Cell—performing some iterations on the PowerPC core would noticeably reduce the overall computation time—but it simplifies comparisons: the speeds reported for the same ECDLP in [3], [5], and [4] also left the central PowerPC core mostly idle.

Our implementation runs independently on each SPE. Each SPE has 256 KB of fast local storage; this storage holds all code and data, including a batch of parallel walks and a table of precomputed points. Communication between the SPE and main memory is very small in this algorithm: a few words of data for every million iterations. Performance is therefore determined almost completely by how well we make use of the computational power of the SPE.

Arithmetic on the SPE. The following description summarizes only the SPE features that are relevant to our implementation. See [13] and [19] for more information about the SPE.

The SPE has a register file consisting of 128 128-bit vector registers. Typical arithmetic instructions are SIMD instructions operating on these registers as vectors of 4 independent 32-bit integers or vectors of 8 independent 16-bit integers. There is a multiplication instruction that multiplies 4 pairs of 16-bit integers in parallel, producing 4 32-bit integers. There is also a fused multiply-add instruction that adds the results into another vector of 4 32-bit integers.

Each SPE cycle carries out at most one of these arithmetic instructions: i.e., 4 32-bit operations or 8 16-bit operations. An algorithm that costs n in the model of Section 4 therefore uses at least $n/4$ cycles on the SPE. There are, however, several reasons that the SPE can take many more cycles than this:

- In-order execution. An arithmetic instruction must wait until 1 cycle after the previous arithmetic instruction in the program.
- Arithmetic latency. An instruction cannot begin until its results are ready. The result of an arithmetic instruction is not ready until several cycles later.
- Load latency. Loads are handled by a separate instruction pipeline but can still delay arithmetic instructions that use the load results.

There are also various function-call overheads, typically consuming 70 cycles per function call. One can eliminate these overheads by inlining and merging functions, but this also increases code size, putting pressure on the SPE's local storage.

Digitsliced multiplication on the SPE. We use 8-way vectorization of our iterations: we repeat our inputs, computations, and outputs 8 independent times. We store 8 independent elements of the ring $\mathbb{Z}/(2^{128} - 3)$ in 10 128-bit vector registers $r_0, \ldots, r_9$, where coefficient i of ring element j is in 16-bit component j of register r_i. We convert each 16-bit operation into a 128-bit vector operation, and we convert each 32-bit operation into two 128-bit vector operations.

Scheduling instructions carefully then works around all arithmetic latency. The multiplication algorithm fits comfortably into the SPE registers: loads and stores are not a bottleneck. (Replacing 8-way vectorization with 16-way vectorization would remove the need for careful instruction scheduling but would put more pressure on registers and, more importantly, would cut in half the number of walks that fit easily into local storage without reshuffling.) The 32-bit results at the end of the algorithm are known to be reduced, so they fit into 16 bits; 10 extra instructions are required to shuffle the results into 10 vectors of 8 16-bit integers, but these extra instructions are handled by the SPE's load/store pipeline and are effectively free when they are interleaved with arithmetic instructions.

Overall our software for vectorized multiplication of 8 pairs of ring elements takes exactly 350 SPE arithmetic instructions, i.e., 43.75 arithmetic instructions per multiplication (1/4 of the cost 175 in the previous section). Our software for vectorized squaring of 8 ring elements uses exactly 254 SPE arithmetic instructions, i.e., 31.75 arithmetic instructions per squaring. Each of our iterations performs 5 multiplications and 1 squaring, in total consuming 250.5 arithmetic instructions, at least 250.5 cycles.

Inversion. To invert modulo $p = (2^{128}-3)/76439$ we simply compute a $(p-2)$nd power. Our addition chain for $p-2$ uses 107 squarings and 32 multiplications. Essentially the same speed would also be achieved by an addition chain for the larger but sparser exponent $2^{128} - 76443 = p - 2 + 76439(p-1)$, using 126 squarings and just 18 multiplications.

Of course, we actually perform 8 independent inversions in parallel, using $8 \cdot 107$ squarings and $8 \cdot 32$ multiplications modulo $2^{128} - 3$. Each inversion uses $107 \cdot 31.75 + 32 \cdot 43.75 = 4797.25$ arithmetic instructions, consuming at least 4797.25 cycles.

Our 8-inversion function actually uses 43293 cycles, i.e., 5411.625 cycles per inversion. The gap is almost entirely explained by the overhead of 64 function calls, half to multiplication and half to an n-squaring function that computes $r \leftarrow r^{2^n}$ for variable n.

To reduce code size we rejected the possibility of more complicated Euclid-type inversion as used in [5]. In context (see below) inversion is already quite fast, only 6.6% of our final iteration cost.

Canonicalizing the y-coordinate. Redundant representations cause trouble for two parts of the algorithm stated in Section 3. First, because $y \in \mathbb{F}_p$ has multiple representations, checking whether $y \in \{0, 2, 4, \ldots, p-1\}$ is not a simple matter of inspecting the bottom bit of the representation of y. Second, because $x \in \mathbb{F}_p$ has multiple representations, finding the hash of x is not a simple matter of extracting bits from the representation of x.

We address both problems by canonicalizing y. We use the canonicalized version of y to decide whether to negate y. Rather than canonicalizing and hashing x, we extract some bits from the canonicalized version of y as a table index. Note that there can be as many as 3 points having the same y-coordinate, but hashing all of those points to the same table index does not merge walks. We also use the canonicalized version of y to define distinguished points.

The most obvious way to canonicalize y is to replace it with $y \bmod p$; but reductions modulo p are expensive. We instead compute $s = 76439y \bmod (2^{128} - 3)$. One can think of this s as a unique representative $y \bmod p$, but represented as $76439(y \bmod p)$. An alternative is to use Montgomery reduction to compute $y \cdot 2^{-16} \pmod p$ with precomputed $2^{-16} \pmod p$, as in [4].

To compute s we multiply y by the cofactor 76439 and then perform a slightly longer reduction chain than the one we use after multiplication. The polynomial-multiplication step here, producing unreduced coefficients $s_0, s_1, \ldots, s_9$, uses only

5 arithmetic instructions (cost 20) instead of 25 (cost 100), because 76439 is represented in just two reduced coefficients: $76439 = 2711 + 9 \cdot 2^{13}$. The usual precomputations of $3g_1$, $2f_1$, etc. also disappear: all we need are the constants $5422 = 2 \cdot 2711$ and $16266 = 6 \cdot 2711$.

To reduce the result we carry $s_6 \to s_7 \to s_8 \to s_9 \to s_0 \to s_1 \to s_2 \to s_3 \to s_4 \to s_5 \to s_6 \to s_7 \to s_8 \to s_9 \to s_0 \to s_1$. This uses 19.75 arithmetic instructions. The overall cost of canonicalization is 24.75 arithmetic instructions per iteration.

We do not claim that the resulting $(s_0, s_1, \ldots, s_9)$ is *always* completely determined by the image of y in $\mathbb{F}_p$. What matters is that the probability of non-uniqueness for random y's is so small that two colliding walks have negligible probability of diverging before they hit a distinguished point. We computer-verified this as explained in Section 6.

Subtraction. The easy part of subtracting two ring elements is performing 10 subtractions of 16-bit integers. The problem is that the output is usually not reduced. Carries would reduce the output but would make subtraction much more expensive.

We use two standard techniques to avoid carries after subtractions. First, we skip unnecessary reductions before multiplications; recall from Section 4 that the multiplication algorithm can safely multiply f by $g - g_2$, where f, g, g_2 are reduced. The elliptic-curve addition formulas involve three subtractions of this type: the denominator $x - x_2$, which is multiplied by a reduced product inside Montgomery's batch-inversion method (except for one product at the beginning of a batch); the numerator $y - y_2$, which is multiplied by the reduced reciprocal of $x - x_2$; and $x_2 - x_3$, which is multiplied by the reduced quotient λ.

Second, we combine multiply-reduce-subtract-reduce into multiply-subtract-reduce. In particular, to compute $\lambda^2 - x - x_2$, we first add x to x_2, and then subtract the sum from λ^2 *before* reducing the coefficients of λ^2. Similarly, we subtract y_2 from $\lambda(x_2 - x_3)$ before reducing the coefficients of $\lambda(x_2 - x_3)$. This makes the subtractions more expensive (32-bit instead of 16-bit), but still much less expensive than extra carries.

Overall these 6 subtractions use 10 arithmetic instructions: 5 instructions for 40 subtractions of 16-bit integers, and another 5 instructions for 20 subtractions of 32-bit integers.

Table lookups. Our iteration function uses a precomputed table of 2048 multiples of P. Each multiple uses 16 bytes for an x-coordinate and 16 bytes for a y-coordinate, for a total of 64 KB of local storage. If we were concerned with defining an iteration function to perform well on many platforms simultaneously then we would use a smaller table, say 16 KB, to avoid L1 cache misses on typical CPUs. However, the analysis in Section 3 shows that this would slightly increase the number of iterations. In this paper our goal is purely to maximize Cell performance, so we keep $r = 2048$.

Normally a precomputed x-coordinate in reduced form $(x_0, x_1, \ldots, x_9)$ would occupy 20 bytes. We instead represent each precomputed x-coordinate in reduced form $(x_0, x_1, \ldots, x_7, 0, 0)$, stored as a contiguous 16-byte vector $(x_0, x_1, \ldots, x_7)$.

Each y-coordinate is stored similarly. The final zeros mean that this representation is limited to approximately 2^{103} different integers; it cannot handle all elements of $\mathbb{Z}/(2^{128}-3)$, or even all elements of $\mathbb{F}_p$. We find 2048 representable multiples of P by generating approximately 400 million random multiples of P; this precomputation has negligible cost.

This table representation also allows lookups with very little arithmetic (but with many load/store instructions, which we carefully interleave with other computations), as explained in the following paragraph. We could further compress the table entries in several ways, but we are not aware of further compression techniques that avoid extra arithmetic. We comment that scaling the same type of precomputation to larger ECDLPs, squeezing the precomputed points as far as we can with a negligible precomputation, reduces the space for rho tables by asymptotically 25%.

The table entry for a point (x, y) has index s_0 mod 2048 where $(s_0, s_1, \ldots, s_9)$ is the canonicalization of y. We perform 8 table lookups in parallel as follows. We perform one arithmetic instruction to obtain 8 parallel table indices; this instruction is a mask of the vector $(2047, 2047, 2047, 2047, 2047, 2047, 2047, 2047)$ with a vector of canonicalizations. We shift the result left by 5 bits (since table entries have 32 bytes) to obtain 8 addresses in memory; this 128-bit shift is handled by the SPE's load/store pipeline. We then shuffle the result into 8 separate registers, perform 8 x-coordinate loads, and perform 8 y-coordinate loads. Finally we use 24 shuffle instructions to convert to digitsliced format.

Batching inversions. We use Montgomery's trick to batch the inversions in 224 independent iterations, replacing them by 669 multiplications and 1 inversion. This batching is on top of the 8-way parallelism of all of our arithmetic operations. Overall the SPE handles $1792 = 224 \cdot 8$ walks at once. At each moment we watch 1/16th of the walks for fruitless cycles.

Each walk uses 70 bytes of storage: 20 for x, 20 for y, 20 for s, 8 for the seed used to start the walk, and 2 for a table index. Each watched walk uses an extra 102 bytes of storage: 20 for the first s to detect a cycle, 60 for the smallest (s, x, y) to escape a cycle, 20 as extra storage needed for conditional doubling, and 2 for a flag indicating whether the walk needs a doubling to escape a cycle. Overall the walks use $1792(70 + 102/16) = 136864$ bytes of local storage.

Overall performance. The most important arithmetic instructions in each iteration are as follows:

- 5 multiplications: 218.75 arithmetic instructions (43.75 each);
- 1 squaring: 31.75 arithmetic instructions;
- 1 canonicalization: 24.75 arithmetic instructions;
- 6 subtractions: 10.00 arithmetic instructions;
- 1/224 inversion minus 3/224 multiplications: ≈ 20.83 arithmetic instructions.

These add up to 306.08 arithmetic instructions per iteration, implying a lower bound of 306.08 cycles per iteration for our software. Our software actually takes 362 cycles per iteration, about 18% more than this lower bound.

Under 4% of the cycles per iteration are spent on operations that can be blamed on negation: specifically, negating s and y, detecting fruitless cycles, and resolving fruitless cycles by doubling. The rest of the gap between 306 and 362 is explained by detection of distinguished points, loop control, and function-call overhead.

There is, outside the iterations, an extra cost for communicating the occasional distinguished points that do appear (and in setting up replacement points). We made no effort to optimize this cost, since it is multiplied by the distinguished-point probability. With the rather large distinguished-point probability used in our experiments, namely 2^{-20}, and with all 6 SPEs running in parallel and competing for communication resources, this cost effectively added 15 cycles to each iteration, slowing the computation down by about 4%. A smaller distinguished-point probability would reduce this penalty below 1%.

Comparison to previous work. Bos, Kaihara, Kleinjung, Lenstra, and Montgomery state in [4, Appendix A] that they use 456 SPE cycles per iteration for the same ECDLP, including 322 cycles for 6 multiplications, 30 cycles for 6 subtractions, 12 cycles for 1/400 inversions, 24 cycles for canonicalization (with Montgomery reduction), and 68 cycles for miscellaneous overhead. Bos, Kaihara, and Montgomery report 453 cycles per iteration in [5, Section 5], with 318 cycles instead of 322 for the multiplications, and 69 cycles instead of 68 for the miscellaneous overhead.

Each of our multiplications is faster than the multiplications in [4] and [5], by a factor of approximately 1.23. This speedup can be traced directly to our use of the non-integer radix $2^{12.8}$, while [4] et al. used the conventional radix 2^{16}. Most of our other operations are also faster than the operations in [4]. We pay a slight penalty for negation but overall gain the same factor of approximately 1.23 in the number of cycles per iteration. Our overall speedup in solving the ECDLP is much larger, because we use far fewer iterations, as discussed in Section 6.

6 Experimental Results and Evaluation

We do not have access to the cluster of 1284 SPEs used for many months by the authors of [3], [4], [5], and [6]. However, a few SPEs at the Jülich and Barcelona supercomputer centers were enough for us to perform some reasonably large discrete-logarithm experiments, demonstrating clearly that our code works and runs at the expected speed. This section presents the details of our experiments.

Scaling elliptic-curve challenges without changing the prime. Our software is dedicated to the prime $p = (2^{128} - 3)/76439$, and is designed to break the ECDLP on the curve secp112r1 over $\mathbb{F}_p$. However, the same software works without modification for points P, Q on any curve of the form $y^2 = x^3 - 3x + b$ over this $\mathbb{F}_p$.

By counting points on $y^2 = x^3 - 3x + b$ for various b we found group orders having many different prime divisors. For example, there are points

- of prime order $1195174242772417 \approx 1.0615 \cdot 2^{50}$ on $y^2 = x^3 - 3x + 238^2$;
- of prime order $36817627222637377 \approx 1.0219 \cdot 2^{55}$ on $y^2 = x^3 - 3x + 372^2$; and
- of prime order $1186848158152955759 \approx 1.0294 \cdot 2^{60}$ on $y^2 = x^3 - 3x + 240^2$.

For each of these prime-order groups we generated a challenge P, Q and then repeatedly solved the challenge, collecting statistics on the distribution of the time needed to solve the challenge.

Distinguished points per second. We used the same distinguished-point property for all of the challenges, inspecting 20 bits of s_4 and s_9. The probability of a point being distinguished is almost exactly 2^{-20}.

We predicted that we would need slightly more than 2^{20} iterations on average to find a distinguished point, for two reasons. The first reason is that a small percentage of the iterations are wasted by fruitless cycles, as discussed in Section 3. This percentage is under 1% and is independent of the group order ℓ.

The second reason is that a walk can enter a long cycle that does not contain a distinguished point. We predicted that this would occur with probability roughly $2^{40}/\ell$ for each seed, i.e., roughly once for every $\ell/2^{40}$ seeds: certainly not an issue for a single-shot experiment with $\ell \approx 2^{112}$, but a serious concern for a careful statistical analysis studying many seeds with $\ell \approx 2^{50}$.

To prevent our software from running forever in case of long cycles, we added a few lines of code to abort each walk after approximately $47 \cdot 2^{18}$ iterations. A walk that is not in a long cycle has probability only about 2^{-17} of surviving for so many iterations and of therefore being aborted. We could also have modified our software to extract discrete logarithms from the long cycles, but there would have been no cryptanalytic benefit from doing so, since long cycles disappear as ℓ grows.

We also reduced our batch size from 224 to 192 (watching 12 instead of 14) to make room in local storage for keeping track of various statistics. This made each iteration slightly slower, 366 cycles instead of 362 cycles. The extra cost of communicating distinguished points adds 15 cycles per iteration as discussed in the previous section, so we predicted that 6 SPEs running in parallel would produce $6 \cdot 3.192 \cdot 10^9/(381 \cdot 2^{20}) \approx 47.94$ distinguished points per second.

We ran 6 SPEs on the original curve secp112r1 and found 48134 distinguished points in 1000 seconds, with no aborted walks. We also ran various experiments on our 50-bit, 55-bit, and 60-bit challenge curves, and in each case found distinguished points at the expected rate. We found 1 aborted walk for every $2^{12.9}$ distinguished points for the 55-bit challenge curve $y^2 = x^3 - 3x + 372^2$ with $\ell \approx 1.0219 \cdot 2^{55}$.

The number of distinguished points needed for a discrete logarithm. We performed the following experiment for our 50-bit challenge. Take a seed, and find the corresponding distinguished point. Take the next seed, and find the corresponding distinguished point. Continue this process until finding two colliding distinguished points. Compute a discrete logarithm from this collision, and verify that it matches the secret scalar used to generate the challenge in the first place.

Our software handles many seeds at once and produces distinguished points out of order, so we sorted the outputs back into the original order of seeds. Skipping this step would have introduced a bias into our experiment, favoring distinguished points that use relatively few iterations.

We then performed this experiment again, starting from the first seed that was not used in the first experiment. We continued in the same way through 32237 experiments, using disjoint seeds for each experiment. We did not verify the discrete logarithms for every experiment, but we verified it for a large random sample of experiments, and encountered no failures. The number of distinguished points used in the experiments was on average $31.526 \approx 1.0789\sqrt{\pi\ell/4}/2^{20}$, with standard deviation $0.558\sqrt{\pi\ell/4}/2^{20}$. The median was $29 \approx 1.0565\sqrt{\ell \log 2}/2^{20}$. A graph of the distribution of the number of distinguished points appears in the full version of this paper.

We then ran 257241 experiments for our 55-bit challenge. The number of distinguished points was on average $163.37 \approx 1.0074\sqrt{\pi\ell/4}/2^{20}$, with standard deviation $0.527\sqrt{\pi\ell/4}/2^{20}$. The median was $152 \approx 0.9977\sqrt{\ell \log 2}/2^{20}$.

We similarly ran 33791 experiments for our 60-bit challenge. The number of distinguished points was on average $920.36 \approx 0.9996\sqrt{\pi\ell/4}/2^{20}$, with standard deviation $0.525\sqrt{\pi\ell/4}/2^{20}$. The median was $864 \approx 0.9989\sqrt{\ell \log 2}/2^{20}$.

Performance extrapolations. On the basis of these experiments we confidently predict that our software would solve the secp112r1 ECDLP in, on average, 37.3 years on a PS3, using $2^{35.71}$ distinguished points, requiring under 1 terabyte of storage. Here $2^{35.71}$ is calculated as $\sqrt{\pi\ell/4}/2^{20}$ with $\ell \approx p \approx 2^{128}/76439$, and 37.3 is calculated as $\sqrt{\pi\ell/4}/(2^{20} \cdot 47.94 \cdot 86400 \cdot 365.25)$.

The software runs in parallel on many PS3s without trouble, and will easily scale beyond the size of the cluster used in [5]. The computation time is inversely proportional to the number of machines, except for a few minutes at the end of the computation (by all machines while the final collision walks towards a distinguished point, and by a central machine recomputing the walks involved in the collision).

One can trivially reduce the storage and communication requirements by, e.g., a factor of 16 by changing the definition of distinguished points to use 24 bits instead of 20. This increases the final few minutes by a factor of 16, but it also saves almost 15 cycles of communication cost for each iteration, as discussed in the previous section, reducing the total time to just 35.6 years on a PS3.

Comparison to previous work. Our speed is directly comparable to, and almost twice as fast as, the speed previously reported by Bos, Kaihara, Kleinjung, Lenstra, and Montgomery.

Specifically, [3, Appendix A] reports an expected number of "$\sqrt{\pi q/2} \approx 8.4 \cdot 10^{16}$" iterations to solve a secp112r1 ECDLP, with each iteration consuming 456 cycles, totalling "about 60 PS3 years". This iteration count (also appearing in [5, Section 5.3]) is slightly too optimistic: the additive walk in [3] uses $r = 16$, creating a noticeable nonrandomness penalty of approximately $1/\sqrt{1 - 1/16}$. The Cell runs at 3.192 GHz, so a better estimate is

$$\frac{456\sqrt{\pi\ell/2}}{6\cdot 3.192\cdot 10^9\cdot 86400\cdot 365.25\cdot\sqrt{1-1/16}}\approx 65.16$$

PS3 years. We have shown how to solve the same ECDLP using just 35.6 PS3 years.

References

[1] Bailey, D.V., Batina, L., Bernstein, D.J., Birkner, P., Bos, J.W., Chen, H.-C., Cheng, C.-M., Van Damme, G., de Meulenaer, G., Dominguez Perez, L.J., Fan, J., Güneysu, T., Gürkaynak, F., Kleinjung, T., Lange, T., Mentens, N., Niederhagen, R., Paar, C., Regazzoni, F., Schwabe, P., Uhsadel, L., Van Herrewege, A., Yang, B.-Y.: Breaking ECC2K-130 (2010), `http://eprint.iacr.org/2009/541/`, Citations in this document: §2, §2

[2] Bernstein, D.J.: Curve25519: new Diffie-Hellman speed records. In: PKC 2006 [24], pp. 207–228 (2006), `http://cr.yp.to/papers.html#curve25519`, Citations in this document: §4

[3] Bos, J.W., Kaihara, M.E., Kleinjung, T., Lenstra, A.K., Montgomery, P.L.: PlayStation 3 computing breaks 2^{60} barrier; 112-bit prime ECDLP solved (2009), `http://lacal.epfl.ch/112bit_prime`, Citations in this document: §1, §1, §4, §4, §5, §6, §6, §6

[4] Bos, J.W., Kaihara, M.E., Kleinjung, T., Lenstra, A.K., Montgomery, P.L.: On the security of 1024-bit RSA and 160-bit elliptic curve cryptography: version 2.1 (2009), `http://eprint.iacr.org/2009/389/`, Citations in this document: §1, §4, §5, §5, §5, §5, §5, §5, §6

[5] Bos, J.W., Kaihara, M.E., Montgomery, P.L.: Pollard rho on the PlayStation 3. In: Workshop record of SHARCS 2009, pp. 35–50 (2009), `http://www.hyperelliptic.org/tanja/SHARCS/record2.pdf`, Citations in this document: §1, §1, §4, §5, §5, §5, §5, §6, §6, §6

[6] Bos, J.W., Kleinjung, T., Lenstra, A.K.: On the use of the negation map in the Pollard rho method. In: ANTS 2010 [11], pp. 66–82 (2010), Citations in this document: §1,§1, §2, §2, §6

[7] Certicom Research, SEC 2: Recommended elliptic curve domain parameters (2000), `http://www.secg.org/collateral/sec2_final.pdf`, Citations in this document: §4

[8] Costigan, N., Schwabe, P.: Fast elliptic-curve cryptography on the Cell Broadband Engine. In: AFRICACRYPT 2009 [18], pp. 368–385 (2009), `http://cryptojedi.org/users/peter/#celldh`, Citations in this document: §4

[9] Duursma, I.M., Gaudry, P., Morain, F.: Speeding up the discrete log computation on curves with automorphisms. In: ASIACRYPT 1999 [14], pp. 103–121 (1999), Citations in this document: §2, §2

[10] Gallant, R.P., Lambert, R.J., Vanstone, S.A.: Improving the parallelized Pollard lambda search on anomalous binary curves. Mathematics of Computation 69, 1699–1705 (2000), Citations in this document: §2

[11] Hanrot, G., Morain, F., Thomé, E. (eds.): Proceedings of Algorithmic Number Theory, 9th International Symposium, ANTS-IX, Nancy, France, July 19-23. LNCS, vol. 6197. Springer, Heidelberg (2010), See [6]

[12] Harley, R.J.: Solution to Certicom's ECC2K-95 problem (email message) (1998), `http://cristal.inria.fr/~harley/ecdl5/ECC2K-95.submission.text`, Citations in this document: §2

[13] IBM DeveloperWorks, Cell Broadband Engine Programming Handbook, Including the PowerXCell 8i Processor (version 1.11) (2008), `http://www-01.ibm.com/chips/techlib/techlib.nsf/techdocs/1741C509C5F64B3300257460006FD68D`, Citations in this document: §5

[14] Lam, K.-Y., Okamoto, E., Xing, C. (eds.): Proceedings of Advances in Cryptology – ASIACRYPT 1999, International Conference on the Theory and Applications of Cryptology and Information Security, Singapore, November 14-18. LNCS, vol. 1716. Springer, Heidelberg (1999), See [9]

[15] Montgomery, P.L.: Speeding the Pollard and elliptic curve methods of factorization. Mathematics of Computation 48, 243–264 (1987), `http://links.jstor.org/sici?sici=0025-5718(198701)48:177<243:STPAEC>2.0.CO;2-3`, ISSN 0025–5718, MR 88e:11130, Citations in this document: §3

[16] van Oorschot, P.C., Wiener, M.: Parallel collision search with cryptanalytic applications. Journal of Cryptology 12, 1–28 (1999), `http://members.rogers.com/paulv/papers/pubs.html`, ISSN 0933–2790, Citations in this document: §2

[17] Pollard, J.M.: Monte Carlo methods for index computation mod p. Mathematics of Computation 32, 918–924 (1978) ISSN 0025–5718, MR 58:10684, Citations in this document: §2

[18] Preneel, B. (ed.): Proceedings of Progress in Cryptology – AFRICACRYPT 2009, Second International Conference on Cryptology in Africa, Gammarth, Tunisia, June 21-25. LNCS, vol. 5580. Springer, Heidelberg (2009), See [8]

[19] Sony Corporation, Cell Broadband Engine architecture, Version 1.01 (2006), `http://cell.scei.co.jp/pdf/CBE_Architecture_v101.pdf`, Citations in this document: §5

[20] Tavares, S., Meijer, H. (eds.): Selected areas of cryptography 1998. LNCS, vol. 1556. Springer, Heidelberg (1999), See [23]

[21] Teske, E.: On random walks for Pollard's rho method. Mathematics of Computation 70, 809–825 (2001), Citations in this document: §2

[22] Wiener, M.J., Zuccherato, R.J.: Faster attacks on elliptic curve cryptosystems (1998); see also newer version [23], `http://grouper.ieee.org/groups/1363/Research/contributions/attackEC.ps`

[23] Wiener, M.J., Zuccherato, R.J.: Faster attacks on elliptic curve cryptosystems. In: SAC 1998 [20], pp. 190–200 (1998); see also older version [22], Citations in this document: §2

[24] Yung, M., Dodis, Y., Kiayias, A., Malkin, T. (eds.): Proceedings of Public Key Cryptography – 9th International Conference on Theory and Practice in Public-Key Cryptography, New York, NY, USA, April 24-26. LNCS, vol. 3958. Springer, Heidelberg (2006), See [2]

Cryptanalysis of the RSA Subgroup Assumption from TCC 2005

Jean-Sébastien Coron[1], Antoine Joux[2,3], Avradip Mandal[1], David Naccache[4], and Mehdi Tibouchi[1,4]

[1] Université du Luxembourg
6, rue Richard Coudenhove-Kalergi
L-1359 Luxembourg, Luxembourg
{jean-sebastien.coron,avradip.mandal}@uni.lu
[2] Direction générale de l'armement (DGA)
[3] Université de Versailles–Saint-Quentin, Laboratoire PRISM
45, avenue des États-Unis, F-78035 Versailles CEDEX, France
antoine.joux@m4x.org
[4] École normale supérieure
Département d'informatique, Groupe de cryptographie
45, rue d'Ulm, F-75230 Paris CEDEX 05, France
{david.naccache,mehdi.tibouchi}@ens.fr

Abstract. At TCC 2005, Groth underlined the usefulness of working in small RSA subgroups of hidden order. In assessing the security of the relevant hard problems, however, the best attack considered for a subgroup of size $2^{2\ell}$ had a complexity of $\mathcal{O}(2^\ell)$. Accordingly, $\ell = 100$ bits was suggested as a concrete parameter.

This paper exhibits an attack with a complexity of roughly $2^{\ell/2}$ operations, suggesting that Groth's original choice of parameters was overly aggressive. It also discusses the practicality of this new attack and various implementation issues.

Keywords: RSA moduli, hidden order, subgroup, cryptanalysis.

1 Introduction

In 2005, Jens Groth [7] proposed a collection of cryptographic primitives based on small RSA subgroups of $\mathbb{Z}_N^*$ of hidden orders. The motivation behind these constructions is improved efficiency and tighter security reductions.

The RSA moduli N used in [7] are of the form:

$$N = p \cdot q = (2p'r + 1) \cdot (2q's + 1)$$

where p, p', q, q' are prime integers and r, s are random integers. Then there exists a unique subgroup $\mathbb{G}$ of $\mathbb{Z}_N^*$ of order $p'q'$. Letting g be a random generator of $\mathbb{G}$, the pair (N, g) is made public whereas everything else including the group order $p'q'$ is kept secret.

D. Catalano et al. (Eds.): PKC 2011, LNCS 6571, pp. 147–155, 2011.

The best attack considered in [7] has complexity $\mathcal{O}(p')$. Therefore, when proposing concrete parameters, the author suggests to take the bit-lengths $\ell_{p'}$ and $\ell_{q'}$ of the primes p' and q' as $\ell_{p'} = \ell_{q'} = 100$.

This paper does not consider any specific scheme from [7]. Instead, it describes an attack that recovers the secret factors of N from the public data (N, g) in $\tilde{\mathcal{O}}(\sqrt{p'})$. This results in a 2^{50} attack making the choice $\ell_{p'} = \ell_{q'} = 100$ potentially insecure. We analyze the practicality of our attack with an implementation, for which we provide the source code in the full version of this paper [5].

Remark 1. In [7], Groth also considers RSA subgroups where r and s are smooth integers (*i.e.* all prime factors of r and s are smaller than some bound B). For this specific variant an attack in complexity $\mathcal{O}(\sqrt{p'})$ is given in [7], and consequently larger parameters ($\ell_{p'} = \ell_{q'} = 160$) are suggested. In this paper we do not consider this variant but directly focus on the general case.

Remark 2. Other works have proposed schemes based on small subgroups of $\mathbb{Z}_n^*$. The attack introduced in this paper applies to some, but not all of them. In particular, the scheme proposed by Damgård *et al.* in [6] uses a subgroup of *prime* order v of $\mathbb{Z}_n^*$, where v is a factor of both $p-1$ and $q-1$ (of around 160 bits). Since the group has the same order modulo p and q, the attack presented herein does not apply to this scheme. On the other hand, it does, in principle, apply to the subgroup variant of the Paillier cryptosystem [11]. The parameter choice from the original paper was more conservative than that of Groth, however (320-bit subgroup), making it out of reach of our new attack.

2 The New Attack

Using the notations above, we factor N in time $\tilde{\mathcal{O}}(\sqrt{p'})$ and memory $\mathcal{O}(\sqrt{p'})$ as follows. Recall that the RSA modulus $N = pq$ is such that:

$$N = p \cdot q = (2p'r + 1) \cdot (2q's + 1)$$

where p' and q' are prime; besides, g is a generator of the subgroup $\mathbb{G}$ of order $p'q'$. From $g^{p'q'} = 1 \mod N$ we get:

$$g^{p'} = 1 \mod p \tag{1}$$

Let ℓ denote the bit-length of p', which we assume is even without loss of generality, and write $\Delta = 2^{\ell/2}$. We then have

$$p' = a + \Delta \cdot b$$

with $0 \le a, b < 2^{\ell/2}$. From (1), we get:

$$g^a = (g^{\Delta})^{-b} \mod p$$

If the prime factor p was known, one could carry out a baby-step giant-step attack by generating the following two lists:

$$L_p = \{g^i \bmod p : 0 < i < 2^{\ell/2}\}$$
$$L'_p = \{(g^{\Delta})^{-j} \bmod p : 0 \le j < 2^{\ell/2}\}$$

and finding a collision between L_p and L'_p, which would reveal a, b and thus p' in total time and space $\mathcal{O}(2^{\ell/2})$.

Obviously p is unknown, so instead of computing L_p and L'_p, we generate the two following lists modulo N:

$$L = \{x_i = g^i \bmod N : 0 < i < 2^{\ell/2}\}$$
$$L' = \{y_j = (g^{\Delta})^{-j} \bmod N : 0 \le j < 2^{\ell/2}\}$$

One could then compute $\gcd(x_i - y_j, N)$ for all $x_i \in L$ and all $y_j \in L'$. Since we have

$$x_a - y_b = 0 \mod p$$

this would reveal the factors of N for $i = a$ and $j = b$. However, the complexity of this naive approach is quadratic in Δ, and will thus require 2^{ℓ} computations, not $2^{\ell/2}$. Hence we proceed as follows instead:

1. Generate the polynomial:
$$f(x) = \prod_{x_i \in L} (x - x_i) \mod N$$
2. For all $y_j \in L'$, evaluate f at y_j and compute $\gcd(f(y_j), N)$.

Since we have

$$f(y_b) = \prod_{x_i \in L} (y_b - x_i) = (y_b - x_a) \cdot R = 0 \mod p$$

computing $\gcd(f(y_j), N)$ reveals the factors of N for $j = b$.

Algorithm 1. Attack overview

```
Let Δ ← 2^{ℓ/2}.
for i = 0 to Δ − 1 do
    x_i ← g^i mod N
    y_i ← (g^Δ)^{−i} mod N
end for
Generate the polynomial
```

$$f(x) \leftarrow \prod_{i=1}^{\Delta-1} (x - x_i) \mod N$$

```
for i = 0 to Δ − 1 do
    Evaluate f(y_i) ∈ Z_N
    Attempt to factor N by computing gcd(f(y_i), N).
end for
```

The attack is summarized in Algorithm 1. In the next section we show that it can be carried out in time quasi-linear in the cardinalities of L and L'.

3 Attack Complexity

This attack involves the computation and evaluation of a polynomial of the form:

$$f(x) = \prod_{i=1}^{d-1} (x - x_i) \mod N$$

with $d = 2^{\ell/2}$. It is a classical fact [1] that the coefficients of such a polynomial can be computed using a product tree, with a total number of operations in $\mathbb{Z}_N$ which is quasilinear in d (namely $\mathcal{O}(M(d) \log d)$, where $M(d)$ is the complexity of the multiplication of two polynomials of degree d). Similarly, with a remainder tree, this polynomial can be evaluated at all points y_j, $0 \leq j < d$ in $\mathcal{O}(M(d) \log d)$ operations.

In our case, however, both (x_i) and (y_j) are geometric progressions, hence even more efficient algorithms exist: the Newton basis conversion algorithms of Bostan and Schost [4] make it possible to compute f using $\mathcal{O}(d)$ precomputations and a single *middle product* of polynomials of degree d, and to evaluate $f(y_j)$ for all j using $\mathcal{O}(d)$ precomputations, a product of polynomials of degree d and a middle product of polynomials of degree d. See the next section for details. This results in an overall complexity of $3M(d) + \mathcal{O}(d)$ for the complete attack, with a small constant in the $\mathcal{O}$. Space requirements are also $\mathcal{O}(d)$, to store a few polynomials of degree d.

Thus, for typical parameter sizes, the attack is *essentially linear in* $\sqrt{p'}$ *both in time and space.*

4 Algorithmic Details

As discussed above, we can break down the attack in two steps: first compute the coefficients of the polynomial $f(x) = \prod_{i=1}^{d-1}(x - x_i) \mod N$, and then evaluate $f \mod N$ at each of the points y_j. Since both (x_i) and (y_j) are geometric progressions, both of these steps reduce to a variant of the discrete Fourier transform, called the "chirp transform" (or its inverse) [12,2]. In our implementation, we carry out these computations using the particularly efficient algorithms of Bostan and Schost [4], as described in [3, §5.5]. More precisely, Bostan gives pseudocode, reproduced as Algorithms 3 and 4 in Appendix A, to compute polynomial interpolation and polynomial evaluation at a geometric progression.

In our case, a number of futher simplifications are possible in the interpolation stage. Indeed, $f(x_i) = 0$ for $1 \leq i \leq d-1$ and $f(1) = \prod_{i=1}^{d-1}(1 - x_i)$, so with the notations of Algorithm 3, $v_0 = (-1)^{n-1} s_{n-1}$ and $v_i = 0$ for $i > 0$. This means in particular that the polynomial multiplication of Algorithm 3, Step 9 reduces to a simple scalar multiplication, and that the computations of Steps 10–12 can

Algorithm 2. Detailed attack

```
function ATTACK(g, n, N)
    p ← 1; q ← 1; s ← 1; u ← 1; z ← 1; w ← 1
    Initialize U, Z, S, W as zero polynomials of degree n − 1
    U_0 ← u; Z_0 ← z; W_0 ← w
    for i = 1 to n − 1 do
        p ← p · g mod N
        q ← q · p mod N
        s ← s · (p − 1) mod N
        u ← u · p/(1 − p) mod N
        z ← (−1)^i u/q mod N
        w ← q/(s · u) mod N
        U_i ← u; Z_i ← z; W_i ← w
    end for
    Z ← (−1)^{n−1} s_{n−1} · Z mod N
    W ← mul^t(n − 1, U, W)
    for i = 0 to n − 1 do
        W_i ← W_i · Z_i mod N
    end for
    g ← 1/(p · g) mod N                                        ▷ g ← g^{−Δ}
    p ← 1; q ← 1; s ← 1; u ← 1; z ← 1; w ← 1
    U ← 0; Z ← 0
    U_0 ← u; Z_0 ← z; S_0 ← s
    for i = 1 to n − 1 do
        p ← p · g mod N
        q ← q · p mod N
        s ← s/(p − 1) mod N
        u ← u · p/(1 − p) mod N
        z ← (−1)^i u/q mod N
        S_i ← s; U_i ← u; Z_i ← z
        W_i ← W_i/z mod N
    end for
    W ← mul^t(n − 1, Z, W)
    for i = 0 to n − 1 do
        W_i ← (−1)^i W_i · U_i mod N
    end for
    W ← W · S
    for i = 0 to n − 1 do
        if gcd(W_i, N) ≠ 1 then return gcd(W_i, N)             ▷ Factor found!
        end if
    end for
end function
```

be carried out in the main loop. We can also have a slightly more conservative memory management, with only 4 polynomials of degree $n-1$ kept in memory for both the interpolation and the evaluation step. Finally, the multiplications by s_i in Algorithm 4, Step 14 can be skipped altogether as we search for a factor of N with GCD computations, since the s_i's are prime to N by construction. We obtain the detailed procedure described in Algorithm 2.

The attack can again be broken down in three stages: interpolation in Steps 2–18, evaluation in Steps 19–36 and factor search in Steps 37–40. Complexity is dominated by the three quasi-linear multiplication steps: the middle products of Steps 15 and 32, and the polynomial multiplication of Step 36.

5 Implementation

We provide the source code of our attack in the full version of this paper [5]. The implementation of polynomial interpolation and evaluation using Newton basis conversions largely follows the pseudocode from [3] (Figure 5.1 and 5.2), implemented in C using the FLINT library [9].

In Table 1, we provide the observed running time of our attack on an Intel Core2 Duo E8500 3.12 GHz, for 1024-bit RSA moduli. The program was linked to the following libraries: FLINT 1.6 (prerelease), MPIR 2.1.3 and MPFR 3.0, and ran on a single CPU core.

Table 1. Experimental attack running times for 1024-bit moduli

$\ell = \lceil \log_2 p' \rceil$	**running time**
26 bits	1.9 seconds
28 bits	4.0 seconds
30 bits	8.1 seconds
32 bits	16.5 seconds
34 bits	33.5 seconds
36 bits	68.9 seconds

From Table 1 we see that, as expected, running times are essentially linear in $\sqrt{p'}$. Direct extrapolation yields the following estimates:

Table 2. Estimated attack running times for 1024-bit moduli

$\ell = \lceil \log_2 p' \rceil$	**running time**	**estimated number of clock cycles**
60 bits	3 days	2^{50}
80 bits	9 years	2^{60}
100 bits	9000 years	2^{70}

Thus, even the parameter $\ell = 100$ suggested in Groth's paper [7] would require a large but not unachievable amount of computation, even by academic standards. As a comparison, the recent factorization of RSA 768 [10] required about 2000 CPU-years.

However, it is not obvious how the algorithm can be efficiently parallelized to distribute the computation. A naive parallelization strategy is to reduce the number of x_i's and increase the number of y_i's by some factor 2^k, but this only reduces time and memory by a factor of about 2^k while requiring 2^{2k} parallel nodes. It would be desirable to be able to distribute the full size computation—both the FFT steps (multiplication and middle product) and the precomputations—but this appears to be nontrivial.

Most importantly, it is difficult to deal with larger parameters because the attack is heavily memory-bound: the $\mathcal{O}(\sqrt{p'})$ memory requirement is a serious hurdle. In experiments, we encountered memory problems as early as $\ell \approx 38$ for a 1024-bit modulus, and even with much more careful memory management and the use of mass storage rather than RAM, it seems unlikely that parameters larger than $\ell \approx 60$ can be attacked unless storage can be efficiently distributed as well.

6 Conclusion

We have described an attack against the RSA subgroup of hidden order described in [7] that works in time $\tilde{\mathcal{O}}(\sqrt{p'})$ while the best attack considered in [7] had complexity $\mathcal{O}(p')$. We have implemented our attack and assessed its practicality. As expected, our attack exhibits a time complexity quasi-linear in $\sqrt{p'}$. In terms of CPU time alone, the parameters suggested in [7] appear to be within reach for a resourceful attacker. However, due to heavy memory requirements and parallelization problems, these parameters may remain unchallenged.

An interesting open question is to decrease the memory requirement: an algorithm similar to Pollard rho or Pollard lambda with constant memory would be the most convenient type of attack on this problem if it exists. If not, a method for distributing the computation and storage efficiently would be the simplest way to make the attack practical for larger parameters.

Acknowledgments. We are grateful to Luca De Feo, Marc Mezzarobba and anonymous referees for useful comments. This work was partly supported by the French ANR-07-TCOM-013-04 PACE Project and by the European Commission through the IST Program under Contract ICT-2007-216646 ECRYPT II.

References

1. Bernstein, D.J.: Fast multiplication and its applications. In: Algorithmic number theory: lattices, number fields, curves and cryptography. MSRI Publications, vol. 44, pp. 325–384. Cambridge University Press, Cambridge (2008)
2. Bluestein, L.I.: A linear filtering approach to the computation of the discrete Fourier transform. IEEE Trans. Electroacoustics 18, 451–455 (1970)
3. Bostan, A.: Algorithmique efficace pour des opérations de base en calcul formel, Ph.D. thesis, École polytechnique (2003) (in English)
4. Bostan, A., Schost, E.: Polynomial evaluation and interpolation on special sets of points. Journal of Complexity 21(4), 420–446 (2005)
5. Coron, J.-S., Joux, A., Mandal, A., Naccache, D., Tibouchi, M.: Cryptanalysis of the RSA subgroup assumption from TCC 2005, Cryptology ePrint Archive, Report 2010/650 (2005), `http://eprint.iacr.org/`, Full version of this paper

6. Damgård, I.B., Geisler, M., Krøigaard, M.: Efficient and Secure Comparison for On-Line Auctions. In: Pieprzyk, J., Ghodosi, H., Dawson, E. (eds.) ACISP 2007. LNCS, vol. 4586, pp. 416–430. Springer, Heidelberg (2007)
7. Groth, J.: Cryptography in Subgroups of Z_n^*. In: Kilian, J. (ed.) TCC 2005. LNCS, vol. 3378, pp. 50–65. Springer, Heidelberg (2005)
8. Hanrot, G., Quercia, M., Zimmermann, P.: The middle product algorithm, I. In: Applicable Algebra in Engineering, Communication and Computing, vol. 14, pp. 415–438. Springer, Heidelberg (2004)
9. Hart, W.B., Harvey, D., et al.: Fast library for number theory, `http://www.flintlib.org`
10. Kleinjung, T., Aoki, K., Franke, J., Lenstra, A.K., Thomé, E., Bos, J.W., Gaudry, P., Kruppa, A., Montgomery, P.L., Osvik, D.A., te Riele, H., Timofeev, A., Zimmermann, P.: Factorization of a 768-Bit RSA Modulus. In: Rabin, T. (ed.) CRYPTO 2010. LNCS, vol. 6223, pp. 333–350. Springer, Heidelberg (2010)
11. Paillier, P., Pointcheval, D.: Efficient Public-Key Cryptosystems Provably Secure Against Active Adversaries. In: Lam, K.-Y., Okamoto, E., Xing, C. (eds.) ASIACRYPT 1999. LNCS, vol. 1716, pp. 165–179. Springer, Heidelberg (1999)
12. Rabiner, L.R., Schafer, R.W., Rader, C.M.: The chirp z-transform algorithm and its applications. Bell System Tech. J. 48, 1249–1292 (1969)
13. Sage Library, `http://www.sagemath.org/`

A Bostan's Algorithms

Algorithm 3. Polynomial interpolation: compute the polynomial F of degree $< n$ such that $F(p_i) = v_i$, where $p_i = q^i$, $0 \le i \le n-1$

function InterpGeom($p_0, \ldots, p_{n-1}; v_0, \ldots, v_{n-1}$)
  $q_0 \leftarrow 1$; $s_0 \leftarrow 1$; $u_0 \leftarrow 1$; $z_0 \leftarrow 1$; $w_0 \leftarrow v_0$
  for $i = 1$ to $n-1$ **do**
    $q_i \leftarrow q_{i-1} \cdot p_i$
    $s_i \leftarrow s_{i-1} \cdot (p_i - 1)$
    $u_i \leftarrow u_{i-1} \cdot p_i / (1 - p_i)$
    $z_i \leftarrow (-1)^i u_i / q_i$
  end for
  $H \leftarrow \left(\sum_{i=0}^{n-1} v_i / s_i x^i\right) \cdot \left(\sum_{i=0}^{n-1} (-x)^i q_i / s_i\right)$
  for $i = 1$ to $n-1$ **do**
    $w_i \leftarrow (-1)^i \, \mathtt{Coeff}(H, i) / u_i$
  end for
  $G \leftarrow \mathrm{mul}^t(n-1, \sum_{i=0}^{n-1} u_i x^i, \sum_{i=0}^{n-1} w_i x^i)$
  return $\sum_{i=0}^{n-1} z_i \, \mathtt{Coeff}(G, i) x^i$
end function

Algorithm 4. Polynomial evaluation: evaluate the polynomial F at all points $p_i = q^i$, $0 \le i \le n-1$

function EVALGEOM($p_0, \ldots, p_{n-1}; F$)
    $q_0 \leftarrow 1$; $s_0 \leftarrow 1$; $u_0 \leftarrow 1$; $z_0 \leftarrow 1$; $g_0 \leftarrow 1$
    for $i = 1$ to $n-1$ **do**
        $q_i \leftarrow q_{i-1} \cdot p_i$
        $s_i \leftarrow s_{i-1} \cdot (p_i - 1)$
        $u_i \leftarrow u_{i-1} \cdot p_i/(1 - p_i)$
        $z_i \leftarrow (-1)^i u_i / q_i$
    end for
    $G \leftarrow \mathrm{mul}^t(n-1, \sum_{i=0}^{n-1} z_i x^i, \sum_{i=0}^{n-1} \mathtt{Coeff}(F, i)/z_i x^i)$
    for $i = 1$ to $n-1$ **do**
        $g_i \leftarrow (-1)^i u_i \, \mathtt{Coeff}(G, i)$
    end for
    $W \leftarrow \left(\sum_{i=0}^{n-1} g_i x^i\right) \cdot \left(\sum_{i=0}^{n-1} s_i^{-1} x^i\right)$
    return $s_0 \, \mathtt{Coeff}(W, 0), \ldots, s_{n-1} \, \mathtt{Coeff}(W, n-1)$
end function

(If) Size Matters: Size-Hiding Private Set Intersection

Giuseppe Ateniese[1], Emiliano De Cristofaro[2], and Gene Tsudik[2]

[1] Computer Science Department, Johns Hopkins University
[2] Computer Science Department, University of California, Irvine

Abstract. Modern society is increasingly dependent on, and fearful of, the availability of electronic information. There are numerous examples of situations where sensitive data must be – sometimes reluctantly – shared between two or more entities without mutual trust. As often happens, the research community has foreseen the need for mechanisms to enable limited (privacy-preserving) sharing of sensitive information and a number of effective solutions have been proposed. Among them, Private Set Intersection (PSI) techniques are particularly appealing for scenarios where two parties wish to compute an intersection of their respective sets of items without revealing to each other *any other information.* Thus far, "any other information" has been interpreted to mean any information about items not in the intersection.

In this paper, we motivate the need for Private Set Intersection with a stronger privacy property of *hiding the size* of the set held by one of the two entities ("client"). We introduce the notion of Size-Hiding Private Set Intersection (SHI-PSI) and propose an efficient construction secure under the RSA assumption in the Random Oracle Model. We also show that input size-hiding is attainable at very low additional cost.

1 Introduction

Operations that involve sharing sensitive or private information are increasingly encountered in everyday life. A typical scenario involves two entities: one that seeks certain information, and the other – that possibly has this information and is either willing or is compelled to share it. At the same time, each entity wants to maximize privacy of its information, beyond the minimum disclosure necessary to complete the required operation. To motivate the problem, we present three concrete (and only slightly contrived) examples illustrating nuanced requirements of such privacy-preserving operations.

Example 1. U.S. Department of Homeland Security (DHS) maintains a dynamic database of suspected terrorists (TWL: Terror Watch List). For every flight, DHS needs to know whether the flight passenger manifest and TWL have any names in common. Airlines are reluctant to unconditionally share passenger information. Some airlines are foreign and some flights might be transit, i.e., they merely fly over, but not land in, United States. At the same time, compliance with DHS is mandatory, meaning that names of any flight passengers that also appear

D. Catalano et al. (Eds.): PKC 2011, LNCS 6571, pp. 156–173, 2011.

in TWL must be supplied to DHS. For its part, DHS treats TWL as secret information and is absolutely unwilling to reveal it to any airline.

Example 2. U.S. Central Intelligence Agency (CIA) has a requirement to periodically (e.g., every year) check whether any of its agents have been arrested or convicted any crimes. It thus needs to approach every state and, ideally, compare its list of employees against the state's list of arrestees and/or convicted offenders. A state (e.g., South Dakota) is unwilling to reveal its information for fear of misuse and legal liability. Whereas, CIA is mandated by law not to reveal the names of *any* of its agents.

Example 3. U.S. Center of Disease Control and Prevention (CDC) maintains a list of people, per city, afflicted by certain contagious diseases, e.g., the H1N1 virus. CDC needs to monitor high or unusual concentrations of infected people in schools, since that might indicate the start of an epidemic. To this end, it periodically needs to cross-check its list with student rosters in each school district. Privacy regulations prevent schools from granting wholesale access to student data. However, information regarding students with highly infectious diseases needs to be disclosed.

1.1 Why Size Matters?

All examples above have some features in common: neither party can reveal its information in its entirety. What they are willing to reveal is limited to common information, i.e., items appearing on both parties' lists. Specifically, our examples involve only *one-way* information sharing, i.e., airlines allow DHS to learn names that appear on both lists, whereas, DHS does not allow airlines to learn the same.

Another important, but more subtle, feature common to our examples is the need to keep client input size secret. Specifically, DHS does not reveal the number of names on the TWL. This list is dynamic (names can be added and removed frequently) and revealing its size would leak sensitive information. Likewise, by law, CIA cannot divulge the number of its agents, for obvious reasons. Finally, the number of infected school-kids in a city (school district) is extremely sensitive: its disclosure can cause wide-spread panic and/or prompt a health insurance rate hikes for that location.

We conclude that there are solid reasons for parties in certain privacy-preserving operations to keep sizes of their inputs secret. The most common reason is that input size represents sensitive information. A related reason is that, given multiple interactions between the same two parties, fluctuations in input size are equally (or even more) sensitive. Another factor motivating input size secrecy is related to the amount of computation imposed on the other party; we discuss it later, in Section 6.

We note, from the outset, that there are limits to input size hiding. For instance, both parties cannot hide their respective input sizes. One obvious reason is that, in all examples above, (at least) one party learns the intersection of its input with that of the other party. The intersection itself is a list (or a set) and its size leaks information about overall input size.

In this paper, we focus on privacy-preserving interactions characterized by the examples above, where one party (client, the one that learns the intersection) aims to keep the size of its input secret. Next, we argue that no current cryptographic primitive, (including generic secure two-party computation [18]) supports input size-hiding for private set intersection. We also discuss the inadequacy of some trivial approaches.

1.2 Size Hiding with Current Tools?

Private Set Intersection (PSI) is a cryptographic primitive, introduced in [13], that allows two parties: server S and client C, to interact on their respective private input sets: $\mathcal{S}$ and $\mathcal{C}$, such that, C learns $(\mathcal{S} \cap \mathcal{C}, |\mathcal{S}|)$ and S learns nothing beyond $|\mathcal{C}|$.[1] Over the last few years, the research community has devised a number of PSI techniques that vary in costs, security assumptions and adversarial models. We discuss prior work on PSI in Section 2. One common feature of all current PSI protocols is that client input size (# of elements in client set) is revealed to server. It is unclear whether they can be extended or amended to support client input size-hiding. Also, generic secure multi-party computation tools [35] are not applicable as they provide all players with the sizes of other players' inputs.

One trivial approach is for client to employ fixed-size input, i.e., pad its input with chaff up to a certain fixed size. However, this has several drawbacks. First and foremost, this always leaks the *upper bound* of input size. Second, if client input is a dynamic set, the fixed size must reflect the maximum possible set size (otherwise, fluctuations would leak information), which entails wasted computation and communication.

1.3 Roadmap and Contributions

In this paper, we introduce the concept of *Size-Hiding Private Set Intersection* (SHI-PSI).We then present the first concrete SHI-PSI construction, offering provable security and efficient operation. Next, we discuss possible extensions to reduce protocol overhead and adapt SHI-PSI to scenarios where client needs to learn data records associated withe each item in the intersection. Finally, we compare costs of our SHI-PSI approach to prior work (on non size-hiding PSI) and show that size-hiding is attainable at very low additional costs.

Notable SHI-PSI features include:

1. It offers a superset of privacy properties of prior PSI protocols: SHI-PSI additionally hides client input size from server. Client input is hidden unconditionally, without relying on any computational assumption – a contribution in its own right.
2. It is secure under the standard RSA assumption in the Random Oracle Model (ROM).

[1] This is sometimes referred to as one-way PSI; a mutual version involves each party learning the intersection.

3. It is very efficient: (1) communication overhead is linear in server input size *only*, (2) server computation complexity is linear in the size of its input *only*, (3) client computation complexity is almost linear – $O(v \cdot \log v)$ – in the size of its input. (We note that this is remarkably low, since the most efficient PSI offers linear complexity; hence, the only "penalty" for size-hiding is a small increase in client computation: $O(v \cdot \log v)$ vs $O(v)$.)
4. It is particularly attractive for scenarios where server is mandated (e.g., by law) to take part in the interaction with client(s). In such scenarios, it makes sense to minimize burden on server, especially, when client input is large. Current PSI schemes involve server computation proportional to client input size. Whereas, in SHI-PSI, server computation depends only on its own input size.

Organization. After reviewing related work in Section 2, Section 3 defines SH-PSI as a concept and specifies its desired properties. Then, Section 4 presents a concrete SHI-PSI protocol and argues its security. Section 5 discusses possible extensions and Section 6 considers performance issues. The paper concludes with a laundry-list of future work items in Section 7.

2 Related Work

We now overview related cryptographic primitives and prior work.

Two-Party Computation (2PC). Private set operations can be performed using secure two-party computation [18,35]. 2PC allows two parties, each with its own private input, to privately evaluate a generic public function, such that nothing else is revealed. However, standard 2PC definitions (see [17]) provide both parties with the length of the other party's input. This contradicts our input size-hiding goal. Furthermore, 2PC incurs several rounds and relatively high computational overhead. Recent techniques, such as [33] and [22], proposed efficient tools for 2PC. However, special-purpose protocols for private set operations are still much faster. For instance, [22] reports the overhead of 12.8 seconds for computing the intersection of two sets with only 100 items using their 2PC-based techniques. In contrast, [11] shows the overhead of only 6 seconds for computing private set intersection for two sets with 5,000 items (on comparable hardware), using specialized PSI tools.

Private Set Intersection (PSI). Freedman, et al. [13] devised a suite of private set operation protocols based on Oblivious Polynomial Evaluations (OPE-s) [31]. The main idea is to represent a set as a polynomial, and set elements – as its roots. Client uses an additively homomorphic cryptosystem (e.g., Paillier [32]) to encrypt coefficients, that are then evaluated by server, such that client learns the intersection (and nothing else) upon decrypting.

Assuming that client and server sets contain v and w items, respectively, their respective computation overheads amount to $O(v + w)$ and $O(w \log \log v)$ exponentiations. The protocol is secure against semi-honest adversaries in the standard model, and against malicious adversaries in ROM (with increased cost).

There are several improvements to this construction against malicious adversaries in the standard model, with linear communication and quadratic computation [7], and linear communication and $O(w \log \log v)$ computation in [20] (all under DDH). Also, [27] extends OPE-s to more than two players, all learning the intersection, with quadratic computational and linear communication complexities.

Other PSI constructs rely on Oblivious Pseudo-Random Functions (OPRF-s) [12]. An OPRF is a two-party protocol that securely evaluates a pseudorandom function $f_k(\cdot)$ on key k contributed by sender and input x contributed by receiver, such that the former learns nothing from the interaction, and the latter only learns $f_k(x)$. OPRF-s can be used to obtain linear complexity PSI, e.g., [19] and [24], secure in the standard model against, respectively, covert [1] and malicious adversaries. Recent PSI results in the Random Oracle Model (ROM) yielded very efficient constructs with linear complexity and short exponentiations. They replace OPRF-s with unpredictable functions [25] and blind signatures [11]. These techniques are secure under *One-More-DH* and *One-More-RSA* assumptions [2], respectively. Very recent work in [10] achieves linear computational and communication complexity (also using short exponents) against malicious adversaries, under the DDH assumption in ROM.

Branching Programs (BP). Ishai and Paskin [23] consider the following problem: given a branching program P (held by server) and encryption c of input x (held by client), is it possible to compute ciphertext c' from which $P(x)$ can be decoded using the secret key? Note that size of c' depends, polynomially, on sizes of x and P. Thus, neither client computation nor protocol communication overhead depends on server input size P, that remains secret to client. Although one can implement PSI with a branching program and thus hide *server* input size (whereas, we focus on hiding client input size), we argue that this generic construction would involve much higher computational overhead – polynomial in the size of inputs. Also, it would require v parallel executions, where v is client input size.

Secure Pattern Matching. Some recent work addressed a somewhat related problem: secure computation of *pattern matching* [19,16,26,21]. One party (P_1) holds a pattern and the other party (P_2) holds a text string. The goal of P_1 is to learn where the pattern appears in the text, without revealing it to P_2 or learning anything else about P_2's input. However, the size of P_1's pattern is always revealed to P_2. [21] sketches a possible way to hide pattern size, however, only by means of random padding. (As discussed earlier, this is an unsatisfying approach that exposes the upper bound.) It also imposes a substantial performance penalty: from linear to quadratic complexity.

Zero-Knowledge Sets. The only cryptographic primitive in the context of which the need for hiding input sizes was discussed is *Zero-Knowledge Sets* [30]. In it, server publishes a short snapshot of its private database, i.e., a commitment. Later, client can request server to prove whether a given item belongs to the committed set. Note that neither the commitment nor the proof reveals server database. However, the problem addressed by ZK-Sets is quite different from (size-hiding) PSI. In fact, in ZK-Sets, client input is not private.

3 Definitions

We now formalize the concept of *Size-Hiding Private Set Intersection* (SHI-PSI). Informally, SHI-PSI extends PSI with an additional privacy feature that client input size must not be revealed to server. Clearly, SHI-PSI implies (one-way) PSI. For ease of presentation, instead of first defining PSI and then adding size-hiding, we begin by defining SHI-PSI directly.

Definition 1. (SHI-PSI.) *A scheme satisfying correctness, server privacy and client privacy (per Definitions 2, 3 and 4 below), involving two parties: C and S, and two components:* {Setup,Interaction}, *where:*

- Setup*: an algorithm that selects all global parameters.*
- Interaction*: a protocol between S and C on respective inputs: $\mathcal{S}=\{s_1,\cdots,s_w\}$ and $\mathcal{C}=\{c_1,\cdots,c_v\}$.*

Definition 2. (Correctness.) *If both parties are honest, at the end of* Interaction, *run on inputs $(\mathcal{S},\mathcal{C})$, S outputs $\perp$, and C outputs $(|\mathcal{S}|,\mathcal{S}\cap\mathcal{C})$, or $|\mathcal{S}|$ the intersection is empty.*

We assume semi-honest parties and use general definitions of secure computation [17]. Specifically, we define SHI-PSI as a secure two-party protocol realizing functionality described above. Our client and server privacy definitions follow from those in related work [28,13,12,19]. In particular, Goldreich ([17], Sec. 7.2.2) states that, in case of semi-honest parties, the general "real-versus-ideal" definition framework is **equivalent** to a much simpler framework that extends the formulation of honest-verifier zero-knowledge. Informally, a protocol privately computes certain functionality if whatever can be obtained from one party's view of a protocol execution can be obtained from input and output of that party. In other words, the view of a semi-honest party (including $\mathcal{C}$ or $\mathcal{S}$, all messages received during execution, and the outcome of that party's internal coin tosses), on each possible input $(\mathcal{C},\mathcal{S})$, can be efficiently simulated considering only that party's own input and output. This is equivalent to the following formulation:

Definition 3. (Client Privacy.) *For every* PPT *S^* that plays the role of S, for every $\mathcal{S}$, and for any client input set $(\mathcal{C}^{(0)},\mathcal{C}^{(1)})$, two views of S^* corresponding to C's inputs: (1) $\mathcal{C}^{(0)}$ and (2) $\mathcal{C}^{(1)}$, are computationally indistinguishable.*

Client privacy is guaranteed if no information is leaked about client input. That is, S^* cannot distinguish between $\mathcal{C}^{(0)}$ and $\mathcal{C}^{(1)}$. S^* cannot even determine whether $|\mathcal{C}^{(0)}|\neq|\mathcal{C}^{(1)}|$. In fact, Definition 3 is strictly stronger than client privacy definition for standard PSI protocols that reveal client input size. In this case, indistinguishability would be relaxed by the constraint $|\mathcal{C}^{(0)}|=|\mathcal{C}^{(1)}|$.

Definition 4. (Server Privacy.) *Let* $\mathsf{View}_C(\mathcal{C},\mathcal{S})$ *be a random variable representing C's view during execution of SHI-PSI with inputs* $\mathcal{C},\mathcal{S}$. *There exists a PPT algorithm* C^* *such that:*

$$\{C^*(\mathcal{C},\mathcal{S}\cap\mathcal{C})\}_{(\mathcal{C},\mathcal{S})} \stackrel{c}{\equiv} \{\mathsf{View}_C(\mathcal{C},\mathcal{S})\}_{(\mathcal{C},\mathcal{S})}$$

In other words, on each possible pair of inputs $(\mathcal{C},\mathcal{S})$, C's view can be efficiently simulated by C^* on input: $\mathcal{C}$ and $\mathcal{S}\cap\mathcal{C}$. Thus, as in [17], we claim that the two distributions implicitly defined above are computationally indistinguishable.

Remark. As mentioned earlier, we consider security in the presence of semi-honest parties, i.e., parties that faithfully follow protocol specifications. However, during or after protocol execution, they might (passively) attempt to infer additional information about the other party's input. Note that this models precisely the class of adversaries considered in our applications. For instance, in our Example 1 in Section 1, DHS and airlines have no incentive to deviate from protocol specifications, because they might be subject to auditing and could face severe penalties for non-compliance. Nonetheless, airline personnel, system administrators, or other malicious insiders might seek to surreptitiously obtain information about contents or size of the DHS Terror Watch List (TWL).

The RSA Assumption on safe moduli. Let τ be a security parameter and let $RSA.Setup(\tau)$ be an algorithm that outputs so-called safe RSA instances, i.e., pairs (N,e), where: (1) $N = pq$ where p and q are random distinct τ-bit primes, such that $p = 2p'+1$ and $q = 2q'+1$ for distinct primes p', q', and (2) $e < \phi(N)$ is a random positive integer, such that $GCD(e,\phi(N)) = 1$. The RSA problem is (τ,t)-hard on 2τ-bit safe RSA moduli, if, for each algorithm $\mathcal{A}$ that runs in time t, it holds that:

$$\Pr[(N,e)\leftarrow_r RSA.Setup(\tau), y\leftarrow_r \mathbb{Z}_N^*{:}\mathcal{A}(N,e,y)=\beta \text{ s.t. } \beta^e = y \bmod N] \leq \mathsf{negl}(\tau).$$

We later assume that y is chosen uniformly at random from QR_N (the set of quadratic residues in $\mathbb{Z}_N^*$). Thus, the order of y is $p'q'$. In this case, we let e be a random integer (chosen independently of y) such that $\gcd(e,p'q') = 1$ with overwhelming probability. If $e = 2^t u$ for an odd integer u, then, if $t \geq 1$, $\mathcal{A}$ would compute square root of y, which is infeasible if the factoring assumption holds. If $t = 0$, then e is odd and $\mathcal{A}$ would solve an instance of the standard RSA problem.

4 SHI-PSI Construction

We now present our SHI-PSI protocol. Its two main building blocks are: (1) tools similar to RSA accumulators [4], and (2) *unpredictable* function $f_{X,p,q}(y) = (X^{1/y}) \bmod N$ (under the RSA assumption on safe moduli).[2]

[2] A function (family) $f_k(\cdot)$ is an (t,q_f,ϵ)−unpredictable if, for any t-time algorithm $\mathcal{A}$ and any auxiliary information z, it holds that: $Pr[(x^*, f_k(x^*) \leftarrow_r \mathcal{A}^{f_k(\cdot)}(z)) \wedge x^* \notin \mathcal{Q}] \leq \epsilon$ where $\mathcal{A}$ makes at most q_f queries to $f_k(\cdot)$, and $\mathcal{Q}$ is the set of queries.

Specifically, client computes a global witness for its input $\mathcal{C} = \{c_1, \cdots, c_v\}$ in the form of an RSA accumulator: $(g^{\prod_{i=1}^{v} H(c_i)}) \bmod N$, where g is a generator of QR_N and $H(\cdot)$ is a full-domain hash function [3]. Then, client securely blinds the accumulator with a random exponent and sends the result (denoted as X) to server. The latter learns no information about client input, not even its size. For its part, for each item $s_j \in \mathcal{S}$, server computes unpredictable function f over client message X. Server then applies a one-way function (in practice, a suitable cryptographic hash function) to each output of f. The results are a set of *tags*, one for each s_j. These tags are then returned to client for matching (details below). We note that the outer hash is crucial, since server privacy is based on the fact that, in ROM, a hash of an unpredictable function is a PRF.

In the above, $H(\cdot)$ is a standard random oracle that does not have to output large primes. Also, we obviate the technical issue of computing the inverse of $H(s_j)$ "in the exponent" by selecting the RSA modulus N as a product of safe primes to ensure that the order of X is itself a product of large and unknown primes (see proof for details).

Client learns the set intersection as well as $|\mathcal{S}|$) since it can only match tags corresponding to the items in the intersection. The intuition is that client computation of $g^{(\prod_{l \neq i} H(c_l))}$ leads to it finding a matching tag only if $c_i \in \mathcal{S} \cap \mathcal{C}$.

Before presenting the actual protocol, we introduce the notation in Table 1.

Table 1. Notation

$a \leftarrow_r \mathcal{A}$	variable a is chosen uniformly at random from set $\mathcal{A}$
τ	security parameter
τ_1, τ_2	security parameters that depend on τ
p, q	two τ-bit safe primes, i.e., $p = 2p' + 1, q = 2q' + 1$
$N = pq, e, d$	RSA modulus, public and private exponents
g	generator of QR_N
$H(\cdot)$	random oracle $H : \{0,1\}^* \rightarrow \{0,1\}^{\tau_1}$
$F(\cdot)$	random oracle $F : \{0,1\}^* \rightarrow \{0,1\}^{\tau_2}$
$\mathcal{C}, \mathcal{S}$	client and server sets, respectively
v, w	sizes of $\mathcal{C}$ and $\mathcal{S}$, respectively
$i \in [1, v]$	indices of elements of $\mathcal{C}$
$j \in [1, w]$	indices of elements of $\mathcal{S}$
c_i, s_j	i-th and j-th elements of $\mathcal{C}$ and $\mathcal{S}$, respectively
hc_i, hs_j	$H(c_i)$ and $H(s_j)$, respectively
π	random permutation

4.1 Protocol Description

Fig. 1 shows our SHI-PSI protocol. Common input is extracted from the output of $RSA.Setup(\tau)$. Primes p' and q' are provided exclusively to server. Client must treat its exponents as large integers. We emphasize that arithmetic operations in the exponent are performed in $\mathbb{Z}^*_{p'q'}$. (In particular, squaring is a permutation of QR_N, in our setting).

Common input: $N, g, H(\cdot), F(\cdot)$

Client C on input:
$\mathcal{C} = \{hc_1, \cdots, hc_v\}$
where $hc_i = H(c_i)$

Server S on input:,
$p', q', \mathcal{S} = \{hs_1, \cdots, hs_w\}$,
where $hs_j = H(s_j)$

For $1 \leq i \leq v$
$PCH_i = \prod_{l=1, l \neq i}^{v} hc_l$
$PCH = \prod_{i=1}^{v} hc_i$
$A = (g^{PCH}) \bmod N$

$R_c \leftarrow_r \{1, \ldots, N^2\}$
$X = (A^{R_c}) \bmod N$

$\xrightarrow{X}$

$R_s \leftarrow_r \{0, \ldots, p'q' - 1\}$
$Z = (g^{R_s}) \bmod N$
For $1 \leq j \leq w$
$K_{s:j} = (X^{R_s \cdot (1/hs_j)}) \bmod N$
$\{t_1, \ldots, t_w\} = \pi(F(K_{s:1}), \ldots, F(K_{s:w}))$

$\xleftarrow{Z, \{t_1, \ldots, t_w\}}$

For $1 \leq i \leq v$
$K_{c:i} = (Z^{R_c \cdot PCH_i}) \bmod N$
$t'_i = F(K_{c:i})$
OUTPUT:
$\{t'_1, \ldots, t'_v\} \cap \{t_1, \ldots, t_w\}$

Fig. 1. SHIPSI protocol secure under RSA assumption in ROM (notation from Table 1)

Theorem 1. *Under the RSA assumption on safe moduli, the protocol in Fig. 1 is a server- and client-private SHI-PSI, satisfying Definitions 1-4. in the Random Oracle Model (ROM).*

Proof. We show that the protocol satisfies *correctness as well as client and server privacy,* defined in Section 3. We assume that all server elements are distinct.[3] Hash functions $H(\cdot)$ and $F(\cdot)$ are modeled as random oracles.

Correctness. We note that: $\forall c_i \in \mathcal{S} \cap \mathcal{C}, \exists\, j\ s.t. c_i = s_j$. Hence, $hc_i = hs_j$ and:

$$K_{c:i} = Z^{R_c \cdot PCH_i} = g^{R_c R_s PCH (1/hs_j)}$$
$$K_{s:j} = X^{R_s \cdot (1/hs_j)} = g^{R_c R_s PCH (1/hs_j)}$$

Consequently, $t'_i = F(K_{c:i}) = F(K_{s:j}) = t_j$; thus, client learns: $c_i \in \mathcal{S} \cap \mathcal{C}$ as well as $|\mathcal{S}| = |\{t_1, \ldots, t_w\}|$.

Client Privacy. Since client's only message to server is $X = g^{(PCH \cdot R_c)} \bmod N$, we claim that the distribution of X is essentially equivalent to that of random elements in QR_N, which is a cyclic group of order $p'q'$. Since PCH and $p'q'$ are

[3] However, we can remove this assumption by adding a counter to the input of $H(\cdot)$.

relatively prime (with overwhelming probability), we assume that $A = g^{PCH} \bmod N$ is a generator of QR_N. Moreover, R_c is chosen uniformly at random from $\{1, \ldots, N^2\}$. Thus, if $R_c = r_1 p'q' + r_2$ with $r_2 \in \{0, \ldots, p'q' - 1\}$, we have that the distribution of r_2 is *statistically* indistinguishable from the uniform distribution on $\{0, \ldots, p'q' - 1\}$ and r_1 and r_2 are essentially independent (see, e.g., [6]). Therefore, $X = A^{R_c} \bmod N$ is essentially distributed as a random quadratic residue independent of PCH *even if factorization of* N *is known.*

Server Privacy. To show that client's view can be efficiently simulated by a PPT algorithm, we follow a hybrid argument: The entire client's view is gradually transformed by replacing values (received by client) that are **outside** the set intersection, with elements chosen uniformly and independently at random. It then suffices to show that this progressive substitution cannot be detected by any efficient algorithm.

Let $I = \mathcal{C} \cap \mathcal{S}$, and $|I| = t$. For any $(\mathcal{C}, \mathcal{S})$, we show that two distributions:

$$\mathcal{D}_0 = \left\{(R_c, T) \;:\; R_c \leftarrow_r \{1, \ldots, N^2\}, T = \pi\left(F(X^{R_s(1/hs_{j1})}), \cdots, F(X^{R_s(1/hs_{jw})})\right)\right\}$$

and

$$\mathcal{D}_{w-t} = \left\{(R_c, T) \;:\; R_c \leftarrow_r \{1, \ldots, N^2\}, T = \pi\left(F(X^{R_s(1/hs_{j1})}), \cdots, r_{t+1}, \cdots, r_w\right)\right\},$$

are computationally indistinguishable, where $(hs_{j1}, \cdots, hs_{jt}) \in I$ and values in $(r_{t+1}, \cdots, r_w)$ are chosen uniformly and independently at random from $\{0,1\}^{\tau_2}$ (i.e., the co-domain of the random oracle $F(\cdot)$).

Our proof follows the standard hybrid argument: Let $z = w - t$. We define a series of intermediate distributions $\mathcal{D}_i$, for $0 < i < z$, where T is constructed by replacing the first i outputs of items NOT in I with random values in the co-domain of $F(\cdot)$.

After fixing index i and probabilistic polynomial-time distinguisher D, we define:

$$\epsilon(\tau) = |\Pr[D = 1|\mathcal{D}_{i+1}] - \Pr[D = 1|\mathcal{D}_i]|$$

Our claim is that $\epsilon(\tau)$ is negligible in τ. Let us assume that this claim is false. The only difference between $\mathcal{D}_i$ and $\mathcal{D}_{i+1}$ is the way T is defined. Specifically, $(i+1)$-st item of T not in I is $F(X^{R_s(1/hs_l)})$ for $\mathcal{D}_i$ and a random value for $\mathcal{D}_{i+1}$.

Since $F(\cdot)$ is a random oracle, distinguisher D must compute $X^{R_s(1/hs_l)} = g^{R_s R_c PCH/hs_l}$ for $hs_l \notin I$. Then, we can build an efficient algorithm $\mathcal{A}$ that, given a challenge (N, e, y), returns $y^{1/e} \bmod N$. (We assume that y is chosen uniformly at random from QR_N. Thus, the order of y is $p'q'$.) The simulation proceeds as follows: First, $\mathcal{A}$ sets $g = y$ and, by programming the random oracle $H(\cdot)$, $\mathcal{A}$ assigns random values to outputs of $H(\cdot)$ and computes $d = \gcd(R_s R_c PCH, hs_l)$, for some integers e and b with $hs_l = ed$ and $R_s R_c PCH = bd$. Since $F(\cdot)$ is a random oracle, $\mathcal{A}$ sees $g^{R_s R_c PCH/hs_l} = g^{b/e}$. Given that $(g^{b/e})^e = g^b$ and $\gcd(e, b) = 1$, $\mathcal{A}$ can use the extended Euclidean algorithm to compute

$g^{1/e} = y^{1/e}$ via the well-known *Shamir's trick*[4]. Thus, under the RSA assumption on safe moduli, formulated for a random exponent, $\epsilon(\tau)$ is negligible in τ.

Remarks. We stress that exponents in our scheme (i.e., outputs of $H(\cdot)$) do not have to be prime, unlike related reductions, such as [34,4,29,15]. This is because client cannot compute $g^{R_s R_c PCH/hs_l}$, for $l \in \{1, \ldots, w\}$, unless $R_c PCH/hs_l$ is an integer. (Recall that R_c is generated honestly). Clearly, if $hs_l \notin I$, $R_c PCH/hs_l$ is, – with negligible probability – an integer as long as hs_l is sufficiently large: random oracles are indeed *division intractable*, as shown in [15,5] (in particular, [5] presents an algorithm for finding division collisions sub-exponential in τ_1, the digest size).

We readily acknowledge that our construction assumes both semi-honest players and the Random Oracle Model. Nevertheless, on a positive side, it is interesting to observe that generic 2PC techniques, following traditional definitions that also apply to malicious adversaries, do not achieve size-hiding of client input. Goldreich [17] remarks that the program of each party (in a protocol for computing the desired functionality) must either depend only on the length of that party's input or obtain information on the counterpart's input length. One intuitive argument against the feasibility of input size-hiding protocols secure in the malicious model is that proving well-formed-ness of client input is only possible by considering each client input set element separately (e.g., via some ZK proofs). Thus, combined proofs would have to reveal at least the upper bound on client input size.

Computational and Communication Complexity. The protocol in Fig. 1 incurs the following computational complexity (in terms of modular exponentiations). In each interaction, server needs to compute $O(w)$ exponentiations, one for each of its items. Whereas, client operations are divided into *off-line* and *on-line* categories. Client *off-line* work amounts to $O(v)$ exponentiations for the computation of $A = g^{PCH} \bmod N$, since PCH is the non-modular product of v values. Additionally, client computes $K_{c:i} = Z^{Rc \cdot PCH_i} \bmod N$ for each item. As each of these operations requires $O(v)$ exponentiations, *on-line* client complexity amounts to $O(v^2)$ exponentiations.[5] Communication complexity in each interaction is dominated by $O(w)$ outputs of $F(\cdot)$ sent from S to C in the second message. (The first message involves the transmission of a single $\log(N)$-bit value).

[4] Note that this is similar to the reduction in [15]. However, in contrast to Theorem 5 in [15], our reduction is not based on the strong RSA assumption, but on the standard RSA assumption in ROM. This is because e is generated independently of base y and, thus, e is effectively provided as input to the adversary. Indeed, the signature scheme in [15] is actually secure under the standard RSA assumption in ROM; this was confirmed via private communication [14].

[5] Note that, if client knew the factorization of N, it could compute PCH and PCH_i's using multiplication $\bmod\, \phi(N)$, thus significantly reducing complexity of each exponentiation. However, as discussed earlier, the fact that client does not know $\phi(N)$ is crucial to server privacy.

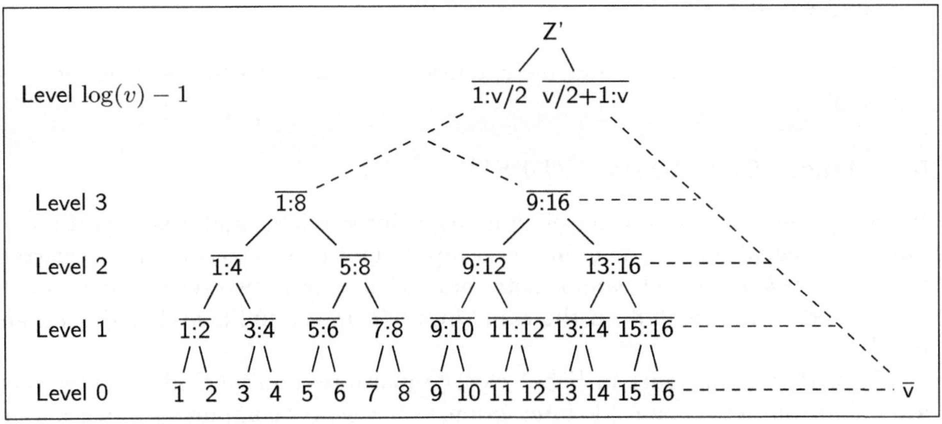

Fig. 2. Tree-based strategy to reduce client computation

4.2 Reducing Client Complexity

We now discuss a simple technique to reduce client computation. Note that the naïve computation of $K_{c:i}$ leads to $O(v^2)$ exponentiations. However, this can be reduced to $O(v \log(v))$ via dynamic programming. Our intuition is as follows: For any (i, j), $K_{c:i}$ and $K_{c:j}$ only differ by one exponent, since $PCH_i = \prod_{l=1,l\neq i}^{v} hc_l$, whereas, $PCH_j = \prod_{l=1,l\neq j}^{v} hc_l$.

We define $Z' = Z^{Rc}$, and $\overline{\text{i:j}} = Z'^{[\prod_{l\notin[i,j]} hc_l]} \bmod N$. We illustrate this technique using a tree in Fig. 2. The leaves contain values $K_{c:i}$, for $1 \leq i \leq v$, e.g., $\bar{i} = Z'^{[\prod_{l\neq i} hc_l]} \bmod N = Z^{Rc \cdot PCH_i} \bmod N = K_{c:i}$.

We now sum up the total number of exponentiations needed to compute all these values. Note that, from a node with value $\overline{\text{i:j}}$, one can obtain the children, $\overline{\text{i:h}}$ and $\overline{\text{h+1:j}}$, as follows:

$$\overline{\text{i:h}} = \left(\overline{\text{i:j}}\right)^{\left(\prod_{l=h+1}^{j} hc_l\right)} (\bmod N)$$
$$\overline{\text{h+1:j}} = \left(\overline{\text{i:j}}\right)^{\left(\prod_{l=i}^{h} hc_l\right)} (\bmod N)$$

For $h = i + (j - i + 1)/2$, each of these operations involves exactly $(j - i + 1)/2$ exponentiations.

At level 0, there are v values, each obtained with a single exponentiation from the parents at level 1. At level 1, there are $v/2$ values, each obtained with 2 exponentiations from nodes at level 2. In general, at level i, there are $v/2^i$ values, each obtained with 2^i exponentiations from nodes at level $i+1$. Thus, client overhead can be estimated as:

$$\#\text{ exponentiations} = \sum_{i=0}^{\log(v)-1} \left(2^i \frac{v}{2^i}\right) = v \log(v).$$

5 Extensions

In this section, we discuss possible extensions to our SHI-PSI construction presented above.

5.1 Linear-Complexity SHI-PSI

In many scenarios, parties engage in multiple interactions, and it is important to hide (from client) any changes in server input. This feature is sometimes referred to as *unlinkability*: client cannot determine whether any two server interactions are related, i.e., executed on the same input (e.g., see unlinkability definitions in [9]).

The SHI-PSI construct in Fig. 1 clearly guarantees unlinkability: server tags are unlinkable across multiple interactions, since server computes a new random R_s and a new $Z \in QR_N$, in each execution. However, although we clearly value unlinkability, it is worth considering scenarios where it might be reasonable to trade off unlinkability for better efficiency.

To this end, we sketch a modified SHI-PSI protocol that reduces the number of client on-line exponentiations to linear. The main intuition is that removing R_s allows client to pre-compute the exponentiations involving (long) PCH_i values.

Common input: $N, g, H(\cdot), F(\cdot)$

Client C on input:
$\mathcal{C} = \{hc_1, \cdots, hc_v\}$
where $hc_i = H(c_i)$

Server S on input:,
$p', q', \mathcal{S} = \{hs_1, \cdots, hs_w\}$,
where $hs_j = H(s_j)$

(*Off-line*)

For $1 \leq i \leq v$
$PCH_i = \prod_{l=1, l \neq i}^{v} hc_l$
$a_i = g^{PCH_i} \bmod N$

$PCH = \prod_{i=1}^{v} hc_i$
$A = (g^{PCH}) \bmod N$

(*On-line*)

Client: $R_c \leftarrow_r \{1, \ldots, N^2\}$
Client: $X = (A^{R_c}) \bmod N$

Client → Server: X

Server: For $1 \leq j \leq w$
$K_{s:j} = (X^{(1/hs_j)}) \bmod N$

Client: For $1 \leq i \leq v$
$K_{c:i} = (a_i^{R_c}) \bmod N$
$t'_i = F(K_{c:i})$

Server: $\{t_1, \cdots, t_w\} = \pi(F(K_{s:1}), \cdots, F(K_{s:w}))$

Server → Client: $\{t_1, \cdots, t_w\}$

OUTPUT:
$\{t'_1, \ldots, t'_v\} \cap \{t_1, \ldots, t_w\}$

Fig. 3. Modified construction of the SHI-PSI protocol to achieve linear client complexity. (Notation is from Table 1).

We illustrate the resulting protocol in Fig. 3. Note that security arguments in Theorem 1 also apply to this protocol variant. Indeed, given assuming semi-honest client, $X = g^{PCH \cdot R_c}$, similar to X^{R_s} in protocol of Fig. 1, is also uniformly distributed in QR_N, for $R_c \leftarrow_r \{1, \ldots, N^2\}$. A more formal treatment of the problem as well as complete formal proofs are deferred to the full version of this paper.

Correctness. Observe that $\forall c_i \in \mathcal{S} \cap \mathcal{C}, \exists\, j\, s.t. c_i = s_j$. Hence, $hc_i = hs_j$ and:

$$K_{s:j} = X^{1/hs_j} = g^{R_c PCH(1/hs_j)} = g^{R_c \cdot PCH_i} = K_{c:i}$$

Consequently, $t'_i = F(K_{c:i}) = F(K_{s:j}) = t_j$, and client learns $c_i \in \mathcal{S} \cap \mathcal{C}$. Also, note that client learns $|\mathcal{S}| = |\{t_1, \ldots, t_w\}|$.

Computational and Communication Complexity. The amended SHI-PSI construct in Fig. 3 incurs the following computational complexity: Server overhead is unaltered from Fig. 1, i.e., $O(w)$ exponentiations. However, client performs $O(v \log(v))$ exponentiations *off-line*, and only $O(v)$ exponentiations *on-line*. Communication overhead is the same as in the protocol of Fig. 1.

5.2 SHI-PSI with Data Transfer

We now show how to extend proposed SHI-PSI constructs to support data transfer. Informally, in *SHI-PSI with Data Transfer*, client additionally obtains data records associated with the items in the intersection. The main idea is to let server encrypt records using a symmetric key (using a secure symmetric cipher, such as AES [8], used with a proper mode of operation to guarantee CPA security) derived from the output of the unpredictable function. For example, keys can be derived by computing a one-way function (e.g., a cryptographic hash) over the unpredictable function output. Correctness and server privacy of SHI-PSI guarantee that client can derive the decryption keys only for items with matching tags, i.e., those in the intersection.

Let $F_{enc}(\cdot)$ be a secure cryptographic hash function (modeled as a random oracle): $F_{enc} : \{0,1\}^* \rightarrow \{0,1\}^{\tau_2}$, chosen at setup. For every j, server computes $k_{s:j} = F_{enc}(K_{s:j})$ and encrypts associated data using $k_{s:j}$. For its part, client, for every i, computes $k_{c:i} = F_{enc}(K_{c:i})$ and decrypts only ciphertexts corresponding to matching tags. (Note that $k_{s:j} = k_{c:i}$ iff $s_j = c_i$ and $t_j = t'_i$). As long as the underlying encryption scheme is CPA-secure, this extension does not affect security or privacy arguments for any protocol discussed thus far. Finally, note that this extension leaves the complexity of both protocols unaltered.

6 Cost of Hiding Size

Although prior work produced a number of PSI protocols with different security assumptions and complexities, we presented the first PSI protocol that hides client input size. Therefore, it seems somewhat counterintuitive to compare our

SHI-PSI constructs with prior PSI protocols. Nonetheless, we provide an estimate of the slow-down incurred by client input size-hiding.

We consider prior PSI techniques and evaluate their asymptotic computation and communication complexities in the RAM computational model (i.e., using a single-processor machine). We estimate computation overhead as the number of *on-line* modular exponentiations performed by server and client. Note that, in order to make the comparison as fair as possible, all protocols are instantiated to provide similar degrees of security in the same model, i.e., semi-honest players and ROM. For instance, we do not count zero-knowledge proofs of protocol compliance in protocols secure against malicious adversaries. Results, reflected in Table 2, summarize: security model (standard or ROM), adversaries (honest-but-curious or malicious), availability of client input size-hiding, communication overhead, number of modular exponentiations by server and client, size of random exponents (i.e., whether they can be selected from subgroups).

Table 2. Performance Comparison of PSI and SHI-PSI constructions

Protocol	Model	Adv	Size Hiding	Comm. Overhead	Server Exp-s	Client Exp-s	Exp-s Length
[13]	Std	HbC	No	$O(w+v)$	$O(w(\log\log v))$	$O(w+v)$	Short
[20]	Std	Mal	No	$O(w+v)$	$O(w(\log\log v))$	$O(w+v)$	Short
[27]	Std	HbC	No	$O(w+v)$	$O(w \cdot v)$	$O(w+v)$	Long
[24]	Std	Mal	No	$O(w+v)$	$O(w+v)$	$O(v)$	Long
[25]	ROM	Mal	No	$O(w+v)$	$O(w+v)$	$O(v)$	Short
[11] (Fig.3)	ROM	HbC	No	$O(w+v)$	$O(w+v)$	$O(v)$	Short
[11] (Fig.4)	ROM	HbC(*)	No	$O(w+v)$	$O(w+v)$	$O(v)$ *mults*	Long
[10]	ROM	Mal	No	$O(w+v)$	$O(w+v)$	$O(v)$	Short
Our Fig.1	ROM	HbC	**Yes**	$O(w)$	$O(w)$	$O(v\log v)$	Long
Our Fig.3	ROM	HbC	**Yes**	$O(w)$	$O(w)$	$O(v)$	Long

(*) This construct actually achieves malicious security with one-sided simulatability.

Observe that the most efficient PSI-s secure in the random oracle model incur linear computational complexities, i.e., $O(w+v)$ for server and $O(v)$ for client, and involve short exponents (e.g., 160-bit) in prime order groups. Whereas, our SHI-PSI protocol (Fig. 1) uses exponents with length close to the RSA modulus (e.g., 1024-bit) and incurs $O(v\log v)$ client complexity. However, such a drawback is experienced by one player – client – that benefits from additional privacy, as its set size is hidden from server. Also, note that our second SHI-PSI construct (Fig. 3) reduces client complexity to $O(v)$.

Finally, we remark that our SHI-PSI constructs achieve better server complexity than PSI-s, in settings where v (the size of client's set) is not negligible. In fact, *server's computational load in SHI-PSI is independent of client's input size*. Also, *protocols not hiding sizes incur higher communication overhead*: client sends a number of values proportional to its set size (as opposed to a single value in SHI-PSI).

7 Conclusions and Future Work

This paper motivated the importance, and introduced the concept, of Size-Hiding Private Set Intersection (SHI-PSI). It also presented two secure and efficient SHI-PSI constructs, Since this work represents an initial foray into SHI-PSI protocols, much remains to be done. Items for future work, include (but are not limited to):

1. Amending proposed SHI-PSI constructs to eliminate the random oracle.
2. Exploring SHI-PSI secure against fully malicious (rather than semi-honest) participants.
3. Investigating SHI-PSI variants that provide authorization of client input, i.e., requiring each item in client set to be pre-authorized by some trusted authority.
4. Extending SHI-PSI to support multiple clients to obtain private computation of size-hiding a *multi-party* set intersection.

Acknowledgements. We are grateful to Stanislaw Jarecki and Adi Shamir for some useful hints in the early stage of SHI-PSI protocol design, as well as the anonymous reviewers for providing valuable feedback. This research was supported by the US Intelligence Advanced Research Projects Activity (IARPA) under grant number FA8750-09-2-0071.

References

1. Aumann, Y., Lindell, Y.: Security Against Covert Adversaries: Efficient Protocols for Realistic Adversaries. In: Vadhan, S.P. (ed.) TCC 2007. LNCS, vol. 4392, pp. 137–156. Springer, Heidelberg (2007)
2. Bellare, M., Namprempre, C., Pointcheval, D., Semanko, M.: The one-more-RSA-inversion problems and the security of Chaum's blind signature scheme. Journal of Cryptology 16(3), 185–215 (2008)
3. Bellare, M., Rogaway, P.: The Exact Security of Digital Signatures - How to Sign with RSA and Rabin. In: Maurer, U.M. (ed.) EUROCRYPT 1996. LNCS, vol. 1070, pp. 399–416. Springer, Heidelberg (1996)
4. Benaloh, J., De Mare, M.: One-way accumulators: A decentralized alternative to digital signatures. In: Helleseth, T. (ed.) EUROCRYPT 1993. LNCS, vol. 765, pp. 274–285. Springer, Heidelberg (1994)
5. Coron, J.-S., Naccache, D.: Security Analysis of the Gennaro-Halevi-Rabin Signature Scheme. In: Preneel, B. (ed.) EUROCRYPT 2000. LNCS, vol. 1807, pp. 91–101. Springer, Heidelberg (2000)
6. Cramer, R., Shoup, V.: Signature schemes based on the strong RSA assumption. ACM Transactions on Information and System Security 3(3), 185 (2000)
7. Dachman-Soled, D., Malkin, T., Raykova, M., Yung, M.: Efficient Robust Private Set Intersection. In: Abdalla, M., Pointcheval, D., Fouque, P.-A., Vergnaud, D. (eds.) ACNS 2009. LNCS, vol. 5536, pp. 125–142. Springer, Heidelberg (2009)
8. Daeman, J., Rijmen, V.: AES proposal: Rijndael (1999)

9. De Cristofaro, E., Jarecki, S., Kim, J., Tsudik, G.: Privacy-Preserving Policy-Based Information Transfer. In: Goldberg, I., Atallah, M.J. (eds.) PETS 2009. LNCS, vol. 5672, pp. 164–184. Springer, Heidelberg (2009)
10. De Cristofaro, E., Kim, J., Tsudik, G.: Linear-Complexity Private Set Intersection Protocols Secure in Malicious Model. In: Abe, M. (ed.) ASIACRYPT 2010. LNCS, vol. 6477, pp. 213–231. Springer, Heidelberg (2010)
11. De Cristofaro, E., Tsudik, G.: Practical Private Set Intersection Protocols with Linear Complexity. In: Sion, R. (ed.) FC 2010. LNCS, vol. 6052, pp. 143–159. Springer, Heidelberg (2010)
12. Freedman, M.J., Ishai, Y., Pinkas, B., Reingold, O.: Keyword search and oblivious pseudorandom functions. In: Kilian, J. (ed.) TCC 2005. LNCS, vol. 3378, pp. 303–324. Springer, Heidelberg (2005)
13. Freedman, M.J., Nissim, K., Pinkas, B.: Efficient private matching and set intersection. In: Cachin, C., Camenisch, J.L. (eds.) EUROCRYPT 2004. LNCS, vol. 3027, pp. 1–19. Springer, Heidelberg (2004)
14. Gennaro, R.: Private Communication (2010)
15. Gennaro, R., Halevi, S., Rabin, T.: Secure hash-and-sign signatures without the random oracle. In: Stern, J. (ed.) EUROCRYPT 1999. LNCS, vol. 1592, pp. 123–139. Springer, Heidelberg (1999)
16. Gennaro, R., Hazay, C., Sorensen, J.S.: Text Search Protocols with Simulation Based Security. In: Nguyen, P.Q., Pointcheval, D. (eds.) PKC 2010. LNCS, vol. 6056, pp. 332–350. Springer, Heidelberg (2010)
17. Goldreich, O.: Foundations of Cryptography, vol. 2. Cambridge Univ. Press, Cambridge (2004)
18. Goldreich, O., Micali, S., Wigderson, A.: How to play any mental game. In: STOC, pp. 218–229 (1987)
19. Hazay, C., Lindell, Y.: Efficient protocols for set intersection and pattern matching with security against malicious and covert adversaries. In: Canetti, R. (ed.) TCC 2008. LNCS, vol. 4948, pp. 155–175. Springer, Heidelberg (2008)
20. Hazay, C., Nissim, K.: Efficient Set Operations in the Presence of Malicious Adversaries. In: Nguyen, P.Q., Pointcheval, D. (eds.) PKC 2010. LNCS, vol. 6056, pp. 312–331. Springer, Heidelberg (2010)
21. Hazay, C., Toft, T.: Computationally secure pattern matching in the presence of malicious adversaries. In: Abe, M. (ed.) ASIACRYPT 2010. LNCS, vol. 6477, pp. 195–212. Springer, Heidelberg (2010)
22. Henecka, W., Koegl, S., Sadeghi, A.-R., Schneider, T., Wehrenberg, I.: TASTY: Tool for Automating Secure Two-partY computations. In: ACM CCS (2010)
23. Ishai, Y., Paskin, A.: Evaluating branching programs on encrypted data. In: Vadhan, S.P. (ed.) TCC 2007. LNCS, vol. 4392, pp. 575–594. Springer, Heidelberg (2007)
24. Jarecki, S., Liu, X.: Efficient Oblivious Pseudorandom Function with Applications to Adaptive OT and Secure Computation of Set Intersection. In: Reingold, O. (ed.) TCC 2009. LNCS, vol. 5444, pp. 577–594. Springer, Heidelberg (2009)
25. Jarecki, S., Liu, X.: Fast Secure Computation of Set Intersection. In: Garay, J.A., De Prisco, R. (eds.) SCN 2010. LNCS, vol. 6280, pp. 418–435. Springer, Heidelberg (2010)
26. Katz, J., Malka, L.: Secure text processing with applications to private dna matching. In: ACM CCS, pp. 485–492 (2010)
27. Kissner, L., Song, D.: Privacy-preserving set operations. In: Shoup, V. (ed.) CRYPTO 2005. LNCS, vol. 3621, pp. 241–257. Springer, Heidelberg (2005)

28. Lindell, Y., Pinkas, B.: Privacy Preserving Data Mining. In: Bellare, M. (ed.) CRYPTO 2000. LNCS, vol. 1880, pp. 36–54. Springer, Heidelberg (2000)
29. Micali, S., Rabin, M., Vadhan, S.: Verifiable random functions. In: FOCS, pp. 120–130 (1999)
30. Micali, S., Rabin, M.O., Kilian, J.: Zero-knowledge sets. In: FOCS (2003)
31. Naor, M., Pinkas, B.: Oblivious polynomial evaluation. SIAM Journal on Computing 1254(5), 1–35 (2006)
32. Paillier, P.: Public-key cryptosystems based on composite degree residuosity classes. In: Stern, J. (ed.) EUROCRYPT 1999. LNCS, vol. 1592, pp. 223–238. Springer, Heidelberg (1999)
33. Pinkas, B., Schneider, T., Smart, N.P., Williams, S.C.: Secure two-party computation is practical. In: Matsui, M. (ed.) ASIACRYPT 2009. LNCS, vol. 5912, pp. 250–267. Springer, Heidelberg (2009)
34. Shamir, A.: On the generation of cryptographically strong pseudorandom sequences. ACM Transactions on Computer Systems 1(1), 38–44 (1983)
35. Yao, A.: Protocols for secure computations. In: FOCS, pp. 160–164 (1982)

Sub-linear, Secure Comparison with Two Non-colluding Parties

Tomas Toft

Dept. of CS, Aarhus University, Denmark
ttoft@cs.au.dk

Abstract. The classic problem in the field of secure computation is Yao's millionaires' problem; we consider two new protocols solving a variation of this: a number of parties, $P_1, \ldots, P_n$, securely hold two ℓ-bit values, x and y – e.g. x and y could be encrypted or secret shared. They wish to obtain a bit stating whether x is greater than y using only secure arithmetic; this should be done without revealing any information, even the output should remain secret. The present setting is special in the sense that it is assumed that two specific parties, referred to as Alice and Bob, are non-colluding. Though this assumption is not satisfied in general, it clearly is for the main example of this work: two-party computation based on Paillier encryption.

The first solution requires $O(\log(\ell)(\kappa + \text{loglog}(\ell)))$ secure arithmetic operations in $O(\log(\ell))$ rounds, where κ is a correctness parameter. The second solution requires only a constant number of rounds, but increases complexity to $O(\sqrt{\ell}(\kappa + \log(\ell)))$ arithmetic operations.

For the motivating setting, each arithmetic operation requires a constant number of Paillier encryptions to be exchanged between Alice and Bob. This implies that both solutions require only a *sub-linear* number of invocations (in the bit-length, ℓ) of the cryptographic primitives. This does not imply sub-linear communication, though, as the size of each encryption transmitted is more than ℓ bits.

Keywords: Secure computation, Yao's Millionaires' problem.

1 Introduction

The concept of secure multiparty computation was introduced by Yao with the now classic millionaires' problem, [Yao82]: two millionaires meet in the street and wish to determine who is richer without revealing their net worths. Since then, this problem – along with variations of it – has received much interest in the research community. Not only is it the original problem, it is also a quite fundamental primitive for secure computation. Auctions are an immediate application, e.g. as considered in [NPS99]. Indeed, they are the motivating examples of many of the papers on secure comparison mentioned below. The problem – and extensions such as determining the minimal of multiple inputs – also plays an important role elsewhere, e.g. in secure data mining such as [LP00] or [JW05] and secure optimization such as [Tof09].

D. Catalano et al. (Eds.): PKC 2011, LNCS 6571, pp. 174–191, 2011.

Secure arithmetic modulo some integer M can be seen as secure integer computation when no overflows modulo M occur. Augmenting such primitives with a protocol for secure comparison provides a setting which allows general integer computation. This approach is often used, when considering applications as those mentioned above. Thus, improving comparison – as the present work – implies improvement of a whole range of applications. These are the first such protocols which invoke the cryptographic primitives $o(\ell)$ times, where ℓ is the bit-length of the inputs.

1.1 Related Work

As the problem of secure comparison has received so much interest, there is a large body of related work, which solves the problem in various settings and with focus on different properties. Some solutions focus on theoretical efficiency – low round complexity (constant or even single round), as well as low communication or computation complexity. Others consider practice, attempting to obtain a fast, real-world solution by balancing the different resources. There are also many variations of the problem. For example: where do the inputs come from; are they known to any of the parties or are they the result of previous computation? Should the result be public, should simple actions be performed based on the outcome, or should the result be private to allow it to be used in arbitrary further computation?

There are essentially two different settings. The two-party case contains Yao circuits, [Yao86], but also includes [Fis01, BK06, DGK07, GSV07, LL07]. The latter are generally not limited to known inputs and public output. The multi-party setting can be split into the cryptographic and the information theoretic (i.t.) settings. [DFK+06, NO07] consider constant rounds solutions in the i.t. setting, while [ST06] considers computationally efficient solutions based on Paillier encryption [Pai99], i.e. the cryptographic setting. Hybrids such as [BCD+09], are also possible: their protocol is based on secret sharing, but the focus is entirely on practice, and primitives guaranteeing only computational security are applied to improve efficiency.

Though radically different, the comparison protocol of Feige et al., [FKN94], is in a sense the closest related work as it uses quadratic residues at its core. However, the basic protocol only allows comparison of integers between zero and two. Though it can be generalized, comparing anything but *very* small values appears highly impractical.

1.2 Contribution

This paper considers a novel approach at secure comparison. The setting consists of n players, $P_1, \ldots, P_n$ that jointly hold two secret, ℓ-bit inputs $x, y \in \mathbb{Z}_M$ to be compared – ordering is obtained by viewing the inputs as ℓ-bit integers. Two of the parties, Alice and Bob, are guaranteed by assumption to be non-colluding. I.e. the adversary cannot corrupt both simultaneously. The desired outcome is for the parties to hold (in a secure fashion) a bit stating whether or not $x > y$. The motivating setting is the two-party setting; for threshold schemes the protocols are limited to the case when the threshold is 1. When there are few parties (say three or four), this is completely acceptable, however, the protocols are not generally applicable.

This clearly solves the "standard" millionaires' problem: the millionaires simply provide their inputs (e.g. encrypt and send them to the relevant parties). Once the result has been determined, it can be revealed (e.g. decrypted). However, having such a comparison protocol realizes the secure integer computation setting from above – the outcome of a comparison could equally well be used in subsequent computation.

Two protocols are presented here, both based on arbitrary secure arithmetic modulo M, where M is either an RSA-modulus (e.g. Paillier encryption) or a prime (e.g. secret sharing over $\mathbb{F}_M$ where M is prime). The first solution requires $O(\log(\ell)(\kappa + \log\log(\ell)))$ arithmetic operations in $O(\log(\ell))$ rounds, where κ is a correctness parameter. The second solution is constant-rounds, but increases complexity to $O(\sqrt{\ell}(\kappa + \log(\ell)))$. With the exception of [FKN94], all previous solutions known to the author utilize access to the binary representation of the inputs (or the binary representation of random values related to the inputs). Hence, this is the first solution that requires less than a linear number of invocations of the primitives in the bit-length of the inputs.

Security against passive adversaries follows almost entirely from the security of the underlying arithmetic. However, this is not the case with respect to active adversaries. The problem is that the protocol specifies Alice and Bob to provide inputs of bounded size, which cannot be verified efficiently using only arithmetic. Fortunately such a primitive exists in many settings including Paillier based ones as well as based on Shamir sharing over a prime field, $\mathbb{F}_M$.

1.3 An Overview of This Paper

Section 2 introduces the setting and primitives in the form of an ideal functionality; Sect. 3 then explains how this can be realized. Secure equality testing is then presented in Sect. 4; this is needed for both solutions. The two contributions are then presented in Sect. 5 and Sect. 6. Finally, two variations are noted in Sect. 7, before the concluding remarks of Sect. 8.

2 Primitives and Notation

The basic setting will consist of an ideal functionality providing the underlying primitives, the main one being secure $\mathbb{Z}_M$ arithmetic, i.e. it is an arithmetic black-box (ABB), [DN03]. This approach ensures that it is possible to ignore the details of the underlying primitives; focus is on the required properties rather than on specific solutions.

2.1 The Arithmetic Black-Box

The arithmetic black-box allows n parties, $P_1, \ldots, P_n$, to securely store and retrieve elements of a ring $\mathbb{Z}_M$. Here, M will be either a prime or an RSA-modulus, i.e. the product of two odd primes.

The secure storage (input/output) can be thought of as secret sharing, and we borrow that notation, writing stored values in square brackets, $[x]$. However, as noted other solutions are also possible, and the notation could equally well represent encryption: secure storage could be obtained by having one or more parties store encryptions under some public key. Input then means "encrypt and send to these parties," while output means "decrypt and broadcast." Naturally the decryption key itself must not be held by anyone also holding the encryptions, however, it could be secret shared or held by a different party. The ABB can also provide an output to only a *single* party. This is always achievable, e.g. by additively masking the output with a random value provided by the receiving party. For specific primitives, other solutions may be preferable, though.

In addition to secure storage, the ideal functionality allows arithmetic computation on the stored values. Such computation is written using infix notation in the "plaintext" space, e.g.

$$[x \cdot y + z] \leftarrow [x] \cdot [y] + [z] .$$

The actual operations to be performed depend on the details of the realization. Presently it does not matter whether an operation is a protocol invocation (e.g. the multiplication protocol of Ben-Or et al. [BGW88]) or simply local computation by one or more parties (e.g. multiplying two Paillier encryptions provides an encryption of the sum of the two plaintexts).

Complexity of a protocol based on the ABB is found by simply counting the number of operations (input/output or arithmetic[1]) performed. It is assumed that the ABB allows operations to be performed concurrently (representing e.g. executing multiple multiplication protocols in parallel), thus, round complexity will be the number of sequential operations performed.

2.2 Required Extensions of the ABB

The ABB-setting considers n parties, $P_1, \ldots, P_n$. The present work requires two of these – denoted Alice and Bob – to be mutually incorruptible, i.e. at least one of them will remain honest. We do not specify the corruption sets further. Indeed, the remaining $n-2$ parties are ignored. The ideas are best explained by focusing on Alice, Bob, and the ABB. Including the "helper parties"[2] would complicate the explanation unnecessarily.

[1] Similarly to other work, we focus on communication complexity and assume that the underlying primitives are additively homomorphic. Thus, only multiplications are counted.

[2] Denoting the remaining $n-2$ parties as "helpers" is not completely fair. They may only be assisting during the present comparison protocols, but could have an equal stake in the overall computation.

As noted above, the present protocols do not provide active security based only on the arithmetic black-box. Even when actively secure arithmetic is given, malicious parties can still deviate. Two extensions to the ABB are therefore needed in order to eliminate these issues. First, given an input, it must be verifiable (using only constant work) that it is less than some public bound. This allows the protocol to specify that an input must come from a small, specified range. Second, it must be verifiable that two held values are indeed equal.

Finally, to improve readability we introduce a bit of syntactic sugar, writing

$$[b]\ ?\ [x] : [y]$$

to denote conditional selection between two secret values, $[x]$ and $[y]$, based on an secret bit, $[b]$. This can be achieved using arithmetic only, and the expression is simply shorthand for

$$[b]\,([x] - [y]) + [y]$$

which clearly equals either x or y depending on $b \in \{0, 1\}$.

3 Realizing the Arithmetic Black-Box

The arithmetic black-box can be realized in different settings and be based on various primitives. However, as the present setting requires two of the parties to be mutually incorruptible, our prime example is two-party computation based on Paillier encryption [Pai99]. The realization is sketched in Sect. 3.1. It is stressed that realizations with more than two parties are also relevant. Two possibilities are presented in Sect. 3.2; both provide security against active adversaries. The realizations of the extensions needed to provide active security below are provided in Sect. 3.3.

3.1 Passively Secure, Two-Party Paillier-Based Arithmetic

Paillier's encryption scheme [Pai99] is a semantically secure, additively homomorphic, public-key cryptosystem over $\mathbb{Z}_N$, where $N = pq$ is an RSA-modulus. In the simple, two-party setting, Bob holds a copy of Alice's public key. To provide an input means to hand Bob a fresh encryption of it. Output is realized by sending the relevant encryption to Alice (rerandomizing it first); she then decrypts and returns the plaintext to Bob. At this point both parties know the value in question.

Regarding the realization of the secure arithmetic, the homomorphic property allows Bob to compute linear combinations of encryptions that he holds. However, he is unable to obtain encryptions of products without the help of Alice. For this he uses the standard protocol seen in Fig. 1, where subtraction simply means "invert and add." (In the ciphertext domain this means "multiply by the multiplicative inverse.") Correctness follows from

$$\begin{aligned} xy &= (xy + r_y x + r_x y + r_x r_y) - r_y x - r_x y - r_x r_y \\ &= (x + r_x)(y + r_y) - r_y x - r_x y - r_x r_y \\ &= x'y' - r_y x - r_x y - r_x r_y. \end{aligned}$$

Alice: sk	$pk = N$	Bob: $[x]$,$[y]$
		$r_x \in_R \mathbb{Z}_N$
		$r_y \in_R \mathbb{Z}_N$
		$[x'] \leftarrow [x] + r_x$
		$[y'] \leftarrow [y] + r_y$
	$\xleftarrow{[x'],[y']}$	
$x' \leftarrow \text{decr}([x'])$		
$y' \leftarrow \text{decr}([y'])$		
	$\xrightarrow{[x'y']}$	
		$[\sigma] \leftarrow r_x[y] + r_y[x] + r_x r_y$
		$[xy] \leftarrow [x'y'] - [\sigma]$

Fig. 1. Multiplication of two encryptions

On the intuitive level, this clearly realizes the arithmetic black-box. For input/output Alice's view consists only of the encryptions of outputs she receives, while Bob only sees the outputs and encryptions under Alice's key. For the multiplication protocol, Alice receives two encryptions of masked (i.e. uniformly random) values, while Bob simply receives two additional encryptions. Neither provides any information.

Formally simulating these views is simple. Sketching the ideas, Alice must be handed fresh encryptions, either of the output to be received (which the simulator knows) or a random value in the case of the multiplication protocol. Bob on the other hand expects to see encryptions of the values in question. The simulator cannot generate these, but from Bob's point of view they are indistinguishable from arbitrary encryptions, as the scheme is semantically secure. Hence any fresh encryptions under Alice's (simulated) key will do.

3.2 The Multiparty Case

The arithmetic black-box can also be realized in a multiparty setting. Note that both of the options mentioned are secure against active adversaries, but may be simplified, if it is assumed that the adversary is honest-but-curious only, i.e. if the corrupt parties follow the protocol as specified.

The first candidate consists of using Shamir's secret sharing scheme over the prime field $\mathbb{F}_M$ along with the protocols of Ben-Or et al. [Sha79, BGW88]. This results in a threshold scheme, which is not applicable in general. However, if the threshold is 1, then at most one party will be corrupt. In that case *any two parties* can take the role of Alice and Bob.

A multiparty solution based on Paillier encryption is also possible using the techniques of Cramer et al. [CDN01]. In that setting, all parties hold all ciphertexts, while the key is secret shared among them. Again, a threshold solution allowing at most one corrupt party is a possible setting. Note that though not presented so, it is possible to use the protocols of [CDN01] with a threshold $t \geq n/2$; this provides an alternative, two-party setting, which is more suited when considering active adversaries.

3.3 Active Security

Both solutions of the previous section realize the arithmetic black-box in the presence of active adversaries. It remains to provide the required extensions, i.e. to describe how to verify that two values are equal, and how to demonstrate that an input is of bounded size.

Equality of two values is easily verified using only arithmetic by outputting their difference. This is of course not secure in general, however, in the present setting, one of the secret values depends only on inputs from one party (who of course knows the actual value). It can therefore be viewed as coming directly from that party. For an honest party, the output is always 0, while for a corrupt party, the adversary will know the output in advance. Hence no new information can be obtained. It is noted that for specific primitives, more direct solutions may be more efficient.

The second problem is to verify that an input provided by some party is indeed of bounded size. This can be done by taking a detour over the integers, as any non-negative integer can be written as the sum of the squares of four integers. For a Paillier based setting, Schoenmakers and Tuyls [ST06] note that this can be achieved using integer commitments [Bou00, Lip03, DJ02]. For the setting based on Shamir sharing, a similar proof that a value is of bounded size can be obtained using linear integer secret sharing, [Tho09]. Sketching the latter solution, the key idea is to first verify that a value (shared over the integers) is of the desired size (i.e. provide the four integers). This sharing is then converted to a Shamir sharing over $\mathbb{F}_M$.

4 Secure Equality Testing

An additional primitive is needed before the comparison protocols can be presented: securely determining a secret bit stating if two values, $[x]$ and $[y]$, are equal. The protocol is a variation of the secure equality testing of secret shared values, [NO07, Tof07]; the latter work notes that the ideas generalize to the case of multiparty computation based on Paillier encryption. Note that in specific settings, specialized variations of the present protocols will most likely be preferable to the general solution presented here.

The main idea is seen in Fig. 2. To test equality of two values, it suffices to test if their difference, $[d]$, is 0. If this is the case, then adding a random square, $[r]^2$, results in a value with Jacobi symbol $j_{d+r^2} = \left(\frac{d+r^2}{M}\right) = 1$. If $[d] \neq 0$, however, then the Jacobi symbol is -1 with probability roughly 1/2,

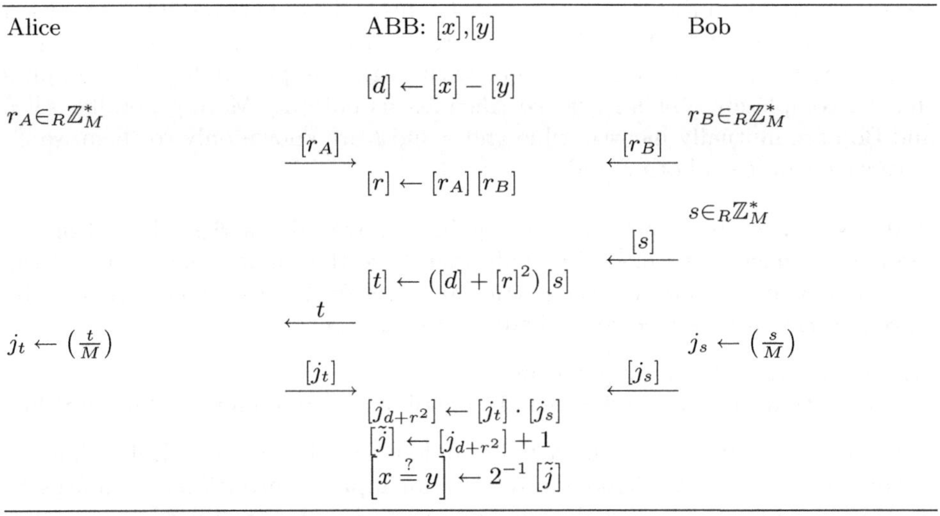

Fig. 2. Testing equality (with error probability $\approx 1/2$)

[Per52, Tof07]. The Jacobi symbol can therefore be used as a test of equality[3]. As the Jacobi symbol of a product is the product of the Jacobi symbols of the factors, and the multiplicative inverse of ± 1 is itself, $[j_{d+r^2}]$ is correctly computed when $[j_s]$ and $[j_t]$ are the Jacobi symbols of $[s]$ and $[t]$ respectively. The final computation simply maps -1 to 0 while preserving a 1.

A false positive – an incorrect 1 – occurs with a probability of roughly 1/2. However, if the protocol is repeated κ times on the same input, the probability that *all* executions provide false positives is negligible. Determining if all invocations have returned 1 can be done by computing the κ-ary fan-in AND of the test results. This can be done in $O(1)$ rounds using $O(\kappa)$ arithmetic operations as described in [DFK$^+$06]. The basic idea is to compute the sum of the bits plus one, $[\sigma]$, and use this as input to the known, $(\kappa + 1)$-degree polynomial mapping $1, \ldots, \kappa$ to 0 and $\kappa + 1$ to 1. The powers of $[\sigma]$ – $[\sigma], [\sigma^2], \ldots, [\sigma^{\kappa+1}]$ – can be computed in constant rounds using $O(\kappa)$ multiplications by utilizing (a simple variation of) the unbounded fan-in multiplication protocol of Bar-Ilan and Beaver, [BB89].

Correctness. Correctness follows by the intuition above: each test always returns an encryption of 1 when $x = y$, and the logical AND of these bits is therefore 1 as well. When $x \neq y$, then except with negligible probability in κ, at least one of the tests will successfully determine this, i.e. return 0. In this case the logical AND is also 0.

[3] As M is either the product of large primes or a large prime itself, $j_{d+r^2} = 0$ occurs with negligible probability. In some settings, the issue can be eliminated using arithmetic, e.g. by considering $\left(\frac{d^2+r^2}{M}\right)$ when $M \equiv 3 \bmod 4$ is prime. There the additive inverse of a quadratic residue will never be a quadratic residue itself.

Passive security. In Fig. 2, Alice receives t. This is the only potential information leak, as the arithmetic black-box is secure by definition. As M only has large prime factors, $d + r^2$ is in $\mathbb{Z}_M^*$, except with negligible probability. This implies that $[s]$ completely blinds it, hence Alice learns nothing. More generally, Alice and Bob are mutually incorruptible and s and t are known only to them, so no adversary will learn both s and t.

Active security. To ensure security against active adversaries, the actions of the parties must be verified. The only issues are the inputs, as security of the arithmetic is immediate. I.e. the parties must verify that the values provided by Alice and Bob are as specified. There are two issues:

1. Demonstrate that an input is in $\mathbb{Z}_M^*$.
2. Verify that an input is the Jacobi symbol of a stored, inputter-known value.

Note that this suffices; $[r]$ is uniformly random as either Alice or Bob is honest.

For Alice or Bob to demonstrate that an input is invertible, it suffices to provide an additional uniformly random invertible value, and have the product of the two revealed. All parties can then verify that that the product is invertible, which implies the same of both factors. This reveals no other information, as the second element masks the real value.

The second issue is slightly more difficult. For Alice to demonstrate that $[j_t]$ is the Jacobi symbol of $[t]$ she provides uniformly random pairs, $([j_i], [t_i])$ for $1 \leq i \leq k$, such that $j_i = \left(\frac{t_i}{M}\right)$. Bob then flips k coins, $b_1, \ldots, b_k$; the ABB is then asked to compute and reveal either the pair $([j_i] \cdot [j_t], [t_i] \cdot [t])$ if b_i is 1 or $([j_i], [t_i])$ otherwise. All parties then verify that the Jacobi symbol of each revealed pair matches the element. To cheat, Alice would have to guess *all* of Bob's coin flips – this can only occur with probability negligible in k. Note that no information is revealed when Alice is honest: the parties merely see random pairs, $([j_i], [t_i])$, or maskings of $[t]$ and its Jacobi symbol. For Bob to demonstrate the same, the roles are simply reversed.

It is often possible to do better than the general solution. When, for example, M is prime, elements with known Jacobi symbol can be efficiently constructed using only arithmetic:

$$[x] \leftarrow [x']^2 \cdot \left(2^{-1}([j_x] + 1) \mathbin{?} 1 : \omega\right),$$

where ω is a fixed element with Jacobi symbol -1, [NO07, Tof07].

A similar calculation is possible when M is the product of two primes both congruent to 3 modulo 4. In this case -1 is a non-residue with Jacobi symbol 1, hence any element $[x]$ can be written as

$$([b] \mathbin{?} 1 : -1)\, [x']^2 \cdot \left(2^{-1}([j_x] + 1) \mathbin{?} 1 : \omega\right)$$

where $[b]$ is a bit. (If Alice cannot compute b, i.e. distinguish quadratic residue from non-residues, she can instead provide a uniformly random value with the same Jacobi symbol. The parties can then reveal the product of this and the actual value to Bob, who verifies that its Jacobi symbol is 1).

Complexity. For passive security, when M is prime, or when M is the product of two primes both congruent to 3 modulo 4, executing κ copies of Fig. 2 in parallel and performing the κ-ary fan-in AND requires $O(\kappa)$ ABB-operations executed in $O(1)$ rounds. In other settings – i.e. when it is not possible to efficiently verify a Jacobi symbol – arithmetic complexity increases by a factor of k. For simplicity we can view this as being incorporated into κ – say $k = \sqrt{\kappa}$ – though strictly speaking it should be treated separately.

5 The log-Rounds Protocol

Based on the arithmetic black-box presented in Sect. 2 – including extensions such as the equality test – the log-rounds protocol can now be explained. In a sense, this paper follows the same overall strategy as many of the previous solutions: transform the problem to a comparison of "known" values and determine their most significant differing bit-position; this provides the result. What differs are the means to achieve this goal.

On the intuitive level, the approach can be viewed as a binary search, though strictly speaking, this is not the case. The full protocol is seen in Fig. 3 and is explained during the argument of correctness. Assume for simplicity that ℓ is a power of two. This can always be ensured by viewing x and y as numbers of larger (but less than double) size. Further, assume that the modulus of the ABB is much larger than than the input size, $M \gg 2^{\ell+\kappa'}$ for security parameter κ'.

Correctness. Correctness of the protocol is quite simple, once the intuition is understood. Therefore, rather than presenting the protocol from beginning to end, we present the ideas. This requires starting at both the end *and* the beginning at the same time, and working our way towards the middle. It explains *why* a given computation is performed by showing *how* it helps solve the original problem.

Initially the arithmetic black-box is used to compute the value $[z]$. As $x \geq y \Leftrightarrow z \geq 2^\ell$, we find that if $[z_\ell]$ really is an encryption of the ℓ'th bit of z, then the result is correct. This is the case if $\tilde{z} = z \bmod 2^\ell$: $z - (z \bmod 2^\ell)$ sets all the ℓ least significant bits of the $(\ell+1)$-bit z to 0 leaving only the top bit. The multiplication by $2^{-\ell}$ (modulo M) simply shifts this down to the desired position.

To perform the modulo reduction of $[z]$, Bob provides a $(2^{\ell+\kappa'})$-bit mask, $[r]$, which is added to $[z]$. The outcome, $[c]$, is then revealed to Alice. As it was assumed that $M \gg 2^{\kappa'+\ell}$, the computation of c can be viewed as occurring over the integers. Therefore

$$z \equiv c - r \mod 2^\ell.$$

Both c and r are easily reduced modulo 2^ℓ. Alice knows the former and Bob the latter, so they may simply supply these values, $[\bar{c}]$ and $[\bar{r}]$. Subtracting the latter from the former provides the desired result, however, this subtraction must occur modulo 2^ℓ, not modulo M.

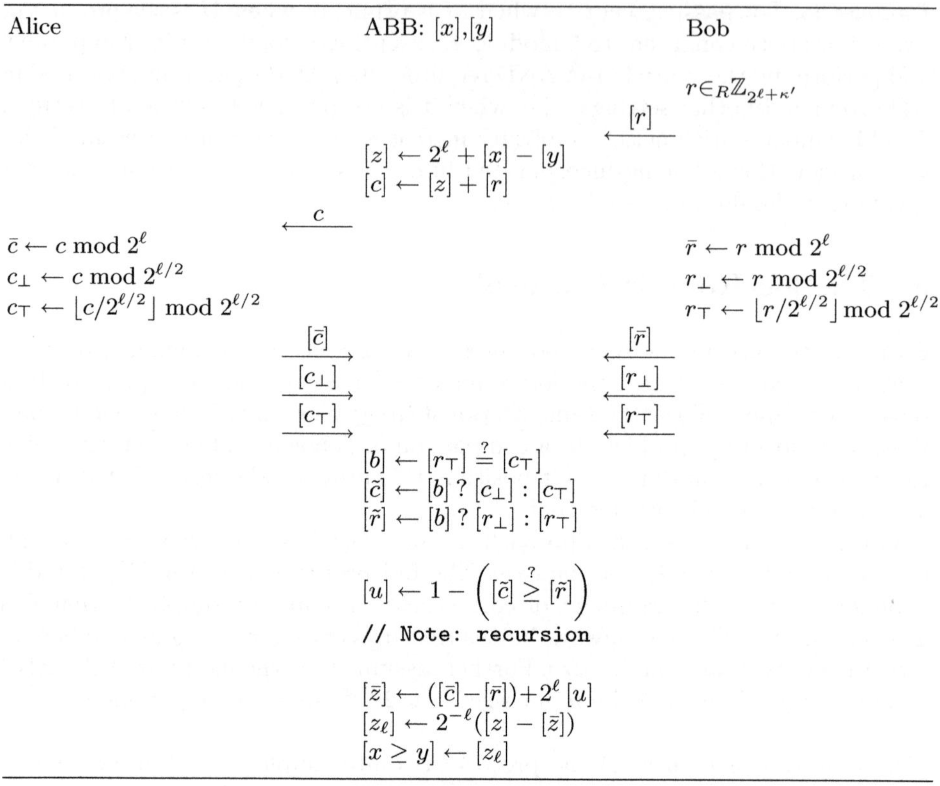

Fig. 3. The $\log(\ell)$-rounds comparison protocol

However, by considering two cases, the desired operation can be simulated with $\mathbb{Z}_M$ arithmetic:

1. If $\bar{c} \geq \bar{r}$, then subtraction modulo M provides the correct result.
2. If $\bar{c} < \bar{r}$, then subtracting $\bar{r}$ results in an underflow modulo M. Adding 2^ℓ at this point provides the correct result.

Unfortunately the parties do not know which case they are in – nor should they learn it – and therefore do not know if they should instruct the ABB to add 2^ℓ.

This leaves us with the problem of comparing $[\bar{c}]$ and $[\bar{r}]$, and while it seems as if we are back at the initial problem, this is not the case, as Alice and Bob each know one of these values. Thus, they can be decomposed into the top and bottom halves – the $\ell/2$ least significant and most significant bits – by having the parties input them. These are denoted $[c_\top]$, $[c_\bot]$, $[r_\top]$, and $[r_\bot]$.

Again there are two cases to consider, and again the parties do not know which is the case:

1. If $c_\top = r_\top$, i.e. if the most significant half of the bits are equal, then it suffices to compare the values representing the least significant bits.

2. If $c_\top \neq r_\top$, i.e. if the top halves differ, then the least significant bits can be ignored, and it suffices to compare $c_\top$ and $r_\top$.

Using the equality test, the parties can determine a secret bit stating which is the case. Based on this, they obliviously choose between the top and bottom halves of $[\bar{c}]$ and $[\bar{r}]$ – i.e. $[r_\top]$, $[c_\top]$ and $[r_\bot]$, $[c_\bot]$ – storing the results as $[\tilde{r}]$ and $[\tilde{c}]$.

The concluding computation consists of determining which of $[\tilde{r}]$ and $[\tilde{c}]$ contains the larger value. This *is* the original problem, however, the bit-length of the numbers is now only $\ell/2$. If $\ell/2 > 1$, then the protocol calls itself recursively. Otherwise – i.e. if the values to be compared are single-bit – it is straightforward to compute an encryption of $\tilde{r} > \tilde{c}$ using only secure arithmetic:

$$(\tilde{c} \oplus \tilde{r})\tilde{r} = \tilde{r} - \tilde{c}\tilde{r}.$$

To see that this is the correct result, note that it is 1 exactly when the bits differ and $\tilde{r}$ is set.

Security. As the arithmetic black-box is secure by definition, an adversary can only obtain information or affect the computation through the input/output. Leakage can only occur through the value, c, received by a corrupt Alice. However, this is statistically indistinguishable (in κ') from a uniformly random $(\ell + \kappa')$-bit value, as $[z]$ is masked by $[r]$ provided by the honest Bob.

Similarly, an adversary can only affect the computation through incorrect inputs. Hence, if it is verified that $[\bar{c}]$, $[c_\top]$, $[c_\bot]$, $[\bar{r}]$, $[r_\top]$, and $[r_\bot]$ are indeed the correct "sub-strings" of $[c]$ and $[r]$ (and that $[r]$ is indeed $\ell + \kappa'$ bits long), then no adversarial behavior is possible.

This verification can be performed by having Alice and Bob provide $[c_I]$ and $[r_I]$, the (ignored) top κ' bits of $[c]$ and $[r]$, as well. Initially it is verified that $[c_\top]$, $[c_\bot]$, $[r_\top]$, and $[r_\bot]$ are all $\ell/2$ bits, and that $[c_I]$ and $[r_I]$ are κ' bits. Then, the parties check that $[\bar{c}] = 2^{\ell/2}[c_\top]+[c_\bot]$, $[\bar{r}] = 2^{\ell/2}[r_\top]+[r_\bot]$, $[c] = 2^{\ell}[c_I]+[\bar{c}]$, and $[r] = 2^{\ell}[r_I]+[\bar{r}]$. At this point it has been ensured that the decomposition of $[c]$ and $[r]$ are correct. Note that there is no need to explicitly verify that $[\bar{c}]$ and $[\bar{r}]$ are ℓ bits, nor that $[r]$ is $\ell + \kappa'$ bits. This has already been verified implicitly.

Complexity. For a reduction of the problem to one of half size – Fig. 3 except for the recursive invocation – one invocation of the equality test and a constant amount of input/output and arithmetic is required. Overall this means $O(\kappa)$ arithmetic and input/output operations are needed, where κ is the correctness parameter for the equality test. These can be performed in $O(1)$ rounds. Each iteration reduces the problem to one of half size, thus only $\log(\ell)$ steps are required until the bit-length reaches 1, at which point the remaining work is constant. This implies that the overall complexity is $O(\kappa \log(\ell))$ operations in $O(\log(\ell))$ rounds.

Note that to ensure correctness of the full protocol, *all* invocations of the equality test must succeed. As the number of tests depends on ℓ, so must κ. Adding $\log\log(\ell)$ provides the desired error probability, as $(1 - 2^{-\kappa})^{\log(\ell)} \approx 1 - \log(\ell)2^{-\kappa}$

(since $2^{-\kappa}$ is negligible)[4]. This implies $O(\log(\ell)(\kappa+\log\log(\ell)))$ operations overall, to ensure correctness except with probability negligible in κ.

Theorem 1. *Given two mutually incorruptible parties, Alice and Bob, and two ℓ-bit values, $[x]$ and $[y]$, stored within an arithmetic black-box providing secure arithmetic in $\mathbb{Z}_M$ (where $M > 2^{\ell+\kappa'+\log\log(\ell)}$ is prime or an RSA-modulus and κ' is a security parameter), the parties may obtain a bit, $[b]$, such that b is 1 iff $x \geq y$ (except with probability negligible in the correctness parameter κ) using $O(\log(\ell)(\kappa + \log\log(\ell)))$ ABB-operations in $O(\log(\ell))$ rounds.*

6 The Constant-Rounds Protocol

Decreasing the round complexity to constant without considering the individual bits is achieved by combining the ideas of the previous section with earlier approaches. Rather than going over *all details*, we present the overall protocol in Fig. 4, and only reference existing sub-protocols in the analysis. This avoids muddling the presentation with details of existing protocols.

Correctness. The protocol starts and ends exactly as the log-rounds solution, hence if $[u]$ correctly states if an underflow occurs in the subtraction, then the correct result is determined. In difference to above, $[\bar{c}]$ and $[\bar{r}]$ are split into $\sqrt{\ell}$ blocks rather than two (the bit-length can be padded to ensure that it is a square). This can be viewed as writing the numbers in $2^{\sqrt{\ell}}$-ary notation. As in the previous section, it suffices to only compare the most significant differing block. There are more of them, but the goal remains the same: the parties must obliviously find and select the block in question using only the ABB.

To do this, an equality test is performed for each block: $[b_i]$ states if the i'th block of the numbers are equal. $\left[\tilde{b}_i\right]$ is the logical AND of the top blocks, i.e. from the most significant one down to the i'th. It is 1 exactly when all these blocks are equal. Thus, the most significant differing block will be the first one containing a 0 (starting with the most significant one). The $\left[\tilde{b}_i\right]$ of the less significant blocks will of course also be 0, as there is a more significant, differing position. The value $[b'_i] = \left[\tilde{b}_{i+1}\right] - \left[\tilde{b}_i\right]$ states if the i'th block is the desired one. For the most significant, differing block it will be $1 = 1 - 0$, while the rest have $b'_i = 0$, either from $1 - 1$ or $0 - 0$. Thus, the $[b'_i]$ that is set can be viewed as a pointer to the correct block position, i, and $\left[r^{(i)}\right]$ and $\left[c^{(i)}\right]$ are selected by the sums[5].

Concluding, the $\sqrt{\ell}$-bit $[\tilde{c}]$ and $[\tilde{r}]$ must be compared. This time, we cannot proceed recursively, as this would not result in a constant-rounds protocol. Instead, the parties transform the problem to one where Alice holds one input and

[4] A similar increase is needed for the security parameter, κ'. This does not change complexity though.

[5] If the inputs are equal this is not true, but as we get $\tilde{c} = \tilde{r} = 0$ which provides the same result, it is acceptable.

Alice		ABB: $[x]$,$[y]$		Bob
				$r \in_R \mathbb{Z}_{2^{\ell+\kappa'}}$
			$\xleftarrow{[r]}$	
		$[z] \leftarrow 2^{\ell} + [x] - [y]$		
		$[c] \leftarrow [z] + [r]$		
	$\xleftarrow{c}$			
$\bar{c} \leftarrow c \bmod 2^{\ell}$				$\bar{r} \leftarrow r \bmod 2^{\ell}$
	$\xrightarrow{[\bar{c}]}$		$\xleftarrow{[\bar{r}]}$	
for $i = 0..\sqrt{\ell} - 1$ **do**				**for** $i = 0..\sqrt{\ell} - 1$ **do**
$\hat{c}^{(i)} \leftarrow \lfloor c/2^{i\sqrt{\ell}} \rfloor$				$\hat{r}^{(i)} \leftarrow \lfloor r/2^{i\sqrt{\ell}} \rfloor$
$c^{(i)} \leftarrow \hat{c}^{(i)} \bmod 2^{\sqrt{\ell}}$				$r^{(i)} \leftarrow \hat{r}^{(i)} \bmod 2^{\sqrt{\ell}}$
	$\xrightarrow{\left[c^{(i)}\right]}$		$\xleftarrow{\left[r^{(i)}\right]}$	
od				**od**
		for $i = 0..\sqrt{\ell} - 1$ **do**		
		$[b_i] \leftarrow [r^{(i)}] \stackrel{?}{=} [c^{(i)}]$		
		od		
		$\left[\tilde{b}_{\sqrt{\ell}}\right] \leftarrow 1$		
		for $i = \sqrt{\ell} - 1..0$ **do**		
		$\left[\tilde{b}_i\right] \leftarrow \bigwedge_{j=i}^{\sqrt{\ell}-1} [b_j]$		
		$[b'_i] \leftarrow \left[\tilde{b}_{i+1}\right] - \left[\tilde{b}_i\right]$		
		od		
		$[\tilde{c}] \leftarrow \sum_{i=0}^{\sqrt{\ell}-1} [b'_i] \left[c^{(i)}\right]$		
		$[\tilde{r}] \leftarrow \sum_{i=0}^{\sqrt{\ell}-1} [b'_i] \left[r^{(i)}\right]$		
		$[u] \leftarrow 1 - \left([\tilde{c}] \stackrel{?}{\geq} [\tilde{r}]\right)$		
		$[\tilde{z}] \leftarrow ([\bar{c}] - [\bar{r}]) + 2^{\ell} [u]$		
		$[z_\ell] \leftarrow 2^{-\ell}([z] - [\tilde{z}])$		
		$[x \geq y] \leftarrow [z_\ell]$		

Fig. 4. The constant-rounds comparison protocol

Bob the other – the initial step of the present protocols. Then, these numbers are bit-decomposed – Alice and Bob provide the binary representation of the numbers – and the problem is brute-forced, e.g. using [DFK+06]. As there are only $\sqrt{\ell}$ bits, linear work is acceptable at this point.

Security. Privacy of the protocol follows by the exact same argument as above: leakage can only happen through c, which is statistically indistinguishable from a uniformly random $(\ell + \kappa')$-bit value. Further, as above, an active adversary can only affect the protocol through inputs. To avoid malicious behavior, it must merely be verified that $[r]$ is indeed of the proper bit-length; that $[r]$ and $[c]$ are decomposed properly into the $\left[c^{(i)}\right]$ and $\left[r^{(i)}\right]$; and that $[\bar{c}]$ and $[\bar{r}]$ are correct. In addition to this, it must be verified that the bit-decomposition needed in the comparison of $[\tilde{c}]$ and $[\tilde{r}]$ is correct. All this can be performed analogously to the constructions of the previous section – the difference is that a $2^{\sqrt{\ell}}$-ary or binary representation is considered rather than a $2^{\ell/2}$-ary one.

Complexity. Initially, Bob provides the mask, $[r]$, after which Alice obtains c, the masking of $[z]$. They decompose these to their $2^{\sqrt{\ell}}$-ary representations and provide these as inputs to the ABB. Complexity of this is clearly $O(\sqrt{\ell})$ input/output operations. Moreover, only three rounds are needed as all elements of the decompositions may be input concurrently.

Regarding the work performed by the ABB, clearly the computation of $[z]$ and $[c]$ is constant. Following this, $\sqrt{\ell}$ equality tests are performed; each of these requires $O(\kappa)$ operations implying $O(\sqrt{\ell} \cdot \kappa)$ operations overall. Round complexity remains constant, though: no test depends on the output of another, so they may be executed in parallel. The next step is to compute the "pointer," $[b'_i]$. The most expensive part of this is the $\sqrt{\ell}$ $\sqrt{\ell}$-ary logical AND's. The naive solution requires $\Omega(\ell)$ operations which is too expensive. However, the AND-gates share the same inputs – overall it is simply a prefix-AND of the b_i. Thus, the computation can be performed with linear work in a constant number of rounds by applying the techniques of [DFK+06]. Concluding the computation are the selection of $[\tilde{r}]$ and $[\tilde{c}]$ each requiring $O(\sqrt{\ell})$ arithmetic operations. This is followed by the brute-force (linear in $\sqrt{\ell}$) comparison and the final, constant-size computation.

The dominating term of all of this is the equality tests, due to their factor of κ. Thus, the overall complexity is $O(\sqrt{\ell}(\kappa+\log(\ell)))$, where the $\log(\ell)$-term is needed to ensure that the error probability remains negligible in κ when performing all $\sqrt{\ell}$ tests.

Theorem 2. *Given two mutually incorruptible parties, Alice and Bob, and two ℓ-bit values, $[x]$ and $[y]$, stored within an arithmetic black-box providing secure arithmetic in $\mathbb{Z}_M$ (where $M > 2^{\ell+\kappa'}$ is prime or an RSA-modulus and κ' is a security parameter), the parties may obtain a bit, $[b]$, such that b is 1 iff $x \geq y$ (except with probability negligible in the correctness parameter κ) using $O(\sqrt{\ell}(\kappa + \log(\ell)))$ ABB-operations in $O(1)$ rounds.*

7 Variations

Handling arbitrary inputs of $\mathbb{Z}_M$. Protocols allowing arbitrary inputs in $\mathbb{Z}_M$ are also possible. This involves comparing both inputs, $[x]$ and $[y]$, as well as $[x - y]$ to $\lfloor M/2 \rfloor$, [NO07]. The answer can be determined using a small arithmetic circuit. Comparison with $\lfloor M/2 \rfloor$ is equivalent to doubling and extracting the least significant bit, which translates to a comparison of values held by Alice and Bob. At this point the present ideas may be applied.

Improved complexity in the constant-rounds case. Complexity of the constant-rounds protocol can be improved slightly. The idea is to split the numbers into $\sqrt{\ell/\kappa}$ blocks of size $\sqrt{\ell\kappa}$. Determining and selecting the relevant block requires only $O(\sqrt{\ell\kappa})$ arithmetic operations, and while the new comparison problem grows to size $\sqrt{\ell\kappa}$, it is still acceptable to brute-force this.

A second approach allows complexity to be reduced further at the cost of extra rounds. Splitting the numbers into $\sqrt[3]{\ell}$ blocks results in a new problem of size $(\sqrt[3]{\ell})^2$, which may be solved with $O(\sqrt[3]{\ell} \cdot \kappa)$ work. This amount is also needed to reduce the size of the problem, thus, it is also the overall complexity. The idea generalizes to $O(c)$-round $O(\sqrt[c]{\ell} \cdot \kappa)$ solutions for any c.

Hybrid protocols for practice. Due to the blowup of κ in the complexity, theoretically worse solutions may be preferable to the log-rounds protocol for small inputs. This suggests that a hybrid approach will be best in practice: Initially the log-rounds solution can be applied repeatedly to reduce the size of the problem, but at some point (when the problem is small enough) another solution will be better. At this point, that solution may be applied instead of continuing with the recursive approach.

8 Concluding Remarks

The protocols presented demonstrate that in order to perform secure comparison, the explicit binary representation does not have to be considered. Rather, by testing equality of "sub-strings" of the full problem (where this test occurs on elements of the ring or field), the size of the problem can be reduced without having to consider each bit-position individually. This implies an improved theoretic complexity over all previous solution (regarding the number of cryptographic invoked).

A sub-linear comparison protocol for the general multiparty setting is left as an open problem. One obvious possibility would be to generate encryptions or sharings of uniformly random values of bounded size, say k-bit, which are unknown to all. It is not clear how to do this without also generating the binary representation as well, though.

The author would like to thank Ivan Damgård, Martin Geisler, and the anonymous referees for their many remarks and suggestions.

References

[BB89] Bar-Ilan, J., Beaver, D.: Non-cryptographic fault-tolerant computing in a constant number of rounds of interaction. In: Rudnicki, P. (ed.) Proceedings of the Eighth Annual ACM Symposium on Principles of Distributed Computing, pp. 201–209. ACM Press, New York (1989)

[BCD+09] Bogetoft, P., Christensen, D.L., Damgård, I., Geisler, M., Jakobsen, T., Krøigaard, M., Nielsen, J.D., Nielsen, J.B., Nielsen, K., Pagter, J., Schwartzbach, M., Toft, T.: Secure multiparty computation goes live. In: Dingledine, R., Golle, P. (eds.) FC 2009. LNCS, vol. 5628, pp. 325–343. Springer, Heidelberg (2009)

[BGW88] Ben-Or, M., Goldwasser, S., Wigderson, A.: Completeness theorems for noncryptographic fault-tolerant distributed computations. In: 20th Annual ACM Symposium on Theory of Computing, pp. 1–10. ACM Press, New York (1988)

[BK06] Blake, I.F., Kolesnikov, V.: Conditional encrypted mapping and comparing encrypted numbers. In: Di Crescenzo, G., Rubin, A. (eds.) FC 2006. LNCS, vol. 4107, pp. 206–220. Springer, Heidelberg (2006)

[Bou00] Boudot, F.: Efficient proofs that a committed number lies in an interval. In: Preneel, B. (ed.) EUROCRYPT 2000. LNCS, vol. 1807, pp. 431–444. Springer, Heidelberg (2000)

[CDN01] Cramer, R., Damgård, I.B., Nielsen, J.B.: Multiparty computation from threshold homomorphic encryption. In: Pfitzmann, B. (ed.) EUROCRYPT 2001. LNCS, vol. 2045, pp. 280–300. Springer, Heidelberg (2001)

[DFK+06] Damgård, I., Fitzi, M., Kiltz, E., Nielsen, J.B., Toft, T.: Unconditionally secure constant-rounds multi-party computation for equality, comparison, bits and exponentiation. In: Halevi, S., Rabin, T. (eds.) TCC 2006. LNCS, vol. 3876, pp. 285–304. Springer, Heidelberg (2006)

[DGK07] Damgård, I., Geisler, M., Krøigaard, M.: Efficient and Secure Comparison for On-Line Auctions. In: Pieprzyk, J., Ghodosi, H., Dawson, E. (eds.) ACISP 2007. LNCS, vol. 4586, pp. 416–430. Springer, Heidelberg (2007)

[DJ02] Damgård, I., Jurik, M.: Client/Server tradeoffs for online elections. In: Naccache, D., Paillier, P. (eds.) PKC 2002. LNCS, vol. 2274, pp. 125–140. Springer, Heidelberg (2002)

[DN03] Damgård, I., Nielsen, J.B.: Universally composable efficient multiparty computation from threshold homomorphic encryption. In: Boneh, D. (ed.) CRYPTO 2003. LNCS, vol. 2729, pp. 247–264. Springer, Heidelberg (2003)

[Fis01] Fischlin, M.: A cost-effective pay-per-multiplication comparison method for millionaires. In: Naccache, D. (ed.) CT-RSA 2001. LNCS, vol. 2020, pp. 457–471. Springer, Heidelberg (2001)

[FKN94] Feige, U., Killian, J., Naor, M.: A minimal model for secure computation (extended abstract). In: STOC 1994: Proceedings of the Twenty-Sixth Annual ACM Symposium on Theory of Computing, pp. 554–563. ACM Press, New York (1994)

[GSV07] Garay, J.A., Schoenmakers, B., Villegas, J.: Practical and secure solutions for integer comparison. In: Okamoto, T., Wang, X. (eds.) PKC 2007. LNCS, vol. 4450, pp. 330–342. Springer, Heidelberg (2007)

[JW05] Jagannathan, G., Wright, R.: Privacy-preserving distributed k-means clustering over arbitrarily partitioned data. In: KDD, pp. 593–599 (2005)

[Lip03] Lipmaa, H.: On diophantine complexity and statistical zero-knowledge arguments. In: Laih, C.-S. (ed.) ASIACRYPT 2003. LNCS, vol. 2894, pp. 398–415. Springer, Heidelberg (2003)

[LL07] Laur, S., Lipmaa, H.: A new protocol for conditional disclosure of secrets and its applications. In: Katz, J., Yung, M. (eds.) ACNS 2007. LNCS, vol. 4521, pp. 207–225. Springer, Heidelberg (2007)

[LP00] Lindell, Y., Pinkas, B.: Privacy preserving data mining. Journal of Cryptology, 36–54 (2000)

[NO07] Nishide, T., Ohta, K.: Multiparty computation for interval, equality, and comparison without bit-decomposition protocol. In: Okamoto, T., Wang, X. (eds.) PKC 2007. LNCS, vol. 4450, pp. 343–360. Springer, Heidelberg (2007)

[NPS99] Naor, M., Pinkas, B., Sumner, R.: Privacy preserving auctions and mechanism design. In: Proceedings of the 1st ACM Conference on Electronic Commerce, pp. 129–139. ACM Press, New York (1999)

[Pai99] Paillier, P.: Public-key cryptosystems based on composite degree residuosity classes. In: Stern, J. (ed.) EUROCRYPT 1999. LNCS, vol. 1592, pp. 223–238. Springer, Heidelberg (1999)

[Per52] Perron, O.: Bemerkungen über die Verteilung der quadratischen Reste. Mathematische Zeitschrift 56(2), 122–130 (1952)

[Sha79] Shamir, A.: How to share a secret. Communications of the ACM 22(11), 612–613 (1979)

[ST06] Schoenmakers, B., Tuyls, P.: Efficient binary conversion for paillier encrypted values. In: Vaudenay, S. (ed.) EUROCRYPT 2006. LNCS, vol. 4004, pp. 522–537. Springer, Heidelberg (2006)

[Tho09] Thorbek, R.: Linear Integer Secret Sharing. PhD thesis, Aarhus University (2009)

[Tof07] Toft, T.: Primitives and Applications for Multi-party Computation. PhD thesis, Aarhus University (March 2007), `http://www.daimi.au.dk/~ttoft/publications/dissertation.pdf`

[Tof09] Toft, T.: Solving linear programs using multiparty computation. In: Dingledine, R., Golle, P. (eds.) FC 2009. LNCS, vol. 5628, pp. 90–107. Springer, Heidelberg (2009)

[Yao82] Yao, A.: Protocols for secure computations (extended abstract). In: 23th Annual Symposium on Foundations of Computer Science (FOCS 1982), pp. 160–164. IEEE Computer Society Press, Los Alamitos (1982)

[Yao86] Yao, A.: How to generate and exchange secrets (extended abstract). In: 27th Annual Symposium on Foundations of Computer Science, pp. 162–167. IEEE Computer Society Press, Los Alamitos (1986)

Oblivious Transfer with Hidden Access Control Policies

Jan Camenisch[1], Maria Dubovitskaya[1], Gregory Neven[1], and Gregory M. Zaverucha[2]

[1] IBM Research - Zurich, Saumerstrasse 4, CH-8803 Ruschlikon Switzerland
[2] Certicom Research, 5520 Explorer Drive Mississauga, ON, L4W 5L1 Canada

Abstract. Consider a database where each record has different access control policies. These policies could be attributes, roles, or rights that the user needs to have in order to access the record. Here we provide a protocol that allows the users to access the database record while: (1) the database does not learn who queries a record; (2) the database does not learn which record is being queried, nor the access control policy of that record; (3) the database does not learn whether a user's attempt to access a record was successful or not; (4) the user can only obtain a single record per query; (5) the user can only access those records for which she has the correct permissions; (6) the user does not learn any other information about the database structure and the access control policies other than whether he was granted access to the queried record, and if so, the content of the record; and (7) the users' credentials can be revoked.

Our scheme builds on the one by Camenisch, Dubovitskaya and Neven (CCS'09), who consider oblivious transfer with access control when the access control policies are public.

Keywords: Privacy, Oblivious Transfer, Anonymous Credentials, Access Control.

1 Introduction

When controlling access to a sensitive resource, it is clear that the applicable access control policies can already reveal too much information about the resource. For example, consider a medical database containing patient records, where the access control policy (ACP) of each record lists the names of the treating doctors. The fact that a patient's record has certain specialists in its ACP leaks information about the patient's disease. Many patients may want to hide, for example, that they are being treated by a plastic surgeon or by a psychiatrist. Also, doctors treating a celebrity may want to remain anonymous to avoid being approached by the press.

As another example, in a multi-user file system, it may be desirable to hide the owner of a file or the groups that have access to it to prevent social engineering attacks, coercion, and bribery. In a military setting, knowing which files are classified "top secret", or even just the percentage of "top secret" files in the system, may help an attacker to focus his attack.

Confidentiality of the stored data and associated ACPs is not the only security concern. Privacy-aware users accessing the database may be worried about malicious database servers prying information from the query traffic. For example, the frequency that a patient's record is accessed gives a good estimate of the seriousness of his

D. Catalano et al. (Eds.): PKC 2011, LNCS 6571, pp. 192–209, 2011.

condition, while the identity of the doctors that access it most frequently may be an indication of the nature of the disorder. Users may therefore prefer to query the database anonymously, i.e., hiding their identity, roles, permissions, etc. from the database server, as well as hiding the index of the queried record. At the same time, the database server wants to rest assured that only permitted users have access to the data, and that they cannot find out who else has access to the data.

1.1 Our Contribution

In this paper we consider access to a database where each record is protected by a (possibly) different access control policy, expressed in terms of the attributes, roles, or rights that a user needs to have to obtain access. To provide the maximal amount of privacy to both users and the database server, we propose a protocol guaranteeing that (1) the database does not learn who queries a record; (2) the database does not learn the index nor the ACP of the queried record; (3) the database does not learn whether a user's attempt to access a record was successful or not; (4) users can only obtain a single record per query, (5) users can only access those records for which they satisfy the ACP; (6) at each query, users learn no more information about the applicable access control policy other than whether they satisfy it or not; and (7) the users' credentials can be revoked. Our ACP structure can be used to implement many practical access control models, including access control matrices, capability lists, role-based access control, and hierarchical access control. This work extends the work of Camenisch, Dubovitskaya, and Neven [8] who consider oblivious transfer with access control when the access control policies are public, i.e., they do not satisfy properties (6) and (7).

1.2 Related Work

Oblivious transfer protocols in their basic form [24,21,11] offer users access to a database without the server learning the contents of the query, but place no restrictions on who can access which records. After Aiello et al. suggested priced oblivious transfer [28], Herranz [18] was the first to add access control restrictions to records, but has users authenticate openly (i.e., non-anonymously) to the server. Later, Coull et al. [9] and Camenisch et al. [8] proposed OT protocols with anonymous access control. In all of these works, however, the access control policies are assumed to be publicly available to all users, and the server notices when a user's attempt to access a record fails.

There is also a line of work devoted to access control with hidden policies and hidden credentials, but none of them consider oblivious access to data, meaning that the server learns which resource is being accessed. In trust negotiation systems [19,26,27], two parties establish trust through iterative disclosure of and requests for credentials. Hidden credentials systems are designed to protect sensitive credentials and policies [4,17]. Neither provide full protection of policies, however, in the sense that the user learns (partial) information about the policy if her credentials satisfy it. The protocol of Frikken et al. [15] does provide full protection, but for arbitrary policies it requires communication exponential in the size of the policies.

Finally, one could always implement a protocol with all desired properties by evaluating an especially designed logical circuit using generic two-party computation techniques [25], but the cost of this approach would be prohibitive. In particular, the computation and communication cost of each record transfer would be linear in the number of records in the database N, whereas the efficiency of our transfer protocol is independent of N.

2 Definition of OT with Hidden Access Control Policies

An oblivious transfer protocol with hidden access control policies (HAC-OT) is run between an issuer, a database, and one or more users. The issuer provides access credentials to users for the data categories that they are entitled to access. The database hosts a list of records and associates to each record an access control policy (ACP). Users can request individual records from the database, and the request will succeed provided they have the necessary credentials. The ACPs are never revealed.

In a nutshell, a HAC-OT protocol works as follows. The issuer generates its key pair for issuing credentials and publishes the public key as a system-wide parameter. The database server initializes a database containing records protected by access control policies. It generates the encrypted database, which also contains the encrypted access control policies, and makes it available to all users, e.g., by posting it on a website or by distributing it on DVDs. Each user contacts the issuer to obtain a credential that lists all data categories that the user is entitled to access. When she wants to access a record in the database, the user proves to the database in zero-knowledge that her credential contains all the data categories required by the access control policy associated to the record. She performs computations on the encrypted access control rule associated to the desired record so that, with the help of the database, she will obtain the record key if and only if she satisfies the (encrypted) ACP. The database learns nothing about the index of the record that is being accessed, nor about the categories in the access control policy. The database does not even learn whether the user's attempt to obtain a record was successful.

2.1 Setting and Procedures

If $\kappa \in \mathbb{N}$, then 1^κ is the string consisting of κ ones. The empty string is denoted ε. If A is a randomized algorithm, then $y \stackrel{\$}{\leftarrow} \mathsf{A}(x)$ denotes the assignment to y of the output of A on input x when run with fresh random coins.

Unless noted, all algorithms are probabilistic polynomial-time (PPT) and we implicitly assume they take an extra parameter 1^κ in their input, where κ is a security parameter. A function $\nu : \mathbb{N} \rightarrow [0, 1]$ is *negligible* if for all $c \in \mathbb{N}$ there exists a $\kappa_c \in \mathbb{N}$ such that $\nu(\kappa) < \kappa^{-c}$ for all $\kappa > \kappa_c$.

We consider a limited universe of data categories $C = \{C_1, \ldots, C_\ell\} \subseteq \{0, 1\}^*$. An access control policy $ACP \subseteq C$ contains those data categories that a user needs to have access to in order to obtain the record. We will usually encode access control policies as vectors $\mathbf{c} = (c_1, \ldots, c_\ell) \in \{0, 1\}^\ell$, where $c_i = 1$ iff $C_i \in ACP$. A database consists of a list of N pairs $((R_1, ACP_1), \ldots, (R_N, ACP_N))$ of records $R_i \in \{0, 1\}^*$ and their associated access control policies $ACP_i \subseteq C$.

Users hold credentials that certify the list of categories that the user is entitled to access. The list is encoded as a vector $\boldsymbol{d} = (d_1, \ldots, d_\ell) \in \{0, 1\}^\ell$, where $d_i = 1$ iff the user is granted access to category C_i. Letting $\boldsymbol{c} \cdot \boldsymbol{d} = \sum_{i=1}^{\ell} c_i d_i$ and $|\boldsymbol{c}| = \sum_{i=1}^{\ell} c_i$, we say that a user's credential $\boldsymbol{d}$ *covers* an access control policy $\mathbf{c}$ iff $\boldsymbol{c} \cdot \boldsymbol{d} = |\mathbf{c}|$. This essentially means that users need to have access to *all* categories in the ACP in order to have access to the record. This fits nicely to a number of real-world access control models. For example, to implement role-based access control, where each database record can be accessed by users with one particular role, one sets ℓ to be the number of roles in the system, one sets $c_i = 1$ for the required role i and $c_j = 0$ for all $j \neq i$, and one sets $d_i = 1$ in the users' credentials for all roles i that a user owns. In a hierarchical access control system, users are granted access if their access level is at least that of the resource. For example, in a military context, the levels may be "top secret", "secret", "restricted", and "declassified", so that someone with "top secret" clearance has access to all records. To implement this in our system, one would set ℓ to be the number of levels, set $c_i = 1$ for the level i of the resource, and set $d_j = 1$ for all levels j lower than or equal to i.

Alternatively, one could use a coverage definition where $\boldsymbol{d}$ covers $\mathbf{c}$ iff $\boldsymbol{c} \cdot \boldsymbol{d} = |\boldsymbol{d}|$, effectively meaning that all of a user's categories have to appear in the ACP in order to be granted access. Our protocol is easily adapted to implement these semantics. This definition of coverage could be useful to implement simple access control matrices: if ℓ is the number of users, then user i would have a credential with $d_i = 1$, and the ACP sets $c_j = 1$ for all users j that are allowed access.

An oblivious transfer protocol with hidden access control policies (HAC-OT) is six polynomial-time algorithms and protocols, i.e., $\mathcal{HAC\text{-}OT} = (\mathsf{ISetup}, \mathsf{Issue}, \mathsf{Revoke}, \mathsf{DBSetup}, \mathsf{DBVerify}, \mathsf{Transfer})$.

- $\mathsf{ISetup}(C) \xrightarrow{\$} (pk_{\mathrm{I}}, sk_{\mathrm{I}}, RL)$. The issuer runs the randomized ISetup algorithm to generate a public key pk_{I}, the corresponding secret key sk_{I}, and an initial revocation list RL for security parameter κ and category universe C. The public key and revocation list are published as system-wide parameters.
- Issue: Common input: pk_{I}, uid, $\boldsymbol{d}$; Issuer input: sk_{I}; User output: $cred_{uid}$ or $\perp$. A user obtains an access credential for a vector of categories $\boldsymbol{d} = (d_1, \ldots, d_\ell) \in \{0,1\}^\ell$ by engaging in the Issue protocol with the issuer. The issuer's public key pk_{I}, the user's identity uid, and the vector $\boldsymbol{d}$ are common inputs. The issuer also uses his secret key sk_{I} as an input. At the end of the protocol, the user obtains the credential $cred_{uid}$.
- $\mathsf{Revoke}(sk_{\mathrm{I}}, RL, uid) \xrightarrow{\$} RL'$. To revoke the credential of user uid, the issuer runs the Revoke algorithm to create an updated revocation list RL' and publishes it as a system-wide parameter.
- $\mathsf{DBSetup}(pk_{\mathrm{I}}, DB = (R_i, ACP_i)_{i=1,\ldots,N}) \xrightarrow{\$} ((pk_{\mathrm{DB}}, ER_1, \ldots, ER_N), sk_{\mathrm{DB}})$. The database server runs the DBSetup algorithm to create a database containing records $R_1, \ldots, R_N$ protected by access control policies $ACP_1, \ldots, ACP_N$. This algorithm generates the encrypted database consisting of a public key pk_{DB} and the encrypted records along with their encrypted access control policies $ER_1, \ldots, ER_N$. The encrypted database is made available to all users, e.g., by posting it on a website.[1] The database server keeps the secret key sk_{DB} for itself.

[1] We assume that each user obtains a copy of the entire encrypted database. It is impossible to obtain our strong privacy requirements with a single database server without running into either computation or communication complexity that is linear in the database size.

• $\mathsf{DBVerify}(pk_{\mathrm{DB}}, EDB) \rightarrow b$. Upon receiving an encrypted database EDB, all users perform a one-time check to test whether EDB is correctly formed ($b = 1$) or not ($b = 0$).

• $\mathsf{Transfer}$: Common input: $pk_{\mathrm{I}}, RL, pk_{\mathrm{DB}}$; User input: $uid, i, ER_i, cred_{uid}$; Database input: sk_{DB}; User output: R_i or $\perp$. When the user wants to access a record in the database, she engages in a $\mathsf{Transfer}$ protocol with the database server. Common inputs are the issuer's public key pk_{I}, the revocation list RL, and the database's public key pk_{DB}. The user has as a secret input her identity uid, her selection index $i \in \{1, \ldots, N\}$, the encrypted record with encrypted ACP, and her credential $cred_{uid}$. The database server uses its secret key sk_{DB} as a private input. At the end of the protocol, the user obtains the database record R_i if her credential satisfies the ACP, or receives $\perp$ if not.

We assume that all communication links are private. We also assume that the communication links between a user and the issuer are authenticated, so that the issuer always knows to which user it is issuing a credential. The communication links between a user and the database are assumed to be anonymous, so that the database does not know which user is making a record query. (Authenticated communication channels between users and the database would obviously ruin the strong anonymity properties of our protocol).

2.2 Security Definitions

We define security of an HAC-OT protocol through indistinguishability of a real-world and an ideal-world experiment as introduced by the UC framework [5,6] and the reactive systems security models [22,23]. The definitions we give, however, do not entail all formalities necessary to fit one of these frameworks; our goal here is solely to prove our scheme secure.

We summarize the ideas underlying these models. In the real world there are a number of players, who run some cryptographic protocols with each other, an adversary A, who controls some of the players, and an environment $\mathcal{E}$. The environment provides the inputs to the honest players and receives their outputs and interacts arbitrarily with the adversary. The dishonest players are subsumed into the adversary.

In the ideal world, we have the same players. However, they do not run any cryptographic protocols but send all their inputs to and receive all their outputs from an ideal all-trusted party T. This party computes the output of the players from their inputs, i.e., applies the functionality that the cryptographic protocol(s) are supposed to realize. The environment again provides the inputs to and receives the output from the honest players, and interacts arbitrarily with the adversary controlling the dishonest players. A set of cryptographic protocols is said to securely implement a functionality if for every real-world adversary A and every environment $\mathcal{E}$ there exists an ideal-world simulator A' controlling the same parties in the ideal world as A does in the real world, such that the environment cannot distinguish whether it is run in the real world interacting with A, or whether it is run in the ideal world interacting with the simulator A'.

Definition 1. *Let* $\mathbf{Real}_{\mathcal{E},\mathsf{A}}(\kappa)$ *denote the probability that* $\mathcal{E}$ *outputs 1 when run in the real world with* A *and let* $\mathbf{Ideal}_{\mathcal{E},\mathsf{A}'}(\kappa)$ *denote the probability that* $\mathcal{E}$ *outputs 1 when* *run in the ideal world with* A'*, then the set of cryptographic protocols is said to securely implement functionality* T *if* $\mathbf{Real}_{\mathcal{E},\mathsf{A}}(\kappa) - \mathbf{Ideal}_{\mathcal{E},\mathsf{A}'}(\kappa)$ *is a negligible function in* κ.

THE REAL WORLD. We first describe how the real-world algorithms presented in §2.1 are orchestrated when all participants are honest, i.e., honest real-world users $\mathsf{U}_1, \dots, \mathsf{U}_M$, an honest issuer I, and an honest database DB. If parties are controlled by the real-world adversary A, they can arbitrarily deviate from the behavior described below.

All begins with I generating a key pair and an initial revocation list $(pk_{\mathrm{I}}, sk_{\mathrm{I}}, RL) \overset{\$}{\leftarrow} \mathsf{ISetup}(C)$ and sending (pk_{I}, RL) to all users $\mathsf{U}_1, \dots, \mathsf{U}_M$ and the database DB.

When the environment $\mathcal{E}$ sends a message $(\mathtt{initdb}, DB = (R_i, ACP_i)_{i=1,\dots,N})$ to the database DB, the latter encrypts DB by running $(EDB, sk_{\mathrm{DB}}) \overset{\$}{\leftarrow} \mathsf{DBSetup}(pk_{\mathrm{I}}, DB)$, and sends the encrypted database $EDB = (pk_{\mathrm{DB}}, ER_1, \dots ER_N)$ to all users $\mathsf{U}_1, \dots, \mathsf{U}_M$. All users execute a $\mathsf{DBVerify}$ protocol with the database and, if it returns 1, return a message $(\mathtt{initdb}, N)$ to the environment.

When $\mathcal{E}$ sends a message $(\mathtt{issue}, \boldsymbol{d})$ to user U_{uid}, she engages in an Issue protocol with I on common input pk_{I}, uid, and a vector $\boldsymbol{d}$ indicating which categories the user is allowed to access, with I using sk_{I} as its secret input. Eventually, U_{uid} obtains the access credential $cred_{uid}$. User U_{uid} returns a message $(\mathtt{issue}, b)$ to the environment indicating whether the issue protocol succeeded ($b = 1$) or failed ($b = 0$). Each user will engage in an Issue protocol only once; if she receives a second message $(\mathtt{issue}, \boldsymbol{d})$ from the environment, she simply returns $(\mathtt{issue}, 0)$.

When $\mathcal{E}$ sends a message $(\mathtt{revoke}, uid)$ to the issuer I, it creates a new revocation list $RL' \overset{\$}{\leftarrow} \mathtt{revoke}(sk_{\mathrm{I}}, RL, uid)$ based on the old revocation list RL and the user identity uid to be revoked. The issuer returns $(\mathtt{revoke}, uid)$ to the environment.

When $\mathcal{E}$ sends a message $(\mathtt{transfer}, i)$ to user U_{uid}, then U_{uid} engages in a $\mathsf{Transfer}$ protocol with DB on common input pk_{I} and pk_{DB}, on U_{uid}'s private input i and her credential $cred_{uid}$, and on DB's private input sk_{DB}. As a result of the protocol U_{uid} obtains the record R_i, or $\perp$ indicating failure. If the transfer succeeded the user returns $(\mathtt{transfer}, i, R_i)$ to the environment; if it failed she returns $(\mathtt{transfer}, i, \perp)$. We note that DB does not return any outputs to the environment.

THE IDEAL WORLD. In the ideal world, all participants communicate through a trusted party T which implements the functionality of our protocol. We describe the behavior of T on the inputs of the ideal-world users $\mathsf{U}'_1, \dots, \mathsf{U}'_M$, the ideal-world issuer I', and the ideal-world database DB'.

The trusted party T maintains an initially empty vector ε for each user U'_{uid}, an initially empty list of revoked users RL, and sets $DB \leftarrow \perp$. It responds to queries from the different parties as follows.

- Upon receiving $(\mathtt{initdb}, N)$ from DB', T sets $DB \leftarrow (\varepsilon_i, \varepsilon_i)_{i=1,\dots,N}$ and sends the message $(\mathtt{initdb}, N)$ to all users. All users send the message $(\mathtt{initdb}, N)$ to the environment.
- Upon receiving $(\mathtt{issue}, \boldsymbol{d})$ from U'_{uid} for the first time, T sends $(\mathtt{issue}, \mathsf{U}'_{uid}, \boldsymbol{d})$ to I' who sends back a bit b. If $b = 1$ then T initializes the category vector $\boldsymbol{d}_{uid} \leftarrow \boldsymbol{d}$ for the user U'_{uid} and sends $(\mathtt{issue}, 1)$ to U'_{uid}; otherwise it simply sends $(\mathtt{issue}, 0)$ to U'_{uid}. T responds to all subsequent messages $(\mathtt{issue}, \boldsymbol{d})$ from the same user U'_{uid} with $(\mathtt{issue}, 0)$.

• Upon receiving ($\texttt{revoke}, uid$) from I', T adds U'_{uid} to the list of revoked users by setting $RL \leftarrow RL \cup \{uid\}$.

• Upon receiving ($\texttt{transfer}, i$) from U'_{uid}, T proceeds as follows. If $DB = \bot$, it sends ($\texttt{transfer}, \bot$) back to U'_{uid}. If $DB = (\varepsilon_i, \varepsilon_i)_{i=1,\ldots,N}$ it sends ($\texttt{transfer}, \varepsilon$) to DB' who sends back a bit b. If $b = 1$ it also sends the whole database $DB = (R_i, ACP_i)_{i=1,\ldots,N}$, and the T sets $DB = (R_i, ACP_i)_{i=1,\ldots,N}$. If the database DB already contained records, T sends $\texttt{transfer}$ to DB', who sends back just a bit b. If $b = 1$, $\boldsymbol{d}_{uid}$ covers ACP_i, and $uid \notin RL$, then it sends ($\texttt{transfer}, R_i$) to U'_{uid}; otherwise, it sends ($\texttt{transfer}, \bot$) to U'_{uid}.

The ideal-world parties $\mathsf{U}'_1, \ldots, \mathsf{U}'_M, \mathsf{I}', \mathsf{DB}'$ simply relay inputs and outputs between the environment $\mathcal{E}$ and the trusted party T.

SECURITY PROPERTIES. It is easy to see that the ideal world definition implies that the users' privacy is protected:

An adversary, controlling all parties except some honest users, cannot tell which of the users access which record nor whether the attempt was successful.

Also, the database is guaranteed that (potentially malicious) users can only access the records for which they were issued credentials and that users do not learn any information about the access control lists apart from the fact whether or not their credentials allow them to access a record. We note that colluding users cannot pool their credentials nor can they obtain access with revoked credentials.

3 Randomizing and Extending Groth-Sahai Proofs

Let $\mathsf{Pg}(1^\kappa)$ be a pairing group generator that on input 1^κ outputs descriptions of multiplicative groups $\mathbb{G}_1, \mathbb{G}_2, \mathbb{G}_\mathrm{T}$ of prime order p where $|p| > \kappa$. Let $\mathsf{Pg}(p)$ be a pairing group generator that on input a prime p outputs descriptions of multiplicative groups $\mathbb{G}_1$, $\mathbb{G}_2$, and $\mathbb{G}_\mathrm{T}$ of order p. Let $\mathbb{G}_1^* = \mathbb{G}_1 \setminus \{1\}, \mathbb{G}_2^* = \mathbb{G}_2 \setminus \{1\}$ and let $g_1 \in \mathbb{G}_1^*, g_2 \in \mathbb{G}_2^*$. The generated groups are such that there exists an admissible bilinear map $e : \mathbb{G}_1 \times \mathbb{G}_2 \to \mathbb{G}_\mathrm{T}$, meaning that (1) for all $a, b \in \mathbb{Z}_p$ it holds that $e(g_1^a, g_2^b) = e(g_1, g_2)^{ab}$; (2) $e(g_1, g_2) \neq 1$; and (3) the bilinear map is efficiently computable. The group setting *GroupSet* is a tuple $(\mathbb{G}_1, \mathbb{G}_2, \mathbb{G}_T, p, e)$.

Groth and Sahai [16] present non-interactive witness-indistinguishable proofs of knowledge for three types of equations involving bilinear groups. These are: (i) pairing product equations, (ii) multi-exponentiation equations, and (iii) quadratic equations modulo the group order. Our protocol primarily uses proofs of the second type of equation, which can be made zero knowledge (ZK).

Belenkiy et al. [3] show that the Groth-Sahai proofs can be randomized such that they still prove the same statement but different proofs for the same statement are indistinguishable. In this paper, we extend these ideas: we take a proof for one statement and transform and randomize it into a proof of a related statement.

Independently of our work, Dodis et al. [12] also suggest to use GS-proofs of an old statement to construct a GS-proof for a modified statement. However, their approach only exploits the additive homomorphism of the GS-proof scheme, i.e., it allows one to modify an old witness by adding a new value to it and construct a proof for the new witness, without knowledge of the old one. We suggest a modification scheme for the GS proof that enables modifying the witness in a more general way using multiplication and addition.

We describe how this is done in the following. We start with the descriptions of the basic algorithms of an instantiation of the Groth-Sahai proof system for multi-exponentiation equations for the prime-order groups in the group setting $GroupSet = (\mathbb{G}_1, \mathbb{G}_2, \mathbb{G}_T, p, e)$, i.e., the proof system

$$\mathsf{NIZKP_{GS}}\{((x_{ij})_{i=1,\dots,M,\ j=1,\dots,\ell}) \ :\ \bigwedge_{i=1}^{M} y_i = \prod_{j=1}^{\ell} g_j^{x_{ij}}\},$$

where the y_i's and g_j's are public group elements of $\mathbb{G}_1$ (cf. [7]). In the following let $stmt = ((x_{ij})_{j=1,\dots,\ell;i=1,\dots,M}) : \bigwedge_{i=1}^{M} y_i = \prod_{j=1}^{\ell} g_j^{x_{ij}}$. The proof system for $GroupSet$ consists of three algorithms GSSetup, GSProve, and GSVerify. A trusted third party generates the common (public) reference string by running $CRS \leftarrow \mathsf{GSSetup}(GroupSet)$. A prover generates a proof as $\pi \leftarrow \mathsf{GSProve}(CRS, stmt, (y_i), (g_j), (x_{ij}))$ and a verifier checks it via $b \leftarrow \mathsf{GSVerify}(CRS, \pi, stmt, (y_i), (g_j))$, where $b = 1$ if π is true w.r.t. $stmt$ and $b = 0$ otherwise. We now present these algorithms in detail, based on the XDDH assumption [16,7]. (For ease of notation, we will denote by (y_i), (g_j), and (x_{ij}) the lists $(y_1, \dots y_M)$, $(g_1, \dots, g_\ell)$, and $(x_{11}, \dots, x_{M\ell})$ whenever the indices are clear from the context).

$\mathsf{GSSetup}(\mathbb{G}_1, \mathbb{G}_2, \mathbb{G}_T, p, e) \xrightarrow{\$} CRS$: Return $CRS = (\chi_1, \chi_2, \gamma_1, \gamma_2) \xleftarrow{\$} \mathbb{G}_2^4$.

$\mathsf{GSProve}(CRS, stmt, (y_i), (g_j), (x_{ij})) \xrightarrow{\$} \pi$:

1. Pick $r_{ij} \xleftarrow{\$} \mathbb{Z}_p$ for $i = 1 \dots M$ and $j = 1, \dots, \ell$.
2. For each x_{ij} in (x_{ij}) compute the set of commitments $C_{ij}^{(1)} \leftarrow \gamma_1^{x_{ij}} \chi_1^{r_{ij}}$; $C_{ij}^{(2)} \leftarrow \gamma_2^{x_{ij}} \chi_2^{r_{ij}}$.
3. For each y_i in (y_i) compute $p_i = \prod_{j=1}^{\ell} g_j^{r_{ij}}$.
4. Return $\pi \leftarrow (p_i, (C_{ij}^{(1)}, C_{ij}^{(2)})_{j=1,\dots\ell})_{i=1\dots M}$.

$\mathsf{GSVerify}(CRS, \pi, stmt, (y_j), (g_i)) \xrightarrow{\$} b$:

1. If for all $i = 1 \dots M$ we have $(\prod_{j=1}^{\ell} e(g_j, C_{ij}^{(1)}) = e(y_i, \gamma_1)e(p_i, \chi_1)) \wedge (\prod_{j=1}^{\ell} e(g_j, C_{ij}^{(2)}) = e(y_i, \gamma_2)e(p_i, \chi_2))$ then return $b \leftarrow 1$, else return $b \leftarrow 0$.

For the security properties of these algorithms we refer to Groth and Sahai [16] and Camenisch et al. [7].

Now, like Belenkiy et al. [3], we extend this basic system with a fourth algorithm GSRand which allows anyone to take a proof π and randomize it to obtain a proof π' for the same statement without knowledge of the witnesses (x_{ij}). Still, the proofs π and π' have the same distribution. This algorithm is as follows.

$\mathsf{GSRand}(CRS, \pi, stmt, (y_i), (g_j)) \xrightarrow{\$} \pi'$:

1. If $0 = \mathsf{GSVerify}(CRS, \pi, stmt, (y_i), (g_j))$ abort.
2. Pick $r'_{ij} \xleftarrow{\$} \mathbb{Z}_p$ for $i = 1 \dots M$ and $j = 1, \dots, \ell$.

3. Re-randomize all commitments:
   For every x_{ij} compute $C'^{(1)}_{ij} = C^{(1)}_{ij}\chi_1^{r'_{ij}}; C'^{(2)}_{ij} = C^{(2)}_{ij}\chi_2^{r'_{ij}}$.
4. Re-randomize p_i (consistent with the new randomness), by computing
   $p_i = p_i \prod_{j=1}^{\ell} g_j^{r'_{ij}}$.
5. Return $\pi' \leftarrow (p'_i, (C'^{(1)}_{ij}, C'^{(2)}_{ij})_{j=1,\ldots\ell})_{i=1\ldots M}$.

It is not hard to see that the proof π' has the same distribution as π and will be accepted by GSVerify. Also note that all the security properties of the proof system are retained (the algorithm essentially only randomizes, cf. [7]).

We now extend the above ideas to a fifth, and new algorithm GSRMod, which allows us not only to re-randomize the proof π for the statement *stmt* but also to extend it to a proof $\hat{\pi}$ for the related statement

$$stmt' = ((\hat{x}_{ij})_{j=1,\ldots,\ell;i=1,\ldots,M}) : \bigwedge_{i=1}^{M} \hat{y}_i = \prod_{j=1}^{\ell} g_j^{\hat{x}_{ij}}$$

where $\hat{y}_i = \prod_{j=1}^{M} y_j^{x'_j} \prod_{j=1}^{\ell} g_j^{x'_{ij}}$. Similarly to just randomizing a proof, it is sufficient to know x'_j and x'_{ij} to do this proof modification, i.e., knowledge of $\hat{x}_{ij}$ is not required. Note that $\hat{x}_{ij} = x'_{ij} + \sum_{k=1}^{M} x_{kj} \cdot x'_k$ will hold w.r.t. the original witnesses x_{ij}.

$\mathsf{GSRMod}(CRS, \pi, stmt', stmt, (\hat{y}_i), (y_i), (g_j), (x'_k)(x'_{ij})) \xrightarrow{\$} \pi'$:
1. If $0 = \mathsf{GSVerify}(CRS, \pi, stmt, (y_i), (g_j))$ abort.
2. Pick $r'_{ij} \xleftarrow{\$} \mathbb{Z}_p$ for $i = 1, \ldots, M$ and $j = 1, \ldots, \ell$.
3. Create commitments for each $\hat{x}_{ij}$ using old commitments:
   $\hat{C}^{(1)}_{ij} = \prod_{k=1}^{M}(C^{(1)}_{kj})^{x'_k}\gamma_1^{x'_{ij}}\chi_1^{r'_{ij}}$; $\hat{C}^{(2)}_{ij} = \prod_{k=1}^{M}(C^{(2)}_{kj})^{x'_k}\gamma_2^{x'_{ij}}\chi_2^{r'_{ij}}$.
4. Re-randomize and modify p_i (consistent with the new witnesses and randomness), by computing
   $\hat{p}_i = \prod_{j=1}^{M}(p_j)^{x'_j} \prod_{j=1}^{\ell} g_j^{r'_{ij}}$.
5. Return $\hat{\pi} \leftarrow (\hat{p}_i, (\hat{C}^{(1)}_{ij}, \hat{C}^{(2)}_{ij})_{j=1,\ldots\ell})_{i=1\ldots M}$.

Again, it is not hard to see that all security properties are retained (cf. [7]).

4 Our Construction

The main ideas underlying our protocol are as follows: the database server starts by encrypting each record with a key that is at the same time a signature on the index of the record and on the access control policy for the record. Furthermore, the server ElGamal-encrypts the access control policy for each record and creates a commitment to the policy. It then provides a non-interactive GS proof that the commitment and the encryptions are consistent. All of these values are then published as the encrypted database.

To be able to access the records, users are issued credentials for the list of data categories they are allowed to access. To revoke a user's credential, we use ideas from the revocation scheme by Nakanishi et al. [20] about revocable group signatures, where the revoked user identities are sorted in lexicographical order and the revocation list contains signatures on each pair of neighboring revoked user identities. To prove that her credential is not revoked, the user needs to show that her identity lies within the open interval defined by one of the signed identity pairs in the current revocation list. For this purpose the issuer signs all possible "distances" within such intervals, called "revocation distances". The user then proves that she possesses a valid signed pair of revoked user identities and valid signatures on the distances to the edges of the revocation interval. We note that to improve efficiency, one could make the maximal interval sizes smaller by revoking a number of dummy user identities by default.

When the user wants to access a record i, she re-encrypts the encrypted ACP for that record under her own freshly generated public key. She also randomizes the commitment to the ACP and then modifies the original GS proof into a new one proving that the new encryptions and the new commitment are consistent. The user also blinds the database server's signature on the ACP. Using the homomorphism of ElGamal encryption, the user computes an encryption of $\delta = \sum c_{ij} d_j - \sum c_{ij}$. Note that δ is 0 if the user is allowed access and non-zero otherwise. Finally, the user sends these values to the database server and proves in zero-knowledge that 1) she computed the encryption of δ correctly from the modified encryptions and w.r.t. the d_j that appear in her credential, that 2) the blinded signature is a valid signature by the database on the ACP values in the randomized commitment (without knowing these values of course), and that 3) her credential was not revoked.

If all of these proofs verify correctly, the database server uses the blinded signature to compute the blinded key of the record and "folds it into" the encryption of δ so that it contains the blinded key if $\delta = 0$ and contains a random plaintext otherwise. The server sends these values to the user and proves that they were computed correctly. Upon receipt, the user decrypts and unblinds the record key, and decrypts the record.

4.1 Issuer Setup

We now describe each step of our scheme in detail. We begin with the setup procedures of the issuer and the database provider. Users do not have their own setup procedure.

To set up its keys, the issuer runs the randomized ISetup algorithm displayed in Figure 1. This will generate groups of prime order p, a public key pk_{I} and corresponding secret key sk_{I} for security parameter κ and category universe C.

The values $g_I, y_I, h_0, \ldots, h_{\ell+1}, u, w$ from the issuer's public key and the corresponding x_I from the secret key are used to issue credentials. The values $\hat{g}, \hat{g}_1, \hat{g}_3, \hat{g}_4, y_r$

$$(\overline{\mathbb{G}}_1, \overline{\mathbb{G}}_2, \overline{\mathbb{G}}_{\mathrm{T}}, p) \xleftarrow{\$} \mathsf{Pg}(1^\kappa) \; ; \; g_I, h_0, \ldots, h_{\ell+1}, u, w, \hat{g}, \hat{g}_1, \hat{g}_2, \hat{g}_3, \hat{g}_4 \xleftarrow{\$} \overline{\mathbb{G}}_1 \; ;$$
$$h_I, h_r, h_\Delta \xleftarrow{\$} \overline{\mathbb{G}}_2 \; ; \; g_t, h_t \xleftarrow{\$} \overline{\mathbb{G}}_{\mathrm{T}} \; ; \; x_I, x_\Delta, x_r \xleftarrow{\$} \mathbb{Z}_p \; ; \; y_I \leftarrow h_I^{x_I} \; ; \; y_\Delta \leftarrow h_\Delta^{x_\Delta} \; ; \; y_r \leftarrow h_r^{x_r} \; ;$$
$$r, s \xleftarrow{\$} \mathbb{Z}_p \; ; \; S \leftarrow (\hat{g}\hat{g}_1^1 \hat{g}_2^0 \hat{g}_3^{\Delta_{\max}+1} \hat{g}_4^r)^{1/(x_r+s)} \quad .$$
$$\text{For } i = 1, \ldots, \Delta_{\max} \text{ do } y_\Delta^{(i)} \leftarrow g_I^{1/(x_\Delta + i)} \quad .$$
$$\text{Return } (sk_{\mathrm{I}} = (x_I, x_\Delta, x_r), pk_{\mathrm{I}} = (g_I, h_0, \ldots, h_{\ell+1}, u, w, \hat{g}, \hat{g}_1, \hat{g}_3, \hat{g}_4, h_I, h_\Delta, h_r, g_t, h_t, y_I,$$
$$y_r, y_\Delta, y_\Delta^{(1)}, \ldots, y_\Delta^{(\Delta_{\max})}), RL = (1, \{0, \Delta_{\max}+1\}, \{(0, \Delta_{\max}+1, S, r, s)\})) \quad .$$

Fig. 1. Issuer Setup algorithm $\mathsf{ISetup}(C)$

and x_r are used to sign pairs of the revoked user identities in the revocation list. And, finally, the issuer uses the Boneh-Boyen signature scheme with secret key x_Δ and public key y_Δ to sign all possible "revocation distances" and makes the set of these signatures $(y_\Delta^{(i)})_{i=1}^{\Delta_{\max}}$ a part of its public key. It will later become apparent what these revocation distances are (users will need them for proving that her credential is not on the list of revoked credentials).

He publishes the public key as a system-wide parameter.

4.2 Issuing Credentials

To be able to make database queries, a user needs to obtain a credential for the categories that she is allowed to access. To this end, the user runs the Issue protocol with the issuer as depicted in Figure 2. How the issuer determines which user has access to which categories is of course out of scope of the scheme.

$\mathsf{U}(pk_\mathrm{I}, uid, \mathbf{d})$:		$\mathsf{I}(pk_\mathrm{I}, uid, \mathbf{d}, sk_\mathrm{I})$:
		$r, s \xleftarrow{\$} \mathbb{Z}_p$
		$A \leftarrow (g_I h_0^{uid} h_1^{d_1} \cdots h_\ell^{d_\ell} h_{\ell+1}^r)^{\frac{1}{x_I+s}}$
	$\xleftarrow{A, r, s}$	
If $\overline{e}(A, h_I^s y_I) = \overline{e}(g_I h_0^{uid} h_1^{d_1} \cdots h_\ell^{d_\ell} h_{\ell+1}^r, h_I)$ then return $cred \leftarrow (\mathbf{d}, A, r, s)$.		

Fig. 2. Issue protocol Issue()

As a result of the issuing protocol, the user will obtain an access credential for the vector $\mathbf{d}$. This credential is a triple (A, r, s), which is a signature on the user's identity uid and the messages $\mathbf{d}$, using the signature scheme proposed and proved secure by Au et al. [1]. It is based on the schemes of Camenisch and Lysyanskaya [10] and of Boneh et al. [2].

4.3 Revoking Credentials

To revoke a user uid's credential, the issuer runs the revocation algorithm Revoke$(pk_\mathrm{I}, sk_\mathrm{I}, RL, uid)$. Recall that in our model each user has only one credential, in which the user's identity uid is embedded. We implement this revocation list with the ideas of Nakanishi et al. [20] for revocation in the context of group signatures. Thus the revocation list $RL = (t, R_1, R_2)$ consists of its current version number t, a set R_1 containing the identities of all revoked users, and a set R_2 of signatures on each pair of neighboring uid's in R_1 (according to their lexicographic order). When the issuer revokes a credential, it takes the latest revocation list $RL = (t, R_1, R_2)$ and constructs the updated revocation list by setting the version number to $t' = t + 1$, by adding the revoked user identity uid to R_1, by sorting this list lexicographically, and by issuing

Parse RL as (t, R_1, R_2) ; $t' \leftarrow t+1$; $R_1' \leftarrow R_1 \cup \{uid\}$; $R_2' \leftarrow \emptyset$.
Sort R_1' lexicographically as $(uid_1, \ldots, uid_n)$.
For $i = 1, \ldots, n-1$ do
$\quad \hat{r}, \hat{s} \xleftarrow{\$} \mathbb{Z}_p$; $S \leftarrow (\hat{g}\hat{g}_1^{t'} \hat{g}_2^{uid_i} \hat{g}_3^{uid_{i+1}} \hat{g}_4^{\hat{r}})^{1/(x_r+\hat{s})}$; $R_2' \leftarrow R_2' \cup \{(t', uid_i, uid_{i+1}, S, \hat{r}, \hat{s})\}$.
Return $RL' = (t', R_1', R_2')$.

Fig. 3. Revocation algorithm $\mathsf{Revoke}(pk_{\mathrm{I}}, sk_{\mathrm{I}}, RL, uid)$

a new set R_2' of signatures on each pair of neighboring revoked user identities in the updated set R_1'. This algorithm is described in detail in Figure 3.

4.4 Database Setup

To set up the database, the database server runs the algorithm shown in Figure 4. That is, it uses the issuer's public key and a pairing group generator to create groups of the same order p and generate keys for encrypting records. First the database provider generates its public and private keys to encrypt records.

Then the database creates a signature σ_i to bind the ACP and index to the encrypted record and "randomizes" it with value v_i. It also computes a commitment V_i to the index, ACP and v_i. The commitment V_i will be used for signature verification.

Next it encrypts each record R_i as (σ_i, F_i), each with its own key σ_i. In fact, these keys are verifiably pseudo random values [13], but are at the same time signatures under the the database provider's secret key (x_{DB}) on the index of the record (i) and the

1. Generate system parameters and keys
   $(\mathbb{G}_1, \mathbb{G}_2, \mathbb{G}_T) \xleftarrow{\$} \mathsf{Pg}(p)$; $g, h, h_{DB} \xleftarrow{\$} \mathbb{G}_1^*$; $g' \xleftarrow{\$} \mathbb{G}_2^*$; $H \leftarrow e(h_{DB}, g')$;
   $x_e, x_{\mathrm{DB}} \xleftarrow{\$} \mathbb{Z}_p$; $y_e \leftarrow g^{x_e}$; $y_{\mathrm{DB}} \leftarrow g^{x_{\mathrm{DB}}}$; $CRS \leftarrow \mathsf{GSSetup}(\mathbb{G}_1, \mathbb{G}_2, \mathbb{G}_T, p, e)$;
   For $i = 1, \ldots, \ell+1$ do $\{x_i \xleftarrow{\$} \mathbb{Z}_p$; $y_i \leftarrow g^{x_i}\}$;
   $sk_{\mathrm{DB}} \leftarrow (h_{DB}, x_e, x_{\mathrm{DB}}, x_1, \ldots, x_{\ell+1})$; $pk_{\mathrm{DB}} \leftarrow (g, g', H, h, y_e, y_{\mathrm{DB}}, y_1, \ldots, y_{\ell+1})$.
2. Create an encrypted database
   For $i = 1, \ldots, N$ do
   2.1 Parse ACP vector $\mathbf{c}_i$ as $(c_{i1}, \ldots c_{i\ell})$;
   2.2 Sign and encrypt records:
   $v_i \xleftarrow{\$} \mathbb{Z}_p$; $V_i \leftarrow g^i y_1^{c_{i1}} \cdots y_\ell^{c_{i\ell}} y_{\ell+1}^{v_i}$; $\sigma_i \leftarrow g'^{\frac{1}{x_{\mathrm{DB}}+i+\sum_{j=1}^{\ell} x_j \cdot c_{ij} + x_{\ell+1} v_i}}$;
   $F_i \leftarrow e(h_{DB}, \sigma_i) \cdot R_i$.
   2.3 Encrypt categories from the record's ACP:
   For each bit c_{ij} generate $r_{ij} \xleftarrow{\$} \mathbb{Z}_p, j = 1 \ldots \ell$;
   $E_{ij}^{(1)} = g^{c_{ij}} y_e^{r_{ij}}$; $E_{ij}^{(2)} = g^{r_{ij}}$; $E_{ij} = (E_{ij}^{(1)}, E_{ij}^{(2)})$
   2.4 Generate GS proof that all keys, signatures and encryptions were computed correctly
   $\pi_i = \mathsf{GSProve}(CRS, stmt_i, ((E_{ij}), V_i), (g, y_e, (y_j)), (v_i, (c_{ij}), (r_{ij})))$
   2.5 $ER_i \leftarrow (\sigma_i, V_i, F_i, (E_{i1}, \ldots, E_{i\ell}), \pi_i)$
3. Publish an encrypted database and public key
   Return $EDB \leftarrow ((pk_{\mathrm{DB}}, ER_1, \ldots, ER_N), sk_{\mathrm{DB}})$

Fig. 4. Database Setup algorithm $\mathsf{DBSetup}(pk_{\mathrm{I}}, DB = (R_i, \mathbf{c}_i)_{i=1,\ldots,N})$

categories defined in the access control policy for the record ($\mathbf{c}_i$). The pairs (σ_i, F_i) can be seen as an ElGamal encryption [14] in $\mathbb{G}_T$ of R_i under the public key H. During the transfer phase, this verifiability allows the database to check that the user is requesting the decryption key for a record with an access control policy satisfied by the user's credential.

To hide the records' ACPs, the database generates ElGamal encryptions of each bit c_{ij} and provides a NIZK GS-proof π_i to prove knowledge of the plaintexts. The statement for this proof for record i is

$$stmt_i = (v_i, c_{ij}, r_{ij}) : V_i = g^i y_1^{c_{i1}} \ldots y_l^{c_{il}} y_{l+1}^{v_i} \bigwedge_{j=1}^{\ell} \left(E_{ij}^{(1)} = g^{c_{ij}} y_e^{r_{ij}} \wedge E_{ij}^{(2)} = g_e^{r_{ij}}\right).$$

4.5 Accessing a Record

After having obtained the encrypted database, the user verifies it for correctness by running for each record R_i the step, denoted as $\mathsf{DBVerify}(pk_{\mathrm{DB}}, EDB)$ and defined as: $\mathsf{GSVerify}(CRS, \pi_i, stmt_i, ((E_{ij}), V_i), (g, y_e, (y_j))) \wedge (e(y_{DB}V_i, \sigma_i) \stackrel{?}{=} e(g, g'))$.

When the user wants to access a record in the database, she engages in a $\mathsf{Transfer}$ protocol (Figure 5) with the database server.

The input of the database server is its secret and public key as well as the public key of the issuer. The input of the user is the public keys of the issuer and the database, the index i of the record she wants to access, her credential, and the encrypted record $ER_i = ((\sigma_i, V_i, F_i, (E_{i1}, \ldots, E_{i\ell}), \pi_i))$.

At a high level, the protocol has three main steps. First, the user takes the encrypted ACP for the record she wants to access and adds a second layer of encryption to it, using a freshly generated key pair (x_u, y_u). Using the homomorphic properties of the encryption scheme, the user's categories in her credential and ACP for the record are compared by constructing a ciphertext $(D_i^{(1)}, D_i^{(2)})$ that encrypts zero if the user's credential satisfies the ACP, and a non-zero value if it does not. Then, the resulting ciphertext is sent to the database together with a proof PK_1 by the user that she constructed it correctly w.r.t. to the credential she possesses and the encrypted database. If that proof is valid, the database removes one layer of encryption and returns the result to the user. Finally, the user removes the remaining layer of encryption to recover the key for the record (or a random value if access is not granted).

Now we describe each step in greater detail. The user takes the ElGamal encryptions of each category bit for the record she wants to access and re-encrypts them with her own key. Then, using these values, she calculates an ElGamal encryption of $\delta = (\sum_{j=1}^{\ell} c_{ij} d_j - \sum_{j=1}^{\ell} c_{ij})$, which will be 0 if and only if $c_{ij} = d_j$. She then re-randomizes and modifies the GS proof π into the new one π' for the statement

$$stmt_i' = (v_i, i, (c_{ij}), (r_{ij}), (r'_{ij}), x_V) : V_i' = g^i y_1^{c_{i1}} \ldots y_\ell^{c_{i\ell}} y_{\ell+1}^{v_i} h^{x_V}$$
$$\bigwedge_{j=1}^{\ell} \left(E'^{(1)}_{ij} = g^{c_{ij}} (y_e y_u)^{r_{ij}+r'_{ij}} \wedge E'^{(2)}_{ij} = g_e^{r_{ij}+r'_{ij}}\right)$$

to prove that the new encryptions are consistent with the new commitment V_i'.

$\mathsf{U}\,(i, cred, pk_{\mathrm{I}}, pk_{\mathrm{DB}}, ER_i, RL)$:

1. Parse the encrypted record: $ER_i \leftarrow (\sigma_i, V_i, F_i, (E_{i1}, \ldots, E_{i\ell}), \pi_i)$;
2. Re-encrypt categories with fresh user's key

$x_u \xleftarrow{\$} \mathbb{Z}_p$; $y_u \leftarrow g^{x_u}$;

For $j = 1, \ldots, \ell$: $r'_{ij} \xleftarrow{\$} \mathbb{Z}_p$; $E'^{(1)}_{ij} \leftarrow E^{(1)}_{ij}(E^{(2)}_{ij})^{x_u}(y_e y_u)^{r'_{ij}}$;

$E'^{(2)}_{ij} \leftarrow E^{(2)}_{ij} g^{r'_{ij}} x_V \xleftarrow{\$} \mathbb{Z}_p$; $V'_i = V_i h^{x_V}$

$\pi'_i \leftarrow \mathsf{GSRMod}(CRS, \pi, stmt', stmt, ((E'^{(1)}_{ij}, E'^{(2)}_{ij}), V'_i),$
$((E^{(1)}_{ij}, E^{(2)}_{ij}), V_i), (g, y_e, y_u, (y_i), h), (x_u, (r_{ij}), x_V))$

3. Create an encryption of $\delta = \sum c_{ij} d_j - \sum c_{ij}$

$$r_d \xleftarrow{\$} \mathbb{Z}_p \; ; \; D^{(1)}_i \leftarrow \frac{\prod_{j=1}^{\ell} (E'^{(1)}_{ij})^{d_j}}{\prod_{j=1}^{\ell} E'^{(1)}_{ij}} y_e^{r_d} y_u^{r_d} \; ;$$

$$D^{(2)}_i \leftarrow \frac{\prod_{j=1}^{\ell} (E'^{(2)}_{ij})^{d_j}}{\prod_{j=1}^{\ell} E'^{(2)}_{ij}} g^{r_d}$$

4. Blind σ_i $k_\sigma \xleftarrow{\$} \mathbb{Z}_p$; $\sigma'_i \leftarrow \sigma_i{}^{k_\sigma}$

$\longrightarrow$ $y_u, (E'^{(1)}_{ij}, E'^{(2)}_{ij})_{j=1,\ldots\ell}, \pi'_i, (D^{(1)}_i, D^{(2)}_i), \sigma'_i$

$\longleftrightarrow$ $\mathsf{PK}_1\{Correct(\sigma'_i, V'_i, D_i, cred)\}$

$\mathsf{DB}(sk_{\mathrm{DB}}, pk_{\mathrm{DB}}, pk_{\mathrm{I}}, RL)$:

1. Verify all proofs: $b \leftarrow \big(\mathsf{PK}_1 \wedge$
$\mathsf{GSVerify}(\pi'_i, stmt', ((E'^{(1)}_{ij}, E'^{(2)}_{ij}), V'_i)),$
$(g, y_e, y_u, (y_i), h))\big)$
2. Remove DB encryption from D_i

$$L^{(1)}_i \leftarrow \frac{D^{(1)}_i}{(D^{(2)}_i)^{x_e}} \; ; \; L^{(2)}_i \leftarrow D^{(2)}_i$$

3. Re-randomize remaining user's encryption L_i

$k_\delta, k_L \xleftarrow{\$} \mathbb{Z}_p$; $L'^{(1)}_i \leftarrow (L^{(1)}_i)^{k_\delta} y_u^{k_L}$;
$L'^{(2)}_i \leftarrow (L^{(2)}_i)^{k_\delta} g^{k_L}$

4. Compute $M \leftarrow e(h_{DB}, \sigma'_i) \cdot e(L'^{(1)}_i, g')$

$\longleftarrow$ $M, L'^{(2)}_i$

$\longleftrightarrow$ $\mathsf{PK}_2\{Correct(M, L'^{(2)}_i)\}$

$R_i \leftarrow F_i / (M \cdot e(L'^{(2)}_i, g')^{-x_u})^{1/k_\sigma}$
Return R_i

Return ε

Fig. 5. The Transfer() protocol. The details of the proof protocols PK_1 and PK_2 are described in the text, as are the definitions of the statements *stmt* and *stmt'* of the GS proofs. The latter essentially say that the encryptions (E_{ij}) and (E'_{ij}) are consistent with V_i and V'_i, respectively.

Then, the user blinds σ_i and sends this blinded version σ'_i to the database server. Note that σ_i is derived from the database provider's secret key, the index of the records, and, most importantly, all the categories associated to the record. Next, the user proves to the database interactively that σ'_i is correctly formed as a randomization of some σ_i for which she possesses all necessary credentials, that V'_i is consistent with σ'_i, and that D_i is is correctly formed from the re-encrypted ACP and her credentials. That is, she executes with the database server the step we refer to as $\mathsf{PK}_1\{Correct(\sigma'_i, V'_i, D_i, cred)\}$ in Figure 5. For this proof, the user first searches in the current revocation list $RL = (t, R_1, R_2)$ for a tuple $(uid_{\mathrm{left}}, uid_{\mathrm{right}}, S, \hat{r}, \hat{s}) \in R_2$ such that $uid_{\mathrm{left}} < uid < uid_{\mathrm{right}}$. Let Δ_{left} and Δ_{right} such that $uid_{\mathrm{left}} + \Delta_{\mathrm{left}} = uid = uid_{\mathrm{right}} - \Delta_{\mathrm{right}}$ and let $y_\Delta^{(\Delta_{\mathrm{left}})}$ and $y_\Delta^{(\Delta_{\mathrm{right}})}$ be the issuer's signatures on these distances. The user next blinds $y_\Delta^{(\Delta_{\mathrm{left}})}$ and $y_\Delta^{(\Delta_{\mathrm{right}})}$ and her credentials, i.e., she computes $t_1, t'_1 \xleftarrow{\$} \mathbb{Z}_p$; $\tilde{A} \leftarrow Au^{t_1}$; $B \leftarrow w^{t_1} u^{t'_1}$, $t_2, t'_2 \xleftarrow{\$} \mathbb{Z}_p$; $\tilde{S} \leftarrow Su^{t_2}$, and $\tilde{S}_1 \leftarrow w^{t_2} u^{t'_2}$ to blind her credentials and computes $t_3, t'_3 \xleftarrow{\$} Z_p; \tilde{Y} \leftarrow (y_\Delta^{(\Delta_{\text{left}})})^{t_3}; \tilde{Y}_1 \leftarrow (y_\Delta^{(\Delta_{\text{right}})})^{t'_3}$ to blind the revocation distance signatures.

The user sends $\tilde{A}, B, \tilde{S}, \tilde{S}_1, \tilde{Y}$, and $\tilde{Y}_1$ to the database server and then executes the following proof of knowledge:

$$\mathsf{PK}_1\{(uid, uid_{\text{left}}, uid_{\text{right}}, k_\sigma, x_V, r_d, r, \alpha, \beta, \alpha_r, \beta_r, t_1, t'_1, t_2, t'_2, t_3, t'_3, d_1, \ldots, d_\ell,$$

$$s, \hat{s}, \hat{r}, c_1, \ldots, c_\ell, v_i) : V'_i = g^i \prod_{i=1}^{\ell} y_i^{c_i} y_{\ell+1}^{v_i} h^{x_V} \wedge 1 = B^{-s} w^\alpha u^\beta \ \wedge B = w^{t_1} u^{t'_1}$$

$$\wedge\, e(y_{DB} V'_i, \sigma'_i) = e(g, g')^{k_\sigma} e(h, \sigma'_i)^{x_V} \wedge \tilde{S}_1 = w^{t_2} u^{t'_2} \wedge 1 = \frac{w^{\alpha_r} u^{\beta_r}}{\tilde{S}_1^{\hat{s}}} \wedge$$

$$\frac{\overline{e}(\tilde{A}, y_I)}{\overline{e}(g_I, h_I)} = \overline{e}(\tilde{A}, h_I)^{-s} \overline{e}(u, y_I^{t_1} h_I^\alpha) \overline{e}(h_0, h_I)^{uid} \overline{e}(h_{\ell+1}, h_I)^r \prod_{k=1}^{\ell} \overline{e}(h_k, h_I)^{d_k} \wedge$$

$$\frac{\overline{e}(\tilde{S}, y_r)}{\overline{e}(\hat{g}, h_r)} = \overline{e}(\tilde{S}, h_r)^{-\hat{s}} \overline{e}(u, y_r^{t_2} h_r^{\alpha_r}) \overline{e}(\hat{g}_1, h_r)^t \overline{e}(\hat{g}_2, h_r)^{uid_{\text{left}}} \overline{e}(\hat{g}_3, h_r)^{uid_{\text{right}}} \overline{e}(\hat{g}_4, h_r)^{\hat{r}}$$

$$\wedge\, \overline{e}(\tilde{Y}, y_\Delta) = \overline{e}(\tilde{Y}, h_\Delta)^{-(uid - uid_{\text{left}})} \overline{e}(g_I, h_\Delta)^{t_3} \wedge$$

$$\overline{e}(\tilde{Y}_1, y_\Delta) = \overline{e}(\tilde{Y}_1, h_\Delta)^{-(uid_{\text{right}} - uid)} \overline{e}(g_I, h_\Delta)^{t'_3} \wedge$$

$$D_i^{(1)} \cdot \prod_{j=1}^{\ell} E_{ij}^{\prime(1)} = \prod_{j=1}^{\ell} (E_{ij}^{\prime(1)})^{d_j} y_e^{r_d} y_u^{r_d} \wedge D_i^{(2)} \cdot \prod_{j=1}^{\ell} E_{ij}^{\prime(2)} = \prod_{j=1}^{\ell} (E_{ij}^{\prime(2)})^{d_j} g^{r_d}\}$$

If the proof is successful, the database removes its layer of encryption from D_i and then randomizes the remaining encryption of δ (using the user's temporary public key y_u). This ensures that if $\delta \neq 0$ holds, then the encryption will contain a random value and is not related to the decryption key for the record. The database then proves to the user that it had computed everything correctly by executing with with her the protocol we referred to as $\mathsf{PK}_2\{\mathit{Correct}(M, L_i^{\prime(2)})\}$ in Figure 5. More precisely, it is the following proof that the values M and $L_i^{\prime(2)}$ were computed correctly (whereby $\gamma = -x_e k_\delta$):

$$\mathsf{PK}\{(h_{DB}, x_e, k_L, k_\delta, \gamma) : H = e(h_{DB}, g') \wedge y_e = g^{x_e} \wedge L_i^{\prime(2)} = (D_i^{(2)})^{k_\delta} g^{k_L} \wedge$$

$$1 = y_e^{k_\delta} g^\gamma \wedge M = e(h_{DB}, \sigma'_i) \cdot e(D_i^{(1)}, g')^{k_\delta} \cdot e(D_i^{(2)}, g')^\gamma e(g, g')^{k_L}\} \ .$$

When the user gets L from the database, removes all randomness and decrypts, the decrypted value R_i is correct if and only if $\delta = 0$.

We finally remark that the database has to calculate encryptions of all ACPs and encrypt all records $(1, \ldots, N)$ only once at the setup phase, and the user has to download and verify the entire encrypted database only once as well. So the communication and computation complexity of the protocol depend on the number of the records in the database only in the setup and verify phases. The other parts of the protocol (issue and transfer) require only $O(\ell)$ group elements to be sent and exponentiations and pairings to be computed, where ℓ is the size of ACP vector.

5 Security Analysis

The security of our protocol is analyzed by proving indistinguishability between adversary actions in the real protocol and in an ideal scenario that is secure by definition.

Given a real-world adversary A, we construct an ideal-world adversary A' such that no environment $\mathcal{E}$ can distinguish whether it is interacting with A or A'. We organize the proof in sub-lemmas according to which subset of parties are corrupted. We do not consider the cases where all parties are honest, where all parties are dishonest, where the issuer is the only honest party, or where the issuer is the only dishonest party, as these cases have no real practical interest.

For each case we prove the indistinguishability between the real and ideal worlds by defining a sequence of hybrid games **Game-0**, ..., **Game-n**. In each game we define a simulator Sim_i that runs A as a subroutine and that provides $\mathcal{E}$'s entire view. We define $\mathbf{Hybrid}_{\mathcal{E},\mathsf{Sim}_i}(\kappa)$ to be the probability that $\mathcal{E}$ outputs 1 when run in the world provided by Sim_i. The games are always constructed such that the first simulator Sim_0 runs A and all honest parties exactly like in the real world, so that $\mathbf{Hybrid}_{\mathcal{E},\mathsf{Sim}_0}(\kappa) = \mathbf{Real}_{\mathcal{E},\mathsf{A}}(\kappa)$, and such that the final simulator Sim_n is easily transformed into an ideal-world adversary A' so that $\mathbf{Hybrid}_{\mathcal{E},\mathsf{Sim}_n}(\kappa) = \mathbf{Ideal}_{\mathcal{E},\mathsf{A}'}(\kappa)$. By upper-bounding and summing the mutual game distances $\mathbf{Hybrid}_{\mathcal{E},\mathsf{Sim}_i}(\kappa) - \mathbf{Hybrid}_{\mathcal{E},\mathsf{Sim}_{i+1}}(\kappa)$ for $i = 0, \ldots, n-1$, we obtain an upper bound for the overall distance $\mathbf{Real}_{\mathcal{E},\mathsf{A}}(\kappa) - \mathbf{Ideal}_{\mathcal{E},\mathsf{A}'}(\kappa)$.

Theorem 2. *If the $(N+2)$-BDHE and XDDH assumptions hold in $\mathbb{G}_1, \mathbb{G}_T$, the $(N+1)$-SDH assumption holds in $\mathbb{G}_1$, and M-SDH assumption holds in $\overline{\mathbb{G}}_1$ then the $\mathcal{HAC}$-$\mathcal{OT}$ protocol in Section 4 securely implements the HAC-OT functionality, where N is the number of database records, and M is the number of the users.*

We prove the theorem by separately proving it for all relevant combinations of corrupted parties in four lemmas.

Due to lack of space, the detailed proof is found in the full version of this paper.

6 Conclusion

We have provided a set of efficient protocols and thereby shown that it is possible to build an access control system for a database with the maximal possible privacy for all involved parties: users can access records they are authorized to access without the server obtaining any information whatsoever about which records they access, which authorizations they users have, or whether the access was successful. Indeed, the database only learns that some user attempted to access the database. At the same time the database server is guaranteed that users can only access a single record and do not get any other information including information about other records or any access control list. Indeed, the user only learn whether or not their current credentials are sufficient to access the record they try to access.

The protocols we provide are fairly practical and we believe applications where records are relatively valuable, e.g., keys to decrypt some media such as movies or a particular DNA-sequence of many people, then our protocol could be used in practice. Still, it remains an open question whether more efficient protocols exist. One way to achieve this, could be using attribute based encryption instead of using anonymous credentials. Our initial investigation of such protocols makes us believe that such an approach would lead to less efficient protocols. Nevertheless, further research along this lines could be fruitful.

Acknowledgements

The authors thank the anonymous referees of CCS 2009 for suggesting the problem of oblivious transfer with hidden access control lists. We are grateful to Robert Enderlein for saving us from a few embarrassing typos. This work was supported by the European Community through the Seventh Framework Programme (FP7/2007-2013) project PrimeLife (grant agreement no. 216483).

References

1. Au, M.H., Susilo, W., Mu, Y.: Constant-size dynamic k-TAA. In: De Prisco, R., Yung, M. (eds.) SCN 2006. LNCS, vol. 4116, pp. 111–125. Springer, Heidelberg (2006)
2. Boneh, D., Boyen, X., Shacham, H.: Short group signatures. In: Franklin, M. (ed.) CRYPTO 2004. LNCS, vol. 3152, pp. 41–55. Springer, Heidelberg (2004)
3. Belenkiy, M., Camenisch, J., Chase, M., Kohlweiss, M., Lysyanskaya, A., Shacham, H.: Randomizable proofs and delegatable anonymous credentials. In: Halevi, S. (ed.) CRYPTO 2009. LNCS, vol. 5677, pp. 108–125. Springer, Heidelberg (2009)
4. Bradshaw, R.W., Holt, J.E., Seamons, K.E.: Concealing complex policies with hidden credentials. In: 11th (CCS 2004), pp. 46–157. ACM Press, New York (2004)
5. Canetti, R.: Studies in Secure Multiparty Computation and Applications. PhD thesis, Weizmann Institute of Science, Rehovot 76100, Israel (June 1995)
6. Canetti, R.: Security and composition of multi-party cryptographic protocols. Journal of Cryptology 13(1), 143–202 (2000)
7. Camenisch, J., Chandran, N., Shoup, V.: A public key encryption scheme secure against key dependent chosen plaintext and adaptive chosen ciphertext attacks. In: Joux, A. (ed.) EUROCRYPT 2009. LNCS, vol. 5479, pp. 351–368. Springer, Heidelberg (2009)
8. Camenisch, J., Dubovitskaya, M., Neven, G.: Oblivious transfer with access control. In: ACM CCS 2009, pp. 131–140. ACM Press, New York (2009)
9. Coull, S., Green, M., Hohenberger, S.: Controlling access to an oblivious database using stateful anonymous credentials. In: Jarecki, S., Tsudik, G. (eds.) PKC 2009. LNCS, vol. 5443, pp. 501–520. Springer, Heidelberg (2009)
10. Camenisch, J., Lysyanskaya, A.: Signature schemes and anonymous credentials from bilinear maps. In: Franklin, M. (ed.) CRYPTO 2004. LNCS, vol. 3152, pp. 56–72. Springer, Heidelberg (2004)
11. Camenisch, J., Neven, G., Shelat, A.: Simulatable adaptive oblivious transfer. In: Naor, M. (ed.) EUROCRYPT 2007. LNCS, vol. 4515, pp. 573–590. Springer, Heidelberg (2007)
12. Dodis, Y., Haralambiev, K., Lopez-Alt, A., Wichs, D.: Cryptography against continuous memory attacks. Cryptology ePrint Archive, Report 2010/196 (2010)
13. Dodis, Y., Yampolskiy, A.: A verifiable random function with short proofs and keys. In: Vaudenay, S. (ed.) PKC 2005. LNCS, vol. 3386, pp. 416–431. Springer, Heidelberg (2005)

14. ElGamal, T.: A public key cryptosystem and a signature scheme based on discrete logarithms. IEEE Transactions on Information Theory 31(4), 469–472 (1985)
15. Frikken, K.B., Atallah, M.J., Li, J.: Attribute-based access control with hidden policies and hidden credentials. IEEE Trans. Computers 55(10), 1259–1270 (2006)
16. Groth, J., Sahai, A.: Efficient non-interactive proof systems for bilinear groups. In: Smart, N.P. (ed.) EUROCRYPT 2008. LNCS, vol. 4965, pp. 415–432. Springer, Heidelberg (2008)
17. Holt, J.E., Bradshaw, R.W., Seamons, K.E., Orman, H.K.: Hidden credentials. In: ACM WPES 2003, USA, pp. 1–8. ACM, New York (2003)
18. Herranz, J.: Restricted adaptive oblivious transfer. Cryptology ePrint Archive, Report 2008/182 (2008)
19. Li, N., Winsborough, W.: Towards practical automated trust negotiation. In: POLICY 2002, Washington, DC, USA, p. 92. IEEE Computer Society, Los Alamitos (2002)
20. Nakanishi, T., Fujii, H., Hira, Y., Funabiki, N.: Revocable group signature schemes with constant costs for signing and verifying. In: Jarecki, S., Tsudik, G. (eds.) PKC 2009. LNCS, vol. 5443, pp. 463–480. Springer, Heidelberg (2009)
21. Naor, M., Pinkas, B.: Oblivious transfer with adaptive queries. In: Wiener, M. (ed.) CRYPTO 1999. LNCS, vol. 1666, pp. 573–590. Springer, Heidelberg (1999)
22. Pfitzmann, B., Waidner, M.: Composition and integrity preservation of secure reactive systems. In: 7th ACM CCS, pp. 245–254. ACM Press, New York (2000)
23. Pfitzmann, B., Waidner, M.: A model for asynchronous reactive systems and its application to secure message transmission. In: Proceedings of the IEEE Symposium on Research in Security and Privacy, pp. 184–200. IEEE Computer Society, IEEE Computer Society Press, Los Alamitos (2001)
24. Rabin, M.O.: How to exchange secrets by oblivious transfer. Technical Report TR-81, Harvard Aiken Computation Laboratory (1981)
25. Yao, A.C.: Protocols for secure computations. In: 23rd FOCS, pp. 160–164. IEEE Computer Society Press, Los Alamitos (1982)
26. Yu, T., Winslett, M.: A unified scheme for resource protection in automated trust negotiation. In: IEEE Symposium on Security and Privacy (S&P 2003), USA, pp. 110–122. IEEE Computer Society, Los Alamitos (2003)
27. Yu, T., Winslett, M., Seamons, K.E.: Interoperable strategies in automated trust negotiation. In: ACM CCS 2001, pp. 146–155. ACM Press, New York (2001)
28. Aiello, W., Ishai, Y., Reingold, O.: Priced oblivious transfer: How to sell digital goods. In: Pfitzmann, B. (ed.) EUROCRYPT 2001. LNCS, vol. 2045, p. 119. Springer, Heidelberg (2001)

Chosen Ciphertext Secure Encryption under Factoring Assumption Revisited*

Qixiang Mei[1,2], Bao Li[1], Xianhui Lu[1], and Dingding Jia[1]

[1] State Key Laboratory of Information Security, Graduate University of Chinese Academy of Sciences, Beijing, 100049, China
[2] School of Information, Guangdong Ocean University, Zhanjiang, 524088, China
nupf@163.com, {lb,xhlu,ddjia}@is.ac.cn

Abstract. In Eurocrypt 2009, Hofheinz and Kiltz proposed a practical chosen ciphertext (CCA) secure public key encryption under factoring assumption based on Rabin trapdoor one-way *permutation*.

We show that when the modulus is special such that Z_N^* has semi-smooth order, the instantiation of Hofheinz-Kiltz 09 scheme (HK09) over a much smaller subgroup of quadratic residue group (Semi-smooth Subgroup) is CCA secure as long as this type of modulus is hard to be factored. Since the exponent domain of this instantiation is much smaller than the original one, the efficiency is substantially improved.

In addition, we show how to construct a practical CCA secure encryption scheme from ElGamal trapdoor one-way *function* under factoring assumption. When instantiated over Semi-smooth Subgroup, this scheme has even better decryption efficiency than HK09 instantiation.

Keywords: public key encryption, chosen ciphertext secure, semi-smooth subgroup, factoring assumption.

1 Introduction

Chosen ciphertext security is now widely accepted as the standard security notion for the public key encryption. The first practical CCA secure public key encryption scheme without random oracle was proposed by Cramer and Shoup [6]. Their construction was later generalized to hash proof system [7]. However, the Cramer-Shoup encryption scheme and all its variants [20,16] inherently rely on decisional assumptions, e.g., the Decisional Diffie-Hellman (DDH) assumption, Decisional Composite Residuosity (DCR) assumption, and Decisional Quadratic Residuosity (DQR) assumption. In [24], Peikert and Waters proposed a general framework of constructing CCA secure encryption from the lossy trapdoor function. In [27], Rosen and Segev proposed a general way under the correlated inputs

* Supported by the National Natural Science Foundation of China (No.60862001, 61070171), the National Basic Research Program of China(973 project) (No.2007CB311201) and the Postdoctoral Science Foundation of China (No.20090460565).

D. Catalano et al. (Eds.): PKC 2011, LNCS 6571, pp. 210–227, 2011.

function. However, all the concrete constructions of lossy trapdoor function and correlated inputs function are also based on decisional assumptions.

It is widely believed that computational assumptions are more standard than their decisional versions. Canetti, Halevi and Katz [3] proposed the first practical public key encryption under a computational assumption, namely the Bilinear Diffie-Hellman assumption. In Eurocrypt 2008, Cash, Kiltz and Shoup [5] (CKS08) proposed a practical CCA secure scheme under the Computational Diffie-Hellman (CDH) assumption. Later in the same year, Hanaoka and Karosawa (HK08) proposed a more efficient CCA secure scheme under the CDH assumption [15]. Very recently, Haralambiev et al.[14], further improved the efficiency of CKS08 and HK08.

Even though the CCA secure schemes under CDH assumption are already fairly practical, since the pseudo-random generator for CDH problem can only extract one bit (or $O(\log(\lambda))$ with loose reduction), the encryption/decryption require $O(\lambda)$ (or $O(\lambda/\log(\lambda))$ respectively) modular exponentiations, where λ denotes the security lever parameter.

Hofheinz and Kiltz proposed a practical CCA secure PKE in Eurocrypt 2009 [17]. The Hofheinz-Kiltz 2009 scheme (HK09) [17] is constructed from the Blum-Goldwasser encryption [2], which itself is based on the Rabin encryption scheme [25] and Blum-Blum-Shub (BBS) generator[1]. The noticeable property of HK09 is that it only add a group element in Z_N^* to BG scheme and can be proved under factoring assumption (instead of the related decisional assumption). Since the BBS generator extracts one bit with only one modular multiplication, both the encryption and decryption require only $O(1)$ modular exponentiations.

However, in original HK09, the exponent is chosen from $[(N-1)/4]$. For the secure level of 80, the bits length of N, ℓ_N, needs to be chosen at least as 1024. For higher security, the length needs to be chosen even larger. A natural problem is that can we choose smaller domain of the exponent to improve the efficiency under factoring assumption?

In HK09, its security proof heavily relies on the fact that Rabin encryption is a trapdoor one-way *permutation*. The same technique seems hard to be directly used to construct CCA encryption from another factoring based encryption, ElGamal encryption over composite modulus, since the latter is only a trapdoor one-way *function*. In TCC 2010, Cramer, Hofheinz and Kiltz obtained an efficient CCA encryption from the ElGamal encryption over composite modulus under RSA assumption [4] (For convenience, throughout this paper, we will refer their scheme under RSA assumption as CHK10). However, the security of CHK10 could not be proved under factoring assumption. In addition, since the authors did not give an efficient pseudo-random bits generator, the encryption/decryption require $O(\lambda)$ modular exponentiations. It is interesting to construct practical CCA encryption such that the encryption/decryption only require $O(1)$ modular exponentiations from ElGamal encryption over composite modulus under factoring assumption.

1.1 Contributions

We present a HK09 instantiation over the much smaller subgroup than QR_N for a special modulus, i,e., Z_N^* has semi-smooth order. We prove that as long as this type of modulus is hard to be factored, this instantiation is CCA secure. Compared to the original HK09, the domain of the exponent is much smaller, thus, the efficiency is substantially improved. More precisely, this type of modulus is of the form $N = PQ = (2pp' + 1)(2qq' + 1)$, where p' and q' are primes large enough but much smaller than P and Q respectively, p and q are product of some distinct odd primes smaller than a low bound B. We instantiate HK09 over the unique subgroup G of QR_N with order $p'q'$. For the convenience, we call this subgroup as Semi-smooth Subgroup throughout this paper. In this instantiation, the domain of the exponent is set as $[2^{\ell_{p'}+\ell_{q'}+\lambda}]$, where a useful property for our proof is that a uniform element has almost the same distribution as the uniform element of $[p'q']$. We prove a simple but crucial lemma: computing the square roots resides of uniformly chosen element in G can be reduce to the factoring assumption about this type of modulus.

Another contribution of ours is that we construct a practical CCA secure encryption from ElGamal encryption over composite modulus under factoring assumption. As in HK09, the ciphertext only consists of two group elements in Z_N^*. Taking account of efficiency, we present the instantiation over Semi-smooth Subgroup. The decryption of this instantiation is more efficient than the decryption of HK09 instantiation. First of all, we need an efficient pseudo-random generator for ElGamal encryption over composite modulus to transform it from one-wayness secure encryption to indistinguishability secure one under factoring assumption. This can be achieved since, adapting the proof technique of [23], we are able to prove that, under factoring assumption, $BBS_r(g^{xy})$ is pseudo-random even given (g^x, g^y). To explain how to transform it into CCA secure, we describe our attempts towards the final scheme step by step. The first attempt is directly applying Kiltz 07 [19] to the composite modulus case: the public key is $(g, X' = g^{\rho'}, X = g^{\rho})$, private key is (ρ', ρ); the ciphertext is $(R = g^{\mu}, S = (X'^t X)^{\mu})$; the encapsulated key is $\mathrm{BBS}_r(X'^{\mu})(= \mathrm{BBS}_r(g^{\rho'\mu}))$. This scheme could not be proved CCA secure under factoring assumption using known techniques since the simulator could not answer the DDH oracle and could not compute exponent inversion modulo unknown order. Inspired by a fact proved in [18], i.e., factoring assumption imply the strong Diffie-Hellman assumption over the signed quadratic residue group, QR_N^+, we make our second attempt by instantiating Kiltz 07 over QR_N^+ so that the simulator is able to answer the DDH oracle. But the simulator still could not compute exponent inversion modulo unknown order. Inspired by the method of HK09, we further modify the scheme as follows: the public key is $(g, X' = g^{\rho'}, X = g^{2^{\nu}\rho})$, private key is (ρ', ρ); the ciphertext is $(R = |g^{\mu 2^{\nu}}|, S = |(X'^t X)^{\mu}|)$; the encapsulated key is $\mathrm{BBS}_r(|(X')^{2^{\nu}\mu}|)(= \mathrm{BBS}_r(|R^{\rho'}|) = \mathrm{BBS}_r(|g^{2^{\nu}\rho'\mu})|)$. Recall that, in HK09, they implicitly used the following fact: Given $A, B \in Z_N^*$, $x, y \in Z$, from the equation $A^x = B^y$, any one can efficiently compute $A^{c/y}$, where $c = \gcd(x, y)$. But when we attempt to directly apply the proof method of HK09 to our case, we find that the simulator does not know one of bases, without of loss generality, we denote it as A. To overcome this problem, we construct another simulator such that he knows both $B' = B^{2^k}$ and $A' = A^{2^k}$ for some suitable $k \in Z^+$. Basing on the underlying fact used in [18] to prove factoring assumption imply strong DH assumption, i.e., if U and V both belong to QR_N^+, then the equation $U = V$ is equivalent to the equation $U^{2^k} = V^{2^k}$ for any $k \in Z^+$, this simulator is able to verify the equivalent equation $A'^x = B'^y$ and then compute the encapsulated key from this equation. In the real proof, instead of black-box reducing the CCA security of the encryption to the pseudo-randomness of the generator, we prove the pseudo-randomness of the generator and the CCA security of the encryption simultaneously.

Both our schemes presented in this paper are actually CCA secure key encapsulation mechanism, from which it is easy to obtain full CCA secure public key encryption [28].

2 Preliminaries

2.1 Key Encapsulation Mechanism

A key encapsulation mechanism consists of three algorithms: Key generation $Gen(1^\lambda)$, Encapsulation $Enc(PK)$, Decapsulation $Dec(SK, C)$.

$Gen(1^\lambda)$: A probabilistic polynomial-time key generation algorithm takes as input a security parameter λ and outputs a public-key PK and secret key SK.

$Enc(PK)$: A probabilistic polynomial-time encryption algorithm takes public-key PK as input, and outputs a pair (K, C), where K is the key and C is a ciphertext.

$Dec(SK, C)$: A decryption algorithm takes a ciphertext C and the secret key SK as input. It returns a key K.

We require that for all (PK, SK) output by $Gen(1^\lambda)$, all (K, C) output by $Enc(PK)$, we have $Dec(SK, C) = K$.

Definition 1. (CCA Secure KEM) *A key encapsulation mechanism is indistinguishable against chosen ciphertext attacks if any PPT adversary* M *has negligible advantage in the game defined between the adversary* M *and the challenger* D *as follows:*

1. When M queries a key generation oracle, D invokes $Gen(1^\lambda)$ to obtain (PK, SK), responds with PK.
2. When M queries a challenge oracle. D invokes $Enc(PK)$ to obtain C^*, K_0, and chooses a random bits string K_1 with the same length as K_0, chooses a random bit b, set $K^* = K_b$, responds with (C^*, K^*).
3. When M makes a sequence of calls to the decryption oracle. For each decryption oracle query, M submits a ciphertext C, and D invokes $Dec(SK, C)$ to obtain K, responds with the K. The only restriction is that the adversary M can not request the decryption of C^*.
4. Finally, the adversary outputs a guess b'.

The adversary's advantage in the above game is

$$\mathrm{Adv}_{\mathrm{KEM,M}}^{\mathrm{CCA}}(\lambda) = |Pr[\mathrm{M}(K_0) = 1] - Pr[\mathrm{M}(K_1) = 1]|$$

2.2 Target Collision Resistant Hash Function

Informally, we say that a function $\mathrm{H} : X \rightarrow Y$ is a target-collision resistant (TCR) hash function, if, given a random pre-image $x \in X$, it is hard to find $x' \neq x$ with $\mathrm{H}(x') = \mathrm{H}(x)$.

Definition 2. *Let* $\mathrm{H} : X \rightarrow Y$ *be a function. For an adversary* M*, define*

$$\mathrm{Adv}_{\mathrm{H,M}}^{TCR}(\lambda) = Pr[x \leftarrow X, x' \leftarrow \mathrm{M}(x, \mathrm{H}) : x' \neq x \wedge \mathrm{H}(x') = \mathrm{H}(x)]$$

We say that H *is target-collision resistant if for any PPT adversary* M, $\mathrm{Adv}_{\mathrm{H,M}}^{TCR}(\lambda)$ *is negligible.*

2.3 Semi-smooth Subgroup

In [13], the author introduced the definition of semi-smooth subgroup.

Let $\mathrm{IGen}(1^\lambda)$ be a probability polynomial-time algorithm such that on input security parameter λ, randomly chooses two $m(\lambda)$-bit primes P and Q satisfying $P = 2p'p + 1$, $Q = 2q'q + 1$, outputs $N = PQ$, where p' and q' are $m'(\lambda)$-bit primes, both p and q are product of some distinct odd primes smaller than a low bound B. We call such integer N as semi-smooth integer.

Definition 3. *Let* $N = (2p'p + 1)(2q'q + 1)$ *be a random output of* $\mathrm{IGen}(1^\lambda)$*, the unique subgroup* G *of order* $p'q'$ *is called the semi-smooth subgroup of* Z_N^*.

Factoring Assumption about Semi-smooth Integer. We assume that there exists no probabilistic polynomial-time algorithm such that given only N, the random output of $\mathrm{IGen}(1^\lambda)$, can factor N with non-negligible probability.

In [13], at secure level of 80, parameters are suggest to be $\ell_{p'} = \ell_{q'} = 160$, $\ell_N = 1024$, and $B = 2^{15}$.

Here, we describe some properties that will be used later.

Property 1. Let h be a uniform element of Z_N^*, $P_B = \prod_{1<p<B,\ \mathrm{p\ is\ prime}} p$, and $g = h^{P_B}$. Then, g is a uniform element of G.

Property 2. With probability $1 - O(2^{-m'(\lambda)})$, a uniform element in G is a generator of G.

Property 3. Any element z of G is a quadratic residue, the unique quadratic residue u such that $u^2 = z$ lies in G.

Property 4. For any element z of G, the unique quadratic residue u such that $z = u^{2^k}$ lies in G for any $k \in Z^+$.

2.4 Signed Quadratic Residues

The signed quadratic residues[18], QR_N^+, are defined as the group $QR_N^+ = \{|x| : x \in QR_N\}$, where $|x|$ is the absolute value when representing elements of Z_N^* as the set $\{-(N-1)/2, \cdots, (N-1)/2\}$, N is a Blum integer. The group operation $\circ$ is defined by $a \circ b = |ab \bmod N|$. For simplification, we denote $|ab|$ instead of $|ab \bmod N|$.

An attractive property is that the membership in QR_N^+ can be efficiently verified since $QR_N^+ = J_N^+ = J_N \bigcap [(N-1)/2]$, where J_N^+ denotes $\{|x| : x \in J_N\}$, and J_N denotes the group of elements with Jacobi symbol 1.

2.5 Some Lemmas

Lemma 1. *Let g be a generator of G, μ is chosen uniformly from $[2^{\ell_{p'}+\ell_{q'}+\lambda}]$, k is any integer, then both $\mu \bmod p'q'$ and $(\mu+k) \bmod p'q'$ are statistically close to the uniform distribution of $[p'q']$, both g^μ and $g^{\mu+k}$ are statistically close to the uniform distribution of G.*

Proof. Write $2^{\ell_{p'}+\ell_{q'}+\lambda}$ as $k_1p'q' + k_2$ over Z, where $0 < k_2 < p'q'$. If μ is uniformly chosen from $[k_1p'q']$, then both $\mu \bmod p'q'$ and $(\mu + k)\ mod\ p'q'$ are uniform in $[p'q']$, and then both g^μ and $g^{\mu+k}$ are uniform in G. But a uniformly chosen element from $2^{\ell_{p'}+\ell_{q'}+\lambda}$ belongs to $[k_1p'q']$ with probability $k_1p'q'/2^{\ell_{p'}+\ell_{q'}+\lambda} = 1 - k_2/2^{\ell_{p'}+\ell_{q'}+\lambda} \geq 1 - O(2^{-\lambda})$.

The following lemma states that computing the square root residue of random element in semi-smooth subgroup can be reduced to the factoring algorithm for the modulus N.

Lemma 2. *Let G be the semi-smooth subgroup of Z_N^*, z is a uniformly chosen element of G, if there exists an adversary* A *can compute the unique quadratic residue u such that $u^2 = z$ with non-negligible probability, then there exists an adversary* C *can factor N with non-negligible probability.*

Proof. Given N, C chooses h uniformly from Z_N^*, set $P'_B = \prod_{2<p<B,\ \text{p is prime}} p$, and $h' = h^2$, $z = h'^{P'_B} (= h^{P_B})$. So z is a uniform element of G. If A can compute u such that $u^2 = z$, then C can compute $\widetilde{h}$ such that $\widetilde{h}^2 = h^2$: compute a, b over Z such that $aP'_B + 2b = gcd(P'_B, 2) = 1$, set $\widetilde{h} = u^a h'^b$. If $\widetilde{h} \neq \pm h$, then C outputs $\gcd(\widetilde{h} - h, N)$. With probability $1/2$, $\widetilde{h} \neq \pm h$, and so $\gcd(\widetilde{h} - h, N)$ is a non-trivial factor of N.

From lemma 2 and the Goldreich-Levin lemma[12], it is easy to see that given $z = u^2$ over G, the Goldreich-Levin predicate, $B_r(u)$, is a hard-core. Using the hybrid argument,we have:

Lemma 3. *Let G be the semi-smooth subgroup of Z_N^*, given a uniform element z of G, then $BBS_r(u)$ is indistinguishable from the uniform bits string U from $[2^{\ell_K}]$ under the assumption factoring N is hard, where u is the unique quadratic residue such that $z = u^{2^{\ell_K}}$, $\mathrm{BBS}_r(u) \overset{\text{def}}{=} (B_r(u), B_r(u^2), \cdots, B_r(u^{2^{\ell_K - 1}}))$, r is a random element with bits-size ℓ_N.*

Lemma 4. *Given A, B in some subgroup F of Z_N^*, along with x, y in Z, such that $A^x = B^y$, $\gcd(x, y) = z$, $\gcd(y, \mathrm{ord}(F)) = 1$. Then one can efficiently compute $A^{z/y}$, where the inversion in the exponent is computed modulo* $\mathrm{ord}(F)$.

Proof. Since $\gcd(x, y) = z$, using the extended Euclidean algorithm, one can compute a, b over Z such that $ax + by = z$. Let $B' = A^b B^a$. Since A, B belongs to F , so B' belongs to F. It is easy to verify that $A^z = (B')^y$. Since $\gcd(y, \mathrm{ord}(F)) = 1$, so $B' = A^{z/y}$.

Lemma 5. *If $A, B \in QR_N^+$, then $A^2 = B^2 \bmod N \Leftrightarrow A = B$. More generally, $A^{2^k} = B^{2^k} \bmod N$ $(k \in Z^+) \Leftrightarrow A = B$.*

Lemma 6. *If $A, B \in QR_N \cup QR_N^+$, then $|AB| = \big||A||B|\big|$.*

3 The Instantiation of HK09 over Semi-smooth Subgroup

3.1 Scheme Description

Gen(1^λ) : Run IGen(1^λ) to get the modulus N. Then, *Gen* chooses a target-collision resistant hash function $\mathrm{H} : Z_N \rightarrow [2^{\ell_\mathrm{H}} - 1]$. Next, *Gen* randomly chooses an element g of G, a bit string r of length ℓ_N, and ρ from $[2^{\ell_{p'}+\ell_{q'}+\lambda}]$. Finally, *Gen* sets $X = g^{\rho 2^\nu}$ $(\nu = \ell_\mathrm{H} + \ell_\mathrm{K})$. The public key is $PK = (N, g, X, r, \mathrm{H})$, and the private key is $SK = \rho$.

Enc(PK) : *Enc* randomly chooses $\mu \in [2^{\ell_{p'}+\ell_{q'}+\lambda}]$, and computes

$$R = g^{\mu 2^\nu}, \qquad t = \mathrm{H}(R), \qquad S = |(g^t X)^\mu|.$$

Set the ciphertext as $C = (R, S)$. Compute the encapsulation key as $K = \mathrm{BBS}_r(g^{\mu 2^{\ell_\mathrm{H}}})$.

Dec(SK, C) : *Dec* writes C as $C = (R, S)$, verifies both R, S belong to $Z_N^* \times (Z_N^* \bigcap [(N-1)/2])$ and rejects it if not. Then *Dec* computes $t = \mathrm{H}(R)$, verifies:

$$(S^2)^{2^\nu} = (R^2)^{t + \rho 2^\nu} \tag{1}$$

Reject it if it not. *Dec* computes $a, b, c \in Z$ such that

$$2^c = gcd(t, 2^\nu) = at + b2^\nu \tag{2}$$

Then *Dec* computes

$$T = ((S^2)^a \cdot (R^2)^{b - a\rho})^{2^{\ell_\mathrm{H} - c - 1}} \tag{3}$$

and $K = \mathrm{BBS}_r(T)$, outputs K

Correctness: Notice that, in *Dec*, even R, S may not sit in the subgroup G, as long as they pass the verification, they must sit in Z_N^*. So from the proof of the original HK09, the computed T is equal to $(R^2)^{1/(2^{\ell_K+1}) \bmod \mathrm{ord}(QR_N)}$. If $R = g^{\mu 2^{\ell_K+\ell_H}} \in G \subset QR_N$, then, we have $(R^2)^{1/(2^{\ell_K+1}) \bmod \mathrm{ord}(QR_N)} = (R^2)^{1/(2^{\ell_K+1}) \bmod \mathrm{ord}(QR_N) \bmod \mathrm{ord}(G)} = (R^2)^{1/(2^{\ell_K+1}) \bmod \mathrm{ord}(G)} = g^{\mu 2^{\ell_H}}$.

Efficiency comparison with original HK09. The encapsulation and decapsulation of both the original HK09 and this variant need $3\ell_{\exp} + \ell_K + 2.5\ell_H$ and $1.5\ell_{\exp} + 4\ell_K + 6.5\ell_H$ multiplications respectively, where both ℓ_K and ℓ_H can be set as λ. The difference is that, in original HK09, $\ell_{\exp}$ equals to ℓ_N, instead, in this variant, $\ell_{\exp}$ equals to $\ell_{p'} + \ell_{q'} + \lambda$. For 80-bits security, $\ell_N = 1024$, $\ell_{p'} = \ell_{q'} = 160$, $\lambda = 80$. In original HK09, the encapsulation requires 3352 multiplications, the decapsulation requires 2376 multiplications. In this variant, the encapsulation requires 1480 multiplications, the decapsulation requires 1440 multiplications.

3.2 Security Proof

Theorem 1. *If factoring the modulus N is hard and* H *is target-collision resistant, then the above key encapsulation mechanism is chosen ciphertext secure.*

Proof. To prove this theorem, from lemma 3, it is enough to reduce the CCA security of this scheme to the pseudo-randomness of the BBS generator over the Semi-smooth Subgroup.

Assume there exists an adversary A on KEM's IND-CCA security. We define an adversary D on the pseudo-randomness of the BBS generator. On input (N, z, V), the goal of D is to distinguish whether V is $\mathrm{BBS}_r(u)$ or a uniform bits string with equal length, where u is the unique quadratic residue in G such that $z = u^{2^{\ell_K}}$, z is a uniform element in G.

Prepare the public key. D chooses a target-collision resistant hash function $\mathrm{H} : Z_N \rightarrow [2^{\ell_H} - 1]$, a bits string r of length ℓ_N, a random element $g \in G$, as well as $\beta \in [2^{\ell_{p'}+\ell_{q'}+\lambda}]$, sets

$$R^* = z, \qquad t^* = \mathrm{H}(R^*), \qquad X = g^{\beta 2^\nu - t^*}.$$

The public key is set as $PK = (N, g, X, r, \mathrm{H})$. The private key is implicitly defined as $\rho = \beta - t^*/2^\nu \bmod p'q'$.

Prepare the challenge cipertext and key. Next, we assume g is a generator of G. So we can write $R^* = g^{\mu^* 2^\nu}$, though μ^* is unknown to D. D defines

$$S^* = |R^{*\beta}| \quad (= |g^{\mu^* \beta 2^\nu}| = |(g^{t^*} X)^{\mu^*}|).$$

The real corresponding key K^* is defined as

$$K^* = \mathrm{BBS}_r(g^{\mu^* 2^{\ell_H}}) = \mathrm{BBS}_r(R^{*\frac{1}{2^{\ell_K}}}) = \mathrm{BBS}_r(z^{\frac{1}{2^{\ell_K}}}) = \mathrm{BBS}_r(u)$$

The challenge ciphertext is $C^* = (R^*, S^*)$, the challenge key is V. Note that, as in the IND-CCA2 game, if V is $\mathrm{BBS}_r(u)$, then V is a real key , else V is a uniform string.

We claim that the distribution of the public key and the challenge ciphertext C^* is almost identical in simulation and IND-CCA game. Firstly, in public key, g, N, r and H are perfectly simulated. From Property 2, with overwhelming probability, g is a generator of G. From Lemma 1, we know that if g is a generator of G, then X in the real game and in simulation are both statistically close to the uniform element in G. So X is simulated perfectly with overwhelming probability. Similarly, with overwhelming, R^* is also perfectly simulated. Conditioned on X, R^*, g, r, N are simulated perfectly, from the simulation, we know that S^* and K^* are also perfectly simulated. As required.

Answer the decryption queries. When A submit a ciphertext (R, S), D does as following.

Check $(R, S) \in Z_N^* \times (Z_N^* \bigcap [(N-1)/2])$, reject if not. Compute $t = \mathrm{H}(R)$. For the case $t \neq t^*$. Verify:

$$(S^2)^{2^\nu} = (R^2)^{t-t^*+\beta 2^\nu} \tag{4}$$

Reject it if it not.

Note that the equation (4) is equivalent to

$$(R^2)^{(t-t^*)} = (R^{-2\beta} S^2)^{2^{\ell_{\mathrm{H}}+\ell_{\mathrm{K}}}}$$

Since R^2 and $R^{-2\beta} S^2$ belong to QR_N, using lemma 4, the simulator can compute $B' = (R^2)^{2^{c'}/2^{\ell_{\mathrm{H}}+\ell_{\mathrm{K}}}}$, where $2^{c'} = \gcd(t - t^*, 2^{\ell_{\mathrm{H}}+\ell_{\mathrm{K}}})$, the inversion in the exponent is computed modulo $\mathrm{ord}(QR_N)$. Then Dec can compute $T = (R^2)^{1/2^{\ell_{\mathrm{K}}+1}} = (B')^{2^{\ell_{\mathrm{H}}-1-c'}}$ since $\ell_{\mathrm{H}} - 1 - c' \geq 0$. If $R = g^{\mu 2^{\ell_{\mathrm{K}}+\ell_{\mathrm{H}}}}$, then $(R^2)^{1/(2^{\ell_{\mathrm{K}}+1}) \bmod \mathrm{ord}(QR_N)} = (R^2)^{1/(2^{\ell_{\mathrm{K}}+1}) \bmod \mathrm{ord}(G)} = g^{\mu 2^{\ell_{\mathrm{H}}}}$. Concretely, the simulator compute $a', b', c' \in Z$ such that

$$2^{c'} = gcd(t - t^*, 2^\nu) = a'(t - t^*) + b' 2^\nu \tag{5}$$

Then compute

$$T = ((S^2)^{a'} \cdot (R^2)^{b' - a'\beta})^{2^{\ell_{\mathrm{H}} - c' - 1}} \tag{6}$$

D answer with $\mathrm{BBS}_r(T)$.

For the case $t = t^*$. If $R = R^*$ and the ciphertext is valid , it will satisfy

$$(S^2) = (R^2)^{(t-t^*)/2^\nu + \beta} = (R^2)^\beta = S^{*2}.$$

Therefore, $S^2 = S^{*2}$. Furthermore, $(R, S) \neq (R^*, S^*)$ implies that $|S| = S \neq S^* = |S^*|$, so that $S \neq \pm S^*$ and $(S + S^*)(S - S^*) = S^2 - S^{*2} = 0 \bmod N$ yields a non-trivial factor of N.

If $t = t^*$ and $R \neq R^*$, then it will contradict the target-collision resistance of H, so D can safely give up this type of ciphertext.

So with overwhelming probability, D perfectly simulates the CCA game.
D outputs what A outputs.
Therefore, D can use A as an oracle to distinguish whether V is $\mathrm{BBS}_r(u)$ or a uniform bits string.

4 Scheme Based on ElGamal Encryption over Composite Modulus

In this section, we show how to construct a practical CCA secure KEM from ElGamal encryption over composite modulus. Taking account of efficiency, we present the instantiation over semi-smooth subgroup. This scheme implicitly uses the signed quadratic residues group [18].

4.1 Scheme Description

$Gen(1^\lambda)$: Run $\mathrm{IGen}(1^\lambda)$ to get the modulus N. Then, choose a target-collision resistant hash function $\mathrm{H} : Z_N \rightarrow [2^{\ell_{\mathrm{H}}} - 1]$. Next, randomly choose an element g of the semi-smooth subgroup G, a bit string r of length $\ell_{(N-1)/2}$, and $\rho, \rho' \in [2^{\ell_{p'}+\ell_{q'}+\lambda}]$. Finally, set $X = g^{\rho 2^\nu}$ $(\nu = \ell_{\mathrm{H}} - 1)$ and $X' = g^{\rho'}$. The public key is $PK = (N, g, X, X', r, \mathrm{H})$, and the private key is $SK = (\rho, \rho')$.

$Enc(PK)$: Enc randomly chooses $\mu \in [2^{\ell_{p'}+\ell_{q'}+\lambda}]$, and computes

$$R = |g^{\mu 2^\nu}|, \qquad t = \mathrm{H}(R), \qquad S = |(X'^t X)^\mu|, \qquad T = |X'^{\mu 2^\nu}|,$$

$$K = \mathrm{BBS}_r^+(T) \stackrel{\text{def}}{=} (B_r(|T|), B_r(|T^2|), \cdots, B_r(|T^{2^{\ell_K - 1}}|)).$$

Set the ciphertext as $C = (R, S)$ and the encapsulation key as K.

$Dec(SK, C)$: Dec writes C as $C = (R, S)$, verifies both R and S belong to $QR_N^+ = J_N \bigcap [(N-1)/2]$. If it holds, then Dec computes $t = \mathrm{H}(R)$, verifies:

$$|S^{2^\nu}| = |R^{\rho' t + \rho 2^\nu}|$$

If it holds, then Dec computes

$$T = |R^{\rho'}|, \quad K = \mathrm{BBS}_r^+(T).$$

Correctness: If R and S are computed according to the encapsulation, then both R and S belong to G^+. Since $G^+ \subseteq QR_N^+$, so $R, S \in QR_N^+$. From lemma 5, we know that $|AB| = ||A||B||$ as long as $A, B \in QR_N \cup QR_N^+$, then

$$|S^{2^\nu}| = ||(X'^t X)^\mu|^{2^\nu}| = |g^{(\rho' t + \rho 2^\nu)\mu 2^\nu}| = |R^{\rho' t + \rho 2^\nu}|.$$

The fact that $|R^{\rho'}|$ equals to $|X'^{\mu 2^\nu}|$ follows from:

$$|X'^{\mu 2^{\nu}}| = |g^{\rho'\mu 2^{\nu}}| = ||g^{\mu 2^{\nu}}|^{\rho'}| = |R^{\rho'}|$$

Efficiency: The ciphertext of this KEM consists of two group element (Notice that for the known CCA secure KEM schemes based on the ElGamal encryption over prime modulus under CDH assumption, the ciphertexts consist of at least three group elements). If we choose $\ell_{q_1} = \ell_{p_1} = 160$, $\lambda = 80$, then $\ell_\rho = \ell_{\rho'} = \ell_{\text{exp}} = 400$. We assume $\ell_{\text{H}} = 80$. As in original HK09, we assume one regular exponentiation with an exponent of length ℓ requires 1.5ℓ modular multiplications and that one squaring takes the same time as one multiplication. Notice that we can compute $g^{\rho'}$ and g^ρ with about 1.2 exponentiations since they share the same base g. The encapsulation requires $4.5\ell_{\text{exp}} + 2.5\ell_{\text{H}} + \ell_{\text{K}} = 2080$ multiplications. The decapsulation requires $1.5 \times 1.2\ell_{\text{exp}} + 2.5\ell_{\text{H}} + \ell_{\text{K}} = 1000$ multiplications.

4.2 Security Proof

Theorem 2. *If factoring the modulus N is hard and* H *is target-collision resistant, then the above key encapsulation mechanism is chosen ciphertext secure.*

High level of the proof: In HK09 instantiation (and the original HK09), the proof consists of two steps: firstly, the pseudo-randomness of the BBS generator $\text{BBS}_r(u)$ is reduced to the factoring assumption; then, the CCA security is black box reduced to the pseudo-randomness of the BBS generator $\text{BBS}_r(u)$. But, in this scheme, if we directly reduce the CCA security to the pseudo-randomness of $\text{BBS}_r^+(g^{\mu\rho'})$ (even g^μ and $g^{\rho'}$ is given), then the simulator could not answer DDH oracle that is needed for the verification and could not compute the inversion modulo the unknown order $p'q'$ which is needed to compute the encapsulated key. Instead, we prove the CCA security of this scheme *and* the pseudo-randomness of $\text{BBS}_r^+(g^{\mu\rho'})$ *simultaneously*. Adapting the proof idea of [23], we firstly reduce the security (both the CCA security of this scheme and the pseudo-randomness of $\text{BBS}_r^+(g^{\mu\rho'})$) to a hardcore distinguisher; next, we reduce the hardcore distinguisher to a hardcore predictor; finally, we reduce the hardcore predictor to a factoring algorithm. In the first step, the distinguisher could compute $\rho' 2^{\ell_{\text{K}}} \bmod p'q'$, so he could compute $|R^{\rho' 2^{\ell_{\text{K}}}}|$. The distinguisher does not directly verify the equation $|S^{2^\nu}| = |R^{\rho' t + \rho 2^\nu}|$, instead, he verify a equivalent equation $S^{2^{\nu+\ell_{\text{K}}}} = R^{\rho' t 2^{\ell_{\text{K}}} + \rho 2^{\nu+\ell_{\text{K}}}}$ (Before this, he should verify both R and S belong to QR_N^+, from lemma 5, we know that the two equations are equivalent. Note that, in [18], this technique has been used for proving the factoring assumption implies the strong DH assumption over QR_N^+). Then, by using lemma 4, the distinguisher is able to efficiently compute the encapsulated key from the latter equation.

Proof. The theorem is the consequence of the following three lemmas.

Reduce to the hard-core distinguisher $\text{D}(v^2, N, r, \alpha)$

Lemma 7. *If there exists a PPT adversary* M *such that* $\mathrm{Adv}^{\mathrm{CCA}}_{\mathrm{KEM,M}}(\lambda)$ *equals to* $\varepsilon(\lambda)$, *then there exists a PPT adversary* $\mathrm{D}(v^2, N, r, \alpha)$ *that distinguishes whether* α *is equal to* $B_r^+(|uw|)$ *or a random bit* b *with advantage* $\varepsilon'(\lambda)$, *where* v^2 *is a uniformly chosen element of* G, u *is the unique square root residue of* v^2, w *is determined by* v^2 *and* D*'s internal coin tosses, and* $\varepsilon'(\lambda) = \frac{\varepsilon(\lambda) - O(2^{-\lambda}) - \mathrm{Adv}^{\mathrm{TCR}}_{\mathrm{H,M}}(\lambda)}{\ell_{\mathrm{K}}}$.

Proof. On input (v^2, N, r, α), D works as follows.

Prepare the public key: Choose a target-collision resistant hash function $\mathrm{H} : Z_N \to [2^{\ell_{\mathrm{H}}} - 1]$. Randomly choose $J = k$ from $\{0, 1, \cdots \ell_{\mathrm{K}} - 1\}$. Select at random bits string $(b_0, b_1, \cdots, b_{k-1})$. Randomly and independently select 2 elements $\xi_i (i = 1, 2)$ from $[2^{\ell_{p'} + \ell_{q'} + \lambda}]$, denote $\overrightarrow{\xi} = (\xi_1, \xi_2)$. Set $s = 2\ell_{\mathrm{K}} - k$, $g = v^{2^s} \bmod N$, and $a_i = (\xi_i + 2^{-\ell_{\mathrm{K}}}) \bmod p'q' \quad (i = 1, 2)$. Set $X' = g^{a_1} = g^{\xi_1 + 2^{-\ell_{\mathrm{K}}} \bmod p'q'}$ (implicitly define $\rho' = (\xi_1 + 2^{-\ell_{\mathrm{K}}}) \bmod p'q'$). Set $B = g^{a_2} = g^{\xi_2 + 2^{-\ell_{\mathrm{K}}} \bmod p'q'}$ (implicitly define $B = g^{\mu^* 2^{\nu}}$). Set $t^* = \mathrm{H}(|B|)$. Randomly choose $\beta \in [2^{\ell_{p'} + \ell_{q'} + \lambda}]$, and set $X = g^{\beta 2^{\nu}} X'^{-t^*}$ (implicitly define $\rho = (\beta - \rho' t^*/2^{\nu}) \bmod p'q'$). The public key is set as $(N, g, X', X, r, \mathrm{H})$.

(D can efficiently compute $X' = g^{a_1} = g^{\xi_1 + 2^{-\ell_{\mathrm{K}}} \bmod p'q'} = v^{2^{(\ell_{\mathrm{K}} - k)}} v^{2^{(2\ell_{\mathrm{K}} - k)} \xi_1}$ since $\ell_{\mathrm{K}} > k$ and v^2 is given. Similar, D can be efficiently computed too. It is easy to see that other elements of the public key can be efficiently simulated by D).

Prepare the challenge ciphertext and key: The challenge ciphertext is set as:

$$R^* = |B| \ \ (= |g^{\mu^* 2^{\nu}}|); \qquad S^* = |R^{*\beta}| \ \ (= |(X'^{t^*} X)^{\mu^*}|)$$

And the challenge key is set as

$$K^* = (b_0, b_1, \cdots, b_{k-1}, \alpha, B_r(|g^{2^{k+1} a_1 a_2}|), \cdots, B_r(|g^{2^{\ell_{\mathrm{K}}-1} a_1 a_2}|))$$

Define w: We define $w = (v^{2^{2\ell_{\mathrm{K}}}})^{\xi_1 \xi_2} (v^{2^{\ell_{\mathrm{K}}}})^{\xi_1 + \xi_2}$. Given the values of ξ_1, ξ_2 and v^2, D is able to efficiently compute w. It is easy to see that, w is a quadratic residue in G (recall that $v^2 \in G$)and is determined by v^2 and D's internal coin tosses.

Claim 1. Let a_1, a_2, g, u, w be defined as above respectively, then $g^{2^k a_1 a_2} = uw$.

Proof.

$$\begin{aligned} g^{2^k a_1 a_2} &= g^{2^k (\xi_1 + 2^{-\ell_{\mathrm{K}}})(\xi_2 + 2^{-\ell_{\mathrm{K}}})} = g^{2^k (\xi_1 \xi_2 + (\xi_1 + \xi_2) 2^{-\ell_{\mathrm{K}}} + 2^{-2\ell_{\mathrm{K}}})} \\ &= v^{2^{2\ell_{\mathrm{K}}} (\xi_1 \xi_2 + (\xi_1 + \xi_2) 2^{-\ell_{\mathrm{K}}} + 2^{-2\ell_{\mathrm{K}}})} = uw. \end{aligned}$$

Claim 2. D is able to compute $g^{2^{k+j} a_1 a_2}$ for $j = 1, \ldots \ell_{\mathrm{K}} - k - 1$.

Proof. From Claim 1, we have $g^{2^k a_1 a_2} = uw$, so each $g^{2^{k+j} a_1 a_2}$ equals to $(uw)^{2^j}$ for $j = 1, \ldots \ell_{\mathrm{K}} - k - 1$. Furthermore, since D knows v^2 and w^2, so he is able to compute $(uw)^{2^j}$ for $j = 1, \ldots \ell_{\mathrm{K}} - k - 1$, as required.

From claim 1 and 2, it is easy to see that, D is able to efficiently prepare the challenge ciphertext and key.

Answer the decryption queries: For the query ciphetext (R, S), D verifies both R and S belong to QR_N^+. If it holds, D computes $t = \mathrm{H}(R)$.

If $t \neq t^*$, D verifies if

$$(S^{2^{\ell_\mathrm{K}}})^{2^\nu} = (R^{(2^{\ell_\mathrm{K}}\xi_1+1)})^{t-t^*}(R^{2^{\ell_\mathrm{K}}})^{\beta 2^\nu} \tag{7}$$

(Note that the right side equals to

$$(R^{(2^{\ell_\mathrm{K}}\rho')})^{t-t^*}(R^{2^{\ell_\mathrm{K}}})^{\beta 2^\nu} = (R^{2^{\ell_\mathrm{K}}})^{\rho' t-\rho' t^*+\beta 2^\nu} = (R^{2^{\ell_\mathrm{K}}})^{\rho' t+\rho 2^\nu}$$

From Lemma 5, we know that verifying $|S^{2^\nu}| = |R^{\rho' t+\rho 2^\nu}|$ is equivalent to verifying $(S^{2^{\ell_\mathrm{K}}})^{2^\nu} = (R^{2^{\ell_\mathrm{K}}})^{\rho' t+\rho 2^\nu}$, as required).

Equation (7) is equivalent to $(R^{(2^{\ell_\mathrm{K}}\xi_1+1)})^{t-t^*} = (SR^{-\beta})^{2^{\ell_\mathrm{K}+\nu}}$. Since $R^{(2^{\ell_\mathrm{K}}\xi_1+1)}$, $SR^{-\beta}$ belong to QR_N^+, using lemma 4, D is able to compute $(R^{(2^{\ell_\mathrm{K}}\xi_1+1)})^{2^{c'}/2^{\ell_\mathrm{K}+\nu}}$, where $2^{c'} = \gcd(t-t^*, 2^{\nu+\ell_\mathrm{K}})$, the inversion in the exponent is computed modulo $\mathrm{ord}(QR_N^+)(= \mathrm{ord}(QR_N))$. Furthermore, since both t and t^* are smaller than 2^{ℓ_H}, then $c' \leq \ell_\mathrm{H} - 1 = \nu$. Therefore, D is able to compute

$$T = |((R^{(2^{\ell_\mathrm{K}}\xi_1+1)})^{2^{c'}/2^{\ell_\mathrm{K}+\nu}})^{2^{\nu-c'}}|(= |(R^{(2^{\ell_\mathrm{K}}\xi_1+1)})^{1/2^{\ell_\mathrm{K}}}| = |R^{\rho'}|)$$

Concretely, if (7) holds, D computes $a', b', c' \in Z$ such that:

$$2^{c'} = \gcd(t - t^*, 2^{\nu+\ell_\mathrm{K}}) = a'(t - t^*) + b' 2^{\nu+\ell_\mathrm{K}}$$

Then D computes

$$T = |((SR^{-\beta})^{a'} R^{b'(2^{\ell_\mathrm{K}}\xi_1+1)})^{2^{\nu-c'}}|$$

Response the oracle with $\mathrm{BBS}_r^+(T)$.

If $t = t^*$, D rejects the query ciphetext (R, S). (If $\mathrm{H}(R) = t = t^* = \mathrm{H}(R^*)$ and $R \neq R^*$, then M has broken the target-collision resistance of H. If $t = t^*$ and $R = R^*$, and the ciphertext is valid, then we have

$$S = |S| = |((R^{(2^{\ell_\mathrm{K}}\xi_1+1)})^{t-t^*}(R^{2^{\ell_\mathrm{K}}})^{\beta 2^\nu})^{1/(2^{\nu+\ell_\mathrm{K}})}| = |R^\beta| = |(R^*)^\beta| = S^*$$

which means that $(R, S) = (R^*, S^*)$, so this query will be rejected, as required).

When M outputs a bit, D outputs the same bit.

The running time of D**:** It is easy to see that D can run in polynomial time.

The success-probability of D**.** To find the success probability of D, we prove that the distribution of the public key, challenge ciphetext, and the decryption in the simulated game is statistically close to that in the real game.

Since v^2 is a uniformly chosen element of G, and squaring is a permutation, so the above defined g is a uniformly distributed element in G. Thus, g is perfectly simulated. Obviously, N, r and H are perfectly simulated. From property 2, we know that with probability $1 - O(2^{-m'(\lambda)}) \geq 1 - O(2^{-\lambda})$, g is a generator. From

Lemma 1, we know that with probability $1 - O(2^{-\lambda})$, X' in simulation and in real game are both statistically close to the uniformly distributed element in G. So, with probability $1 - O(2^{-\lambda})$, X' is perfectly simulated. With the same analysis, with probability $1 - O(2^{-\lambda})$, X and R^* are perfectly simulated.

Therefore, the statistical distance between distribution of the public key in the simulated game and that in the real game is $O(2^{-\lambda})$.

Note that, conditioned on the public key is simulated perfectly, the challenge ciphertext is perfectly simulated, and the decryption oracle is simulated perfectly except the case that M finds a target collision, which occurs with negligible probability $\mathrm{Adv}^{\mathrm{TCR}}_{\mathrm{H,M}}(\lambda)$.

For convenience, we denote some hybrid experiments $H^J (J = 0, \cdots, \ell_{\mathrm{K}})$ the same as the real game except the way the challenge key is responded with: the first J bits are chosen randomly, while the other $\ell_{\mathrm{K}} - J$ bits are computed as in K_0. So in the experiment H^0, the distribution of the key that the adversary sees is the same as K_0, whereas in the experiment $H^{\ell_{\mathrm{K}}}$, the distribution of the key is the same as K_1.

From Claim 1, we know that, if $\alpha = B_r^+(|uw|)$, then the distribution that M sees is the simulated H^J , while if α is a random bit b, then the distribution that M sees is the simulated H^{J+1}. We denote the simulated H^k as H^k_S for each $k \in \{0, 1, \cdots, \ell_{\mathrm{K}}\}$, and still denote real H^k as H^k. So the advantage of D is:

$$
\begin{aligned}
&|Pr[\mathrm{D}(B_r^+(|uw|)) = 1] - Pr[\mathrm{D}(b) = 1]| \\
&= \tfrac{1}{\ell_{\mathrm{K}}} |\textstyle\sum_{j=0}^{\ell_{\mathrm{K}}-1} \{Pr[\mathrm{D}(B_r^+(|uw|)) = 1 | J = j] - Pr[\mathrm{D}(b) = 1 | J = j]\}| \\
&= \tfrac{1}{\ell_{\mathrm{K}}} |\textstyle\sum_{j=0}^{\ell_{\mathrm{K}}-1} \{Pr[\mathrm{M}(H^j_S) = 1] - Pr[\mathrm{M}(H^{j+1}_S) = 1]\}| \\
&= \tfrac{1}{\ell_{\mathrm{K}}} |Pr[\mathrm{M}(H^0_S) = 1] - Pr[\mathrm{M}(H^{\ell_{\mathrm{K}}}_S) = 1]| \\
&\geq \tfrac{1}{\ell_{\mathrm{K}}} \{|Pr[\mathrm{M}(H^0) = 1] - Pr[\mathrm{M}(H^{\ell_{\mathrm{K}}}) = 1]| - O(2^{-\lambda}) - \mathrm{Adv}^{\mathrm{TCR}}_{\mathrm{H,M}}(\lambda)\} \\
&= \frac{\varepsilon(\lambda) - O(2^{-\lambda}) - \mathrm{Adv}^{\mathrm{TCR}}_{\mathrm{H,M}}(\lambda)}{\ell_{\mathrm{K}}}
\end{aligned}
$$

This completes the proof of Lemma 7.

Reduce to the hard-core predictor $\mathrm{D}'_{N,\xi_1,\xi_2,v^2}$

Since D defined in Lemma 7 chooses ξ_1, ξ_2 itself and w depends on v^2 and ξ_1, ξ_2 , then the value of w potentially changes each time D is invoked. Furthermore, D is not a predictor for $B_r^+(|uw|)$ but rather a distinguisher. So D is not suitable to be used as an oracle for the Goldreich-Levin reconstruction algorithm [12]. The first problem can be solved by fixing the value ξ_1, ξ_2 in advance. The second problem can be addressed by reducing the hard-core distinguisher to a suitable hard-core predictor. On input $< r >$, the hard-core predictor $\mathrm{D}'_{N,\xi_1,\xi_2,v^2}$ is defined as follows:

1. Uniformly choose random bits α and β .
2. Invoke D on input $< v^2, N, r, \alpha >$, and feed it with ξ_1, ξ_2 .
3. If D outputs 1, then output α, else if D outputs 0, then output β.

Note that now, the value of w does not change with the invoking of $\mathrm{D}'_{N,\xi_1,\xi_2,v^2}$. So it is possible to use $\mathrm{D}'_{N,\xi_1,\xi_2,v^2}$ as an oracle to reconstruct $|uw|$.

Lemma 8. *If there exists a PPT adversary* M *such that* $\mathrm{Adv}^{\mathrm{CCA}}_{\mathrm{KEM,M}}(\lambda)$ *equals to* $\varepsilon(\lambda)$*, then there exists a PPT hard-core predictor,* $\mathrm{D}'_{N,\xi_1,\xi_2,v^2}$*, with the probability* $\varepsilon'(\lambda)/2$ *(over the choice of* N, v^2*, and* ξ_1, ξ_2*), can predict the value of* $B^+_r(|uw|)$ *with advantage* $\varepsilon'(\lambda)/4$*, where* u *is the unique square root residue of* v^2*,* w *is determined by* v^2 *and* ξ_1, ξ_2*, and* $\varepsilon'(\lambda) = \frac{\varepsilon(\lambda)-O(2^{-\lambda})-\mathrm{Adv}^{\mathrm{TCR}}_{\mathrm{H,M}}(\lambda)}{\ell_{\mathrm{K}}}$.

Proof. By Lemma 7, M has $\varepsilon'(\lambda)$-advantage in distinguishing $B^+_r(|uw|)$ from a random bit b. Then for at least $\varepsilon'(\lambda)/2$ fraction of the choices of N, v^2, and ξ_1, ξ_2, M has $\varepsilon'(\lambda)/2$-advantage in distinguishing $B^+_r(|uw|)$ from the random bit b . So it is straightforward that $\mathrm{D}'_{N,\xi_1,\xi_2,v^2}$ can predict the value of $B^+_r(|uw|)$ with advantage $\varepsilon'(\lambda)/4$.

Reduce to the factoring algorithm A(N)

Lemma 9. *If there exists a PPT adversary* M *such that* $\mathrm{Adv}^{\mathrm{CCA}}_{\mathrm{KEM,M}}(\lambda)$ *equals to* $\varepsilon(\lambda)$*, then there exists a PPT algorithm* A *factoring* N *with success probability* $\Omega(\varepsilon'(\lambda)^2)$*, where* $\varepsilon'(\lambda)$ *equals to* $\varepsilon'(\lambda) = \frac{\varepsilon(\lambda)-O(2^{-\lambda})-\mathrm{Adv}^{\mathrm{TCR}}_{\mathrm{H,M}}(\lambda)}{\ell_{\mathrm{K}}}$.

Proof. From lemma 2, it is enough to prove that there exists another adversary $\mathrm{A}'(N, v^2)$, using M as oracle, is able to compute the unique quadratic residue u such that $u^2 = v^2$ with probability $\Omega(\varepsilon'(\lambda)^2)$, where v^2 is a uniform element of G.

On input (N, v^2), A′ is defined as follows:

1. Choose ξ_1, ξ_2 uniformly from $[2^{\ell_{p'}+\ell_{q'}+\lambda}]$.
2. Compute $w = (v^{2^{2\ell_{\mathrm{K}}}})^{\xi_1\xi_2}(v^{2^{\ell_{\mathrm{K}}}})^{\xi_1+\xi_2}$.
3. Invoke the Goldreich-Levin reconstruction algorithm, $R(1^\lambda)$:
   (a) Whenever asked for $B_{r_i}(z)$, invoke $\mathrm{D}'_{N,\xi_1,\xi_2,v^2}$ on input $< r_i >$, and give its output as an answer. (Recall that $\mathrm{D}'_{N,\xi_1,\xi_1,v^2}$ invokes M and answers its queries).
   (b) Denote by z the output of R.
4. Compute $u' = zw^{-1}$. Given that R outputs the correct value, i.e., $z = |uw|$, then $z = uw$ or $z = -uw$.

The successful probability of A′**:** Since with the probability $\varepsilon'(\lambda)/2$, $\mathrm{D}'_{N,\xi_1,\xi_2,v^2}$ predicts the value of $B^+_r(|uw|)$ with advantage $\varepsilon'(\lambda)/4$, then by Goldreich-Levin theorem [12], we have that R retrives the value of $|uw|$ with probability at least $\Omega(\varepsilon'(n)^2)$. Given that R outputs the correct value, with probability 1/2, $u' = u$. Therefore, A′ compute u with probability $\Omega(\varepsilon'(n)^2)$, as required.

References

1. Blum, L., Blum, M., Shub, M.: A simple unpredictable pseudo-random number generator. SIAM Journal on Computing 15(2), 364–383 (1986)
2. Blum, M., Goldwasser, S.: An probabilistic public key encryption scheme which hides all partial information. In: Blakely, G.R., Chaum, D. (eds.) CRYPTO 1984. LNCS, vol. 196, pp. 289–299. Springer, Heidelberg (1985)

3. Canetti, R., Halevi, S., Katz, J.: Chosen-Ciphertext Security from Identity-Based Encryption. In: Cachin, C., Camenisch, J.L. (eds.) EUROCRYPT 2004. LNCS, vol. 3027, pp. 207–222. Springer, Heidelberg (2004)
4. Cramer, R., Hofheinz, D., Kiltz, E.: A Twist on the Naor-Yung Paradigm and Its Application to Efficient CCA-Secure Encryption from Hard Search Problems. In: Micciancio, D. (ed.) TCC 2010. LNCS, vol. 5978, pp. 146–164. Springer, Heidelberg (2010)
5. Cash, D.M., Kiltz, E., Shoup, V.: The twin diffie-hellman problem and applications. In: Smart, N.P. (ed.) EUROCRYPT 2008. LNCS, vol. 4965, pp. 127–145. Springer, Heidelberg (2008)
6. Cramer, R., Shoup, V.: A practical public key cryptosystem provably secure against adaptive chosen ciphertext attack. In: Krawczyk, H. (ed.) CRYPTO 1998. LNCS, vol. 1462, pp. 13–25. Springer, Heidelberg (1998)
7. Cramer, R., Shoup, V.: Universal hash proofs and a paradigm for adaptive chosen ciphertext secure public-key encryption. In: Knudsen, L.R. (ed.) EUROCRYPT 2002. LNCS, vol. 2332, pp. 45–64. Springer, Heidelberg (2002)
8. Cramer, R., Shoup, V.: Design and analysis of practical public-key encryption schemes secure against adaptive chosen ciphertext attack. SIAM Journal on Computing 33(1), 167–226 (2003)
9. Dolev, D., Dwork, C., Naor, M.: Non-malleable cryptography. In: Proceedings of the 23rd ACM Symposium on Theory of Computing, pp. 542–552. IEEE Computer Society Press, Los Alamitos (1991)
10. Diffie, W., Hellman, M.: New directions in cryptography. IEEE Transactions on Information Theory 22, 644–654 (1976)
11. ElGama, T.: A public key cryptosystem and a signature scheme based on discrete loga- rithms. IEEE Transactions on Information Theory 31(4), 469–472 (1985)
12. Goldreich, O., Levin, L.A.: A hard-core predicate for all one-way functions. In: Proceedings of the 21st Annual ACM Symposium on Theory of Computing, pp. 25–32. ACM Press, New York (1989)
13. Groth, J.: Cryptography in subgroups of Z_n^*. In: Kilian, J. (ed.) TCC 2005. LNCS, vol. 3378, pp. 50–65. Springer, Heidelberg (2005)
14. Haralambiev, K., Jager, T., Kiltz, E., Shoup, V.: Simple and efficient public-key encryption from computational Diffie-Hellman in the standard model. In: Nguyen, P.Q., Pointcheval, D. (eds.) PKC 2010. LNCS, vol. 6056, pp. 1–18. Springer, Heidelberg (2010)
15. Hanaoka, G., Kurosawa, K.: Efficient chosen ciphertext secure public key encryption under the computational Diffie-Hellman assumption. In: Pieprzyk, J. (ed.) ASIACRYPT 2008. LNCS, vol. 5350, pp. 308–325. Springer, Heidelberg (2008)
16. Hofheinz, D., Kiltz, E.: Secure Hybrid Encryption from Weakened Key Encapsulation. In: Menezes, A. (ed.) CRYPTO 2007. LNCS, vol. 4622, pp. 553–571. Springer, Heidelberg (2007)
17. Hofheinz, D., Kiltz, E.: Practical Chosen Ciphertext Secure Encryption from Factoring. In: Joux, A. (ed.) EUROCRYPT 2009. LNCS, vol. 5479, pp. 313–332. Springer, Heidelberg (2009)
18. Hofheinz, D., Kiltz, E.: The group of signed quadratic residues and applications. In: Halevi, S. (ed.) CRYPTO 2009. LNCS, vol. 5677, pp. 637–653. Springer, Heidelberg (2009)
19. Kiltz, E.: Chosen-Ciphertext Secure Key-Encapsulation Based on Gap Hashed Diffie-Hellman. In: Okamoto, T., Wang, X. (eds.) PKC 2007. LNCS, vol. 4450, pp. 282–297. Springer, Heidelberg (2007)

20. Kurosawa, K., Desmedt, Y.: A New Paradigm of Hybrid Encryption Scheme. In: Franklin, M. (ed.) CRYPTO 2004. LNCS, vol. 3152, pp. 426–442. Springer, Heidelberg (2004)
21. Yehuda Lindell, A.: Simpler Construction of CCA2-Secure Public-Key Encryption under General Assumptions. In: Biham, E. (ed.) EUROCRYPT 2003. LNCS, vol. 2656, pp. 241–254. Springer, Heidelberg (2003)
22. McCurley, K.: A Key Distribution System Equivalent to Factoring. Journal of Cryptology 1(2), 95–105 (1988)
23. Naor, M., Reingold, O., Rosen, A.: Pseudo-random functions and factoring. SIAM Journal on Computing 31(5), 1383–1404 (2002)
24. Peikert, C., Waters, B.: Lossy trapdoor functions and their applications. In: Proceedings of the 40th Annual ACM Symposium on Theory of Computing 2008, pp. 187–196. ACM, New York (2008)
25. Rabin, M.O.: Digital signatures and public key functions as intractable as factorization. Technical Report MIT/LCS/TR-212, Massachusetts Institute of Technology (January 1979)
26. Rackoff, C., Simon, D.R.: Non-interactive zero-knowledge proof of knowledge and chosen ciphertext attack. In: Feigenbaum, J. (ed.) CRYPTO 1991. LNCS, vol. 576, pp. 433–444. Springer, Heidelberg (1992)
27. Rosen, A., Segev, G.: Chosen-Ciphertext Security via Correlated Products. In: Reingold, O. (ed.) TCC 2009. LNCS, vol. 5444, pp. 419–436. Springer, Heidelberg (2009)
28. Shoup, V.: Using Hash Functions as a Hedge against Chosen Ciphertext Attack. In: Preneel, B. (ed.) EUROCRYPT 2000. LNCS, vol. 1807, pp. 275–288. Springer, Heidelberg (2000)

A Proof of Some Properties and Lemmas

Property 1. Let h be a uniform element of Z_N^*, $P_B = \prod_{1<p<B,\ \text{p is prime}} p$, and $g = h^{P_B}$. Then, g is a uniform element of G.

Proof. Notice that order of h is one of the factors of $2pqp'q'$, and $2pq|P_B$, $\gcd(p'q', P_B) = 1$. Then the order of g must be one of the factors of $p'q'$. Thus g lies in the unique subgroup of order $p'q'$, G. On the other hand, for every element g of G, there must exists an element h belongs to Z_N^*, such that $g = h^{P_B}$ (Reason: since $\gcd(p'q', P_B) = 1$, then there exists $a, b \in Z$ such that $aP_B + bp'q' = 1$. Then $g = g^{aP_B + bp'q'} = (g^a)^{P_B}$). Therefore, $G = \{g | g = h^{P_B},\ h \in Z_N^*\}$. Observe that the mapping $f(x) = x^{P_B}$ is a $4pq$ to 1 mapping from Z_N^* to G, that is, for every z in G, there exists exactly $4pq$ solutions in Z_N^* such that $z = x^{P_B}$ (Reason: We firstly consider the set $X_I = \{x | x \in Z_N^*,\ x^{P_B} = 1\}$. Let the number of X_I,$|X_I|$, be m. Let y' be an element of G, x' be an element of Z_N^* such that $y' = (x')^{P_B}$. For every element x of X_I, it must be that $y' = (x'x)^{P_B}$. For every x does not belong to X_I, it must be that $y' \neq (x'x)^{P_B}$. So it must be that for every z of G, the equation $z = x^{P_B}$ has exactly m solutions in Z_N^*. Since the number of G ,$|G|$, equals to $p'q'$. So it must be $mp'q' = 4pqp'q'$. So m equals to $4p'q'$). When x is chosen uniformly from Z_N^*, $z = x^{P_B}$ is uniformly distributed in G. So g is a uniformly random element of G.

Property 2. With probability $1 - O(2^{-m'(\lambda)})$, a uniform element in G is a generator of G.

Proof. The order of G is $p'q'$, there are $(p'-1)(q'-1)$ elements of order $p'q'$. So with probability $1-(p'-1)(q'-1)/p'q' = 1 - O(2^{-m'(\lambda)})$,a uniform element in G is a generator of G.

Property 3. Any element z of G is a quadratic residue, the unique quadratic residue u such that $u^2 = z$ lies in G.

Proof. From property 1 and 2, with overwhelming probability, $g = h^{P_B} = (h^{P'_B})^2$ is a generator of G, where $P'_B = \prod_{2<p<B,\text{ p is prime}} p$. Obviously, g is a quadratic residue. So any element of $G =< g >$ is a quadratic residue. Since N is a Blum integer, then the equation $u^2 = z$ has unique solution in QR_N. Furthermore, the order of G, $p'q'$, is odd, then $gcd(2, p'q') = 1$, so $2^{-1}\text{mod } p'q'$ exists. Since z lies in G, then $z^{2^{-1}\text{mod } p'q'}$ lies in G. Finally, since $(z^{2^{-1}\text{mod } p'q'})^2 = z$, then $z^{2^{-1}\text{mod } p'q'}$ which lies in G is the unique solution of the equation $u^2 = z$.

Property 4. For any element z of G, then the unique quadratic residue u such that $z = u^{2^k}$ lies in G for any $k \in Z^+$.

Proof. Since $gcd(2, p'q') = 1$ and so $gcd(2^k, p'q') = 1$, then $2^{-k}\text{mod } p'q'$ exists, thus $z^{2^{-k}\text{mod } p'q'}$ lies in G and is a quadratic residue. Since N is a Blum integer, then squaring in quadratic residue group, QR_N, is a permutation. Then $u^2 = z$ has unique solution in QR_N. By induction, $z = u^{2^k}$ has unique solution in QR_N. Since $(z^{2^{-k}\text{mod } p'q'})^{2^k} = z$, then $u = z^{2^{-k}\text{mod } p'q'}$ is the unique quadratic residue satisfies $z = u^{2^k}$ and lies in G.

Lemma 5. If $A, B \in QR_N^+$, then $A^2 = B^2 \text{ mod } N \Leftrightarrow A = B$. More generally, $A^{2^k} = B^{2^k} \text{ mod } N\ (k \in Z^+) \Leftrightarrow A = B$.

Proof. The necessity is obvious. We only prove the sufficiency.

Since $A \in QR_N^+$, then there exists $u \in Z_N^*$ such that $A = u^2$ if $0 \leq u^2 < N/2$ or else $A = -u^2$. Similarly, there exists $v \in Z_N^*$ such that $B = v^2$ if $0 \leq v^2 < N/2$ or else $B = -v^2$. Now

$$A^2 = B^2 \text{ mod } N \Rightarrow u^4 = v^4 \text{ mod } N$$

From the uniqueness of square quadratic root (recall that N is a Blum integer), we have $u^2 = v^2 \text{ mod } N$.

So if $0 \leq u^2 < N/2$ then $A = u^2 = v^2 = B$; else if $-N/2 < u^2 < 0$ then $A = -u^2 = -v^2 = B$.

The general case can be proved by induction.

Lemma 6. If $A, B \in QR_N \cup QR_N^+$, then $|AB| = \big||A||B|\big|$.

Proof. If $A \in QR_N \cup QR_N^+$, then exists u such that $A = u^2$ or $A = -u^2$. Similarly, if $B \in QR_N \cup QR_N^+$, then exists v such that $B = v^2$ or $B = -v^2$. On one hand, we have $|AB| = |u^2v^2|$ or $|AB| = |-u^2v^2|$. So, we have $|AB| = |\pm u^2v^2| = |u^2v^2|$. On the other hand, we also have $||A||B|| = |\pm u^2v^2| = |u^2v^2|$ since $|A|$ equals to u^2 or $-u^2$ and $|B|$ equals to v^2 or $-v^2$. The Lemma follows since both $|AB|$ and $||A||B||$ equal to $|u^2v^2|$.

Chameleon All-But-One TDFs and Their Application to Chosen-Ciphertext Security

Junzuo Lai[1], Robert H. Deng[1], and Shengli Liu[2]

[1] School of Information Systems,
Singapore Management University, Singapore 178902
{junzuolai,robertdeng}@smu.edu.sg
[2] Department of Computer Science and Engineering
Shanghai Jiao Tong University, Shanghai 200240, China
slliu@sjtu.edu.cn

Abstract. In STOC'08, Peikert and Waters introduced a new powerful primitive called lossy trapdoor functions (LTDFs) and a richer abstraction called all-but-one trapdoor functions (ABO-TDFs). They also presented a black-box construction of CCA-secure PKE from an LTDF and an ABO-TDF. An important component of their construction is the use of a strongly unforgeable one-time signature scheme for CCA-security.

In this paper, we introduce the notion of chameleon ABO-TDFs, which is a special kind of ABO-TDFs. We give a generic as well as a concrete construction of chameleon ABO-TDFs. Based on an LTDF and a chameleon ABO-TDF, we presented a black-box construction, free of one-time signature, of variant of the CCA secure PKE proposed by Peikert and Waters.

Keywords: Chosen Ciphertext Security, Lossy Trapdoor Functions, Chameleon All-But-One Trapdoor Functions.

1 Introduction

Chosen-ciphertext security (CCA-security, for short) [33,14] is now considered as a standard notion of security for public key encryption (PKE) in practice. Numerous CCA-secure PKE schemes in the standard model, under both specific hardness assumption and general assumption, have been constructed over the years following several structural approaches.

The first approach for constructing CCA-secure PKE schemes was put forward by Naor and Yung [28]. As explained in [17], the approach employs a "two key" construction, where the well-formedness of a ciphertext is guaranteed by a (simulation-sound) non-interactive zero knowledge (NIZK) proof. The two-key/NIZK paradigm has led to CCA-secure PKE schemes based on general assumption [14], such as trapdoor permutations, and efficient schemes based on specific number theoretic assumptions [12,13], such as the decisional Diffie-Hellman (DDH) and composite residuosity assumptions.

Canetti, Halevi, and Katz [10] presented another approach for constructing CCA-secure PKE schemes using identity-based encryption (IBE) as a building

D. Catalano et al. (Eds.): PKC 2011, LNCS 6571, pp. 228–245, 2011.

block. The idea is to use, for each encryption, a fresh random verification key of a strongly unforgeable one-time signature scheme as the "identity" for IBE encryption. In order to tie the IBE ciphertext to this verification key, the ciphertext is signed using the corresponding signing key. Boneh and Katz [7] improved the efficiency of the scheme by using a MAC instead of a strongly unforgeable one-time signature. Some other efforts [8,25] further improved the efficiency.

The PKE schemes in [10,7,8,25] follow a similar method in the proof simulation. After the setup phase there is a certain set of well-formed ciphertexts that the simulator can decrypt corresponding to "identities" that the simulator knows the private keys. The remaining well-formed ciphertexts, that the simulator cannot decrypt corresponding to "identities" for which the simulator does not know the private keys, can be used as challenge ciphertexts in the simulation.

Recently, Peikert and Waters [31] introduced a new primitive called lossy trapdoor functions (LTDFs) and a richer abstraction called all-but-one trapdoor functions (ABO-TDFs). Peikert and Waters [31] constructed an elegant CCA-secure PKE scheme in a black-box manner based on an LTDF and an ABO-TDF. The scheme can be viewed as an application of the two-key paradigm [28], and the proof of security is similar to that of the IBE-based schemes [10]. An important component of their construction is the use of a strongly unforgeable one-time signature scheme for CCA-security, similar to that of Canetti, Halevi, and Katz [10]. This paper is motivated by improving the CCA-secure PKE construction of Peikert and Waters [31].

1.1 Our Contributions

Chameleon all-but-one TDFs. We introduce the notion of chameleon all-but-one TDFs (ABO-TDFs), which is a special kind of ABO-TDFs.

In an ABO-TDFs collection [31], each function has several *branches*. Almost all the branches are injective trapdoor functions, except for *one* branch which is lossy. Freeman et al. [18] generalized the definition of ABO trapdoor functions by allowing possibly *many* lossy branches (other than *one*).

As for chameleon ABO-TDFs, each function has many lossy branches just as the generalized definition in [18], but each branch is now represented by a pair (a, b). The "chameleon" property requires that for each a, it is easy to determine a unique b to come up with a lossy branch (a, b) with a trapdoor, while it is computationally indistinguishable to tell a lossy branch (a, b_0) from an injective branch (a, b_1) without the trapdoor.

Based on any CPA-secure PKE scheme with some additional property (mostly additively homomorphism), we propose a generic construction of chameleon ABO-TDFs. We can construct a chameleon ABO-TDF from any ABO-TDF in the sense of [31] and a chameleon hash function [24] targeting to the branch set. Yet the properties of the chameleon hash are a bit overkill for what we need and we build the needed properties directly into the constructions for better efficiency. In our construction, each chameleon ABO-TDF takes as input $((a, b), x)$, and outputs a ciphertext, which is an encryption of $x(ax_a + bx_b + x_d)$, where x_a, x_b, x_d are the trapdoor and the encryptions of x_a, x_b, x_d using the CPA-secure

PKE scheme are the public parameters of the function. Note that, due to the homomorphism of the CPA-secure PKE scheme, the chameleon ABO-TDF can be computed publicly. If (a, b) is a lossy branch, the chameleon ABO-TDF outputs an encryption of 0. Given a, with the trapdoor x_a, x_b, x_d, one can compute $b = (-ax_a - x_d) \cdot {x_b}^{-1}$, where (a, b) is a lossy branch. This computation requires that the message space of the PKE scheme is a finite field. In previous constructions of ABO-TDFs [31,18], each ABO-TDF takes as input (b, x), and outputs a ciphertext, which is an encryption of $x(b - b^*)$, where an encryption of b^* is the public parameter. The only lossy branch is $b = b^*$, and the ABO-TDF outputs an encryption of 0.

We also show how to instantiate the generic construction based on the Damgård-Jurik encryption scheme [15]. In fact, it is easy to transform the DDH-based all-but-one trapdoor function proposed by Freeman et al. [18] into a chameleon ABO-TDF using the same technique in our generic construction of chameleon ABO-TDFs.

CCA-SECURE PKE. We present a black-box construction of CCA-secure PKE based on an LTDF and a chameleon ABO-TDF. Our construction does not require strongly unforgeable one-time signature scheme, but a collision-resistant hash function, making it more efficient than that of Peikert and Waters [31].

The security proof of the construction does not rely on random oracles (RO) [5]. We follow a similar method of Peikert and Waters [31] in the proof simulation. In the security proof of Peikert and Waters's construction, when the adversary issues decryption queries, with overwhelming probability, the ABO-TDF works as an injective trapdoor function and the simulator uses the corresponding trapdoor to respond. In the challenge phase, the ABO-TDF works in lossy branch and the encrypted message is information hidden from the adversary. Because the ABO-TDF of Peikert and Waters has only *one* lossy branch, the simulator needs to know the lossy branch before the challenge phase and it resorts to strongly unforgeable one-time signature scheme.

In our CCA-secure PKE construction, the ABO-TDF is replaced by a chameleon ABO-TDF. Each chameleon ABO-TDF has many lossy branches (other than *one*) and each branch is represented by a pair (a, b). In our scheme, the first component of a branch a is correlated with the encrypted message, but b is independent of the message. Now, in the proof simulation, when the adversary submits the challenge messages, the simulator first computes a^*, which is correlated with the challenge messages, and computes b^* with the trapdoor of the chameleon ABO-TDF to make the branch (a^*, b^*) lossy. In other words, the simulator does not need to know the lossy branch before the challenge phase and it can generate a lossy branch after the adversary submits the challenge messages, which allows us to remove the requirement of strongly unforgeable one-time signature scheme. In our proof simulation, the simulator uses the same method of Peikert and Waters [31] to answer the adversary's decryption queries.

1.2 Related Work

Lossy trapdoor functions (LTDFs) were introduced by Peikert and Waters in [31]. Since their introduction, LTDFs have found many uses in cryptography. In particular, Peikert and Waters showed that any LTDF with enough lossiness can be used to construct an ABO-TDF, which can then be used to achieve CCA-security. In addition to CCA-secure encryption, LTDFs have been used in achieving deterministic encryption [3], lossy encryption [30], hedged public key encryption [2], security against selective opening attacks [4].

Peikert and Waters [31] presented constructions of LTDFs from the Decisional Diffie-Hellman (DDH) assumption and lattice assumptions. Boldyreva et al. [6] and Freeman et al. [18] gave (identical) efficient constructions of LTDFs from Paillier's decisional composite residuosity (DCR) assumption [29]. Freeman et al. [18] also gave efficient constructions of LTDFs based on composite residuosity assumption and d-Linear assumption [19,34]. Hemenway and Ostrovsky [20] showed that smooth homomorphic hash proof systems imply LTDFs. Kiltz et al. [23] showed that the RSA trapdoor function is lossy under the Φ-Hiding assumption of Cachin et al. [11]. Recently, Boyen and Waters [9] proposed two new discrete-log-type LTDFs, which are more efficient than earlier comparable constructions.

Rosen and Segev [32] showed that any collection of injective trapdoor functions that is secure under very natural correlated products can be used to construct a CCA-secure PKE scheme, and demonstrated that any collection of LTDFs with sufficient lossiness yields a collection of injective trapdoor functions that is secure under natural correlated products.

Mol and Yilek [27] extended the results of [31] and [32] and showed that only a non-negligible fraction of a single bit of lossiness is sufficient for building CCA-secure PKE schemes.

Hemenway and Ostrovsky [21] studied under which condition a homomorphic encryption implies CCA. They showed that a homomorphic encryption with cyclic plaintexts implies a family of LTDFs, and henceforth a CCA-secure encryption using the results of Peikert and Waters [31]. Our paper focuses on efficient construction of CCA-secure systems from families of LTDFs, compared with the construction of Peikert and Waters [31]. Our result is that we can do that with a special kind of LTDFs, namely, chameleon ABO-TDFs. A homomorphic encryption with cyclic plaintexts is not enough to construct a family of chameleon ABO-TDFs.

Recently, Kiltz et al. [22] introduced the notion of adaptive trapdoor functions (ATDFs) and a natural generalization they called tag-based adaptive trapdoor functions (TB-ATDFs). They showed that ATDFs and TB-ATDFs can be constructed directly using lossy+ABO-TDFs. They also showed that ATDFs and TB-ATDFs are strictly weaker than correlated-product trapdoor functions [32] and LTDFs [31]. They gave black-box constructions of CCA-secure PKE from both ATDFs and TB-ATDFs. The construction of CCA-secure PKE from TB-ATDFs is similar to the construction of Peikert and Waters [31]. But, compared with [31], one-time signature can be replaced by a MAC using the transform of

Boneh et al. [7]. They used the hardcore bit of the ATDF to construction one-bit PKE. But, if the given ATDF is a permutation or has linearly many simultaneous hardcore bits, they can use the ATDF as a key-encapsulation mechanism (KEM) for an CCA-secure symmetric encryption scheme to construction a CCA-secure PKE. In their construction of ATDF from lossy+ABO-TDF, the branch of ABO-TDF is the output of a target collision-resistant hash function, which takes the output of LTDF as input. The branch of ABO-TDF in their construction of TB-ATDF from lossy+ABO-TDF is a tag chosen randomly. As opposed to their construction, the first component of a branch (a, b) of chameleon ABO-TDF in our scheme a is chosen randomly, and the second component b is the output of a collision-resistant hash function, which takes the output of LTDF and $h(x) \oplus m$ as inputs, where m is the encrypted message. (Note that, our construction needs a fully collision resistant hash as opposed to target collision resistant as in [22].) This technique, which has already used in the RO model to construct CCA-secure PKE schemes [1], allows us to avoid using one-time signature, MAC, or other symmetric-key primitives.

Our works are also related to chameleon hash function [24]. Roughly speaking, chameleon hash functions are randomized collision-resistant hash functions with the additional property that given a trapdoor, one can efficiently generate collisions. Mohassel showed in [26] how to construct one-time signature from chameleon hash functions, which can be used in the construction of Peikert and Waters [31].

1.3 Organization

The rest of the paper is organized as follows. In Section 2, we review some standard notations and cryptographic definitions. We introduce the notion of chameleon ABO-TDFs and present a generic construction and a concrete construction in Section 3. In Section 4, we present black-box constructions of CCA-secure PKE based on an LTDF and a chameleon ABO-TDF. Finally, we state our conclusion in Section 5.

2 Preliminaries

If S is a set, then $|S|$ denotes its size and $s \stackrel{\$}{\leftarrow} S$ denotes the operation of picking an element s uniformly at random from S . Let $\mathbb{N}$ denote the natural numbers. If $\lambda \in \mathbb{N}$ then 1^λ denotes the string of λ ones. Let $z \leftarrow \mathsf{A}(x, y, \ldots)$ denote the operation of running an algorithm A with inputs $(x, y, \ldots)$ and output z. Let U_ℓ denote the uniform distribution on ℓ-bit binary strings. A function $f(\lambda)$ is *negligible* if for every $c > 0$ there exists an λ_c such that $f(\lambda) < 1/\lambda^c$ for all $\lambda > \lambda_c$.

2.1 Hashing

Formally, a function $H : X \rightarrow Y$ is a collision-resistant (CR) hash function, if for all probabilistic polynomial-time (PPT) algorithm $\mathcal{A}$, $\mathsf{Adv}_{\mathcal{A}}^{\mathsf{CR}}(\lambda)$ is negligible in λ, where

$$\mathsf{Adv}_{\mathcal{A}}^{\mathsf{CR}}(\lambda) = \mathsf{Pr}[x, x' \leftarrow \mathcal{A}(H) : x' \neq x \wedge H(x') = H(x)].$$

A family of function $\mathcal{H} = \{h_i : X \to Y\}$ is pairwise independent, if for every distinct $x, x' \in X$ and every $y, y' \in Y$, $\mathsf{Pr}_{h \xleftarrow{\$} \mathcal{H}}[h(x) = y \text{ and } h(x') = y'] = 1/|Y|^2$.

2.2 Extracting Randomness

The *min-entropy* $\mathbf{H}_\infty(X)$ of a random variable X is $-\log(\max_x \mathsf{Pr}(X = x))$. Dodis, Reyzin and Smith [16] defined *average min-entropy* of X given Y to be the logarithm of the average probability of the most likely value of X given Y: $\tilde{\mathbf{H}}_\infty(X|Y) = -\log\left(\mathrm{E}_{y \leftarrow Y}\left[2^{-\mathbf{H}_\infty(X|Y=y)}\right]\right)$. They proved that if Y has 2^ℓ possible values and Z is any random variable, then $\tilde{\mathbf{H}}_\infty(X|(Y,Z)) \geq \mathbf{H}_\infty(X|Z) - \ell$.

The *statistical distance* between two probability distributions X and Y is $\mathbf{SD}(X, Y) = \frac{1}{2}\sum_v |\mathsf{Pr}(X = v) - \mathsf{Pr}(Y = v)|$. Dodis, Reyzin and Smith [16] proved that if X, Y are random variables such that $X \in \{0,1\}^n$ and $\tilde{\mathbf{H}}_\infty(X|Y) \geq k$, and $\mathcal{H}$ is a family of pairwise independent hash functions from $\{0,1\}^n$ to $\{0,1\}^\ell$, then for $h \xleftarrow{\$} \mathcal{H}$, $\mathbf{SD}((Y, h, h(X)), (Y, h, U_\ell)) \leq \epsilon$ as long as $\ell \leq k - 2\log(1/\epsilon)$.

2.3 Public Key Encryption

A PKE scheme is a tuple of algorithms described as follows:

$\mathsf{Kg}(\lambda)$ takes as input a security parameter λ. It outputs a public/private key pair $(\mathsf{PK}, \mathsf{SK})$.

$\mathsf{Enc}(\mathsf{PK}, m)$ takes as input a public key PK and a message m. It outputs a ciphertext.

$\mathsf{Dec}(\mathsf{SK}, c)$ takes as input a private key SK and a ciphertext c. It outputs a plaintext message or the special symbol $\perp$ meaning that the ciphertext is invalid.

We insist that all public key encryption schemes satisfy the obvious correctness condition (that decryption "undoes" encryption).

The strongest and commonly accepted notion of security for a PKE scheme is that of indistinguishability against an adaptive chosen ciphertext attack (CCA). It is defined using the following game between an adversary $\mathcal{A}$ and a challenger.

Setup. The challenger runs $\mathsf{Kg}(\lambda)$ to obtain a public/private key pair $(\mathsf{PK}, \mathsf{SK})$. It gives the public key PK to the adversary.

Query phase 1. The adversary $\mathcal{A}$ adaptively issues decryption queries c. The challenger responds with $\mathsf{Dec}(\mathsf{SK}, c)$.

Challenge. The adversary $\mathcal{A}$ submits two (equal length) messages m_0, m_1. The challenger selects a random bit $\beta \in \{0,1\}$, sets $c^* = \mathsf{Enc}(\mathsf{PK}, m_\beta)$ and sends c^* to the adversary as its challenge ciphertext.

Query phase 2. The adversary continues to adaptively issue decryption queries c, as in Query phase 1, but with the natural constraint that the adversary does not request the decryption of c^*.

Guess. The adversary $\mathcal{A}$ outputs its guess $\beta' \in \{0,1\}$ for β and wins the game if $\beta = \beta'$.

We define $\mathcal{A}$'s advantage in attacking the public key encryption scheme PKE with the security parameter λ as $\mathsf{Adv}_{\mathcal{A}}^{\mathsf{PKE}}(\lambda) = |\Pr[\beta = \beta'] - \frac{1}{2}|$.

Definition 1. *A public key encryption scheme PKE is CCA secure, if for all polynomial-time adversary $\mathcal{A}$, the advantage $\mathsf{Adv}_{\mathcal{A}}^{\mathsf{PKE}}(\lambda)$ is negligible.*

The chosen-plaintext security CPA for a public key encryption scheme can be defined as the preceding game, except that adversary $\mathcal{A}$ is disallowed to issue any decryption query.

2.4 Lossy Trapdoor Functions

Informally, a collection of LTDFs is a collection of functions with two computationally indistinguishable branches: an injective branch with a trapdoor and a lossy branch losing information about its input.

Definition 2 (Lossy Trapdoor Functions). A collection of (n, k)-lossy trapdoor functions is a 3-tuple of (possibly probabilistic) polynomial-time algorithms $(\mathsf{G}, \mathsf{F}, \mathsf{F}^{-1})$ such that:

1. **Sampling an injective function:** $\mathsf{G}(1^\lambda, \mathsf{injective})$ outputs (s, td) where s is a function index and td is its trapdoor. The algorithm $\mathsf{F}(s, \cdot)$ computes a (deterministic) injective function $f_s(\cdot)$ over the domain $\{0,1\}^n$, and $\mathsf{F}^{-1}(s, td, \cdot)$ computes $f_s^{-1}(\cdot)$.
2. **Sampling a lossy function:** $\mathsf{G}(1^\lambda, \mathsf{lossy})$ outputs s where s is a function index. The algorithm $\mathsf{F}(s, \cdot)$ computes a (deterministic) function $f_s(\cdot)$ over the domain $\{0,1\}^n$ whose image has size at most 2^{n-k}.
3. **Hard to distinguish injective from lossy:** The ensembles $\{s : (s, td) \leftarrow \mathsf{G}(1^\lambda, \mathsf{injective})\}_{\lambda \in \mathbb{N}}$ and $\{s : s \leftarrow \mathsf{G}(1^\lambda, \mathsf{lossy})\}_{\lambda \in \mathbb{N}}$ are computationally indistinguishable.

3 Chameleon ABO-TDFs and Its Constructions

In this section, we first introduce our notion of chameleon ABO-TDFs. Then, based on a CPA-secure pubic key encryption scheme with some additional property, we propose a generic construction of chameleon ABO-TDFs. Finally, we instantiate the generic construction using the Damgård-Jurik encryption scheme [15].

3.1 Chameleon ABO-TDFs

The notion of ABO-TDFs, introduced by Peikert and Waters [31], is a richer abstraction of LTDFs. In an ABO collection, each function has several *branches*. Almost all the branches are injective trapdoor functions, except for *one* branch which is lossy. Freeman et al. [18] generalized the definition of ABO-TDFs by allowing possibly *many* lossy branches (other than *one*). Let $\mathbb{B} = \{B_\lambda\}_{\lambda\in\mathbb{N}}$ be a collection of sets whose elements represent the branches, and we recall the definition of ABO-TDFs [18].

Definition 3(All-But-One Trapdoor Functions). A collection of (n,k)-all-but-one trapdoor functions is a 3-tuple of (possibly probabilistic) polynomial-time algorithms $(\mathsf{G}_{abo}, \mathsf{F}_{abo}, \mathsf{F}_{abo}^{-1})$ such that:

1. **Sampling a function:** For any $\lambda \in \mathbb{N}$ and $b^* \in B_\lambda$, $\mathsf{G}_{abo}(1^\lambda, b^*)$ outputs $(s, td, \tilde{S})$ where s is a function index, td is its trapdoor and $\tilde{S}$ is a set of lossy branches with $b^* \in \tilde{S} \subset B_\lambda$.
2. **Evaluation of injective functions:** For any $b \in B_\lambda$, if $b \notin \tilde{S}$ where $(s, td, \tilde{S}) \leftarrow \mathsf{G}_{abo}(1^\lambda, b^*)$, then $\mathsf{F}_{abo}(s, b, \cdot)$ computes a (deterministic) injective function $f_{s,b}(\cdot)$ over the domain $\{0,1\}^n$, and $\mathsf{F}_{abo}^{-1}(s, td, b, \cdot)$ computes $f_{s,b}^{-1}(\cdot)$.
3. **Evaluation of lossy functions:** For any $b \in B_\lambda$, if $b \in \tilde{S}$ where $(s, td, \tilde{S}) \leftarrow \mathsf{G}_{abo}(1^\lambda, b^*)$, then $\mathsf{F}_{abo}(s, b, \cdot)$ computes a (deterministic) function $f_{s,b}(\cdot)$ over the domain $\{0,1\}^n$ whose image has size at most 2^{n-k}.
4. **Security:** The ensembles $\{s : (s, td, \tilde{S}) \leftarrow \mathsf{G}_{abo}(1^\lambda, b_1^*)\}_{\lambda\in\mathbb{N}, b_1^*\in B_\lambda}$ and $\{s : (s, td, \tilde{S}) \leftarrow \mathsf{G}_{abo}(1^\lambda, b_2^*)\}_{\lambda\in\mathbb{N}, b_2^*\in B_\lambda}$ are computationally indistinguishable.
5. **Hard to find one-more lossy branch:** Any probabilistic polynomial-time algorithm $\mathcal{A}$ that receives (s, b) as input, where $(s, td, \tilde{S}) \leftarrow \mathsf{G}_{abo}(1^\lambda, b^*)$ and $b \xleftarrow{\$} \tilde{S}$, has only a negligible probability of outputting an element $b' \in \tilde{S}\backslash\{b\}$.

We are now ready to introduce the notion of chameleon ABO-TDFs, which is a specific kind of ABO-TDFs with two variable (a, b) as a branch. The property we require is that given any a, it is easy to determine a unique lossy branch (a, b) with the help of a trapdoor, while (a, b_0) from a lossy branch family is computationally indistinguishable from (a, b_1) from an injective branch family with the trapdoor. We can construct a chameleon ABO-TDF from any ABO-TDF in the sense of [31] and a chameleon hash function [24] targeting to the branch set. Yet the properties of the chameleon hash are a bit overkill for what we need and we build the needed properties directly into the constructions for better efficiency.

Let $\mathbb{A} \times \mathbb{B} = \{A_\lambda \times B_\lambda\}_{\lambda\in\mathbb{N}}$ be a collection of sets whose elements represent the branches.

Definition 4 (Chameleon All-But-One Trapdoor Functions). A collection of (n,k)-chameleon all-but-one trapdoor functions is a 4-tuple of (possibly probabilistic) polynomial-time algorithms $(\mathsf{G}_{ch}, \mathsf{F}_{ch}, \mathsf{F}_{ch}^{-1}, \mathsf{CLB}_{ch})$ such that:

1. **Sampling a function:** For any $\lambda \in \mathbb{N}$, $\mathsf{G}_{ch}(1^\lambda)$ outputs $(s, td, \tilde{S})$ where s is a function index, td is its trapdoor and $\tilde{S} \subset A_\lambda \times B_\lambda$ is a set of lossy branches.

   Note that, a lossy branch is specified as a parameter to the function sampler of an ABO collection, but we have no such requirement.

2. **Evaluation of injective functions:** For any $(a, b) \in A_\lambda \times B_\lambda$, if $(a, b) \notin \tilde{S}$ where $(s, td, \tilde{S}) \leftarrow \mathsf{G}_{ch}(1^\lambda)$, then $\mathsf{F}_{ch}(s, a, b, \cdot)$ computes a (deterministic) injective function $g_{s,a,b}(\cdot)$ over the domain $\{0,1\}^n$, and $\mathsf{F}_{ch}^{-1}(s, td, a, b, \cdot)$ computes $g_{s,a,b}^{-1}(\cdot)$.
3. **Evaluation of lossy functions:** For any $(a, b) \in A_\lambda \times B_\lambda$, if $(a, b) \in \tilde{S}$ where $(s, td, \tilde{S}) \leftarrow \mathsf{G}_{ch}(1^\lambda)$, then $\mathsf{F}_{ch}(s, a, b, \cdot)$ computes a (deterministic) function $g_{s,a,b}(\cdot)$ over the domain $\{0,1\}^n$ whose image has size at most 2^{n-k}.
4. **Chameleon property:**
   (a) **Computing a lossy branch:** For any $a \in A_\lambda$, $\mathsf{CLB}_{ch}(s, td, a)$ computes a unique $b \in B_\lambda$ to result in a lossy branch (a, b). The uniqueness of b for a given a implies that any randomly chosen branch from $A_\lambda \times B_\lambda$ is injective with overwhelming probability.
   (b) **Hard to distinguish a lossy branch from an injective branch:** Any probabilistic polynomial-time algorithm $\mathcal{A}$ that receives s as input, where $(s, td, \tilde{S}) \leftarrow \mathsf{G}_{ch}(1^\lambda)$, has only a negligible probability of distinguishing a pair $(a, b_0) \in \tilde{S}$ from $(a, b_1) \notin \tilde{S}$, even a is chosen by $\mathcal{A}$. Formally, Let $\mathcal{A} = (\mathcal{A}_1, \mathcal{A}_2)$ be a CH-LI distinguisher and define its advantage as

$$\mathsf{Adv}_{\mathcal{A}}^{\mathsf{CH\text{-}LI}}(\lambda) = \left| \mathsf{Pr}\left[\beta = \beta' : \begin{array}{c} (s, td, \tilde{S}) \leftarrow \mathsf{G}_{ch}(1^\lambda); a \leftarrow \mathcal{A}_1(s); \\ b_0 = \mathsf{CLB}_{ch}(s, td, a); b_1 \xleftarrow{\$} B_\lambda; \\ \beta \xleftarrow{\$} \{0,1\}; \beta' \leftarrow \mathcal{A}_2(s, a, b_\beta) \end{array} \right] - \frac{1}{2} \right|.$$

   Given a collection of chameleon all-but-one trapdoor functions, it is hard to distinguish a lossy branch from an injective branch, if $\mathsf{Adv}_{\mathcal{A}}^{\mathsf{CH\text{-}LI}}(\cdot)$ is negligible for every PPT distinguisher $\mathcal{A}$. This property implies that given a, without the trapdoor, the component b of the lossy branch (a, b) is distributed uniformly.
5. **Hard to find one-more lossy branch:** Any probabilistic polynomial-time algorithm $\mathcal{A}$ that receives (s, a, b) as input, where $(s, td, \tilde{S}) \leftarrow \mathsf{G}_{ch}(1^\lambda)$ and $(a, b) \xleftarrow{\$} \tilde{S}$, has only a negligible probability of outputting a pair $(a', b') \in \tilde{S} \backslash \{(a, b)\}$. This property implies that the size of $\tilde{S}$ should not be too small.

In [31], Peikert and Waters also introduced a slightly relaxed definition of LTDFs, which they called *almost-always* LTDFs. Namely, there is only a negligible probability that $f_s(\cdot)$ is not injective or that $\mathsf{F}^{-1}(s, td, \cdot)$ incorrectly computes $f_s^{-1}(\cdot)$ for some input.

Similarly, we define *almost-always* chameleon ABO-TDFs. In a *almost-always* chameleon ABO-TDFs, *with overwhelming probability*, $\mathsf{F}_{ch}^{-1}(s, td, a, b, \cdot)$ inverts correctly on all values in the image of $g_{s,a,b}(\cdot)$ if $(a, b) \notin \tilde{S}$, and $\mathsf{CLB}_{ch}(s, td, a)$ outputs b such that $(a, b) \in \tilde{S}$.

3.2 Generic Construction

Let $(\mathsf{Kg}, \mathsf{Enc}, \mathsf{Dec})$ be a CPA-secure PKE scheme, which is additively homomorphic. For the PKE scheme, we also assume that

1. $\mathcal{M}$ is its message space and $\mathcal{R}$ is its randomness space. Both spaces are large enough and $|\mathcal{M}| > |\mathcal{R}|$.
2. $\mathcal{M}$ is a finite field. For constructing *almost-always* chameleon ABO-TDFs, we only require that, $\mathcal{M}$ is a commutative ring with multiplicative identity and, *with overwhelming probability*, each element in $\mathcal{M}$ has multiplicative inverse (See the concrete construction in Section 3.3.).
3. $\mathsf{Enc}(\mathsf{PK}, m) \odot \mathsf{Enc}(\mathsf{PK}, m') = \mathsf{Enc}(\mathsf{PK}, m + m')$, where $(\mathsf{PK}, \mathsf{SK}) \leftarrow \mathsf{Kg}(\lambda), m, m' \in \mathcal{M}$, and $\odot$ denotes coordinate-wise multiplication of ciphertexts.
4. For $a, m \in \mathcal{M}$, $(\mathsf{Enc}(\mathsf{PK}, m))^a = \mathsf{Enc}(\mathsf{PK}, am)$, where exponentiation of a ciphertext is also coordinate-wise.

Now, we define a 4-tuple algorithms $(\mathsf{G}_{ch}, \mathsf{F}_{ch}, \mathsf{F}_{ch}^{-1}, \mathsf{CLB}_{ch})$ as follows:

1. **Sampling a function:** G_{ch} takes as input 1^λ, where λ is a security parameter. It first generates a keypair for the public key encryption scheme: $(\mathsf{PK}, \mathsf{SK}) \leftarrow \mathsf{Kg}(\lambda)$. It then chooses $x_a, x_b, x_d \xleftarrow{\$} \mathcal{M}$ and computes
$$c_a = \mathsf{Enc}(\mathsf{PK}, x_a),\ c_b = \mathsf{Enc}(\mathsf{PK}, x_b),\ c_d = \mathsf{Enc}(\mathsf{PK}, x_d).$$
The function index is $s = (\mathsf{PK}, c_a, c_b, c_d)$, the trapdoor is $td = (\mathsf{SK}, x_a, x_b, x_d)$ and the set of lossy branches $\tilde{S}$ is all pairs $(a, b) \in \mathcal{M} \times \mathcal{M}$ such that $ax_a + bx_b + x_d = 0$.
2. **Evaluating a function:** F_{ch} takes as input (s, a, b, x), where $s = (\mathsf{PK}, c_a, c_b, c_d)$ is a function index and $x \in \mathcal{M}$. It computes $y = \left((c_a)^a \odot (c_b)^b \odot c_d\right)^x$, and outputs y.
3. **Inverting an injective function:** F_{ch}^{-1} takes as input (s, td, a, b, y), where $s = (\mathsf{PK}, c_a, c_b, c_d)$ is a function index, $td = (\mathsf{SK}, x_a, x_b, x_d)$ is the trapdoor and $(a, b) \notin \tilde{S}$. It computes $x = \mathsf{Dec}(\mathsf{SK}, y) \cdot (ax_a + bx_b + x_d)^{-1}$, and outputs x.
4. **Computing a lossy branch:** CLB_{ch} takes as input (s, td, a), where $s = (\mathsf{PK}, c_a, c_b, c_d)$ is a function index and $td = (\mathsf{SK}, x_a, x_b, x_d)$ is the trapdoor. It computes $b = (-ax_a - x_d) \cdot {x_b}^{-1}$, and outputs b.

Theorem 1. *The algorithms described above give a collection of* $(\log|\mathcal{M}|, \log|\mathcal{M}| - \log|\mathcal{R}|)$*-chameleon all-but-one trapdoor functions.*

Proof. We observe that, if (a, b) is not a lossy branch, namely $b \neq \mathsf{CLB}_{ch}(s, td, a) = (-ax_a - x_d) \cdot {x_b}^{-1}$, then $\mathsf{F}_{ch}(s, a, b, x)$ computes

$$\begin{aligned} y &= \left((c_a)^a \odot (c_b)^b \odot c_d\right)^x = \left((\mathsf{Enc}(\mathsf{PK}, x_a))^a \odot (\mathsf{Enc}(\mathsf{PK}, x_b))^b \odot \mathsf{Enc}(\mathsf{PK}, x_d)\right)^x \\ &= \mathsf{Enc}(\mathsf{PK}, x(ax_a + bx_b + x_d)), \end{aligned}$$

and $\mathsf{F}_{ch}^{-1}(s, td, a, b, y)$ computes

$$\begin{aligned} \mathsf{Dec}(\mathsf{SK}, y) \cdot (ax_a + bx_b + x_d)^{-1} &= \mathsf{Dec}(\mathsf{SK}, \mathsf{Enc}(\mathsf{PK}, x(ax_a + bx_b + x_d))) \\ &\quad \cdot (ax_a + bx_b + x_d)^{-1} \\ &= x(ax_a + bx_b + x_d) \cdot (ax_a + bx_b + x_d)^{-1} = x. \end{aligned}$$

So, we have shown invertibility for injective functions via the trapdoor information. Next, we show that if (a, b) is a lossy branch, namely $b = \mathsf{CLB}_{ch}(s, td, a) = (-ax_a - x_d) \cdot x_b^{-1}$, then F_{ch} evaluates a lossy function. In this case, F_{ch} computes

$$\left((c_a)^a \odot (c_b)^b \odot c_d\right)^x = \left((\mathsf{Enc}(\mathsf{PK}, x_a))^a \odot (\mathsf{Enc}(\mathsf{PK}, x_b))^b \odot \mathsf{Enc}(\mathsf{PK}, x_d)\right)^x$$
$$= \mathsf{Enc}(\mathsf{PK}, 0),$$

and most of the information on the input is lost. The function $\mathsf{F}_{ch}(s, a, b, \cdot)$ is defined over the domain $\mathcal{M}$, and if $(a, b) \in \tilde{S}$, F_{ch} is a lossy function and the image size is at most $|\mathcal{R}|$. Therefore the amount of lossiness is at least $\log|\mathcal{M}| - \log|\mathcal{R}|$.

Given the public key encryption scheme (Kg,Enc,Dec) is CPA secure, it is easy to see that any probabilistic polynomial-time algorithm $\mathcal{A}$ has only a negligible probability of distinguishing a pair $(a, b_0) \in \tilde{S}$ from $(a, b_1) \notin \tilde{S}$, even a is chosen by $\mathcal{A}$.

Finally, we show that any probabilistic polynomial-time algorithm $\mathcal{A}$ that receives (s, a, b) as input, where $(a, b) \in \tilde{S}$, has only a negligible probability of outputting a pair $(a', b') \in \tilde{S} \backslash \{(a, b)\}$. To see this, observe that the values x_a, x_b and x_d are initially hidden by the CPA secure public key encryption scheme. $\mathcal{A}$ could obtain the information that $ax_a + bx_b + x_d = 0$. However, there are exactly $|\mathcal{M}|^2$ pairs that satisfy this equation and each of them are equally likely. Thus, the adversary can output a pair (a', b') satisfying $(a', b') \neq (a, b)$ and $a'x_a + b'x_b + x_d = 0$ with negligible probability.

The formal proofs of the hardness of distinguishing a lossy branch from an injective branch of the chameleon ABO-TDFs, which can be reduced to the CPA security of the PKE scheme, and the hardness of finding one-more lossy branch, which can be reduced to the one-wayness of the PKE scheme, will be given in the full version of the paper.

3.3 A Concrete Construction

Based on the Damgård-Jurik encryption scheme [15], which is additively homomorphic, we present a concrete construction of *almost-always* chameleon ABO-TDFs by instantiating the generic construction. We begin with a brief description of the Damgård-Jurik encryption scheme [15], and then describe our construction.

Consider a modulus $N = PQ$, where P and Q are odd primes and $\gcd(N, \phi(N)) = 1$. Such an N is called *admissible* by Damgård and Jurik [15]. The following theorem was proved in [15]:

Theorem 2. *For any admissible N and a natural number $\ell < P, Q$, the map $\psi_\ell : \mathbb{Z}_{N^\ell} \times \mathbb{Z}_N^* \to \mathbb{Z}_{N^{\ell+1}}^*$ defined by $\psi_\ell(x, r) = (1+N)^x r^{N^\ell} \mod N^{\ell+1}$ is an isomorphism, where*

$$\psi_\ell(x_1 + x_2 \mod N^\ell, r_1 r_2 \mod N) = \psi_\ell(x_1, r_1) \cdot \psi_\ell(x_2, r_2) \mod N^{\ell+1}.$$

Moreover, it can be inverted in polynomial time given lcm$(P-1, Q-1)$.

The following describes the Damgård-Jurik encryption scheme [15].

$\mathsf{DJ.Kg}(\lambda)$ Given the security parameter λ, choose an admissible modulus $N = PQ$ and a natural number $\ell < P, Q$. The published public key is $\mathsf{PK} = (N, \ell)$, and the private key is $\mathsf{SK} = \mathrm{lcm}(P-1, Q-1)$.

$\mathsf{DJ.Enc}(\mathsf{PK}, m)$ Given PK and a message $m \in \mathbb{Z}_{N^\ell}$, choose $r \xleftarrow{\$} \mathbb{Z}_N^*$ and output $c = (1+N)^m r^{N^\ell} \mod N^{\ell+1}$.

$\mathsf{DJ.Dec}(\mathsf{SK}, c)$ Given $\mathsf{SK} = \mathrm{lcm}(P-1, Q-1)$ and a ciphertext c, apply the inversion algorithm provided by Theorem 2 to compute $(m, r) = \psi_\ell^{-1}(\mathsf{SK}, c)$, and output m.

Damgård and Jurik [15] also proved that based on decisional composite residuosity assumption, the encryption scheme described above is CPA secure.

Now, given a Damgård and Jurik encryption scheme with algorithms DJ.Kg, DJ.Enc and DJ.Dec, we define a 4-tuple algorithms $(\mathsf{G}_{ch}, \mathsf{F}_{ch}, \mathsf{F}_{ch}^{-1}, \mathsf{CLB}_{ch})$ as follows:

1. **Sampling a function:** G_{ch} takes as input 1^λ, where λ is a security parameter. It runs $(\mathsf{PK}, \mathsf{SK}) \leftarrow \mathsf{DJ.Kg}(\lambda)$, where $\mathsf{PK} = (N, \ell)$, and chooses $x_a, x_b, x_d \xleftarrow{\$} \mathbb{Z}_{N^\ell}$. Next, it computes
$$c_a = \mathsf{DJ.Enc}(\mathsf{PK}, x_a),\ c_b = \mathsf{DJ.Enc}(\mathsf{PK}, x_b),\ c_d = \mathsf{DJ.Enc}(\mathsf{PK}, x_d).$$
The function index is $s = (\mathsf{PK}, c_a, c_b, c_d)$, the trapdoor is $td = (\mathsf{SK}, x_a, x_b, x_d)$ and the set of lossy branches $\tilde{S}$ is all pairs $(a, b) \in \mathbb{Z}_{N^\ell} \times \mathbb{Z}_{N^\ell}$ such that $ax_a + bx_b + x_d = 0 \mod N^\ell$.
2. **Evaluating a function:** F_{ch} takes as input (s, a, b, x), where $s = (N, \ell, c_a, c_b, c_d)$, $a, b, x \in \mathbb{Z}_{N^\ell}$. It computes $y = \left((c_a)^a \cdot (c_b)^b \cdot c_d\right)^x \mod N^{\ell+1}$, and outputs y.
3. **Inverting an injective function:** F_{ch}^{-1} takes as input (s, td, a, b, y), where $s = (N, \ell, c_a, c_b, c_d)$, $td = (\mathsf{SK}, x_a, x_b, x_d)$ and $(a, b) \notin \tilde{S}$. It computes
$$x' = \mathsf{DJ.Dec}(\mathsf{SK}, y),$$
and outputs $x = x' \cdot (ax_a + bx_b + x_d)^{-1} \mod N^\ell$.
Note that, *with overwhelming probability*, $(ax_a + bx_b + x_d) \mod N^\ell$ has multiplicative inverse.
4. **Computing a lossy branch:** CLB_{ch} takes as input (s, td, a), where $s = (N, \ell, c_a, c_b, c_d)$, $td = (\mathsf{SK}, x_a, x_b, x_d)$ and $a \in \mathbb{Z}_{N^\ell}$. It computes
$$b = (-ax_a - x_d) \cdot {x_b}^{-1} \mod N^\ell,$$
and outputs b.

Theorem 3. *Under the composite residuosity assumption, the algorithms described above give a collection of* $((n-1)\ell, (n-1)\ell - n)$*-almost-always chameleon all-but-one trapdoor functions.*

Proof. The CPA security of the Damgård-Jurik encryption scheme and Theorem 1 guarantee that the algorithms described above give a collection of *almost-always* chameleon ABO-TDFs. Thus, it only remains to bound the amount of lossiness.

The function $\mathsf{F}_{ch}(s, a, b, \cdot)$ is defined over the domain $\{0,1\}^{(n-1)\ell}$, and if $(a, b) \in \tilde{S}$, F_{ch} is a lossy function and the image size is at most 2^n. Therefore the amount of lossiness is at least $(n-1)\ell - n$.

4 CCA-Secure PKE Scheme

Let $(\mathsf{G}, \mathsf{F}, \mathsf{F}^{-1})$ be a collection of (n, k_1)-lossy trapdoor functions, and let $(\mathsf{G}_{ch}, \mathsf{F}_{ch}, \mathsf{F}_{ch}^{-1}, \mathsf{CLB}_{ch})$ be a collection of (n, k_2)-chameleon all-but-one trapdoor functions having branches $\mathbb{A} \times \mathbb{B} = \{A_\lambda \times B_\lambda\}_{\lambda \in \mathbb{N}}$. Let $\mathcal{H}$ be a family of pairwise independent hash functions from $\{0,1\}^n$ to $\{0,1\}^\ell$.

We assume that the public key encryption scheme has message space $\{0,1\}^\ell$. We also require that $k_1 + k_2 - n \geq k$ for some $k = \omega(\log n)$, and $\ell \leq k - 2\log(1/\epsilon)$ for some negligible ϵ (in λ).

Our PKE scheme consists of the following algorithms:

- $\mathsf{Kg}(\lambda)$ Given the security parameter λ, generate an injective trapdoor function: $(s, td) \leftarrow \mathsf{G}(1^\lambda, \mathsf{injective})$. Then generate a chameleon all-but-one trapdoor function:$(s', td') \leftarrow \mathsf{G}_{ch}(1^\lambda)$. Finally, choose a collision-resistant hash function $H : \{0,1\}^* \rightarrow A_\lambda$ and $h \stackrel{\$}{\leftarrow} \mathcal{H}$. The published public key is $\mathsf{PK} = (s, s', H, h)$, and the private key is $\mathsf{SK} = (td, td')$.
- $\mathsf{Enc}(\mathsf{PK}, m)$ Given PK and a message $m \in \{0,1\}^\ell$, choose $x \stackrel{\$}{\leftarrow} \{0,1\}^n, r \stackrel{\$}{\leftarrow} B_\lambda$ and compute $c_0 = h(x) \oplus m$, $c_1 = \mathsf{F}(s, x)$, $c_2 = \mathsf{F}_{ch}(s', t, r, x)$, where $t = H(c_0, c_1)$. Finally, output the ciphertext $c = (c_0, c_1, c_2, r)$.
- $\mathsf{Dec}(\mathsf{SK}, c)$ Given $\mathsf{SK} = (td, td')$ and a ciphertext $c = (c_0, c_1, c_2, r)$, compute $x = \mathsf{F}^{-1}(s, td, c_1)$ and $t = H(c_0, c_1)$. Then check whether

  $$c_1 = \mathsf{F}(s, x) \text{ and } c_2 = \mathsf{F}_{ch}(s', t, r, x).$$

  If not, output $\perp$, else output $m = c_0 \oplus h(x)$.

It is clear that the above construction satisfies *correctness*. Our construction does not require strongly unforgeable one-time signature scheme. Compared with the scheme of Peikert and Waters [31], the ciphertext is compact without attached signature and decryption does not require performing signature verification. We now turn to security.

Theorem 4. *The algorithms (Kg, Enc, Dec) described above are a public key encryption scheme secure against adaptive chosen ciphertext attack.*

Proof. The proof is a sequence of games [35], $\text{Game}_0, \text{Game}_1, \ldots, \text{Game}_5$, where Game_0 is the original adaptive chosen ciphertext attack game. Then we show that for all $i = 0, \ldots, 4$, Game_i and Game_{i+1} are (computationally) indistinguishable. Finally, we make an unconditional argument that an adversary must have negligible advantage in $\mathbf{Game}_5$. It follows that the public key encryption scheme is CCA-secure.

Game$_1$. We modify the way that the challenger computes the challenge ciphertext $c^* = (c_0^*, c_1^*, c_2^*, r^*)$ as

$$c_0^* = h(x^*) \oplus m_\beta,\ c_1^* = \mathsf{F}(s, x^*),\ c_2^* = \mathsf{F}_{ch}(s', t^*, r^*, x^*),$$

where $x^* \stackrel{\$}{\leftarrow} \{0,1\}^n$, $t^* = H(c_0^*, c_1^*)$ and $r^* = \mathsf{CLB}_{ch}(s', td', t^*)$.

Game$_2$. We modify the decryption oracle so that it rejects all ciphertexts $c = (c_0, c_1, c_2, r)$, such that $r = r^*$ and $t = H(c_0, c_1) = t^*$.

Game$_3$. We modify the decryption oracle so that it applies the following *special rejection rule*: if the adversary submits a ciphertext $c = (c_0, c_1, c_2, r)$ for decryption, such that $r = \mathsf{CLB}_{ch}(s', td', t)$, where $t = H(c_0, c_1)$, then the decryption oracle immediately outputs reject and halts.

Game$_4$. This game is identical to **Game$_3$**, except for a small modification to the decryption oracle. When the adversary submits a ciphertext $c = (c_0, c_1, c_2, r)$ for decryption, the challenger computes $x = \mathsf{F}_{ch}^{-1}(s', td', c_2)$ and $t = H(c_0, c_1)$. Then it checks whether

$$c_1 = \mathsf{F}(s, x) \text{ and } c_2 = \mathsf{F}_{ch}(s', t, r, x).$$

If not, it outputs $\perp$, else outputs $m = c_0 \oplus h(x)$.

Game$_5$. In this game, we replace the injective function with a lossy one. Formally, in the **Setup** phase, we replace $(s, td) \leftarrow \mathsf{G}(1^\lambda, \mathsf{injective})$ with $s \leftarrow \mathsf{G}(1^\lambda, \mathsf{lossy})$.

Claim 1. *Game$_0$ and Game$_1$ are computationally indistinguishable, given the hardness of distinguishing a lossy branch from an injective branch of the chameleon all-but-one trapdoor functions collection.*

Proof. We prove this claim by describing a CH-LI distinguisher algorithm $\mathcal{A} = (\mathcal{A}_1, \mathcal{A}_2)$ that receives s' as input where $(s', td') \leftarrow \mathsf{G}_{ch}(1^\lambda)$. The distinguisher $\mathcal{A}$ interacts with the adversary as follows.

In the **Setup** phase, $\mathcal{A}$ runs $(s, td) \leftarrow \mathsf{G}(1^\lambda, \mathsf{injective})$, and chooses a collision-resistant hash function H and $h \stackrel{\$}{\leftarrow} \mathcal{H}$. The public key is $\mathsf{PK} = (s, s', H, h)$. We point out that $\mathcal{A}$ knows the injective trapdoor td, but does not know the trapdoor td' corresponding to s'.

When the adversary issues decryption queries, $\mathcal{A}$ responds using the injective trapdoor td. Note that, the only secret information that the decryption oracle needs to operate is td, which $\mathcal{A}$ knows.

When the adversary submits two (equal length) messages m_0, m_1, $\mathcal{A}$ flips a fair coin $\beta \in \{0,1\}$ and constructs the challenge ciphertext as follows:

1. It chooses $x^* \stackrel{\$}{\leftarrow} \{0,1\}^n$ and computes

$$c_0^* = h(x^*) \oplus m_\beta,\ c_1^* = \mathsf{F}(s, x^*),\ t^* = H(c_0^*, c_1^*).$$

2. Next, $\mathcal{A}_1$ submits t^* and sets the response as r^*. Then, it computes $c_2^* = \mathsf{F}_{ch}(s', t^*, r^*, x^*)$.
3. Finally, it outputs the ciphertext $c^* = (c_0^*, c_1^*, c_2^*, r^*)$.

When $r^* = \mathsf{CLB}_{ch}(s', td', t^*)$, $\mathcal{A}$ simulates Game$_1$ perfectly; otherwise, it simulates Game$_0$. Therefore, any difference in behavior between Game$_0$ and Game$_1$ immediately breaks the hardness of distinguishing a lossy branch from an injective branch of the chameleon all-but-one trapdoor functions collection.

Claim 2. *Game$_1$ and Game$_2$ are computationally indistinguishable, given the collision-resistant property of the hash function H.*

Proof. We observe that Game$_1$ and Game$_2$ behave equivalently unless an event E happens, which is that the adversary makes a legal (i.e., not equal to c^*) decryption query of the form $c = (c_0, c_1, c_2, r = r^*)$, where $t^* = H(c_0, c_1)$. We show that event E happens with negligible probability.

If event E happens, then because $c \neq c^*$ we must have $(c_0, c_1) \neq (c_0^*, c_1^*)$ and $H(c_0, c_1) = H(c_0^*, c_1^*) = t^*$. Therefore, we find a collision of the hash function H.

Because the hash function H is collision-resistant, we conclude that E happens with negligible probability.

Claim 3. *Game$_2$ and Game$_3$ are computationally indistinguishable, given the hardness of finding one-more lossy branch of the chameleon all-but-one trapdoor functions collection.*

Proof. We define the event F to be the event that the adversary makes a legal (i.e., not equal to c^*) decryption query of the form $c = (c_0, c_1, c_2, r)$, such that $r = \mathsf{CLB}_{ch}(s', td', t)$, where $t = H(c_0, c_1)$. It is clear that Game$_2$ and Game$_3$ proceed identically until event F occurs. We show that event F happens with negligible probability.

Note that, if $c \neq c^*$ and $(t, r) = (t^*, r^*)$, the decryption oracle rejects the ciphertext in both games. Therefore, if even F happens, then (t, r) is a new lossy branch of the chameleon all-but-one trapdoor function.

Because of the hardness of finding one-more lossy branch of the chameleon all-but-one trapdoor functions collection, we conclude that F happens with negligible probability.

Claim 4. *Game$_3$ and Game$_4$ are equivalent.*

Proof. The only difference between Game$_3$ and Game$_4$ is in the implementation of decryption oracle. We show that decryption oracle is equivalent in the two games.

In both games, when the adversary makes a legal (i.e., not equal to c^*) decryption query of the form $c = (c_0, c_1, c_2, r)$, where $t = H(c_0, c_1)$, the challenger checks that $c_1 = \mathsf{F}(s, x)$ and $c_2 = \mathsf{F}_{ch}(s', t, r, x)$ for some x that they compute (in different ways), and outputs $\perp$ if not.

Note that, if $r = \mathsf{CLB}_{ch}(s', td', t)$, the decryption oracle outputs rejects and halts in both games. Therefore, $\mathsf{F}(s, \cdot)$ and $\mathsf{F}_{ch}(s', t, r, \cdot)$ are both injective, and there is a unique x such that $(c_1, c_2) = (\mathsf{F}(s, x), \mathsf{F}_{ch}(s', t, r, x))$. Game$_3$ finds that x by computing $\mathsf{F}^{-1}(s, td, c_1)$, while Game$_4$ finds it by computing $\mathsf{F}_{ch}^{-1}(s', td', t, r, c_2)$.

Claim 5. *$Game_4$ and $Game_5$ are computationally indistinguishable, given the hardness of distinguishing injective functions from lossy functions of the lossy trapdoor functions collection.*

Proof. The only difference between Game_4 and Game_5 is in the **Setup** phase. In the **Setup** phase of Game_4, the challenger proceeds as in the original CCA game, outputting the public key $\mathsf{PK} = (s, s', H, h)$ where $(s, td) \leftarrow \mathsf{G}(1^\lambda, \mathsf{injective})$ and $(s', td') \leftarrow \mathsf{G}_{ch}(1^\lambda)$. In Game_5, **Setup** generates a lossy function instead, outputting $\mathsf{PK} = (s, s', H, h)$ where $s \leftarrow \mathsf{G}(1^\lambda, \mathsf{lossy})$ and $(s', td') \leftarrow \mathsf{G}_{ch}(1^\lambda)$.

We point out that the challenger knows the trapdoor td' of the chameleon all-but-one trapdoor function, but does not know the trapdoor td corresponding to s (if it even exists). The only secret information that the decryption oracle needs to operate is td', which the challenger knows.

It is straightforward to show that the adversary's views in the two games are indistinguishable, using the indistinguishability of injective and lossy functions.

Claim 6. *No (even unbounded) adversary has more than a negligible advantage in $Game_5$.*

We prove this claim by showing the fact that the value $h(x^*)$ is a nearly uniform and independent "one-time pad", and therefore the adversary has negligible advantage in guessing which message was encrypted.

Note that, in Game_5, both $\mathsf{F}(s, \cdot)$ and $\mathsf{F}_{ch}(s', t^*, r^*, \cdot)$ are lossy functions, and its image size is at most 2^{n-k_1} and 2^{n-k_2}, respectively. By the hypothesis that $k_1 + k_2 - n \geq k$, we have that the random variable $(c_1^*, c_2^*) = (\mathsf{F}(s, x^*), \mathsf{F}_{ch}(s', t^*, r^*, x^*))$ can take at most $2^{2n-k_1-k_2} \leq 2^{n-k}$ values.

Because x^* and h are independent, We also have $\tilde{\mathbf{H}}_\infty(x^* | (c_1^*, c_2^*, h)) \geq \mathbf{H}_\infty(x^* | h) - (n - k) = k$. Therefore, we have that $(c_1^*, c_2^*, h, h(x^*))$ *and*$(c_1^*, c_2^*, h, U_\ell)$ are within ϵ in statistical distance by our requirement that $\ell \leq k - 2\log(1/\epsilon)$, and we are done.

5 Conclusions

We introduced a new primitive called chameleon ABO-TDFs, which is a special kind of ABO-LTDFs. Given a CPA-secure public key encryption scheme with some additional property (mostly additively homomorphism), we also gave a generic and concrete construction of chameleon ABO-TDFs. Based on an LTDF and a chameleon ABO-TDF, we proposed a black-box construction of CCA-secure PKE which is more efficient than that of Peikert and Waters [31]. A future direction is to find other constructions and applications of chameleon ABO-TDFs.

Acknowledgement

We are grateful to the anonymous reviewers for their helpful comments. Our special thanks go to Adam O'Neill for helpful discussions and comments that improved the presentation of our paper. This work is partially funded by National Natural Science Foundation of China (No. 60873229) and Shanghai Rising-star Program (No. 09QA1403000), and also supported in part by A*Star SERC Grant No. 102 101 0027 in Singapore.

References

1. Abe, M., Kiltz, E., Okamoto, T.: Chosen Ciphertext Security with Optimal Ciphertext Overhead. In: Pieprzyk, J. (ed.) ASIACRYPT 2008. LNCS, vol. 5350, pp. 355–371. Springer, Heidelberg (2008)
2. Bellare, M., Brakerski, Z., Naor, M., Ristenpart, T., Segev, G., Shacham, H., Yilek, S.: Hedged Public-Key Encryption: How to Protect Against Bad Randomness. In: Matsui, M. (ed.) ASIACRYPT 2009. LNCS, vol. 5912, pp. 232–249. Springer, Heidelberg (2009)
3. Bellare, M., Fischlin, M., O'Neill, A., Ristenpart, T.: Deterministic encryption: Definitional equivalences and constructions without random oracles. In: Wagner, D. (ed.) CRYPTO 2008. LNCS, vol. 5157, pp. 360–378. Springer, Heidelberg (2008)
4. Bellare, M., Hofheinz, D., Yilek, S.: Possibility and impossibility results for encryption and commitment secure under selective opening. In: Joux, A. (ed.) EUROCRYPT 2009. LNCS, vol. 5479, pp. 1–35. Springer, Heidelberg (2009)
5. Bellare, M., Rogaway, P.: Random oracles are practical: a paradigm for designing efficient protocols. In: Proc. of ACM CCS 1993, pp. 62–73. ACM Press, New York (1993)
6. Boldyreva, A., Fehr, S., O'Neill, A.: On notions of security for deterministic encryption, and efficient constructions without random oracles. In: Wagner, D. (ed.) CRYPTO 2008. LNCS, vol. 5157, pp. 335–359. Springer, Heidelberg (2008)
7. Boneh, D., Katz, J.: Improved efficiency for CCA-secure cryptosystems built using identity-based encryption. In: Menezes, A. (ed.) CT-RSA 2005. LNCS, vol. 3376, pp. 87–103. Springer, Heidelberg (2005)
8. Boyen, X., Mei, Q., Waters, B.: Direct chosen ciphertext security from identity-based techniques. In: Proc. of ACM CCS 2005, pp. 320–329. ACM Press, New York (2005)
9. Boyen, X., Waters, B.: Shrinking the Keys of Discrete-Log-Type Lossy Trapdoor Functions. In: Zhou, J., Yung, M. (eds.) ACNS 2010. LNCS, vol. 6123, pp. 35–52. Springer, Heidelberg (2010)
10. Canetti, R., Halevi, S., Katz, J.: Chosen-ciphertext security from identity-based encryption. In: Cachin, C., Camenisch, J.L. (eds.) EUROCRYPT 2004. LNCS, vol. 3027, pp. 207–222. Springer, Heidelberg (2004)
11. Cachin, C., Micali, S., Stadler, M.: Computationally private information retrieval with polylogarithmic communication. In: Stern, J. (ed.) EUROCRYPT 1999. LNCS, vol. 1592, pp. 402–414. Springer, Heidelberg (1999)
12. Cramer, R., Shoup, V.: A practical public key cryptosystem provably secure against adaptive chosen ciphertext attack. In: Krawczyk, H. (ed.) CRYPTO 1998. LNCS, vol. 1462, pp. 13–25. Springer, Heidelberg (1998)
13. Cramer, R., Shoup, V.: Universal hash proofs and a paradigm for adaptive chosen ciphertext secure public-key encryption. In: Knudsen, L.R. (ed.) EUROCRYPT 2002. LNCS, vol. 2332, pp. 45–64. Springer, Heidelberg (2002)
14. Dolev, D., Dwork, C., Naor, M.: Nonmalleable cryptography. SIAM J. Comput. 30(2), 391–437 (2000); Preliminary version in STOC 1991
15. Damgård, I., Jurik, M.: A generalisation, a simplification and some applications of Paillier's probabilistic public-key system. In: Kim, K.-c. (ed.) PKC 2001. LNCS, vol. 1992, pp. 119–136. Springer, Heidelberg (2001) (Full version with additional co-author J. B. Nielsen)
16. Dodis, Y., Reyzin, L., Smith, A.: Fuzzy extractors: How to generate strong keys from biometrics and other noisy data. In: Cachin, C., Camenisch, J.L. (eds.) EUROCRYPT 2004. LNCS, vol. 3027, pp. 523–540. Springer, Heidelberg (2004)

17. Elkind, E., Sahai, A.: A unified methodology for constructing public-key encryption schemes secure against adaptive chosen-ciphertext attack. Cryptology ePrint Archive, Report 2002/042 (2002), `http://eprint.iacr.org/`
18. Freeman, D.M., Goldreich, O., Kiltz, E., Rosen, A., Segev, G.: More constructions of lossy and correlation-secure trapdoor functions. In: Nguyen, P.Q., Pointcheval, D. (eds.) PKC 2010. LNCS, vol. 6056, pp. 279–295. Springer, Heidelberg (2010)
19. Hofheinz, D., Kiltz, E.: Secure hybrid encryption from weakened key encapsulation. In: Menezes, A. (ed.) CRYPTO 2007. LNCS, vol. 4622, pp. 553–571. Springer, Heidelberg (2007)
20. Hemenway, B., Ostrovsky, R.: Lossy trapdoor functions from smooth homomorphic hash proof systems. In: ECCC, vol. 16(127) (2009)
21. Hemenway, B., Ostrovsky, R.: Homomorphic Encryption Over Cyclic Groups Implies Chosen-Ciphertext Security. Cryptology ePrint Archive, Report 2010/099 (2010)
22. Kiltz, E., Mohassel, P., O'Neill, A.: Adaptive trapdoor functions and chosen-ciphertext security. In: Gilbert, H. (ed.) EUROCRYPT 2010. LNCS, vol. 6110, pp. 673–692. Springer, Heidelberg (2010)
23. Kiltz, E., O'Neill, A., Smith, A.: Lossiness of RSA and the chosen-plaintext security of OAEP without random oracles (2009) (manuscript)
24. Krawczyk, H., Rabin, T.: Chameleon signatures. In: NDSS 2000. The Internet Society, San Diego (2000)
25. Lai, J., Deng, R.H., Liu, S., Kou, W.: Efficient CCA-Secure PKE from Identity-Based Techniques. In: Pieprzyk, J. (ed.) CT-RSA 2010. LNCS, vol. 5985, pp. 132–147. Springer, Heidelberg (2010)
26. Mohassel, P.: One-time Signatures and Chameleon Hash Functions. To appear in Proc. of SAC 2010. Springer, Heidelberg (2010)
27. Mol, P., Yilek, S.: Chosen-ciphertext security from slightly lossy trapdoor functions. Cryptology ePrint Archive, Report 2009/524 (2009)
28. Naor, M., Yung, M.: Public-key cryptosystems provably secure against chosen ciphertext attacks. In: STOC 1990, pp. 427–437. ACM, New York (1990)
29. Paillier, P.: Public-key cryptosystems based on composite degree residuosity classes. In: Stern, J. (ed.) EUROCRYPT 1999. LNCS, vol. 1592, pp. 223–238. Springer, Heidelberg (1999)
30. Peikert, C., Vaikuntanathan, V., Waters, B.: A framework for efficient and composable oblivious transfer. In: Wagner, D. (ed.) CRYPTO 2008. LNCS, vol. 5157, pp. 554–571. Springer, Heidelberg (2008)
31. Peikert, C., Waters, B.: Lossy Trapdoor Functions and Their Applications. In: STOC 2008, pp. 187–196. ACM, New York (2008)
32. Rosen, A., Segev, G.: Chosen-ciphertext security via correlated products. In: Reingold, O. (ed.) TCC 2009. LNCS, vol. 5444, pp. 419–436. Springer, Heidelberg (2009)
33. Rackoff, C., Simon, D.R.: Non-interactive zero-knowledge proof of knowledge and chosen ciphertext attack. In: Feigenbaum, J. (ed.) CRYPTO 1991. LNCS, vol. 576, pp. 433–444. Springer, Heidelberg (1992)
34. Shacham, H.: A Cramer-Shoup encryption scheme from the Linear assumption and from progressively weaker Linear variants. Cryptology ePrint Archive, Report 2007/074 (2007)
35. Shoup, V.: Sequences of Games: A Tool for Taming Complexity in Security Proofs. Cryptology ePrint Archive: Report 2004/332

Parallel Decryption Queries in Bounded Chosen Ciphertext Attacks

Takahiro Matsuda* and Kanta Matsuura

The University of Tokyo, Japan
{tmatsuda,kanta}@iis.u-tokyo.ac.jp

Abstract. Whether it is possible to construct a chosen ciphertext secure (CCA secure) public key encryption (PKE) scheme only from a chosen plaintext secure (CPA secure) one is a fundamental open problem, and the best known positive results regarding this problem are the constructions of so-called *bounded CCA* secure schemes. Since we can achieve the best possible security in the bounded CCA security notions, in order to further tackle the problem, we would need other new security notions that capture intermediate security notions that lie between CPA and CCA security. Motivated by this situation, we focus on "parallel" decryption queries (originally introduced by Bellare and Sahai) for the extension of bounded CCA security, and introduce a new security notion which we call *mixed CCA* security. It captures security against adversaries that make single and parallel decryption queries in a predetermined order, where each parallel query can contain *unboundedly* many ciphertexts. Moreover, how the decryption oracle is available before and after the challenge is also taken into account in this new security definition, which enables us to capture existing major security notions that lie between CPA and CCA security in a unified security notion. We investigate the relations among mixed CCA security notions, and show a necessary and sufficient condition of implications/separations between any two notions in mixed CCA security. We also show two black-box constructions of PKE schemes with improved security only using CPA secure schemes as building blocks.

Keywords: public key encryption, bounded CCA security, parallel decryption query, relations among security notions, black-box construction.

1 Introduction

Background. Studies on constructing and understanding public key encryption (PKE) schemes that satisfy security against chosen ciphertext attacks (CCA) [22,27], which is nowadays considered as a standard security notion needed in most practical applications/situations where PKE schemes are used, are important research topics in the area of cryptography. We can roughly categorize the

* Takahiro Matsuda is supported by JSPS Research Fellowships for Young Scientists.

D. Catalano et al. (Eds.): PKC 2011, LNCS 6571, pp. 246–264, 2011.

approaches for constructing CCA secure PKE schemes into two types: Constructions from specific number-theoretic assumptions and constructions from general assumptions. (In the following, we write IND-CCA1 to denote non-adaptive CCA security [22] and IND-CCA2 to denote adaptive CCA security [27]).

The approaches of the first type have been successful so far from both theoretical and practical points of view. After the first novel practical scheme based on the decisional Diffie-Hellman (DDH) assumption by Cramer and Shoup [10], many practical IND-CCA2 secure PKE schemes that pursue smaller ciphertext size and/or base security on weaker assumptions have been constructed so far, e.g. [20,6,17,7,14,18,15,30].

The approaches of the second type have also been successful, mainly from a theoretical point of view. Especially, it is known that if there exists an (enhanced) trapdoor permutation, which is one of the most fundamental primitives, then we can construct an IND-CCA2 secure PKE schemes generically [22,4,11]. There are also several elegant generic constructions of IND-CCA2 secure PKE schemes from primitives with some "stronger" functionality and/or security, such as constructions from identity-based encryption [5], and from special types of injective trapdoor functions and trapdoor relations [25,29,19,30].

However, one of the most fundamental problems still remains open: *Is it possible to generically construct a CCA (*IND-CCA1 or IND-CCA2*) secure PKE scheme from any semantically secure [13] (i.e.* IND-CPA *secure) one?*

So far, there are several negative and positive results related to this problem. Gertner et al. [12] showed that constructing an IND-CCA1 secure PKE scheme only from IND-CPA secure PKE schemes in a black-box manner is impossible, if the construction satisfies the property called *shielding*, where a PKE-to-PKE construction is said to be shielding if the decryption algorithm of the construction does not call the encryption algorithm of the building block PKE scheme.

Pass et al. [23] showed how to construct a PKE scheme that is non-malleable against chosen plaintext attacks (NM-CPA) from any IND-CPA secure PKE scheme. Their construction uses a certain class of NIZK proofs and was non-black-box.

Cramer et al. [9] introduced the notion of *bounded CCA* security which is defined in exactly the same way as ordinary IND-CCA2 security, except that the number of decryption oracle queries that an adversary can ask is bounded by some predetermined value (say, q) that is known a priori (we denote this notion by q-CCA2). Then they showed that for any polynomial q it is possible to construct an IND-q-CCA2 secure PKE scheme from any IND-CPA secure one in a black-box and shielding manner. They furthermore showed that for any polynomial q it is possible to construct a PKE scheme that satisfies non-malleability against q-bounded CCA (NM-q-CCA2) in a non-black-box manner.

Recently, Choi et al. [8] showed the constructions of PKE schemes from any IND-CPA secure scheme both in a black-box and shielding manner. Their first construction achieves NM-CPA security, and their second construction, which is essentially the same as the first construction but needs larger parameters, can achieve NM-q-CCA2 security.

These previous results show that we can achieve the best possible security notion (NM-q-CCA2) in the bounded CCA framework. This suggests that in order to proceed from the current situation, we would need new security notions which are intermediate between CPA and CCA security in a different sense from bounded CCA security. The motivation of this paper is to introduce and study such intermediate security notions as an extension of the bounded CCA security as a foundation for tackling the above fundamental problem.

Extending Bounded CCA Security with Parallel Decryption Queries. For the purpose mentioned above, we focus on and use the concept of the *parallel chosen ciphertext attacks* which is originally introduced by Bellare and Sahai [3] in the context of non-malleability [11] for PKE schemes, and consider *parallel* queries in the bounded CCA security framework. More specifically, as an extension of bounded CCA security, we introduce a new security notion, which we call *mixed CCA security*, that captures security against adversaries that make single (i.e. ordinary) decryption queries and parallel decryption queries in a predetermined order, where each parallel query can contain *unboundedly* many ciphertexts. (The name "mixed" is because we consider a mix of single and parallel queries). Moreover, the difference among decryption queries that are only allowed to make before/after the challenge and those that are allowed to make both before and after the challenge (an adversary can decide "flexibly" how to issue queries as long as it follows the predetermined order of queries and types) is also taken into account in our definition, which enables us to capture existing major security notions that lie between CPA and CCA security, including slightly complex notions such as non-malleability against bounded CCA (NM-q-CCA2) that considers "stage-specific" decryption queries, in a unified security notion. As a natural and interesting special class of mixed CCA security, we also introduce the notion of *bounded parallel CCA* security. For more details, see Section 3. We believe that the mixed CCA security provides a theoretical foundation for discussion of the problem of whether constructing (unbounded) CCA secure PKE schemes from any CPA secure PKE schemes is possible or not, and for intermediate results towards the problem.

1.1 Our Contributions

Relations among Mixed CCA Security Notions. We investigate the relations among mixed CCA security notions for PKE schemes and for key encapsulation mechanisms (KEMs) in Section 4. As one of the main results, we show necessary and sufficient conditions for implications/separations between any two notions in mixed CCA security. Interestingly and perhaps somewhat surprisingly, *the relations for PKE schemes differ depending on its plaintext space size.* More specifically, the relations among security notions for PKE schemes with super-polynomially large plaintext space size and those with polynomially bounded plaintext space size are different. Therefore, this difference suggests that when we consider the relations among security notions for PKE schemes, we have to be also careful about the plaintext space size, though seemingly unrelated.

The relations for KEMs are the same as those of PKE schemes with polynomially bounded plaintext space size.

Black-Box Feasibility Results from CPA-Security. Using the notion of mixed CCA security, in Section 5, we show two new black-box constructions of PKE schemes (which can encrypt plaintexts of polynomial length, and thus have exponentially large plaintext space size) from an IND-CPA secure PKE scheme. The first one is constructed based on the construction by Choi et al. [8] which is NM-q-CCA2 secure, and achieves slightly but strictly stronger security notion than NM-q-CCA2. Our approach for the first construction is to use the Choi et al. scheme as a KEM and combine it with an IND-CCA2 secure data encapsulation mechanism (DEM), and thus is a very simple extension. In order for this simple approach to work, we show some implication result for mixed CCA security of KEMs (and PKE schemes with polynomially bounded plaintext space size). The second one is constructed based on the above result and the construction of PKE scheme by Cramer et al. [9], and achieves yet another security notion which cannot be directly compared with the security notion achieved by our other constructions and with NM-q-CCA2 security.

As will be explained later, one of the important and interesting observations that our results suggests, combined with previously known results, is that *the difficulty of constructing an* IND-CCA1 *secure PKE scheme only from* IND-CPA *secure one lies not in whether the number of decryption results that an adversary can see is bounded or not, but in whether the number of the adversary's "adaptive" decryption queries is bounded.* To the best of our knowledge, this observation has not been explicitly stated before.

2 Preliminaries

In this section, we review the basic notations and definitions of primitives used in this paper. Due to space limitation, the definitions for key encapsulation mechanism (KEM) and data encapsulation mechanism (DEM) are omitted and are provided in the full version. (They can also be found in [10,16], for example).

Basic Notations. If q is a natural number, then $[q] = \{1, \dots, q\}$. "$x \leftarrow y$" denotes that x is chosen uniformly at random from y if y is a finite set, x is output from y if y is a function or an algorithm, or y is assigned to x otherwise. "$|x|$" denotes the size of the set if x is a finite set or bit length of x if x is an element of some set. "PPTA" denotes a *probabilistic polynomial time algorithm.* $\mathcal{A}^{\mathcal{O}}$ denotes an algorithm $\mathcal{A}$ with oracle access to $\mathcal{O}$. Unless otherwise stated, k denotes the security parameter. A function $f : \mathbb{N} \rightarrow [0, 1]$ is said to be *negligible* if for any positive polynomial $p(k)$ and for all sufficiently large k, we have $f(k) < \frac{1}{p(k)}$.

Public Key Encryption. A PKE scheme Π consists of the following three PPTAs $(\mathsf{PKG}, \mathsf{PEnc}, \mathsf{PDec})$: A key generation algorithm PKG takes 1^k (security parameter k) as input, and outputs a public/private key pair (pk, sk); An encryption algorithm PEnc takes pk and a plaintext $m \in \mathcal{M}_{\Pi}$ (the plaintext space of Π) as input, and outputs a ciphertext c; A deterministic decryption algorithm PDec takes sk and c as input, and outputs a plaintext m (or a symbol $\perp$ meaning "decryption error"). As a correctness requirement, we require $\mathsf{PDec}(sk, \mathsf{PEnc}(pk, m)) = m$ for all (pk, sk) output from PKG and all $m \in \mathcal{M}_\Pi$.

Conventional Security Notions. The security notions for PKE schemes are expressed by a combination of a `GOAL` and an attack type (`ATK`) of an adversary. For conventional security notions for PKE schemes, we consider *indistinguishability* (`IND`) and *non-malleability* (`NM`) for security goals and *chosen plaintext attacks* (`CPA`), *non-adaptive chosen ciphertext attacks* (`CCA1`), *adaptive chosen ciphertext attacks* (`CCA2`), and *q-bounded chosen ciphertext attacks* (q-`CCA2`) [9] for attack types of an adversary. Non-malleability for PKE schemes we treat in this paper is the so-called *parallel chosen-ciphertext attack* based definition [3], which is equivalent to the indistinguishability based definition used in [23,24][1]. Since these conventional `GOAL-ATK` security notions can be expressed as special cases of *mixed CCA* security defined in Section 3, here we omit the definitions.

Implications and Separations of Security Notions. We will show several implications and separations of security notions, and thus we recall here. Though we write only the definition for PKE schemes, the same is defined for KEMs.

Definition 1. *Let* `X` *and* `Y` *be security notions for PKE schemes. We say that* `X` *security implies* `Y` *security if any* `X` *secure PKE scheme is also* `Y` *secure. We say that* `X` *security does not imply* `Y` *security if, under the assumption that* `X` *secure PKE schemes exist, there exists a PKE scheme which is* `X` *secure but is not* `Y` *secure. We say that* `X` *security and* `Y` *security are equivalent if we have implications for both directions (i.e. from* `X` *to* `Y` *and from* `Y` *to* `X`*).*

Shielding Black-Box Constructions. We briefly recall the definition of a shielding black-box construction of an `X` secure PKE scheme from a `Y` secure scheme. The notion of black-box constructions we mention in this paper is classified as *fully-black-box* ones [28], but specified for PKE-to-PKE constructions. (for details, see [28]). The notion of the *shielding* constructions is from [12].

Definition 2. *Let* `X` *and* `Y` *be security notions for PKE schemes. We say that there exists a shielding black-box construction of an* `X` *secure PKE scheme from a* `Y` *secure one, if there exist oracle PPTAs* $\Pi = (\mathsf{PKG}, \mathsf{PEnc}, \mathsf{PDec})$ *and* $\mathcal{B}$ *with the following properties. For all algorithms* $\pi = (\mathsf{G}, \mathsf{E}, \mathsf{D})$ *and* $\mathcal{A}$ *(each algorithm can be of arbitrary complexity), the following two conditions are satisfied:*

(Correctness:) *If* $\pi = (\mathsf{G}, \mathsf{E}, \mathsf{D})$ *satisfies correctness as a PKE scheme, so does* $\Pi^{\mathsf{G,E,D}} = (\mathsf{PKG}^{\mathsf{G,E,D}}, \mathsf{PEnc}^{\mathsf{G,E,D}}, \mathsf{PDec}^{\mathsf{G,D}})$.
(Security:) *If* $\mathcal{A}$ *breaks* `X` *security of* $\Pi^{\mathsf{G,E,D}} = (\mathsf{PKG}^{\mathsf{G,E,D}}, \mathsf{PEnc}^{\mathsf{G,E,D}}, \mathsf{PDec}^{\mathsf{G,D}})$ *then* $\mathcal{B}^{\mathcal{A},\mathsf{G,E,D}}$ *breaks* `Y` *security of* π.

(Note that PDec *does not have access to* E*).*

[1] Pass et al. [24] prove that *many-message* (indistinguishability-based) non-malleability, which considers multiple challenge messages, and single-message non-malleability, considered in this paper, are equivalent.

3 Extending Bounded CCA: Mixed CCA Security

In order to deal with and discuss existing security notions for PKE schemes and KEMs that lie between IND-CPA and IND-CCA2 security in a unified way, in this section we introduce an extension of conventional bounded CCA security [9], which we call security against *mixed chosen ciphertext attacks* (mixed CCA security), where the decryption oracle in the security experiment accepts both single decryption queries and *parallel* decryption queries in a predetermined order, and "how" the decryption oracle is available before/after the challenge is also taken into account.

Preliminary Definitions. We first formally define the notion of a parallel query to an oracle.

Definition 3. *Let $\mathcal{O} : \{0,1\}^* \to \{0,1\}^*$ be an oracle. A parallel query to $\mathcal{O}$ is a vector $\overrightarrow{x} = (x_1, x_2, \dots)$ of inputs for $\mathcal{O}$, where the size of the vector $\overrightarrow{x}$ is not predetermined, and a response to the parallel query $\overrightarrow{x}$ is a vector of the output values $\overrightarrow{y} = (y_1, y_2, \dots)$ where $y_i = \mathcal{O}(x_i)$ for every $1 \le i \le |\overrightarrow{x}|$.*

We stress that the number of inputs in each parallel query $\overrightarrow{x}$ is unbounded and can be dependent only on an algorithm that uses the oracle.

To define mixed CCA security, we need to introduce several notations. The symbols "s" and "p" denote one *single query* and one *parallel query*, respectively. Let $q \ge 0$ be an integer. "s^q" and "p^q" denote q single queries and q parallel queries, respectively. We define $\mathsf{s}^0 = \mathsf{p}^0 = \emptyset$.

If we write "$(\mathsf{s}^{q_1}\mathsf{p}^{q_2}\dots)$" with some integers $q_1, q_2, \dots \ge 0$, then it denotes a *query sequence*. This query sequence will define how the decryption oracle in the mixed CCA experiment accepts the queries. For example, $(\mathsf{s}^2\mathsf{p}^3)$ denotes two single decryption queries followed by three parallel decryption queries. We denote by "unbound" a special sequence that indicates "unboundedly" many single queries, i.e. $\mathsf{unbound} = \mathsf{s}^\infty$.

"$\mathcal{QS}$" denotes a set consisting of all possible query sequences with the restriction that the total number of queries in each sequence is bounded to be polynomial (in the security parameter). We furthermore define $\mathcal{QS}^* = \mathcal{QS} \cup \{\mathsf{unbound}\}$. We refer to queries following the query sequence $\mathsf{seq} \in \mathcal{QS}^*$ as "seq-queries".

If $\mathsf{seq} \in \mathcal{QS}$, then we denote by "$|\mathsf{seq}|$" the length of the query sequence. For example, if $\mathsf{seq} = (\mathsf{s}^2\mathsf{p})$ then $|\mathsf{seq}| = 3$. We define $|\mathsf{unbound}| = \infty$.

We define a concatenation operation "$||$" for query sequences naturally. For example, if $\mathsf{seq}_1 = (\mathsf{s}^2\mathsf{p})$ and $\mathsf{seq}_2 = (\mathsf{p}^2\mathsf{s}^3)$, then $(\mathsf{seq}_1||\mathsf{seq}_2) = (\mathsf{s}^2\mathsf{p}\mathsf{p}^2\mathsf{s}^3) = (\mathsf{s}^2\mathsf{p}^3\mathsf{s}^3)$. For any $\mathsf{seq} \in \mathcal{QS}^*$, we define $(\mathsf{seq}||\emptyset) = (\emptyset||\mathsf{seq}) = \mathsf{seq}$ and $(\mathsf{seq}||\mathsf{unbound}) = (\mathsf{unbound}||\mathsf{seq}) = \mathsf{unbound}$.

3.1 Definition of Mixed CCA Security

Now we define mixed CCA security for a PKE scheme $\Pi = (\mathsf{PKG}, \mathsf{PEnc}, \mathsf{PDec})$ as IND-ATK-like security parameterized by three query sequences $\mathsf{B}, \mathsf{F}, \mathsf{A} \in \mathcal{QS}^*$, denoted by $\langle \mathsf{B} : \mathsf{F} : \mathsf{A} \rangle$-mCCA security, via the $\langle \mathsf{B} : \mathsf{F} : \mathsf{A} \rangle$-mCCA experiment $\mathsf{Expt}_{\Pi,\mathcal{A}}^{\langle \mathsf{B}:\mathsf{F}:\mathsf{A} \rangle\text{-mCCA}}(k)$ that an adversary $\mathcal{A} = (\mathcal{A}_1, \mathcal{A}_2)$ runs in:

$$\mathsf{Expt}_{\Pi,\mathcal{A}}^{\langle \mathsf{B}:\mathsf{F}:\mathsf{A}\rangle\text{-mCCA}}(k) : [(pk, sk) \leftarrow \mathsf{PKG}(1^k); (m_0, m_1, \mathsf{st}) \leftarrow \mathcal{A}_1^{\mathcal{O}}(pk); b \leftarrow \{0,1\};$$
$$c^* \leftarrow \mathsf{PEnc}(pk, m_b); b' \leftarrow \mathcal{A}_2^{\mathcal{O}}(c^*, \mathsf{st}) : \texttt{If } b' = b \texttt{ then return } 1 \texttt{ else return } 0]$$

where $\mathcal{O}(\cdot) = \mathsf{PDec}(sk, \cdot)$ is a decryption oracle. However, how the decryption oracle is available is determined depending on the query sequences B, F, A in the following way: $\mathcal{A}_1$ can issue decryption queries following the sequence B, and after all B-queries are completed, $\mathcal{A}_1$ can further issue decryption queries following the sequence F. However, $\mathcal{A}_1$ need not complete all F-queries, and the ability to issue F-queries can be "shared" with $\mathcal{A}_2$. That is, as long as the order, the types, and the number of queries are maintained, the F-queries that $\mathcal{A}_1$ has not been completed can be taken over by $\mathcal{A}_2$. $\mathcal{A}_2$ can issue the remaining F-queries that $\mathcal{A}_1$ has left for $\mathcal{A}_2$, and after all F-queries are completed, $\mathcal{A}_2$ can further issue decryption queries following the sequence A.[2] Moreover, as usual, $\mathcal{A}_2$'s queries must not contain the challenge ciphertext c^*.

We refer to B, F, and A as "**B**efore-challenge" queries, "**F**lexible" queries (in the sense that $\mathcal{A}$ can "flexibly" decide how it issues queries before/after the challenge), and "**A**fter-challenge" queries, respectively. For notational convenience, if $\mathsf{F} = \emptyset$ then we write $\langle \mathsf{B} :: \mathsf{A}\rangle$-mCCA, instead of $\langle \mathsf{B} : \emptyset : \mathsf{A}\rangle$-mCCA.

Definition 4. *Let* $\mathsf{B}, \mathsf{F}, \mathsf{A} \in \mathcal{QS}^*$. *We say that a PKE scheme* Π *is* $\langle \mathsf{B} : \mathsf{F} : \mathsf{A}\rangle$-mCCA *secure if* $\mathsf{Adv}_{\Pi,\mathcal{A}}^{\langle \mathsf{B}:\mathsf{F}:\mathsf{A}\rangle\text{-mCCA}}(k) = |\Pr[\mathsf{Expt}_{\Pi,\mathcal{A}}^{\langle \mathsf{B}:\mathsf{F}:\mathsf{A}\rangle\text{-mCCA}}(k) = 1] - \frac{1}{2}|$ *is negligible for any PPTA* $\mathcal{A}$.

We define mixed CCA security for KEMs in exactly the same way as above.

With the mixed CCA security notions, we can express all the security notions mentioned in Section 2. These are summarized in Table 1. As noted earlier, for non-malleability, we adopt the characterization using a parallel query by Bellare and Sahai [3]. (In the table, we also include the *bounded parallel CCA* security notions defined below).

We remark that we can also define a parallel decryption query in mixed CCA security experiment (i.e. the $\langle \mathsf{B} : \mathsf{F} : \mathsf{A}\rangle$-mCCA experiment) so that the number of ciphertexts contained in each parallel query is also bounded to be some predetermined value (say, t). However, such security definition is implied by $(|(\mathsf{B}||\mathsf{F}||\mathsf{A})| \cdot t)$-Bounded CCA security, which is already achieved by the existing PKE schemes that are constructed only from IND-CPA secure schemes by the previous results [9,8]. Therefore, we think that studying security with such limitation is less interesting than studying mixed CCA security defined in this section, and is not treated in this paper.

Previously to this paper, Phan and Pointcheval [26] defined a similar notion which they call (i,j)-IND security and (i,j)-NM security, which are equivalent to $\langle \mathsf{s}^i :: \mathsf{s}^j\rangle$-mCCA security and $\langle \mathsf{s}^i :: \mathsf{s}^j\mathsf{p}\rangle$-mCCA security in our definition, respectively (for NM, we interpret it with parallel CCA-based characterization in [3]). They did not consider the "flexible" F-queries.

[2] In other words, in the $\langle \mathsf{B} : \mathsf{F} : \mathsf{A}\rangle$-mCCA experiment, $\mathcal{A}_1$ can issue $(\mathsf{B}||\mathsf{F}_1)$-queries, and $\mathcal{A}_2$ can issue $(\mathsf{F}_2||\mathsf{A})$-queries, for any pair of query sequences $(\mathsf{F}_1, \mathsf{F}_2)$ satisfying $(\mathsf{F}_1||\mathsf{F}_2) = \mathsf{F}$, and how F is split into F_1 and F_2 can be decided adaptively by $\mathcal{A}$ in the experiment.

Table 1. Compatibility with Existing Security Notions

Existing Notions	Notation in Mixed CCA Security
IND-CPA	$\langle\emptyset :: \emptyset\rangle$-mCCA
NM-CPA	$\langle\emptyset :: \mathsf{p}\rangle$-mCCA
IND-q-CCA2	$\langle\emptyset : \mathsf{s}^q : \emptyset\rangle$-mCCA
NM-q-CCA2	$\langle\emptyset : \mathsf{s}^q : \mathsf{p}\rangle$-mCCA
IND-q-pCCA1	$\langle\mathsf{p}^q :: \emptyset\rangle$-mCCA
NM-q-pCCA1	$\langle\mathsf{p}^q :: \mathsf{p}\rangle$-mCCA
IND-q-pCCA2	$\langle\emptyset : \mathsf{p}^q : \emptyset\rangle$-mCCA
NM-q-pCCA2	$\langle\emptyset : \mathsf{p}^q : \mathsf{p}\rangle$-mCCA
IND-CCA1	$\langle\mathsf{unbound} :: \emptyset\rangle$-mCCA
NM-CCA1	$\langle\mathsf{unbound} :: \mathsf{p}\rangle$-mCCA
IND-CCA2	$\langle\mathsf{unbound} :: \mathsf{unbound}\rangle$-mCCA

Bounded Parallel CCA Security. Here, we define a natural and interesting special class of mixed CCA security which we call *bounded parallel CCA* security. This captures security against adversaries whose decryption queries are always parallel, and is a natural extension from the original bounded CCA security [9].

Depending on how the decryption oracle is available for an adversary, we define pCCA1 and pCCA2 as natural analogue of CCA1 and CCA2, respectively. Moreover, as is similar to the existing security notions, we define indistinguishability (IND) and non-malleability (NM).

Definition 5. *Let $q \geq 0$ be an integer. We say that a PKE scheme is* IND-q-pCCA1 *(resp.* IND-q-pCCA2, NM-q-pCCA1, *and* NM-q-pCCA2*) secure if it is* $\langle\mathsf{p}^q :: \emptyset\rangle$-mCCA *(resp.* $\langle\emptyset : \mathsf{p}^q : \emptyset\rangle$-mCCA, $\langle\mathsf{p}^q :: \mathsf{p}\rangle$-mCCA, *and* $\langle\emptyset : \mathsf{p}^q : \mathsf{p}\rangle$-mCCA*) secure.*

We define the bounded parallel CCA security notions for KEMs in the same way.

3.2 General Properties of Mixed CCA Security

Here, we show two general implication results about the mixed CCA security notions. (In this section, we always assume $\mathsf{B}, \mathsf{F}, \mathsf{A} \in \mathcal{QS}^*$, and do not write it explicitly).

Firstly, by noticing the property of the "flexible" queries F, we obtain the following.

Theorem 1. *For both PKE schemes and KEMs,* $\langle\mathsf{B} : \mathsf{F} : \mathsf{A}\rangle$-mCCA *security and "the combination of all security notions of the form* $\langle(\mathsf{B}||\mathsf{F}_1) :: (\mathsf{F}_2||\mathsf{A})\rangle$-mCCA *satisfying* $(\mathsf{F}_1||\mathsf{F}_2) = \mathsf{F}$*" are equivalent.*

The implication from the former to the latter is immediate by definition. Since the proof for the other direction is almost trivial, we omit the proof and only mention the intuition using the simplest case $\mathsf{F} = \mathsf{s}$. It is easy to see that $\langle\mathsf{B} : \mathsf{s} : \mathsf{A}\rangle$-mCCA adversary can be divided into two types: The first type that makes $(\mathsf{B}||\mathsf{s})$-queries before the challenge, and A-queries after the challenge, and the second type that makes B-queries before, and (s||A)-queries after the challenge. Then, the experiment for the first type can be simulated by a ⟨(B||s) :: A⟩-mCCA adversary while that for the second type can be simulated by a ⟨B :: (s||A)⟩-mCCA adversary. This is easily extended to any $\mathsf{F} \in \mathcal{QS}$ case. Note that if F = unbound, then the statement is again trivial because we can have $\mathsf{F}_1 = \mathsf{F}_2$ = unbound (since unbound = (unbound||unbound)), and thus in this case ⟨(B||F_1) :: (F_2||A)⟩-mCCA security is equivalent to ⟨unbound :: unbound⟩-mCCA = IND-CCA2 security, which implies all the mixed CCA security notions.

Next, we show that for PKE schemes with polynomially bounded plaintext space size and for KEMs, the A-queries, which is intended to be only available after the challenge, can actually be issued "flexibly" without destroying security.

Theorem 2. *For PKE schemes with polynomially bounded plaintext space size and for KEMs,* ⟨B : F : A⟩-mCCA *security and* ⟨B : (F||A) : ∅⟩-mCCA *security are equivalent.*

The implication from the latter notion to the former is immediate by definition. The proof for the other direction is given in the full version. Very roughly, showing the implication from the former notion to the latter is possible because the challenge ciphertext can be made "in advance" for PKE schemes with polynomially bounded plaintext space size and for KEMs. (In particular, for the PKE case, since the plaintext space size is polynomially bounded, the adversary's two challenge plaintexts can be guessed correctly with probability $1/poly(k)$). Therefore, we can construct a reduction algorithm $\mathcal{B}$ that can successfully attack ⟨B : F : A⟩-mCCA security using a successful ⟨B : (F||A) : ∅⟩-mCCA adversary $\mathcal{A}$. Actually, in showing the proof, we have to be careful about the situation in which some of $\mathcal{A}$'s flexible decryption queries (i.e. (F||A)-queries) issued by $\mathcal{A}$ before "$\mathcal{A}$'s" challenge contains $\mathcal{B}$'s challenge ciphertext (which will be later used as $\mathcal{A}$'s challenge). However, the statistical property of PKE schemes and KEMs called *smoothness*, formalized in [2], guarantees that the probability of such a problematic situation occurring is negligible. For more details, we refer the reader to the full version.

4 Relations among Mixed CCA Security Notions

Due to its stage-specific queries and the difference between single and parallel queries, given two mixed CCA security notions, it is not always easy to tell if one notion implies the other. Therefore, a natural and yet non-trivial question is: *given two mixed CCA security notions* ⟨B : F : A⟩-mCCA *and* ⟨$\widetilde{\mathsf{B}}$: $\widetilde{\mathsf{F}}$: $\widetilde{\mathsf{A}}$⟩-mCCA*, under what conditions on* B, F, A, $\widetilde{\mathsf{B}}, \widetilde{\mathsf{F}}, \widetilde{\mathsf{A}}$ *are there implications/separations?*

In this section, we fully answer this question and show a necessary and sufficient condition for implications/separations between any two mixed CCA security notions. Interestingly, it turns out that *for PKE schemes, the relations among security notions are different depending on its plaintext space size.* The relations among mixed CCA security notions for PKE schemes with polynomially bounded plaintext space size and those for KEMs are always the same.

The rest of this section is organized as follows: In Section 4.1 we introduce a relation over query sequences which plays a key role for our results. Then in Sections 4.2 and 4.3 we show separation results and implication results, respectively. Finally in Section 4.4, we summarize the results by showing the necessary and sufficient conditions. For notational convenience, throughout this section we always assume $\mathsf{B}, \mathsf{F}, \mathsf{A}, \widetilde{\mathsf{B}}, \widetilde{\mathsf{F}}, \widetilde{\mathsf{A}} \in \mathcal{QS}^*$. Due to space limitation, most of the proofs in this section are omitted and are given in the full version, and for theorems whose proofs are omitted, we provide some ideas for the proofs.

4.1 "is-Simulatable-by" Relation for Query Sequences

We first introduce the following relation over symbols.

Definition 6. *We define a partial order "$\subseteq_1$" over symbols $\{\mathsf{s}, \mathsf{p}\}$ by $\mathsf{s} \subseteq_1 \mathsf{s}$, $\mathsf{s} \subseteq_1 \mathsf{p}$, and $\mathsf{p} \subseteq_1 \mathsf{p}$.*

Intuitively, the meaning of "$\subseteq_1$" is that the former type oracle query "is-simulatable-by" the latter type of oracle query. The subscript "1" of "$\subseteq_1$" denotes that it is a relation for one symbol, and it should not be mixed up with the relation for query sequences below (although the meaning is essentially the same).

Now, we extend the "is-simulatable-by" relation to query sequences:

Definition 7. *Let $\mathsf{seq}, \widetilde{\mathsf{seq}} \in \mathcal{QS}^*$. We define a binary relation "$\subseteq_{qs}$" over $\mathcal{QS}^*$ as follows. "$\widetilde{\mathsf{seq}} \subseteq_{qs} \mathsf{seq}$" if and only if one of the following is satisfied:*

- $\mathsf{seq} = \mathsf{unbound}$ *or* $\widetilde{\mathsf{seq}} = \emptyset$
- $\mathsf{seq} = (a_1 \dots a_m), \widetilde{\mathsf{seq}} = (b_1 \dots b_n) \in \mathcal{QS}\backslash\{\emptyset\}$ *where* $a_i, b_j \in \{\mathsf{s}, \mathsf{p}\}$ *for each* $i \in [m]$, $j \in [n]$, *and there exists a strictly increasing function* $f : [n] \to [m]$ *such that* $b_j \subseteq_1 a_{f(j)}$ *holds for all* $j \in [n]$.

If seq *and* $\widetilde{\mathsf{seq}}$ *do not satisfy the above, we write "*$\widetilde{\mathsf{seq}} \not\subseteq_{qs} \mathsf{seq}$*".*

The subscript "qs" of $\subseteq_{qs}$ stands for *query sequence*. It is easy to see that the above relation "$\subseteq_{qs}$" is a natural extension from $\subseteq_1$. Suppose $\widetilde{\mathsf{seq}} \subseteq_{qs} \mathsf{seq}$. Consider two PPTA adversaries $\mathcal{A}$ and $\mathcal{B}$ attacking a same PKE scheme, where $\mathcal{A}$ makes seq-queries and $\mathcal{B}$ makes $\widetilde{\mathsf{seq}}$-queries, and a situation in which $\mathcal{A}$ simulates the experiment for $\mathcal{B}$. If $\widetilde{\mathsf{seq}} = \emptyset$, then $\mathcal{B}$ makes no query. If $\mathsf{seq} = \mathsf{unbound}$, then $\mathcal{A}$ can use unbounded oracle access, and thus $\mathcal{B}$'s decryption oracle can be simulated. Otherwise, (i.e. $\mathsf{seq}, \widetilde{\mathsf{seq}} \in \mathcal{QS}\backslash\{\emptyset\}$), then i-th query from $\mathcal{B}$ can be simulated by $\mathcal{A}$'s $f(i)$-th query (where f is a strictly increasing function guaranteed to exist by definition) for all $i \in [n]$.

Now, given two sequences $\mathsf{seq}, \widetilde{\mathsf{seq}} \in \mathcal{QS}^*$ we can tell if $\widetilde{\mathsf{seq}} \subseteq_{qs} \mathsf{seq}$ or $\widetilde{\mathsf{seq}} \not\subseteq_{qs} \mathsf{seq}$.[3] For example, $(\mathsf{s}^2\mathsf{p}) \subseteq_{qs} (\mathsf{psps})$; $(\mathsf{sp}^2) \not\subseteq_{qs} (\mathsf{s}^2\mathsf{ps}^2)$; $\mathsf{s}^r \subseteq_{qs} \mathsf{s}^q$ iff $q \geq r$.

---

[3] Note that "$\subseteq_{qs}$" forms a partial order over $\mathcal{QS}^*$. However, it is not a total order. For example, we have both $(\mathsf{sp}) \not\subseteq_{qs} (\mathsf{ps})$ and $(\mathsf{ps}) \not\subseteq_{qs} (\mathsf{sp})$.

4.2 Separation Results

A common approach for showing a separation of a security notion $\mathtt{X}$ from a security notion $\mathtt{Y}$ for PKE schemes is to construct a "separating" PKE scheme from a building block $\mathtt{X}$-secure PKE scheme: the decryption algorithm of the separating scheme typically has some "backdoor" mechanism, which leads to some "critical information" v (e.g. secret key for the building block PKE scheme) for breaking $\mathtt{Y}$ security so that $\mathtt{Y}$-adversary can, by using a decryption oracle, reach for v and break $\mathtt{Y}$-security of the separating PKE scheme while an $\mathtt{X}$-adversary cannot reach for v or simply v is useless for breaking $\mathtt{X}$-security of the scheme. We also follow this approach.

Useful Tool for Separation: Backdoor-Sequence Scheme. In order for the above approach to work, what to use as the critical information and how to implement such backdoor mechanism are the main issues. We wish to implement a backdoor mechanism so that given two sequences $\mathsf{seq}, \widetilde{\mathsf{seq}} \in \mathcal{QS}^*$, if $\widetilde{\mathsf{seq}} \not\sqsubseteq_{qs} \mathsf{seq}$,[4]. then an adversary making $\widetilde{\mathsf{seq}}$-queries can finally reach for a critical information (and break some security of a separating PKE scheme) while an adversary making seq-queries cannot. This is indeed possible. We can implement such backdoor mechanism as a *sequence of backdoor information* $(u_1, \dots, u_{|\widetilde{\mathsf{seq}}|+1})$ and a *strategy for "how to release next backdoor information"*, based on seq, $\widetilde{\mathsf{seq}}$, and the critical information v. Specifically, let $\widetilde{\mathsf{seq}} = (b_1 \dots b_n)$ such that $b_i \in \{\mathsf{s}, \mathsf{p}\}$ for $i \in [n]$.

- The sequence of backdoor information $(u_1, \dots, u_{n+1})$ is set up so that $u_1 = 1^k$ (any publicly known value will do), $u_2, \dots, u_n$ are random values (which must be hard to guess), and u_{n+1} is the critical information v.
- The strategy for "how to release next backdoor information", depending on $\widetilde{\mathsf{seq}} = (b_1 \dots b_n)$, is set up so that: If $b_i = \mathsf{s}$, this "release strategy" on input u_i outputs u_{i+1} directly; If $b_i = \mathsf{p}$, this "release strategy" on input u_i (together with some index j) outputs a (j-th) "secret-share" of u_{i+1}, so that if we collect the shares more than a threshold which is set to be a value greater than $|\mathsf{seq}|$, we can reconstruct u_{i+1}.

(Intuitively, such a release strategy is implemented into a decryption algorithm of a separating PKE scheme so that if the decryption algorithm takes some special information indicating "backdoor mode" as input, the output of the release strategy is used instead of decrypting as a ciphertext). Constructed as above, an adversary making $\widetilde{\mathsf{seq}}$-queries to the release strategy can finally obtain $u_{n+1} = v$. In particular, if $b_i = \mathsf{p}$ then an adversary can make a parallel query to the release strategy to obtain all the share of u_{i+1} at once, and thus can reconstruct u_{i+1}. It is actually possible to show that *if* $\widetilde{\mathsf{seq}} \not\sqsubseteq_{qs} \mathsf{seq}$, *then no adversary who is only allowed to make* seq*-queries to the release strategy can reach for* $u_{n+1} = v$, and thus we can make a difference in the information available for an adversary making $\widetilde{\mathsf{seq}}$-queries and that making seq-queries.

---

[4] As we have seen in Section 4.1, if $\widetilde{\mathsf{seq}} \sqsubseteq_{qs} \mathsf{seq}$, then any information available for adversaries making $\widetilde{\mathsf{seq}}$-queries is also available for those making seq-queries.

In order to make it easier to analyze PKE schemes used to show separations, in the full version we formalize this "backdoor mechanism" as a "stand alone" primitive. We call it a *backdoor-sequence scheme*, and use it as one of main building blocks for constructing the separating schemes that are used to establish the separations in the following paragraphs in this subsection.

Separation by Total Query Sequence.

Theorem 3. *For both PKE schemes and KEMs, if* $(\widetilde{\mathsf{B}}||\widetilde{\mathsf{F}}||\widetilde{\mathsf{A}}) \not\leq_{qs} (\mathsf{B}||\mathsf{F}||\mathsf{A})$*, then* $\langle \mathsf{B} : \mathsf{F} : \mathsf{A} \rangle$-$\mathtt{mCCA}$ *security does not imply* $\langle \widetilde{\mathsf{B}} : \widetilde{\mathsf{F}} : \widetilde{\mathsf{A}} \rangle$-$\mathtt{mCCA}$ *security.*

The idea for building the separating PKE scheme for showing Theorem 3 is straightforward. We use the secret key *sk* for the building block scheme of the separating PKE scheme as a critical information, and setup the backdoor-sequence scheme appropriately. That is, a $\langle \widetilde{\mathsf{B}} : \widetilde{\mathsf{F}} : \widetilde{\mathsf{A}} \rangle$-$\mathtt{mCCA}$ adversary who can (in total) make $(\widetilde{\mathsf{B}}||\widetilde{\mathsf{F}}||\widetilde{\mathsf{A}})$-queries can finally reach for *sk* and decrypt the challenge ciphertext. However, since $(\widetilde{\mathsf{B}}||\widetilde{\mathsf{F}}||\widetilde{\mathsf{A}}) \not\leq_{qs} (\mathsf{B}||\mathsf{F}||\mathsf{A})$, the property of the backdoor-sequence scheme guarantees that a $\langle \mathsf{B} : \mathsf{F} : \mathsf{A} \rangle$-$\mathtt{mCCA}$ adversary who is only allowed to make $(\mathsf{B}||\mathsf{F}||\mathsf{A})$-queries in total cannot reach for it, and thus the separating PKE scheme remains $\langle \mathsf{B} : \mathsf{F} : \mathsf{A} \rangle$-$\mathtt{mCCA}$ secure. The same proof strategy works for the KEM case.

Separation by After-challenge Queries.

Theorem 4. *For both PKE schemes and KEMs, if* $(\widetilde{\mathsf{F}}||\widetilde{\mathsf{A}}) \not\leq_{qs} (\mathsf{F}||\mathsf{A})$*, then* $\langle \mathsf{B} : \mathsf{F} : \mathsf{A} \rangle$-$\mathtt{mCCA}$ *security does not imply* $\langle \widetilde{\mathsf{B}} : \widetilde{\mathsf{F}} : \widetilde{\mathsf{A}} \rangle$-$\mathtt{mCCA}$ *security.*

For an explanation here, consider the extreme case: the separation of $\langle \emptyset :: \widetilde{\mathsf{A}} \rangle$-$\mathtt{mCCA}$ security from $\langle \mathsf{unbound} :: \mathsf{A} \rangle$-$\mathtt{mCCA}$ security under the condition $\widetilde{\mathsf{A}} \not\leq_{qs} \mathsf{A}$. Note that a $\langle \mathsf{unbound} :: \mathsf{A} \rangle$-$\mathtt{mCCA}$ adversary can make unbounded single queries before the challenge while a $\langle \emptyset :: \widetilde{\mathsf{A}} \rangle$-$\mathtt{mCCA}$ adversary can make no query. Therefore, the critical information for breaking $\langle \emptyset :: \widetilde{\mathsf{A}} \rangle$-$\mathtt{mCCA}$ security must be something that is useful and available only after the challenge. We set the critical information to be the decryption result of a ciphertext (which can be a challenge ciphertext) that is input together with the backdoor information into the decryption algorithm of the separating scheme, and use a pseudorandom function F to realize the separating PKE scheme that has a "ciphertext-dependent" backdoor sequence. More specifically, a seed K for F is picked as a part of a secret key of the separating PKE scheme. The decryption algorithm of the separating scheme, on input a ciphertext c together with backdoor and some information that indicates "backdoor mode", derives a pseudorandom value $R = F_K(c)$ and use this R as a randomness for deriving the sequence of backdoors, and then outputs a corresponding "next backdoor". Since the backdoor-sequence scheme is set up so that an adversary that can make $\widetilde{\mathsf{A}}$-queries can finally reach for the critical information (decryption of any ciphertext), a $\langle \emptyset :: \widetilde{\mathsf{A}} \rangle$-$\mathtt{mCCA}$ adversary can finally reach for the decryption result of the challenge ciphertext by appropriately making decryption queries after the challenge. However, since $\widetilde{\mathsf{A}} \not\leq_{qs} \mathsf{A}$, the same does not apply to a $\langle \mathsf{unbound} :: \mathsf{A} \rangle$-$\mathtt{mCCA}$ adversary that can make only

A-queries after the challenge. Therefore, to break $\langle \mathsf{unbound} :: \mathsf{A} \rangle$-mCCA security of the separating scheme, the adversary has to essentially break $\langle \mathsf{unbound} :: \mathsf{A} \rangle$-mCCA security of the building block scheme (unless it breaks security of the pseudorandom function or the backdoor-sequence scheme). Essentially the same proof strategy works for proving the KEM case. Using a pseudorandom function to set up "ciphertext-dependent" backdoor information was previously used in [1] to separate NM-CCA2 = IND-CCA2 security from NM-CCA1 security.

Separation by Before-challenge Queries.

Theorem 5. *For PKE schemes with superpolynomially large plaintext space size, if* $(\widetilde{\mathsf{B}}||\widetilde{\mathsf{F}}) \not\subseteq_{qs} (\mathsf{B}||\mathsf{F})$*, then* $\langle \mathsf{B} : \mathsf{F} : \mathsf{A} \rangle$-mCCA *security does not imply* $\langle \widetilde{\mathsf{B}} : \widetilde{\mathsf{F}} : \widetilde{\mathsf{A}} \rangle$-mCCA *security.*

Note that this theorem is only true for PKE schemes with superpolynomially large plaintext space size. For an explanation here, consider the extreme case: the separation of $\langle \widetilde{\mathsf{B}} :: \emptyset \rangle$-mCCA security from $\langle \mathsf{B} :: \mathsf{unbound} \rangle$-mCCA security under the condition $\widetilde{\mathsf{B}} \not\subseteq_{qs} \mathsf{B}$. This time, the critical information for breaking $\langle \widetilde{\mathsf{B}} :: \emptyset \rangle$-mCCA security must be something that is useful only before the challenge, because $\langle \widetilde{\mathsf{B}} :: \emptyset \rangle$-mCCA adversary can make no query after the challenge. We use a one-way function f to construct the separating PKE scheme so that it has "weak" plaintexts, which are not encrypted at all by the encryption algorithm of the separating PKE scheme. (Similar ideas are used in [26,24,2]). A public key of the separating PKE scheme contains $V = f(m^*)$ for some random element m^* chosen from the plaintext space of the underlying PKE scheme, and weak plaintexts m are the ones satisfying $f(m) = V$. We set the critical information to be m^* itself, where the backdoor-sequence scheme will finally release m^* if $\widetilde{\mathsf{B}}$-queries are appropriately performed, and thus m^* can be used as one of two challenge plaintexts to break $\langle \widetilde{\mathsf{B}} :: \emptyset \rangle$-mCCA security. However, since $\widetilde{\mathsf{B}} \not\subseteq_{qs} \mathsf{B}$, the property of the backdoor-sequence scheme guarantees that a $\langle \mathsf{B} :: \mathsf{unbound} \rangle$-mCCA adversary cannot reach for m^* before the challenge (unless it break one-wayness of f). Moreover, the weak plaintext m^* is useless even if it is found after the challenge. Therefore, in order to break $\langle \mathsf{B} :: \mathsf{unbound} \rangle$-mCCA security of the separating scheme, the adversary essentially has to attack $\langle \mathsf{B} :: \mathsf{unbound} \rangle$-mCCA security of the building block scheme.

4.3 Implication Results

A combination of Theorems 3, 4, and 5 shows that given two mixed CCA security notions $\langle \mathsf{B} : \mathsf{F} : \mathsf{A} \rangle$-mCCA and $\langle \widetilde{\mathsf{B}} : \widetilde{\mathsf{F}} : \widetilde{\mathsf{A}} \rangle$-mCCA, the latter notion is separated from the former if $(\widetilde{\mathsf{B}}||\widetilde{\mathsf{F}}||\widetilde{\mathsf{A}}) \not\subseteq_{qs} (\mathsf{B}||\mathsf{F}||\mathsf{A})$, $(\widetilde{\mathsf{B}}||\widetilde{\mathsf{F}}) \not\subseteq_{qs} (\mathsf{B}||\mathsf{F})$, or $(\widetilde{\mathsf{F}}||\widetilde{\mathsf{A}}) \not\subseteq_{qs} (\mathsf{F}||\mathsf{A})$ holds for PKE schemes with superpolynomially large plaintext space. We show that if none of the above conditions are satisfied, then we actually have an implication from the former notion to the latter, where this implication is also true for all PKE schemes and KEMs.

Theorem 6. *For both PKE schemes and KEMs, if* $(\widetilde{\mathsf{B}}||\widetilde{\mathsf{F}}||\widetilde{\mathsf{A}}) \subseteq_{qs} (\mathsf{B}||\mathsf{F}||\mathsf{A})$, $(\widetilde{\mathsf{B}}||\widetilde{\mathsf{F}}) \subseteq_{qs} (\mathsf{B}||\mathsf{F})$*, and* $(\widetilde{\mathsf{F}}||\widetilde{\mathsf{A}}) \subseteq_{qs} (\mathsf{F}||\mathsf{A})$ *hold simultaneously, then* $\langle \mathsf{B} : \mathsf{F} : \mathsf{A} \rangle$-mCCA *security implies* $\langle \widetilde{\mathsf{B}} : \widetilde{\mathsf{F}} : \widetilde{\mathsf{A}} \rangle$-mCCA *security.*

This theorem holds because it can be shown that if the three conditions regarding query sequences are satisfied, then whatever strategy regarding the "flexible" queries an $\langle \widetilde{\mathsf{B}} : \widetilde{\mathsf{F}} : \widetilde{\mathsf{A}} \rangle$-$\mathtt{mCCA}$ adversary may take, the $\langle \widetilde{\mathsf{B}} : \widetilde{\mathsf{F}} : \widetilde{\mathsf{A}} \rangle$-$\mathtt{mCCA}$ experiment can be perfectly simulated by an $\langle \mathsf{B} : \mathsf{F} : \mathsf{A} \rangle$-$\mathtt{mCCA}$ adversary.

Combining Theorem 6 with Theorem 2, we obtain the following corollary.

Corollary 1. *For PKE schemes with polynomially bounded plaintext space size and for KEMs, if* $(\widetilde{\mathsf{B}}||\widetilde{\mathsf{F}}||\widetilde{\mathsf{A}}) \subseteq_{qs} (\mathsf{B}||\mathsf{F}||\mathsf{A})$ *and* $(\widetilde{\mathsf{F}}||\widetilde{\mathsf{A}}) \subseteq_{qs} (\mathsf{F}||\mathsf{A})$ *hold simultaneously, then* $\langle \mathsf{B} : \mathsf{F} : \mathsf{A} \rangle$-$\mathtt{mCCA}$ *security implies* $\langle \widetilde{\mathsf{B}} : \widetilde{\mathsf{F}} : \widetilde{\mathsf{A}} \rangle$-$\mathtt{mCCA}$ *security.*

Proof. By Theorem 2, we know that for PKE schemes with polynomially bounded plaintext space size and for KEMs, $\langle \mathsf{B} : \mathsf{F} : \mathsf{A} \rangle$-$\mathtt{mCCA}$ security implies $\langle \widetilde{\mathsf{B}} : \widetilde{\mathsf{F}} : \widetilde{\mathsf{A}} \rangle$-$\mathtt{mCCA}$ security if and only if $\langle \mathsf{B} : (\mathsf{F}||\mathsf{A}) : \emptyset \rangle$-$\mathtt{mCCA}$ security implies $\langle \widetilde{\mathsf{B}} : (\widetilde{\mathsf{F}}||\widetilde{\mathsf{A}}) : \emptyset \rangle$-$\mathtt{mCCA}$ security. Then, Theorem 6 tells us that the sufficient condition of the implication from the former notion to the latter is: "$(\widetilde{\mathsf{B}}||(\widetilde{\mathsf{F}}||\widetilde{\mathsf{A}})||\emptyset) \subseteq_{qs} (\mathsf{B}||(\mathsf{F}||\mathsf{A})||\emptyset)$, $(\widetilde{\mathsf{B}}||(\widetilde{\mathsf{F}}||\widetilde{\mathsf{A}})) \subseteq_{qs} (\mathsf{B}||(\mathsf{F}||\mathsf{A}))$, and $((\widetilde{\mathsf{F}}||\widetilde{\mathsf{A}})||\emptyset) \subseteq_{qs} ((\mathsf{F}||\mathsf{A})||\emptyset)$ hold simultaneously." Simplifying this condition yields Corollary 1. □

4.4 Necessary and Sufficient Conditions for Implication/Separation

As a summarization of the results in this section, we show the following necessary and sufficient conditions for implication/separation among mixed CCA security.

Theorem 7. *For PKE schemes with superpolynomially large plaintext space size,* $\langle \mathsf{B} : \mathsf{F} : \mathsf{A} \rangle$-$\mathtt{mCCA}$ *security implies* $\langle \widetilde{\mathsf{B}} : \widetilde{\mathsf{F}} : \widetilde{\mathsf{A}} \rangle$-$\mathtt{mCCA}$ *security if and only if* $(\widetilde{\mathsf{B}}||\widetilde{\mathsf{F}}||\widetilde{\mathsf{A}}) \subseteq_{qs} (\mathsf{B}||\mathsf{F}||\mathsf{A})$, $(\widetilde{\mathsf{B}}||\widetilde{\mathsf{F}}) \subseteq_{qs} (\mathsf{B}||\mathsf{F})$, *and* $(\widetilde{\mathsf{F}}||\widetilde{\mathsf{A}}) \subseteq_{qs} (\mathsf{F}||\mathsf{A})$ *hold simultaneously.*

Proof. This follows from a combination of Theorems 3, 4, 5, and 6. □

Theorem 8. *For PKE schemes with polynomially bounded plaintext space size and for KEMs,* $\langle \mathsf{B} : \mathsf{F} : \mathsf{A} \rangle$-$\mathtt{mCCA}$ *security implies* $\langle \widetilde{\mathsf{B}} : \widetilde{\mathsf{F}} : \widetilde{\mathsf{A}} \rangle$-$\mathtt{mCCA}$ *security if and only if* $(\widetilde{\mathsf{B}}||\widetilde{\mathsf{F}}||\widetilde{\mathsf{A}}) \subseteq_{qs} (\mathsf{B}||\mathsf{F}||\mathsf{A})$ *and* $(\widetilde{\mathsf{F}}||\widetilde{\mathsf{A}}) \subseteq_{qs} (\mathsf{F}||\mathsf{A})$ *hold simultaneously.*

Proof. This follows from a combination of Theorems 3, 4, and Corollary 1. □

We believe the relations among security notions shown in this section are useful for future studies on PKE schemes and KEMs whose security notions can be expressed in mixed CCA security notions. For example, by utilizing the above theorems, we can fully establish the relations among bounded parallel CCA security and other existing security notions in Table 1. We also note that the previously established relations among security notions [1,9,16] can be re-proved as corollaries from the above theorems.

Importance of Plaintext Space Size in Relations among Security Notions for PKE Schemes. As our results in this section have clarified, it is important to care about the size of the plaintext space size for relations among security notions for PKE schemes. Specifically, Theorems 7 and 8 tell us that given $\langle \mathsf{B} : \mathsf{F} : \mathsf{A} \rangle$-$\mathtt{mCCA}$ and $\langle \widetilde{\mathsf{B}} : \widetilde{\mathsf{F}} : \widetilde{\mathsf{A}} \rangle$-$\mathtt{mCCA}$ security notions, the implication/separation from the former notion to the latter notion differs if $(\widetilde{\mathsf{B}}||\widetilde{\mathsf{F}}||\widetilde{\mathsf{A}}) \subseteq_{qs} (\mathsf{B}||\mathsf{F}||\mathsf{A})$, $(\widetilde{\mathsf{F}}||\widetilde{\mathsf{A}}) \subseteq_{qs} (\mathsf{F}||\mathsf{A})$, and $(\widetilde{\mathsf{B}}||\widetilde{\mathsf{F}}) \not\subseteq_{qs} (\mathsf{B}||\mathsf{F})$ hold simultaneously.

5 Feasibility Results from IND-CPA Secure PKE Schemes

By adopting the notion of mixed CCA security, in this section we show two black-box constructions of PKE schemes, which can encrypt plaintexts of polynomial length (thus, exponentially large plaintext space), from IND-CPA secure schemes.

The first result is the following.

Theorem 9. *For any polynomial $q \geq 0$, there exists a shielding black-box construction of a $\langle \emptyset : \mathsf{s}^q\mathsf{p} : \emptyset \rangle$-mCCA secure PKE scheme which can encrypt plaintexts of polynomial length from an* IND-CPA *secure PKE scheme.*

Proof. This theorem is proved by combining the existing results and Theorem 2 in Section 3.2. The following statement is due to the result by Choi et al. [8].

Lemma 1. *[8] For any polynomial $q \geq 0$, there exists a shielding black-box construction of an* NM-q-CCA2 *secure PKE scheme which can encrypt plaintexts of polynomial length from an* IND-CPA *secure PKE scheme.*

Recall that NM-q-CCA2 $= \langle \emptyset : \mathsf{s}^q : \mathsf{p} \rangle$-mCCA (see Table 1). Since any $\langle \mathsf{B} : \mathsf{F} : \mathsf{A} \rangle$-mCCA secure PKE scheme can be trivially used as a KEM with the same security by encrypting a uniformly random string K and using it as a session-key, Lemma 1 implies that we can construct a $\langle \emptyset : \mathsf{s}^q : \mathsf{p} \rangle$-mCCA secure KEM (we call it the *CDMW KEM*) from any IND-CPA secure PKE scheme in a black-box and shielding manner. Then, by Theorem 2 for KEMs, we can immediately say that the CDMW KEM is $\langle \emptyset : \mathsf{s}^q\mathsf{p} : \emptyset \rangle$-mCCA secure. Finally, by combining the CDMW KEM with an IND-CCA2 secure DEM, we obtain the desired result. (It is implicit from the works by Cramer and Shoup [10] and by Herranz et al. [16] that if we combine a $\langle \mathsf{B} : \mathsf{F} : \mathsf{A} \rangle$-mCCA secure KEM and an IND-CCA2 secure DEM in a straightforward manner, we can obtain a $\langle \mathsf{B} : \mathsf{F} : \mathsf{A} \rangle$-mCCA secure PKE scheme). Note that we can construct an IND-CCA2 secure DEM even without any computational assumption (see e.g. [10, Section 7.2.2]). Moreover, the "shielding" and "black-box" properties are trivially preserved by our construction. This completes the proof of Theorem 9. □

$\langle \emptyset : \mathsf{s}^q\mathsf{p} : \emptyset \rangle$-mCCA security implies NM-q-CCA2 $= \langle \emptyset : s^q : \mathsf{p} \rangle$-mCCA security by definition, while by Theorem 7 we know that for PKE schemes with superpolynomially large plaintext space size, $\langle \emptyset : \mathsf{s}^q : \mathsf{p} \rangle$-mCCA security does not imply $\langle \emptyset : \mathsf{s}^q\mathsf{p} : \emptyset \rangle$-mCCA security. Therefore, for these types of PKE schemes, $\langle \emptyset : \mathsf{s}^q\mathsf{p} : \emptyset \rangle$-mCCA security is strictly stronger than NM-q-CCA2 security.

We remark that Theorem 2 actually implies that the original CDMW PKE scheme [8] already achieves the shielding black-box construction of $\langle \emptyset : \mathsf{s}^q\mathsf{p} : \emptyset \rangle$-mCCA secure PKE schemes *if it is used with short plaintexts (so that the plaintext space size is bounded to be polynomial)*. However, the Choi et al. result itself does not imply Theorem 9, because it is not obvious how to construct $\langle \emptyset : \mathsf{s}^q\mathsf{p} : \emptyset \rangle$-mCCA secure PKE schemes which can encrypt plaintexts of polynomially length from PKE schemes that satisfies the same security but has only polynomially bounded plaintext space size, in a black-box and shielding manner[5].

We also remark that the original CDMW PKE scheme [8] might be shown to be $\langle\emptyset : \mathsf{s}^q\mathsf{p} : \emptyset\rangle$-mCCA secure as it is for large plaintext space size, using the same assumptions used to show its NM-q-CCA2 security. However, our main purpose here is to show the improved feasibility rather than the concrete construction and efficiency, and thus we did not try proving directly that the CDMW PKE is $\langle\emptyset : \mathsf{s}^q\mathsf{p} : \emptyset\rangle$-mCCA secure.

Theorem 9 implies the following corollary.

Corollary 2. *There exists a shielding black-box construction of an* IND-1-pCCA2 *secure PKE scheme which can encrypt plaintexts of polynomial length from an* IND-CPA *secure PKE scheme.*

Our second result on black-box constructions is the following.

Theorem 10. *For any polynomials $q, q' \geq 0$, there exists a shielding black-box construction of a $\langle\mathsf{s}^q\mathsf{p} : \mathsf{s}^{q'} : \emptyset\rangle$-mCCA secure PKE scheme which can encrypt plaintexts of polynomial length from an* IND-CPA *secure PKE scheme.*

Proof. To prove this theorem, we will use the following result which is implicit from [9, Lemma 1][6]:

Lemma 2. *(Implicit from [9]). For any $\mathsf{B} \in \mathcal{QS}^*$ and any polynomial $q' \geq 0$, there exists a shielding black-box construction of a $\langle\mathsf{B} : s^{q'} : \emptyset\rangle$-mCCA secure PKE scheme from a $\langle\mathsf{B} :: \emptyset\rangle$-mCCA secure PKE scheme.*

We call the construction by Cramer et al. [9] the *CHH+ PKE scheme.* Due to Theorem 9 above, for any polynomial $q \geq 0$, we can construct a $\langle\emptyset : \mathsf{s}^q\mathsf{p} : \emptyset\rangle$-mCCA secure PKE scheme, which is also $\langle\mathsf{s}^q\mathsf{p} :: \emptyset\rangle$-mCCA secure, from any IND-CPA secure PKE scheme. Then, by using this PKE scheme as a building block of the CHH+ PKE scheme, due to Lemma 2, we have a PKE scheme which satisfies the claimed security. The CHH+ PKE construction is shielding and black-box. Since the construction of the PKE scheme in Theorem 9 is also shielding and black-box, so is the construction as a whole. The size of the plaintext space is maintained as well. This completes the proof of Theorem 10. □

We note that by Theorem 7, for PKE schemes with superpolynomially large plaintext space size, $\langle\mathsf{s}^q\mathsf{p} : \mathsf{s}^{q'} : \emptyset\rangle$-mCCA security achieved in Theorem 10 cannot be directly compared with the notion achieved in Theorem 9 (actually even with NM-CPA$=\langle\emptyset :: \mathsf{p}\rangle$-mCCA security). However, the security achieved in Theorem 10

[5] Recently, Myers and Shelat [21] showed a black-box construction of multi-bit IND-CCA2 secure PKE schemes from 1-bit IND-CCA2 secure PKE schemes. Though it seems that their result can be extended (with some modification) to any mixed CCA security, we remark that their construction is non-shielding.

[6] The original statement of Lemma 1 in [9] shows a special case of Lemma 2 in which $\mathsf{B} = \emptyset$. Moreover, the special case of Lemma 2 in which $\mathsf{B} = \mathsf{unbound}$ is also mentioned in [9]. See Remark 2 after the proof of Lemma 1 in [9].

allows the bounded number of "flexible" single queries before and after the challenge, after the parallel query in the first stage, while the security achieved by Theorem 9 does not allow any query after one parallel query for an adversary. Thus we believe that Theorem 10 is also interesting as a feasibility result.

Handling Decryption of Unboundedly Many Ciphertexts before the Challenge. Previous to our work, none of the constructions of PKE schemes that use only IND-CPA secure ones have achieved security against adversaries that can observe unboundedly many decryption results (via the decryption oracle) in the first stage, i.e., before choosing two challenge plaintexts, regardless of whether the construction is black-box or non-black-box. On the other hand, the constructions in Theorems 9 and 10 (and also the combination of [8] and Theorem 2) achieve security against adversaries that can observe unboundedly many decryption results by one parallel decryption query before the challenge.

Thus, due to the results in this section, it has been clarified that *the difficulty of constructing an* IND-CCA1 *secure PKE scheme only from* IND-CPA *secure ones lies not in whether the number of decryption results that the adversary can see before the challenge is bounded or not, but in whether the number of the adversary's "adaptive" decryption queries is bounded.*

6 Open Problems

Constructions Secure against Two or More Parallel Queries. None of our feasibility results achieves mixed CCA security in which we can handle more than one parallel decryption query, and whether we can construct a PKE scheme with such security only using IND-CPA secure schemes is still unclear. Therefore, we would like to leave it as an open problem. Since any (unbounded) CCA secure PKE construction from IND-CPA secure ones must first be secure against adversaries who make two or more parallel decryption queries, we believe that overcoming this barrier of "two parallel queries" is worth tackling.

We notice that if we can, by only using an IND-CPA secure PKE scheme as a building block, construct a (strong) *designated verifier* (DV) NIZK proof system [23,9] for any NP language with q-bounded "*parallel*" strong soundness, which is a natural extension of a (strong) DV-NIZK with q-bounded strong soundness [9] in the soundness experiment of which an adversary can ask verification of many theorem/proof pairs in a parallel manner, then by using the DV-NIZK proof system in the Dolev-Dwork-Naor construction [11,23,9] (resp. the Naor-Yung construction [22]) we will be able to construct an IND-$(q+1)$-pCCA2 (resp. IND-$(q+1)$-pCCA1) secure PKE scheme. However, how to construct such a DV-NIZK proof system only from IND-CPA secure PKE schemes is not known so far. This might be worth looking at towards the next step from our results.

Stronger Black-Box Impossibility Results. Since the constructions in Theorems 9 and 10 are shielding and black-box, according to the impossibility result of [12] and the transitivity of black-box constructions, we have that there exists no shielding black-box construction of an IND-CCA1 secure PKE scheme from PKE schemes which satisfy any security notion achieved in Theorems 9 and 10.

It would also be interesting to clarify if we can show a stronger impossibility result than [12] such that constructing IND-q-pCCA1 secure PKE schemes from IND-CPA secure one in a shielding and black-box manner for some $q > 1$ is impossible. (Or more generally, we can also consider the impossibility of some of mixed CCA security notion). Note that this strengthening of the impossibility result of [12] can make sense only if we consider parallel decryption queries, because the result by Choi et al. [8] already shows that it is possible to achieve the strongest form of (ordinary) bounded CCA security, namely, NM-q-CCA2, in a black-box and shielding manner.

Acknowledgement

The authors would like to thank Jacob Schuldt for his helpful comments and suggestions. The authors also would like to thank anonymous reviewers of PKC 2011 for their invaluable comments.

References

1. Bellare, M., Desai, A., Pointcheval, D., Rogaway, P.: Relations among notions of security for public-key encryption schemes. In: Krawczyk, H. (ed.) CRYPTO 1998. LNCS, vol. 1462, pp. 26–45. Springer, Heidelberg (1998)
2. Bellare, M., Hofheinz, D., Kiltz, E.: Subtleties in the definition of IND-CCA: When and how should challenge-decryption be disallowed?, Cryptology ePrint Archive: Report 2009/418 (2009)
3. Bellare, M., Sahai, A.: Non-malleable encryption: Equivalence between two notions, and an indistinguishability-based characterization. In: Wiener, M. (ed.) CRYPTO 1999. LNCS, vol. 1666, pp. 519–536. Springer, Heidelberg (1999); The revised version is available in Cryptology ePrint Archive (Report 2006/228)
4. Bellare, M., Yung, M.: Certifying permutations: Noninteractive zero-knowledge based on any trapdoor permutation. J. Cryptology 9(3), 149–166 (1996)
5. Boneh, D., Canetti, R., Halevi, S., Katz, J.: Chosen-ciphertext security from identity-based encryption. SIAM J. Comput. 36(5), 1301–1328 (2007)
6. Boyen, X., Mei, Q., Waters, B.: Direct chosen ciphertext security from identity-based techniques. In: CCS 2005, pp. 320–329 (2005)
7. Cash, D., Kiltz, E., Shoup, V.: The twin diffie-hellman problem and applications. In: Smart, N.P. (ed.) EUROCRYPT 2008. LNCS, vol. 4965, pp. 127–145. Springer, Heidelberg (2008)
8. Choi, S.G., Dachman-Soled, D., Malkin, T., Wee, H.: Black-box construction of a non-malleable encryption scheme from any semantically secure one. In: Canetti, R. (ed.) TCC 2008. LNCS, vol. 4948, pp. 427–444. Springer, Heidelberg (2008)
9. Cramer, R., Hanaoka, G., Hofheinz, D., Imai, H., Kiltz, E., Pass, R., Shelat, A., Vaikuntanathan, V.: Bounded CCA2-Secure Encryption. In: Kurosawa, K. (ed.) ASIACRYPT 2007. LNCS, vol. 4833, pp. 502–518. Springer, Heidelberg (2007)
10. Cramer, R., Shoup, V.: Design and analysis of practical public-key encryption schemes secure against adaptive chosen ciphertext attack. SIAM J. Comput. 33(1), 167–226 (2003)
11. Dolev, D., Dwork, C., Naor, M.: Non-malleable cryptography. SIAM J. Comput. 30(2), 391–437 (2000)

12. Gertner, Y., Malkin, T., Myers, S.: Towards a separation of semantic and CCA security for public key encryption. In: Vadhan, S.P. (ed.) TCC 2007. LNCS, vol. 4392, pp. 434–455. Springer, Heidelberg (2007)
13. Goldwasser, S., Micali, S.: Probabilistic encryption. J. Comput. Syst. Sci. 28(2), 270–299 (1984)
14. Hanaoka, G., Kurosawa, K.: Efficient chosen ciphertext secure public key encryption under the computational diffie-hellman assumption. In: Pieprzyk, J. (ed.) ASIACRYPT 2008. LNCS, vol. 5350, pp. 308–325. Springer, Heidelberg (2008)
15. Haralambiev, K., Jager, T., Kiltz, E., Shoup, V.: Simple and efficient public-key encryption from computational diffie-hellman in the standard model. In: Nguyen, P.Q., Pointcheval, D. (eds.) PKC 2010. LNCS, vol. 6056, pp. 1–18. Springer, Heidelberg (2010)
16. Herranz, J., Hofheinz, D., Kiltz, E.: Some (in)sufficient conditions for secure hybrid encryption. Inf. Comput. 208(11), 1243–1257 (2010)
17. Hofheinz, D., Kiltz, E.: Secure hybrid encryption from weakened key encapsulation. In: Menezes, A. (ed.) CRYPTO 2007. LNCS, vol. 4622, pp. 553–571. Springer, Heidelberg (2007)
18. Hofheinz, D., Kiltz, E.: Practical chosen ciphertext secure encryption from factoring. In: Joux, A. (ed.) EUROCRYPT 2009. LNCS, vol. 5479, pp. 313–332. Springer, Heidelberg (2009)
19. Kiltz, E., Mohassel, P., O'Neill, A.: Adaptive trapdoor functions and chosen-ciphertext security. In: Gilbert, H. (ed.) EUROCRYPT 2010. LNCS, vol. 6110, pp. 673–692. Springer, Heidelberg (2010)
20. Kurosawa, K., Desmedt, Y.: A new paradigm of hybrid encryption scheme. In: Franklin, M. (ed.) CRYPTO 2004. LNCS, vol. 3152, pp. 426–442. Springer, Heidelberg (2004)
21. Myers, S., Shelat, A.: Bit encryption is complete. In: FOCS 2009, pp. 607–616 (2009)
22. Naor, M., Yung, M.: Public-key cryptosystems provably secure against chosen ciphertext attacks. In: STOC 1990, pp. 427–437 (1990)
23. Pass, R., Shelat, A., Vaikuntanathan, V.: Construction of a Non-malleable Encryption Scheme from Any Semantically Secure One. In: Dwork, C. (ed.) CRYPTO 2006. LNCS, vol. 4117, pp. 271–289. Springer, Heidelberg (2006)
24. Pass, R., Shelat, A., Vaikuntanathan, V.: Relations Among Notions of Non-malleability for Encryption. In: Kurosawa, K. (ed.) ASIACRYPT 2007. LNCS, vol. 4833, pp. 519–535. Springer, Heidelberg (2007)
25. Peikert, C., Waters, B.: Lossy trapdoor functions and their applications. In: STOC 2008, pp. 187–196 (2008)
26. Phan, D.H., Pointcheval, D.: On the security notions for public-key encryption schemes. In: Blundo, C., Cimato, S. (eds.) SCN 2004. LNCS, vol. 3352, pp. 33–46. Springer, Heidelberg (2005)
27. Rackoff, C., Simon, D.R.: Non-interactive zero-knowledge proof of knowledge and chosen ciphertext attack. In: Feigenbaum, J. (ed.) CRYPTO 1991. LNCS, vol. 576, pp. 433–444. Springer, Heidelberg (1992)
28. Reingold, O., Trevisan, L., Vadhan, S.: Notions of reducibility between cryptographic primitives. In: Naor, M. (ed.) TCC 2004. LNCS, vol. 2951, pp. 1–20. Springer, Heidelberg (2004)
29. Rosen, A., Segev, G.: Chosen-ciphertext security via correlated products. In: Reingold, O. (ed.) TCC 2009. LNCS, vol. 5444, pp. 419–436. Springer, Heidelberg (2009)
30. Wee, H.: Efficient chosen-ciphertext security via extractable hash proofs. In: Rabin, T. (ed.) CRYPTO 2010. LNCS, vol. 6223, pp. 314–332. Springer, Heidelberg (2010)

Secure Blind Decryption*

Matthew Green

Johns Hopkins University
Information Security Institute
3400 N. Charles Street; Baltimore, MD 21218, USA
mgreen@cs.jhu.edu

Abstract. In this work we construct public key encryption schemes that admit a protocol for *blindly* decrypting ciphertexts. In a blind decryption protocol, a user with a ciphertext interacts with a secret keyholder such that the user obtains the decryption of the ciphertext and the keyholder learns nothing about what it decrypted. While we are not the first to consider this problem, previous works provided only weak security guarantees against malicious users. We provide, to our knowledge, the first practical blind decryption schemes that are secure under a strong CCA security definition. We prove our construction secure in the standard model under simple, well-studied assumptions in bilinear groups. To motivate the usefulness of this primitive we discuss several applications including privacy-preserving distributed file systems and Oblivious Transfer schemes that admit *public* contribution.

1 Introduction

The past several years have seen a trend towards outsourcing data storage to remote data stores and cloud-based services. While much attention has been paid to securing this data, relatively little has been given to the problem of securing the data's access pattern. This is a real problem for some systems where users' access histories are more sensitive than the data itself, for example patent databases. Even in business there are many practical applications where users' access history is sensitive. For example, the data access patterns of a major corporation's executives could be worth millions of dollars to the right person, particularly in advance of a merger or acquisition.

To address these concerns, many recent works have proposed tools that allow users to transact online without sacrificing their privacy. These tools include (but are not limited to) efficient adaptive oblivious transfer protocols [15, 28, 29, 44], anonymous credential schemes [13, 4], and group signature schemes [16, 7]. One recent application for these tools is to the construction of *oblivious databases* that provide strong access control while preventing the operator from learning

* This work was supported in part by NSF Grant CNS-1010928 and by HHS Grant number 90TR0003/01. The contents of this publication are solely the responsibility of the authors and do not necessarily represent the official views of HHS or any other sponsor.

D. Catalano et al. (Eds.): PKC 2011, LNCS 6571, pp. 265–282, 2011.

which records its users access [20, 12]. Despite this progress, there are still many primitives that we do not know how to implement efficiently using the techniques available to us.

Blind Decryption. In this work we consider one such primitive, which we refer to as *blind decryption*. A blind decryption scheme is a public-key encryption (PKE) scheme that admits an efficient protocol for obliviously decrypting ciphertexts. In this protocol a User who possesses a ciphertext interacts with a Decryptor who holds the necessary secret key. At the conclusion of the protocol, the User obtains the plaintext while the Decryptor learns nothing about what it decrypted. Given that the fundamental purpose of a blind decryption protocol is to decrypt ciphertexts, it seems reasonable to analyze any such protocol with malicious adversaries in mind. Specifically, since such an adversary can implicitly use the blind decryption protocol to decrypt chosen ciphertexts, we will restrict out investigation to secure blind decryption schemes that retain their security even under (adaptive) chosen ciphertext attack.

Blind decryption has many applications to privacy-preserving protocols and systems. For example, blind decryption implies k-out-of-N oblivious transfer [11], which is important theoretically as well as practically for its applications to the construction of oblivious databases [15, 20, 12]. Moreover, blind decryption has practical applications to distributed cryptographic filesystems and for supporting rapid deletion [43].

We are not the first to consider the problem of constructing blind decryption schemes. The primitive was originally formalized by Sakura and Yamane [45] in the mid-1990s, but folklore solutions are thought to have predated that work by more than a decade. Despite an abundance of research in this area, most proposed constructions are insecure under adaptive chosen ciphertext attack [24, 49,41,23,47,42]. Several protocols have recently been proposed containing "blind decryption-like" techniques (see *e.g.*, the simulatable oblivious transfer protocols of [15, 29, 44, 30, 33]). However, these protocols use symmetric (or at least, non-public) encryption procedures, and it does not seem easy to adapt them to the public-key model.

Of course, blind decryption is an instance of secure multi-party computation (MPC) and can be achieved by applying general techniques (*e.g.*, [50, 26, 34]) to the decryption algorithm of a CCA-secure PKE scheme. However, the protocols yielded by this approach are likely to be quite inefficient, making them impractical for real-world applications.

Our Contributions. In this paper we present what is, to our knowledge, the first practical blind decryption scheme that is IND-CCA2-secure in the standard model. We prove our scheme secure under reasonable assumptions in bilinear groups. At the cost of introducing an optional Common Reference String, the protocol can be conducted in a single communication round.

To motivate the usefulness of this new primitive we consider several applications. Chief among these is the construction of privacy-preserving encrypted filesystems (and databases), where a central authority manages the decryption of many ciphertexts without learning users' access patterns. This is important in situations where the access pattern might leak critical information about the information being accessed. Unlike previous attempts to solve this problem [15, 20, 12], our encryption algorithm is *public*, *i.e.*, users can encrypt new messages offline without assistance from a trusted party. By combining blind decryption with the new oblivious access control techiques of [20, 12] (which use anonymous credentials to enforce complex access control policies) we can achieve strong proactive access control without sacrificing privacy.

Of potential theoretical interest, blind decryption can be used as a building block in constructing adaptive k-out-of-N Oblivious Transfer protocols [15, 29, 44, 30, 33, 37]. In fact, it is possible to achieve a multi-party primitive that is more flexible than traditional OT, in that *any* party can commit messages to the message database (rather than just the Sender). We refer to this enhanced primitive as Oblivious Transfer with Public Contribution (OTPC). We discuss these applications in Section 5.

1.1 Related Work

The first blind decryption protocol is generally attributed Chaum [19], who proposed a technique for blinding an RSA ciphertext in order to obtain its decryption c^d *mod* N. Since traditional RSA ciphertexts are malleable and hence vulnerable to chosen ciphertext attack, this approach does not lead to a secure blind decryption scheme. Furthermore, standard encryption padding techniques [5] do not seem helpful.

Subsequent works [45,24,49] adapted Chaum's approach to other CPA-secure cryptosystems such as Elgamal. These constructions were employed within various protocols, including a 1-out-of-N Oblivious Transfer scheme due to Dodis *et al.* [24]. Unfortunately, since the cryptosystems underlying these protocols are not CCA-secure, security analyses of those protocols frequently required strong assumptions such as honest-but-curious adversaries[1]. Mambo, Sakurai and Okamoto [41] proposed to address chosen ciphertext attacks by signing the ciphertexts to prevent an adversary from mauling them. Their *transformable signature* could be blinded in tandem with the ciphertext. The trouble with this approach and other related approaches [15, 29, 30, 33, 44] is that the encryption scheme is no longer a PKE, since encryption now requires a knowledge of a secret signing key (furthermore, these transformable signatures were successfully cryptanalyzed [23]). Schnorr and Jakobsson [47] proposed a scheme secure under the weaker *one-more decryption attack* and used this to construct a PIR protocol. Unfortunately, their protocol is secure only for *random* messages, and furthermore cannot be extended to construct stronger primitives such as simulatable OT [15].

Recently, Green and Hohenberger [28] proposed a technique for blindly extracting decryption keys in an Identity-Based Encryption scheme. Subsequently, Ogata and Le Trieu [42] used this tool to obtain a weak blind decryption scheme

[1] For example, Dodis *et al.* [24] analyzed their 1-out-of-N oblivious transfer construction in the honest-but-curious model.

(by encrypting ciphertexts under a random identity, then blindly extracting the appropriate secret key). The resulting protocol is efficient, but the ciphertexts are malleable and thus vulnerable to adaptive chosen ciphertext attack.

1.2 Intuition

Ideally the development of a blind decryption scheme would begin with an existing CCA-secure PKE, and would only require us to develop an efficient two-party protocol for computing the decryption algorithm. Indeed, the literature provides us with many candidate PKE constructions that can be so adapted if we are willing to accept the costs associated with general multi-party computation techniques [50, 26, 34].

However, in this work we are interested in protocols that are both secure and *practical.* This rules out inefficient gate-by-gate decryption protocols, limiting us to a relatively small collection of techniques that can be used to build efficient protocols. This toolbox includes primitives such as homomorphic commitment schemes, which we might combine with zero knowledge proofs for statements involving algebraic relations among cyclic group elements, *e.g.,* [46, 31]. While these techniques have been deployed successfully to construct other privacy-preserving protocols, there are strict limitations on what they can accomplish.

To illustrate this point, let us review several of the most popular encryption techniques in the literature. Random oracle paradigms such as OAEP [5] and Fujisaki-Okamoto [25] seem fundamentally difficult to adapt, since these approaches require the decryptor to evaluate an ideal hash function on a partially-decrypted value *prior* to outputting a result. Even the more efficient standard-model CCA-secure paradigms such as Cramer-Shoup [22] and recent bilinear constructions (*e.g.,* [8, 10, 35]) require components that we cannot efficiently adapt. For example, when implemented in a group $\mathbb{G}$ of order p, the Cramer-Shoup scheme assumes a collision-resistant mapping $H : \mathbb{G} \times \mathbb{G} \times \mathbb{G} \rightarrow \mathbb{Z}_p$. We know of no efficient two-party technique for evaluating such a function[2].

Our approach. Rather than adapt an existing scheme, we set out to design a new one. Our approach is based on the TBE-to-PKE paradigm proposed independently by Canetti *et al.* [18] and MacKenzie *et al.* [40]. This technique converts a Tag-Based Encryption (TBE) scheme into a CCA-secure public PKE with the assistance of a strongly unforgeable one-time signature (OTS). In this generic transform, encryption is conducted by first generating a keypair (vk, sk) for the OTS, encrypting the message using the TBE with vk as the tag, then signing the resulting ciphertext with sk. Intuitively the presence of the signature (which is verified at decryption time) prevents an adversary from mauling the ciphertext.

To blindly decrypt such a ciphertext, we propose the following approach: the User first commits to the ciphertext and vk using a homomorphic commitment or encryption scheme. She then efficiently proves knowledge of the associated signature for these committed values. If this proof verifies, the Decryptor

[2] Conceivably it might be possible to develop one, however it might be tied to the specific construction of $\mathbb{G}$ and thus be be quite inflexible.

may then apply the TBE decryption algorithm to the (homomorphically) committed ciphertext, secure in the knowledge that the commitment contains an appropriately-distributed value. Finally, the result can be opened by the User.

For this protocol to be efficient, we must choose our underlying primitives with care. Specifically, we must ensure that (1) the OTS verification key maps to the tag-space of the TBE, (2) and the TBE ciphertext maps to the message space of the OTS. Of course, the easiest way to achieve these goals is to use an OTS that directly signs the TBE ciphertext space, with a TBE whose tag-space includes the OTS verification keyspace. These primitives must admit efficient protocols for the operations we will conduct with them. Finally, we would like to avoid relying on complex or novel complexity assumptions in order to achieve these goals.

Our proposed construction is based on a variant of Cramer-Shoup that was adapted by Shacham [48] for security in bilinear groups. We first modify Shacham's construction into a TBE with the following ciphertext structure. Let $\alpha \in \mathbb{Z}_p^*$ be an arbitrary ciphertext tag and $m \in \mathbb{G}$ a message to be encrypted. Given a public key $g, g_1, g_2, g_3, h_1, h_2, c_1, c_2, d_1, d_2 \in \mathbb{G}$ an encryptor selects random elements $r_1, r_2 \in \mathbb{Z}_p^*$ and outputs the ciphertext:

$$(u_1, u_2, u_3, e, v, vk) = (g_1^{r_1}, g_2^{r_2}, g_3^{r_1+r_2}, m \cdot h_1^{r_1} h_2^{r_2}, (c_1 d_1^{\alpha})^{r_1} \cdot (c_2 d_2^{\alpha})^{r_2}, g^{\alpha})$$

An important feature of this construction is that the decryptor does *not* need to know the tag value α[3]. Therefore, in constructing our PKE we can "dual-purpose" α as both the ciphertext tag *and* as the secret key of a one-time signature (OTS) scheme. Specifically, our encryption process will select a random α, encrypt the message using the TBE with α as the tag, and finally sign the resulting elements (u_1, u_2, u_3, e, v) under α. The resulting ciphertext contains $(u_1, u_2, u_3, e, v, vk)$ along with the signature on those values.

The remaining challenge is therefore to construct an efficient OTS that can sign multiple bilinear group elements, yet admits an efficient proof-of-knowledge for a signature on committed elements. To address this we propose a new multi-block one-time "F-signature" that we believe may be of independent interest[4]. Interestingly, our signing algorithm does not actually operate on elements of $\mathbb{G}$, but rather signs message vectors of the form $(m_1, \ldots, m_n) \in \mathbb{Z}_p^{*n}$ (for some arbitrary vector length n). Once a message is signed, however, the signature can be *verified* given the tuple $(g^{m_1}, \ldots, g^{m_n}) \in \mathbb{G}^n$, rather than the original message

[3] This differs from many other candidate TBE and IBE schemes, *e.g.,* Boneh and Boyen's IBE [6] and Kiltz's TBE [35] where the tag/identity is an element of $\mathbb{Z}_p^*$ and *must* be provided at decryption time (or in the case of IBE, when a secret key is extracted). This requirement stems from the nature of those schemes' security proofs.

[4] F-signature is a contraction of F-unforgeable signature, which is a concept proposed by Belinkiy *et al.* [4], and later developed by Green and Hohenberger [30]. In this paradigm, the signing algorithm operates on a message m, but there exists a signature verification algorithm that can operate given only $F(m)$ for some one-way function F.

vector. Strictly speaking, this construction does not meet our requirements—an encryptor won't always know the discrete logarithm base g of (u_1, u_2, u_3, e, v). Our key insight is to show that encryptors can produce an identically distributed "workalike" signature even when the discrete logarithms are not known. We prove that, in the context of our encryption scheme, no adversary can forge these workalike signatures. Our signature construction is presented independently in Appendix 2.4.

2 Technical Preliminaries

2.1 Bilinear Groups and Cryptographic Assumptions

Let λ be a security parameter. We define BMsetup as an algorithm that, on input 1^λ, outputs the parameters for a bilinear mapping as $\gamma = (p, \mathbb{G}, \mathbb{G}_T, e, g \in \mathbb{G})$, where g generates $\mathbb{G}$, the groups $\mathbb{G}, \mathbb{G}_T$ each have prime order p, and $e : \mathbb{G} \times \mathbb{G} \to \mathbb{G}_T$. For $\langle g \rangle = \langle h \rangle = \mathbb{G}$ the efficiently-computable mapping e must be both *non-degenerate* ($\langle e(g, h) \rangle = \mathbb{G}_T$) and *bilinear* (for $a, b \in \mathbb{Z}_p^*$, $e(g^a, h^b) = e(g, h)^{ab}$).

The Decision Linear Assumption (DLIN) [7]. Let $\mathbb{G}$ be a group of prime order $p \in \Theta(2^\lambda)$. For all p.p.t. adversaries $\mathcal{A}$, the following probability is 1/2 plus an amount negligible in λ: $\Pr[f, g, h, z_0 \xleftarrow{R} \mathbb{G}; a, b \xleftarrow{R} \mathbb{Z}_p^*; z_1 \leftarrow h^{a+b}; d \xleftarrow{R} \{0, 1\}; d' \leftarrow \mathcal{A}(f, g, h, f^a, g^b, z_d) : d = d']$.

The Flexible Diffie-Hellman Assumption (FDH) [36,30]. Let $\mathbb{G}$ be a group of prime order $p \in \Theta(2^\lambda)$. For all p.p.t. adversaries $\mathcal{A}$, the following probability is negligible in λ: $\Pr[g, g^a, g^b; a, b \xleftarrow{R} \mathbb{Z}_p^*; (w, w') \leftarrow \mathcal{A}(g, g^a, g^b) : w \neq 1 \wedge w' = w^{ab}]$.

This assumption was previously described as the 2-out-of-3 CDH assumption by Kunz-Jacques and Pointcheval [36]. We adopt the name Flexible Diffie-Hellman for consistency with recent work [39,30]. To instill confidence in this assumption, Green and Hohenberger [30] showed that a solver for the Flexible Diffie-Hellman problem implies a solver for a related decisional problem, the Decisional 3-Party Diffie-Hellman assumption (3DDH) which has been used several times in the literature [38, 9, 32, 30].

2.2 Proofs of Knowledge

We use several standard results for proving statements about the satisfiability of one or more pairing-product equations. For variables $\{\mathcal{X}\}_{1...n} \in \mathbb{G}$ and constants $\{\mathcal{A}\}_{1...n} \in \mathbb{G}, a_{i,j} \in \mathbb{Z}_p^*$, and $t_T \in \mathbb{G}_T$, these equations have the form:

$$\prod_{i=1}^{n} e(\mathcal{A}_i, \mathcal{X}_i) \prod_{i=1}^{n} \prod_{j=1}^{n} e(\mathcal{X}_i, \mathcal{X}_j)^{a_{i,j}} = t_T$$

The proof-of-knowledge protocols in this work can be instantiated using one of two approaches. The first approach is to use the interactive zero-knowledge proof technique of Schnorr [46], with extensions due to *e.g.*, [21, 14, 17, 2, 15]. Note that this may require that the proofs be executed sequentially (indeed, this requirement is explicit in our security definitions). For details, see the work of Adida *et al.* [2], which provides a taxonomy of interactive proof techniques for pairing-based statements.

Alternatively, the proofs can be instantiated using the Groth-Sahai proof system [31] which permits efficient non-interactive proofs of the satisfiability of multiple pairing product equations. In the general case these proofs are witness indistinguishable. However a subset of special cases (including where $t_T = 1$) may be conducted in zero-knowledge[5]. The Groth-Sahai system can be instantiated under the Decision Linear assumption in the Common Reference String model.

We refer the reader to the cited works for formal security definitions of ZK and WI proof systems. In our security analysis we will assume some generic instantiation Π_{ZK} that is secure under the Decision Linear assumption in $\mathbb{G}$. Either of the techniques mentioned above can satisfy this requirement. When referring to WI and ZK proofs we will use the notation of Camenisch and Stadler [16]. For instance, $\mathsf{WIPoK}\{(g, h) : e(g, h) = T \ \wedge \ e(g, v) = 1\}$ denotes a witness indistinguishable proof of knowledge of elements g and h that satisfy both $e(g, h) = T$ and $e(g, v) = 1$. All values not in enclosed in ()'s are assumed to be known to the verifier.

2.3 Linear Encryption

Our blind decryption protocol employs a multiplicatively homomorphic scheme that encrypts elements of $\mathbb{G}$. We instantiate this scheme with the Linear Encryption scheme of Boneh, Boyen and Shacham [7] which is semantically secure under the Decision Linear assumption. Ciphertexts in this scheme have the form $(c_1, c_2, c_3) \in \mathbb{G}^3$, and the homomorphic operation is simple pairwise multiplication. Exponentiation by a scalar z can be performed as c_1^z, c_2^z, c_3^z. To re-randomize a ciphertext one multiplies it by $\mathsf{LE.Enc}(pk, 1)$. Our protocols also require an efficient ZK proof-of-knowledge of the plaintext m underlying a ciphertext C, which we denote by $\mathsf{ZKPoK}\{(m) : C \in \mathsf{LE.Enc}(pk, m)\}$. We refer the reader to the full version [27] for formal algorithm descriptions.

2.4 A One-Time F-Signature on Multiblock Messages

Our constructions require a strongly unforgeable one-time F-signature scheme that signs messages of the form $(m_1, \ldots, m_N) \in \mathbb{Z}_p^{*n}$ (for arbitrary values of n), but can *verify* signatures given only a function of the messages, specifically, $(g_1^{m_1}, \ldots, g_n^{m_n}) \in \mathbb{G}^n$ for fixed $g_1, \ldots, g_n \in \mathbb{G}$. Note that $g_1, \ldots, g_n$ need not be distinct.

[5] In many cases it is easy to re-write pairing products equation as a composition of multiple distinct equations having $t_T = 1$ (see [31]). Although we do not explicitly perform this translation in our protocols, we note that it can be applied to all of the ZKPoKs used in our constructions.

To construct FS, we adapt a weakly-unforgeable signature due to Green and Hohenberger [30] to admit multi-block messages, while simplifying the scheme into a one-time signature. The latter modification has the incidental effect of strengthening the signature to be strongly unforgeable. Let us now describe FS:

FS.KG. On input group parameters γ, a vector length n, select $g, g_1, \ldots, g_n, v, d, u_1, \ldots, u_n \xleftarrow{R} \mathbb{G}$ and $a \xleftarrow{R} \mathbb{Z}_p^*$. Output $vk = (\gamma, g, g^a, v, d, g_1, \ldots, g_n, u_1, \ldots, u_n, n)$ and $sk = (vk, a)$.

FS.Sign. Given sk and a message vector $(m_1, \ldots, m_n) \in \mathbb{Z}_p^{*n}$, first select $r \xleftarrow{R} \mathbb{Z}_p^*$ and output the signature $\sigma = ((\prod_{i=1}^n u_i^{m_i} \cdot v^r\ d)^a, g_1^{am_1}, \ldots, g_n^{am_n}, u_1^{m_1}, \ldots, u_n^{m_n}, r)$.

FS.Verify. Given $pk, (g_1^{m_1}, \ldots, g_n^{m_n})$, parse $\sigma = (\sigma_1, e_1, \ldots, e_n, f_1, \ldots, f_n, r)$, output 1 if the following check holds: $e(\sigma_1, g) = e(\prod_{i=1}^n f_i \cdot v^r d, g^a) \ \wedge \ \{e(g_i^{m_i}, g^a) = e(e_i, g) \ \wedge \ e(g_i^{m_i}, u_i) = e(g_i, f_i)\}_{i \in [1,n]}$.

Note that verification is a pairing product equation. Thus we can efficiently prove knowledge of a signature using the techniques described in Section 2.2. We denote such a proof by *e.g.*, $\mathsf{WIPoK}\{(\sigma) : \mathsf{Verify}(vk, (g^{m_1}, \ldots, g^{m_n}), \sigma) = 1\}$. Note that vk or the messages may reside within a commitment. In the full version of this paper [27] we provide details on these proofs of knowledge, as well as definitions of security and a proof that FS is strongly unforgeable under the Flexible Diffie-Hellman assumption.

Workalike signatures. Our blind decryption constructions make use of the "workalike" algorithms (WAKG, WASign). While the public outputs of these algorithms are identically distributed those of KG and Sign, the WASign algorithm operates on messages of the form $(g_1, \ldots, g_n) \in \mathbb{G}^n$. We stress that (WAKG, WASign, Verify) is *not* a secure signature scheme on arbitrary group elements, but can be used securely under the special conditions of our constructions..

FS.WAKG. Select $x_1, \ldots, x_n \xleftarrow{R} \mathbb{Z}_p^*$ and set $(u_1, \ldots, u_n) = (g^{x_1}, \ldots, g^{x_n})$. Compute the remaining elements as in KG and set $sk = (vk, a, x_1, \ldots, x_n)$.

FS.WASign. Given a message vector $(h_1, \ldots, h_n) \in \mathbb{G}^n$, first select $r \xleftarrow{R} \mathbb{Z}_p^*$ and output the signature $\sigma = ((\prod_{i=1}^n h_i^{x_i} \cdot v^r d)^a, h_1^a, \ldots, h_n^a, h_1^{x_1}, \ldots, h_n^{x_n}, r)$.

3 Definitions

Notation: Let $\mathcal{M}$ be the message space and $\mathcal{C}$ be the ciphertext space. We write $P(\mathcal{A}(a), \mathcal{B}(b)) \to (c, d)$ to indicate the protocol P is between parties $\mathcal{A}$ and $\mathcal{B}$, where a is $\mathcal{A}$'s input, c is $\mathcal{A}$'s output, b is $\mathcal{B}$'s input and d is $\mathcal{B}$'s output. We will define $\nu(\cdot)$ as a negligible function.

Definition 1 (Blind Decryption Scheme). A public-key blind decryption scheme consists of a tuple of algorithms (KG, Enc, Dec) and a protocol BlindDec.

$\mathsf{KG}(1^\lambda)$. On input a security parameter λ, the key generation algorithm KG outputs a public key pk and a secret key sk.

$\mathsf{Enc}(pk, m)$. On input a public key pk and a message m, Enc outputs a ciphertext C.

$\mathsf{Dec}(pk, sk, C)$. On input pk, sk and a ciphertext C, Dec outputs a message m or the error symbol $\perp$.

The two-party protocol BlindDec is conducted between a user $\mathcal{U}$ and a decryptor $\mathcal{D}$:

$\mathsf{BlindDec}(\{\mathcal{U}(pk, C)\}, \{\mathcal{D}(pk, sk)\}) \rightarrow (m, \mathsf{nothing})$. On input pk and a ciphertext C, an honest user $\mathcal{U}$ outputs the decryption m or the error symbol $\perp$. The decryptor $\mathcal{D}$ outputs nothing or an error message.

We now present the standard definition of adaptive chosen ciphertext security for public key encryption.

Definition 2 (IND-CCA2)**.** A public key encryption scheme $\Pi = (\mathsf{KG}, \mathsf{Enc}, \mathsf{Dec})$ is IND-CCA2 secure if every *p.p.t.* adversary $\mathcal{A} = (\mathcal{A}_1, \mathcal{A}_2)$ has advantage $\leq \nu(\lambda)$ in the following experiment.

$$
\begin{array}{l}
\mathsf{IND\text{-}CCA2}(\Pi, \mathcal{A}, \lambda) \\
\quad (pk, sk) \leftarrow \mathsf{KG}(1^\lambda) \\
\quad (m_0, m_1, z) \leftarrow \mathcal{A}_1^{\mathcal{O}_{dec}(pk, sk, \cdot)}(pk) \text{ s.t. } m_0, m_1 \in \mathcal{M} \\
\quad b \leftarrow \{0,1\}; c^* \leftarrow \mathsf{Enc}(pk, m_b) \\
\quad b' \leftarrow \mathcal{A}_2^{\mathcal{O}'_{dec}(pk, sk, \cdot)}(c^*, z) \\
\quad \text{Output } b'
\end{array}
$$

Where $\mathcal{O}_{dec}$ is an oracle that, on input a ciphertext c, returns $\mathsf{Dec}(pk, sk, c)$ and $\mathcal{O}'_{dec}$ operates identically but returns $\perp$ whenever $c = c^*$. We define $\mathcal{A}$'s advantage in the above game by:

$$|\mathbf{Pr}\left[\, b = b' \,\right] - 1/2|$$

Additional security properties. A secure blind decryption scheme must possess the additional properties of *leak-freeness* and *blindness.* Intuitively, leak-freeness [28] ensures that an adversarial User gains no more information from the blind decryption protocol than she would from access to a standard decryption oracle. Blindness prevents a malicious Decryptor from learning *which* ciphertext a User is attempting to decrypt. Let us now formally state these properties.

Definition 3 (Leak-Freeness [28]). A protocol BlindDec associated with a PKE scheme $\Pi = (\mathsf{KG}, \mathsf{Enc}, \mathsf{Dec})$ is *leak free* if for all *p.p.t.* adversaries $\mathcal{A}$, there exists an efficient simulator $\mathcal{S}$ such that for every value λ, no *p.p.t.* distinguisher D can distinguish the output of Game Real from Game Ideal with non-negligible advantage:

Game Real: Run $(pk, sk) \leftarrow \mathsf{KG}(1^\lambda)$ and publish pk. As many times as D wants, $\mathcal{A}$ chooses a ciphertext C and atomically executes the BlindDec protocol with $\mathcal{D}$:
$\mathsf{BlindDec}(\{\mathcal{U}(pk, C)\}, \{\mathcal{D}(pk, sk)\})$. $\mathcal{A}$'s output (which is the output of the game) includes the list of ciphertexts and decrypted plaintexts.

Game Ideal: A trusted party runs $(pk, sk) \leftarrow \mathsf{KG}(1^\lambda)$ and publishes pk. As many times as D wants, $\mathcal{S}$ chooses a ciphertext C and queries the trusted party to obtain the output of $\mathsf{Dec}(pk, sk, C)$, if $C \in \mathcal{C}$ and $\perp$ otherwise. $\mathcal{S}$'s output (which is the output of the game) includes the list of ciphertexts and decrypted plaintexts.

In the games above, BlindDec and Dec are treated as atomic operations. Hence D and $\mathcal{A}$ (or $\mathcal{S}$) may communicate at any time except during the execution of those protocols. Additionally, while we do not explicitly specify that auxiliary information is given to the parties, this information must be provided in order to achieve a sequential composition property.

Definition 4 (Ciphertext Blindness). Let $\mathcal{O}_\mathcal{U}(pk, C)$ be an oracle that, on input a public key and ciphertext, initiates the User's portion of the BlindDec protocol, interacting with an adversary. A protocol $\mathsf{BlindDec}(\mathcal{U}(\cdot,\cdot)\ \mathcal{A}(\cdot,\cdot))$ is Blind secure if every *p.p.t.* adversary $\mathcal{A} = (\mathcal{A}_1, \mathcal{A}_2)$ has advantage $\leq \nu(\lambda)$ in the following game.

$$\begin{array}{l}\mathsf{Blind}(\mathsf{BlindDec}, \mathcal{A}, \lambda)\\ \quad (pk, C_0, C_1, z) \leftarrow \mathcal{A}_1(1^\lambda)\\ \quad b \leftarrow \{0,1\}; b' \leftarrow \mathcal{A}_2^{\mathcal{O}_\mathcal{U}(pk, C_b), \mathcal{O}_\mathcal{U}(pk, C_{b-1})}(z)\end{array}$$

We define $\mathcal{A}$'s advantage in the above game as: $|\Pr[b' = b] - 1/2|$. Note that a stronger notion of blindness is *selective-failure* blindness, which was proposed by Camenisch *et al.* [15]. While our constructions do not natively achieve this definition, in section 4.1 we discuss techniques for achieving this stronger definition.

Definition 5 (CCA2-secure Blind Decryption). A blind decryption scheme Π = (KG, Enc, Dec, BlindDec) is IND-CCA2-secure if and only if: (1) (KG, Enc, Dec) is IND-CCA2-secure, (2) BlindDec is leak free, and (3) BlindDec possesses the property of ciphertext blindness.

4 Constructions

We now present a blind decryption scheme BCS that is secure under the Decision Linear and Flexible Diffie-Hellman assumptions. BCS is based on a variant of Cramer-Shoup that was proposed by Shacham [48], with significant extensions to permit blind decryption.

The core algorithms. We now describe the algorithms (KG, Enc, Dec), which are responsible for key generation, encryption and decryption respectively. BCS encrypts elements of $\mathbb{G}$, which may necessitate an encoding scheme from other message spaces (see *e.g.*, [3]).

$\mathsf{BCS.KG}(1^\lambda)$. First sample $\gamma = (p, \mathbb{G}, \mathbb{G}_T, \hat{e}, g \in \mathbb{G}) \leftarrow \mathsf{BMsetup}(1^\lambda)$. Choose $g, g_1, g_2, g_3, v', d', u'_1, \ldots, u'_5 \xleftarrow{R} \mathbb{G}$, and $x_1, x_2, x_3, y_1, y_2, y_3, z_1, z_2, z_3 \xleftarrow{R} \mathbb{Z}_p^*$ and compute:

$$c_1 \leftarrow g_1^{x_1} g_3^{x_3} \qquad d_1 \leftarrow g_1^{y_1} g_3^{y_3} \qquad h_1 \leftarrow g_1^{z_1} g_3^{z_3}$$
$$c_2 \leftarrow g_2^{x_2} g_3^{x_3} \qquad d_2 \leftarrow g_2^{y_2} g_3^{y_3} \qquad h_2 \leftarrow g_2^{z_2} g_3^{z_3}$$

Output $pk = (\gamma, g, g_1, g_2, g_3, c_1, c_2, d_1, d_2, h_1, h_2, v', d', u'_1, \ldots, u'_5)$, $sk = (x_1, x_2, x_3, y_1, y_2, y_3, z_1, z_2, z_3)$.

$\mathsf{BCS.Enc}(pk, m \in \mathbb{G})$. Select $\alpha, r_1, r_2, c, \psi \xleftarrow{R} \mathbb{Z}_p^*$. Construct a FS keypair (vk_1, sk_1) and a second "workalike" keypair (vk_2, sk_2) as follows:

$$vk_1 \leftarrow (\gamma, g, g^\alpha, v', d', g_1, g_2, g_3, g, g, u'_1, \ldots, u'_5, 5) \qquad vk_2 = (\gamma, g, g^\psi, v', d', g, g^c, 1)$$
$$sk_1 \leftarrow (vk_1, \alpha) \qquad sk_2 = (vk_2, \psi, c)$$

Next, compute the ciphertext $C = (u_1, u_2, u_3, e, v, vk, e_1, e_2, e_3, f_1, f_2, \sigma_1, \sigma_2)$ as:

$$u_1 \leftarrow g_1^{r_1} \quad u_2 \leftarrow g_2^{r_2} \quad u_3 \leftarrow g_3^{r_1+r_2} \quad e \leftarrow m \cdot h_1^{r_1} h_2^{r_2} \quad v \leftarrow (c_1 d_1^\alpha)^{r_1} \cdot (c_2 d_2^\alpha)^{r_2}$$
$$vk \leftarrow g^\alpha \quad e_1 \leftarrow u_1^\alpha \quad e_2 \leftarrow u_2^\alpha \quad e_3 \leftarrow u_3^\alpha \quad f_1 \leftarrow g^c \quad f_2 \leftarrow g^\psi$$
$$\sigma_1 \leftarrow \mathsf{FS.Sign}\left(sk_1, (r1, r2, r1 + r2, c, \psi)\right) \quad \sigma_2 \leftarrow \mathsf{FS.WASign}\left(sk_2, e\right)$$

$\mathsf{BCS.Dec}(pk, sk, C)$. Parse sk and C as above. Assemble $vk_1 \leftarrow (\gamma, g, vk, v', d', g_1, g_2, g_3, g, g, u'_1, \ldots, u'_5, 5)$ and $vk_2 \leftarrow (\gamma, g, f_2, v', d', g, f_1, 1)$. Now, verify the relations:

$$\begin{aligned} &\{\hat{e}(vk, u_i) = \hat{e}(e_i, g)\}_{i \in [1,3]} \wedge \\ \mathsf{FS.Verify}\left(vk_1, (u_1, u_2, u_3, f_1, f_2), \sigma_1\right) = 1 \wedge &\ \mathsf{FS.Verify}\left(vk_2, (e), \sigma_2\right) = 1\} \end{aligned} \tag{1}$$

If this check fails, output $\perp$. Otherwise, parse $sk = (x_1, x_2, x_3, y_1, y_2, y_3, z_1, z_2, z_3)$ and select $z \xleftarrow{R} \mathbb{Z}_p^*$. Compute the decryption m' as:

$$m' = e \cdot \frac{(u_1^{x_1} e_1^{y_1} \cdot u_2^{x_2} e_2^{y_2} \cdot u_3^{x_3} e_3^{y_3})^z}{u_1^{z_1} u_2^{z_2} u_3^{z_3} \cdot v^z} \tag{2}$$

Ciphertexts consist of approximately 25 elements of $\mathbb{G}$ and two element of $\mathbb{Z}_p^*$. While at first glance these ciphertexts may seem large, note that the scheme can be instantiated in asymmetric bilinear settings such as the MNT group of elliptic curves, where group elements can be represented in as little as 170 bits at the 80-bit security level. In this setting we are able to achieve a relatively ciphertext size of approximately 5100 bits. While this is large compared to RSA, a 640-byte per file overhead is quite reasonable for many practical applications. Also note that in our description the KG algorithm samples a unique set of bilinear group parameters γ for each key; however, it is perfectly acceptable for many keyholders to share the same group parameters.

The Blind Decryption Protocol. The blind decryption protocol $\mathsf{BlindDec}$ with respect to BCS is shown in Figure 1. The protocol requires a multiplicatively homomorphic IND-CPA-secure encryption scheme, which we instantiate using the Linear Encryption scheme (LE) of Boneh *et al.* [7].[6]

The protocol employs the homomorphic property of LE to construct a a two-party implementation of the Dec algorithm, with ZKPoKs used to ensure that both the User and Decryptor's contributions are correctly formed. Note that for security reasons it is critical that the Decryptor *re-randomize* the ciphertext that it sends back to the User in its portion of the protocol. In the LE scheme this can be accomplished by multiplying a ciphertext with a fresh encryption of the identity element.

SECURITY. Let Π_{ZK} be a zero-knowledge (and, implicitly, witness indistinguishable) proof system secure under the Decision Linear assumption (possibly in the Common Reference String model). In the following theorems we will show that if the Decision Linear and Flexible Diffie-Hellman assumptions hold in $\mathbb{G}$ then $\mathsf{BCS} = (\mathsf{KG}, \mathsf{Enc}, \mathsf{Dec}, \mathsf{BlindDec})$ implemented with Π_{ZK} is a secure blind decryption scheme in the sense of Definition 5. To accomplish this we must show that: (1) the algorithms (KG, Enc, Dec) comprise an IND-CCA2-secure encryption scheme, (2) the BlindDec protocol is leak-free, and (3) BlindDec achieves ciphertext blindness.

Theorem 1. *If the Decision Linear and Flexible Diffie-Hellman assumptions hold in* $\mathbb{G}$*, then* (BCS.KG, BCS.Enc, BCS.Dec) *comprise an* IND-CCA2-*secure public-key encryption scheme secure in the standard model.*

Due to space concerns, we must leave a full proof of Theorem 1 to the full version of this work [27]. Here we will sketch the intuition behind the proof, which employs techniques from the Cramer-Shoup variant proposed by Shacham [48]. As in that scheme, our simulator knows the scheme's secret key, and can use it to answer decryption queries. The exceptions to this rule are certain queries related to the challenge ciphertext. Specifically, we must be careful with queries that are (*a*) "malformed", *i.e.*, the queried value $v \neq u_1^{x_1} e_1^{y_1} \cdot u_2^{x_2} e_2^{y_2} \cdot u_3^{x_3} e_3^{y_3}$, or that (*b*) embed the value vk^* from the challenge ciphertext.

Note that equation (2) of the Dec algorithm ensures that malformed ciphertexts decrypt to a random element of $\mathbb{G}$, so the first case is easily dealt with in our simulation. The adversary cannot maul the ciphertext due to the presence of the checksum v. Thus it remains to consider well-formed ciphertexts with $vk = vk^*$. We argue that the challenge ciphertext itself is the only ciphertext that will pass all of our checks.

Intuitively our simulation accomplishes this by setting $vk = g^{\alpha^*}$ as the public key of a strongly unforgeable OTS which is secure under the Flexible Diffie-Hellman assumption. In principle we use this key to sign the challenge ciphertext components $(u_1^*, u_2^*, u_3^*, e^*)$, which produces all of the remaining components of the ciphertext. When the adversary submits a decryption query with $vk = vk^*$

[6] In asymmetric bilinear groups where the Decisional Diffie-Hellman problem is hard, this can easily be replaced with Elgamal encryption, resulting in a significant efficiency improvement.

$\mathcal{U}(pk, C)$	$\mathcal{D}(pk, sk)$

1. Parse C as $(u_1, u_2, u_3, e, v, vk, e_1, e_2, e_3, f_1, f_2, \sigma_1, \sigma_2)$, and parse $pk = (\gamma, g, g_1, g_2, g_3, c_1, c_2, d_1, d_2, h_1, h_2, v', d', u'_1, \ldots, u'_5)$. Verify that C satisfies equation (1) of the Dec algorithm. If not, abort and output $\perp$.
2. Generate $(pk_{\mathcal{U}}, sk_{\mathcal{U}}) \leftarrow \mathsf{LE.KG}(\gamma)$ and select $\bar{z} \xleftarrow{R} \mathbb{Z}_p^*$. Compute:
   $\mathbf{c}_1 \leftarrow \mathsf{LE.Enc}(pk_{\mathcal{U}}, u_1^{\bar{z}})$, $\mathbf{c}_2 \leftarrow \mathsf{LE.Enc}(pk_{\mathcal{U}}, u_2^{\bar{z}})$, $\mathbf{c}_3 \leftarrow \mathsf{LE.Enc}(pk_{\mathcal{U}}, u_3^{\bar{z}})$,
   $\mathbf{c}_4 \leftarrow \mathsf{LE.Enc}(pk_{\mathcal{U}}, e_1^{\bar{z}})$, $\mathbf{c}_5 \leftarrow \mathsf{LE.Enc}(pk_{\mathcal{U}}, e_2^{\bar{z}})$, $\mathbf{c}_6 \leftarrow \mathsf{LE.Enc}(pk_{\mathcal{U}}, e_3^{\bar{z}})$,
   $\mathbf{c}_7 \leftarrow \mathsf{LE.Enc}(pk_{\mathcal{U}}, v^{\bar{z}})$ and set $vk_1 \leftarrow (\gamma, g, vk, v', d', g_1, g_2, g_3, g, g, u'_1, \ldots, u'_5, 5)$,
   $vk_2 \leftarrow (\gamma, g, f_2, v', d', g, f_1, 1)$
3. Send $pk_{\mathcal{U}}, \mathbf{c}_1, \ldots, \mathbf{c}_7$ to $\mathcal{D}$ and conduct the following PoK with $\mathcal{D}$:
   $\mathsf{WIPoK}\{(u_1, u_2, u_3, v, vk, e_1, e_2, e_3, f_1, f_2, \sigma_1, \sigma_2, vk_1, vk_2, \bar{z})$:
   $\mathbf{c}_1 = \mathsf{LE.Enc}(pk_{\mathcal{U}}, u_1^{\bar{z}}) \wedge \mathbf{c}_2 = \mathsf{LE.Enc}(pk_{\mathcal{U}}, u_2^{\bar{z}}) \wedge \mathbf{c}_3 = \mathsf{LE.Enc}(pk_{\mathcal{U}}, u_3^{\bar{z}}) \wedge$
   $\mathbf{c}_4 = \mathsf{LE.Enc}(pk_{\mathcal{U}}, e_1^{\bar{z}}) \wedge \mathbf{c}_5 = \mathsf{LE.Enc}(pk_{\mathcal{U}}, e_2^{\bar{z}}) \wedge \mathbf{c}_6 = \mathsf{LE.Enc}(pk_{\mathcal{U}}, e_3^{\bar{z}}) \wedge$
   $\mathbf{c}_7 = \mathsf{LE.Enc}(pk_{\mathcal{U}}, v^{\bar{z}}) \ \wedge\ \hat{e}(vk, u_i) = \hat{e}(e_i, g)\}_{i \in [1,3]} \ \wedge$
   $\mathsf{FS.Verify}(vk_1, (u_1, u_2, u_3, f_1, f_2), \sigma_1) = 1 \ \wedge\ \mathsf{FS.Verify}(vk_2, (e), \sigma_2) = 1\}$
4. If the proof does not verify, abort.
5. Compute $\mathbf{c}' = \mathsf{LE.Enc}(pk_{\mathcal{U}}, 1)$, $\bar{z}' \xleftarrow{R} \mathbb{Z}_p^*$.
6. Using the homomorphic property of LE,
   compute: $\mathbf{c}'' \leftarrow \frac{(\mathbf{c}_1^{x_1}\mathbf{c}_4^{y_1} \cdot \mathbf{c}_2^{x_2}\mathbf{c}_5^{y_2} \cdot \mathbf{c}_3^{x_3}\mathbf{c}_6^{y_3})^{\bar{z}'}}{\mathbf{c}_1^{z_1}\mathbf{c}_2^{z_2}\mathbf{c}_3^{z_3} \cdot \mathbf{c}_7^{\bar{z}'}} \cdot \mathbf{c}'$.
7. Return $\mathbf{c}''$ and conduct the following proof:
   $ZKPoK\{(x_1, x_2, x_3, y_1, y_2, y_3, z_1, z_2, z_3, \bar{z}', \mathbf{c}')$:
   $\mathbf{c}' = \mathsf{LE.Enc}(pk, 1) \wedge$
   $\mathbf{c}'' = \frac{(\mathbf{c}_1^{x_1}\mathbf{c}_4^{y_1} \cdot \mathbf{c}_2^{x_2}\mathbf{c}_5^{y_2} \cdot \mathbf{c}_3^{x_3}\mathbf{c}_6^{y_3})^{\bar{z}'}}{\mathbf{c}_1^{z_1}\mathbf{c}_2^{z_2}\mathbf{c}_3^{z_3} \cdot \mathbf{c}_7^{\bar{z}'}} \cdot \mathbf{c}'\}$
8. If the proof does not verify, abort and return $\perp$.
9. Compute $m' = e \cdot (\mathsf{LE.Dec}(sk, \mathbf{c}''))^{1/\bar{z}}$.

Output m'.	Output nothing.

Fig. 1. The Blind Decryption protocol $\mathsf{BlindDec}(\mathcal{U}(pk, C), \mathcal{D}(pk, sk)) \rightarrow (m', \mathsf{nothing})$. For compactness of notation we represent the homomorphic operation on two LE ciphertexts $\mathbf{c}_1, \mathbf{c}_2$ using simple multiplicative notation $(\mathbf{c}_1\mathbf{c}_2)$, and exponentiation by a scalar z as $\mathbf{c}_1^{z}$.

we can be assured that the query is identical to the challenge ciphertext, as any other result would require the adversary to forge the OTS.

It remains to separately argue that the signatures σ_1, σ_2 are unforgeable. This is non-trivial, since the OTS operates on messages of the form $m_1, \ldots, m_n \in \mathbb{Z}_p^*$. In a separate simulation we could select the elements u_1^*, u_2^*, u_3^* such the simulator knows their discrete logarithm base g. Unfortunately, even this is not sufficient, since our simulator cannot always know the discrete logarithm of the value e^* which is based on a message chosen by the adversary. The core intuition of our proof is to give two separate simulations: in one the signing key α^* is known and we can *simulate* the signature, producing a correctly-distributed (but not unforgeable) signature over arbitrary group elements. In the second simulation the signing key is unknown: the simulator chooses $(u_1^*, u_2^*, u_3^*, e^*)$ at random such that it knows the discrete logarithm (base g) of each value. Although the resulting ciphertext does not encrypt either m_0 *or* m_1, an adversary is unable to detect this condition under the Decision Linear assumption.

Theorem 2. *If the Decision Linear assumption holds in $\mathbb{G}$ and Π_{ZK} is secure under the Decision Linear assumption, then the* BCS *protocol* BlindDec *is leak-free.*

We present a proof sketch of Theorem 2 in the full version of this work [27]. Intuitively this proof is quite simple: we show that for any real-world adversary $\mathcal{A}$ we can construct an ideal-world adversary $\mathcal{S}$ that, whenever $\mathcal{A}$ initiates the BlindDec protocol, operates as follows: (1) $\mathcal{S}$ uses the extractor for the PoK system to obtain $\mathcal{A}$'s requested ciphertext, (2) queries this result to the trusted decryption oracle, (3) re-blinds and returns the correctly formulated result to the adversary, simulating the necessary ZK proofs. We show that under the Decision Linear assumption no *p.p.t.* distinguisher can differentiate the output of $\mathcal{S}$ playing the Ideal-World game from the output of $\mathcal{A}$ in the Real-World game except with negligible probability.

Theorem 3. *If the Decision Linear assumption holds in $\mathbb{G}$ and Π_{ZK} is secure under the Decision Linear assumption, then the* BCS *protocol* BlindDec *satisfies the property of Ciphertext Blindness (*Blind*).*

We sketch a proof of Theorem 3 in the full version of this work [27]. Intuitively, we show that an adversarial Decryptor who distinguishes the User's exection of the blind decryption protocol on two distinct (and adversarially-chosen) ciphertexts C_0 and C_1 must imply a distinguisher for the witness indistinguishable proof system, or a CPA adversary against the LE encryption scheme.

4.1 Extensions

Tag-Based Encryption. Tag-Based Encryption (TBE) allows encryptors to apply a tag (label) to each ciphertext. This tag is used during the decryption process. The BCS construction is in fact natively based on a TBE scheme, but this functionality is lost as part of the TBE-to-PKE transform we use. In the full version of this work [27] we show that with some minor extensions it is possible to retain the scheme's full TBE functionality.

Selective-failure blindness. Camenisch *et al.* [15] propose a stronger definition of blindness (for signature schemes) that they refer to as "selective-failure" blindness. Intuitively, this definition captures the notion that an adversarial Decryptor might attempt to induce failures in the protocol (*e.g.*, by generating malformed ciphertexts) in order to deprive the User of privacy. Unfortunately our protocols do not natively achieve this definition because the Decryptor can create ciphertexts with an improperly formed check value v. Unfortunately, due to the nature of our scheme this check cannot be verified independently by the user. One potential solution to this problem is to add to each ciphertext a non-interactive proof that v is correctly formed. Such a proof could be constructed using the Fiat-Shamir heuristic in the random oracle model, or using the Groth-Sahai system in the Common Reference String model. Note that this approach would not require any changes to the blind decryption protocol. We elaborate on this approach in the full version of this work [27].

5 Applications

Blind decryption has applications to a number of privacy-preserving protocols. Several applications have already been proposed in the literature, *e.g.,* [43, 24]. Below we propose two specific applications motivated by our construction.

Privacy-preserving Distributed Filesystems. Many organizations are responding to the difficulty of securing data in a distributed network, where storage locations can include semi-trusted file servers, desktop computers and mobile devices. An increasingly popular approach is to employ cryptographic access control to restrict and monitor file access in these environments. In this approach (*e.g.,* [1]), access control is performed by encrypting files at rest; authorized users contact a centralized server in order to decrypt them when necessary.

A concern with this approach is that the server gains a great deal of information regarding users' access patterns. In some cases, knowing *which* content a user is accessing may by itself leak confidential information. For example, the pattern of file accesses by executives during a corporate merger might have enormous financial value to an investor. While it is desirable to centralize access control, it may also be important to restrict this centralized party from learning which information is being managed. While these goals seems contradictory, Coull *et al.* [20] and Camenisch *et al.* [12] recently showed how to construct sophisticated access control mechanisms using anonymous credentials. In these protocols a server provides strong, and even *history-dependent* access control without ever learning user's access pattern. Our blind decryption protocols are amenable to integration with these access control techniques. In particular, by extending BCS to include *encryption tags* as in Section 4.1, data can be explicitly categorized and policies can be defined around these categories.

Oblivious Transfer with Public Contribution. In an adaptively-secure k-out-of-N Oblivious Transfer protocol ($\mathsf{OT}^N_{k\times 1}$) a Receiver obtains up to k items from a Sender's N-item database, without revealing to the Sender *which* messages were transferred. There has been much recent interest in $\mathsf{OT}^N_{k\times 1}$ [15, 28, 29, 33, 44, 20, 12], as it is particularly well suited for constructing privacy-preserving databases in which the user's query pattern is cryptographically protected (this is critical in *e.g.,* patent and medical databases).

For practical reasons, there are situations in which it is desirable to distribute the authorship of records, particularly when database updates are performed offline. Unfortunately, existing $\mathsf{OT}^N_{k\times 1}$ protocols seem fundamentally incapable of supporting message contributions by third parties without the explicit cooperation of the Sender. Our blind decryption constructions admit new $\mathsf{OT}^N_{k\times 1}$

protocols that support *public contribution*. Intuitively, contributors simply encrypt their messages using the Enc algorithm under the Sender's public key and send the resulting ciphertexts directly to the Receiver. The Receiver can then obtain up to k decryptions by running BlindDec with the Sender. Proving this intuitive protocol secure under a strong simulation-based definition [15, 28] requires some additional components that are easily achieved using the techniques available to us.

Acknowledgements. The author would like to thank Susan Hohenberger for her helpful comments.

References

1. Eruces Tricryption, `http://www.eruces.com/index.php`
2. Adida, B., Hohenberger, S., Rivest, R.L.: Ad-hoc group signatures from hijacked keypairs. In: DIMACS Workshop on Theft in E-Commerce (preliminary version) (April 2005)
3. Ateniese, G., Camenisch, J., de Medeiros, B.: Untraceable RFID tags via insubvertible encryption. In: CCS 2005, pp. 92–101. ACM Press, New York (2005)
4. Belenkiy, M., Chase, M., Kohlweiss, M., Lysyanskaya, A.: P-signatures and Noninteractive Anonymous Credentials. In: Canetti, R. (ed.) TCC 2008. LNCS, vol. 4948, pp. 356–374. Springer, Heidelberg (2008)
5. Bellare, M., Rogaway, P.: Optimal asymmetric encryption padding — how to encrypt with rsa. In: De Santis, A. (ed.) EUROCRYPT 1994. LNCS, vol. 950, pp. 92–111. Springer, Heidelberg (1995)
6. Boneh, D., Boyen, X.: Efficient Selective-ID Secure Identity-Based Encryption Without Random Oracles. In: Cachin, C., Camenisch, J.L. (eds.) EUROCRYPT 2004. LNCS, vol. 3027, pp. 223–238. Springer, Heidelberg (2004)
7. Boneh, D., Boyen, X., Shacham, H.: Short Group Signatures. In: Franklin, M. (ed.) CRYPTO 2004. LNCS, vol. 3152, pp. 41–55. Springer, Heidelberg (2004)
8. Boneh, D., Katz, J.: Improved Efficiency for CCA-Secure Cryptosystems Built Using Identity-Based Encryption. In: Menezes, A. (ed.) CT-RSA 2005. LNCS, vol. 3376, pp. 87–103. Springer, Heidelberg (2005)
9. Boneh, D., Sahai, A., Waters, B.: Fully Collusion Resistant Traitor Tracing with Short Ciphertexts and Private Keys. In: Vaudenay, S. (ed.) EUROCRYPT 2006. LNCS, vol. 4004, pp. 573–592. Springer, Heidelberg (2006)
10. Boyen, X., Mei, Q., Waters, B.: Direct chosen ciphertext security from identity-based techniques. In: CCS 2005, pp. 320–329. ACM Press, New York (2005)
11. Brassard, G., Crépeau, C., Robert, J.M.: All-or-Nothing Disclosure of Secrets. In: Odlyzko, A.M. (ed.) CRYPTO 1986. LNCS, vol. 263, pp. 234–238. Springer, Heidelberg (1987)
12. Camenisch, J., Dubovitskaya, M., Neven, G.: Oblivious transfer with access control. In: CCS 2009, pp. 131–140. ACM, New York (2009)
13. Camenisch, J., Lysyanskaya, A.: Signature Schemes and Anonymous Credentials from Bilinear Maps. In: Franklin, M. (ed.) CRYPTO 2004. LNCS, vol. 3152, pp. 56–72. Springer, Heidelberg (2004)
14. Camenisch, J., Michels, M.: Proving in zero-knowledge that a number n is the product of two safe primes. In: Stern, J. (ed.) EUROCRYPT 1999. LNCS, vol. 1592, pp. 107–122. Springer, Heidelberg (1999)

15. Camenisch, J., Neven, G., Shelat, A.: Simulatable Adaptive Oblivious Transfer. In: Naor, M. (ed.) EUROCRYPT 2007. LNCS, vol. 4515, pp. 573–590. Springer, Heidelberg (2007)
16. Camenisch, J., Stadler, M.: Efficient Group Signature Schemes for Large Groups. In: Kaliski Jr., B.S. (ed.) CRYPTO 1997. LNCS, vol. 1294, pp. 410–424. Springer, Heidelberg (1997)
17. Camenisch, J.L.: Group Signature Schemes and Payment Systems Based on the Discrete Logarithm Problem. PhD thesis, ETH Zürich (1998)
18. Canetti, R., Halevi, S., Katz, J.: A forward-secure public-key encryption scheme. In: Biham, E. (ed.) EUROCRYPT 2003. LNCS, vol. 2656, pp. 255–271. Springer, Heidelberg (2003)
19. Chaum, D.: Blind signatures for untraceable payments. In: CRYPTO 1982, pp. 199–203. Plenum Press, New York (1982)
20. Coull, S., Green, M., Hohenberger, S.: Controlling Access to an Oblivious Database Using Stateful Anonymous Credentials. In: Jarecki, S., Tsudik, G. (eds.) PKC 2009. LNCS, vol. 5443, pp. 501–520. Springer, Heidelberg (2009)
21. Cramer, R., Damgård, I., Schoenmakers, B.: Proof of Partial Knowledge and Simplified Design of Witness Hiding Protocols. In: Desmedt, Y.G. (ed.) CRYPTO 1994. LNCS, vol. 839, pp. 174–187. Springer, Heidelberg (1994)
22. Cramer, R., Shoup, V.: A Practical Public Key Cryptosystem Provably Secure against Adaptive Chosen Ciphertext Attack. In: Krawczyk, H. (ed.) CRYPTO 1998. LNCS, vol. 1462, pp. 13–25. Springer, Heidelberg (1998)
23. Damgård, I., Mambo, M., Okamoto, E.: Further study on the transformability of digital signatures and the blind decryption. In: SCIS 1997-33B (1997)
24. Dodis, Y., Halevi, S., Rabin, T.: A Cryptographic Solution to a Game Theoretic Problem. In: Bellare, M. (ed.) CRYPTO 2000. LNCS, vol. 1880, pp. 112–130. Springer, Heidelberg (2000)
25. Fujisaki, E., Okamoto, T.: Secure Integration of Asymmetric and Symmetric Encryption Schemes. In: Wiener, M. (ed.) CRYPTO 1999. LNCS, vol. 1666, pp. 537–554. Springer, Heidelberg (1999)
26. Goldreich, O., Micali, S., Wigderson, A.: How to play any mental game or a completeness theorem for protocols with honest majority. In: STOC 1987, pp. 218–229 (1987)
27. Green, M.: Secure blind decryption (full version) (December 2010), `http://eprint.iacr.org`, `http://spar.isi.jhu.edu/~mgreen/SecureBlindDecryption.pdf`
28. Green, M., Hohenberger, S.: Blind Identity-Based Encryption and Simulatable Oblivious Transfer. In: Kurosawa, K. (ed.) ASIACRYPT 2007. LNCS, vol. 4833, pp. 265–282. Springer, Heidelberg (2007)
29. Green, M., Hohenberger, S.: Universally Composable Adaptive Oblivious Transfer. In: Pieprzyk, J. (ed.) ASIACRYPT 2008. LNCS, vol. 5350, pp. 179–197. Springer, Heidelberg (2008)
30. Green, M., Hohenberger, S.: Practical adaptive oblivious transfer from a simple assumption. In: TCC 2011. LNCS, vol. 6597. Springer, Heidelberg (to appear), `http://eprint.iacr.org/2010/109`
31. Groth, J., Sahai, A.: Efficient Non-interactive Proof Systems for Bilinear Groups. In: Smart, N.P. (ed.) EUROCRYPT 2008. LNCS, vol. 4965, pp. 415–432. Springer, Heidelberg (2008)
32. Hohenberger, S., Rothblum, G.N., Shelat, A., Vaikuntanathan, V.: Securely Obfuscating Re-encryption. In: Vadhan, S.P. (ed.) TCC 2007. LNCS, vol. 4392, pp. 233–252. Springer, Heidelberg (2007)

33. Jarecki, S., Liu, X.: Efficient Oblivious Pseudorandom Function with Applications to Adaptive OT and Secure Computation of Set Intersection. In: Reingold, O. (ed.) TCC 2009. LNCS, vol. 5444, pp. 577–594. Springer, Heidelberg (2009)
34. Kilian, J.: Founding cryptography on oblivious transfer. In: STOC 1988, pp. 20–31 (1988)
35. Kiltz, E.: Chosen-ciphertext security from tag-based encryption. In: Halevi, S., Rabin, T. (eds.) TCC 2006. LNCS, vol. 3876, pp. 581–600. Springer, Heidelberg (2006)
36. Kunz-Jacques, S., Pointcheval, D.: About the Security of MTI/C0 and MQV. In: De Prisco, R., Yung, M. (eds.) SCN 2006. LNCS, vol. 4116, pp. 156–172. Springer, Heidelberg (2006)
37. Kurosawa, K., Nojima, R.: Simple Adaptive Oblivious Transfer without Random Oracle. In: Matsui, M. (ed.) ASIACRYPT 2009. LNCS, vol. 5912, pp. 334–346. Springer, Heidelberg (2009)
38. Laguillaumie, F., Paillier, P., Vergnaud, D.: Universally Convertible Directed Signatures. In: Roy, B. (ed.) ASIACRYPT 2005. LNCS, vol. 3788, pp. 682–701. Springer, Heidelberg (2005)
39. Libert, B., Vergnaud, D.: Multi-use unidirectional proxy re-signatures. In: CCS 2008, pp. 511–520. ACM, New York (2008)
40. MacKenzie, P.D., Reiter, M.K., Yang, K.: Alternatives to non-malleability: Definitions, constructions, and applications (extended abstract). In: Naor, M. (ed.) TCC 2004. LNCS, vol. 2951, pp. 171–190. Springer, Heidelberg (2004)
41. Mambo, M., Sakurai, K., Okamoto, E.: How to utilize the transformability of digital signatures for solving the oracle problem. In: Kim, K.-c., Matsumoto, T. (eds.) ASIACRYPT 1996. LNCS, vol. 1163, pp. 322–333. Springer, Heidelberg (1996)
42. Ogata, W., Le Trieu, P.: Blind HIBE and its application to blind decryption. In: SCIS, pp. 4D1–2 (2008)
43. Radia, P.: The ephemerizer: Making data disappear. Journal of Information System Security 1(1), 51–68 (2005)
44. Rial, A., Kohlweiss, M., Preneel, B.: Universally Composable Adaptive Priced Oblivious Transfer. In: Shacham, H., Waters, B. (eds.) Pairing 2009. LNCS, vol. 5671, pp. 231–247. Springer, Heidelberg (2009)
45. Sakurai, K., Yamane, Y.: Blind decoding, blind undeniable signatures, and their applications to privacy protection. In: Anderson, R. (ed.) IH 1996. LNCS, vol. 1174, pp. 257–264. Springer, Heidelberg (1996)
46. Schnorr, C.-P.: Efficient signature generation for smart cards. Journal of Cryptology 4(3), 239–252 (1991)
47. Schnorr, C.-P., Jakobsson, M.: Security of Signed ElGamal Encryption. In: Okamoto, T. (ed.) ASIACRYPT 2000. LNCS, vol. 1976, pp. 73–89. Springer, Heidelberg (2000)
48. Shacham, H.: A Cramer-Shoup encryption scheme from the linear assumption and from progressively weaker linear variants. Cryptology ePrint Archive, Report 2007/074 (2007), `http://eprint.iacr.org/`
49. Teague, V.: Selecting Correlated Random Actions. In: Juels, A. (ed.) FC 2004. LNCS, vol. 3110, pp. 181–195. Springer, Heidelberg (2004)
50. Yao, A.: How to generate and exchange secrets. In: FOCS 1986, pp. 162–167 (1986)

New Developments in Leakage-Resilient Cryptography

Vinod Vaikuntanathan

Microsoft Research
vinodv@alum.mit.edu

Abstract. Much of modern cryptography is predicated on the assumption that users have secrets which are generated using perfect randomness, and kept perfectly secret from an attacker. The attacker is then constrained to black-box (input/output) access to the user's program. In reality, neither assumption holds, as evidenced by numerous side-channel attacks that have surfaced over the last few decades.

This leads naturally to the question – is it possible to secure cryptography against general types of information leakage at a fundamental, algorithmic level (as opposed to, say, solutions for specific attacks)? This is the goal of leakage-resilient cryptography.

In this talk, we will survey recent developments in leakage-resilient cryptography, including definitions and constructions of various cryptographic primitives secure against general forms of leakage. We will place particular emphasis on the new tools and techniques that we have developed to handle information leakage, as well as the relation between leakage-resilience and other questions in cryptography.

D. Catalano et al. (Eds.): PKC 2011, LNCS 6571, p. 283, 2011.

On the Security of a Bidirectional Proxy Re-encryption Scheme from PKC 2010

Jian Weng[1,2,3], Yunlei Zhao[4,*], and Goichiro Hanaoka[5]

[1] Department of Computer Science, Jinan University, Guangzhou, China
[2] State Key Laboratory of Networking and Switching Technology
Beijing University of Posts and Telecommunications, Beijing, China
[3] State Key Laboratory of Information Security
Institute of Software, Chinese Academy of Sciences, Beijing, China
[4] Software School, Fudan University, Shanghai, China
[5] National Institute of Advanced Industrial Science and Technology, Tokyo, Japan
cryptjweng@gmail.com, yunleizhao@gmail.com, hanaoka-goichiro@aist.go.jp

Abstract. In ACM CCS 2007, Canetti and Hohenberger left an interesting open problem of how to construct a chosen-ciphertext secure proxy re-encryption (PRE) scheme without bilinear maps. This is a rather interesting problem and has attracted great interest in recent years. In PKC 2010, Matsuda, Nishimaki and Tanaka introduced a novel primitive named re-applicable lossy trapdoor function, and then used it to construct a PRE scheme without bilinear maps. Their scheme is claimed to be chosen-ciphertext secure in the standard model. In this paper, we make a careful observation on their PRE scheme, and indicate that their scheme does not satisfy chosen-ciphertext security. The purpose of this paper is to clarify the fact that, it is still an open problem to come up with a chosen-ciphertext secure PRE scheme without bilinear maps in the standard model.

Keywords: bilinear map, proxy re-encryption, chosen-ciphertext security, standard model.

1 Introduction

Proxy re-encryption (PRE), introduced by Blaze, Bleumer and Strauss [3] in Eurocrypt'98, allows a semi-trust proxy to translate a ciphertext intended for Alice into another ciphertext intended for Bob. The proxy, however, can not learn anything about the underlying messages. According to the direction of transformation, PRE can be categorized into *bidirectional* PRE, in which the proxy can transform ciphertexts from Alice to Bob and vice versa, and *unidirectional* PRE, in which the proxy cannot transform ciphertexts in the opposite direction. PRE can also be categorized into *multi-hop* PRE, in which the ciphertexts can be transformed from Alice to Bob and then to Charlie and so on, and *single-hop* PRE, in which the ciphertexts can only be transformed once.

* Corresponding author.

D. Catalano et al. (Eds.): PKC 2011, LNCS 6571, pp. 284–295, 2011.

In their seminal paper, Blaze *et al.* [3] proposed the first bidirectional PRE scheme. Ateniese *et al.* [1, 2] presented unidirectional PRE schemes from bilinear maps. All of these schemes are only secure against chosen-plaintext attacks (CPA). However, applications often require security against chosen-ciphertext attacks (CCA).

To fill this gap, Canetti and Hohenberger [7] presented the first CCA-secure bidirectional multi-hop PRE scheme in the standard model. Libert and Vergnaud [14, 13] proposed a unidirectional single-hop PRE scheme, which is replayable CCA-secure [8] in the standard model. Recently, Weng et al. [18] presented a unidirectional single-hop PRE scheme, which is CCA-secure against adaptive corruption of users in the standard model. These schemes rely on bilinear maps. In spite of the recent advances in implementation technique, compared with standard operations such as modular exponentiation in finite fields, the bilinear map computation is still considered as a rather expensive operation. It would be desirable for cryptosystems to be constructed without relying on pairings, especially in computation resource limited settings. Thus, in ACM CCS'07, Canetti and Hohenberger [7] left an open problem of how to construct a CCA-secure PRE scheme without bilinear maps.

Deng et al. [10, 19] presented a bidirectional single-hop PRE scheme without bilinear maps, and proved its CCA-security in the random oracle model. Shao et al. [17] presented a unidirectional single-hop PRE scheme without bilinear maps in the random oracle model, but their scheme was later identified a security flaw in [9]. Sherman et al. [9] presented a CCA-secure unidirectional single-hop PRE scheme without bilinear maps, again in the random oracle model. It is well-known [5, 6] that a proof in the random oracle model can only serve as an argument, which does not imply the security for real implementations. Thus, it is more desirable to come up with a CCA-secure PRE scheme without bilinear maps in the standard model.

In PKC 2010, Matsuda, Nishimaki and Tanaka made an important step and tried to construct such a scheme. They first introduced a new cryptographic primitive named re-applicable lossy trapdoor functions (re-applicable LTDFs), which are specialized lossy trapdoor functions [16, 11, 4, 12], and then used this primitive to construct a PRE scheme without bilinear maps. They claimed that their scheme is CCA-secure in the standard model. However, in this paper, we present a concrete attack, and indicate that their PRE scheme does not achieve the CCA-security. However, we stress that Matsuda et al.'s work is still considered as an important step in this research area. Namely, due to their scheme, we can figure out one of main difficulties for constructing CCA-secure PRE without using bilinear maps, and this would enable us to further design novel schemes which overcome the same problem.

2 Preliminaries

The Matsuda-Nishimaki-Tanaka PRE scheme involves the primitives of all-but-one trapdoor function and re-applicable (n, k) lossy trapdoor functions (LTDFs).

Thus in this section, we shall review the definitions of these two primitives (for more details, the reader is referred to [15] and [16]). We shall also review the definition and security notion for bidirectional multi-hop PRE.

2.1 All-but-One Trapdoor Function

Let $B = \{B_\lambda\}_{\lambda \in \mathbb{N}}$ be a collection of sets whose elements represents the branches. A collection of (n, k)-all-but-one trapdoor functions is a tuple of probabilistic polynomial time (PPT) algorithms $(\mathsf{G}_{\text{abo}}, \mathsf{F}_{\text{abo}}, \mathsf{F}_{\text{abo}}^{-1})$ having the following properties:

- **All-but-one property:** Given a lossy branch $b^* \in B_\lambda$, algorithm $\mathsf{G}_{\text{abo}}(1^\lambda, b^*)$ outputs a pair (s, td), where s is a function index and td is its trapdoor. For any $b \in B_\lambda \backslash \{b^*\}$, the algorithm $\mathsf{F}_{\text{abo}}(s, b, \cdot)$ computes an injective function $f_{s,b}(\cdot)$ over $\{0,1\}^n$, and $\mathsf{F}_{\text{abo}}^{-1}(td, b, \cdot)$ computes $f_{s,b}^{-1}(\cdot)$. For the lossy branch b^*, $\mathsf{F}_{\text{abo}}(s, b^*, \cdot)$ computes a lossy function $f_{s,b^*}(\cdot)$ over the domain $\{0,1\}^n$, where $|f_{s,b^*}(\{0,1\}^n)| \leq 2^{n-k}$.
- **Indistinguishability:** For every b_1^* and $b_2^* \in B_\lambda$, the first output s_0 of $\mathsf{G}_{\text{abo}}(1^\lambda, b_0^*)$ and the first output s_1 of $\mathsf{G}_{\text{abo}}(1^\lambda, b_1^*)$ are computationally indistinguishable.

2.2 Re-applicable (n, k)-Lossy Trapdoor Functions

A collection of re-applicable (n, k)-lossy trapdoor functions (LTDFs) with respect to function indices is a tuple of PPT algorithms (ParGen, LossyGen, LossyEval, LossyInv, ReIndex, ReEval, PrivReEval, Trans, FakeKey) such that:

Injectivity: For every public parameter $\mathsf{par} \leftarrow \mathsf{ParGen}(1^\lambda)$ and every tag $\tau \in \mathcal{T} \backslash \{\tau_{\text{los}}\}$, $\mathsf{LossyGen}(\tau)$ outputs a pair of a function index and its trapdoor (s, td), $\mathsf{LossyEval}(s, \cdot)$ computes an injective function $f_{s,\tau}(\cdot)$ over $\{0,1\}^n$, and $\mathsf{LossyInv}(td, \tau, \cdot)$ computes $f_{s,\tau}^{-1}(\cdot)$. (We represent the function $f_{s,\tau}$, not f_s, in order to clarify a tag τ. If we do not need to clarify a tag, we represent a function as $f_{s,\star}$).

Lossiness: For every public parameter $\mathsf{par} \leftarrow \mathsf{ParGen}(1^\lambda)$, $\mathsf{LossyGen}(\tau_{\text{los}})$ outputs $(s, \perp)$ and $\mathsf{LossyEval}(s, \cdot)$ computes a function $f_{s,\tau_{\text{los}}}(\cdot)$ over $\{0,1\}^n$, where $|f_{s,\tau_{\text{los}}}(\{0,1\}^n)| \leq 2^{n-k}$.

Indistinguishability between injective and lossy indices: Let X_λ denote the distribution of $(\mathsf{par}, s_{\text{inj}}, \tau)$, and Y_λ denote the distribution of $(par, s_{\text{los}}, \tau')$, where par is a public parameter from $\mathsf{ParGen}(1^\lambda)$, τ and τ' are random elements in $\mathcal{T}$, and the function indices s_{inj} and s_{los} are the first element output from $\mathsf{LossyGen}(\tau)$ and $\mathsf{LossyGen}(\tau_{\text{los}})$ respectively. Then, $\{X_\lambda\}$ and $\{Y_\lambda\}$ are computationally indistinguishable.

Re-applying with respect to function indices: Let τ_i and τ_j be any tags with $\tau_i \neq \tau_{\text{los}}$ and $\tau_j \neq \tau_{\text{los}}$. The algorithm $\mathsf{ReIndex}(td_i, td_j)$ outputs $s_{i \leftrightarrow j}$, where td_i and td_j are the second elements of $\mathsf{LossyGen}(\tau_i)$ and $\mathsf{LossyGen}(\tau_j)$. Then, for any $x \in \{0,1\}^n$, $x = \mathsf{LossyInv}(td_j, \tau_i, \mathsf{ReEval}(s_{i \leftrightarrow j}, \mathsf{LossyEval}(s_i, x)))$. Note that $\mathsf{LossyInv}$ takes τ_i as one of the inputs, not τ_j.

Generating proper outputs: Let c be an output from $\mathsf{ReEval}(s_{i\leftrightarrow j}, \mathsf{LossyEval}(s_i, x))$, where $s_{i\leftrightarrow j}$ and s_i have the same meaning as that in the above paragraph. Then, $\mathsf{PrivReEval}(x, \tau_i, \tau_j, s_j)$ outputs the same c, where x, τ_i, τ_j, and s_j have the same meaning as that in the above paragraph. That is, $\mathsf{ReEval}(s_{i\leftrightarrow j}, \mathsf{LossyEval}(s_i, \cdot))$ and $\mathsf{PrivReEval}(\cdot, \tau_i, \tau_j, s_j)$ are equivalent as a function (i.e. any output of $\mathsf{ReEval}(s_{i\leftrightarrow j}, \mathsf{LossyEval}(s_i, \cdot))$ is independent of s_i).

Transitivity: Let (s_i, td_i), (s_j, td_j) and (s_k, td_k) be outputs from $\mathsf{LossyGen}(\tau_i)$, $\mathsf{LossyGen}(\tau_j)$, and $\mathsf{LossyGen}(\tau_k)$, and let $s_{i\leftrightarrow j}$ and $s_{i\leftrightarrow k}$ be the outputs from $\mathsf{ReIndex}(td_i, td_j)$ and $\mathsf{ReIndex}(td_i, td_k)$, respectively. Then, $\mathsf{Trans}(s_{i\leftrightarrow j}, s_{i\leftrightarrow k})$ outputs $s_{j\leftrightarrow k}$ which is the same output from $\mathsf{ReIndex}(td_j, td_k)$.

Fake key statistical indistinguishability: The algorithm $\mathsf{FakeKey}(s_i, \tau_i)$ outputs $(s'_j, s'_{i\leftrightarrow j}, \tau'_j)$, where s_i is the first element of an output from $\mathsf{LossyGen}(\tau_i)$. Let X_λ denote the distribution of $(\mathsf{par}, s_i, s_j, s_{i\leftrightarrow j}, \tau_i, \tau_j)$, and let Y_λ denote the distribution of $(\mathsf{par}, s_i, s'_j, s'_{i\leftrightarrow j}, \tau_i, \tau'_j)$, where each par, s_j, $s_{i\leftrightarrow j}$, and τ_j has the same meaning as that in the above paragraph. Then, $\{X_\lambda\}$ and $\{Y_\lambda\}$ are statistically indistinguishable.

Generation of injective functions from lossy functions: Let s be the first element of an output from $\mathsf{FakeKey}(s_{\mathrm{los}}, \tau)$, where τ is a tag and s_{los} is the first element of an output from $\mathsf{LossyGen}(\tau_{\mathrm{los}})$. Then, for every τ, $\mathsf{LossyEval}(s, \cdot)$ represents an injective function $f_{s,\star}$ with overwhelming probability, where a random variable is the randomness of $\mathsf{FakeKey}(s_{\mathrm{los}}, \tau)$. (We do not require other properties of index s if $f_{s,\star}$ is injective. The function $f_{s,\star}$ cannot have any trapdoor information).

2.3 Realization of Re-applicable LTDFs

Based on Peikert and Waters' LTDFs [16], Matsuda, Nishimaki and Tanaka [15] gave a realization of re-applicable LTDFs, which is specified as below (for more details, the reader is referred to [15]):

ParGen: This algorithm first generates a cyclic group $\mathbb{G}$ with prime order p, and then chooses a random generator $g \in \mathbb{G}$. Next, it selects random numbers $r_1, \cdots, r_n \in_R \mathbb{Z}_p$, and outputs the public parameters $\mathbf{C}_1$ as

$$\mathbf{C}_1 = \begin{pmatrix} c_1 \\ \vdots \\ c_n \end{pmatrix} = \begin{pmatrix} g^{r_1} \\ \vdots \\ g^{r_n} \end{pmatrix}.$$

LossyGen: Taking as input the public parameter $\mathbf{C}_1$ and a tag $\tau \in \mathbb{G}$ (note that if τ is the identity element e of $\mathbb{G}$, it means execution of the lossy mode; otherwise, execution of the injective mode), this algorithm first selects random elements $z_1, z_2, \cdots, z_n \in_R \mathbb{Z}_p$, and then computes a function index as

$$\mathbf{C}_2 = \begin{pmatrix} c_{1,1} & \cdots & c_{1,n} \\ \vdots & \ddots & \vdots \\ c_{n,1} & \cdots & c_{n,n} \end{pmatrix} = \begin{pmatrix} c_1^{z_1} \cdot \tau & \cdots & c_1^{z_n} \\ \vdots & \ddots & \vdots \\ c_n^{z_1} & \cdots & c_n^{z_n} \cdot \tau \end{pmatrix} = \begin{cases} c_{i,j} = c_i^{z_j} \cdot \tau, & \text{if } i = j; \\ c_{i,j} = c_i^{z_j}, & \text{otherwise.} \end{cases}$$

Finally, it outputs the function index $s = (\mathbf{C}_1, \mathbf{C}_2)$ and the trapdoor $td = z = (z_1, \cdots, z_n)$.

LossyEval: Taking as input a function index $s = (C_1, C_2)$ and an n-bit input $\mathbf{x} = (x_1, \cdots, x_n) \in \{0,1\}^n$, this algorithm outputs $(y_1, \mathbf{y}_2)$ such that

$$y_1 = \mathbf{x}\mathbf{C}_1 = \prod_{i=1}^{n} c_i^{x_i},$$

$$\mathbf{y}_2 = \mathbf{x}\mathbf{C}_2 = \left(\prod_{i=1}^{n} c_{i,1}^{x_i}, \cdots, \prod_{i=1}^{n} c_{i,n}^{x_i}\right) = \left((\prod_{i=1}^{n} c_i^{z_1 x_i})\tau^{x_1}, \cdots, (\prod_{i=1}^{n} c_i^{z_n x_i})\tau^{x_n}\right).$$

LossyInv: Taking as input $(td, \tau, (y_1, \mathbf{y}_2))$, where the trapdoor information td consists of $z = (z_1, \cdots, z_n)$, the tag τ is an element in $\mathbb{G}\backslash\{e\}$, and $\mathbf{y}_2 = (y_{2,1}, \cdots, y_{2,n}) \in \mathbb{G}^{1\times n}$, this algorithm computes $\mathbf{w} = (y_{2,1} \cdot y_1^{-z_1}, y_{2,2} \cdot y_1^{-z_2}, \cdots, y_{2,n} \cdot y_1^{-z_n})$. Then, if j-th element of $\mathbf{w}$ is the identity element of $\mathbb{G}$, then it sets $x_j = 0$; else if the j-th element of $\mathbf{w}$ is τ then it sets $x_j = 1$; otherwise, it outputs $\perp$. Finally, it outputs $x = (x_1, \cdots, x_n)$.

ReIndex: Taking as input trapdoors $td_i = (z_1, \cdots, z_n)$ and $td_j = (z_1', \cdots, z_n')$, this algorithm outputs $s_{i\leftrightarrow j} = td_j - td_i = (z_1' - z_1, \cdots, z_n' - z_n) = (z_{1,i\leftrightarrow j}, \cdots, z_{n,i\leftrightarrow j})$.

ReEval: On input $(s_{i\leftrightarrow j}, (y_1, \mathbf{y}_2))$, where $s_{i\leftrightarrow j} = (z_{1,i\leftrightarrow j}, z_{2,i\leftrightarrow j}, \cdots, z_{n,i\leftrightarrow j})$ and $(y_1, \mathbf{y}_2) = (y_1, (y_{2,1}, y_{2,2}, \cdots, y_{2,n}))$, this algorithm computes $\mathbf{y}_2' = (y_{2,1}', y_{2,2}', \cdots, y_{2,n}') = (y_{2,1} \cdot y_1^{z_{1,i\leftrightarrow j}}, y_{2,2} \cdot y_1^{z_{2,i\leftrightarrow j}}, \cdots, y_{2,n} \cdot y_1^{z_{n,i\leftrightarrow j}})$. Then it outputs $(y_1, \mathbf{y}_2')$.

PrivReEval: Taking as input $\mathbf{x}, \tau_i, \tau_j$ and s_j, where $\mathbf{x} = (x_1, \cdots, x_n)$ is n-bit input, this algorithm first computes $(\hat{y}_1, \hat{\mathbf{y}}_2) \leftarrow \mathsf{LossyEval}(s_j, \mathbf{x})$. Next, it makes $\hat{\mathbf{y}}_2'$ from $\hat{\mathbf{y}}_2$ in the following process: for each $i \in [1, n]$, if $x_i = 1$ then $\hat{y}_{2,i}' = \hat{y}_{2,i}\tau_j^{-1}\tau_i$; else $\hat{y}_{2,i}' = \hat{y}_{2,i}$, where $\hat{y}_{2,i}$ and $\hat{y}_{2,i}'$ are the i-th elements of $\hat{\mathbf{y}}_2$ and $\hat{\mathbf{y}}_2'$ respectively. Finally, it outputs $(\hat{y}_1, \hat{\mathbf{y}}_2')$.

Trans: Taking as input $s_{i\leftrightarrow j}$ and $s_{i\leftrightarrow k}$, this algorithm outputs $s_{i\leftrightarrow k} - s_{i\leftrightarrow j} = (td_k - td_i) - (td_j - td_i) = td_k - td_j = s_{i\leftrightarrow k}$.

FakeKey: Taking as input a function index $s_i = (\mathbf{C}_1, \mathbf{C}_2)$ and a tag $\tau_i \in \mathbb{G}$, this algorithm first chooses a random element $t \in \mathbb{G}$. Next, it chooses random numbers $s_{i\leftrightarrow j} = (z_{1,i\leftrightarrow j}, \cdots, z_{n,i\leftrightarrow j}) \in_R \mathbb{Z}_p^n$. Then it makes a new matrix $\mathbf{C}_2'$ as follows:

$$\mathbf{C}_2' = \begin{pmatrix} c_{1,1} \cdot c_1^{z_{1,i\leftrightarrow j}} \cdot t & \cdots & c_{1,n} \cdot c_1^{z_{n,i\leftrightarrow j}} \\ \vdots & \ddots & \vdots \\ c_{n,1} \cdot c_n^{z_{1,i\leftrightarrow j}} & \cdots & c_{n,n} \cdot c_n^{z_{n,i\leftrightarrow j}} \cdot t \end{pmatrix} = \begin{cases} c_{k,\ell}' = c_{k,\ell} \cdot c_k^{z_{\ell,i\leftrightarrow j}} \cdot t, & \text{if } k = \ell; \\ c_{k,\ell}' = c_{k,\ell} \cdot c_k^{z_{\ell,i\leftrightarrow j}}, & \text{otherwise,} \end{cases}$$

where c_k is the k entry of $\mathbf{C}_1$, and $c_{k,\ell}$ is the (k,ℓ) entry of $\mathbf{C}_2$. Finally, it outputs $s_j = (\mathbf{C}_1, \mathbf{C}_2')$, $s_{i\leftrightarrow j} = (z_{1,i\leftrightarrow j}, \cdots, z_{n,i\leftrightarrow j})$ and $\tau_j = \tau_i \cdot t$.

3 Bidirectional Multi-hop PRE

A bidirectional PRE scheme Π = ($\mathsf{Setup}, \mathsf{KeyGen}, \mathsf{Enc}, \mathsf{ReKeyGen}, \mathsf{ReEnc}, \mathsf{Dec}$) consists of the following six algorithms:

- $\mathsf{Setup}(1^\lambda)$: Given a security parameter 1^λ, this setup algorithm outputs a public parameter PP. Denote this by $PP \leftarrow \mathsf{Setup}(1^\lambda)$.
- $\mathsf{KeyGen}(PP)$: Given the public parameter PP, this key generation algorithm outputs a public key pk and a secret key sk. Denote this by $(pk, sk) \leftarrow \mathsf{KeyGen}(PP)$.
- $\mathsf{Enc}(PP, pk, m)$: Given the public parameter PP, a public key pk and a message m in the message space $\mathcal{M}$, this encryption algorithm outputs a ciphertext C. Denote this by $C \leftarrow \mathsf{Enc}(PP, pk, m)$.
- $\mathsf{ReKeyGen}(PP, sk_i, sk_j)$: Given the public parameter PP, a pair of secret keys sk_i and sk_j where $i \neq j$, this re-encryption key generation algorithm outputs a re-encryption key $rk_{i \leftrightarrow j}$. Denote this by $rk_{i \leftrightarrow j} \leftarrow \mathsf{ReKeyGen}(PP, sk_i, sk_j)$.
- $\mathsf{ReEnc}(PP, rk_{i \leftrightarrow j}, C_i)$: Given the public parameter PP, a re-encryption key $rk_{i \leftrightarrow j}$ and a ciphertext C_i intended for user i, this re-encryption algorithm outputs another ciphertext C_j for user j or the error symbol $\perp$. Denote this by $C_j \leftarrow \mathsf{ReEnc}(PP, rk_{i \leftrightarrow j}, C_i)$.
- $\mathsf{Dec}(PP, sk, C)$: Given the public parameter PP, a public key sk and a ciphertext C, this decryption algorithm outputs a message m or the error symbol $\perp$.

Next, we review the definition of chosen-ciphertext security for bidirectional multi-hop PRE scheme as defined in [15, 7]. Let λ be the security parameter, $\mathcal{A}$ be an oracle TM, representing the adversary, and Γ_U and Γ_C be two lists which are initially empty. The game consists of an execution of $\mathcal{A}$ with the following oracles, which can be invoked multiple times in any order, subject to the constraints specified as below:

Setup Oracle: This oracle can be queried first in the game only once. This oracle generates the public parameters $PP \longleftarrow \mathsf{Setup}(1^\lambda)$, and gives PP to $\mathcal{A}$.

Uncorrupted key generation: This oracle first generates a new key pair by running $(pk, sk) \leftarrow \mathsf{KeyGen}(PP)$. Next, it adds pk in Γ_U, and gives pk to $\mathcal{A}$.

Corrupted key generation: This oracle generates a new key pair by running $(pk, sk) \leftarrow \mathsf{KeyGen}(PP)$. Next, it adds pk in Γ_C, and gives (pk, sk) to $\mathcal{A}$.

Challenge oracle: This oracle can be queried only once. On input (pk_{i^*}, m_0, m_1), this oracle randomly chooses a bit $b \in \{0, 1\}$ and gives $C_{i^*} = \mathsf{Enc}(PP, pk_{i^*}, m_b)$ to $\mathcal{A}$. Here it is required that $pk_{i^*} \in \Gamma_U$. We call pk_{i^*} the challenge key and C_{i^*} the challenge ciphertext.

Re-encryption key generation: On input (pk_i, pk_j) from the adversary, this oracle gives the re-encryption key $rk_{i \leftrightarrow j} = \mathsf{ReKeyGen}(PP, sk_i, sk_j)$ to $\mathcal{A}$, where sk_i and sk_j are the secret keys corresponding to pk_i and pk_j, respectively. Here it is required that pk_i and pk_j are both in Γ_C, or alternatively are both in Γ_U.

Re-encryption oracle: On input (pk_i, pk_j, C_i), if $pk_j \in \Gamma_C$ and (pk_i, C_i) is a *derivative* of (pk_{i^*}, C_{i^*}), this oracle give $\mathcal{A}$ a special symbol $\perp$, which is not in the domain of messages or ciphertext. Otherwise, it gives the re-encrypted ciphertext $C_j = \mathsf{ReEnc}(PP, \mathsf{ReKeyGen}(PP, sk_i, sk_j), C_i)$ to $\mathcal{A}$. Derivatives of (pk_{i^*}, C_{i^*}) are defined inductively as follows:
- (pk_{i^*}, C_{i^*}) is a derivative of itself.
- If (pk, C) is a derivative of (pk_{i^*}, C_{i^*}), and (pk', C') is a derivative of (pk, C), then (pk', C') is a derivative of (pk_{i^*}, C_{i^*}).
- If $\mathcal{A}$ has queried the re-encryption oracle on input (pk, pk', C) and obtained the response C', then (pk', C') is a derivative of (pk, C).
- If $\mathcal{A}$ has queried the re-encryption key generation oracle on input (pk, pk') or (pk', pk), and $C' = \mathsf{ReEnc}(PP, \mathsf{ReKeyGen}(PP, sk, sk'), C)$, then (pk', C') is a derivative of (pk, C), where sk and sk' are the secret keys corresponding to pk and pk', respectively.

Decryption oracle: On input (pk, C), if the pair (pk, C) is a derivative of the challenge key and ciphertext (pk_{i^*}, C_{i^*}), or pk is not in $\Gamma_U \cup \Gamma_C$, this oracle returns the special symbol $\perp$ to $\mathcal{A}$. Otherwise, it returns the result of $\mathsf{Dec}(PP, sk, C)$ to $\mathcal{A}$, where sk is the secret key with respect to pk.

Decision oracle: This oracle can be queried at the end of the game. On input b', if $b' = b$ and the challenge key $pk_{i^*} \in \Gamma_U$, this algorithm output 1; else output 0.

We describe the output of the decision oracle in the above CCA-security definitional game as $\mathsf{Expt}_{\Pi,\mathcal{A}}^{\text{bid-PRE-CCA}}(\lambda) = b$ for an adversary A and a scheme Π. We define the advantage of adversary $\mathcal{A}$ as

$$\mathsf{Adv}_{\Pi,\mathcal{A}}^{\text{bid-PRE-CCA}}(\lambda) \stackrel{\text{def}}{=} \left| \Pr[\mathsf{Expt}_{\Pi,\mathcal{A}}^{\text{bid-PRE-CCA}}(\lambda) = 1] - \frac{1}{2} \right|,$$

where the probability is over the random choices of $\mathcal{A}$ and oracles. We say that the scheme Π is secure under the bidirectional PRE-CCA attack, if for any PPT adversary A, his advantage $\mathsf{Adv}_{\Pi,\mathcal{A}}^{\text{bid-PRE-CCA}}(\lambda)$ is negligible in the security parameter λ (for sufficiently large λ).

4 Review of the Matsuda-Nishimaki-Tanaka PRE Scheme

In this section, we shall review the Matsuda-Nishimaki-Tanaka bidirectional multi-hop PRE scheme.

Let λ be the security parameter, and let n, k, k', k'' and v be parameters depended on λ. Let ($\mathsf{SigGen}, \mathsf{SigSign}, \mathsf{SigVer}$) be a strongly unforgeable one-time signature scheme where the verification keys are in $\{0,1\}^v$. Let ($\mathsf{ParGen}, \mathsf{LossyGen}, \mathsf{LossyEval}, \mathsf{LossyInv}, \mathsf{ReEval}, \mathsf{PrivReEval}, \mathsf{Trans}, \mathsf{FakeKey}$) be a collection of reapplicable (n, k)-LTDFs and $\mathcal{T}$ be a set of tags. Let $(\mathsf{G}_{\text{abo}}, \mathsf{F}_{\text{abo}}, \mathsf{F}_{\text{abo}}^{-1})$ be a collection of (n, k')-ABO trapdoor functions with branches $B_\lambda = \{0,1\}^v$, which contains the set of signature verification keys. Let $\mathcal{H}$ be a family of pairwise independent hash functions from $\{0,1\}^n$ to $\{0,1\}^{k''}$. It is required that the above parameters satisfy $(k+k')-(k''+n) \geq \delta = \delta_1+\delta_2$ for some $\delta_1 = \omega(\log \lambda)$ and $\delta_2 = \omega(\log \lambda)$. The message space of the system is $\{0,1\}^{k''}$. The Matsuda-Nishimaki-Tanaka PRE scheme [15] is specified by the following algorithms:

Setup(1^λ)**:** This algorithm first generates an index of all-but-one trapdoor functions with lossy branch 0^v: $(s_{\text{abo}}, td_{\text{abo}}) \leftarrow \mathsf{G}_{\text{abo}}(1^\lambda, 0^v)$. Then, it generates a public parameter of re-applicable LTDFs: $\mathsf{par} \leftarrow \mathsf{ParGen}(1^\lambda)$. Next, it chooses a hash function h from $\mathcal{H}$. Finally, it outputs a public parameter as $PP = (s_{\text{abo}}, \mathsf{par}, h)$.

Note that the algorithm Setup erases the trapdoor td_{abo} because the following algorithms do not use td_{abo}.

KeyGen(PP)**:** Taking as input the pubic parameters $PP = (s_{\text{abo}}, \mathsf{par}, h)$, this algorithm first chooses a tag $\tau \in \mathcal{T}\backslash\{\tau_{\text{los}}\}$ and generates an injective index of re-applicable LTDFs: $(s_{\text{rltdf}}, td_{\text{rltdf}}) \leftarrow \mathsf{LossyGen}(\tau)$. Finally, it outputs the public key $pk = (s_{\text{rltdf}}, \tau)$ and the secret key $sk = (td_{\text{rltdf}}, s_{\text{rltdf}}, \tau)$.

Enc(PP, pk, m)**:** Taking as input the public parameters $PP = (s_{\text{abo}}, \mathsf{par}, h)$, public key $pk = (s_{\text{rltdf}}, \tau)$ and a message $m \in \{0,1\}^{k''}$, this encryption algorithm first chooses $x \in \{0,1\}^n$ uniformly at random. Next it generates a key-pair for the one-time signature scheme: $(vk, sk_\sigma) \leftarrow \mathsf{SigGen}(1^\lambda)$, and computes

$$c_1 = \mathsf{LossyEval}(s_{\text{rltdf}}, x), c_2 = \mathsf{F}_{\text{abo}}(s_{\text{abo}}, vk, x), c_3 = h(x) \oplus m.$$

Then it signs a tuple (c_2, c_3, τ) as $\sigma \leftarrow \mathsf{SigSign}(sk_\sigma, (c2, c3, \tau))$. Finally, it outputs the ciphertext $C = (vk, c_1, c_2, c_3, \tau, \sigma)$.

ReKeyGen(PP, sk_i, sk_j)**:** On public parameter $PP = (s_{\text{abo}}, \mathsf{par}, h)$, the secret keys $sk_i = (td_i, s_i, \tau_i)$ and $sk_j = (td_j, s_j, \tau_j)$, this algorithm computes $s_{i\leftrightarrow j} \leftarrow \mathsf{ReIndex}(td_i, td_j)$, and then outputs a re-encryption key $rk_{i\leftrightarrow j} = s_{i\leftrightarrow j}$.

ReEnc$(PP, rk_{i\leftrightarrow j}, C_i)$**:** Taking as input the public parameter $PP = (s_{\text{abo}}, \mathsf{par}, h)$, the re-encryption key $rk_{i\leftrightarrow j} = s_{i\leftrightarrow j}$ and a ciphertext $C_i = (vk, c_{1,i}, c_2, c_3, \tau, \sigma)$, this algorithm computes $c_{1,j} \leftarrow \mathsf{ReEval}(s_{i\leftrightarrow j}, c_{1,i})$. It then outputs $C_j = (vk, c_{1,j}, c_2, c_3, \tau, \sigma)$ as a new ciphertext for the user with sk_j.

Dec(PP, sk, C)**:** Taking as input the public parameter $PP = (s_{\text{abo}}, \mathsf{par}, h)$, a secret key $sk = (td_{\text{rltdf}}, s_{\text{rltdf}}, \tau)$ and a ciphertext $C = (vk, c_1, c_2, c_3, \tau', \sigma)$, this decryption algorithm acts as follows:

1. Check whether $\mathsf{SigVer}(vk, (c_2, c_3, \tau'), \sigma) = 1$ holds. If not, output $\perp$.
2. Compute $x = \mathsf{LossyInv}(td_{\text{rltdf}}, \tau', c_1)$. If $\tau = \tau'$, it checks $\mathsf{LossyEval}(s_{\text{rltdf}}, x) = c_1$; else it checks $\mathsf{PrivReEval}(x, \tau', \tau, s_{\text{rltdf}}) = c_1$. If not, it outputs $\perp$. It also checks $\mathsf{F}_{\text{abo}}(s_{\text{abo}}, vk, x) = c_2$. If not, it outputs $\perp$.
3. Finally, output $m = c_3 \oplus h(x)$.

5 Security Analysis

In this section, we shall present a concrete attack against the Matsuda-Nishimaki-Tanaka PRE scheme. Before presenting its details, we first identify the potential weakness in their scheme: for a ciphertext $C_i = (vk, c_{1,i}, c_2, c_3, \tau, \sigma)$, their ReEnc algorithm simply transforms the ciphertext component $c_{1,i}$ into $c_{1,j}$, without verifying the validity of $c_{1,i}$. Then there might exist an adversary who can break the CCA-security of their scheme as follows: Given the challenge ciphertext $C_{i^*} = (vk, c_{1,i^*}, c_2, c_3, \tau, \sigma)$, the adversary can first modify the ciphertext component c_{1,i^*} to obtain a new (ill-formed) ciphertext C'_{i^*} and then ask the re-encryption oracle to re-encrypt C'_{i^*} into another ciphertext C'_j for a *corrupted* user j (note that according to the security model, it is legal for the adversary to issue such a query); next, the adversary can modify C'_j to obtain the *right* re-encrypted ciphertext C_j of the challenge ciphertext, and thus he can derive the underlying plaintext by decrypting C_j with user j's secret key.

Below we give the attack details. For an easy explanation of how the adversary can modify C'_j to obtain the right transformed ciphertext C_j, when describing the underlying re-applicable LTDFs we shall take Matsuda et al.'s concrete realization (recalled in Section 2.3) as the example. Concretely, the adversary works as follows:

1. The adversary first obtains the public parameters PP from the setup oracle.
2. The adversary obtains a public key pk_{i^*} from the uncorrupted key generation oracle. Note that pk_{i^*} will be added in Γ_U by the oracle.
3. The adversary obtains a public/secret key pair (pk_j, sk_j) from the corrupted key generation oracle. Note that pk_j will be added in Γ_C by the oracle.
4. The adversary submits (pk_{i^*}, m_0, m_1) to the challenge oracle, and then is given the challenge ciphertext $C_{i^*} = (vk^*, c_{1,i^*}, c_2^*, c_3^*, \tau^*, \sigma^*)$, where c_{1,i^*} is the output of function LossyEval. Here we use Matsuda et al.'s concrete realization of LossyEval as an example. Wlog, suppose $c_{1,i^*} = (y_1, \mathbf{y}_2) = (y_1, (y_{2,1}, \cdots, y_{2,n}))$.
5. The adversary first randomly picks $\tilde{y}_{2,1}, \cdots, \tilde{y}_{2,n}$ from $\mathbb{G}$, and modifies the challenge ciphertext to obtain a new (ill-formed) ciphertext $C'_{i^*} = (vk^*, c'_{1,i^*}, c_2^*, c_3^*, \tau^*, \sigma^*)$, where $c'_{1,i^*} = (y_1, (\tilde{y}_{2,1}, \cdots, \tilde{y}_{2,n}))$. Then, the adversary submits $(pk_{i^*}, pk_j, C'_{i^*})$ to the re-encryption oracle. Note that, although $pk_j \in \Gamma_C$, it is legal for the adversary to issue this query, since (pk_{i^*}, C'_{i^*}) is *not* a derivative of (pk_{i^*}, C_{i^*}). Note that the re-encryption algorithm ReEnc cannot check the validity of the ciphertext component c'_{1,i^*}. So, it will return the re-encrypted ciphertext $C'_j = \mathsf{ReEnc}(PP, \mathsf{ReKeyGen}(PP, sk_{i^*}, sk_j), C'_{i^*})$ to the adversary.

   According to the re-encryption algorithm, we get $C'_j = (vk^*, c'_{1,j}, c_2^*, c_3^*, \tau^*, \sigma^*)$, where $c'_{1,j} = \mathsf{ReEval}(s_{i^* \leftrightarrow j}, c'_{1,i^*})$. According to Matsuda et al.'s concrete realization of ReEval, we have

$$c'_{1,j} = (y_1, (\tilde{y}'_{2,1}, \cdots, \tilde{y}'_{2,n})) = \left(y_1, \left(\tilde{y}_{2,1} \cdot y_1^{z_{1,i^* \leftrightarrow j}}, \cdots, \tilde{y}_{2,n} \cdot y_1^{z_{n,i^* \leftrightarrow j}}\right)\right).$$

   Now, from $c'_{1,j} = (y_1, (\tilde{y}'_{2,1}, \cdots, \tilde{y}'_{2,n}))$, the adversary can compute the following

$$c_{1,j} = \left(y_1, (\frac{\tilde{y}'_{2,1} y_{2,1}}{\tilde{y}_{2,1}}, \cdots, \frac{\tilde{y}'_{2,n} y_{2,n}}{\tilde{y}_{2,n}}) \right)$$
$$= \left(y_1, (\frac{\tilde{y}_{2,1} \cdot y_1^{z_{1,i^*\leftrightarrow j}} y_{2,1}}{\tilde{y}_{2,1}}, \cdots, \frac{\tilde{y}_{2,n} \cdot y_1^{z_{n,i^*\leftrightarrow j}} y_{2,n}}{\tilde{y}_{2,n}}) \right)$$
$$= \left(y_1, (y_{2,1} \cdot y_1^{z_{1,i^*\leftrightarrow j}}, \cdots, y_{2,n} \cdot y_1^{z_{n,i^*\leftrightarrow j}}) \right) .$$

Observe that $c_{1,j}$ is indeed equivalent to the result of $\mathsf{ReEval}(s_{i^*\leftrightarrow j}, c_{1,i^*})$. Thus, we have $C_j = (vk^*, c_{1,j}, c_2^*, c_3^*, \tau^*, \sigma^*)$ is indeed the result of $\mathsf{ReEnc}(PP, \mathsf{ReKeyGen}(PP, sk_{i^*}, sk_j), C_{i^*})$, which is an encryption of m_b. Now, the adversary can obtain the underlying plaintext m_b by decrypting the re-encrypted ciphertext C_j using the secret key sk_j, and obviously can break CCA-security of the Matsuda-Nishimaki-Tanaka PRE scheme.

The above attack can also be simply extended to the case that the user j is uncorrupted. In this case, the adversary $\mathcal{A}$ directly request (pk_j, C_j) to the decryption oracle, which will return the plaintext m_b to $\mathcal{A}$.

6 Discussions and Conclusion

The authors constructed 11 games, Game-0 to Game-10, to prove the CCA-security of the PRE scheme developed in [15], where Game-0 is just the CCA definitional game of PRE (recalled in Section 3) and Game-10 is a game which any adversary can win with only probability 1/2. They also discussed that difference of advantage between any two successive games is negligible, and hence any adversary cannot win Game-0 with a better probability than 1/2 plus a negligible value. However, as we observed in the previous section, there exists an adversary which can always win Game-0, and this implies that for at least one of pairs of two successive games, difference of advantage between them is non-negligible. If we correctly understand the security proof in [15], such two games seem Game-7 and 8. This is basically due to the fact that until Game-7, the challenger generates all re-encryption keys for all users (including both uncorrupted and corrupted users), and by using these re-encryption keys, the challenger simulates both the re-encryption key generation oracle and the re-encryption oracle. In contrast, in Game-8, the challenger generates re-encryption keys among uncorrupted users in a specific manner without knowing these users' secret keys (see [15] for detail), and therefore, the same strategy for simulating these two oracles as in Game-7 cannot be immediately applied to this game. Namely, we see that in Game-8, it is still straightforward to generate re-encryption keys among uncorrupted users or those among corrupted users, but it seems hard to generate any re-encryption key between an uncorrupted user and a corrupted user. This also implies that the re-encryption key generation oracle can be still simulated, but the re-encryption oracle can not as long as the same simulation technique as in Game-7 is used.

The PRE scheme developed in [15] is based upon the CCA-secure public-key encryption (PKE) scheme of Peikert and Waters (that is in turn based upon LTDFs) [16], which can be viewed as an extension of the Peikert-Waters PKE scheme into the proxy re-encryption setting. One key difference between the Peikert-Waters PKE construction and the PRE construction of [15] is that: all components in the ciphertext of the Peikert-Waters PKE are signed by the one-time signature (under the verification key vk), but the key component c_1 is not signed in the ciphertext of the PRE of [15]. Of course, signing c_1 can prevent our concrete attack, but the resultant scheme is not a PRE scheme any longer (particularly, the proxy cannot translate ciphertexts among players, as the underlying signing key w.r.t. vk is unknown to the proxy). From our view, constructing CCA-secure proxy re-encryption without bilinear maps in the standard model may need significantly new ideas and techniques. It is still an open problem to come up with a (bidirectional or unidirectional) proxy re-encryption scheme without bilinear maps in the standard model.

Acknowledgements

The first author is supported by the National Science Foundation of China under Grant Nos. 60903178 and 61005049, and it also supported by the Fundamental Research Funds for the Central Universities. The second author is supported in part by a grant from the Major State Basic Research Development (973) Program of China (No. 2007CB807901), and grants from the National Natural Science Foundation of China (Nos. 60703091 and 61070248) and the QiMingXing Program of Shanghai, and Young Faculty Fund of Computer Science School of Fudan University.

References

1. Ateniese, G., Fu, K., Green, M., Hohenberger, S.: Improved Proxy Re-Encryption Schemes with Applications to Secure Distributed Storage. In: NDSS. The Internet Society, San Diego (2005)
2. Ateniese, G., Fu, K., Green, M., Hohenberger, S.: Improved Proxy Re-Encryption Schemes with Applications to Secure Distributed Storage. ACM Trans. Inf. Syst. Secur. 9(1), 1–30 (2006)
3. Blaze, M., Bleumer, G., Strauss, M.: Divertible Protocols and Atomic Proxy Cryptography. In: Nyberg, K. (ed.) EUROCRYPT 1998. LNCS, vol. 1403, pp. 127–144. Springer, Heidelberg (1998)
4. Boyen, X., Waters, B.: Shrinking the keys of discrete-log-type lossy trapdoor functions. In: Zhou, J., Yung, M. (eds.) ACNS 2010. LNCS, vol. 6123, pp. 35–52. Springer, Heidelberg (2010)
5. Canetti, R., Goldreich, O., Halevi, S.: The Random Oracle Methodology, Revisited (Preliminary Version). In: STOC, pp. 209–218 (1998)
6. Canetti, R., Goldreich, O., Halevi, S.: The random oracle methodology, revisited. J. ACM 51(4), 557–594 (2004)

7. Canetti, R., Hohenberger, S.: Chosen-Ciphertext Secure Proxy Re-Encryption. In: Ning, P., di Vimercati, S.D.C., Syverson, P.F. (eds.) ACM Conference on Computer and Communications Security, pp. 185–194. ACM, New York (2007)
8. Canetti, R., Krawczyk, H., Nielsen, J.B.: Relaxing Chosen-Ciphertext Security. In: Boneh, D. (ed.) CRYPTO 2003. LNCS, vol. 2729, pp. 565–582. Springer, Heidelberg (2003)
9. Chow, S.S.M., Weng, J., Yang, Y., Deng, R.H.: Efficient Unidirectional Proxy Re-Encryption. In: Bernstein, D.J., Lange, T. (eds.) AFRICACRYPT 2010. LNCS, vol. 6055, pp. 316–332. Springer, Heidelberg (2010)
10. Deng, R.H., Weng, J., Liu, S., Chen, K.: Chosen-Ciphertext Secure Proxy Re-encryption without Pairings. In: Franklin, M.K., Hui, L.C.K., Wong, D.S. (eds.) CANS 2008. LNCS, vol. 5339, pp. 1–17. Springer, Heidelberg (2008)
11. Freeman, D.M., Goldreich, O., Kiltz, E., Rosen, A., Segev, G.: More constructions of lossy and correlation-secure trapdoor functions. In: Nguyen, P.Q., Pointcheval, D. (eds.) PKC 2010. LNCS, vol. 6056, pp. 279–295. Springer, Heidelberg (2010)
12. Hemenway, B., Ostrovsky, R.: Lossy trapdoor functions from smooth homomorphic hash proof systems (2010) (manuscript), `http://www.math.ucla.edu/~bretth/papers/uhp_ltdf.pdf`
13. Libert, B., Vergnaud, D.: Unidirectional Chosen-Ciphertext Secure Proxy Re-Encryption, `http://hal.inria.fr/inria-00339530/en/`; This is the extended version [14]
14. Libert, B., Vergnaud, D.: Unidirectional Chosen-Ciphertext Secure Proxy Re-encryption. In: Cramer, R. (ed.) PKC 2008. LNCS, vol. 4939, pp. 360–379. Springer, Heidelberg (2008)
15. Matsuda, T., Nishimaki, R., Tanaka, K.: CCA Proxy Re-Encryption without Bilinear Maps in the Standard Model. In: Nguyen, P., Pointcheval, D. (eds.) PKC 2010. LNCS, vol. 6056, pp. 261–278. Springer, Heidelberg (2010)
16. Peikert, C., Waters, B.: Lossy trapdoor functions and their applications. In: Ladner, R.E., Dwork, C. (eds.) STOC, pp. 187–196. ACM, New York (2008)
17. Shao, J., Cao, Z.: CCA-Secure Proxy Re-encryption without Pairings. In: Jarecki, S., Tsudik, G. (eds.) PKC 2009. LNCS, vol. 5443, pp. 357–376. Springer, Heidelberg (2009)
18. Weng, J., Chen, M.-R., Yang, Y., Deng, R.H., Chen, K., Bao, F.: CCA-Secure Unidirectional Proxy Re-Encryption in the Adaptive Corruption Model without Random Oracles. Science China: Information Science 53(3), 593–606 (2010)
19. Weng, J., Deng, R.H., Liu, S., Chen, K.: Chosen-ciphertext secure bidirectional proxy re-encryption schemes without pairings. Inf. Sci. 180(24), 5077–5089 (2010)

Fully Secure Accountable-Authority Identity-Based Encryption

Amit Sahai and Hakan Seyalioglu

The University of California - Los Angeles
sahai@cs.ucla.edu, hseyalioglu@ucla.edu

Abstract. The problem of trust is one of the biggest concerns in any identity-based infrastructure where the key-generation authority (called the PKG) must choose secret keys for participants and therefore be highly trusted by all parties. While some abilities of the PKG are intrinsic to this setting, reducing this trust as much as possible is beneficial to both user and authority as the less trust is placed in it, the less an honest authority can be accused of abusing that trust. Goyal (CRYPTO 2007) defined the notion of Accountable-Authority IBE in which a dishonest PKG who had leaked a user's private key could be proven guilty. Later, Goyal et al. (CCS 2008) asked whether it would be possible to implicate a PKG who produced an unauthorized decoder box, enabling decryption with a noticeable probability but which may not actually grant access to a well-formed key. Formally, would it be possible for a tracing algorithm to implicate a dishonest PKG given only black-box access to such a decoder? Goyal et al. could only provide such a scheme in the weaker setting of selective security, where an adversary must declare at the start of the game which identity it intends to target. In this work, we provide the first fully secure accountable-authority IBE scheme. We prove security from the standard DBDH assumption while losing none of the functionality or security of the original proposal.

Keywords: Identity-Based Encryption, Accountable Authority, Tracing.

1 Introduction

Since its introduction by Shamir [24] and first constructions by Boneh and Franklin [4] and Cocks [7], identity-based encryption (IBE) has been one of the most active areas of cryptographic research, with numerous applications to computer security and privacy (e.g. [3,2,9,25]). Many concepts of independent interest sprang from this topic such as Fuzzy IBE and Attribute Based Encryption [23,14,12,22] which allow encryption to groups of users whose credentials (a.k.a. attributes) satisfy a given access policy, and Hierarchical IBE [10,11,17] which allows key generation in a leveled fashion.

Despite its applications and practicality, the security of an IBE scheme relies heavily on trusting the key generation authority. Despite the tremendous practical security benefits of not having to manage a public-key infrastructure (PKI)

D. Catalano et al. (Eds.): PKC 2011, LNCS 6571, pp. 296–316, 2011.

based on individual public keys, trusting any single authority is very troubling. It is a natural and important question to ask how one can reduce the trust one must have in the private key generator. One proposed method to reduce this trust is to employ multiple PKGs such that no single entity can compromise security [2]; however, the question of widespread collusion naturally remained.

This left open the problem of reducing trust in a single central authority. Note that such a central authority can clearly decrypt all messages in the system. However, in many practical situations, users may not be worried that the authority would have the inclination or resources to specifically eavesdrop on their communications. On the other hand, users would want some assurance that the authority doesn't issue their secret keys to *other potentially malicious* users, who might have a lower profile and have less to lose if caught. This is precisely the problem proposed and addressed by the work of Goyal [13] by introducing the notion of Accountable-Authority IBE (A-IBE). If a user finds a key for his identity being disseminated without permission (e.g. the user finds that his secret key is being put up for auction on EBay), Goyal proposed a system where the user can prove that it would have been computationally impossible for the user to have created this compromised key. Since the only other entity with access to keys to the user's identity is the PKG, this allowed the user to successfully implicate a malicious PKG of malfeasance.

Goyal's first construction relied heavily on seeing the *actual* compromised key, and as-such this first construction is called only *white-box* secure. Such white-box security leaves a great deal to be desired, for instance, while able to still decrypt, the key may have been manipulated in some way. A natural extension is to ask whether it would be possible to implicate a PKG who has disseminated a decoder "box" which allows decryption of ciphertexts encrypted to some identity rather than simply leaking a key itself in the clear (e.g. selling such a box or a heuristically obfuscated piece of decryption software). Since the exact information used to create this decoder box may be hidden, this is called *black box* security and is the strongest current proposed notion of A-IBE. If we can trace a PKG who issues decoder boxes, it forces a user attempting to dishonestly acquire decryptions to continually interact with the PKG (who may, for example, be acting as a decryption oracle). Such a setting is much riskier for the dishonest PKG since it requires a much greater level of communication with the dishonest user.

Giving a black-box secure A-IBE scheme was specifically labeled the most important problem left open in Goyal's original work. The first work made some progress towards obtaining black-box security, which was later refined [15,19] (the latter of which proves security if a cheating authority is denied a decryption oracle); however, major shortcomings remained. The only known *black-box* secure constructions are in a very weak model of security called "selective security" where the adversary must designate the identity which it will create a decoder box for before even seeing the public parameters. While selective security was useful for exploring these concepts, it is a completely unreasonable requirement for security since real-world adversaries have the ability to pick targets adaptively.

In fact, selective security is especially troubling in the A-IBE setting since selective security is only a reasonable benchmark if there are only a few high value targets whom an adversary would be interested in attacking (making the prospect of picking a target adaptively during the attack less appealing). However, in a subscription service handled through decoder boxes (one of the most natural applications for a black-box A-IBE scheme), there would be many targets of equal value, making selective-security particularly troubling.

1.1 Our Contributions

In this work, we provide the first fully secure black-box accountable-authority IBE scheme. We are able to prove security based on the standard DBDH number-theoretic intractability assumption while losing none of the functionality or security of the original proposal of Goyal [13].

A limitation of previous work [15] which obtained selective black-box security was that it obtained functionality from established Attribute-Based Encryption schemes without modifying the scheme to better suit the A-IBE framework. A main contribution of this work is that we can obtain A-IBE by only moving "halfway" between basic IBE and Attribute-Based Encryption, in a setting where identities do have attributes, but these attributes are assigned randomly. We call such a scheme a *Dummy-IBE* scheme and use it to construct black-box A-IBE. We then give a construction of a *Dummy-IBE* scheme by combining mechanics of Waters' IBE scheme [25] and the Attribute-Based Encryption scheme due to Sahai and Waters [23]. We note that a fully secure ABE scheme would not directly suffice to make the previous construction work, in particular, ciphertext and key-sanity checks (to be defined later) would also be necessary. Indeed in contrast to recent findings in fully secure ABE [18,21], we are able to follow the framework of previous *selectively secure* ABE constructions [23] to achieve *full* black-box security, and as a result inherit these schemes' improved efficiency and analytic simplicity.

2 Preliminaries

We will often use the notation $[a, b]$ to denote all integers between a and b, inclusive and bold font (e.g. $\mathbf{x}$, $\mathbf{A}$) to denote vectors. A bracketed index (e.g. $\mathbf{x}[j]$, $\mathbf{A}[i]$) will denote the value at that index of the relevant vector. A negligible value is one that grows slower than any inverse polynomial of a certain parameter (usually the security parameter) and a probability is said to be overwhelming if it is within an additive negligible probability of 1.

2.1 Bilinear Maps

Let $\mathbb{G}, \mathbb{G}_T$ be two multiplicative groups of prime order p. Let g be a generator of $\mathbb{G}$ and $e : \mathbb{G} \times \mathbb{G} \to \mathbb{G}_T$. We assume e satisfies bilinearity (for all $u, v \in \mathbb{G}_1$ and $a, b \in \mathbb{Z}_p$, we have $e(u^a, v^b) = e(u, v)^{ab}$) and non-degeneracy ($e(g, g) \neq 1$).

2.2 Security Assumption

Definition 1 (Decisional Bilinear Diffie-Hellman (DBDH) Assumption) *Suppose the values $a, b, c, z \leftarrow \mathbb{Z}_p, \beta \leftarrow \{0,1\}$ are chosen at random. The DBDH assumption states that no probabilistic polynomial-time algorithm can distinguish the tuple $(g, A = g^a, B = g^b, C = g^c, e(g,g)^{abc})$ from the tuple $(g, A = g^a, B = g^b, C = g^c, e(g,g)^z)$ with more than a negligible advantage.*

2.3 Fully-Simulatable Oblivious Transfer

Informally, a k-out-of-n oblivious transfer protocol [8] is a two-party protocol in which n strings are stored by one party and at the end, the other party has received exactly k of them with the requirements that the receiving party gains no more information than the k strings it received and the sending party gains no information about which k the receiving party acquired.

Formally, by a fully-simulatable OT protocol, we will refer to one that meets Canetti's requirements for universal composability [6]. If an adversary $\mathcal{A}'$ is interacting with a challenger and has non-negligible success probability ϵ in a game which includes some Fully-Simulatable OT protocol, there also exists an adversary $\mathcal{A}$ with probability of success within a negligible factor of ϵ which instead uses the ideal-world OT protocol where $\mathcal{A}$ directly queries the indices of the k values it will receive from the challenger. We will often transfer our adversaries into ones which engage in the ideal-world OT protocol. For examples of fully simulatable Oblivious Transfer using only the DDH and DBDH assumptions please refer to the following papers: [20,16,5].

2.4 Accountable Authority IBE (A-IBE)

We now recall what it means for an IBE scheme to be A-IBE secure. First, the scheme must function as a good IBE scheme. Traditionally, it is assumed that the Key Generation algorithm simply outputs a key, however, since the key received by each party should not be available to the central authority, we instead assume it is generated through an interactive protocol.

Definition 2 (Identity Based Encryption (IBE) Scheme). *An* Identity Based Encryption Scheme *consists of the following four probabilistic polynomial time algorithms:*

- **Setup**$(1^\lambda) \rightarrow$ PK, MK. The setup takes as input 1^λ with λ the security parameter and outputs public parameters PK and master secret key MK.
- **KeyGen**(PK, MK,ID)$(\rightarrow)$ $\mathrm{K_{ID}}$. The user key generation algorithm takes as inputs PK, MK and an identity ID and engages in an interactive protocol with the recipient. At the end, the recipient receives a key for ID, $\mathrm{K_{ID}}$. We use the notation $(\rightarrow)$ to highlight the fact that **KeyGen** may not know exactly which key the user receives (for example, if **KeyGen** is implemented using an oblivious transfer protocol).

– **Encrypt**(M, PK, ID)→ C_{ID}. The encryption algorithm takes as input PK, ID and a message M and outputs a ciphertext C_{ID}.
– **Decrypt**(C_{ID}, K_{ID}, PK) → M. The decryption algorithm outputs the original message with overwhelming probability assuming it is given as input a key to the identity ID.

We will assume that ID is included in the plain as part of C_{ID} and K_{ID} during our analysis for notational simplicity.

Definition 3 (ϵ-Useful Decoder Box). *For non-negligible ϵ, a probabilistic polynomial time algorithm $\mathcal{D}_{ID}$ is an ϵ-**useful decoder box** for the identity ID if:*

$$\Pr[M \leftarrow \mathcal{M} : \mathcal{D}_{\mathrm{ID}}(\mathbf{Encrypt}(M, PK, ID) = M] \geq \epsilon$$

The main additional functionality required of an A-IBE scheme follows:

– $\mathbf{Trace}^{\mathcal{D}_{\mathrm{ID}}}$(PK, ID, K_{ID}) outputs either User or PKG. Given oracle access to an ϵ-useful decoder box $\mathcal{D}_{ID}$, PK, ID and a key K_{ID}, the tracing algorithm will decide if $\mathcal{D}_{ID}$ was created by the trusted authority or by the user who supplied K_{ID}.

Intuitively the tracing algorithm allows a user who has recovered a decoder box for its identity to reveal its secret key to prove that it could not have generated this decoder box. This would be impossible if all keys could decrypt perfectly, which is why we allow decryption a small probability of failure. Additionally, it has proven very useful to include two more algorithms in most A-IBE schemes for use in tracing. A term is said to be valid if it is a possible output of the relevant generation algorithm (in other words, a key K_{ID} is valid if it is a possible output of the key generation algorithm on inputs PK, MK and a ciphertext is valid if it is the encryption of some M to the relevant identity):

– **KeySanity**(K_{ID},PK) tests if K_{ID} is a proper key for identity ID and outputs Valid or Not-Valid.
– **CiphertextSanity**(C_{ID},PK) tests if C_{ID} is a proper ciphertext for ID and outputs Valid or Not-Valid.

2.5 Security Requirements

An A-IBE scheme is called black-box secure if it satisfies the following requirements. First, it must satisfy IND-ID-CPA security. Second, if a decoder box $\mathcal{D}_{ID}$ was created by the central authority, the tracing algorithm should implicate the PKG and when it was created by the colluding users, it should implicate the users. This security requirement is captured in the following games.

Definition 4 (IND-ID-CPA Security). *An A-IBE scheme is* IND-ID-CPA *secure if the advantage of any probabilistic polynomial time adversary $\mathcal{B}$ is negligible in the game below:*

1. **Setup.** The challenger runs the **Setup** algorithm and sends PK to $\mathcal{B}$.
2. **Phase 1.** $\mathcal{B}$ engages in a **KeyGen** protocol with the challenger with adaptively chosen ID_i and receives K_{ID_i}. This may be repeated multiple times for different identities.
3. **Challenge.** $\mathcal{B}$ submits two messages M_0, M_1 and an identity ID which it did not query during Phase 1. The challenger flips a coin b and encrypts M_b to the identity ID and sends the encryption C to $\mathcal{B}$.
4. **Phase 2.** Same as Phase 1 except $\mathcal{B}$ may not query a decryption key for ID.
5. **Guess.** The adversary outputs $b' \in \{0, 1\}$.

The advantage of $\mathcal{B}$ is defined as $\Pr[b' = b] - \frac{1}{2}$.

Definition 5 (Dishonest-User Security). *An A-IBE scheme is Dishonest-User secure if the advantage of any probabilistic polynomial time adversary $\mathcal{B}$ is negligible in the game below:*

1. **Setup.** The challenger runs the **Setup** algorithm and sends **PK** to $\mathcal{B}$.
2. **Key Generation.** The adversary adaptively queries keys for distinct ID_i, and receives K_{ID_i}. This may be repeated multiple times for different identities.
3. **Create Decoder Box.** The adversary outputs an ϵ-useful decoder box $\mathcal{D}_{\text{ID}}$ and K'_{ID} for some identity ID queried during the Key-Generation phase.
4. **Tracing Failure.** Finally, success is defined as the event that: $\textbf{Trace}^{\mathcal{D}_{\text{ID}}}(\text{PK}, \text{ID}, K'_{\text{ID}}) = \mathsf{PKG}$.

At this point it may seem artificial to have the requirement that ID was queried once in the Key Generation phase of the game, however since the scheme is required to be IND-ID-CPA secure, outputting a key for an identity which has not been queried would contradict the previous security requirement and therefore adding this additional requirement does not weaken security.

Definition 6 (Dishonest-PKG Security). *An A-IBE scheme is Dishonest-PKG secure if the advantage of any probabilistic polynomial time adversary $\mathcal{B}$ is negligible in the game below:*

1. **Setup.** The adversary $\mathcal{B}$ generates and passes public parameters PK and ID to the challenger. The challenger checks PK and ID are well-formed, aborts if not.
2. **Key Generation.** The challenger and $\mathcal{B}$ engage in the key generation protocol for an identity ID. If neither party aborts, the challenger receives K_{ID} as output.
3. **Decryption.** The adversary adaptively queries ciphertexts C_i to the challenger and the challenger replies with the decryption under K_{ID}. This may be repeated multiple times for different ciphertexts.
4. **Create Decoder Box..** The adversary outputs an ϵ-useful decoder box $\mathcal{D}_{\text{ID}}$.
5. **Tracing Failure.** Finally, success is defined as the event that $\textbf{Trace}^{\mathcal{D}_{\text{ID}}}(\text{PK}, \text{ID}, \text{K}_{\text{ID}}) = \mathsf{USER}$.

If our scheme has a **CiphertextSanity** check we can always assume decryption is preceded with verification that the ciphertext is well-formed and if it is not well-formed, the decryption process can just output $\perp$. Informally, this will allow us to argue that the PKG gains no information from its decryption queries since the decryption oracle will only decrypt well-formed ciphertexts, which the PKG could have decrypted without using the oracle (since the PKG can already generate a key for any identity).

3 Preliminary Reduction

The first key step in our proof is realizing that A-IBE can be built from an encryption scheme which falls somewhere between usual IBE and Attribute-Based Encryption. For this, we introduce the notion of *Dummy-IBE* in which every user is assigned a set of attributes which restricts the ciphertexts that can be decrypted but the user has no control over which attributes are assigned. We stress again that because of the importance of the sanity checks in our setting, a fully secure attribute based encryption scheme by itself would not suffice for our purposes.

3.1 Dummy Identity-Based Encryption

The intuition for *Dummy IBE* (*D-IBE*) comes from previous work [15] which achieves full *Dishonest-PKG* security (but not full *Dishonest-User* security). Keys for identities and ciphertexts will have k attributes and decryption will only succeed if the encryption was to the target identity and the attribute sets overlap in at least d indices.

Definition 7 (Dummy Identity-Based Encryption (D-IBE) Scheme). *A D-IBE Encryption scheme D with parameters $d \leq k \leq n \in \Theta(\lambda)$ consists of the following four poly-time algorithms:*

- **Setup**$(1^\lambda, d, k, n) \rightarrow$ PK, MK public and master keys.
- **KeyGen**(PK, MK) $\rightarrow K_{\mathrm{ID}}(S)$ The key generation algorithm selects $S \subset [1, n]$ a random subset of size k, generates[1] K_{ID}^{α} for all $\alpha \in [1, n]$ and outputs $K_{\mathrm{ID}}(S) = \{K_{\mathrm{ID}}^{\alpha} \mid \alpha \in S\}$.
- **Encrypt**(M, PK, ID, S)$\rightarrow C_{\mathrm{ID}}(S)$.
- **Decrypt**($\mathrm{C_{ID}}(S)$,$\mathrm{K_{ID}}(S')$,PK)$\rightarrow M$ where $C_{\mathrm{ID}}(S)$ is an encryption of M if $|S \cap S'| \geq d$.
- **KeySanity**$(K_{\mathrm{ID}}^{*}(S), PK)$ outputs Valid or Not-Valid depending on whether the key is a valid key for the implied identity and attribute set.

[1] The fact that the key generation algorithm is able to generate all key components will be important for our reduction. To formalize the above notion, we could have the key generation algorithm output all key components and only send the ones corresponding to members in S to the user but we present it as above for notational simplicity.

– **CiphertextSanity**$(C^*_{\mathrm{ID}}(S), PK)$ outputs either Valid or Not-Valid depending on whether the ciphertext is a valid encryption of some message for the implied identity and attribute set.

We will assume that correctly formatted keys and ciphertexts will include a description of the relevant identity and dummy-attribute set. In the above notation α are called the dummy attributes and S is called a dummy attribute set. Correctness of a Dummy-IBE scheme is as expected, decryption should succeed with overwhelming probability if the identities of the key and ciphertext match and the dummy attribute sets overlap in at least d indices.

Definition 8 (Dummy-IBE Security). *A Dummy-IBE scheme is said to be* D-IBE *secure if the advantage of any probabilistic polynomial time adversary* $\mathcal{B}$ *is negligible in the game below:*

1. **Setup.** The challenger runs the **Setup** algorithm and sends PK to $\mathcal{B}$.
2. **Queries.** $\mathcal{B}$ adaptively queries keys for distinct ID_i, the challenger returns $K_{\mathrm{ID}_i}(S_i)$. This may be repeated multiple times for different identities.
3. **Challenge.** The adversary specifies M_0, M_1 and an identity ID_j which has been queried during the **Queries** phase and a dummy attribute set S such $|S \cap S_j| < d$ where S_j is from the **Queries** phase. The challenger picks $b \in \{0, 1\}$ at random and sends $\mathcal{B}$, **Encrypt**(M_b, PK, ID_j, S).
4. **Guess.** The adversary outputs a guess b' for b.

3.2 D-IBE Implies Dishonest-User Security

We describe how to transform a Dummy-IBE scheme D into a *Dishonest-User* Secure A-IBE scheme A. The overall structure of our construction can be considered a generalization of the construction in [15] where the role of our Dummy-IBE scheme is instead replaced by a ABE scheme which is not able to satisfy all the required notions of security. The similarities in structure will allow us to reuse an information theoretic result for *Dishonest-PKG* security from previous work [15]. We will from now on assume that the message space is a group of size superpolynomial in the security parameter, as is the case for our construction (notice that making a ϵ-useful decoder box is trivial if the size is not superpolynomial).

Let $d \leq k \leq n \in \Theta(\lambda)$, we give concrete bounds k, d and n should satisfy at the end of this section.

– **Setup**$(1^\lambda) = D.$**Setup**$(1^\lambda, n, k, d)$
– **KeyGen**(MK, PK, ID) Using MK, PK, ID, generate $K = \{K^\alpha_{\mathrm{ID}} : \alpha \in [1, n]\}$ and randomly permute them (call this permutation σ) and engage in a non-adaptive k-out-of-n oblivious transfer protocol with the user querying the permuted set as the set of possible values to be transferred. After the OT protocol concludes send the user σ in the clear. The user now has access to $K_{\mathrm{ID}}(S)$ for some random $S \subset [1, n]$ and runs a **KeySanity** check on $K_{\mathrm{ID}}(S)$ and terminates the protocol if it fails.

- **Encryption**(M, PK, ID) Pick $S \subset [1, n]$ at random of size k, output D.**Encrypt**(M, PK, ID, S).
- **Decrypt**($C_{ID}(S)$, PK, K_{ID}(S')) If the ciphertext sanity check of $C_{\rm ID}(S)$ passes output:
  D.**Decrypt**($C_{ID}(S)$,$K_{ID}(S')$,PK), else output $\perp$.

Notice that at the moment we only have a guarantee that the decryption algorithm will decrypt if $|S \cap S'| \geq d$. Therefore, pick k and d such that given some $S \subset [1, n], |S| = k$ (the decryption dummy-atributes) a random set $S' \subset [1, n]$ (the dummy-atributes used during encryption),

$$Pr[|S \cap S'| \geq d] > 1 - \mu(\lambda)$$

for some negligible function μ. If we pick $k = cn$, $d = (c^2 - \delta)n$ for some constant $c \in (0, 1)$ and $\delta \in (0, c^2)$ by Chernoff bounds, two sets of size k will intersect in at least d locations with overwhelming probability. From this point, assume k and d are initialized thusly and $c < 1/4$.

3.3 Tracing Algorithm

The overarching idea is that if a message has been encrypted to identity ID under dummy attribute set A, it should only be decryptable by someone holding a key $K_{ID}(A')$ where $|A \cap A'| \geq d$. Therefore, the tracing algorithm, which has oracle access to $\mathcal{D}_{\rm ID}$ will repeatedly query ciphertexts the user ID should not be able to decrypt given the attribute set that was assigned during the key generation phase. If the decoder box decrypts such a ciphertext, we will be able to conclude that the decoder box was not created by the user, proving malfeasance on the part of the PKG.

On input $K_{\rm ID}(S)$ with oracle access to $\mathcal{D}_{\rm ID}$, the tracing algorithm will run a **KeySanity** check to verify the validity of the input key. Then, it will repeat the following experiment $\nu(\epsilon) \in poly(\lambda)$ times (we will choose $\nu(\epsilon)$ after the analysis):

- Select a dummy attribute set S_i with $|S_i \cap S| < d$.
- Select a random message M and encrypt M using S_i as the dummy attributes.
- The decoder box outputs some $M' = \mathcal{D}_{\rm ID}(C_{\rm ID}(S_i))$.

If for any iteration $M' = M$, implicate the PKG otherwise, implicate the User.

Theorem 1. *If D is a secure Dummy-IBE scheme, the A-IBE construction A is Dishonest-User secure.*

Proof. Assume p.p.t. $\mathcal{A}'$ succeeds in the Dishonest-User game above with non-negligible probability δ. Since **KeyGen** is implemented with a fully-simulatable k-out-of-n oblivious transfer scheme, we can work in the OT-hybrid model by Canetti's [6] theorem on composability, which implies there exists p.p.t. $\mathcal{A}$ which also succeeds in the game with non-negligible probability such that $\mathcal{A}$ queries indices during the OT protocol directly from the challenger and the challenger simply sends the values stored in these indices to $\mathcal{A}$. With this new $\mathcal{A}$ which simply requests indices for messages to receive from the challenger as in the OT-Hybrid model, the reduction is as follows:

- **Setup.** Send the public key of the D-IBE scheme to the adversary $\mathcal{A}$.
- **Key Generation.** On each of the adversary's queries for an identity to ID_j and indices (for the OT protocol) $I \subset [1,n]$, $|I| = k$, query the D-IBE scheme for a key on identity ID_j and receive $K_{\text{ID}}(S_j) = \{K_{\text{ID}}^{\alpha_i}\}_{i\in[1,k]}$. Now pick a random σ such that $\sigma(I) = S_j$. Send $\mathcal{A}$: $\{K_{\text{ID}}^{\alpha}, \alpha \in S_j\}$ and σ.
- **Create Decoder Box.** The adversary outputs a decoder box $\mathcal{D}_{\text{ID}_j}$ along with a key $K_{ID_j}(S)$ with an identity ID_j queried at some point in the key generation phase. Run a **KeySanity** check to make sure $K_{ID_j}(S)$ is a valid key. If $S \neq S_j$ run **New Attributes** phase, otherwise proceed to the **Tracing Failure** phase.
- **New Attributes.** Since $S \neq S_j$, pick a d element subset S^* of S which does not overlap with S_j in more than $d-1$ indices (recall both are of size k) and $k-d$ element subset $A \subset [1,n] \setminus (S \cup S_j)$ (this is why we assume $d \leq k < n/4$) and initiate the challenge query with $M_0, M_1 \leftarrow \mathcal{M}$ with identity ID_j and attribute set $S^* \cup A$. Since the challenger has a key which overlaps the challenge attribute set in more than $d-1$ indices with the same identity, it can trivially decrypt and output b. This is a valid challenge since $|(S^* \cup A) \cap S_j| < d$.
- **Tracing Failure.** Pick some $d \in [1, \nu(\epsilon)]$ at random. Then, run the tracing algorithm normally $d-1$ times. For the dth iteration, choose M_0, M_1 at random and initiate the challenge phase of the D-IBE scheme by outputting both messages, the challenge identity ID and a random dummy attribute set S' such that $|S' \cap S| < d$. Receive C from the D-IBE scheme, either an encryption of M_0 or M_1 under identity ID under the attribute set A and compute $\mathcal{D}_{\text{ID}}(C_{\text{ID}}(S'))$. If $\mathcal{D}_{\text{ID}}(C_{\text{ID}}(S')) = M_0$ output 0, if $\mathcal{D}_{\text{ID}}(C_{\text{ID}}(S')) = M_1$ output 1, otherwise, guess randomly.

Now, notice that if the **New-Attributes** phase is initiated, the challenger decrypts and breaks the Dummy-IBE security trivially. Also, since the tracing algorithm only ever queries messages encrypted under dummy-attribute sets which do not overlap with the key dummy-atribute set in more than $d-1$ indices, if the tracing algorithm outputs PKG with non-negligible probability, for some $d \in [1, \nu(\epsilon)]$ it must succeed in correctly decrypting with non-negligible probability. Since the challenge ciphertext the challenger returns to it is from the exact distribution that it expects, with non-negligible probability (since $\nu(\epsilon)$ is polynomial in the security parameter) it will succeed in decrypting the challenge ciphertext, breaking semantic security. Since $\mathcal{D}_{\text{ID}}$ is oblivious to the second message in the challenge phase, with probability within a negligible function of 1, whenever we do not guess randomly, we will be correct. Since we have at least a $1/\nu(\epsilon)$ probability of not guessing randomly if the decoder decrypts on a single query and we assume the decoder will decrypt some query with non-negligible probability, we succeed in breaking semantic security.

3.4 Running in Parallel for PKG Security

To achieve Dishonest-PKG security, we actually will have to modify our construction slightly by running it in parallel m times where $m \in \Theta(\lambda)$. In this setting, if we call the A-IBE candidate above A, we will call our new scheme **A**, where **A**[1], ..., **A**[m] denote the m different instantiations of A. The public key and master secret key are simply the collection of public and master secret keys of each of the individual instantiations.

A private key is obtained for identity ID by running **KeyGen** in parallel m times and giving the output of each to the user. The user's dummy attribute set is now $\mathbf{S}$, with each $\mathbf{S}[i]$ a random k element subset of $[1, n]$. Encryption is handled by splitting M into m shares, $M = M_1 \oplus M_2 \oplus \ldots \oplus M_m$ where $M_1, \ldots M_{m-1}$ are chosen are random and each M_i is encrypted under $\mathbf{A}[i]$ with a dummy attribute set $\mathbf{S}'[i]$, with each $\mathbf{S}'[i]$ a k element subset of $[1, n]$. Now $C_{\mathrm{ID}}(\mathbf{S}')$ should be decrypted only by keys $K_{\mathrm{ID}}(\mathbf{S})$ where $|\mathbf{S}[i] \cap \mathbf{S}'[i]| \geq d$ for each $i \in [1, m]$. **KeySanity** and **CiphertextSanity** checks of **A** work by checking each of the individual components with the existing sanity checks for A. We now describe the tracing algorithm:

3.5 Tracing Algorithm for the Parallel Scheme

On input $K_{\mathrm{ID}}(\mathbf{S})$, the tracing algorithm will run a **KeySanity** check to verify the validity of the input key. Then, it will then repeat the following experiment $\nu \in poly_m(\lambda)$ times:

- Pick $j \in [1, m]$ at random,
- If $i \neq j$, choose $\mathbf{S}'[i]$ a k element subset of $[1, n]$,
- If $i = j$ include the additional restriction that $|\mathbf{S}'[i] \cap \mathbf{S}[i]| < d$,
- Select a random message M and encrypt M using $\mathbf{S}'$ as the dummy attributes,
- The decoder box outputs some $M' = \mathcal{D}_{\mathrm{ID}}(C_{\mathrm{ID}}(\mathbf{S}'))$.

If for any iteration $M' = M$ output PKG otherwise, output User.

Theorem 2. *If D is a secure Dummy-IBE scheme,* **A** *is Dishonest-User secure.*

Proof. The proof for this is identical to the proof for the proof of security for A, first pick $i \in [1, m]$ at random and for all $j \neq i$, the challenger will instantiate $\mathbf{A}[j]$ itself and use D, the Dummy-IBE scheme its trying to break to instantiate the jth index. Since there are only polynomially many choices for j, with non-negligible probability, the adversary which outputs $K_{\mathrm{ID}}(\mathbf{S}')$ and $\mathcal{D}_{\mathrm{ID}}$ will either have $\mathbf{S}'[j]$ overlap the queried keys $\mathbf{S}[j]$ less than d locations or, the tracing algorithm will decrypt a message whose dummy attribute set $\mathbf{S}^*$ has $|\mathbf{S}^*[j] \cap \mathbf{S}[j]| < d$.

Theorem 3. *If D is a secure Dummy-IBE scheme,* **A** *is Dishonest-PKG secure.*

Proof. Since the setup of **A** is very closely related to the construction in Goyal et al.[15], (which is fully secure in the *Dishonest-PKG* game) we will be able to reuse their combinatorial results for the security of **A**. They also compose m sets of dummy-attributes in parallel and use fully-simulatable OT during the key generation phase in the same manner we do. In their result, by using the ciphertext and key sanity checks they are able to prove a purely information theoretic bound on the fraction of dummy-attribute sets that will implicate the PKG with access to a decoder box $\mathcal{D}_{\text{ID}}$. They show that as long as **KeyGen** does not terminate with non-negligible probability, with overwhelming probability the PKG will not query a ciphertext the user is unable to decrypt. Since all ciphertexts which the user can decrypt decrypt to the same value by the ciphertext sanity check and the PKG can decrypt to this value by itself, this allows analysis in the Dishonest-PKG game where the PKG has no decryption capabilities.

Their result (Lemma 5 in the original paper)[2] gives us precisely the second information theoretic guarantee we need, namely that any tracing algorithm with an ϵ-useful decoder box will output PKG for all but a negligible fraction of possible dummy attributes. We refer the reader to the original paper for the proof of the lemma below. Assume m is super-logarithmic and n is linear in the security parameter λ, $k = c_1 n$, $d = c_2 n$ positive constants less than 1 such that $c_2 < c_1^2$ (which ensures decryption with overwhelming probability the Chernoff bound).

Lemma 1. *Let ϵ be non-negligible and $\mathcal{D}_{ID}$ an ϵ-useful decoder box. Let* **S** *be a dummy attribute set for the user and consider the following experiment:*

- *Select a dummy attribute set* $\mathbf{S}'$ *such that* $|\mathbf{S}'[i] \cap \mathbf{S}[i]| < d$ *for some* $i \in [1, m]$
- *Select a random message M and encrypt it using* $\mathbf{S}'$ *as the dummy attributes and outputs the ciphertext C*
- *The decoder box outputs* $M' = \mathcal{D}(C)$
- *Output* PKG *if* $M' = M$

  Then, for all but a negligible fraction of choices of **S**, *the above experiment has probability of outputting* PKG *greater than* $\epsilon/(24m)$.

By the above lemma, if we consider ideal OT functionality and reduce to the case where the PKG has no decryption capabilities, the tracing algorithm will output PKG with overwhelming probability by only running in polynomial time in the *Dishonest-PKG* game as desired as long as **S** does not come from a exceptional set which forms a negligible fraction of all possible **S**. Since full *Dishonest-PKG* security is obtained information theoretically for an identical abstract construction in [15] we refer to their paper for a rigorous treatment of the above argument.

3.6 A Modification for IND-ID-CPA Security

After we have a Dishonest-User and Dishonest-PKG secure A-IBE scheme, we may use the same method as the previous selective secure construction [15]. We will achieve IND-ID-CPA security by using the above A-IBE scheme and a secure

[2] Note that the proof of **Lemma 5** is not in the original, but instead in the full version of the cited paper.

IBE scheme together. For any M, we will secret share M into $M_1 \oplus M_2$ where M_1 was chosen at random and encrypt M_1 with the A-IBE scheme and M_2 with the IBE scheme. Since Waters' IBE scheme [25] achieves IND-ID-CPA security using the DBDH assumption, we may combine it with our Dummy-IBE scheme to achieve full A-IBE with no additional security assumptions.

4 A Dummy-IBE Scheme

The main technical contribution of this paper is the construction of a secure *Dummy-IBE* scheme, which as described, provides a fully secure *A-IBE* scheme. For our construction, we will reuse many of the methods introduced by Waters [25]. As such, we will use Waters' hash extensively. Given an identity ID $\in \{0,1\}^n$ and $\mathbf{u}[i] \in \mathbb{G}$ for $i \in [0,n]$ where ID$[j]$ is the jth bit of ID:

$$H(\mathbf{u}, \text{ID}) = \mathbf{u}[0] \prod_{j \in [1,n]} \mathbf{u}[j]^{\text{ID}[j]}$$

We use $x \xleftarrow{\$} X$ to denote x being picked at random from a set X and $S \subset_\$ X$ a set S being picked at random as a subset of X. We describe the scheme below with $|p|, n, d, k \in \Theta(\lambda)$ and $d \le k \le n/4$. The terms $e, \mathbb{G}_1, \mathbb{G}_2, g, p$ are parameters of the common bilinear map.

Setup (1^λ). Pick $a \xleftarrow{\$} \mathbb{Z}_p$, $g_1 = g^a$. $g_2 \xleftarrow{\$} \mathbb{G}$, $\mathbf{T}[i] \xleftarrow{\$} \mathbb{G}, \mathbf{u}[i] \xleftarrow{\$} \mathbb{G}$ for $i \in [0,n]$. The public key is $(g_1, g_2, \mathbf{u}, \mathbf{T})$. The master secret key is g_2^a.

KeyGen. (PK, ID) Pick $S \subset_\$ [1,n]$ of size k, $\forall\, i \in S$, $r_i \xleftarrow{\$} \mathbb{Z}_p$ and output[3]:

$$K_{\text{ID}}(S) = \left(\text{ID}, S, (g_2^a \left[H(\mathbf{u}, \text{ID})T_i\right]^{r_i}, g^{r_i})_{i \in S}\right).$$

Encrypt. (M, ID, S) Pick $c \xleftarrow{\$} \mathbb{Z}_p$. Select a $d-1$-degree polynomial q in $\mathbb{Z}_p$ at random with $q(0) = c$,

$$C_{\text{ID}}(S) = \left(\text{ID}, S, Me(g_1, g_2)^c, \left(g^{q(i)}, [H(\mathbf{u}, \text{ID})T_i]^{q(i)}\right)_{i \in S}\right).$$

Where the entries in the third component are given in ascending order of $i \in [1,n]$ to avoid ambiguity.

Decrypt($C_{\text{ID}}(S')$,$K_{ID}(S)$). Take $I \subset S' \cap S$ of size d. For all $i \in I$, compute:

$$\frac{e([H(\mathbf{u}, \text{ID})T_i]^{q(i)}, g^{r_i})}{e(g_2^a \left[H(\mathbf{u}, \text{ID})T_i\right]^{r_i}, g^{q(i)})} = \frac{1}{e(g, g_2^a)^{q(i)}} = \frac{1}{e(g_1, g_2)^{q(i)}},$$

$$Me(g_1, g_2)^c \prod_{i \in I} \left(\frac{1}{e(g_1, g_2)^{q(i)}}\right)^{\Delta_{i,I}(0)} = M.$$

Where $\Delta_{i,I}(j)$ is the Lagrange coefficient ($\Delta_{i,I}(j) = \prod_{k \in I \setminus \{i\}} \frac{k-j}{i-j}$).

[3] Notice it is possible to generate K_{ID}^α for all $\alpha \in [1,n]$ as necessary for the previous reduction.

CiphertextSanity($C_{\text{ID}}(S)$, PK). First check the input is formatted as a valid ciphertext ($|S| = k$):

$$\left(\text{ID}, S, C, \left(C_i^{(1)}, C_i^{(2)}\right)_{i \in S}\right)$$

Assuming $H(\mathbf{u}, ID)T_i \neq 1 \in \mathbb{G}$, since $\mathbb{G}, \mathbb{G}_T$ are of prime order, write the last two components for a single $i \in S$ as $g^{w_i}, [H(\mathbf{u}, ID)T_i]^{y_i}$ for some constants $w_i, y_i \in \mathbb{Z}_p$. If $H(\mathbf{u}, ID)T_i = 1 \in \mathbb{G}$, unless the second component above is also 1 output Not-Valid. The ciphertext is well formed if for all components with second entry not 1 the $w_i = y_i$ and the w_i values fall on the same $d-1$ degree polynomial q (note that the value $q(0)$ is actually completely irrelevant to the ciphertext sanity check since as long as all the w_i are on the same $d-1$ degree polynomial, if $q(0) = c$, it will be a valid encryption of $Ce(g,g)^{-c}$). To check that $w_i = y_i$, notice:

$$e\left(H(\mathbf{u}, \text{ID})T_j), C_i^{(1)}\right) \stackrel{?}{=} e\left(C_i^{(2)}, g\right) \Leftrightarrow$$

$$e\left((H(\mathbf{u}, \text{ID})T_j)^{w_i}, g\right) \stackrel{?}{=} e\left(H(\mathbf{u}, \text{ID})T_j, g^{y_i}\right) \Leftrightarrow$$

$$e\left(H(\mathbf{u}, \text{ID})T_j), g\right)^{w_i} \stackrel{?}{=} e\left(H(\mathbf{u}, \text{ID})T_j, g\right)^{y_i} \Leftrightarrow w_i \stackrel{?}{=} y_i.$$

The final implication follows from the fact that both numbers being paired are not 1 in their original group of prime order and so the target group element is not 1 with non-trivial pairing. Now, to check that all the w_i lie on the same $d-1$ degree polynomial, pick a subset $I \subset S$ with $|J| = d$. Then, interpolate $C_i^{(1)}$ to all the other $x \in S \setminus I$ values. In other words, for all $x \in S \setminus I$, check that:

$$\prod_{i \in I} C_i^{(1)\,\Delta_{i,I}(x)} = \prod_{i \in I} g^{w_i \Delta_{i,I}(x)} \stackrel{?}{=} g^{w_x} = C_x^{(1)}.$$

It's now clear that this process will only accept if all w_i lie on some polynomial q of degree not exceeding $d-1$. This suffices to show that the interpolation to $q(0)$ is unique no matter which d points are picked, showing this is a valid ciphertext (can only be decrypted to one value).

KeySanity($K_{\text{ID}}(S)$, PK). First check that the key is formatted correctly. If $H(\mathbf{u}, ID)T_i = 1 \in \mathbb{G}$, checking this component is valid is trivial (the user then should have access to the master secret key and can check this with one pairing from the PK). We can now write each key component as:

$$(H(\mathbf{u}, ID)T_i)^r\, g_2^a, g^{r'}.$$

This will now be a valid key component if $r = r'$ (for elements in $\mathbb{Z}_p$ we write equality as elements in this group). Check (recall $g_1 = g^a$):

$$e\left[(H(\mathbf{u}, ID)T_i)^r\, g_2^a, g\right] \stackrel{?}{=} e\left(H(\mathbf{u}, ID)T_i, g^{r'}\right) e(g_2, g_1) \Leftrightarrow$$

$$e\left[(H(\mathbf{u}, ID)T_i), g\right]^r \stackrel{?}{=} e\left[(H(\mathbf{u}, ID)T_i), g\right]^{r'}.$$

Which is equivalent with $r \stackrel{?}{=} r'$ since g, $H(\mathbf{u}, ID)T_i$ are not 1.

4.1 Proof of Security

We will now prove the *Dummy-IBE* security of the given construction. Let q be an upper bound on the number of queries an adversary $\mathcal{A}$ makes to succeed in the *Dummy-IBE* security game with non-negligible probability ϵ. We now describe how to use this simulator to succeed in the DBDH game with non-negligible probability. Below m will be a value polynomial in an upper bound of the number of queries the adversary makes that will be initialized later in the analysis. We now describe the simulator $\mathcal{S}$ that will be given as input (A, B, C, Z) where $(g, g_1, g_2, C) = (g, g^a, g^b, g^c)$ and $Z = e(g,g)^{abc}$ or $Z \xleftarrow{\$} \mathbb{G}_T$. Generate Φ a random k element subset of $[1, n]$, which will serve as the dummy attribute set of the challenge query and:

$$\mathbf{x}[i] \xleftarrow{\$} [0, m-1] \text{ for } i \in [1, n]\ ,\ \mathbf{x}[0] \xleftarrow{\$} [-n(m-1), 0]$$

$$\mathbf{y}[i] \xleftarrow{\$} \mathbb{Z}_p \text{ for } i \in [0, m-1].$$

Define the following two functions:

$$F'(\mathbf{x}, \mathrm{ID}) = \mathbf{x}[0] + \sum_{i=1}^{n} \mathbf{x}[i]\mathrm{ID}[i]$$

$$G'(\mathbf{y}, I) = \mathbf{y}[0] + \sum_{i=1}^{n} \mathbf{y}[i]\mathrm{ID}[i].$$

Our simulator $\mathcal{S}$ will abort unless for exactly one key query $\mathbf{ID}[i]$, $F'(\mathbf{x}, \mathbf{ID}[i]) = 0$ and for all others $F'(\mathbf{x}, \mathbf{ID}[j]) \notin \{0, 1\}$. Additionally, $\mathcal{S}$ will abort and guess randomly unless $\mathbf{ID}[i]$ as specified above is not the challenge identity (recall that by the definition of *Dummy-IBE* security, we may assume that the challenge identity has been queried once during the key queries phase and that the challenge dummy-attribute set overlaps the dummy-attribute set the adversary received during the key generation phase in no more than $d-1$ indices). We now describe how $\mathcal{S}$ will generate the public key and answer key queries.

Simulated Public Key Generation. $\mathcal{S}$.**Setup**(1^λ) is defined as follows: For $i \in [1, n]$, $\mathbf{B}[i] \xleftarrow{\$} \mathbb{Z}_p$ and,

$$\mathbf{T}[j] = \begin{cases} g_2 g^{\mathbf{B}[j]}, & \text{if } j \in \Phi; \\ g^{\mathbf{B}[j]}, & \text{otherwise.} \end{cases}$$

Related to $\mathbf{T}$, define $G(\mathbf{y}, \mathrm{ID}, j) = G'(\mathbf{y}, \mathrm{ID}) + \mathbf{B}[j]$ and,

$$F(\mathbf{x}, \mathrm{ID}, \Phi, j) = \begin{cases} F'(\mathbf{x}, \mathrm{ID}) + 1, & \text{if } j \in \Phi; \\ F'(\mathbf{x}, \mathrm{ID}), & \text{otherwise.} \end{cases}$$

Output PK $= (g_1, g_2, \mathbf{u}, \mathbf{T})$. Notice if $\mathbf{u}[i] = g_2^{\mathbf{x}[i]} g^{\mathbf{y}[i]}$ then, $H(\mathbf{u}, \mathrm{ID})\mathbf{T}[i] = g_2^{F(\mathbf{x}, \mathrm{ID}, \Phi, i)} g^{G(\mathbf{y}, \mathrm{ID}, j)}$. Observe the output is completely independent of Φ and the output distribution is precisely that of the true public key generation algorithm.

Simulated KeyGen Queries. If $F'(\mathbf{x}, \mathrm{ID}) \notin \{-1, 0\}$, $\mathcal{S}$.**KeyGen**(ID) is defined as follows: Pick $S \subset_\$ [1, n]$ of size k. For $j \in S$: $r \stackrel{\$}{\leftarrow} \mathbb{Z}_p$ and let $w = F(\mathbf{x}, \mathrm{ID}, \Phi, j)^{-1}$ (which exists since $F'(\mathbf{x}, \mathrm{ID}) \neq 0$). Set:

$$(K_j^{(1)}, K_j^{(2)}) = (g_2^{F(\mathbf{x},\mathrm{ID},\Phi,j)\cdot r} g^{G(\mathbf{y},\mathrm{ID},\Phi)\cdot r} g_1^{-G(\mathbf{y},\mathrm{ID},\Phi)\cdot w}, g^r g_1^{-w})$$

For all $j \in S$. Output $K_{\mathrm{ID}}(S) = \left(\mathrm{ID}, S, (K_j^{(1)}, K_j^{(2)})_{j \in S}\right)$. Notice for $j \in S$:

$$\begin{aligned} g_2^{F(\mathbf{x},\mathrm{ID},\Phi,j)\cdot r} g^{G(\mathbf{y},\mathrm{ID},\Phi)\cdot r} g_1^{-G(\mathbf{y},\mathrm{ID},\Phi)\cdot w} &= g_2^a (g_2^{F(\mathbf{x},\mathrm{ID},\Phi,j)} g^{G(\mathbf{y},\mathrm{ID},\Phi)})^{-a\cdot w} (g_2^{F(\mathbf{x},\mathrm{ID},\Phi,j)} g^{G(\mathbf{y},\mathrm{ID},\Phi)})^r \\ &= g_2^a (g_2^{F(\mathbf{x},\mathrm{ID},\Phi,j)} g^{G(\mathbf{y},\mathrm{ID},\Phi)})^{r-a\cdot w}. \end{aligned}$$

And similarly $g^r g_1^{-w} = g^{r-a\cdot w}$. Therefore, this is a valid key component with randomness $r - a \cdot w$, which is also distributed uniformly over $\mathbb{Z}_p$. So, every key component is generated from the correct distribution if $F(\mathbf{x}, \mathrm{ID}) \notin \{-1, 0\}$. This is where the main divergence with Waters' proof occurs, we must be able to handle one query to the identity that will later be challenged on. For this, we notice that if $j \in \Phi$, and $F'(\mathbf{x}, \mathrm{ID}) = 0$ then, $F(\mathbf{x}, \mathrm{ID}, \Phi, j) \neq 0$ and we can repeat the above process by taking $S = \Phi$.

If $F'(\mathbf{x}, \mathrm{ID}) = 0$, $\mathcal{S}$.**KeyGen**(ID) is the same as the above except instead of $S \subset_\$ [1, n]$, take $S = \Phi$.

Recall that $\mathcal{S}$ aborts if there is ever more than one query with $F'(\mathbf{x}, \mathrm{ID}) = 0$ and that the output of the simulated public key is independent of Φ. Therefore, if Φ is only used once during the key queries phase, as the dummy-attribute set of a single query, the view of the adversary interacting with $\mathcal{S}$ is still identical to the view of the adversary interacting with the true *Dummy-IBE* scheme.

Simulated Challenge Ciphertext. Assume $F'(\mathbf{x}, \mathrm{ID}) = 0$ and $|S \cap \Phi| < d$ then,

$\mathcal{S}$.**Challenge**$(M_0, M_1, \mathrm{ID}, S)$ is defined as follows: Pick $\gamma \stackrel{\$}{\leftarrow} \{0, 1\}$ and let $C_1 = Z \cdot M_\gamma$ and choose $S \supset K \supset S \cap \Phi$, with $|K| = d - 1$ and:

For $i \in K$:

$$r_i \stackrel{\$}{\leftarrow} \mathbb{Z}_p\,,\; C_i^{(1)} = g^{r_i}\,,\; C_i^{(2)} = \left(g_2^{F(\mathbf{x},\mathrm{ID},\Phi,i)} g^{G(\mathbf{y},\mathrm{ID},i)}\right)^{r_i}.$$

For $i \notin K$:

$$C_i^{(1)} = (g^c)^{\Delta_{0,K}(i)} \prod_{j \in K} (g^{r_j})^{\Delta_{j,K}(i)}, C_i^{(2)} = \left(C_i^{(1)}\right)^{G(\mathbf{y},\mathrm{ID},i)}.$$

Respond to the query with:

$$C_{\mathrm{ID}}(S) = \left(\mathrm{ID}, S, C_1, \left(C_i^{(1)}, C_i^{(2)}\right)_{i \in S}\right).$$

Remark (1). Notice the encryption is valid with the implied polynomial $q(x)$ in the exponent being defined by $q(i) = r_i \xleftarrow{\$} \mathbb{Z}_p$ for all $i \in K$ and $q(0) = c$ where $C = g^c$. Since choosing a $d-1$ degree polynomial at random is equivalent to specifying its value randomly at d points, the encryption is valid and uniform (since r_i are generated at random during the simulated challenge and the view of the adversary has been independent of c which was chosen at random in the DBDH initialization). This shows that the $C_i^{(1)}$ components are correctly generated from the stated distribution for $i \in K$.

Remark (2). Additionally, since $F'(\mathbf{x}, \mathrm{ID}) = 0$, we have that for all $i \notin K (\Rightarrow i \notin \Phi$ if $i \in S)$, $F(\mathbf{x}, \mathrm{ID}, \Phi, i) = 0$ and therefore $H(\mathbf{u}, \mathrm{ID})T_i = g^{G(\mathbf{y}, \mathrm{ID}, i)}$ which shows the second component is also generated correctly with implied polynomial q. Therefore, as long as $F'(\mathbf{x}, ID) = 0$ the $C_i^{(b)}$ values are drawn from the correct distribution.

Therefore, if Z is $e(g_1, g_2)^c$ we have the above is a uniform encryption of M_γ. Otherwise, Z is a random element of $\mathbb{G}_T$ and the ciphertext gives no information on the choice of γ.

To remind the reader of the relevant security game, the adversary $\mathcal{A}$ will make q distinct key queries $\mathbf{ID}[i]$ to which the simulator will respond with keys $K_{\mathbf{ID}[i]}(\mathbf{S}[i])$ where $\mathbf{S}[i]$ is a k element subset of $[1, n]$ consisting of the dummy attributes of the key. Finally, the adversary will make a challenge query with $M_0, M_1, S^*, \mathbf{ID}[j]$ where $\mathbf{ID}[j]$ with $j \in [1, q]$ was queried before and $|S^* \cap \mathbf{S}[j]| < d$. Our simulator will not abort if and only if:

1. $F'(\mathbf{x}, \mathbf{ID}[i]) \notin \{0, 1\}$ for all $i \in [1, q] \setminus \{j\}$ for some $j \in [1, q]$,
2. $F'(\mathbf{x}, \mathbf{ID}[j]) = 0$,
3. The challenge identity is $\mathbf{ID}[j]$.

This is very similar to the requirement in Waters' IBE, where it is required that $F'(\mathbf{x}, \mathbf{ID}[i]) \neq 0$ for all key queries and $F'(\mathbf{x}, \mathbf{ID}[j]) = 0$ for the challenge identity $\mathbf{ID}[j]$. Here, we require that the challenge identity be queried once before, with the small caveat that for most key queries we are aborting on two values instead of one, but the impact of this on analysis is minimal. Notice that if our algorithm does not abort, it is drawing from the same distribution as the real *Dummy-IBE* scheme since while Φ is used as the dummy attribute set for a single query, the view of the adversary otherwise is independent of Φ (the requirement that $|\Phi \cap S| < d$ is automatically fulfilled since $\mathcal{A}$ was given Φ as the dummy attribute set of $\mathbf{ID}[j]$ and therefore, by the rules of the *Dummy-IBE* security game, the dummy attribute set of the challenge must overlap the dummy attribute set received during the key generation phase in less than d indices). It only remains to bound below the probability of aborting naturally and introduce an artifical abort step to make sure the success probability of the simulator is not correlated with the probability of aborting. We defer bounding the abord probability to **Appendix A**. If for any sequence of queries, our simulation has a non-negligible probability of perfectly simulating the behavior of *Dummy-IBE* security game in the case where the fourth member of the DBDH tuple is $Z = e(g, g)^{abc}$ and in the other case, all information on the challenge message is lost, we can use the advantage of a distinguishing adversary to discern which of the above two possibilities the fourth member of the DBDH tuple is, since its advantage can only be maintained (information theoretically) in the case where $Z = e(g,g)^{abc}$.

5 Conclusion

We have demonstrated the first provably secure black-box accountable authority IBE scheme, answering the primary question posed in multiple previous works [13,15] using the standard DBDH assumption. To achieve this goal, we introduced the notion of *Dummy-IBE* encryption, a hybrid between usual IBE and *Attribute-Based Encryption* where the exact attributes given to an identity are not important but which should exist for some added functionality (in this case tracing), which may be of independent interest.

References

1. Bellare, M., Ristenpart, T.: Simulation without the Artificial Abort: Simplified Proof and Improved Concrete Security for Waters' IBE Scheme. In: Joux, A. (ed.) EUROCRYPT 2009. LNCS, vol. 5479, pp. 407–424. Springer, Heidelberg (2009)
2. Boneh, D., Franklin, M.: Identity-based encryption from the Weil pairing. SIAM Journal on Computing 32(3), 586–615 (2003)
3. Boneh, D., Boyen, X.: Secure Identity Based Encryption Without Random Oracles. In: Franklin, M.K. (ed.) CRYPTO 2004. LNCS, vol. 3152, pp. 443–459. Springer, Heidelberg (2004)
4. Boneh, D., Franklin, M.K.: Identity-Based Encryption from the Weil Pairing. In: Kilian, J. (ed.) CRYPTO 2001. LNCS, vol. 2139, pp. 213–229. Springer, Heidelberg (2001)
5. Camenisch, J., Neven, G., Shelat, A.: Simulatable Adaptive Oblivious Transfer. In: Naor, M. (ed.) EUROCRYPT 2007. LNCS, vol. 4515, pp. 573–590. Springer, Heidelberg (2007)
6. Canetti, R.: Security and composition of multiparty cryptographic protocols. Journal of Cryptology 13(1), 143–202 (2000)
7. Cocks, C.: An Identity Based Encryption Scheme Based on Quadratic Residues. In: Honary, B. (ed.) Cryptography and Coding 2001. LNCS, vol. 2260, pp. 360–363. Springer, Heidelberg (2001)
8. Even, S., Goldreich, O., Lempel, A.: A randomized protocol for signing contracts. Communications of the ACM 28(6), 647 (1985)
9. Gentry, C.: Practical identity-based encryption without random oracles. In: Vaudenay, S. (ed.) EUROCRYPT 2006. LNCS, vol. 4004, pp. 445–464. Springer, Heidelberg (2006)
10. Gentry, C., Halevi, S.: Hierarchical Identity Based Encryption with Polynomially Many Levels. In: Reingold, O. (ed.) TCC 2009. LNCS, vol. 5444, pp. 437–456. Springer, Heidelberg (2009)
11. Gentry, C., Silverberg, A.: Hierarchical ID-Based Cryptography. In: Zheng, Y. (ed.) ASIACRYPT 2002. LNCS, vol. 2501, pp. 548–566. Springer, Heidelberg (2002)

12. Goyal, V., Pandey, O., Sahai, A., Waters, B.: Attribute-based encryption for fine-grained access control of encrypted data. In: Proceedings of the 13th ACM Conference on Computer and Communications Security, p. 98. ACM, New York (2006)
13. Goyal, V.: Reducing Trust in the PKG in Identity Based Cryptosystems. In: Menezes, A. (ed.) CRYPTO 2007. LNCS, vol. 4622, pp. 430–447. Springer, Heidelberg (2007)
14. Goyal, V., Jain, A., Pandey, O., Sahai, A.: Bounded Ciphertext Policy Attribute Based Encryption. In: Aceto, L., Damgård, I., Goldberg, L.A., Halldórsson, M.M., Ingólfsdóttir, A., Walukiewicz, I. (eds.) ICALP 2008, Part II. LNCS, vol. 5126, pp. 579–591. Springer, Heidelberg (2008)
15. Goyal, V., Lu, S., Sahai, A., Waters, B.: Black-box accountable authority identity-based encryption. In: Ning, P., Syverson, P.F., Jha, S. (eds.) ACM Conference on Computer and Communications Security, pp. 427–436. ACM, New York (2008)
16. Green, M., Hohenberger, S.: Blind identity-based encryption and simulatable oblivious transfer. In: Kurosawa, K. (ed.) ASIACRYPT 2007. LNCS, vol. 4833, pp. 265–282. Springer, Heidelberg (2007)
17. Horwitz, J., Lynn, B.: Toward Hierarchical Identity-Based Encryption. In: Knudsen, L.R. (ed.) EUROCRYPT 2002. LNCS, vol. 2332, pp. 466–481. Springer, Heidelberg (2002)
18. Lewko, A., Okamoto, T., Sahai, A., Takashima, K., Waters, B.: Fully Secure Functional Encryption: Attribute-Based Encryption and (Hierarchical) Inner Product Encryption. In: Gilbert, H. (ed.) EUROCRYPT 2010. LNCS, vol. 6110, pp. 62–91. Springer, Heidelberg (2010)
19. Libert, B., Vergnaud, D.: Towards Black-Box Accountable Authority IBE with Short Ciphertexts and Private Keys. In: Jarecki, S., Tsudik, G. (eds.) PKC 2009. LNCS, vol. 5443, pp. 235–255. Springer, Heidelberg (2009)
20. Lindell, A.Y.: Efficient Fully-Simulatable Oblivious Transfer. In: Malkin, T. (ed.) CT-RSA 2008. LNCS, vol. 4964, pp. 52–70. Springer, Heidelberg (2008)
21. Okamoto, T., Takashima, K.: Fully Secure Functional Encryption with General Relations from the Decisional Linear Assumption. In: Rabin, T. (ed.) CRYPTO 2010. LNCS, vol. 6223, pp. 191–208. Springer, Heidelberg (2010)
22. Ostrovsky, R., Sahai, A., Waters, B.: Attribute-based encryption with non-monotonic access structures. In: Proceedings of the 14th ACM Conference on Computer and Communications Security, p. 203. ACM, New York (2007)
23. Sahai, A., Waters, B.: Fuzzy identity-based encryption. In: Cramer, R. (ed.) EUROCRYPT 2005. LNCS, vol. 3494, pp. 457–473. Springer, Heidelberg (2005)
24. Shamir, A.: Identity-based cryptosystems and signature schemes. In: Blakely, G.R., Chaum, D. (eds.) CRYPTO 1984. LNCS, vol. 196, pp. 47–53. Springer, Heidelberg (1985)
25. Waters, B.: Efficient identity-based encryption without random oracles. In: Cramer, R. (ed.) EUROCRYPT 2005. LNCS, vol. 3494, pp. 114–127. Springer, Heidelberg (2005)

A Bounding the Abort Probability

Artificial Abort. Additionally to the probability of aborting naturally, we must make sure that if the simulator $\mathcal{S}$'s probability of success is not correlated with its probability of aborting. Since this method is standard by now, we refer the reader to Waters' original work [25] for more in depth analysis of the necessity of this step. Our application differs little from the analysis in [25] except for a factor of 2 due to the fact that we have two restricted values but include the analysis for completeness[4].

As in Waters' scheme we can now define a second simulator which first generates the secret key g_2^a before initializing $\mathbf{x}$ and $\mathbf{y}$. With the master secret key, it is now able to perfectly respond to all key generation queries. Now, this second simulator will abort and guess randomly unless $\mathbf{ID}[j]$, the challenge identity (which has been queried once before) satisfies $F'(\mathbf{x}, \mathbf{ID}[j]) = 0$ and for all other identities keys are queried for $\mathbf{ID}[i] \neq \mathbf{ID}[j]$, $F'(\mathbf{x}, \mathbf{ID}[j]) \notin \{0, 1\}$.

As long as we can show the second simulator has a non-negligible probability of not aborting for any set of queries, by using artificial aborts, we can get a near-uniform, non-negligible probability to abort and guess randomly for any set of identity queries. Then, our first simulator works as a distinguisher in the DBDH game since if for a given set of queries our simulator has not aborted, the adversary will have a non-negligible advantage of guessing the correct b if $Z = e(g, g)^{abc}$ and no advantage in guessing b if Z is random since b is information theoretically masked in the ciphertext.

Analysis of Abort Probability: We now give a lower bound on the probability for a set of identities $\{\mathbf{ID}[i]\}_{i \in [1,q]}$ with $\mathbf{ID}[j]$, $j \in [1, q]$ the challenge identity that the second simulator does not abort.

$$
\begin{aligned}
\Pr[\overline{\texttt{abort}}] &= \Pr[\bigwedge_{i \in [1,q] \setminus \{j\}} F'(\mathbf{x}, \mathbf{ID}[i]) \notin \{0,1\}] \wedge F'(\mathbf{x}, \mathbf{ID}[j]) = 0 \\
&= (1 - \Pr[\bigvee_{i \in [1,q] \setminus \{j\}} F'(\mathbf{x}, \mathbf{ID}[i]) \in \{0,1\}]) \times \Pr[F'(\mathbf{x}, \mathbf{ID}[j]) = 0 | \bigwedge_{i \in [1,q] \setminus \{j\}} F'(\mathbf{x}, \mathbf{ID}[i]) \notin \{0,1\}] \\
&\geq (1 - \sum_{i \in [1,q] \setminus \{j\}} \Pr[F'(\mathbf{x}, \mathbf{ID}[i]) \in \{0,1\}]) \times \Pr[F'(\mathbf{x}, \mathbf{ID}[j]) = 0 | \bigwedge_{i \in [1,q] \setminus \{j\}} F'(\mathbf{x}, \mathbf{ID}[i]) \notin \{0,1\}] \\
&\geq (1 - \frac{2q}{(n+1)m}) \Pr[F'(\mathbf{x}, \mathbf{ID}[j]) = 0 | \bigwedge_{i \in [1,q] \setminus \{j\}} F'(\mathbf{x}, \mathbf{ID}[i]) \notin \{0,1\}]
\end{aligned}
$$

As in lines (1e) through (1i) in Waters' derivation [25] we can simplify the probability on the right to:

$$
= \frac{\Pr[F'(\mathbf{x}, \mathbf{ID}[j]) = 0]}{\Pr[\bigwedge_{i \in [1,q] \setminus \{j\}} F'(\mathbf{x}, \mathbf{ID}[i]) \notin \{0,1\}]} \times \left(1 - \Pr[\bigvee_{i \in [1,q] \setminus \{j\}} F'(\mathbf{x}, \mathbf{ID}[i]) \in \{0,1\} | F'(\mathbf{x}, \mathbf{ID}[j]) = 0]\right)
$$

Which can be bounded below by:

$$
\frac{1}{(n+1)m}(1 - \sum_{i \in [1,q] \setminus \{j\}} \Pr[F'(\mathbf{x}, \mathbf{ID}[i]) \in \{0,1\} | F'(\mathbf{x}, \mathbf{ID}[j]) = 0]).
$$

[4] Note that it is possible to prove our main result in a fashion similar to Bellare and Ristenpart's recent simplification [1] of Waters' proof using game playing techniques but we include this proof due to its transparency. A version of this paper which includes a game playing proof of security is available from the authors.

Notice we can replace $F'(\mathbf{x}, \mathbf{ID}[i]) \in \{0, 1\}$ with $F'(\mathbf{x}, \mathbf{ID}[i]) \in \{0, 1\} \bmod m$ to get a lower bound. Now, for each $\mathbf{ID}[i]$ with $i \neq j$, $F'(\mathbf{x}, \mathbf{ID}[i])$ modulo m is independent from $F'(\mathbf{x}, \mathbf{ID}[j])$ by the properties of the Waters hash, and therefore, using this substitution the above is bounded below by:

$$\frac{1}{(n+1)m}\left(1 - \frac{2q}{m}\right).$$

Substituting back to the original inequality yields:

$$\left(1 - \frac{2q}{(n+1)m}\right)\frac{1}{(n+1)m}\left(1 - \frac{2q}{m}\right) \geq \frac{1}{(n+1)m}\left(1 - \frac{4q}{m}\right).$$

Which can be taken to be non-negligible with $m = 8q$. This shows for any sequence of queries, the probability that the simulator does not abort can be bounded below by a non-negligible amount. As discussed before, using artificial aborts, this suffices to prove security.

One-Pass HMQV and Asymmetric Key-Wrapping

Shai Halevi and Hugo Krawczyk

IBM Research

Abstract. Consider the task of asymmetric key-wrapping, where a key-management server encrypts a cryptographic key under the public key of a client. When used in storage and access-control systems, it is often the case that the server has no knowledge about the client (beyond its public key) and no means of coordinating with it. For example, a wrapped key used to encrypt a backup tape may be needed many years after wrapping, when the server is no longer available, key-wrapping standards have changed, and even the security requirements of the client might have changed. Hence we need a flexible mechanism that seamlessly supports different options depending on what the original server was using and the current standards and requirements.

We show that one-pass HMQV (which we call HOMQV) is a perfect fit for this type of applications in terms of security, efficiency and flexibility. It offers server authentication if the server has its own public key, and degenerates down to the standardized DHIES encryption scheme if the server does not have a public key. The performance difference between the unauthenticated DHIES and the authenticated HOMQV is very minimal (essentially for free for the server and only 1/2 exponentiation for the client). We provide a formal analysis of the protocol's security showing many desirable properties such as sender's forward-secrecy and resilience to compromise of ephemeral data. When adding a DEM part (as needed for key-wrapping) it yields a secure signcryption scheme (equivalently a UC-secure messaging protocol).

The combination of security, flexibility, and efficiency, makes HOMQV a very desirable protocol for asymmetric key wrapping, one that we believe should be incorporated into implementations and standards.

1 Introduction

Key management is an essential component of secure systems, it is used in many systems both for confidentiality and for enforcing access-control policies. In this work we deal with settings where we have a key-server that needs to securely send a symmetric key to a client. The symmetric key could be a freshly generated one (to be used later by Alice) or it could be a pre-set key (e.g., to let Alice decrypt a previously encrypted file). We call the server Bob (or the sender) and call the client Alice (or the receiver). We specifically focus on settings where the protocol must be one-way, with the server sending a single message to the client.

D. Catalano et al. (Eds.): PKC 2011, LNCS 6571, pp. 317–334, 2011.

For example, consider a typical tape-encryption setting, where backup tapes are encrypted and then stored for future need by a potential client Alice who may need them one day. To enable decryption, Bob "wraps" the encryption key under some public key and stores the wrapped key to the tape itself. Note that Alice is off-line (or perhaps does not even exist) when the encryption key is wrapped, so no interaction can take place. Later, when Alice comes to need the backup tape, she asks her key-management module for the private key corresponding to the public key used by Bob, and then she can unwrap the decryption key and decrypt the backup tape. Such one-way communication can be viewed as *one-pass key-exchange protocols* with implicit client authentication (and in some cases also server authentication). In the case that a pre-set key is transmitted (as in the tape encryption example), the operation is referred to as *key wrapping*.

Key-wrapping (and key-management in general) is a topic of intensive activity in the industry, with many competing services and many standardization efforts (e.g., PKCS #11, FIPS SP 800-57 and 800-130, IETF KeyProv, OASIS KMIP, IEEE 1619.3, IEEE P1363, etc). Symmetric key-wrapping was addressed in the work of Rogaway and Shrimpton [16] and later Gennaro and Halevi [8], with a focus on using deterministic encryption for this purpose. In this work, however, we focus on asymmetric key-wrapping, where Bob uses Alice's public key to wrap the symmetric key. In this context the added complexity of using randomization is insignificant in comparison to the public-key operations that are needed. Hence we just use standard encryption, and view key-wrapping as a target application rather than a separate security goal.

Many real-world implementations of asymmetric key wrapping are based on RSA, but the increase in the use of elliptic-curve cryptosystems suggest that they will be useful for key wrapping as well. The leading mechanisms in this respect is the "elliptic-curve integrated encryption scheme" (ECIES) [12], which is based on the "Diffie-Hellman integrated encryption scheme" (DHIES) encryption scheme of Abdalla et al. [1]. DHIES is an Elgamal-based encryption scheme, proven CCA secure under the "hashed Diffie-Hellman" assumption. (Alternatively, under the Gap Diffie-Hellman assumption in the random-oracle model).

On a high level, in DHIES Alice has a secret exponent a, and the corresponding public key is the group element $A = g^a$ (where g is a generator in a prime-order group). To encrypt a message M, Bob chooses a random exponent y and computes $Y = g^y$, then computes the Diffie-Hellman value $\sigma = A^y$ and uses $K = H(\sigma, Y)$ as a key in a symmetric-key authenticated-encryption scheme to encrypt M. Note that DHIES is an instance of the KEM/DEM paradigm [17].

When used for key-wrapping (i.e., when M is a cryptographic key), this encryption scheme can also be viewed as a key-exchange protocol where only the client is implicitly authenticated. However, there are applications where the server too should be authenticated. For example, consider the tape-backup application from above where Alice is a third party that provides backup/restore services. When Alice gets a tape with a wrapped key on it, she wants to consult the policy of the original server Bob to know if decryption is permitted. So in particular Alice needs to be able to authenticate the source of the wrapped key.

A natural solution is for Bob to sign the ciphertext (e.g. with ECDSA signature if we want an elliptic-curve scheme), but this solution has some drawbacks. For one thing, it adds non-trivial complexity. We need to implement the signature scheme and use additional bandwidth to communicate the signatures. Also, in some cases this does not even provide adequate authentication, since it allows an adversary Charlie to strip the original Bob signature from the wrapped key and replace it with a signature by Charlie. (This may enable Charlie to later ask Alice to unwrap the key and decrypt the tape for him).

Instead, one would like a solution that (a) ensures that the identity of Bob cannot be stripped from the key, and (b) remains as close as possible to the base DHIES scheme. A good candidate to achieve this goal has been suggested in [13], namely the one-pass HMQV protocol. However, while a main appeal of HMQV is its provable security, the one-pass variant of HMQV has not been proven. Here we give a full specification of a one-pass HMQV protocol, with a full proof of security. We call this protocol HOMQV (for **H**ashed-**O**ne-pass-**MQV**, pronounced "Home-Queue-Vee"). We note that HOMQV is different than the one-pass HMQV protocol in [13, Sec. 9] in that HOMQV hashes the session identifier while deriving the session key (as suggested in [14]).

Roughly speaking, the only difference between HOMQV and (the KEM part of) DHIES is the way the Diffie-Hellman value σ is computed. Whereas in DHIES we set $\sigma = A^y = Y^a$, here Bob also has public and secret keys ($B = g^b$) so we use them in the computation. Roughly, we compute a half-size exponent $e = \bar{H}(Y)$ and then set $\sigma = (YB^e)^a = A^{y+be}$ and $K = H(\sigma, Y)$. (In the actual protocol we also add the identities of Alice and Bob inside the hashing and use cofactor exponentiation, see Table 1 for an overview and Section 3 for a precise definition of HOMQV).

We also show that slight variations of the same base protocol can handle a large variety of scenarios in the key transmission and key wrapping settings. For example, a server Bob that does not have a public key can just use the dummy public key $B = 1$, and then the protocol reverts to (the KEM part of) the underlying DHIES. Other variants of the protocol offer increased security at a minimal cost in computation and communication. In all cases the schemes provide forward secrecy for the server, namely the compromise of the sender's private key (if any) does not expose past transmissions[1]. This property has particular importance for our setting, since in key-management and storage applications some of the keys may have a very long lifetime. In some variants this property is achieved in a weak sense, namely, as long as the attacker was passive during these transmissions, but we also show inexpensive variants that provide full sender forward secrecy, even against active attacks.

Remarkably, all the variants except the (degenerate) case of DHIES enjoy the security property that leakage of the ephemeral Diffie-Hellman exponent y does not compromise security; the function of this exponent is to provide forward-secrecy capabilities. We call this property *y-security*. The significance of this

[1] On the other hand, it is easy to see that one-pass protocols inherently cannot offer PFS for the receiver.

Table 1. One-pass modes. In all $Y = g^y$, $\sigma = (YB^e)^a = A^{y+be}$, $e = \bar{H}(\hat{A}, Y)$, $K = H(\sigma, \hat{B}, \hat{A}, Y)$, $\hat{B}, \hat{A}$ represent the identities of sender and receiver. In DHIES we use $B = 1, b = 0$. Last column counts number of exponentiations per sender/receiver.

KEM modes	$\hat{B}$ sends to $\hat{A}$	Key-derivation	impl.auth.	y-secure	Sender FS	#exp (s/r)
DHIES KEM	Y	$SK = K$	$\hat{A}$ only	No	No	2/1
HOMQV	Y	$SK = K$	$\hat{B}, \hat{A}$	Yes	weak	2/1.5
HOMQV + key-conf.	Y, $\text{MAC}_{K_a}(1)$	$SK = PRF_K(0)$ $K_a = PRF_K(1)$	$\hat{B}, \hat{A}$	Yes	full	2/1.5

ENC. modes	$\hat{B}$ sends to $\hat{A}$	Key-derivation	DEM	security	#exp (s/r)
DHIES	Y, C, T	$K_a = PRF_K(1)$ $K_e = PRF_K(2)$	$C = \text{ENC}_{K_e}(M)$ $T = \text{MAC}_{K_a}(C)$	CCA	2/1
HOMQV + DEM	Y, C, T	$K_a = PRF_K(1)$ $K_e = PRF_K(2)$	$C = \text{ENC}_{K_e}(M)$ $T = \text{MAC}_{K_a}(C)$	signcryption	2/1.5

property is that it allows pre-computing (y, g^y) pairs, even if they are stored in less secure media.

We have three variants for the basic setting of a one-pass KEM-only protocol: From the "degenerate" DHIES (that offers only CCA security), via plain HOMQV (that offers server-authentication and server-PFS against passive attacks), to a variant with key-confirmation that adds server-PFS also against active attacks. When adding the DEM part, we get two variants of encryption (or key-wrapping): Without server authentication we only get the CCA-secure DHIES encryption, and with server authentication we get a secure signcryption scheme (equivalently, UC-secure replayable message transmission). These variants (with a slight simplification) are summarized in Table 1, detailed description are found in Sections 3, 4, and 5.4.

We note that the precise connection between key-exchange and signcryption was established in the work of Gorantla et al. [10], and a comprehensive theory of KEM/DEM for signcryptions was developed by Dent [5,6,7]. See Section 4.

As all these schemes are one-pass protocols, they are all inherently open to replay attacks. This can be addressed by having the sender and receiver maintain synchronized state via counters or timestamps, or having the receiver store all past communication (or a combination of both). Alternatively, in Section 6 we show that HOMQV extends smoothly to the interactive setting, where we can prevent replay and offer full PFS for both sides.

In this paper we analyze all these variants and prove their security in the Canetti-Krawczyk model of key exchange [3] (specialized to the case of one-pass protocols). We prove both "basic security" of these variants as well as the additional properties of forward security and y-security (i.e., resilience to leakage of the ephemeral Diffie-Hellman exponents). Importantly, using these additional features we can appeal to the connections proven in [10] to conclude that HOMQV+DEM is a secure signcryption scheme (which is also a secure UC realization of the functionality of replayable message transmission, see [9]).

We highlight some advantages of HOMQV that make it attractive for applications such as key-wrapping for storage. One advantage is its very minimal overhead as compared to DHIES. For the sender Bob there is essentially no added cost, the only change is that it computes the Diffie-Hellman value as $\sigma = A^{y+be}$ rather than $\sigma = A^y$. This only entails an additional hashing (to compute e) and an addition and multiplication modulo the group order, but no additional exponentiations. For the receiver there is an additional 1/2 exponentiation, since it now computes the Diffie-Hellman value as $\sigma = (YB^e)^a$ rather than $\sigma = Y^a$. We mention that as described in Table 1, HOMQV and DHIES require a check that the elements belong to a prime-order sub-group (which can be as costly as one full exponentiation). In Section 3 we show how this check can often be avoided.

2 Security Model for One-Pass Key-Exchange Protocols

We specialize the Canetti-Krawczyk (CK) security model [3] to one-pass key-exchange protocols. A key-exchange (KE) protocol is run in a network of connected parties, where each party can be activated to run an instance of the protocol called a session. During the session, a party creates and maintains a session state, may send and receive messages, and eventually completes the session by outputting a session key and erasing the session state. A session may also be aborted without generating a session key. A KE session is associated with its holder or owner (the party at which the session exists), a peer (the party with which the session key is intended to be established), and a session identifier. In one-pass protocols a peer to a session is either a sender or a receiver.

For simplicity we assume below that a session is always activated with the name of the intended peer (this is called "pre-specified peer" in [4]) and the session identifier is a triple $(\hat{B}, \hat{A}, Y)$ where $\hat{B}$ is the sender identity, $\hat{A}$ the receiver identity, and Y is the message sent in the protocol. Two sessions with the same identifier are called matching. Matching sessions play a fundamental role in the definition of security.

Each party owns a long-term pair of private and public keys, and other parties can verify the binding between an identity and a public key (e.g., using a CA or manually, the exact mechanism is outside the scope of this paper). A corrupted party can choose at any point to "register" any public key of its choice, including public keys equal or related to keys of other parties in the system (and there is no requirement that it knows the corresponding private key). In this paper, a public key will always be a group element, and the private key its secret exponent.

Notations and identities. A "hat" on top of a capital letter denotes an identity; without the hat the letter denotes the public key of that party, and the same letter in lower case denotes a private key. For example, Alice has identity $\hat{A}$ and a public key $A = g^a$ with a as the private key[2].

[2] This notation assumes that there is *a unique public key* associated with each identity. In the real world, where a party may have more than one public key, the symbol $\hat{A}$ is assumed to include an indication of a unique public key.

Attacker Model. The attacker, denoted $\mathcal{M}$, is an active "Man-in-the-Middle" adversary with full control of the communication links between parties. $\mathcal{M}$ can read, modify, inject, delete, or delay messages at will. (Formally, it is $\mathcal{M}$ to whom parties hand their outgoing messages for delivery). $\mathcal{M}$ also schedules all session activations and session-message delivery. In addition, in order to model potential disclosure of secret information, the attacker is allowed access to secret information via session exposure attacks of three types: state-reveal queries, session-key queries, and party corruption. A state-reveal query is directed at a single session while still incomplete (i.e., before outputting the session key) and its result is that the attacker learns the session state for that particular session (such as the secret exponent of an ephemeral public DH value). A session-key query can be performed against a single session after completion and the result is that the attacker learns the corresponding session-key (this models leakage on the session key either via usage of the key by applications, cryptanalysis, break-ins, known-key attacks, etc).. Finally, party corruption means that the attacker learns *all* information in the memory of that party (including the long-term private key of the party as well all session states and session keys stored at the party); in addition, from the moment a party is corrupted all its actions are controlled by the attacker. Indeed, note that the knowledge of the private key allows the attacker to impersonate the party at will.

Sessions against which any one of the above attacks is performed (including sessions compromised via party corruption) are called exposed. In addition, a session is also exposed if the matching session has been exposed (since matching sessions must output the same session key, the compromise of one inevitably implies the compromise of the other).

Note on state-reveal queries. "State-reveal queries" in the CK model is meant to capture the distinction between secrets whose exposure will derail security "forever" (such as long-term keys, or the ephemeral exponent in a DSA signature) and secrets whose exposure only compromises the current session. Data that can be accessed via such state-reveal queries is thought of as "less secret", its exposure only compromises the current session (and thus is can perhaps be stored in less secure memory in a real implementation). Data not accessible via state-reveal is assumed to get the same protection of long-term secrets, such data is revealed to the adversary only upon full compromise of a party.

Basic security. The security of session keys generated in unexposed sessions is captured via the inability of the attacker $\mathcal{M}$ to distinguish the session key of a test session (chosen by $\mathcal{M}$) from a random value. When $\mathcal{M}$ chooses the test session, a random bit b is tossed and $\mathcal{M}$ gets either the real value of the session key (if $b = 1$), or an unrelated random value (if $b = 0$). The attacker can continue with the regular actions against the protocol also after the test session; at the end of its run $\mathcal{M}$ outputs a guess b' for the value of b. The attacker succeeds in its distinguishing attack if (1) the test session is not exposed, and (2) it guessed correctly, $b = b'$. The protocol satisfies the basic notion of security if feasible attackers cannot succeed with probability significantly better than 1/2.

Definition 1 ([3]). *A polynomial-time attacker with the above capabilities is called a* KE-attacker. *A key-exchange protocol π is called* secure *if for all KE-attackers $\mathcal{M}$ running against π it holds:*

1. *If two uncorrupted parties complete matching sessions in a run of protocol π under attacker $\mathcal{M}$ then they output the same key (except for a negligible probability).*
2. *$\mathcal{M}$ succeeds (in its test-session distinguishing attack) with probability not more that 1/2 plus a negligible fraction.*

Sender's forward secrecy. An important property that is not captured by basic security is perfect forward secrecy (PFS) [15], namely the assurance that a session key which is erased from memory cannot be learned by the attacker even if the long-term key of that party is later exposed. This is captured formally in [3] via the notion of session-key expiration (which represents the erasure of a session key from memory). A key-exchange protocol is secure with PFS if Definition 1 holds even when the attacker is allowed to corrupt a peer to the test session *after the test-session key expired* at that peer. In the case of one-pass protocols, forward secrecy cannot be provided in general (since exposing the receiver's private key clearly lets the attacker decrypt all incoming traffic), but the sender can still enjoy forward secrecy. As we will see, the basic HOMQV protocol provides sender's forward secrecy against passive attackers, and adding key-confirmation provides forward secrecy also against active attackers.

Replay attacks. A one-pass key-exchange protocol where parties do not maintain evolving state between sessions is always open to replay attacks, where the attacker forces the establishment of the same session at a receiver by replaying old incoming messages to that party. The model deals with this issue by defining all instances of such replayed sessions to be matching. This means that all these sessions have the same session key and also that for the attacker to be considered successful in attacking a session it should have not exposed any one of the session copies.

3 The Basic HOMQV Protocol

On groups and supergroups. The protocol uses a cyclic group G of prime order q generated by a given generator g. There is no particular requirement from the cyclic group $G = \langle g \rangle$ except for its prime order. However, in cases where G is a subgroup of another group G' and testing membership in G' is easier than testing membership in G, we will exploit this property in the protocol. For this we define f as the "cofactor" $f = |G'|/|G|$. For example, if g is an element of Z_p^* of prime order q, then G' is Z_p^* and $f = (p-1)/q$. If g is an element in an elliptic curve group E then $f = |E|/q$. We always assume that f, q are co-primes.

Using the cofactor, we describe a protocol where one only needs to verify that the various elements are in G' (an efficient test in the above two examples). Note that it can be the case that $f = 1$. The property of the cofactor f that we use is that $X^f \in G$ for any $X \in G'$.

Hash functions. The protocol uses two hash functions (which are viewed as two independent random oracles): one, denoted $\bar{H}$, hashes strings into $\{0, 1, \ldots, \sqrt{q}\}$ and is used to compute exponents of size $|q|/2$, the other, denoted H, hashes into $\{0, 1\}^k$ where k is the length of the output key K.

Protocol HOMQV. Let $\hat{A}, \hat{B}$ be two *different* parties with keys $A = g^a$, $B = g^b$, respectively. To exchange a key with $\hat{A}$, sender $\hat{B}$ checks that $A \in G'$ (if not it aborts), then chooses random $y \in_{\mathrm{R}} Z_q$, and sends $Y = g^y$ to $\hat{A}$. $\hat{B}$ computes the key as $H(\sigma, \hat{B}, \hat{A}, Y)$ where $\sigma = A^{f \cdot (y+eb)}$ and $e = \bar{H}(Y, \hat{A})$. On incoming value Y and peer's identity $\hat{B}$, $\hat{A}$ checks that Y and $\hat{B}$'s public key B are in G' (if not, it aborts) and then computes the session key as $H(\sigma', \hat{B}, \hat{A}, Y)$ where $\sigma' = (YB^e)^{f \cdot a}$. Both parties compute the same session key since $\sigma = \sigma'$, and the session-id is the triple $(\hat{B}, \hat{A}, Y)$.

Performance. For the sender, the protocol requires just two exponentiations and a membership test in G' (the latter has negligible cost for the typical Z_p^* and elliptic curve cases). The exponentiations are for computing $Y = g^y$ and for computing $\sigma = A^{f \cdot (y+eb)}$ (In the latter case we first set $t = f \cdot (y+eb) \bmod \mathsf{ord}(G')$ and then σ is set to A^t. If $\mathsf{ord}(G')$ is not much larger than q — as in the case of elliptic curves — then the cofactor exponentiation is essentially for free). The group element Y can be computed offline, even before knowing the identity of the peer.

The cost for the receiver $\hat{A}$ is 1.5 on-line exponentiations: a half exponentiation when computing (YB^e), and a full exponentiation for raising it to the power fa. (As before, computing $t' = fa \bmod \mathsf{ord}(G')$ is essentially for free when $\mathsf{ord}(G') \approx q$). Comparing it with the cost of the unauthenticated DHIES, we see that *authentication is for free for the sender, and only costs 1/2 exponentiation for the receiver!*

Note, in particular, that there are no hidden costs in the protocol such as the need to test group elements for membership in the prime order subgroup G, only inexpensive G'-membership tests are needed. Also in contrast to some other protocols, HOMQV does not need the CA to checks that parties know the secret key for the public key that they want to register, thus avoiding complex proof-of-possession protocols.

A variant without the cofactor. In cases where $\mathsf{ord}(G') \gg q$ (as typically happens when working over Z_p^* with $q|p-1$), using the cofactor as above result in having to compute long multiplications in G', which could be much more expensive than exponentiations in the subgroup G. In these cases it may be better to use a variant without cofactor, and instead directly verify that the different elements are in the subgroup G: The sender $\hat{B}$ must verify that the receiver's public key A is in the subgroup G while the receiver $\hat{A}$ needs to test $YB^e \in G$ (there is no need to check Y and B separately). We also remark that verification of the receiver's public key A by the sender is only needed to get y-security; if we assume that the y value is never exposed then one can prove security even without that verification step, see more discussion in the full version [11].

3.1 Notes on Security

Session-state exposure. For the HOMQV protocol, one can readily see that disclosure of both the Diffie-Hellman value σ as well as of the exponent y from the same session leads to the exposure of the value $A^b = B^a$, which suffices for impersonating $\hat{B}$ to $\hat{A}$ and vice versa. So clearly, these two pieces of session-specific data should not be stored together in less secure memory.

How about disclosing any one of these values but not the other? For reasons similar to the above, if one discloses σ at the receiver then an attacker can do the following: chooses y, sends $Y = g^y$ to $\hat{A}$ purporting to be $\hat{B}$ (an honest player), then it finds σ at $\hat{A}$ and derives from it A^b which as said allows to impersonate $\hat{A}$ and $\hat{B}$ to each other. As for the secret exponent y, we show in Section 5.3 that its disclosure does not even compromise the session in which it is used. To have a negative effect, the attacker needs to find either the σ value of that session or the private key of the party that generated this exponent. This is an important property of HOMQV, since it means that the pair $(y, Y = g^y)$ can be generated in an off-line phase (even before knowing the identity of $\hat{A}$) and stored for later use. An exposure of such pair does not cause any real damage as long as the long-term key is not disclosed.

Replay. Being a one-pass protocol, HOMQV is obviously open to replay. This can be prevented by including synchronized timestamps (or a shared increasing counter). The timestamp or counter becomes part of the session-id together with $(\hat{B}, \hat{A}, Y)$ and hence included under H when computing the session key K. Another option is to use an interactive protocol in which $\hat{A}$ sends to $\hat{B}$ a nonce, which is then included in the computation of e. Note that the resulting interactive protocol is weaker than HMQV (e.g., it does not provide receiver PFS), but it is cheaper and prevents replay. See Section 6.

On Forward Secrecy. The protocol HOMQV as above provides forward security for the sender, namely, the guarantee that the disclosure of the private key of the sender $\hat{B}$ does not expose past sessions created by $\hat{B}$. However, an active attacker can choose $Y = g^y$ and send it to a receiver $\hat{A}$, purporting to be $\hat{B}$. Later, if the attacker find $\hat{B}$'s private key then it can compute the session. Hence forward secrecy is only ensured against passive attackers. We can get full sender's forward secrecy by adding a key confirmation field in $\hat{B}$'s message, see Section 5.4. Receiver's forward secrecy is impossible in one-pass key exchange, since the receiver does not contribute any session-specific values.

KCI attacks. In a Key Compromise Impersonation attack (KCI), knowing the private key of a party $\hat{A}$ allows the attacker to impersonate other parties to $\hat{A}$. This is obviously possible in HOMQV.

4 Using HOMQV for Encryption and Key-Wrapping

On its own, HOMQV is used to establish a new secret key between the client and server. To get encryption or key-wrapping we need to add a "data encapsulation module" (DEM), where the new key is used in a symmetric encryption mode in order to encrypt (or wrap) the transmitted message/key. Specifically, we use the DEM part of DHIES, where the new key from HOMQV is expanded into encryption and authentication keys, which are then used in an Encrypt-then-Authenticate mode to send the message.

Namely, after computing $K = H(\sigma, \hat{B}, \hat{A}, Y)$ we set $K_a = PRF_K(1)$ and $K_e = PRF_K(2)$ (where 1,2 can be replaced by any two different fixed messages that are publicly determined by the protocol flow). Then to encrypt a message M we compute $C = \text{ENC}_{K_e}(M)$ where ENC is a CPA-secure symmetric encryption scheme and then $T = \text{MAC}_{K_a}(C)$ where MAC is a secure message-authentication code. The DEM part (C, T) — which is an authenticated encryption of M under the shared key (K_e, K_a) — is then added to the HOMQV flow to make the composite ciphertext (Y, C, T).

Regarding security, if we use the "degenerate case" of HOMQV where the public key B is set to 1 then we get exactly the DHIES encryption scheme, which was proven CCA secure by Abdalla et al. [1]. When adding the DEM part to the HOMQV scheme from Section 3 (with server authentication) we get a signcryption scheme [18], whose security can be argued as follows:

Step 1: signcryption-KEM. Gorantla et al. proved in [10, Thm 3] that a one-pass key-exchange protocol Π which is secure in the Canetti-Krawczyk model and offers sender forward secrecy, is also a signcryption-KEM scheme secure in the sense of insider confidentiality and outsider unforgeability in the multi-user setting. See [10] for the definitions of these notions of security. Roughly speaking this means that CCA-security of the encryption part holds even against an attacker that knows the sender's private key, while unforgeability of the signature holds only against an attacker that does not know the receiver's private key.

Moreover, one can verify that the proof from [10] works even if the protocol Π offers sender forward secrecy only against passive attackers. (Roughly speaking, this is because this property is only used with respect to the challenge ciphertext in the KEM-CCA game, which is generated honestly according to the signcryption algorithm). Since we prove in Theorem 2 and Lemma 3 that HOMQV is secure in the Canetti-Krawczyk model and it offers sender forward secrecy against passive attackers, it follows that it is also a signcryption-KEM scheme secure in the sense of insider confidentiality and outsider unforgeability[3].

Step 2: adding the DEM. A signcryption-KEM secure as above can be transformed into regular signcryption (secure in the same sense) by adding an authenticated-encryption DEM. Such results were proven by Dent [5] in slightly different models, and the same proofs work for our case as well. Namely, we have:

[3] Gorantla et al. already suggested using one-pass HMQV to get a secure signcryption KEM, but did not prove security for the variant of one-pass HMQV that they used.

Lemma 1. *The combination of a signcryption KEM with insider confidentiality and outsider unforgeability in the multi-user setting, together with an authenticated-encryption DEM, yields a signcryption scheme with insider confidentiality and outsider unforgeability in the multi-user setting.*

Proof. (sketch) The confidentiality part against insider attacks is identical to the case of standard hybrid encryption. The integrity part against outsider attacks follows roughly from these arguments: (i) An outside attacker cannot generate a valid KEM for any symmetric key except those that were included in one of the ciphertexts that the attacker obtained from its oracle, due to the KEM integrity against outside attackers; (ii) The KEM confidentiality implies that the keys included in the ciphertext from the oracle are indistinguishable from random; and (iii) The integrity-of-ciphertext property of the DEM implies that the attacker cannot generate new valid DEM ciphertexts under these pseudo-random keys.

It therefore follows that augmenting HOMQV with DEM as above gives a signcryption scheme secure in the sense of insider confidentiality and outsider unforgeability in the multi-user setting. These arguments can be made formal via a standard sequence-of-games proof. □

Gjøsteen and Kråkmo proved in [9] that a signcryption scheme secure as above can be used in conjunction with PKI to realize a secure messaging functionality in the UC framework. (The messaging functionality is similar to Canetti's secure-message-transmission functionality from [2], except that signcryption does not prevent replay). We thus conclude that when used in conjunction with PKI, the protocol HOMQV with the DEM part can UC-realize secure messaging. This UC-secure-messaging, in turn, can be used to transport keys from server to client, providing as much security as can be obtained from a one-pass protocol.

Why use HOMQV for Key-Wrapping? The main advantage of HOMQV over the underlying DHIES is server authentication: We let the client verify that wrapped keys indeed arrive from the correct server without adding much complexity or making any changes in the data path. Another advantage is the "y-security" of HOMQV. Recall that HOMQV remains secure even if the ephemeral secret exponent y is exposed, as long as the long-term secret key of $\hat{B}$ is protected. On the other hand, DHIES is clearly broken in this case. This provides a significant line of defense in settings where exposure of y is a real concern, such as when pairs $(y, Y = g^y)$ are precomputed and stored in less secure memory.

At the same time, the fact that HOMQV degenerates back to DHIES by using server public key $B = 1$ means that a HOMQV client can be used to unwrap legacy DHIES-wrapped keys with very little added complexity (if at all). In this case of course we would have to fall back to out-of-band authentication, but otherwise the legacy support is nearly transparent.

5 Security Analysis of HOMQV

5.1 XCR Signatures and Gap-DH

XCR signatures are challenge-response signature where only the challenger can verify the signature. Specifically, the signer $\hat{B}$ has a private key $b \in_{\mathrm{R}} Z_q$ and a public key $B = g^b$. On input a message m and "challenge" $X = g^x$, the signature of $\hat{B}$ on m and challenge X is a pair $(Y = g^y, \sigma)$ where y is chosen by $\hat{B}$ at random in Z_q and σ is defined as $X^{f\cdot(y+\bar{H}(Y,m)b)}$. The challenger, who knows x, verifies the signature by checking that Y is in G' and that $\sigma = (YB^{\bar{H}(Y,m)})^{f\cdot x}$.

Security of XCR signatures. XCR is called secure if it is secure under a chosen-message (and chosen-challenge) attack in the following game between a forger $\mathcal{F}$ and a signing oracle $\hat{B}$. The input to $\mathcal{F}$ is $\hat{B}$'s public key $B \in_{\mathrm{R}} G$ and a challenge $X_0 \in_{\mathrm{R}} G$. $\mathcal{F}$ provides queries (X, m) to $\hat{B}$ which responds with a valid XCR signature (Y, σ). After polynomially many adaptive queries the forger wins if it outputs a valid signature (Y_0, σ_0) on any message m_0 using the (input) challenge X_0. The only requirement is "strong" existential unforgeability, namely, that either $\hat{B}$ was not queried for a signature on message m_0 or, if it was queried on m_0 it output a signature different than the pair (Y_0, σ_0).

Theorem 1 ([13]). *Under the CDH assumption, XCR signatures are secure in the random oracle model.*

The Gap-DH Assumption. The security of XCR is proved under the Computational Diffie-Hellman (CDH) Assumption over G, namely, given $U, V \in_{\mathrm{R}} G = \langle g \rangle$ it is infeasible to compute $DH_g(U, V)$ (the Diffie-Hellman function, with generator g, applied to U and V). To prove HOMQV we need the stronger Gap-DH assumption. We say that a decision algorithm $\mathcal{O}$ is a *Decisional Diffie-Hellman (DDH) Oracle* for a group G and generator g if on input a triple (U, V, W), for $U, V \in G$, oracle $\mathcal{O}$ outputs 1 if and only if $W = DH_g(U, V)$. We say that G satisfies the Gap-Diffie-Hellman (GDH) assumption if no feasible algorithm exists to solve the CDH problem, even when the algorithm is provided with a DDH-oracle for G.

5.2 Proof of Basic Security

Theorem 2. *In the random oracle model and under the Gap-DH assumption, HOMQV is a secure one-pass key-exchange protocol as per Definition 1.*

Proof. We need to prove that (1) Any two matching sessions with honest peers output the same session key; and (2) A KE-attacker has negligible advantage in winning a test-session experiment. Since sessions are matching if and only if they have the same session id $(\hat{B}, \hat{A}, Y)$, condition (1) follows from the fact that these three values determine a unique value for σ and hence for the session key.

The proof of (2) works by reduction to the security of XCR signatures, assuming a decisional Diffie-Hellman oracle as in the Gap-DH assumption. That is, given an assumed KE-attacker $\mathcal{M}$ against HOMQV and a decisional DH oracle we build a (0-message) forger $\mathcal{F}$ against XCR.

Let N be such that the adversary $\mathcal{M}$ has non-negligible advantage in attacking the KE-security of HOMQV when used with N honest parties. $\mathcal{F}$ gets as input a public key B and challenge X_0 (both random elements in G). It starts by simulating a run of HOMQV with N honest parties, and selects two different parties at random among these N honest parties which we denote by $\hat{A}$ and $\hat{B}$; this is $\mathcal{F}$'s guess for the receiver and sender, respectively, of the test session to be chosen by $\mathcal{M}$. Using its inputs B and X_0, $\mathcal{F}$ sets $\hat{A}$'s public key to $A = X_0$ and $\hat{B}$'s public key to B (note that $\mathcal{F}$ does not know the private keys corresponding to these public keys). For all other honest parties, $\mathcal{F}$ chooses their private and public keys. (Corrupted parties have their PKs chosen by the attacker $\mathcal{M}$ who also manages all their actions). $\mathcal{M}$ controls all activations, traffic and delivery of messages. Invocations of the random oracle ($\bar{H}$ and H) are usually answered by $\mathcal{F}$ by choosing a random response from the appropriate range, except as described in the special cases below. (If the same query is made more then once, then it gets the same answer every time). In any case in which a corruption query is issued by $\mathcal{M}$ against $\hat{A}$ or $\hat{B}$, or if $\mathcal{M}$ chooses a test session with peers other than $\hat{A}$ as receiver and $\hat{B}$ as sender, $\mathcal{F}$ aborts its run (in both cases $\mathcal{F}$ failed to guess correctly the peers to the test session).

Any query (activation) of an honest party other than $\hat{A}$ or $\hat{B}$ is answered by $\mathcal{F}$ just as in the protocol, which $\mathcal{F}$ can do since it knows the secret keys of all these honest parties. Similarly, corruption of honest parties other than $\hat{A}$ or $\hat{B}$ are answered by $\mathcal{F}$ using the secret keys that it knows. It is left to show how to simulate queries to $\hat{A}$ and $\hat{B}$.

Send-query (potentially followed later by a session-key reveal query). Assume that the sender is $\hat{B}$, the case where the sender is $\hat{A}$ is handled in exactly the symmetric way. Denote the session's receiver by $\hat{C}$ and its public key by C. If $\hat{C} = \hat{B}$ of if $C \notin G'$ then $\mathcal{F}$ aborts the session. Otherwise $\mathcal{F}$ chooses y at random and sets $Y = g^y$ as the outgoing value to be sent.

If the receiver $\hat{C}$ is an honest party other than $\hat{A}$ then $\mathcal{F}$ computes the session key using $\hat{C}$'s secret key (as done in the protocol), and uses that value to answer secret-key reveal query against this session of $\hat{B}$ (if any). Otherwise (i.e., $\hat{C}$ is either $\hat{A}$ or a corrupted party), $\mathcal{F}$ chooses at random $K \in_{\mathrm{R}} \{0,1\}^k$ and uses this K for future secret-key reveal queries against this session of $\hat{B}$. From this point on, $\mathcal{F}$ will use its DDH oracle to check for each new H-query of the form $(Q, \hat{B}, \hat{C}, Y)$ (for some Q), whether $Q = DH_g(YB^{\bar{H}(Y,\hat{C})}, C^f)$. If so, $\mathcal{F}$ responds to that random-oracle query with the value K.

Note that due to the hashing of the session id in the computation of a session key, the value K is independent of any other session key except for the possible matching session held in this case by a corrupted party or $\hat{A}$. Thus, F's responses to other session-key queries (or corruption queries) are independent of K except if $\hat{C} = \hat{A}$ in which case the matching session at $\hat{A}$ is set to K as well[4].

Receive-query (potentially followed later by a session-key reveal query). Assume that the receiver is $\hat{A}$, the case where the receiver is $\hat{B}$ is handled in exactly the symmetric way. Denote the session's sender by $\hat{C}$ and its public key by C, and denote the incoming value in this session by Z.

If $\hat{C} = \hat{A}$ or if C or Z are not in G' then $\mathcal{F}$ aborts the session. If (C, Z) was used before to activate $\hat{A}$ as receiver (because replay is possible), $\mathcal{F}$ sets the session key to its previous value. If $\hat{C}$ is an honest party other than $\hat{B}$, and Z was previously generated by an instance of this honest $\hat{C}$ in the role of a sender (so $\mathcal{F}$ knows z s.t. $Z = g^z$), then $\mathcal{F}$ computes the session key using this z together with $\hat{C}$'s secret key as done in the protocol, and uses that session key to answer secret-key reveal query against this session of $\hat{A}$ (if any).

Otherwise, we know that $\hat{C}$ is either $\hat{B}$ or a corrupted party, or $\hat{C}$ is honest but Z was never generated by an instance of $\hat{C}$ in the role of a sender. Then for each H-query of the form $(Q, \hat{C}, \hat{A}, Z)$ (for some Q) ever issued by $\mathcal{M}$, $\mathcal{F}$ uses its DDH oracle to check whether $Q = DH_g((ZC^{\bar{H}(Z,\hat{A})})^f, A)$. If such Q is found then $\mathcal{F}$ sets the session key to $K = H(Q, \hat{C}, \hat{A}, Z)$. Else, if $\hat{C} = \hat{B}$ and Z was indeed sent by a simulated session of $\hat{B}$ then $\mathcal{F}$ sets the session key to be the same value K that it chose when processing that previous Send query at $\hat{B}$. Else $\mathcal{F}$ sets the session key K as a new random ℓ-bit string. Thereafter, each time a session key reveal query is performed at a session $(\hat{C}, \hat{A}, Z)$, $\mathcal{F}$ responds with K. (As said, there may be more than one because of replay). From now on, $\mathcal{F}$ uses its DDH oracle to check for any H-query of the form $(Q, \hat{C}, \hat{A}, Z)$ (for any Q) whether $Q = DH_g((ZC^{\bar{H}(Z,\hat{A})})^f, A)$ and if so it answers that query with K.

Test session. In any case where the test session is chosen with peers other than $\hat{A}$ as receiver and $\hat{B}$ as sender, $\mathcal{F}$ aborts its run. When the test session is chosen with $\hat{A}$ as receiver and $\hat{B}$ as sender, $\mathcal{F}$ returns to $\mathcal{M}$ with probability 1/2 the same answer K that it would have returned in a session-key reveal query (say, against the session in $\hat{B}$), and with probability 1/2 $\mathcal{F}$ returns just an independent random value K'.

As long as $\mathcal{M}$ does not query $H(\sigma, \hat{B}, \hat{A}, Y)$ with the values σ, Y corresponding to the test session, it has only probability 1/2 to answer correctly. This is because no session $(\hat{B}, \hat{A}, Y)$ was compromised and all other sessions have keys that are independent of K (since they were generated via H calls with different inputs). Therefore both K and K' are random values independent of $\mathcal{M}$'s view. By assumption $\mathcal{M}$ succeeds with non-negligible advantage, thus implying that $\mathcal{M}$ computes the correct σ with non-negligible probability.

Finally, whenever $\mathcal{F}$ guesses correctly the peers to the test session and $\mathcal{M}$ successfully generates the σ value corresponding to the test session ($\mathcal{F}$ identifies it via the decisional oracle), $\mathcal{F}$ stops its run and outputs the pair (Y, σ) where

4 Without hashing the session id, a UKS attack is possible where different, non-matching, sessions have the same key.

Y is the test session's outgoing value and $\sigma = A^{y+be}$ with $e = \bar{H}(Y, \hat{A})$. But this pair (Y, σ) is a *valid* XCR signature of $\hat{B}$ on message $\hat{A}$ and challenge $A = X_0$; since $\mathcal{F}$ has never queried $\hat{B}$ for any signature, $\mathcal{F}$ wins the forgery game. Since this happens with non-negligible probability, we obtain a contradiction to the security of XCR signatures, thus proving the theorem. □

Note regarding the Gap-DH queries. A valid query to a decisional oracle is one where the two inputs are elements of G. In the above proof, this is indeed the case for the $Q \stackrel{?}{=} DH_g(YB^{\bar{H}(Y,\hat{C})}, C^f)$ query since Y, B and C^f are all in G (Y by $\mathcal{F}$'s choice, B as an input to $\mathcal{F}$, and C^f since $\mathcal{F}$ checks that $C \in G'$). The same is the case for the query $Q \stackrel{?}{=} DH_g((ZC^{\bar{H}(Z,\hat{A})})^f, A)$ since both Z and C are checked to be in G' thus the value $(ZC^{\bar{H}(Z,\hat{A})})^f$ is necessarily in G. Note that if the cofactor f is not used in the protocol then the sender needs to check explicitly that the receiver's public key C is in G (i.e., an element of G' of order q) and the receiver needs to check that the value $ZC^{\bar{H}(Z,\hat{A})}$ is in G.

Beyond basic security. Below we show that HOMQV is strongly resilient to the disclosure of ephemeral DH exponents and also that it provides sender's forward secrecy in case of the disclosure of the private key of a party.

5.3 Resilience of HOMQV to Disclosure of Ephemeral Exponents

We prove that HOMQV has maximal resilience to the leakage of the ephemeral DH exponents y used by a sender to produce outgoing DH values $Y = g^y$ (we referred to this property as y-security). Indeed knowledge of these y's by an attacker (even ahead of their use) does not even compromise the security of the sessions where they are used, except of course if the attacker also learns the private key of the sender. The practical meaning is that it is as safe as possible to compute pairs $(y, Y = g^y)$ offline and store them for later use even if this storage is less protected than the more sensitive long-term private key.

Lemma 2. *Protocol HOMQV remains secure even when making the ephemeral DH exponents y available to the attacker via state reveal queries. Moreover, even sessions for which y is known remain secure (formally, the attacker is allowed to choose a test session for which it knows y).*

Proof. We show an even stronger property: Let $\hat{B}$ be an honest party and assume that $\mathcal{M}$ gets to see all pairs $\{(y_i, Y_i = g^{y_i})\}$ to be used by $\hat{B}$ for all its sessions at the onset of the protocol run. We claim that all of $\hat{B}$'s unexposed sessions (where learning the y_i exponent is *not* considered exposure) are still secure. In other words, even if the test session chosen by $\mathcal{M}$ is one of these outgoing sessions of $\hat{B}$ (or one of the matching sessions at the peer) the attacker cannot win the distinguishing game with non-negligible advantage. The proof is essentially the same as the one for Theorem 2; all we need to observe is that in the simulation in that proof the forger $\mathcal{F}$ chooses (and hence knows) all exponents y corresponding to outgoing DH values Y at $\hat{B}$. Hence, $\mathcal{F}$ can provide these values to $\mathcal{M}$, even before they are used.

For this, the description of $\mathcal{F}$ when simulating a Send query to $\hat{A}$ or $\hat{B}$ needs to be slightly changed. Specifically, it needs to use its DDH oracle to test all previous query to H, just as it is done for a Receive query. This is needed here since $\mathcal{M}$, knowing y from the start, could have made a query that depends on this value even before it was used by $\hat{B}$. □

5.4 Sender's Forward Secrecy

We show that HOMQV ensures *sender's forward secrecy* against passive attackers. Specifically, we show that the exposure of the private key of a party does not compromise any of the past sessions where that party acted as sender. Moreover, this is the case even for sessions established after the disclosure of the private key, provided that the attacker does not control the session's Y or knows its exponent y. This property, however, does not guarantee forward secrecy for sessions where the attacker actively chose the value Y prior to the private key exposure. For example, an attacker can choose y, set $Y = g^y$, and activate party $\hat{A}$ as receiver with incoming Y and sender's identity $\hat{B}$. Later, if the attacker finds $\hat{B}$'s private key, it can compute the value of the session key generated by $\hat{A}$ using Y. In other words, the protocol provides *weak forward secrecy* against passive attackers, but not full PFS against active attacks. Fortunately, we will see below that one can slightly change HOMQV by adding a (key confirmation) field to the message from $\hat{B}$ to $\hat{A}$ and then achieve full sender's forward secrecy.

Lemma 3. *HOMQV enjoys sender's forward secrecy against passive attackers.*

Proof. (sketch) Let $\hat{B}$ be an honest party and assume that at some point $\mathcal{M}$ learns $\hat{B}$'s private key b. We claim that even in this case $\mathcal{M}$ cannot win the test session experiment for any session in which $\hat{B}$ acted as sender provided that the outgoing value Y for the session was indeed generated by $\hat{B}$ and that the session was not exposed via a session-key query or via a state-reveal query. We informally outline the proof of this property.

For contradiction we assume an attacker $\mathcal{M}$ that wins the test-session experiment with non-negligible advantage in a session as above after learning the sender's private key. We use $\mathcal{M}$ to build an algorithm $\mathcal{S}$ that solves CDH given a decisional oracle as in the Gap-DH setting. Let $X, Y \in_{\mathrm{R}} G$ be an instance of the CDH problem. $\mathcal{S}$ will run a simulation similar to the one $\mathcal{F}$ runs in the proof of Theorem 2; in particular, it will try to guess the peers to the test session (call $\hat{B}$ the guess for sender and $\hat{A}$ the guess for receiver) as well as which of the sessions between $\hat{B}$ and $\hat{A}$ will be chosen as the test. $\mathcal{S}$ chooses public keys for all honest parties *including* $\hat{B}$ *but excluding* $\hat{A}$. For $\hat{A}$ it sets the public key to X. The rest of the simulation is similar except that $\hat{B}$'s private key is known (chosen by $\mathcal{S}$) so $\hat{B}$ is treated as any other uncorrupted party. The simulation of $\hat{A}$ for whom the private key is unknown is same as in the proof of Theorem 2 (in particular, it uses the decisional oracles). When the guessed test session is activated at $\hat{B}$, $\mathcal{S}$ will use the value Y from the CDH input as the outgoing value from $\hat{B}$. When $\mathcal{M}$ corrupts $\hat{B}$, $\mathcal{S}$ provides it with $\hat{B}$'s private key b which $\mathcal{S}$ knows. As before, to win the test experiment $\mathcal{M}$ needs to be able to compute with non-negligible probability the value $\sigma = (YB^e)^a$ which $\mathcal{S}$ learns from the H-queries. From σ, $\mathcal{S}$ computes $Z = Y^a$ (where $A = g^a$) by setting $Z = \sigma/A^{eb}$. Since $Z = DH_g(X, Y)$, $\mathcal{S}$ solved the CDH challenge contradicting the Gap-DH assumption. □

Full Sender's Forward Secrecy. As said, HOMQV does not achieve forward secrecy for sessions activated at a receiver $\hat{A}$ where the incoming value Y was chosen by the attacker $\mathcal{M}$. To fix this we need to assure that $\hat{A}$ will not accept an incoming Y that has not been generated by the purported sender. For this, add to the protocol's message *key confirmation* field. First, we define a key derivation step that from the key K, as defined in HOMQV, derives two keys K^*, K_a. The former is output as the session key while K_a is used as a key to a MAC function. The key confirmation field (included in the message sent from $\hat{B}$ to $\hat{A}$) is a tag $\text{MAC}_{K_a}(1)$ (where 1 can be replaced by any fixed message that is publicly determined by the protocol flows). It is not hard to see that in this way a session at a honest receiver will only be established if the MAC was successful hence proving that Y was generated by the claimed sender. (Formally, one shows that if a MAC verification succeeds for a value not generated by the claimed sender then the security of the original HOMQV protocol is broken, or more specifically, that a successful XCR forgery occurs). We have:

Lemma 4. *The modified HOMQV protocol with $\hat{B}$'s key confirmation provides full sender's forward secrecy (against active attacks).*

6 Extensions in the Interactive Setting

In settings where interaction between server and client is possible, the HOMQV protocol smoothly "extends up" to provide better security at a very low cost. Replay attacks can be prevented simply by having $\hat{A}$ send a nonce to $\hat{B}$ in the first flow, and then incorporate that nonce in the final hash calculation, setting $K = H(\sigma, \hat{B}, \hat{A}, Y, n)$ (with n the nonce sent by $\hat{A}$). This simple variant does not offer receiver forward secrecy or protection against KCI attacks, but it does prevent replay attacks without incurring any additional computational cost. Note also that by making the nonce n default to null in an implementation of this variant, we get "transparent" support also for the non-interactive HOMQV.

To get also receiver forward secrecy and protection against KCI attacks, we can move up to two- or three-pass HMQV, where $\hat{A}$ sends $X = g^x$ to $\hat{B}$ and σ is computed as $\sigma = (XA^d)^{f(y+eb)} = (YB^e)^{f(x+da)} = g^{f(x+da)(y+eb)}$, with $d = \bar{H}(X, \hat{B})$ and $e = \bar{H}(Y, \hat{A})$. The price we pay for this added security over the previous variant is another 1/2 exponentiation for the sender (to compute (XA^d)) and another full exponentiation for the receiver (to compute $X = g^x$).

References

1. Abdalla, M., Bellare, M., Rogaway, P.: The oracle diffie-hellman assumptions and an analysis of DHIES. In: Naccache, D. (ed.) CT-RSA 2001. LNCS, vol. 2020, pp. 143–158. Springer, Heidelberg (2001)

2. Canetti, R.: Universally Composable Security: A New paradigm for Cryptographic Protocols. In: 42nd Annual Symposium on Foundations of Computer Science FOCS 2001, pp. 136–145. IEEE, Los Alamitos (2001)
3. Canetti, R., Krawczyk, H.: Analysis of Key-Exchange Protocols and Their Use for Building Secure Channels. In: Pfitzmann, B. (ed.) EUROCRYPT 2001. LNCS, vol. 2045, pp. 453–474. Springer, Heidelberg (2001)
4. Canetti, R., Krawczyk, H.: Security Analysis of IKE's Signature-Based Key-Exchange Protocol. In: Yung, M. (ed.) CRYPTO 2002. LNCS, vol. 2442, pp. 143–161. Springer, Heidelberg (2002)
5. Dent, A.W.: Hybrid Cryptography. ePrint archive 2004/210 (2004), `http://eprint.iacr.org/`
6. Dent, A.W.: Hybrid Signcryption Schemes with Insider Security. In: Boyd, C., González Nieto, J.M. (eds.) ACISP 2005. LNCS, vol. 3574, pp. 253–266. Springer, Heidelberg (2005)
7. Dent, A.W.: Hybrid Signcryption Schemes with Outsider Security. In: Zhou, J., López, J., Deng, R.H., Bao, F. (eds.) ISC 2005. LNCS, vol. 3650, pp. 203–217. Springer, Heidelberg (2005)
8. Gennaro, R., Halevi, S.: More on Key Wrapping. In: Jacobson Jr., M.J., Rijmen, V., Safavi-Naini, R. (eds.) SAC 2009. LNCS, vol. 5867, pp. 53–70. Springer, Heidelberg (2009)
9. Gjøsteen, K., Kråkmo, L.: Universally Composable Signcryption. In: López, J., Samarati, P., Ferrer, J. (eds.) EuroPKI 2007. LNCS, vol. 4582, pp. 346–353. Springer, Heidelberg (2007)
10. Gorantla, M., Boyd, C., González Nieto, J.: On the Connection Between Signcryption and One-Pass Key Establishment. In: Galbraith, S. (ed.) Cryptography and Coding 2007. LNCS, vol. 4887, pp. 277–301. Springer, Heidelberg (2007)
11. Halevi, S., Krawczyk, H.: One-pass HMQV and asymmetric key-wrapping. Cryptology ePrint Archive, Report 2010/638 (2010), `http://eprint.iacr.org/`
12. IEEE 1363a-2004: Standard Specifications for Public Key Cryptography
13. Krawczyk, H.: HMQV: A High-Performance Secure Diffie-Hellman Protocol. In: Shoup, V. (ed.) CRYPTO 2005. LNCS, vol. 3621, pp. 546–566. Springer, Heidelberg (2005), `http://eprint.iacr.org/`
14. Menezes, A.: Another Look at HMQV (2005), `http://eprint.iacr.org/2005/205`
15. Menezes, A., Van Oorschot, P., Vanstone, S.: Handbook of Applied Cryptography. CRC Press, Boca Raton (1996)
16. Rogaway, P., Shrimpton, T.: A provable-security treatment of the key-wrap problem. In: Vaudenay, S. (ed.) EUROCRYPT 2006. LNCS, vol. 4004, pp. 373–390. Springer, Heidelberg (2006)
17. Shoup, V.: ISO 18033-2: An emerging standard for public-key encryption, `http://shoup.net/iso/`
18. Zheng, Y.: Digital signcryption or how to achieve cost (Signature & encryption) $<<$ cost(Signature) + cost(Encryption). In: Kaliski Jr., B.S. (ed.) CRYPTO 1997. LNCS, vol. 1294, pp. 165–179. Springer, Heidelberg (1997)

Linear Recurring Sequences for the UOV Key Generation

Albrecht Petzoldt[1,2], Stanislav Bulygin[1,2], and Johannes Buchmann[1,2]

[1] Technische Universität Darmstadt, Department of Computer Science
Hochschulstraße 10, 64289 Darmstadt, Germany
{apetzoldt,buchmann}@cdc.informatik.tu-darmstadt.de
[2] Center for Advanced Security Research Darmstadt - CASED
Mornewegstraße 32, 64293 Darmstadt, Germany
{johannes.buchmann,Stanislav.Bulygin}@cased.de

Abstract. Multivariate public key cryptography is one of the main approaches to guarantee the security of communication in the post-quantum world. Due to its high efficiency and modest computational requirements, multivariate cryptography seems especially appropriate for signature schemes on low cost devices. However, multivariate schemes are not much used yet, mainly because of the large size of their public keys. In [PB10] Petzoldt et al. presented an idea how to create a multivariate signature scheme with a partially cyclic public key based on the UOV scheme of Kipnis and Patarin [KP99]. In this paper we use their idea to create a multivariate signature scheme whose public key is mainly given by a linear recurring sequence (LRS). By doing so, we are able to reduce the size of the public key by up to 86 %. Moreover, we get a public key with good statistical properties.

Keywords: Multivariate Cryptography, UOV Signature Scheme, Key Size Reduction, Linear Recurring Sequences.

1 Introduction

When quantum computers arrive, cryptosystems based on number theoretic problems such as integer factoring or discrete logarithms will become insecure, since such problems can be efficiently solved via Shor's algorithm [Sh97] [BB08]. So, to guarantee the security of communication in the post-quantum world, alternatives to classical public key schemes are needed. Besides lattice-, code- and hash-based cryptosystems, multivariate public key cryptography [DG06] is one of the main approaches to achieve this goal. Since they require only modest computational resources, multivariate schemes seem to be appropriate for the use on low cost devices like RFID's and smartcards. However, these schemes are not widely used yet, mainly because of the large size of their public and private keys.

The basic idea behind multivariate cryptography is to choose a system $\mathcal{Q}$ of m quadratic polynomials in n variables which can be easily inverted (central map).

D. Catalano et al. (Eds.): PKC 2011, LNCS 6571, pp. 335–350, 2011.

After that one chooses two affine invertible maps $\mathcal{S}$ and $\mathcal{T}$ to hide the structure of the central map. The public key of the cryptosystem is the composed quadratic map $\mathcal{P} = \mathcal{S} \circ \mathcal{Q} \circ \mathcal{T}$ which should be difficult to invert. The private key consists of $\mathcal{S}$, $\mathcal{Q}$ and $\mathcal{T}$ and therefore allows to invert $\mathcal{P}$.

In the last years, a lot of work has been done to find ways how to reduce the key size of multivariate schemes. Thereby, most researchers concentrated on reducing the size of the private key. One way to achieve this is by choosing the coefficients of the private maps out of smaller fields (e.g. $GF(16)$ instead of $GF(256)$). However, this increases the signature length [CC08]. Another way to reduce the size of the private key is by using sparse central polynomials, which is done for example in the TTS schemes of Yang and Chen [YC05]. By using a strategy called "similar keys" Hu et al. [HW05] produced interesting results in this direction, too.

In [PB10] Petzoldt et al. presented an idea how to reduce the public key size of the UOV signature scheme of Kipnis and Patarin [KP99]. They achieved this by inserting a partially circulant matrix into the coefficient matrix of the public key polynomials. By doing so, they were able to reduce the public key size of the standard UOV scheme by a large factor.

In this paper we use their idea to create a multivariate signature scheme whose public key is mainly given by a linear recurring sequence (LRS). Despite of the fact that until now no attack against the partially cyclic scheme is known, we aim at replacing the partially cyclic key by a key which is statistically more random (see Subsection 4.2) without increasing the key size. So, it should become more difficult to develop a dedicated attack against the scheme. We also get closer to the "provably secure" UOV scheme of [BP10].

As in [PB10], we are not able to create a scheme whose public key is completely given by an LRS. So, we will have $M_P = (B|E)$, where B is generated by an LRS and E is a matrix with no apparent structure. Thus we have to store only the parameters of the LRS and the matrix E, which reduces the size of the public key by up to 86 %.

The rest of the paper is organized as follows:

In Section 2 we describe the Unbalanced Oil and Vinegar (UOV) signature scheme, which is the basis of our construction. Section 3 reviews the approach of [PB10] to create a UOV-based scheme with a partially cyclic public key. In Section 4 we repeat results from the theory of linear recurring sequences (LRS's) needed in the following sections and make some remarks about randomness measurements of sequences. Section 5 describes the construction and presents our new scheme in detail. In Section 6 we answer the question how to choose the parameters of the LRS, whereas Section 7 studies the security of our scheme under known attacks. Parameter proposals for our scheme can be found in Section 8, and Section 9 concludes the paper.

2 The (Unbalanced) Oil and Vinegar Signature Scheme

One way to create easily invertible multivariate quadratic systems is the principle of Oil and Vinegar, which was first proposed by J. Patarin in [Pa97].

Let $\mathbb{F}_q$ be a finite field. Let o and v be two integers and set $n = o + v$. We set $V = \{1, \ldots, v\}$ and $O = \{v+1, \ldots, n\}$. Of the n variables $x_1, \ldots, x_n$ we call $x_1, \ldots, x_v$ the Vinegar variables and $x_{v+1}, \ldots, x_n$ Oil variables. We define o quadratic polynomials $q^{(k)}(\mathbf{x}) = q^{(k)}(x_1, \ldots, x_n)$ by

$$q^{(k)}(\mathbf{x}) = \sum_{i \in V,\, j \in O} \alpha_{ij}^{(k)} x_i x_j + \sum_{i,j \in V,\, i \leq j} \beta_{ij}^{(k)} x_i x_j + \sum_{i \in V \cup O} \gamma_i^{(k)} x_i + \eta^{(k)} \ (1 \leq k \leq o)$$

Note that Oil and Vinegar variables are not fully mixed, just like oil and vinegar in a salad dressing.

The map $\mathcal{Q} = (q^{(1)}(\mathbf{x}), \ldots, q^{(o)}(\mathbf{x}))$ can be easily inverted. First, we choose the values of the v Vinegar variables $x_1, \ldots, x_v$ at random. Therefore we get a system of o linear equations in the o Oil variables $x_{v+1}, \ldots, x_n$ which can be solved by Gaussian Elimination. (If the system does not have a solution, one has to choose other values of $x_1, \ldots, x_v$ and try again).

To hide the structure of $\mathcal{Q}$ in the public key one concatenates it with an affine invertible map $\mathcal{T} : \mathbb{F}^n \rightarrow \mathbb{F}^n$. So, the public key of the UOV signature scheme is given as

$$\mathcal{P} = \mathcal{Q} \circ \mathcal{T} \tag{1}$$

Remark 1: In opposite to other multivariate schemes the second affine map $\mathcal{S}$ is not needed for the security of UOV. So it can be dropped.

Signature generation and verification. To *sign* a message with a hash value $h \in \mathbb{F}_q^o$, one computes recursively $y = \mathcal{Q}^{-1}(h)$ and $z = \mathcal{T}^{-1}(y)$. The signature of the message is $z \in \mathbb{F}_q^n$. Here $\mathcal{Q}^{-1}(h)$ means finding one pre-image of $h \in \mathbb{F}_q^o$ under $\mathcal{Q}$, which we get by choosing the Vinegar variables at random and solving the resulting linear system for the Oil variables. To *verify* a signature $z \in \mathbb{F}_q^n$, one computes $w = \mathcal{P}(z) \in \mathbb{F}_q^o$. If $w = h$ holds, the signature is accepted, otherwise rejected.

In the original paper [Pa97], Patarin suggested to use $o = v$ (Balanced Oil and Vinegar (OV)). After this scheme was broken by Kipnis and Shamir in [KS98], it was suggested in [KP99] to use $v > o$ (Unbalanced Oil and Vinegar (UOV)).

The UOV signature scheme over $GF(2^8)$ is commonly believed to be secure for $o \geq 26$ equations and $v = 2 \cdot o$ Vinegar variables [BF08].

3 The Approach of [PB10]

In this section we review the approach of [PB10] to create a UOV-based scheme with a partially cyclic public key.

Remember that, in the case of the Unbalanced Oil and Vinegar signature scheme [KP99], the public key $\mathcal{P}$ is given as the concatenation of the central UOV-map $\mathcal{Q}$ and an affine invertible map $\mathcal{T} = ((t_{ij})_{i,j=1}^n, c_T)$, i.e.

$$\mathcal{P} = \mathcal{Q} \circ \mathcal{T}. \tag{2}$$

The authors of [PB10] observed, that this equation (after fixing the affine map $\mathcal{T}$), leads to a linear relation between the coefficients of the quadratic monomials of $\mathcal{P}$ and $\mathcal{Q}$ of the form

$$p_{ij}^{(k)} = \sum_{i=1}^{n} \sum_{j=i}^{n} \alpha_{ij}^{rs} \cdot q_{rs}^{(k)}, \tag{3}$$

where $p_{ij}^{(k)}$ and $q_{ij}^{(k)}$ are the coefficients of $x_i x_j$ in the k-th component of $\mathcal{P}$ and $\mathcal{Q}$ respectively and the α_{ij}^{rs} are given as

$$\alpha_{ij}^{rs} = \begin{cases} t_{ri} \cdot t_{si} & (i = j) \\ t_{ri} \cdot t_{sj} + t_{rj} \cdot t_{si} & \text{otherwise} \end{cases}. \tag{4}$$

Let $D := \frac{v \cdot (v+1)}{2} + o \cdot v$ be the number of non-zero quadratic terms in any component of $\mathcal{Q}$ and $D' := \frac{n \cdot (n+1)}{2}$ be the number of quadratic terms in the public polynomials. Let M_P and M_Q be the Macaulay matrices of $\mathcal{P}$ and $\mathcal{Q}$ respectively (in graded lexicographical order). The matrices M_P and M_Q are divided into submatrices as shown in Figure 1. Note that, due to the absence of oil $\times$ oil terms in the central polynomials, we have a block of zeros in the middle of M_Q.

Furthermore, the authors of [PB10] defined the so called transformation matrix $A \in \mathbb{F}_q^{D \times D}$ containing the coefficients α_{ij}^{rs} of equation (3)

$$A = \left(\alpha_{ij}^{rs}\right) \ (1 \leq r \leq v, r \leq s \leq n \text{ for the rows, } 1 \leq i \leq v, i \leq j \leq n$$
$$\text{for the columns), i.e.}$$

$$A = \begin{pmatrix} \alpha_{11}^{11} & \alpha_{12}^{11} & \dots & \alpha_{vn}^{11} \\ \alpha_{11}^{12} & \alpha_{12}^{12} & \dots & \alpha_{vn}^{12} \\ \vdots & & & \vdots \\ \alpha_{11}^{vn} & \alpha_{12}^{vn} & \dots & \alpha_{vn}^{vn} \end{pmatrix}. \tag{5}$$

With this notation, equation (3) yields

$$B = Q \cdot A \tag{6}$$

If the matrix A is invertible, this relation becomes bijective.

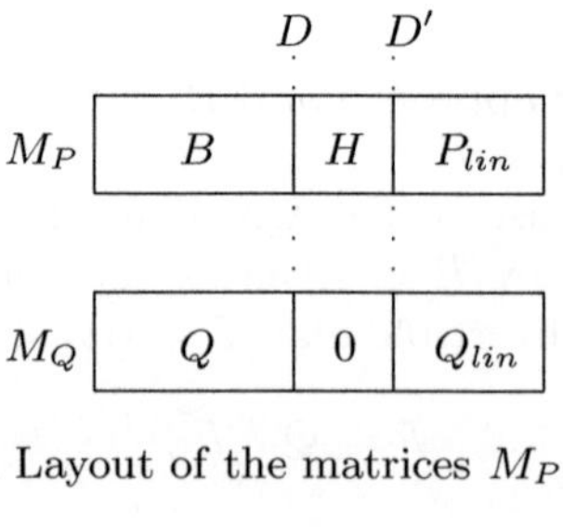

Fig. 1. Layout of the matrices M_P and M_Q

By solving equation (6) for Q, the authors of [PB10] were able to insert a partially circulant matrix into the UOV public key. By doing so, they reduced the public key size of the scheme by a large factor.

4 Preliminaries

4.1 Linear Recurring Sequences (LRS)

In this subsection we repeat briefly results from the theory of linear recurring sequences (LRS's) needed in the following sections. For a more detailed introduction and the proofs we refer to [LN86].

Definition 1. *Let L be a positive integer and $\gamma_1, \ldots, \gamma_L$ be given elements of a finite field $\mathbb{F}_q$. A linear recurring sequence (LRS) of length L is a sequence $\{s_1, s_2, \ldots\}$ of $\mathbb{F}_q$-elements satisfying the relation*

$$s_j = \gamma_1 \cdot s_{j-1} + \gamma_2 \cdot s_{j-2} + \cdots + \gamma_L \cdot s_{j-L} = \sum_{i=1}^{L} \gamma_i \cdot s_{j-i} \quad (\forall j > L). \tag{7}$$

The values $s_1, \ldots, s_L$ are called the initial values of the LRS.

Definition 2. *The connection polynomial of an LRS is defined as*

$$C(x) = \gamma_L x^L + \gamma_{L-1} x^{L-1} + \cdots + \gamma_1 \cdot X + 1 = \sum_{i=1}^{L} \gamma_i X^i + 1.$$

The LRS S is uniquely determined by its initial values $s_1, \ldots, s_L$ and the connection polynomial C (due to equation (7)). Therefore we denote the LRS by $S(s_1, \ldots, s_L, C)$.

Definition 3. *An irreducible polynomial $f(x) \in \mathbb{F}_q[x]$ of degree d is called a primitive polynomial if one of the roots of $f(x)$ is a generator of $\mathbb{F}^{\star}_{q^d}$, the multiplicative group of all the non-zero elements of $\mathbb{F}_{q^d}$.*

Lemma 1. *The irreducible polynomial $f(x) \in \mathbb{F}_q[x]$ of degree d is a primitive polynomial if and only if $f(x)$ divides $x^k - 1$ for $k = q^d - 1$ and for no smaller positive integer k.*

Definition 4. *A sequence $\{\sigma_1, \sigma_2, \ldots\}$ of $\mathbb{F}_q$-elements is said to be periodic with minimal period k, if k is the smallest integer such that $\sigma_i = \sigma_{i+t \cdot k}$ $(\forall i, t \in \mathbb{N})$.*

Lemma 2. *An LRS of length L with primitive connection polynomial $C(x) \in \mathbb{F}_q[x]$ and $(s_1, \ldots, s_L) \in \mathbb{F}_q^L \setminus \{\mathbf{0}\}$ is periodic with minimal period $q^L - 1$.*

Definition 5. *An LRS as in Lemma 2 is called an m-sequence.*

Definition 6. *Let $\Sigma = \{\sigma_1, \sigma_2, \ldots\}$ be a (finite or infinite) sequence of $\mathbb{F}_q$-elements. The linear complexity $\mathrm{LC}(\Sigma)$ is defined as the length of the shortest LRS S such that $\sigma_i = s_i$ $\forall i$.*

Lemma 3. *Let $S = S(s_1, \ldots, s_L, C)$ be an LRS of length L with irreducible connection polynomial C. Then, the linear complexity of S is equal to L.*

4.2 Golomb's Randomness Postulates [Go67]

In this subsection we look at sequences over a finite field $\mathbb{F}_q$. We cite from [GG05] some criteria a sequence Σ must fulfill to be considered a random sequence.

Definition 7. *Let $\lambda, \eta, \zeta \in \mathbb{F}_q$ with $\lambda \neq \eta$ and $\lambda \neq \zeta$. A subsequence $\bar{\sigma}$ of $\Sigma = \{\sigma_1, \sigma_2, \dots\}$ of the form*

$$\eta, \underbrace{\lambda, \dots, \lambda}_{k-\text{times}}, \zeta$$

is called a run of λ of length k.

Definition 8. *The auto-correlation function of a sequence $\Sigma = \{\sigma_1, \sigma_2, \dots\}$ with period $q^n - 1$ is defined as*

$$\mathrm{AC}_\Sigma(\tau) = \sum_{i=0}^{q^n-2} \chi(\sigma_i) \cdot \overline{\chi(\sigma_{i+\tau})} \; (0 \leq \tau \leq q^n - 2),$$

where χ is given by

$$\chi(x) = e^{2\pi i \,\mathrm{Tr}(x)/p}$$

with Tr *being the standard trace function between $\mathbb{F}_q$ and its prime field $\mathbb{F}_p$.*

Golomb formulated three postulates a sequence must fulfill to be considered a random sequence. Let Σ be a sequence with period $q^n - 1$.

R-1. In every period, every non-zero element occurs q^{n-1} times and the zero element occurs $q^{n-1} - 1$ times.

R-2. In every period,
1. for $1 \leq k \leq n-2$, the runs of each element of length k occur $(q-1)^2 \cdot q^{n-k-2}$ times.
2. the runs of each non-zero element of $\mathbb{F}_q$ of length $n-1$ occur $q-2$ times.
3. the runs of the zero element of length $n-1$ occurs $q-1$ times.
4. the run of every non-zero element of length n occurs once.

R-3. The auto-correlation function AC_Σ is two valued with

$$\mathrm{AC}_\Sigma(\tau) = \begin{cases} q^n - 1 & \text{if } \tau \equiv 0 \bmod (q^n - 1) \\ -1 & \text{if } \tau \not\equiv 0 \bmod (q^n - 1) \end{cases}$$

Remark 2: The auto-correlation function AC_Σ measures the amount of similarity between the sequence Σ and its shift by τ positions. Postulate R-3 states that for $\tau \geq 1$ the value $AC_\Sigma(\tau)$ should be quite small.

Postulate R-1 can be extended as follows

R-4. In every period, each non-zero n-tuple $(\lambda_1, \dots, \lambda_n) \in \mathbb{F}_q^n$ appears exactly once.

Lemma 4. *Any m-sequence fulfills the postulates R-1 to R-4 (for $n = L$).*

Remark 3: In the partially cyclic approach of [PB10], the rows of the matrix B are given as $\mathbf{b}^{(i)} = \mathcal{R}^{i-1}(\mathbf{b})$ $(i = 1, \ldots, o)$, where $\mathcal{R}$ is the cyclic right shift and $\mathbf{b}$ is a randomly chosen vector. The sequence obtained by this construction clearly doesn't fulfill these postulates. For example, for most of the $\lambda \in \mathbb{F}_q$ the 2-run (λ, λ) does not appear in such a sequence (contradiction to postulate R-2).

Remark 4: Because of the good statistical properties of m-sequences, linear recurring sequences are used to bring randomness into a large number of areas, for example digital broadcasting and the Global Positioning System (GPS). However, an m-sequence can't be said to be a truely random sequence. For example, the linear complexity of an m-sequence obtained by an LRS of length L is L, whereas the linear complexity of a random sequence of length N should be about $N/2$. Therefore, the elements of an m-sequence are easily predictable. Hence, for cryptographic applications like stream ciphers, one has to add some non-linearity features.

5 Description of the Scheme

In this section we deal with the construction of our scheme and describe it in detail.

Additionally to the matrix A defined in Section 3 we define a matrix $A' \in \mathbb{F}_q^{D \times D'}$ by

$$A' = \left(\alpha_{ij}^{rs}\right) \ (1 \leq r \leq v,\ r \leq s \leq n \text{ for the rows, } 1 \leq i \leq j \leq n \text{ for the columns}), \quad (8)$$

whose entries α_{ij}^{rs} are given by equation (4). The order in which the α_{ij}^{rs} appear in A' is given by the graded lexicographical ordering (for both rows and columns). Note that the matrix A (as defined in Section 3) is a submatrix of A'.

5.1 Construction

At the beginning of our construction we choose randomly an affine invertible map $\mathcal{T}$ (given as a matrix $M_T = (t_{ij})_{i,j=1}^n$ and an $n-$vector c_T) and compute the corresponding transformation matrix A (using equations (4) and (5)). Furthermore, we choose an LRS of length L with initial values $s_1, \ldots, s_L$ and primitive connection polynomial $C(X) = \sum_{i=1}^L \gamma_i X^i + 1$ and compute its first $o \cdot D$ elements (using equation (7)).

We define the $o \times D$-matrix B (see Figure 1) as

$$B = (b_{ij}) \text{ with } b_{ij} = s_{D \cdot (i-1)+j} \ (i = 1, \ldots o,\ j = 1, \ldots, D) \quad (9)$$

As in [PB10] equation (3) yields

$$B = Q \cdot A \quad (10)$$

and we get

$$(B|H) = Q \cdot A' \quad (11)$$

Under the assumption of A being invertible we can invert equation (10) and compute the homogeneous quadratic part of the central map.

Remark 5: To justify the assumption of A being invertible, we carried out a number of experiments. For different values of o and v we created 1000 matrices M_T each time and tested, how many of the corresponding matrices A were invertible. Table 1 shows the results.

Table 1. Percentage of the matrices A being invertible

(2^8,o,v)	(2,4)	(5,10)	(10,20)	(15,30)	(20,40)
% invertible	99.3	99.6	99.7	99.5	99.4

As the table shows, the condition of A being invertible is nearly always complied.

5.2 The Scheme

Key Generation

1. Choose randomly a vector $(s_1, \ldots, s_L) \in \mathbb{F}_q^L \setminus \{\mathbf{0}\}$ and a primitive connection polynomial $C(X) = \sum_{i=1}^{L} \gamma_i X^i + 1$.
2. Compute the first $o \cdot D$ elements of the LRS $S = S(s_1, \ldots, s_L, C)$ using equation (7).
3. Compute the matrix B using equation (9).
4. Choose an affine map $\mathcal{T} = (M_T, c_T) : \mathbb{F}_q^n \to \mathbb{F}_q^n$ at random. If M_T is not invertible, choose again.
5. Compute for T the corresponding transformation matrix A (using equations (4) and (5)). If A is not invertible, go back to step 4.
6. Solve the linear systems given by equation (10) to get the matrix Q and therewith the homogeneous quadratic part of the central map $\mathcal{Q}$.
7. Choose the coefficients of the linear terms of the central polynomials at random.
8. Compute the public key as $\mathcal{P} = \mathcal{Q} \circ \mathcal{T}$.

Signature generation and verification work as in the case of the standard UOV scheme (see Section 2).

The *public key* consists of the last $\left(\frac{o\cdot(o+1)}{2} + o + v + 1\right)$ columns of the matrix M_P, the initial values $s_1, \ldots, s_L$ and the connection polynomial C of the LRS.

The *private key* consists of the maps $\mathcal{Q}$ and $\mathcal{T}$.

The *size of the public key* is given as

$$o \cdot \left(\frac{o \cdot (o+1)}{2} + o + v + 1\right) + 2 \cdot L \text{ field elements,}$$

the *size of the private key* is

$$o \cdot \left(\frac{v \cdot (v+1)}{2} + o \cdot v + o + v + 1\right) \text{ field elements.}$$

We denote the scheme by UOVLRS(q, o, v, L), where q is the cardinality of the underlying field.

6 Choice of the Parameter L

In this section we look at the question how to choose the length of the LRS.

6.1 General Remarks

Proposition 1. *Let the $o \times D$ matrix B be generated by an LRS of length $L \leq o$ (as described in Subsection 5.2). Then we have* $\text{rank}(B) \leq L$.

Proof. We denote B' to be the upper left $L \times L$ submatrix of B. Its rows we denote by $\mathbf{b}'^{(1)}, \ldots, \mathbf{b}'^{(L)}$.

Let's assume that we have already found L linear independent rows of B. W.l.o.g. these are the first L rows of B, namely $\mathbf{b}^{(1)}, \ldots, \mathbf{b}^{(L)}$. Due to Lemma 5 B' is then invertible. We have to show, that all the other rows $\mathbf{b}^{(L+1)}, \ldots, \mathbf{b}^{(o)}$ can be written as linear combinations of the rows $\mathbf{b}^{(1)}, \ldots, \mathbf{b}^{(L)}$.

Let $L < i \leq o$. First we show that the vector $\mathbf{b}'^{(i)}$ (consisting of the first L elements of $\mathbf{b}^{(i)}$) can be written as a linear combination of the vectors $\mathbf{b}'^{(1)}, \ldots, \mathbf{b}'^{(L)}$. In other words, we need to find a vector $\beta^{(i)} \in \mathbb{F}_q^L$ such that $\mathbf{b}'^{(i)} = B' \cdot \beta^{(i)}$. Since B' is invertible, $\beta^{(i)}$ can be computed by $\beta^{(i)} = B'^{-1} \cdot \mathbf{b}'(i)$.

To finish the proof, it remains to show that the vector $\beta^{(i)}$ fulfills the relation $\mathbf{b}^{(i)} = \sum_{j=1}^{L} \beta_j^{(i)} \cdot \mathbf{b}^{(j)}$. Remember that $\beta^{(i)}$ was chosen in such a way that the relation is fulfilled for the first L elements of $\mathbf{b}^{(i)}$. In the following we show this equality for every element $\mathbf{b}_r^{(i)}$ with $(L < r \leq D)$ by induction. Note that, due to the recurrence relation (7), we have $\mathbf{b}_r^{(i)} = \sum_{j=1}^{L} \gamma_j \mathbf{b}_{r-j}^{(i)}$ $(i = 1, \ldots, o,\ r > L)$.
$r = L + 1$:

$$\mathbf{b}_{L+1}^{(i)} = \sum_{j=1}^{L} \gamma_j \cdot \mathbf{b}_{L+1-j}^{(i)} = \sum_{j=1}^{L} \gamma_j \cdot \left(\sum_{l=1}^{L} \beta_l^{(i)} \cdot \mathbf{b}_{L+1-j}^{(l)} \right)$$
$$= \sum_{l=1}^{L} \beta_l^{(i)} \cdot \left(\sum_{j=1}^{L} \gamma_j \cdot \mathbf{b}_{L+1-j}^{(l)} \right) = \sum_{l=1}^{L} \beta_l^{(i)} \cdot \mathbf{b}_{L+1}^{l}$$

$r \leftarrow r + 1$:

$$\mathbf{b}_r^{(i)} = \sum_{j=1}^{L} \gamma_j \cdot \mathbf{b}_{r-j}^{(i)} = \sum_{j=1}^{L} \gamma_j \cdot \left(\sum_{l=1}^{L} \beta_l^{(i)} \cdot \mathbf{b}_{r-j}^{(l)} \right)$$
$$= \sum_{l=1}^{L} \beta_l^{(i)} \cdot \left(\sum_{j=1}^{L} \gamma_j \cdot \mathbf{b}_{r-j}^{(l)} \right) = \sum_{l=1}^{L} \beta_l^{(i)} \cdot \mathbf{b}_r^{l} \qquad \square$$

Lemma 5. *If the first L rows of B are linearly independent, then the matrix B' is invertible.*

Proof. Let's assume that B' doesn't have full rank. Then there exists a linear relation of the form $\sum_{i=1}^{L} \beta_i \cdot B'_i = \mathbf{0}$, where B'_i $(i = 1, \ldots, L)$ are the columns of B'. In other words, there exists an index $j \in \{1, \ldots, L\}$ such that $B'_j = \sum_{i=1,\ i \neq j}^{L} \delta_i \cdot B'_i$.

Let $L+1 \leq k \leq D$. If we denote by B''_k the k-th column of the matrix B'', which we define to be given by the first L rows of B, we get due to the recurrence relation (7)

$$B''_k = \sum_{i=1}^{L} \eta_i \cdot B'_i = \sum_{i=1,\ i \neq j}^{L} \eta_i \cdot B'_i + \eta_j \cdot \sum_{i=1,\ i \neq j}^{L} \delta_i \cdot B'_i = \sum_{i=1,\ i \neq j}^{L} (\eta_i + \eta_j \cdot \delta_i) \cdot B'_i.$$

Therefore, the rank of the matrix B'' would be less than L and the vectors $\mathbf{b}^{(1)}, \ldots, \mathbf{b}^{(L)}$ would be linearly dependent. □

Remark 6: To study the question, which values of rank(B) occur in practice, we carried out a number of experiments. For different parameter sets we created 10000 matrices B and computed their rank. Table 2 shows the results.

Table 2. Number of matrices B with rank L

(o, v, L)		GF(2)	GF(3)	GF(4)	GF(5)	GF(7)	GF(8)	GF(16)	GF(31)	GF(256)
$(8, 16, 5)$	rank$(B) = L$	9037	9118	9722	9871	9974	10000	10000	10000	10000
	B' invertible	9037	9118	9722	9871	9974	10000	10000	10000	10000
$(8, 16, 8)$	rank$(B) = L$	9973	9980	9983	9984	9995	10000	10000	10000	10000
	B' invertible	9973	9980	9983	9984	9995	10000	10000	10000	10000
$(20, 40, 15)$	rank$(B) = L$	9981	9998	10000	10000	10000	10000	10000	10000	10000
	B' invertible	9981	9998	10000	10000	10000	10000	10000	10000	10000
$(20, 40, 20)$	rank$(B) = L$	9962	9983	9995	10000	10000	10000	10000	10000	10000
	B' invertible	9962	9983	9995	10000	10000	10000	10000	10000	10000

The experiments seem to show that for fields of cardinality ≥ 8 the rank of the matrix B is always equal to L. Furthermore, the matrix B' was always invertible for these fields.

Proposition 2. *Let $(\mathcal{P}, \mathcal{Q}, \mathcal{T})$ be a UOV-scheme, whose public key is generated by an LRS of length $L \leq o$. Then we have* rank$(Q) =$ rank$(B) \leq L$.

Proof. According to our assumption the matrix A is invertible. Therefore, the proposition follows directly from equation (10). □

Remark 7: Despite of the relation between the rows of Q, there is no obvious relation between the columns of Q. In particular, there exists no LRS of small length which creates Q.

Theorem 1. *Let $(\mathcal{P}, \mathcal{Q}, \mathcal{T})$ be a UOV scheme generated by an LRS of length $L \leq o$. Then we have* rank$(B|H) =$ rank$(B) \leq L$.

Proof. Since B is a submatrix of $(B|H)$, the rank of $(B|H)$ can't be less than that of B. But, according to equation (11), the rank of $(B|H)$ can't be larger than rank$(Q) =$ rank(B), too. □

Theorem 1 states that for $L < o$ the homogeneous quadratic parts of the public polynomials are linearly dependent. In particular, of the o quadratic polynomials of a public key generated by an LRS of length $L < o$, only L have linear independent homogeneous quadratic parts. So, solving the equation $\mathcal{P}(\mathbf{x}) = h$ (o equations) is only as difficult as solving a system of L quadratic equations. As a consequence of this, to achieve the maximal possible security, we should choose the length of the LRS at least o[1].

To check the correctness of these theoretical considerations, we carried out a number of experiments with MAGMA [BC97]. For different parameter sets $(2^8, o, v, L)$ we created instances of our scheme and solved the corresponding systems using the MAGMA command `GroebnerBasis`. Table 3 shows the results.

Table 3. Running time of the direct attack for different values of L

$(2^8, o, v)$	$(10, 20)$	$(11, 22)$	$(12, 24)$	$(13, 26)$	$(14, 28)$
$L = o$	67 s	384 s	3071 s	23528 s	186382 s
$L = o - 1$	8.3 s	68 s	395 s	3215 s	24652 s
$L = o - 2$	1.7 s	8.4 s	67 s	408 s	3249 s

As the table shows, solving a UOV system with o equations generated by an LRS of length $L < o$ is only as difficult as solving a system with L equations.

6.2 Choice of L for Smaller Fields

For small fields (e.g. $GF(16)$) it might be useful to choose $L < o$. The reason for this is that for small fields the needed number of equations is determined by the length of the hash value and not by attacks against the scheme itself. For example, for $GF(16)$ one needs 40 equations to achieve a hash length of 160 bit. However, only 30 equations are needed to defend the scheme against direct attacks [BF08]. So, it might be useful to choose the homogeneous quadratic part of the last 10 public equations to be a linear combination of the quadratic parts of the previous ones. The fact that the linear part of the public equations is independent of the homogeneous quadratic part, guarantees the functionality of the scheme. This strategy decreases the sizes of both public and private key by about 25 %. Furthermore, key generation and signature generation/ verification become faster.

We plan to study this idea (especially its effects on the security of the scheme) further.

7 Security

In this section we look at known attacks against the UOV signature scheme and study the effect of the special structure of our public key.

1 For smaller fields (e.g. $GF(2^4)$) it might be useful to choose $L < o$. (see Subsection 6.2).

7.1 Direct Attacks

The most straightforward method to forge a signature for a message h is by trying to solve the system $\mathcal{P}(\mathbf{x}) = h$ directly, i.e. by an equation solver like XL or a Gröbner Basis method like Buchbergers algorithm or Faugère's F_4/F_5. We carried out a number of experiments with MAGMA, which contains an efficient implementation of the F_4 algorithm [Fa99]. Before using the MAGMA command `GroebnerBasis`, we had to fix some of the variables to create a determined system. Since the number of solutions of an underdetermined UOV system is approximately q^v, it can be expected that, after fixing v of the variables, the determined system has a solution. Table 4 shows the results of our experiments on UOV-like schemes and random systems.

Table 4. Results of our experiments with direct attacks

$(2^8, o, v, L)$	$(10, 20, 10)$	$(11, 22, 11)$	$(12, 24, 12)$	$(13, 26, 13)$	$(14, 28, 14)$
UOVLRS	67 s	384 s	3071 s	23528 s	186382 s
UOV	68 s	386 s	3068 s	23677 s	186425 s
random system	68 s	386 s	3072 s	23725 s	186483 s

As the table shows, the running time of direct attacks against our scheme is nearly the same as for the standard UOV scheme and for random systems. So, for $o \geq 26$ equations [BF08] our scheme seems to be secure against direct attacks.

Definition 9. *Let $p(\mathbf{x}) = p(x_1, \ldots, x_n)$ be a quadratic multivariate polynomial and*

$$dp(\mathbf{x}, \mathbf{c}) = p(\mathbf{x} + \mathbf{c}) - p(\mathbf{x}) - p(\mathbf{c}) + p(\mathbf{0})$$

its discrete differential. We define H_p to be the symmetric matrix such that

$$dp = \mathbf{x}^T \cdot H_p \cdot \mathbf{c}$$

For the matrix H_{p_i} representing the quadratic part of the i-th public polynomial we write in short H_i. Analogous, we denote the symmetric matrix representing the homogeneous quadratic part of the i-th central polynomial by Q_i $(i = 1, \ldots, o)$.

7.2 UOV-Reconciliation

The goal of the UOV-Reconciliation attack is to find a change of variables which brings the matrices H_i into UOV-form, which means that the lower right $o \times o$ submatrix is the zeromatrix. By doing so, the attacker creates an equivalent private key and therefore is able to forge signatures for arbitrary messages.

To achieve this goal, the attacker has to solve several multivariate quadratic systems. The complexity of the attack is mainly determined by the complexity

Table 5. Running time of the UOV-Reconciliation attack

(2^8,o,v,L)	(10,20,10)	(11,22,11)	(12,24,12)	(13,26,13)	(14,28,14)
UOVLRS	66 s	385 s	3072 s	23526 s	186380 s
UOV	68 s	384 s	3074 s	23534 s	186423 s

of the first step which is the solving of a quadratic system of o equations in v variables. Table 5 shows the time MAGMA needs for solving this initial system for our scheme and the standard UOV scheme.

As the table shows, the special structure of our public key has only a negligible effect on the running time of the UOV-Reconciliation attack.

Since, for the parameters proposed in Section 2, the UOV scheme is believed to be secure against the UOV-Reconciliation attack, we can assume the same for our scheme.

7.3 Rank Attacks

In this paragraph we look at the behavior of Rank attacks against the standard UOV and our scheme. To do this, we carried out experiments with 10000 instances of our scheme for different parameters $(2^8, o, v, o)$. We observed that, just as in the case of the standard UOV scheme, all the matrices Q_i representing the homogeneous quadratic parts of the central equations have full rank n. This prevents the MinRank attack. Furthermore, all the variables $x_1, \ldots, x_n$ appear in every of the o central equations, which prevents HighRank attacks.

7.4 UOV Attack [KP99]

The goal of this attack is to find the pre-image of the oil subspace $\mathcal{O} = \{x \in K^n : x_1 = \cdots = x_v = 0\}$ under the affine invertible transformation $\mathcal{T}$. To achieve this, one forms a random linear combination $P = \sum_{j=1}^{o} \beta_j H_j$, multiplies it with the inverse of one of the H_i and looks for invariant subspaces of this matrix. For each parameter set $(2^8, o, v, L)$ listed in the table we created 100 instances of both schemes. Then we attacked these instances by the UOV-attack to find out the number of trials we need to find a basis of $\mathcal{T}^{-1}(\mathcal{O})$. Table 6 shows the results.

Table 6. Average number of trials in the UOV-attack

(2^8,o,v,L)	(5,7,5)	(8,11,8)	(12, 15,12)	(15, 18,15)
UOVLRS	1725	530826	851836	1178392
UOV	1734	531768	852738	1183621

As the table shows, there is only a negligible difference between the number of trials we need between our scheme and the standard UOV. Since for the parameters proposed in Section 2 UOV is believed to be secure against this attack, we can say the same for our scheme.

7.5 Summary

As the previous four subsections showed, known attacks against the UOV signature scheme do not work significantly better in our case, which means that they can not use the special structure of our public key. So, in this sense our scheme seems to be secure and we do not have to adapt our parameter sets.

However, in the future we are going to study the security of our scheme under other attacks, e.g. decomposition attacks [FP09]. It might also be possible that dedicated attacks against our scheme exist. Still, since the statistical properties of the public key are rather strong due to the use of m-sequences, we believe that the development of such an attack is a hard task.

8 Parameters

Based on our security analysis (see previous section) we propose for our scheme the same parameters as for the standard UOV signature scheme (see Section 2). According to our considerations in Section 6, the length of the LRS should be at least o. Such we get

$$q = 2^8, \ o = 26, \ v = 52, \ L = 26.$$

Table 7 compares our scheme with the scheme of [PB10] and the standard UOV for this and a more conservative parameter set.

Table 7. Comparison of different UOV based schemes

	public key size (kB)	private key size (kB)	hash size (bit)	signature size (bit)	reduction factor (%)
UOV(2^8,26,52)	80.2	71.3	208	624	-
cyclicUOV(2^8,26,52)	13.6	71.3	208	624	83.0
UOVLRS(2^8,26,52,26)	11.0	71.3	208	624	86.3
UOV(2^8,28,56)	99.9	88.8	224	672	-
cyclicUOV(2^8,28,56)	16.5	88.8	224	672	83.4
UOVLRS(2^8,28,56,28)	13.5	88.8	224	672	86.4

As the table shows, the public key size of our scheme is only slightly smaller than that of the cyclicUOV scheme of [PB10]. However, due to the good statistical properties of our public key, we believe our scheme to be more secure.

9 Conclusion

In this paper we proposed a multivariate signature scheme whose public key is mainly generated by a linear recurring sequence (LRS). By doing so, we were able to reduce the public key size of the standard UOV scheme by up to 86 %.

Moreover, the so obtained public keys have good statistical properties, which makes it difficult to develop dedicated attacks against our scheme. We think that our approach is an interesting idea on reducing the key size of multivariate schemes. Points of research we want to address in the future include

- Exhaustive security analysis (including decomposition attacks).
- Extension of the strategy to other underlying fields. While in this paper we have concentrated on an underlying field of 256 elements, we are planning to use our strategy for other fields (especially GF(16) or GF(31)). Here, the main points of our construction stay the same, while one has to study the invertibility of the matrix A and the security of the scheme. Furthermore, we are going to study the impact of the idea mentioned in Subsection 6.2.
- Use of pseudo-random number generators (PRNG's) for generating the public key. By the use of PRNG's (for example AES in the OFB mode) we will get public keys with even better statistical properties. Moreover, we hope that this will bring us closer to the "provably secure" UOV scheme of [BP10].

Acknowledgements

The first author is supported by the Horst Görtz Foundation within the project "Efficient and practically applicable multivariate-based schemes with estimated secure parameters for now and the future" run by the second author as a principal investigator. The second author is partially supported by the German Science Foundation (DFG) grant BU 630/22-1. Furthermore, we want to thank the anonymous reviewers for their valuable comments which helped to improve the paper.

References

[BB08] Bernstein, D.J., Buchmann, J., Dahmen, E. (eds.): Post-Quantum Cryptography. Springer, Heidelberg (2009)

[BC97] Bosma, W., Cannon, J., Playoust, C.: The Magma algebra system. I. The user language. J. Symbolic Comput. 24(3-4), 235–265 (1997)

[BF08] Bettale, L., Faugére, J.C., Perret, L.: Hybrid approach for solving multivariate systems over finite fields. Journal Math. Crypt. 2, 1–22 (2008)

[BP10] Bulygin, S., Petzoldt, A., Buchmann, J.: Towards provable security of the UOV Signature Scheme under direct attacks. In: Gong, G., Gupta, K.C. (eds.) INDOCRYPT 2010. LNCS, vol. 6498, pp. 17–32. Springer, Heidelberg (2010)

[CC08] Chen, A.I.-T., Chen, C.-H.O., Chen, M.-S., Cheng, C.M., Yang, B.-Y.: Practical-Sized Instances of Multivariate PKCs: Rainbow, TTS, and ℓIC-Derivatives. In: Buchmann, J., Ding, J. (eds.) PQCrypto 2008. LNCS, vol. 5299, pp. 95–108. Springer, Heidelberg (2008)

[DG06] Ding, J., Gower, J.E., Schmidt, D.: Multivariate Public Key Cryptosystems. Springer, Heidelberg (2006)

[Fa99] Faugére, J.C.: A new efficient algorithm for computing Groebner bases (F4). Journal of Pure and Applied Algebra 139, 61–88 (1999)

[FP09] Faugére, J.C., Perret, L.: An efficient algorithm for decomposing multivariate polynomials and its applications to cryptography. Journal of Symbolic Computation 44(12), 1676–1689 (2009)

[Go67] Golomb, S.W.: Shift Register Sequences. Holden Day, San Francisco (1967)

[GG05] Golomb, S.W., Gong, G.: Signal Design for Good Correlation. Cambridge University Press, New York (2005)

[HW05] Hu, Y.-H., Wang, L.-C., Chou, C.-Y., Lai, F.: Similar Keys of Multivariate Quadratic Public Key Cryptosystems. In: Desmedt, Y.G., Wang, H., Mu, Y., Li, Y. (eds.) CANS 2005. LNCS, vol. 3810, pp. 211–222. Springer, Heidelberg (2005)

[KP99] Kipnis, A., Patarin, J., Goubin, L.: Unbalanced Oil and Vinegar Signature Schemes. In: Stern, J. (ed.) EUROCRYPT 1999. LNCS, vol. 1592, pp. 206–222. Springer, Heidelberg (1999)

[KS98] Kipnis, A., Shamir, A.: Cryptanalysis of the Oil & Vinegar Signature Scheme. In: Krawczyk, H. (ed.) CRYPTO 1998. LNCS, vol. 1462, pp. 257–266. Springer, Heidelberg (1998)

[LN86] Lidl, R., Niederreiter, H.: Introduction to finite fields and their applications. Cambridge University Press, Cambridge (1986)

[Pa97] Patarin, J.: The oil and vinegar signature scheme, presented at the Dagstuhl Workshop on Cryptography (September 1997)

[PB10] Petzoldt, A., Bulygin, S., Buchmann, J.: A Multivariate Signature Scheme with a partially cyclic public key. In: Proceedings of SCC, pp. 229–235 (2010)

[Sh97] Shor, P.: Polynomial-Time Algorithms for Prime Factorization and Discrete Logarithms on a Quantum Computer. SIAM J. Comput. 26(5), 1484–1509

[YC05] Yang, B.-Y., Chen, J.-M.: Building secure tame-like multivariate public-key cryptosystems: The new TTS. In: Boyd, C., González Nieto, J.M. (eds.) ACISP 2005. LNCS, vol. 3574, pp. 518–531. Springer, Heidelberg (2005)

On the Impossibility of Instantiating PSS in the Standard Model

Rishiraj Bhattacharyya[1] and Avradip Mandal[2]

[1] Cryptology Research Group, Applied Statistics Unit,
Indian Statistical Institute, Kolkata
rishi_r@isical.ac.in
[2] Université du Luxembourg, Luxembourg
avradip.mandal@uni.lu

Abstract. In this paper we consider the problem of securely instantiating Probabilistic Signature Scheme (PSS) in the standard model. PSS, proposed by Bellare and Rogaway [3] is a widely deployed randomized signature scheme, provably secure (*unforgeable under adaptively chosen message attacks*) in Random Oracle Model.

Our main result is a black-box impossibility result showing that one can not prove unforgeability of PSS against chosen message attacks using blackbox techniques even assuming existence of *ideal trapdoor permutations* (a strong abstraction of trapdoor permutations which inherits all security properties of a random permutation, introduced by Kiltz and Pietrzak in Eurocrypt 2009) or the recently proposed *lossy trapdoor permutations* [20]. Moreover, we show *onewayness*, the most common security property of a trapdoor permutation does not suffice to prove even the weakest security criteria, namely *unforgeability under zero message attack*. Our negative results can easily be extended to any randomized signature scheme where one can recover the random string from a valid signature.

Keywords: PSS, Blackbox Reductions, Randomized Signature, Standard Model.

1 Introduction

Probabilistic Signature Scheme (PSS) is one of the most known and widely deployed provably secure randomized signature schemes. It was designed by Bellare and Rogaway [3] as a generic scheme based on a trapdoor permutation (like RSA). In [3], Bellare and Rogaway showed the scheme is secure in Random Oracle (RO) Model [2]. Coron improved the previous security bound in [7]. Recently in [8], PSS is proven secure even against fault attacks exploiting the Chinese Remainder Theorem (CRT) implementation of RSA . However, all the previous security proofs are valid only in RO model, where one assumes the existence of ideal, truly random hash functions. Unfortunately truly random functions do not exist and in practice, the "ideal" functions are instantiated with some efficient hash functions. Hence it is important that the proofs are valid while replacing

D. Catalano et al. (Eds.): PKC 2011, LNCS 6571, pp. 351–368, 2011.

random oracles by a standard hash functions. Otherwise such proofs merely provide heuristic evidence that breaking the scheme may be hard (or there is no generic attack against the scheme).

A number of papers [6,9,14,18], starting from famous results of Canetti et. al. [6], showed that there are schemes secure in the Random Oracle model, which are uninstantiable under standard model. Naturally, these results raise concerns about the soundness of the schemes proven secure in random oracle model. Particularly for widely deployed scheme like PSS, it is especially important to have an secure instantiation by a standard, efficiently computable hash function so that we do not build our technology in vacuum. In this paper, we ask essentially this particular question about PSS: *Whether it is possible, to securely instantiate PSS based on reasonable assumptions to the underlying trapdoor permutation.*

1.1 Our Results

Our main result is a general negative result to the above question. Roughly, we extend *all* the negative results by Dodis et. al. [9] for Full Domain Hash (FDH) to PSS. Specifically, we show the following

- There is no instantiation of PSS such that, *unforgeability under chosen message attack* can be reduced to any security property of a random permutation using *black-box* reduction techniques. As a random permutation satisfies almost all reasonable security notions, our result covers many of the standard security notions, like inverting trapdoor permutation on a random point (one-way), finding some bits of pre-image of a random point (partial domain one-wayness), finding correlated inputs etc. Our result is perfectly valid even if the hash functions used in PSS can query the trapdoor permutation and digests are arbitrarily related to the responses.
- We also rule out any black box reduction from recently proposed Lossy Trapdoor Permutations [20]. In Crypto 2010, Kiltz et. al. [17] has proven IND-CPA security of OAEP based on Lossy Trapdoor Permutation. Hence it is important to analyze whether positive result could be possible for PSS.
- We also show that even the weakest security criteria , namely unforgeability under *no message attack* cannot be black-box reduced to the one-wayness of the trapdoor permutation if the randomness space in PSS is "super-polynomial" in security parameter.
- All our results can easily be extended to the scenario when the adversary can invert some points of his choice (with some restrictions) for a fixed bounded number of times.

We would like to mention that our results does not completely rule out the possibility of instantiating PSS in standard model. A "whitebox" reduction, using the code of the adversary, may still exist. On the other hand, it may be possible to show a reduction from other cryptographic functions like homomorphic encryption. Still, we believe our result is important from theoretical point of view as it shows PSS requires special property of underlying trapdoor permutation as opposed to "Only randomness of hash is sufficient" notion of random oracle model.

1.2 Overview of Our Technique

We use the technique of two oracles due to Hsio and Reyzin [15] for our separation results. We construct two oracles T and G such that T implements a ideal trapdoor permutation and G can be used to forge the PSS scheme. However, G does not help the attacker to break any security property of the ideal trapdoor permutation. Informally, this ensures that a black-box security proof cannot exist as any such proof should be valid against our T and G.

On a very high level our technique can be seen as an extension of the technique of Dodis et. al. [9] to rule out black box reduction of FDH. Separation from a random permutation is achieved in two steps. As the first step, we instantiate T by permutation chosen uniformly at random from the set of exponentially many permutations. Intuitively, G, the main forger oracle, should output a forgery after checking whether the adversary truly has access to a signer by sending polynomially many challenge messages. However the reduction could design the underlying hash function in such a way, so that the digests of the messages either collide with each other (hence reducing the number of points on which inversion is needed) or the digest is the result one of the evaluation queries made to the trapdoor permutation (hence the reduction can get the signature from the corresponding query by evaluating the hash function). For this reason we define G to output the forgery only if the adversary can produce distinct signatures, which were not a query to the trapdoor permutation during the computation of digests, for all the challenge messages.

In the second step we show that a reduction algorithm (which does not have access to inversion oracle) can not produce valid, signature meeting both the conditions with non-negligible probability. Hence to win any hard game, G is of no use to the adversary. However, we construct an efficient adversary with an access to a valid sign oracle (available in an unforgeability game) that can either find a forgery on its own or can construct signatures satisfying all the conditions of G. We stress that the efficient algorithm in [9], which precomputes all the hash values to check for the conditions, does not work efficiently when the signature scheme is randomized. Specifically, when the random strings are of super-logarithmic length, it is no longer possible for a polynomial time algorithm to compute all possible hash values for even a single message. It might very well happen that the computed digests meet the conditions but the digests on which signer generated the signature do not meet the condition. To solve this problem we use an elegant adaptive "evaluate on the fly" technique where we sample polynomially many random strings and check for the conditions. If the conditions are satisfied for the sampled digests, we repeatedly query the signer with fresh random coins for multiple signatures of same message. We show that, with probability exponentially close to 1, one gets either a set of valid signatures maintaining the conditions from the signer or could find a forgery during the sampling stage.

1.3 Previous Results

A rich body of work [11,12,13,14,16,21] on blackbox separation exists in the literature starting from the seminal work of Impagliazzo and Rudich [16]. Regarding the separation of random oracle from the standard model, the first result was due to Canetti, Goldreich and Halevi [5,6] who showed an artificial albeit valid signature scheme that can not be securely instantiated by standard hash functions. Many such results [9,10,14,18] were subsequently published. To obtain our separation results we use the two oracle technique of Hsiao and Reyzin[15]. The most relevant results to our paper is the work of Dodis et. al. [9] and of Kiltz and Pietrzak [18]. In [9], Dodis, Oliviera and Pietrzak showed that the popular Full Domain Hash (FDH) signature scheme can not be instantiated (using blackbox technique) in standard model by a ideal trapdoor permutation. Kiltz and Pietrzak [18] established that there is no blackbox reduction of any padding based CCA secure encryption scheme from ideal trapdoor permutations. In [19], Paillier showed impossibility of reduction of many RSA based signatures including PSS from different security assumptions of RSA. However, their result is based on an additional assumption (namely, instance non-malleability) of RSA. In comparison, our result is more generic as we rule out blackbox reduction from any property of random permutation.

1.4 Differences from Dodis et al's Crypto'05 Paper [9]

Although our definition of oracles are quite similar to that in [9], difference comes in when finally implementing a forgery. The technique of [9] is not readily applicable for randomized signatures. Specifically in case of PSS the forger cannot force the signer to choose any particular random string. On the other hand, if the randomness space is super-polynomial the forger cannot pre-compute all the possible value of the hashes of any message. As a result the forger, as defined in [9], cannot output a forgery when G aborts. Our contribution is in constructing adaptive forger that can forge PSS with overwhelming probability even when the randomness space is super-polynomial. Moreover, our technique to rule out black-box reduction to one way trapdoor permutation is completely different. Looking ahead, we show that when the randomness space is of super-polynomial size, no Probabilistic Polynomial-Time Turing Machine (PPTM) can use a random signature (over the choice of random string during signing) of any fixed message to invert the one way trapdoor permutation.

2 Preliminaries

2.1 Notations

Throughout the paper, if x is a string, $|x|$ denotes the length of the string. 1^n denotes the string of n many 1s. If S is a set $|S|$ denotes the cardinality of the set. We use $negl(n)$ to denote any function $\gamma : \mathbb{N} \rightarrow [0,1]$ where for any constant $c > 0$ there exist n_0 such that for all $n > n_0$; $\gamma(n) < 1/n^c$. We call a function $f(n)$ to be super-polynomial if for any constant $c > 0$, there exists n_0 such that for all $n > n_0$, $f(n) > n^c$.

2.2 Trapdoor Permutations (TDPs)

Definition 1. *A trapdoor permutation family is a triplet of PPTM (Tdg, F, F^{-1}). Tdg is probabilistic and on input 1^n outputs a key-pair $(pk, td) \leftarrow_R Tdg(1^n)$. $F(pk, .)$ implements a permutation f_{pk} over $\{0,1\}^n$ and $F^{-1}(td, .)$ implements the corresponding inverse f_{pk}^{-1}.*

The most standard security property of TDP is *one-wayness* which says that it is hard to invert a random element without knowing the trapdoor. Formally, for any PPTM A

$$Pr[(pk, td) \leftarrow_R Tdg(1^n), x \leftarrow_R \{0,1\}^n : A(f_{pk}(x)) = x] \leq negl(n).$$

Many other security notion for Trapdoor Permutations are known. Like [9,18], we consider a wide class of security properties using the notion of *δ-hard games.*

2.3 Hard Games

A cryptographic game consists of two PPTMs C (Challenger) and A (Prover) who can interact over a shared tape. After the interaction, C finally outputs a bit d. We say, A wins the game if $d = 1$ and denote it, following [9], by $\langle C, A \rangle = 1$.

Definition 2. *A game defined as above is called δ-hard game if for all PPT A (in the security parameter n) the probability of win , when both C and A has oracle access to t uniform random permutations $\pi_1, \pi_2, \cdots, \pi_t$ over $\{0,1\}^n$, is at most negligible more than δ. Formally C is a* δ-hard *game if for all PPTM A*

$$Adv_C(A, n) = Pr[\langle C^{\pi_1, \pi_2, \cdots, \pi_t}, A^{\pi_1, \pi_2, \cdots, \pi_t} \rangle = 1] \leq \delta + negl(n)$$

The hardness of the game C (denoted by $\delta(C)$) is the minimum δ such that C is δ-hard.

For cryptographic games like one-wayness, partial one-wayness, claw-freeness; $\delta = 0$. For the game of pseudo-randomness $\delta = 1/2$. The notion of δ-hard game was considered in [18] as a generalization of hard games considered in [9]. It was pointed out in [18] that the result of [9] can easily be extended to this notion.

2.4 Ideal Trapdoor Permutations

The notion of Ideal Trapdoor permutation was coined in [18]. To remain consistent with literature, we follow the same notion.

Let $TDP = (Tdg, F, F^{-1})$ be a trapdoor permutation. We say that TDP is secure for δ-hard game C if for all PPTM A, $\text{Adv}_C(A, n) - \delta(C)$ is negligible even when the random permutations in the definition of hard game is replaced by TDP. Formally, TDP is secure iff,

$$Pr[\langle C^{F(pk_1),F(pk_2),\cdots,F(pk_t)}, A^{F(pk_1),F(pk_2),\cdots,F(pk_t)}\rangle = 1] \leq \delta + negl(n),$$

where $(pk_i, td_i) \leftarrow_R Tdg(1^n)$ *for* $i = 1, \cdots t$.

Definition 3. *TDP is said to be an ideal trapdoor permutation if it is secure for any δ-hard game C.*

We stress that, ideal trapdoor permutation does not exist (see [9] for proof). However as we are proving negative result, showing that PSS cannot be reduced to ideal trapdoor permutation (hence to any hard game) makes our result stronger. This implies that PSS cannot be black-box reduced to security notions like collision resistant hashing, pseudo-random functions, IND-CCA secure public key encryption schemes etc.

2.5 Lossy Trapdoor Permutations(LTDPs)

Lossy Trapdoor Functions were introduced by Peikert et. al. in [20]. In this paper we consider a straightforward generalization to permutations. A family of (n, l) Lossy Trapdoor Permutations (LTDPs) is given by a tuple $(\mathcal{S}, \mathcal{F}, \mathcal{F}')$ of PPTMs. $\mathcal{S}$ is a sampling algorithm which on input 1 invokes $\mathcal{F}$ and on input 0 invokes $\mathcal{F}'$. $\mathcal{F}$ (called "Injective Mode") describes a usual trapdoor permutation; i.e. it outputs (f, f^{-1}) where f is a permutation over $\{0,1\}^n$ and f^-1 is the corresponding inverse. $\mathcal{F}'$ (called "Lossy Mode") outputs a function f' on $\{0,1\}^n$ with range size at most 2^l. For any distinguisher D, *LTDP-Advantage* is defined as

$$\mathrm{Adv}^{ltdp}_{(\mathcal{F},\mathcal{F}'),D} = \left| Pr[D^f(.) = 1 : (f, f^{-1}) \leftarrow^R \mathcal{F}] - Pr[D^{f'}(.) = 1 : f' \leftarrow^R \mathcal{F}'] \right|.$$

We call $\mathcal{F}$ "lossy" if it is the first component of some lossy LTDP.

3 Signature Schemes

A signature scheme $(\texttt{Gen}, \texttt{Sign}, \texttt{Verify})$ is defined as follows:

- The *key generation algorithm* `Gen` is a probabilistic algorithm which given 1^k, outputs a pair of matching public and private keys, (pk, td).

- The *signing algorithm* `Sign` takes the message M to be signed, the public key *pk* and the private key *td*, and returns a signature $\sigma = \texttt{Sign}_{td}(M)$. The signing algorithm may be probabilistic.

- The *verification algorithm* `Verify` takes a message M, a candidate signature σ' and pk. It returns a bit $\texttt{Verify}_{pk}(M, \sigma')$, equal to one if the signature is accepted, and zero otherwise. We require that if $\sigma \leftarrow \texttt{Sign}_{td}(M)$, then $\texttt{Verify}_{pk}(M, \sigma) = 1$.

3.1 Security of a Signature Scheme

In the *existential unforgeability under an adaptive chosen message attack* scenario, the forger can dynamically obtain signatures of messages of his choice and attempts to output a valid forgery. A *valid forgery* is a message/signature pair (M, x) such that $\texttt{Verify}_{pk}(M, x) = 1$ whereas the signature of M was never requested by the forger.

3.2 Probabilistic Signature Scheme(PSS)

Let $TDP = (Tdg, F, F^{-1})$ be a trapdoor permutation. PSS uses a triplet $H = (h, g_1, g_2)$ of hash functions such that, $h : \{0,1\}^* \to \{0,1\}^{k_1}$, $g_1 : \{0,1\}^{k_1} \to \{0,1\}^{k_0}$ and $g_2 : \{0,1\}^{k_1} \to \{0,1\}^{k-k_0-k_1-1}$, where k, k_0 and k_1 are parameters.

$\texttt{Gen}(1^k)$

1. Return $(pk, td) = Tdg(1^k)$

$\texttt{Sign}_{td}(m)$

1. $r \leftarrow \{0,1\}^{k_0}$
2. $\omega \leftarrow h(m\|r)$
3. $r^* \leftarrow g_1(\omega) \oplus r$
4. $y \leftarrow 0\|\omega\|r^*\|g_2(\omega)$
5. Return $\sigma = F^{-1}(td, y)$.

$\texttt{Verify}_{pk}(m, \sigma)$

1. Let $y = F(pk, \sigma)$
2. Parse y as $0\|\omega\|r^*\|\gamma$. If the parsing fails return 0.
3. $r \leftarrow r^* \oplus g_1(\omega)$
4. If $h(m\|r) = \omega$ and $g_2(\omega) = \gamma$ return 1.
5. else return 0.

Any PSS signature scheme can be instantiated by specifying the triplet of hash functions $H = (h, g_1, g_2)$ and the trapdoor permutation TDP. PSS_H^{TDP} be the PSS signature scheme instantiated by H and TDP. For any $H = (h, g_1, g_2)$, the PSS transformation described above is defined as

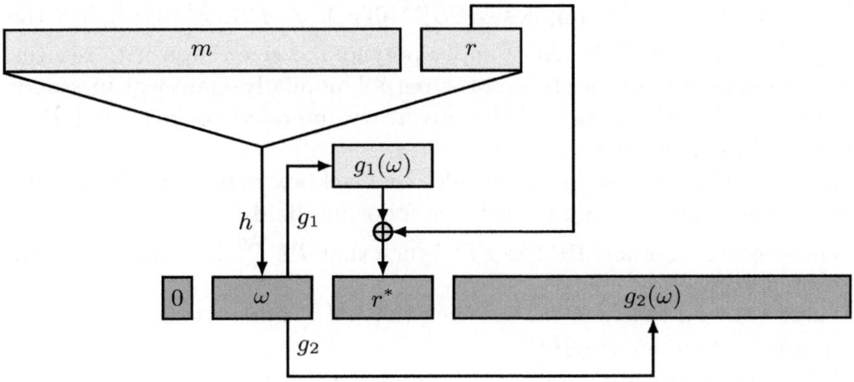

Fig. 1. PSS_H^{TDP}: The components of the image $y = 0\|\omega\|r^*\|g_2(\omega)$ are darkened. The signature of m is $F^{-1}(td, y)$.

$$PSS_H^{f_{pk}}(m\|r) = 0\|h(m\|r)\|(r \oplus g_1(h(m\|r))\|g_2(h(m\|r)).$$

$PSS_H^{f_{pk}}(m\|r)$ is in fact the darkened area in Figure 1, $y = 0\|\omega\|r^*\|g_2(\omega)$. *Note, h, g_1, g_2 can be oracle circuits with oracle access to f_{pk}.* For the rest of the paper, PSS_H^{TDP} denotes the signature scheme, where as $PSS_H^{f_{pk}}(\cdot)$ is the PSS transformation during `Sign` procedure before applying the trapdoor permutation. From the context these two notations are easily distinguishable.

The following observation is very important to our technique.

Observation 1. *A collision after the PSS transformation implies collision in the random space. In other words,*

$$PSS_H^{f_{pk}}(M_1\|r_1) = PSS_H^{f_{pk}}(M_2\|r_2)$$

implies $r_1 = r_2$.

As both the digests are same, $\omega_1\|r_1^*\|\gamma_1 = \omega_2\|r_2^*\|\gamma_2$; we have $\omega_1 = \omega_2$ and $r_1^* = r_2^*$. This leads to $r_1 = r_2$. So for two distinct random strings r_1 and r_2, the digests of $PSS_H^{f_{pk}}$ and hence the signatures are always different (irrespective of whether the messages are same or not)! As a side note, all our results are valid not only for PSS, but for any randomized signature scheme where the randomness is recoverable and hence the Observation 1 holds true.

4 No Blackbox Reduction from One Way Trapdoor Permutations

One-wayness is the most common security property of a trapdoor permutation. All the previous security proofs of PSS in Random Oracle model are based on one wayness of underlying trapdoor permutation (specifically RSA). In this section we consider the possibility of reducing security of PSS from one-wayness of a trapdoor permutation, but in standard model. We show that when $k_0 = \omega(\log n)$, one cannot prove PSS secure via a blackbox reduction from one way trapdoor permutation even if the forger is never allowed to query the signer.

Recall that, $r_1 \neq r_2$ implies $PSS_H^{f_{pk}}(0\|r_1) \neq PSS_H^{f_{pk}}(0\|r_2)$. So the set $\{PSS_H^{f_{pk}}(0\|r)|r \in \{0,1\}^{k_0}\}$ is of super-polynomial size. Even if G returns one random signature (from a choice of superpolynomially many) of message 0, it is unlikely to be of any use of the adversary intended to invert TDP^T on a uniformly chosen element z.

Following [15], Proposition 1, to rule out blackbox reductions, it is enough to construct two oracles T and G such the following holds:

- There exists an oracle PPTM TDP such that TDP^T implements a trapdoor permutation.
- There exist an oracle PPTM A such that $A^{T,G}$ finds a forgery under chosen message attack for $PSS_H^{TDP^T}$.
- TDP^T is an one-way trapdoor permutation relative to the oracles T and G. That is, TDP^T is an one-way permutation even if the adversary is given oracle access to T and G.

Definition of T. For any $n \in \mathbb{N}$, Choose $2^n + 1$ permutations $f_0, f_1, f_2, \cdots,$ f_{2^n-1} and g uniformly at random from the set of all permutations over $\{0,1\}^n$. Now the oracle T is defined as follows:

- $T_1(td) \rightarrow g(td)$ (generate public key from the trapdoor)
- $T_2(pk, y) \rightarrow f_{pk}(y)$ (evaluate)
- $T_3(td, z) \rightarrow f^{-1}_{g(td)}(z)$ (inversion)

Implementing TDP^T. We use $T = (T_1, T_2, T_3)$ in the following way to construct (in the functional sense) the trapdoor permutation $TDP^T = (Tdg, F_T, F_T^{-1})$.

- $Tdg(1^n)$ chooses a uniform random $td \leftarrow \{0,1\}^n$ and computes the corresponding public key as $pk = T_1(td)$ and outputs (td, pk).
- $F_T(pk, y)$ returns $T_2(pk, y)$.
- $F_T^{-1}(td, z)$ returns $T_3(td, z)$.

It is easy to check that as TDP^T implements a trapdoor permutation, as $g(td) = pk$.

Description of G. The oracle G takes as input $k \in \mathbb{N}$ and $H \in \{0,1\}^*$. G selects an $r \in_R \{0,1\}^{k_0}$ and returns $f^{-1}_{pk}(PSS_H^{f_{pk}}(0\|r))$. Here pk is the public key generated by $Tdg(1^k)$.

As G always outputs a forgery for message 0, we get the following result.

Lemma 1. *There is a PPTM A such that A^G outputs a forgery for PSS signature scheme.*

G Does Not Break Security of TDP^T

Next we shall prove that TDP^T is one way, even relative to G. This is not at all obvious as G always provides forgery of the form $f^{-1}_{pk}(PSS_H^{f_{pk}}(.))$ for a H of our choice! But we note that $G(.)$ samples one z' from a set of superpolynomial size and outputs $f^{-1}_{pk}(z)$. Even if the adversary sets $PSS_H^{f_{pk}}(0, r)$ for one r to be the challenge z she received, probability that $f_{pk}(G(.)) = z$ is negligible. On the other hand if $f_{pk}(G(.)) \neq z$, then knowledge of inverse of some other point does not help the adversary to find $f^{-1}_{pk}(z)$ with significant probability for a pseudorandom f_{pk}. Following the above discussion we have Lemma 2, whose detailed proof is given in the full version [4].

Lemma 2. *A random permutation $\pi : \{0,1\}^n \rightarrow \{0,1\}^n$ is one way even if adversary is allowed to make one inverse query on any input except the challenge.*

Now, we can claim that TDP^T is one way even relative to G.

Lemma 3

$$Pr[A^{T,G}(pk, z) = x : F_T(pk, x) = z] \leq negl(n),$$

where $x \leftarrow_R \{0,1\}^n$ *and* $(pk, td) \leftarrow Tdg(1^n)$.

For a proof of the above lemma, see [4].

We get the main result of this section as follows

Theorem 1. *There is no blackbox reduction of Security under no message attack of Probabilistic Signature Scheme with superpolynomial randomness space from Oneway Trapdoor Permutations.*

5 No Blackbox Reduction from an Ideal Trapdoor Permutation

The following theorem states that there is no adversary that can break the security of the TDP^T using any adversary (in black-box way) breaking $PSS_H^{TDP^T}$ by chosen message attack when TDP^T is an ideal permutation.

Theorem 2. *There is no black-box reduction from a family of ideal trapdoor permutations to the existential unforgeability against chosen message attack of the PSS signature scheme.*

Like the previous section, we shall construct a oracle G such that there exist a PPTM B such that B^G can forge PSS although TDP^T is secure even relative to G. We define, T and TDP^T as in section 4.

Definition of *G*. The oracle G works as follows. On input the description of the hash function triplet $H = (h, g_1, g_2)$, and the security parameter n, it selects $t = \max(|H|, n)$ messages $m_1, m_2, \cdots m_t$ uniformly at random from $\{0,1\}^* \setminus \{0\}$ and outputs them as a set of challenge messages. G expects valid and *distinct* signatures of all the messages. G also keeps a list (initially empty) of description of input hash functions, the challenge messages and the forgery it returns. If the description of the hash matches then G outputs the same challenge messages. If it gets valid signatures (as described below) then it outputs the previously returned forgery from the list.

Once it receives the messages and the signatures $(m_1, m_2, \cdots, m_t, \sigma_1, \sigma_2, \cdots, \sigma_t)$, G checks for the following conditions.

1. $\sigma_1, \cdots, \sigma_t$ are valid signatures for $m_1, \cdots, m_t$. Recover $r_1, \cdots, r_t$ such that,
$$PSS_H^{f_{pk}}(m_1 \| r_1) = f_{pk}(\sigma_1), \cdots, PSS_H^{f_{pk}}(m_t \| r_t) = f_{pk}(\sigma_t).$$
2. $\sigma_i \neq \sigma_j$ (or equivalently $PSS_H^{f_{pk}}(m_i \| r_i) \neq PSS_H^{f_{pk}}(m_j \| r_j)$) for all $1 \le i < j \le t$.
3. $\{PSS_H^{f_{pk}}(m_1 \| r_1), \cdots, PSS_H^{f_{pk}}(m_t \| r_t)\} \cap Y_{f_{pk}}^{PSS_H}(r_1, \cdots, r_t) = \emptyset$ where
$$Y_{f_{pk}}^{PSS_H}(r_1, \cdots, r_t) = \{f_{pk}(x) | \exists i,\ 1 \le i \le t, \\ PSS_H^{f_{pk}}(m_i \| r_i) \text{ makes the oracle query } x\}.$$

If all the above conditions are satisfied then G chooses one r uniformly at random from $\{0,1\}^{k_0}$ and returns $f_{pk}^{-1}(PSS_H^{f_{pk}}(0 \| r))$. Here pk is the public key generated by $Tdg(1^k)$.

G Breaks the Security of $PSS_H^{TDP^T}$

Lemma 4. *There is a PPTM B^G that can mount existential forgery by chosen message attack on PSS with overwhelming probability.*

Proof. The goal of B^G is to either generate a forgery on it's own or use the `sign` oracle to get signatures of $m_1, \cdots, m_t$ such that Condition 1, Condition 2 and 3 get satisfied. Then B^G can use output of G to produce forgery for the message 0. We describe two constructions of B^G depending on size of the randomness space or k_0.

Case I: $k_0 = \mathcal{O}(\log n)$. In this case B^G precomputes $PSS_H^{f_{pk}}(m_i\|r)$ for all $r \in \{0,1\}^{k_0}$ and $i = 1 \cdots t$ and checks whether the Condition 3 from Section 5 would get satisfied or not for any possible choice of r by the `Sign` oracle. If not B can find some m_i, m_j, r_i, r_j, x such that

$$PSS_H^{f_{pk}}(m_i\|r_i) = f_{pk}(x),$$

where $PSS_H^{f_{pk}}(m_j\|r_j)$ makes the oracle query $f_{pk}(x)$. In this case B can easily produce the forged signature x for the message m_i.

Otherwise to take care of Condition 2, B^G calls the `Sign` oracle to get valid signatures for message m_i's one by one for $i = 1$ to t. After receiving the ith signature σ_i it always recovers the randomness r_i and checks whether

$$PSS_H^{f_{pk}}(m_i\|r_i) = PSS_H^{f_{pk}}(m_j\|r_i)$$

for some $i < j \leq t$. Because of Observation 1 it is sufficient to check with the fixed r_i for collision detection purposes. If the above condition gets true again B^G can readily output a forged signature for message m_j as σ_i. Otherwise, B^G ends up with $\sigma_1, \cdots, \sigma_t$ such that all the three conditions in Section 5 are satisfied. So B^G can easily use G to produce a forgery for the message 0. Hence B^G succeeds to forge PSS with probability 1.

Case II: $k_0 = \omega(\log n)$. In this case the randomness space is of superpolynomial size, hence B^G cannot precompute all the possible outputs of $PSS_H^{f_{pk}}(m\|\cdot)$ even for a single message m. However, we observe that the "no collision" requirement or Condition 2 can easily be taken care of by a technique similar to the previous one. To take care of Condition 3, we adopt a sampling procedure. B^G works in two phases. In *Phase-I*, B samples some random r's from $\{0,1\}^{k_0}$ uniformly and simulate the signing procedure by the real `Sign` oracle that would be queried in *Phase-II.* Then the probabilities that Condition 3 gets satisfied in Phase-I or in Phase-II are essentially the same. We set our parameters such a way, with high probability either Condition 3 does not hold in Phase I (hence direct forgery) or it holds in Phase-II (forgery via oracle G, provided Condition 2 holds).

Algorithm 1. B^G : Phase-I

```
r_i^k ←_R {0,1}^{k_0} : 1 ≤ i ≤ t, 1 ≤ k ≤ t
V = {PSS_H^{f_pk}(m_i || r_i^k) : 1 ≤ i ≤ t, 1 ≤ k ≤ t}
Y = {f_pk(x) | ∃i, k, 1 ≤ i ≤ t, 1 ≤ k ≤ t
        s. t. PSS_H^{f_pk}(m_i, r_i^k) makes oracle query f_pk(x)}
if V ∩ Y ≠ ∅ then
    Output Direct Forgery
end if
```

Success Probability of B^G in Case II. In Line 21 of Algorithm 2, Condition 1 and Condition 2 are always satisfied. So B^G can abort only in two ways.

1. In Line 16 of Algorithm 2, Σ_i becomes empty for some i, $1 \leq i \leq t$.
2. In Line 21 of Algorithm 2, Condition 3 gets violated. $r_1, \cdots, r_t$ be the random strings recovered from $\sigma_1, \cdots, \sigma_t$. Violation of Condition 3 over here implies there exists some i, j, $1 \leq i, j \leq t$, $i \neq j$ such that

$$PSS_H^{f_{pk}}(m_i \| r_i) = f_{pk}(x),$$

where $PSS_H^{f_{pk}}(m_j \| r_j)$ makes the oracle query x.

Moreover, in both the cases no forgery was found in Algorithm 1.

Let us consider the case where for some i, Σ_i is empty. It implies for some i, for all $k = 1, \cdots, t$, $\sigma_i^k \in X_{i,k}$ and hence was removed from Σ_i . Fix some i . Let us call the set of r for which $PSS_H^{f_{pk}}(m_i, r) = f_{pk}(x)$ and x was queried while computing $PSS_H^{f_{pk}}(m_i, r)$ as BAD. Suppose

$$Pr_r[r \in \text{BAD}] = \theta.$$

Now the event $\Sigma_i = \emptyset$ and no forgery was found in Phase-I implies that the random strings $r^{(i)}$, sampled in Phase 1 were not from the BAD set and all of $r_i^1, r_i^2, \cdots, r_i^t$ was from BAD. As `Sign` and B samples independently, probability of $\Sigma_i = \emptyset$ is $\theta^t(1-\theta)^t \leq 2^{-t}$. Taking union bound over all i, the probability that for some i, Σ_i is empty is at most $t/2^t$.

For the second case, the chosen σ_is were not queried while computing them; rather one σ_i was queried while computing some other σ_j. Recall that maximum number of f_{pk} queries (made by $PSS_H^{f_{pk}}$) while computing one signature is $|H|$. As, for any j $\Sigma_j \leq t$, for each $j = 1, 2, \cdots, t$; $j \neq i$, maximum number of f_{pk} queries made while computing Σ_j is at most $t|H|$. So overall, for all $j \neq i$, total number of f_{pk} queries made by the $PSS_H^{f_{pk}}$ was $t^2|H|$. As, there are $2^{|r|}$ choices of random string, implying $2^{|r|}$ choices for each σ_i^k, and `Sign` runs each time with independent random coins, probability that at least one σ_i^k was from those $t^2|H|$ many f_{pk} queries is at most $\frac{t^4}{2^{k_0}}$.

Algorithm 2. B^G : Phase-II

for $i = 1$ to t **do**
    $\sigma_i^1 \leftarrow \mathtt{Sign}(m_i), \cdots, \sigma_i^t \leftarrow \mathtt{Sign}(m_i)$
    $\Sigma_i = \{\sigma_i^1, \cdots, \sigma_i^t\}$
    Recover $r_i^1, \cdots, r_i^t$ from $\sigma_i^1, \cdots \sigma_i^t$ using `Verify`.
    for $j = i + 1$ to t **do**
        if $PSS_H^{f_{pk}}(m_i \| r_i^k) == PSS_H^{f_{pk}}(m_j \| r_i^k)$ *for some* $1 \le k \le t$ **then**
            Output Direct Forgery (m_j, σ_i^k)
        end if
    end for
    for $k = 1$ to t **do**
        $X_{i,k} \leftarrow \{x | PSS_H^{f_{pk}}(m_i \| r_i^k) \;\; \textit{makes oracle query} \;\; f_{pk}(x)\}$
        if $\sigma_i^k \in X_{i,k}$ **then**
            $\Sigma_i \leftarrow \Sigma_i \setminus \{\sigma_i^k\}$
        end if
    end for
    if $\Sigma_i = \emptyset$ **then**
        Output $\perp$
    end if
    Pick any $\sigma_i \in \Sigma_i$
end for
if $\sigma_1, \cdots, \sigma_t$ satisfy Condition 1, Condition 2 and Condition 3 from Section 5 **then**
    Output forgery via G
else
    Output $\perp$
end if

Hence we get that

$$\begin{aligned} &\Pr[B^G \rightarrow \perp] \\ &\le \Pr[\exists i; \Sigma_i = \emptyset] + \Pr[\exists i, j; \sigma_i \in \{\; f_{pk} \text{ queries made while computing } \sigma_j \;\}] \\ &\le \frac{t}{2^t} + \frac{t^4|H|}{2^{|r|}} \end{aligned}$$

Putting $t = max(|H|, n), |r| = \omega(\log n)$ and $|H| \le n^c$ for some constant c, $\Pr[B^G \rightarrow \perp]$ is $negl(n)$. □

G Does Not Break the Security of TDP^T

Lemma 5. *For any oracle PPTM B and any δ-hard game C (with $t = t(n)$ implicitly defined by C),*

$$Pr[\langle C^{F(pk_1),F(pk_2),\cdots,F(pk_t)}, A^{F(pk_1),F(pk_2),\cdots,F(pk_t),G} \rangle = 1] \le \delta + negl(n),$$

where $(pk_i, td_i) \leftarrow_R Tdg(1^n)$ for $i = 1, \cdots t$.

Proof. The proof of the above lemma is essentially same as in proof of Lemma 2 in [9], where one argues in the absence of oracle G the claim holds because of computational indistinguishability of f_{pk} from a random permutation. Moreover, Lemma 6 below states the accepting condition of oracle G can only be satisfied with a negligible probability. □

Lemma 6. *Let f be a random permutation on $\{0,1\}^n$ and $c \geq 1$ be a constant, $m_1, \cdots, m_t$ be n-bit values with $t = \max(|H|, n)$. For any oracle TM A which makes at most n^c oracle queries, we have (the probability is over randomness of f)*

$$\Pr[A^f \to (H, x_1, \cdots, x_t)] = \boldsymbol{negl}(n)$$

where, $|H| \leq n^c$ and the output satisfies the following conditions for some k_0-bit $r_1, \cdots, r_t$

1. $f^{-1}(PSS_H^f(m_1\|r_1)) = x_1, \cdots, f^{-1}(PSS_H^f(m_t\|r_t)) = x_t$.
2. $f^{-1}(PSS_H^f(m_i\|r_i)) \neq f^{-1}(PSS_H^f(m_j\|r_j))$ *for all* $1 \leq i < j \leq t$.
3. $\{PSS_H^f(m_1\|r_1), \cdots, PSS_H^f(m_t\|r_t)\} \cap Y_f^{PSS_H}(r_1, \cdots, r_t) = \emptyset$, *where*

$$\begin{aligned} Y_f^{PSS_H}(r_1, \cdots, r_t) = &\{f(x) | \exists i, 1 \leq i \leq t, \\ &PSS_H^f(m_i\|r_i) \text{ makes the oracle query } x\}. \end{aligned}$$

Lemma 6 can proved following the same technique of Lemma 3 of [9]. For a proof, we refer the reader to the full version of this paper [4].

6 No Reduction from Lossy Trapdoor Permutations

Lossy Trapdoor Functions, introduced by Peikert et. al. has gained considerable attention in recent years. In a recent work [17], has proven IND-CPA security of OAEP under Lossy Trapdoor Permutation. Moreover different constructions like IND-CCA secure encryption, which cannot be reduced to standard trapdoor permutation using blackbox techniques, were proven reducible to Lossy Trapdoor Permutations. In this section we show that there is no blackbox reduction of existential unforgeability of PSS against chosen message attack from Lossy Trapdoor Permutations as well. Specifically, Let $LTDP = (S, F, F')$ be a family of Lossy Trapdoor Permutation. We define the output of PSS based on LTDP as $\sigma = f^{-1}(PSS_H(m\|r))$ where $(f, f^{-1}) \in F$. Note that, while instantiating PSS by a lossy TDP, we consider the trapdoor permutation to be the injective mode of the TDP.

Theorem 3. *There is no blackbox reduction of existential unforgeability against chosen message attack of Probabilistic Signature Scheme from Lossy Trapdoor Permutations.*

Proof. To prove Theorem 3, we need new definitions of the oracles.

Definition of $\mathcal{T}$. $\mathcal{T}$ is defined as a pair (T, T'). Choose $2^n + 1$ permutations $f_0, \cdots, f_{2^n-1}$ and g uniformly at random from the set of all permutations over $\{0,1\}^n$. Moreover choose 2^n functions $e_0, \cdots, e_{2^n-1}$ uniformly at random from the set of all functions from $\{0,1\}^n$ to $\{0,1\}^l$.

Oracle T works as follows:

- $T_1(td) \rightarrow g(td)$ (generate public key from the trapdoor)
- $T_2(pk, y) \rightarrow f_{pk}(y)$ (evaluate)
- $T_3(td, z) \rightarrow f^{-1}_{g(td)}(z)$ (inversion)

On the other hand T' is defined as follows

- $T'(pk, x) = f_{pk}(1^{n-l} || e_{pk}(x))$

Now we define the $LTDP^{T,T'} = (S, (F, F^{-1}), F')$ as follows

- $S(b)$ If $b = 1$, choose a uniform random $td \leftarrow \{0,1\}^n$ computed $pk = T_1(td)$ and return (pk,td), otherwise choose a uniform random $pk \leftarrow \{0,1\}^n$ and return $(pk, \perp)$.
- $F(pk, y)$ returns $T_2(pk, y)$.
- $F^{-1}(td, z)$ returns $T_3(td, z)$.
- $F'(pk, y)$ returns $T'(pk, x)$.

Lemma 7. *$LTDP^{T,T'}$ implements a secure (n, l) Lossy Trapdoor Permutation when $l = \mathcal{O}(n^{\frac{1}{c}})$ for a positive constant c.*

Proof. Recall that, to show the security of $LTDP^{T,T'}$, we need to argue that for any efficient distinguisher D, $|Pr[D^F = 1] - Pr[D^{F'} = 1]|$ is negligible. Consider a random function $e' : \{0,1\}^n \rightarrow \{0,1\}^l$ and a random permutation $\pi : \{0,1\}^n \rightarrow \{0,1\}^n$. It is easy to check that $\pi(1^{n-l}||e'())$ has the same distribution of a random permutation until a collision in e'. e' being a random function, the collision probability is $q^2/2^l$, which is negligible for $q = \mathcal{O}(n^{c_1}))$ for some constant $c_1 > 0$.

Now using the fact that a function (permutation) chosen uniformly at random from the set of exponentially many functions (permutations) is indistinguishable form a random function (permutation), the lemma follows. □

Definition of G. Informally, G will work exactly the same way as in the previous case when the underlying permutation is in injective mode. When the permutation is lossy G can abort instead of returning a forgery. So effectively, when instantiated by the lossy mode G always aborts and in injective mode G aborts if the conditions are not satisfied.

In more detail, G works in the following way. On input the description of the hash functions h, g_1 and g_2, it selects t (to be fixed later) messages $m_1, m_2, \cdots m_t$ uniformly at random from $\{0,1\}^* \setminus 0$ and outputs them as a set of challenge messages. G expects valid and *distinct* signatures of all the messages. G also keeps a list (initially empty) of description of input hash functions, the challenge messages and the forgery it returns. If the description of the hash matches then G outputs the same challenge messages. If it gets valid signatures (as described below) then it outputs the same forgery from the list.

Once it receives the messages and the signatures $(m_1, m_2, \cdots, m_t, \sigma_1, \sigma_2, \cdots, \sigma_t)$, G first checks whether the signatures are valid and distinct.

- $F(pk, \sigma_i) = PSS_H^{f_{pk}}(m_i \| r)$ for some r. This signature verification is to make sure that that calling algorithm has access to signing oracle.
- $\sigma_i \neq \sigma_j$ for all $i \neq j$.

If the above two conditions are satisfied then G finds the random strings used in the signatures. Let $r_1, r_2, \cdots, r_t$ be the random strings

- $\{F(pk, \sigma_1), F(pk, \sigma_2), \cdots, F(pk, \sigma_t)\} \cap Y_T = \emptyset$ where

$$Y_T = \{F(pk, x) | \exists i, 1 \leq i \leq t,\ PSS_H^{f_{pk}}(m_i \| r_i) \text{ queries } F(pk, x)\}.$$

Finally G checks whether F is the lossy mode[1], if yes it aborts; otherwise G chooses one r uniformly at random from $\{0,1\}^{k_0}$ and computes the PSS hash of $0\|r$ as $y = 0\|h(0\|r)\|g_1(h(0\|r)) \oplus r\|g_2(h(0\|r))$. Finally it returns the forgery as $(0, F^{-1}(td, y))$.

In order to use G to distinguish the lossy and the injective mode, any distinguisher has to construct a satisfying assignment of G in injective mode. By Lemma 6, it happens with negligible probability and we get the following result.

Lemma 8. *Suppose $k = \mathcal{O}(n^{\frac{1}{c}})$ for a positive constant c. $LTDP^{T,T'}$ implements a secure (n, k) Lossy Trapdoor Permutation even relative to G.*

Existence of a forger B^G for PSS using the injective mode of the LTDP is satisfied by Lemma 4. This completes the proof of Theorem 3. □

7 No Reduction from Hard Games with Inversion

Like [9], our result can also be extended to the hard games with inversions. Informally, in a hard game with bounded inversion $\mathcal{C}$, the adversary is allowed to make polynomial $q(n)$ many inversion queries except on some points defined in the game (for one way game adversary is not allowed to make inversion queries on the challenge she received). Following [9], if we modify G to ask for signatures of $|H| + q(n)$ messages and modify Lemma 6 accordingly, we get the following two theorems.

Theorem 4. *There is no blackbox reduction of security against existential forgery under chosen message attack for PSS from any hard game with polynomial number of inversion queries.*

Theorem 5. *There is no blackbox reduction of security against existential forgery against zero message attack for PSS from an oneway trapdoor permutation, even with polynomial number of inversion queries.*

---

[1] As description of F can be hardwired in G, G can easily check the mode of F by finding the possible inverses.

8 Conclusion

Following the negative results, on generic insecurity of FDH by Dodis et. al [9] and of OAEP by Kiltz and Pietrzak [18] in standard model, we show security of PSS also can not be black box reduced to any property of an ideal trapdoor permutation. Moreover, we also show one can not even hope to achieve security of PSS based on Lossy Trapdoor Permutations. On the contrary recently a secure instantiation of OAEP has been realized based on Lossy Trapdoor Permutations [17].

Acknowledgements

We sincerely thank Palash Sarkar for helpful discussions.

References

1. An, J.H., Dodis, Y., Rabin, T.: On the security of joint signature and encryption. In: Knudsen, L.R. (ed.) EUROCRYPT 2002. LNCS, vol. 2332, pp. 83–107. Springer, Heidelberg (2002)
2. Bellare, M., Rogaway, P.: Random oracles are practical: A paradigm for designing efficient protocols. In: ACM Conference on Computer and Communications Security, pp. 62–73 (1993)
3. Bellare, M., Rogaway, P.: The exact security of digital signatures - how to sign with RSA and rabin. In: Maurer, U.M. (ed.) EUROCRYPT 1996. LNCS, vol. 1070, pp. 399–416. Springer, Heidelberg (1996)
4. Bhattacharyya, R., Mandal, A.: On the impossibility of instantiating pss in the standard model: Full version of this paper. Cryptology ePrint Archive, Report 2010/651 (2010)
5. Canetti, R., Goldreich, O., Halevi, S.: The random oracle methodology, revisited. In: STOC, pp. 209–218 (1998)
6. Canetti, R., Goldreich, O., Halevi, S.: The random oracle methodology, revisited. J. ACM 51(4), 557–594 (2004)
7. Coron, J.-S.: Optimal security proofs for PSS and other signature schemes. In: Knudsen, L.R. (ed.) EUROCRYPT 2002. LNCS, vol. 2332, pp. 272–287. Springer, Heidelberg (2002)
8. Coron, J.-S., Mandal, A.: PSS is secure against random fault attacks. In: Matsui, M. (ed.) ASIACRYPT 2009. LNCS, vol. 5912, pp. 653–666. Springer, Heidelberg (2009)
9. Dodis, Y., Oliveira, R., Pietrzak, K.: On the generic insecurity of the full domain hash. In: Shoup, V. (ed.) CRYPTO 2005. LNCS, vol. 3621, pp. 449–466. Springer, Heidelberg (2005)
10. Fischlin, M., Schröder, D.: On the impossibility of three-move blind signature schemes. In: Gilbert, H. (ed.) EUROCRYPT 2010. LNCS, vol. 6110, pp. 197–215. Springer, Heidelberg (2010)
11. Gennaro, R., Trevisan, L.: Lower bounds on the efficiency of generic cryptographic constructions. In: FOCS, pp. 305–313 (2000)
12. Gertner, Y., Kannan, S., Malkin, T., Reingold, O., Viswanathan, M.: The relationship between public key encryption and oblivious transfer. In: FOCS, pp. 325–335 (2000)

13. Gertner, Y., Malkin, T., Reingold, O.: On the impossibility of basing trapdoor functions on trapdoor predicates. In: FOCS, pp. 126–135 (2001)
14. Goldwasser, S., Kalai, Y.T.: On the (in)security of the fiat-shamir paradigm. In: FOCS, pp. 102–113 (2003)
15. Hsiao, C.-Y., Reyzin, L.: Finding collisions on a public road, or do secure hash functions need secret coins? In: Franklin, M. (ed.) CRYPTO 2004. LNCS, vol. 3152, pp. 92–105. Springer, Heidelberg (2004)
16. Impagliazzo, R., Rudich, S.: Limits on the provable consequences of one-way permutations. In: STOC, pp. 44–61 (1989)
17. Kiltz, E., O'Neill, A., Smith, A.: Instantiability of RSA-OAEP under chosen-plaintext attack. In: Rabin, T. (ed.) CRYPTO 2010. LNCS, vol. 6223, pp. 295–313. Springer, Heidelberg (2010)
18. Kiltz, E., Pietrzak, K.: On the security of padding-based encryption schemes – or – why we cannot prove OAEP secure in the standard model. In: Joux, A. (ed.) EUROCRYPT 2009. LNCS, vol. 5479, pp. 389–406. Springer, Heidelberg (2009)
19. Paillier, P.: Impossibility proofs for RSA signatures in the standard model. In: Abe, M. (ed.) CT-RSA 2007. LNCS, vol. 4377, pp. 31–48. Springer, Heidelberg (2006)
20. Peikert, C., Waters, B.: Lossy trapdoor functions and their applications. In: STOC, pp. 187–196 (2008)
21. Simon, D.R.: Findings Collisions on a One-Way Street: Can Secure Hash Functions Be Based on General Assumptions? In: Nyberg, K. (ed.) EUROCRYPT 1998. LNCS, vol. 1403, pp. 334–345. Springer, Heidelberg (1998)

On-line Non-transferable Signatures Revisited*

Jacob C.N. Schuldt[1,**] and Kanta Matsuura[2]

[1] Research Center for Information Security, AIST, Japan
jacob.schuldt@aist.go.jp
[2] Institute of Industrial Science, The University of Tokyo, Japan
kanta@iis.u-tokyo.ac.jp

Abstract. Undeniable signatures, introduced by Chaum and van Antwerpen, and designated confirmer signatures, introduced by Chaum, allow a signer to control the verifiability of his signatures by requiring a verifier to interact with the signer to verify a signature. An important security requirement for these types of signature schemes is *non-transferability* which informally guarantees that even though a verifier has confirmed the validity of a signature by interacting with the signer, he cannot prove this knowledge to a third party. Recently Liskov and Micali pointed out that the commonly used notion of non-transferability only guarantees security against an off-line attacker which cannot influence the verifier while he interacts with the signer, and that almost all previous schemes relying on interactive protocols are vulnerable to on-line attacks. To address this, Liskov and Micali formalized on-line non-transferable signatures which are resistant to on-line attacks, and proposed a generic construction based on a standard signature scheme and an encryption scheme. In this paper, we revisit on-line non-transferable signatures. Firstly, we extend the security model of Liskov and Micali to cover not only the sign protocol, but also the confirm and disavow protocols executed by the confirmer. Our security model furthermore considers the use of multiple (potentially corrupted or malicious) confirmers, and guarantees security against attacks related to the use of signer specific confirmer keys. We then present a new approach to the construction of on-line non-transferable signatures, and propose a new concrete construction which is provably secure in the standard model. Unlike the construction by Liskov and Micali, our construction does not require the signer to issue "fake" signatures to maintain security, and allows the confirmer to both confirm and disavow signatures. Lastly, our construction provides noticeably shorter signatures than the construction by Liskov and Micali.

Keywords: signatures, on-line non-transferability, standard model.

1 Introduction

An ordinary signature scheme provides public verifiability i.e. anyone is able to verify the validity of a given signature using the public key of the signer.

* The full version of this paper is available at http://eprint.iacr.org/2009/406

** This author is supported by a JSPS scholarship.

D. Catalano et al. (Eds.): PKC 2011, LNCS 6571, pp. 369–386, 2011.

While this property is useful in many scenarios, it might not always be desirable. For example, a signer who signs a sensitive message might prefer to be able to control who can verify the validity of his signature. Chaum *et al.* [5] addressed this problem with their proposal of undeniable signatures in which a verifier is required to interact with the signer to verify a signature. Furthermore, to preserve non-repudiation, the signer is also able to prove invalidity of a signature through a disavow protocol. Hence, in a dispute, the signer will either be able to confirm or disavow a purported signature. However, in some scenarios, a signer might become unavailable or might refuse to cooperate with a verifier, in which case the validity of a signature cannot be determined. To address this, Chaum [4] introduced designated confirmer signatures in which a third party, the confirmer, can interact with a verifier to confirm or disavow a signature on behalf of the signer. Furthermore, the confirmer can, in the case of a dispute, extract a publicly verifiable signature (of the signer) from a valid designated confirmer signature. Since their introduction, a number of undeniable schemes and designated confirmer schemes have been proposed, e.g. see [3,7,12,14,2,8].

Off-line and On-line Non-transferability. An important security notion for these types of signature schemes is non-transferability. Intuitively, non-transferability guarantees that once a verifier has verified a signature and is convinced about its validity, he cannot transfer this conviction to a third party. This is achieved by ensuring that a verifier is able to simulate a transcript of the interaction with the signer/confirmer i.e. any "evidence" of validity obtained through the interaction, could have been generated by the verifier himself. A scheme providing this property is said to be *off-line non-transferable.* However, Liskov and Micali [13] pointed out that almost all[1] previous schemes relying on interactive protocols to provide off-line non-transferability are vulnerable to *on-line* attacks, i.e. an attacker who is present *while* the verifier interacts with a signer/confirmer might be able to determine the validity of a signature by influencing messages sent by the verifier. A scheme preventing these types of attacks is said to be *on-line non-transferable* and is constructed by enabling the verifier to *interactively* simulate the interaction with a signer/confirmer. To preserve soundness of the scheme, only the verifier should be able to simulate a proof, and to facilitate this, Liskov and Micali [13] assumed the verifier holds a public/private key pair i.e. to simulate the interaction between the signer/confirmer and a verifier, the private key of the verifier is required. Note that this approach to on-line non-transferability requires that the verifier knows the private key corresponding to his public key to maintain security. More specifically, if it is possible for a verifier to convince a third party that he does not know his private key (e.g. by generating his public key by applying a hash function to a random seed $pk_V = H(x)$, and then presenting x to the third party), the scheme will no longer provide on-line non-transferability. To prevent this type of malicious behavior, verifier key registration is required i.e. a verifier should prove knowledge of his private key when registering his public key (see [13] for further discussion of this). In this

[1] See *Related Work* below for a few exceptions in the random oracle model.

paper, we adopt the same general approach as [13], assume verifiers are equipped with public/private key pairs, and will furthermore explicitly model verifier key registration in our security model[2].

In [13], Liskov and Micali illustrated the feasibility of constructing an on-line non-transferable signature scheme under the above described assumption of verifier key registration. More specifically, they proposed a generic construction based on `ind-cpa` secure public key encryption and `uf-cma` secure signatures. The resulting scheme provides on-line non-transferability of an interactive sign protocol through which the signer both constructs and proves validity of a signature. Furthermore, the scheme supports the use of confirmers and is proved secure in the standard model. However, to achieve on-line non-transferability, a signer has to be willing to issue "fake" signatures to anyone requesting them. This is essential since a verifier will not be able to simulate the sign protocol without the ability to ask the signer for fake signatures. This drawback limits the practical applicability of the scheme. Furthermore, the functionality and security guarantees of the confirmer are somewhat limited. More specifically, a confirmer can disavow but not confirm the validity of a signature, and neither off-line nor on-line non-transferability are considered for the disavow protocol[3].

Our Contribution. In this paper, we address many of the limitations of the approach by Liskov and Micali. Firstly, we extend the security model to model not only the on-line non-transferability of the sign protocol, but also of the confirm and disavow protocols executed by the confirmer. Furthermore, we introduce two additional security notions, confirmer soundness and key unforgeability, required by the added ability of the confirmer to confirm signatures and to prevent attacks related to the forgery of signer specific confirmer keys which are used both in our construction and in [13] (see Section 3 for details). Unlike [13], our security model also allows the signer to make use of multiple confirmers and ensures unforgeability even against malicious confirmers, which will guarantee security in a more realistic usage scenario.

We then propose a new general approach to the construction of on-line non-transferable signatures. More specifically, we show how a simple *core confirmer signature scheme*, which essentially implements the non-interactive functionality of an on-line non-transferable signature scheme, can be extended to a fully secure scheme with the additional use of ordinary signatures, sigma protocols, and trapdoor commitments with an enhanced binding property. Based on this approach, we propose a concrete instantiation which is provably secure in the standard model assuming the computational Diffie-Hellman problem and the decisional linear problem are hard.

Compared to the approach taken by Liskov and Micali, our scheme has several advantages. Besides implementing additional confirmer functionality and

---

[2] Note that while the security definitions in [13] does not explicitly describe verifier key registration, this *is* a requirement to ensure basic security, and we argue that our security models are fundamentally the same.

[3] The defined disavow protocol in [13] is non-interactive and provides a publicly verifiable proof of invalidity.

providing security in our extended security model, our scheme allows a verifier to independently simulate the sign, confirm and disavow protocols, and does not require the signer to issue "fake" signatures to maintain security. Lastly, our concrete instantiation provides efficient protocols and short signatures consisting of four group elements and an integer, whereas the scheme by Liskov and Micali requires signatures consisting of more than $3k$ encryptions, where k is the security parameter. However, we note that our concrete scheme requires large public keys due to the use of the techniques by Waters [19].

Related Work. Jakobsson *et al.* [10] introduced an alternative approach to limiting the verifiability of signatures with their proposal of designated verifier signatures in which only a specific verifier chosen by the signer will be convinced about the validity of a signature. This concept was extended by Steinfeld *et al.* [17] who introduced universal designated verifier signatures which allow any user to convert a publicly verifiable signature into a designated verifier signature for a chosen verifier. Since this type of schemes do not rely on interactive protocols for signature confirmation, on-line attacks are not a concern. However, these schemes do not provide a mechanism to determine the validity of a (converted) signature in a dispute, and most of the recently proposed schemes will in fact enable the designated verifier to construct signatures which are perfectly indistinguishable from signatures constructed by the signer. Hence, unlike undeniable and designated confirmer signatures, non-repudiation cannot be enforced in these schemes which make them unsuitable for a number of applications.

A few existing schemes, which are provably secure in the random oracle model, implicitly provide protection against on-line attacks. For example, the undeniable signature schemes by Kudla *et al.* [11] and Huang *et al.* [9] provide non-interactive proofs which are simulatable by the verifier, and hence avoid the problem of on-line attacks. Furthermore, Monnerat *et al.* [15] proposed an undeniable signature scheme with interactive 2-move confirm and disavow protocols. While the used definition of non-transferability in [15] only guarantees transcript simulatability (i.e. defines off-line non-transferability), the concrete scheme allows a verifier to use his private key to simulate proofs interactively, and hence the scheme provides on-line non-transferability. However, we emphasize that all of the above schemes are only provable secure in the random oracle model, and that the schemes furthermore do not support the use of confirmers to ensure non-repudiation if the signer becomes off-line or refuses to cooperate.

2 On-line Non-transferable Signatures

An on-line non-transferable signature (ONS) scheme involves a signer S, a confirmer C, and a verifier V, and is given by the following probabilistic polynomial time (PPT) algorithms:

- `Setup` which, given a security parameter 1^k, returns the public parameters *par*.

- $\mathtt{KeyGen}_S$, $\mathtt{KeyGen}_C$, and $\mathtt{KeyGen}_V$ which, given *par*, return public/private key pairs (pk_S, sk_S), (pk_C, sk_C), and (pk_V, sk_V) for a signer, a confirmer, and a verifier, respectively.
- `CSetup` which on input *par*, sk_C and pk_S, returns a signer specific public/private confirmer key pair $(pk_{C,S}, sk_{C,S})$. This algorithm is run once by the confirmer for each signer S, and the confirmer stores $sk_{C,S}$ for later use. The public key $pk_{C,S}$ is given to the signer who is to use this when constructing signatures with confirmer C.
- `CKeyValid` which, on input *par*, pk_S, pk_C, and $pk_{C,S}$, outputs either `accept` or `reject`.
- (`Sign`, `Receive`) which is a pair of interactive algorithms with common input $(par, pk_S, pk_C, pk_{C,S}, pk_V, m)$. `Sign` is run by the signer and is given sk_S as private input, and `Receive` is run by the verifier. At the end of the interaction, both `Sign` and `Receive` will output a signature, σ_S and σ_R, and `Receive` will in addition output either `accept` or `reject`.
- `Convert` which, on input *par*, pk_S, m, σ, and $sk_{C,S}$, returns a verification token tk_σ.
- `TkVerify` which, on input *par*, pk_S, pk_C, $pk_{C,S}$, m, σ, and tk_σ, returns either `accept` or `reject`.
- ($\mathtt{Confirm}, \mathtt{V}_C$) which is a pair of interactive algorithms with common input $(par, pk_S, pk_C, pk_{C,S}, pk_V, m, \sigma)$. `Confirm` is run by the confirmer and is given $sk_{C,S}$ as private input, and $\mathtt{V}_C$ is run by the verifier. At the end of the interaction, $\mathtt{V}_C$ outputs either `accept` or `reject`.
- ($\mathtt{Disavow}, \mathtt{V}_D$) which is also a pair of interactive algorithms. Input for `Disavow` and $\mathtt{V}_D$ is exactly as in ($\mathtt{Confirm}, \mathtt{V}_C$) above, and the output of $\mathtt{V}_D$ is either `accept` or `reject`.

Like Liskov and Micali [13], we require that before signer S makes use of a confirmer C, he will approach C to obtain a signer specific confirmer key $pk_{C,S}$ which C generates by running `CSetup`. This process can be seen as a registration procedure in which the confirmer agrees to act as a confirmer for this specific signer. Note that this does not require a confidential channel between the signer and confirmer. Our definition differs slightly from that of [13] in that we explicitly define a key validation algorithm `CKeyValid` for signer specific confirmer keys[4], and introduce ($\mathtt{Confirm}, V_C$) to allow C to confirm signatures. Furthermore, we do not include a fake signature algorithm which is required to maintain the security of the scheme in [13].

Using the above defined algorithms, a confirmer can verify a signature by first computing a verification token using `Convert` and then verifying the signature using `TkVerify`. To simplify notation, we define an algorithm `Valid` which performs these two steps:

- `Valid`: given the input $(par, pk_S, pk_C, pk_{C,S}, m, \sigma, sk_{C,S})$, compute the verification token $tk_\sigma \leftarrow \mathtt{Convert}(par, pk_S, m, \sigma, sk_{C,S})$ and return the output of $\mathtt{TkVerify}(par, pk_S, pk_C, pk_{C,S}, m, \sigma, tk_\sigma)$.

[4] This algorithm is required by our extended security model. More specifically, it is required to define key unforgeability (see Section 3).

We will use the notation $\{\texttt{Sign}(sk_S) \leftrightarrow \texttt{Receive}\}(par, pk_S, pk_C, pk_{C,S}, pk_V, m)$ to denote the interaction between Sign and Receive on the common input $(par, pk_S, pk_C, pk_{C,S}, pk_V, m)$ and private input sk_S to the Sign algorithm. To shorten this notation, we will sometimes use $PK = (pk_S, pk_C, pk_{C,S}, pk_V)$ to represent the public keys. We furthermore use $(\sigma_S, (\sigma_R, z)) \leftarrow \{\texttt{Sign}(sk_S) \leftrightarrow \texttt{Receive}\}(par, PK, m)$ to denote the output of Sign and Receive, respectively, and use $(\sigma_R, z) \leftarrow_2 \{\texttt{Sign}(sk_S) \leftrightarrow \texttt{Receive}\}(par, PK, m)$ when we are only considering the output of Receive. Similar notation is used for the confirm and disavow protocols.

Correctness. We require that for all honestly generated public parameters par and key pairs (pk_S, sk_C), (pk_C, sk_C), (pk_V, sk_V), and $(pk_{C,S}, sk_{C,S})$, that $z \leftarrow \texttt{CKeyValid}(par, pk_C, pk_S, pk_{C,S})$ yields $z = \texttt{accept}$ and that, for all messages m, the signature generation $(\sigma_S, (\sigma_R, z_R)) \leftarrow \{\texttt{Sign}(sk_S) \leftrightarrow \texttt{Receive}\}(par, PK, m)$, where $PK \leftarrow (pk_S, pk_C, pk_{C,S}, pk_V)$, yields that $z_R = \texttt{accept}$ and $\sigma_R = \sigma_S$, that $\texttt{Valid}(par, pk_S, pk_C, pk_{C,S}, sk_{C,S}, m, \sigma) = \texttt{true}$ and that $z_C \leftarrow \{\texttt{Confirm}(sk_{C,S}) \leftrightarrow \texttt{V}_C\}(par, PK, m, \sigma)$ results in $z_C = \texttt{accept}$. Furthermore, for all (m', σ') such that $\texttt{Valid}(par, pk_S, pk_C, pk_{C,S}, sk_{C,S}, m', \sigma') = \texttt{false}$, we require that $z_D \leftarrow_2 \{\texttt{Disavow}(sk_{C,S}) \leftrightarrow \texttt{V}_D\}(par, PK, m', \sigma')$ yields $z_D = \texttt{accept}$.

3 Security Model

An ONS scheme is required to satisfy the security notions *unforgeability*, *key unforgeability*, *soundness*, *non-repudiation* and *non-transferability* to be considered secure. However, before we can formally define these security notions, we require a scheme to define the verifier simulation algorithms $\texttt{SimSign}(par, PK, m, sk_V)$, $\texttt{SimCon}(par, PK, m, \sigma, sk_V)$ and $\texttt{SimDis}(par, PK, m, \sigma, sk_V)$ which simulates the Sign, Confirm and Disavow algorithms, respectively. While these algorithms are not part of the basic functionality of an ONS scheme, they must be defined to ensure that a verifier can simulate the interactive protocols of the scheme as required by the non-transferability notion defined below. Furthermore, since an adversary might observe the execution of these algorithms while attempting to mount attacks against other security properties of the scheme, we must provide the adversary with oracle access to these algorithms in the relevant security definitions.

Unforgeability. Our notion of unforgeability requires that, even for a maliciously chosen confirmer key, an adversary with oracle access to an honest signer cannot produce a new message/signature pair and convince a verifier about the validity of this pair, either by interacting with the verifier in the confirm protocol or by producing a token such that TkVerify outputs accept. Our definition allows the adversary to obtain signatures using any confirmer key, and thereby ensures security in a scenario where a signer makes use of multiple potentially malicious confirmers. In comparison, the unforgeability notion defined by Liskov and Micali only considers a signer using a single honest confirmer. Formally, we define unforgeability of an ONS scheme N via the experiment $\texttt{Exp}^{\texttt{uf-cma}}_{N,\mathcal{A}}$ shown in Figure 1. In the experiment, $\mathcal{A}$ has access to the oracles $\mathcal{O} = \{\mathcal{O}_{VKeyReg}, \mathcal{O}_{Sign},$

$\mathcal{O}_{SimSign}, \mathcal{O}_{SimCon}, \mathcal{O}_{SimDis}\}$ which are defined below. The oracle $\mathcal{O}_{VKeyReg}$ implements verifier key registration and maintains a list, $L_{VKeyReg}$, of registered keys. It is assumed that it can be verified that a key pair (pk_V, sk_V) is valid i.e. that (pk_V, sk_V) lies in the range of $\texttt{KeyGen}_V$.

- $\mathcal{O}_{VKeyReg}$: given (pk_V, sk_V), this oracle stores (pk_V, sk_V) in the list $L_{VKeyReg}$ and returns $\top$ to $\mathcal{A}$ if (pk_V, sk_V) is a valid key pair. Otherwise, the oracle returns $\perp$ to $\mathcal{A}$. In the following, if a query to an oracle involves a verifier key pk_V, it is assumed that $\mathcal{A}$ has previously submitted pk_V to this oracle as part of a valid key pair. If this is not the case, the relevant oracle will return $\perp$ to $\mathcal{A}$.
- $\mathcal{O}_{Sign}$: given input $(pk_C, pk_{C,S}, pk_V, m)$, this oracle interacts with $\mathcal{A}$ by running $\texttt{Sign}$ with common input $(par, pk_S, pk_C, pk_{C,S}, pk_V, m)$ and secret input sk_S. Local output of $\texttt{Sign}$ will be a signature σ, and $(pk_C, pk_{C,S}, m, \sigma)$ is added to L_{Sign}.
- $\mathcal{O}_{SimSign}$: given input pk_S, pk_C, $pk_{C,S}$, and m, this oracle interact with $\mathcal{A}$ by running the simulation algorithm $\texttt{SimSign}(par, pk_S, pk_C, pk_{C,S}, m, sk_V)$.
- $\mathcal{O}_{SimCon}$: given input pk_S, pk_C, $pk_{C,S}$, m and σ, this oracle interacts with $\mathcal{A}$ by running the algorithm $\texttt{SimCon}(par, pk_S, pk_C, pk_{C,S}, pk_V, m, \sigma, sk_V)$.
- $\mathcal{O}_{SimDis}$: given the same input as $\mathcal{O}_{SimCon}$, this oracle interacts with $\mathcal{A}$ by running the algorithm $\texttt{SimDis}(par, pk_S, pk_C, pk_{C,S}, pk_V, m, \sigma, sk_V)$.

Definition 1. *An ONS scheme N is said to be* unforgeable, *if no PPT algorithm $\mathcal{A}$ with non-negligible advantage* $\texttt{Adv}^{\textit{uf-cma}}_{N,\mathcal{A}}(k) = \Pr[\texttt{Exp}^{\textit{uf-cma}}_{N,\mathcal{A}}(1^k) = 1]$ *exists.*

Key unforgeability. The use of the confirmer setup, $\texttt{CSetup}$, warrants additional security requirements. Key unforgeability requires that an adversary without access to the private confirmer key, cannot produce a new valid signer specific confirmer key i.e. a new key which is accepted by $\texttt{CKeyValid}$. The security model in [13] does not have a similar security requirement and does in fact not rule out the possibility that a signer is able to forge a signer specific confirmer key and then use this forged key in the sign protocol. This would leave the confirmer unable to either confirm, disavow or convert the signature. However, such concerns are eliminated by explicitly requiring key unforgeability. Formally, key unforgeability of an ONS scheme N is defined via the experiment $\texttt{Exp}^{\texttt{key-uf}}_{N,\mathcal{A}}(1^k)$ shown in Figure 1. In the experiment, $\mathcal{A}$ has access to the oracles $\mathcal{O} = \{\mathcal{O}_{VKeyReg}, \mathcal{O}_{CSetup}, \mathcal{O}_{Convert}, \mathcal{O}_{Con}, \mathcal{O}_{Dis}\}$ where $\mathcal{O}_{VKeyReg}$ is defined as above, and the remaining oracles are defined as follows:

- $\mathcal{O}_{CSetup}$: given pk_S, this oracle runs $(pk_{C,S}, sk_{C,S}) \leftarrow \texttt{CSetup}(par, pk_S, sk_C)$, stores $(pk_S, pk_{C,S}, sk_{C,S})$ in L_{CSetup}, and returns $pk_{C,S}$ to $\mathcal{A}$.
- $\mathcal{O}_{Convert}$: given pk_S, $pk_{C,S}$, m, and σ, this oracle searches for a matching tuple $(pk_S, pk_{C,S}, sk_{C,S})$ in L_{CSetup}, and returns $\perp$ if no such tuple is found. Otherwise, the oracle returns $tk_\sigma \leftarrow \texttt{Convert}(par, pk_S, m, \sigma, sk_{C,S})$.
- $\mathcal{O}_{Con}$: given pk_S, $pk_{C,S}$, pk_V, m, and σ, the oracle searches for a tuple $(pk_S, pk_{C,S}, sk_{C,S})$ in L_{CSetup}. If no such tuple is found the oracle returns $\perp$. Otherwise, the oracle interacts with $\mathcal{A}$ by running $\texttt{Confirm}$ with common input $(par, pk_S, pk_C, pk_{C,S}, pk_V, m, \sigma)$ and secret input $sk_{C,S}$.

$\mathtt{Exp}^{\mathtt{uf\text{-}cma}}_{\mathcal{S},\mathcal{A}}(1^k)$
  $L_{Sign} \leftarrow \{\}$; $L_{VKeyReg} \leftarrow \{\}$
  $par \leftarrow \mathtt{Setup}(1^k)$
  $(pk_S, sk_S) \leftarrow \mathtt{KeyGen}_S(par)$
  $(pk_V, sk_V) \leftarrow \mathtt{KeyGen}_V(par)$
  $(pk_C, pk_{C,S}, m, \sigma, tk_\sigma, st)$
    $\leftarrow \mathcal{A}^{\mathcal{O}}(par, pk_S, pk_V)$
  $PK \leftarrow (pk_S, pk_C, pk_{C,S}, pk_V)$
  $z \leftarrow_2 \{\mathcal{A}^{\mathcal{O}}(st) \leftrightarrow V_C(par, PK, m, \sigma)\}$
  $z' \leftarrow \mathtt{TkVer}(par, pk_S, pk_C, pk_{C,S}, m, \sigma, tk_\sigma)$
  if $(pk_C, pk_{C,S}, m, \sigma) \notin L_{Sign} \wedge$
    $(z = \mathtt{accept} \vee z' = \mathtt{accept})$
    output 1
  else output 0

$\mathtt{Exp}^{\mathtt{key\text{-}uf}}_{\mathcal{S},\mathcal{A}}(1^k)$
  $L_{CSetup} \leftarrow \{\}$; $L_{VKeyReg} \leftarrow \{\}$
  $par \leftarrow \mathtt{Setup}(1^k)$
  $(pk_C, sk_C) \leftarrow \mathtt{KeyGen}_S(par)$
  $(pk_S, pk_{C,S}) \leftarrow \mathcal{A}^{\mathcal{O}}(par, pk_C)$
  $z \leftarrow \mathtt{CKeyValid}(par, pk_S, pk_C, pk_{C,S})$
  if $(pk_S, pk_{C,S}, *) \notin L_{CSetup} \wedge$
    $z = \mathtt{true}$
    output 1
  else output 0

$\mathtt{Exp}^{\mathtt{non\text{-}rep}}_{\mathcal{S},\mathcal{A}}(1^k)$
  $par \leftarrow \mathtt{Setup}(1^k)$
  $(pk_V, sk_V) \leftarrow \mathtt{KeyGen}_V(par)$
  $(pk_S, pk_C, pk_{C,S}, m, st) \leftarrow \mathcal{A}^{\mathcal{O}}(par, pk_V)$
  $PK \leftarrow (pk_S, pk_C, pk_{C,S}, pk_V)$
  $(st', (\sigma, z_1)) \leftarrow \{\mathcal{A}^{\mathcal{O}}(st) \leftrightarrow \mathtt{Receive}(par, PK, m)\}$
  $z_2 \leftarrow_2 \{\mathcal{A}^{\mathcal{O}}(st') \leftrightarrow \mathtt{V}_D(par, PK, m, \sigma)\}$
  if $z_1 = z_2 = \mathtt{accept}$ output 1
  else output 0

Fig. 1. Unforgeability, key unforgeability and non-repudiation security experiments

- $\mathcal{O}_{Dis}$: given the same input as $\mathcal{O}_{Con}$, this oracle returns $\perp$ to $\mathcal{A}$ if there is no tuple $(pk_S, pk_{C,S}, sk_{C,S})$ in L_{CSetup}. Otherwise, the oracle interacts with $\mathcal{A}$ by running `Disavow` with common input $(par, pk_S, pk_C, pk_{C,S}, pk_V, m, \sigma)$ and secret input $sk_{C,S}$.

Definition 2. *An ONS scheme N is said to be* key unforgeable *if no PPT algorithm $\mathcal{A}$ with non-negligible advantage $\mathtt{Adv}^{key\text{-}uf}_{N,\mathcal{A}}(k) = \Pr[\mathtt{Exp}^{key\text{-}uf}_{N,\mathcal{A}}(1^k) = 1]$ exists.*

Non-repudiation. Informally, non-repudiation requires that, even if a malicious signer and confirmer collude, it is not possible for the signer to make an honest verifier accept a message/signature pair as valid in the sign protocol, while the confirmer is able to disavow the validity of the message/signature pair. Our definition of non-repudiation is slightly weaker than the definition given in [13] in that we allow the adversary a negligible success probability whereas [13] requires the success probability to be zero. However, we highlight that [13] makes use of a non-interactive disavow protocol which is both off-line and on-line transferable which allows the slightly stronger non-repudiation property, whereas our constructions will rely on interactive non-transferable protocols with negligible soundness error. We define non-repudiation of an ONS scheme N via the experiment $\mathtt{Exp}^{\mathtt{non\text{-}rep}}_{N,\mathcal{A}}(1^k)$ shown in Figure 1. In the experiment, $\mathcal{A}$ has access to the oracles $\mathcal{O} = \{\mathcal{O}_{SimSign}, \mathcal{O}_{SimCon}, \mathcal{O}_{SimDis}\}$ which are defined as above.

$\mathtt{Exp}^{\mathtt{snd\text{-}sign}}_{N,\mathcal{A}}(1^k)$
$par \leftarrow \mathtt{Setup}(1^k)$; $(pk_C, sk_C) \leftarrow \mathtt{KeyGen}_C(par)$
$(pk_V, sk_V) \leftarrow \mathtt{KeyGen}_V(par)$
$(pk_S, m, st) \leftarrow \mathcal{A}^{\mathcal{O}}(par, pk_C, sk_C, pk_V)$
$(pk_{C,S}, sk_{C,S}) \leftarrow \mathtt{CSetup}(par, pk_S, sk_C)$
$PK \leftarrow (pk_S, pk_C, pk_{C,S}, pk_V)$
$(\sigma, z_1) \leftarrow_2 \{\mathcal{A}^{\mathcal{O}}(st, pk_{C,S}, sk_{C,S}) \leftrightarrow \mathtt{Receive}(par, PK, m)\}$
$z_2 \leftarrow \mathtt{Valid}(par, pk_S, pk_C, pk_{C,S}, m, \sigma, sk_{C,S})$
if $z_1 = \mathtt{accept} \wedge z_2 = \mathtt{reject}$ output 1
else output 0

$\mathtt{Exp}^{\mathtt{snd\text{-}conf}}_{N,\mathcal{A}}(1^k)$
$par \leftarrow \mathtt{Setup}(1^k)$; $(pk_V, sk_V) \leftarrow \mathtt{KeyGen}_V(par)$
$(pk_S, pk_C, pk_{C,S}, m, \sigma, tk_\sigma, st) \leftarrow \mathcal{A}^{\mathcal{O}}(par, pk_V)$
$PK \leftarrow (pk_S, pk_C, pk_{C,S}, pk_V)$
$z_1 \leftarrow_2 \{\mathcal{A}^{\mathcal{O}}(st) \leftrightarrow \mathtt{V}_D(par, PK, m, \sigma)\}$
$z_2 \leftarrow_2 \{\mathcal{A}^{\mathcal{O}}(st) \leftrightarrow \mathtt{V}_C(par, PK, m, \sigma)\}$
$z_3 \leftarrow \mathtt{TkVerify}(par, pk_S, pk_C, pk_{C,S}, m, \sigma, tk_\sigma)$
if $z_1 = \mathtt{accept} \wedge (z_2 = \mathtt{accept} \vee z_3 = \mathtt{accept})$ output 1
else output 0

Fig. 2. Soundness security experiments

Definition 3. *An ONS scheme N is said to provide* non-repudiation *if no PPT algorithm $\mathcal{A}$ with non-negligible advantage* $\mathtt{Adv}^{non\text{-}rep}_{N,\mathcal{A}}(k) = \Pr[\mathtt{Exp}^{non\text{-}rep}_{N,\mathcal{A}}(1^k) = 1]$ *exists.*

Soundness. We consider two soundness notions – *signer soundness* and *confirmer soundness*. The first notion, signer soundness, guarantees that a signer cannot make a verifier accept a message/signature pair as valid through the sign protocol without the confirmer being able to confirm the validity of this pair as well as compute a verification token showing validity. Our definition guarantees signer soundness even if the confirmer is corrupted since the adversary is allowed to access the private confirmer key, and implies the soundness notion by Liskov and Micali which only grants the adversary access to the public confirmer key. Formally, we define signer soundness of an ONS scheme N via the experiment $\mathtt{Exp}^{\mathtt{snd\text{-}sign}}_{N,\mathcal{A}}(1^k)$ shown in Figure 2. In the experiment, $\mathcal{A}$ will have access to the oracles $\mathcal{O} = \{\mathcal{O}_{SimSign}, \mathcal{O}_{SimCon}, \mathcal{O}_{SimDis}\}$ defined as above.

Definition 4. *An ONS scheme N is said to provide* signer soundness *if no PPT algorithm $\mathcal{A}$ with non-negligible advantage* $\mathtt{Adv}^{snd\text{-}sign}_{N,\mathcal{A}}(k) = \Pr[\mathtt{Exp}^{snd\text{-}sign}_{N,\mathcal{A}}(1^k) = 1]$ *exists.*

The second notion, confirmer soundness, guarantees that even if both signer and confirmer key is maliciously generated, a confirmer cannot produce a message/signature pair which he can successfully disavow by completing the disavow protocol, while still being able to confirm validity of this pair, either by completing the confirm protocol or by producing a token that will make `TkVerify` output `accept`. Since Liskov and Micali do not consider the confirmer's ability to confirm a signature, an equivalent security notion is not defined in [13]. We define confirmer soundness of an ONS scheme N via the experiment $\mathtt{Exp}_{N,\mathcal{A}}^{\mathtt{snd\text{-}conf}}(1^k)$ shown in Figure 2. In the experiment, $\mathcal{A}$ has access to the oracles $\mathcal{O} = \{\mathcal{O}_{SimSign}, \mathcal{O}_{SimCon}, \mathcal{O}_{SimDis}\}$ defined as above.

Definition 5. *An ONS scheme is said to provide* confirmer soundness *if no PPT algorithm $\mathcal{A}$ with non-negligible advantage* $\mathtt{Adv}_{N,\mathcal{A}}^{\mathtt{snd\text{-}conf}}(k) = \Pr[\mathtt{Exp}_{N,\mathcal{A}}^{\mathtt{snd\text{-}conf}}(1^k) = 1]$ *exists.*

On-line Non-transferability. Intuitively, on-line non-transferability of a protocol requires that an adversary cannot distinguish between a real execution of the protocol and a simulated execution by the verifier. Note that since we are considering the on-line non-transferability of both the sign, confirm and disavow protocols, a verifier must be able to provide a consistent response, even if the adversary first obtains a signature through the (simulated) sign protocol and later try to re-confirm the validity through the (simulated) confirm protocol. We define a single non-transferability notion covering the non-transferability of all three interactive protocols. More specifically, we require that an adversary cannot distinguish between a scenario in which he obtains a valid signature through the sign protocol, confirms the validity through the confirm protocol, and then interacts in the simulated disavow protocol, and a scenario in which he obtains a signature through the simulated sign protocol, confirms the validity through the simulated confirm protocol, and then interacts in the disavow protocol. Our non-transferability notion implies a similar type of non-transferability of the sign protocol as defined by Liskov and Micali, but does not involve fake signature generation. Formally, we define on-line non-transferability of a ONS scheme N via the experiment $\mathtt{Exp}_{N,\mathcal{A}}^{\mathtt{non\text{-}trans}}(1^k)$ shown in Figure 3. In the experiment, $\mathcal{A}$ has

$\mathtt{Exp}_{N,\mathcal{A}}^{\mathtt{non\text{-}trans}}(1^k)$
  $L_{VKeyReg} \leftarrow \{\}$; $par \leftarrow \mathtt{Setup}(1^k)$; $(pk_S, sk_S) \leftarrow \mathtt{KeyGen}_S(par)$
  $(pk_C, sk_C) \leftarrow \mathtt{KeyGen}_C(par)$; $(pk_{C,S}, sk_{C,S}) \leftarrow \mathtt{CSetup}(par, pk_S, sk_C)$
  $(pk_V, sk_V) \leftarrow \mathtt{KeyGen}_V(par)$; $PK \leftarrow (pk_S, pk_C, pk_{C,S}, pk_V)$
  $(m, st) \leftarrow \mathcal{A}^{\mathcal{O}}(par, PK, sk_V)$
  $b \leftarrow \{0,1\}$
  if $b = 0$
    $(\sigma, st') \leftarrow \{\mathtt{Sign}(par, PK, m, sk_S) \leftrightarrow \mathcal{A}(st)\}$
    $st'' \leftarrow_2 \{\mathtt{Confirm}(par, PK, m, \sigma^*, sk_{C,S}) \leftrightarrow \mathcal{A}(st')\}$
    $st''' \leftarrow_2 \{\mathtt{SimDis}(par, PK, m, \sigma, sk_V) \leftrightarrow \mathcal{A}(st'')\}$
  else $(b = 1)$
    $(\sigma^*, st') \leftarrow \{\mathtt{SimSign}(par, PK, m, sk_V) \leftrightarrow \mathcal{A}(st)\}$
    $st'' \leftarrow_2 \{\mathtt{SimCon}(par, PK, m, \sigma, sk_V) \leftrightarrow \mathcal{A}(st')\}$
    $st''' \leftarrow_2 \{\mathtt{Disavow}(par, PK, m, \sigma, sk_{C,S}) \leftrightarrow \mathcal{A}(st'')\}$
  $b' \leftarrow \mathcal{A}^{\mathcal{O}'}(st''')$
  if $b = b'$ output 1
  else output 0

Fig. 3. On-line non-transferability security experiment

access to the oracles $\mathcal{O} = \{\mathcal{O}_{VKeyReg}, \mathcal{O}_{CSetup}, \mathcal{O}_{Sign}, \mathcal{O}_{Convert}, \mathcal{O}_{Con}, \mathcal{O}_{Dis}\}$ defined as in the unforgeability and the key unforgeability experiments. The oracles $\mathcal{O}'$ are defined exactly as $\mathcal{O}$, except that $\mathcal{O}_{Convert}$ will not respond to the query consisting of the challenge $(pk_S, pk_{C,S})$, m^* and σ^*, and $\mathcal{O}_{Con}$ and $\mathcal{O}_{Dis}$ will not respond to queries on the challenge $(pk_S, pk_{C,S})$, m^*, σ^* and any pk_V. Note that $\mathcal{O}_{Sign}$ allows the adversary to obtain signatures under any confirmer key, and that $\mathcal{O}_{Convert}$, $\mathcal{O}_{Con}$ and $\mathcal{O}_{Dis}$ accepts any signer key. This ensures security in a scenario where multiple confirmers service multiple signers. Note also that the adversary is given the private key of the verifier. This will ensure that even if the verifier is compromised, the non-transferability is still maintained.

Definition 6. *An ONS scheme N is said to be* on-line non-transferable *if no PPT algorithm $\mathcal{A}$ with non-negligible advantage* $\mathtt{Adv}_{N,\mathcal{A}}^{non-trans}(k) = |\Pr[\mathtt{Exp}_{N,\mathcal{A}}^{non-trans}(1^k) = 1] - \frac{1}{2}|$ *exits .*

4 Construction of an ONS Scheme

In this section we will present a construction of an ONS scheme based on four simpler building blocks: a standard signature scheme, a core confirmer signature scheme, sigma protocols, and a trapdoor commitment scheme with an enhanced binding property. In the following, we will formally define a core confirmer signature scheme as well as the needed security requirements, motivate and define the enhanced binding property of a trapdoor commitment scheme, and finally show how the above mentioned primitives can be combined into a secure ONS scheme. Formal definitions of standard signatures, sigma protocols and trapdoor commitments can be found in the full version.

4.1 Core Confirmer Signature Scheme

A core confirmer signature scheme is essentially an ONS scheme without any of the interactive algorithms. More specifically, a core confirmer signature scheme is defined by $CS = \{$`CS.Setup`, `CS.KeyGen`$_S$, `CS.KeyGen`$_C$, `CS.Sign`, `CS.Convert`, `CS.TkVerify`$\}$ where the algorithms `CS.Setup`, `CS.KeyGen`$_S$, and `CS.KeyGen`$_C$ are defined as in a full ONS scheme, and the `CS.Sign`, `CS.Convert` and `CS.TkVerify` algorithms are defined as follows:

- `CS.Sign`: given par, pk_C, m, and sk_S, this algorithm returns a signature σ.
- `CS.Convert`: given par, pk_S, m, σ and sk_C, this algorithm returns a verification token tk_σ.
- `CS.TkVerify`: given par, pk_S, pk_C, m, σ and tk_σ, this algorithm returns either `accept` or `reject`.

Both `CS.Convert` and `CS.TkVerify` are assumed to be deterministic. Note that all algorithms are non-interactive and that no specific confirmer keys $pk_{C,S}$ or verifier keys pk_V are required. Like for an ONS scheme, we define an algorithm `CS.Valid` as

– CS.Valid: given par, pk_S, pk_C, m, σ and sk_C, this algorithms computes the verification token $tk_\sigma \leftarrow$ CS.Convert$(par, pk_S, m, \sigma, sk_C)$ and returns the output of CS.TkVerify$(par, pk_S, pk_C, m, \sigma, tk_\sigma)$.

We require that a scheme is *correct* i.e. for all $par \leftarrow$ CS.Setup(1^k), $(pk_S, sk_S) \leftarrow$ CS.KeyGen$_S(par)$, $(pk_C, sk_C) \leftarrow$ CS.KeyGen$_V(par)$, all messages m and all $\sigma \leftarrow$ CS.Sign(par, pk_C, m, sk_S), we require that $z \leftarrow$ CS.Valid$(par, pk_S, pk_C, m, \sigma, sk_C)$ yields $z =$ accept. Furthermore, we require that a core confirmer signature scheme has *unique* private confirmer keys i.e. for any pk_C, there exists at most one sk_C such that $(pk_C, sk_C) \in \{$CS.KeyGen$_C(par)\}$ where $\{$CS.KeyGen$_C(par)\}$ denotes the set of all possible confirmer key pairs generated by CS.KeyGen$_C$[5].

Security Requirements. For a core confirmer signature scheme to be secure, we require that the scheme provides *unforgeability*, *invisibility* and *token soundness.* However, due to the reduced functionality of a core confirmer signature scheme, these definitions will be much simpler compared to the security definitions of a full ONS scheme.

We define unforgeability of a core confirmer signature scheme CS via the experiment $\mathtt{Exp}^{\mathtt{cs\text{-}uf\text{-}cma}}_{CS,\mathcal{A}}(1^k)$ shown in Figure 4. In the experiment, $\mathcal{A}$ has access to the oracle $\mathcal{O}_{Sign}$ defined as follows:

– $\mathcal{O}_{Sign}$: given pk_C, and m, this oracle computes $\sigma \leftarrow$ CS.Sign(par, pk_C, m, sk_S), adds (pk_C, m, σ) to L_{CSSign}, and returns σ.

Definition 7. *A core confirmer signature scheme CS is said to be* unforgeable *if no PPT algorithm $\mathcal{A}$ with non-negligible advantage* $\mathtt{Adv}^{\mathtt{cs\text{-}uf\text{-}cma}}_{CS,\mathcal{A}}(k) = \Pr[\mathtt{Exp}^{\mathtt{cs\text{-}uf\text{-}cma}}_{CS,\mathcal{A}}(1^k) = 1]$ *exists.*

Invisibility of a core confirmer signature scheme CS, which captures the property that valid signatures cannot be distinguished from random elements of the signature space, is defined via the experiment $\mathtt{Exp}^{\mathtt{cs\text{-}inv\text{-}cma}}_{CS,\mathcal{A}}$ shown in Figure 4. In the experiment, $\mathcal{S}$ denotes the signature space of the scheme, and $\mathcal{A}$ has access to the oracles $\mathcal{O} = \{\mathcal{O}_{Sign}, \mathcal{O}_{Convert}\}$ where $\mathcal{O}_{Sign}$ is defined as in the above unforgeability experiment, and $\mathcal{O}_{Convert}$ is defined as follows:

– $\mathcal{O}_{Convert}$: given m and σ, this oracle returns the verification token $tk_\sigma \leftarrow$ CS.Convert$(par, pk_S, pk_C, m, \sigma, sk_C)$.

Note that $\mathcal{O}_{Convert}$ will only convert signatures from the signer pk_S and is not required to work for maliciously generated public signer keys for which the adversary might know the corresponding private key. Hence, intuitively, this security requirement only requires the scheme to be secure in a "single user" setting in which a confirmer only services a single signer. This weaker requirement is important for the security proof of our concrete construction.

Definition 8. *A core confirmer signature scheme CS is said to be* invisible *if there exists no PPT algorithm $\mathcal{A}$ with non-negligible advantage* $\mathtt{Adv}^{\mathtt{cs\text{-}inv\text{-}cma}}_{CS,\mathcal{A}}(k) = |\Pr[\mathtt{Exp}^{\mathtt{cs\text{-}inv\text{-}cma}}_{CS,\mathcal{A}}(1^k) = 1] - \frac{1}{2}|$ *exists.*

[5] This property is needed to prove confirmer soundness (Theorem 15) of the construction presented in Section 4.3.

$\mathtt{Exp}^{\mathtt{cs\text{-}uf\text{-}cma}}_{CS,\mathcal{A}}(1^k)$
  $L_{CSSign} \leftarrow \{\}$
  $par \leftarrow \mathtt{CS.Setup}(1^k)$
  $(pk_S, sk_S) \leftarrow \mathtt{CS.KeyGen}_S(par)$
  $(pk^*_C, m^*, \sigma^*, tk^*_\sigma) \leftarrow \mathcal{A}^{\mathcal{O}}(par, pk_S)$
  $z \leftarrow \mathtt{CS.TkVerify}(par, pk_S, pk^*_C, \sigma^*, m^*, tk^*_\sigma)$
  if $(pk^*_C, m^*, \sigma^*) \notin L_{CSSign} \wedge z = \mathtt{accept}$
    output 1
  else output 0

$\mathtt{Exp}^{\mathtt{cs\text{-}inv\text{-}cma}}_{CS,\mathcal{A}}(1^k)$
  $par \leftarrow \mathtt{CS.Setup}(1^k)$
  $(pk_S, sk_S) \leftarrow \mathtt{CS.KeyGen}_S(par)$
  $(pk_C, sk_C) \leftarrow \mathtt{CS.KeyGen}_C(par)$
  $(m^*, st) \leftarrow \mathcal{A}^{\mathcal{O}}(par, pk_S, pk_C)$
  $b \leftarrow \{0, 1\}$
  if $b = 0$ $\sigma^* \leftarrow \mathcal{S}$
  else $\sigma^* \leftarrow \mathtt{CS.Sign}(par, pk_C, m, sk_S)$
  $b' \leftarrow \mathcal{A}^{\mathcal{O}}(st, \sigma^*)$
  if $b = b'$ output 1
  else output 0

$\mathtt{Exp}^{\mathtt{cs\text{-}tk\text{-}snd}}_{CS,\mathcal{A}}(1^k)$
  $par \leftarrow \mathtt{CS.Setup}(1^k)$
  $(pk^*_S, pk^*_C, sk^*_C, m^*, \sigma^*, tk^*_\sigma) \leftarrow \mathcal{A}(par)$
  $z_1 \leftarrow \mathtt{CS.TkVerify}(par, pk^*_S, pk^*_C, \sigma^*, m^*, tk^*_\sigma)$
  $z_2 \leftarrow \mathtt{CS.Valid}(par, pk^*_S, pk^*_C, \sigma^*, m^*, sk^*_C)$
  if $(pk^*_C, sk^*_C) \in \{\mathtt{CS.KeyGen}_C(par)\} \wedge z_1 = \mathtt{accept} \wedge z_2 = \mathtt{reject}$
    output 1
  else output 0

Fig. 4. Unforgeability, invisibility and token soundness experiments for a core confirmer signature scheme

Lastly, we consider token soundness which intuitively captures the property that an accepting verification token cannot be constructed for an invalid signature. Formally, we define token soundness of a scheme CS via the experiment $\mathtt{Exp}^{\mathtt{cs\text{-}tk\text{-}snd}}_{CS,\mathcal{A}}$ shown in Figure 4. In the figure, $\{\mathtt{CS.KeyGen}_C(par)\}$ denotes the set of all possible key pairs generated by $\mathtt{CS.KeyGen}_C$.

Definition 9. *A core confirmer signature scheme CS is said to provide* token soundness *if there exists no PPT algorithm $\mathcal{A}$ with non-negligible advantage $\mathtt{Adv}^{cs\text{-}tk\text{-}snd}_{CS,\mathcal{A}}(k) = \Pr[\mathtt{Exp}^{cs\text{-}tk\text{-}snd}_{CS,\mathcal{A}}(1^k) = 1]$.*

Compatible Sigma Protocols. In our full ONS scheme, we will base the construction of on-line non-transferable protocols on sigma protocols. For this purpose, we require that a set of sigma protocols compatible with the core confirmer signature scheme exists. More specifically, we say that a triple of sigma protocols, Σ_S, Σ_C and $\overline{\Sigma}_C$, and a core confirmer signature scheme CS are compatible if the sigma protocols are defined for the common input $x = (par, pk_S, pk_C, m, \sigma)$ and the following relations.

$$\Sigma_S\{(x, (sk_S, r)) : (pk_S, sk_S) \in \{\mathtt{CS.KeyGen}_S(par)\} \wedge \sigma = \mathtt{CS.Sign}(par, pk_C, m, sk_S; r)\}$$

$$\Sigma_C\{(x, sk_C) : (pk_C, sk_C) \in \{\mathtt{CS.KeyGen}_C(par)\} \wedge \mathtt{CS.Valid}(par, pk_S, pk_C, m, \sigma, sk_C) = \mathtt{accept}\}$$

$$\overline{\Sigma}_C\{(x, sk_C) : (pk_C, sk_C) \in \{\texttt{CS.KeyGen}_C(par)\} \wedge \\ \texttt{CS.Valid}(par, pk_S, pk_C, m, \sigma, sk_C) = \texttt{false}\}$$

In the above, we use the notation $\Sigma\{(x, w) : R(x, w) = 1\}$ to denote the sigma protocol Σ for relation R with common input x and witness w. For simplicity, we assume that the challenge space of Σ_S, Σ_C and $\overline{\Sigma}_C$ are of the same size.

4.2 On-line Non-transferable Protocols

Our construction of on-line non-transferable protocols is based on a simple and intuitive approach inspired by the construction of designated verifier proofs by Jakobsson *et al.* [10] and is furthermore closely related to the construction of efficient zero-knowledge proofs in the auxiliary string model [6]. More specifically, we modify the compatible sigma protocols using a trapdoor commitment scheme, and let a prover and a verifier interact as follows:

1. The prover computes the first message a of the sigma protocol and the commitment $com \leftarrow \texttt{Comm}(ck, a, r)$ for random r, and sends com to the verifier.
2. The verifier then sends a random challenge c to the prover.
3. The prover computes the last message z of the sigma protocol and sends the opening (a, r) together with z to the verifier. The verifier checks that $com = \texttt{Comm}(ck, a, r)$, and accepts if (a, c, z) is an accepting transcript of the sigma protocol.

The commitment key and trapdoor (ck, td) will be used as the public/private key pair (pk_V, sk_V) of the verifier. Hence, using sk_V, the verifier will be able to open com to any message a of his choice, and can therefor postpone generating a until after the challenge c is revealed which allows him to simulate the proof interactively. We use the notation $\texttt{NT}(ck)\text{-}\Sigma$ to denote the non-transferable protocol obtained by modifying the sigma protocol Σ as described above using the commitment key ck.

However, the above approach is not sufficient for proving our constructions secure. Essentially the problem is that an adversary can request to interact with SimSign, SimCon and SimDis in many of the security notions defined in Section 3, choosing any message or message/signature pair (valid or invalid) as input. This type of query can be difficult to handle for a simulator not knowing the trapdoor of the commitment scheme, whereas a simulator knowing the trapdoor might not gain sufficient information from an adversary breaking the security of the scheme. To address this problem, we introduce a commitment scheme with a stronger binding property. Specifically, we consider the advantage of an adversary $\mathcal{A}$ against a scheme T defined by

$$\texttt{Adv}^{bind}_{T,\mathcal{A}}(k) = \Pr[(ck, td) \leftarrow \texttt{G}(1^k); (w, r, w', r') \leftarrow \mathcal{A}^{\mathcal{O}_c, \mathcal{O}_o}(ck) : \\ w \neq w' \wedge \texttt{Comm}(ck, w, r) = \texttt{Comm}(ck, w', r')]$$

where $\mathcal{A}$ has access to a commit and an open oracle, $\mathcal{O}_c$ and $\mathcal{O}_o$, which behave as follows: upon request, $\mathcal{O}_c$ computes $(com, aux) \leftarrow \texttt{TdComm}$, stores aux and returns com to $\mathcal{A}$. Given a commitment com returned by $\mathcal{O}_c$ and a value w, $\mathcal{O}_o$ retrieves the corresponding aux and returns $r \leftarrow \texttt{TdOpen}(aux, w, td)$ such that $com = \texttt{Comm}(ck, w, r)$. The adversary is only allowed to query $\mathcal{O}_o$ with a commitment com obtained from O_c, and is not allowed to make more than one query to $\mathcal{O}_o$ for a given commitment com.

Definition 10. *A trapdoor commitment scheme T is said to be* binding under selective trapdoor openings *if no PPT algorithm $\mathcal{A}$ with non-negligible advantage $\texttt{Adv}^{bind}_{T,\mathcal{A}}(k)$ exists.*

Pedersen's commitment scheme [16] can be shown to be binding under selective trapdoor openings assuming the one more discrete logarithm problem is hard. However, to obtain a commitment scheme which can be shown secure only assuming the ordinary discrete logarithm problem is hard, we can make use of the "double trapdoor" extension also used to strengthen the security of ordinary signatures [18]. In the full version, we recall this scheme and prove it binding under selective trapdoor openings.

4.3 Combined Scheme

We now show how to combine the above mentioned primitives into a full ONS scheme. More specifically, let $CS = \{\texttt{CS.Setup}, \texttt{CS.KeyGen}_S, \texttt{CS.KeyGen}_C, \texttt{CS.Sign}, \texttt{CS.Convert}, \texttt{CS.TkVerify}\}$ be a core confirmer signature scheme with compatible sigma protocols Σ_S, Σ_C and $\overline{\Sigma}_C$, let $T = \{\texttt{T.G}, \texttt{T.Comm}, \texttt{T.TdComm}, \texttt{T.TdOpen}\}$ be a trapdoor commitment scheme, and let $S = \{\texttt{S.Setup}, \texttt{S.KeyGen}, \texttt{S.Sign}, \texttt{S.Verify}\}$ be an ordinary signature scheme. We construct an ONS scheme N as follows:

- $\texttt{Setup}(1^k)$: Compute $par_S \leftarrow \texttt{S.Setup}(1^k)$ and $par_{CS} \leftarrow \texttt{CS.Setup}(1^k)$, and return the parameters $par \leftarrow (par_S, par_{CS})$. It is assumed that par_{CS} include a description of the randomness space $\mathcal{R}$ used by the $\texttt{CS.Sign}$ algorithm.
- $\texttt{KeyGen}_S(par)$: Return $(pk_S, sk_S) \leftarrow \texttt{CS.KeyGen}_S(par_{CS})$.
- $\texttt{KeyGen}_C(par)$: Return $(pk_C, sk_C) \leftarrow \texttt{S.KeyGen}(par_S)$.
- $\texttt{KeyGen}_V(par)$: Return $(pk_V, sk_V) \leftarrow \texttt{T.G}(1^k)$.
- $\texttt{CSetup}(par, pk_S, sk_C)$: Compute the key pair $(pk'_C, sk'_C) \leftarrow \texttt{CS.KeyGen}_C(par_{CS})$ and the signature $\delta \leftarrow \texttt{S.Sign}(par_S, \text{“}pk_S||pk'_C\text{”}, sk_C)$, and return $pk_{C,S} \leftarrow (pk'_C, \delta)$ and $sk_{C,S} \leftarrow sk'_C$.
- $\texttt{CKeyValid}(par, pk_S, pk_C, pk_{C,S})$ Let $pk_{C,S} = (pk'_C, \delta)$ and return the result of the verification $\texttt{S.Verify}(par_S, pk_C, \text{“}pk_S||pk'_C\text{”}, \delta)$.
- $(\texttt{Sign}, \texttt{Receive})$: The common input is $(par, pk_S, pk_C, pk_{C,S}, pk_V, m)$ where $pk_{C,S} = (pk'_C, \delta)$ and the signer is given sk_S as private input. The signer picks $r \in \mathcal{R}$ and computes $\sigma \leftarrow \texttt{CS.Sign}(par_{CS}, pk'_C, sk_S, pk_C||pk_{C,S}||m; r)$. Then the signer sends σ to the verifier, and interacts with the verifier in the protocol $\texttt{NT}(pk_V)\text{-}\Sigma_S$ using $(par, pk_S, pk'_C, pk_C||pk_{C,S}||m, \sigma)$ as common input and (sk_S, r) as secret input[6].

[6] Note that this construction of the sign protocol is slightly more flexible than required by the definition in Section 2 in that a signer is able to re-confirm a signature by running $\texttt{NT}(pk_V)\text{-}\Sigma_S$. This, however, requires the signer to remember the randomness used to construct the signature.

– $\texttt{Convert}(par, pk_S, m, \sigma, sk_{C,S})$: Return the core confirmer signature verification token $tk_\sigma \leftarrow \texttt{CS.Convert}(par_{CS}, pk_S, m, \sigma, sk_{C,S})$.
– $\texttt{TkVerify}(par, pk_S, pk_C, pk_{C,S}, m, \sigma, tk_\sigma)$: Firstly, verify the validity of $pk_{C,S}$ by computing $z \leftarrow \texttt{CKeyValid}(par, pk_S, pk_C, pk_{C,S})$, and return $\texttt{reject}$ if $z = \texttt{reject}$. Otherwise, let $pk_{C,S} = (pk'_C, \delta)$ and return the output of $\texttt{CS.TkVerify}(par_{CS}, pk_S, pk'_C, pk_C || pk_{C,S} || m, \sigma, tk_\sigma)$.
– $(\texttt{Confirm}, V_C)$: The common input is given by $(par, pk_S, pk_C, pk_{C,S}, pk_V, m, \sigma)$ where $pk_{C,S} = (pk'_C, \delta)$ and the signer is given $sk_{C,S}$ as private input. Firstly, the verifier checks validity of $pk_{C,S}$ by running $\texttt{CKeyValid}(par, pk_S, pk_C, pk_{C,S})$, and aborts if the output is $\texttt{reject}$. The confirmer then interacts with the verifier in the protocol $\texttt{NT}(pk_V)\text{-}\Sigma_C$ with private input $sk_{C,S}$ and common input $(par_S, pk_S, pk'_C, pk_C || pk_{C,S} || m, \sigma)$.
– $(\texttt{Disavow}, V_D)$: Having the same input as in $(\texttt{Confirm}, V_C)$, the verifier firstly checks if $\texttt{CKeyValid}(par, pk_S, pk_C, pk_{C,S}) = \texttt{accept}$, and aborts if this is not the case. The verifier and signer then interact in the protocol $\texttt{NT}(pk_V)\text{-}\overline{\Sigma}_C$ with common input $(par_S, pk_S, pk'_C, pk_C || pk_{C,S} || m, \sigma)$ and private input $sk_{C,S}$ to the confirmer.

Security. We will now state the theorems showing that the above constructed ONS scheme satisfies the security definitions given in Section 3 assuming the underlying primitives are secure. Due to space limitation, the proofs are not included here, but can be found in the full version.

Theorem 11. *Assume that CS is unforgeable, that Σ_S is honest verifier zero-knowledge and has special soundness, and that T is perfectly hiding and binding under selective trapdoor openings. Then the above ONS scheme N is unforgeable.*

Theorem 12. *Assume that S is strongly unforgeable. Then the above ONS scheme N has key unforgeability.*

Theorem 13. *Assume that Σ_S and $\overline{\Sigma}_C$ have special soundness, and that T is binding under selective trapdoor openings. Then the above ONS scheme N provides non-repudiation.*

Theorem 14. *Assume that Σ_S have special soundness, and that T is binding under selective trapdoor openings. Then the above ONS scheme N provides signer soundness.*

Theorem 15. *Assume that CS has unique private confirmer keys and provides token soundness, that Σ_C and $\overline{\Sigma}_C$ have special soundness, and that T is binding under selective trapdoor openings. Then the above ONS scheme N provides confirmer soundness.*

Theorem 16. *Assume that CS is invisible, that Σ_S, Σ_C and $\overline{\Sigma}_C$ are honest verifier zero-knowledge, and that T provides a perfect trapdoor property. Then the above ONS scheme N provides on-line non-transferability.*

4.4 Concrete Instantiation

To instantiate the above construction, we can use the strongly unforgeable signature scheme by Boneh *et al.* [1] and the double trapdoor Pedersen commitment scheme mentioned above. In the full version, we define and prove secure a core confirmer signature scheme and compatible sigma protocols to complete the instantiation. The scheme is essentially based on a linear encryption of the first component of a Waters signature [19] combined with the technique of Boneh *et al.* [1] to obtain a strongly unforgeable scheme. More specifically, the public/private key pairs of the signer and confirmer is given by $(pk_S, sk_S) = ((g^\alpha, g_2, h, F), \alpha)$ and $(pk_C, sk_C) = ((u, v), (x, y))$ where F is a Waters hash function and $u^x = v^y = g$, and a signature is of the form $\sigma = (u^a, v^b, g^{a+b+r}, g_2^\alpha H(M)^r, s)$ where $M = g^t h^s$, $t = H(pk_C||u^a||v^b||g^{a+b+r}||m)$, and H is a collision resistant hash function. The scheme is invisible assuming the decisional linear problem is hard, and is strongly unforgeable assume the discrete logarithm problem is hard, H is collision resistant and Waters signatures are (weakly) unforgeable. Furthermore, the structure of the scheme allows the compatible sigma protocols to be implemented using well-know techniques for proving equality and inequality of discrete logarithms. We refer the reader to the full version for the details.

5 Comparison

The generic construction by Liskov and Micali [13] provides many instantiation options due to their use of standard primitives, whereas our approach relies on special building blocks. However, our concrete instantiation has several advantages compared to any instantiation of the scheme by Liskov and Micali, both in terms of functionality, efficiency and security. More specifically, our scheme allows a confirmer to both confirm and disavow signatures, provides short signatures compared to the $\mathcal{O}(k)$ size signatures of [13] which furthermore allows a more efficient sign protocol, and lastly, all interactive protocols are on-line non-transferable, security is guaranteed when multiple (potentially malicious) confirmers are used, and the signer is not required to engage in any "fake" signing protocols to maintain security. As mentioned in the introduction, these security properties are not enjoyed by [13]. We note, however, that our scheme requires large public keys whereas [13] can be instantiated to provide compact public keys, and that the non-interactive disavow protocol of [13] allows perfect non-repudiation whereas our scheme only achieves computational non-repudiation.

References

1. Boneh, D., Shen, E., Waters, B.: Strongly Unforgeable Signatures Based on Computational Diffie-Hellman. In: Yung, M., Dodis, Y., Kiayias, A., Malkin, T.G. (eds.) PKC 2006. LNCS, vol. 3958, pp. 229–240. Springer, Heidelberg (2006)

2. Camenisch, J., Michels, M.: Confirmer signature schemes secure against adaptive adversaries. In: Preneel, B. (ed.) EUROCRYPT 2000. LNCS, vol. 1807, pp. 243–258. Springer, Heidelberg (2000)
3. Chaum, D.: Zero-knowledge undeniable signatures. In: Damgård, I.B. (ed.) EUROCRYPT 1990. LNCS, vol. 473, pp. 458–464. Springer, Heidelberg (1991)
4. Chaum, D.: Designated Confirmer Signatures. In: De Santis, A. (ed.) EUROCRYPT 1994. LNCS, vol. 950, pp. 86–91. Springer, Heidelberg (1995)
5. Chaum, D., van Antwerpen, H.: Undeniable Signatures. In: Brassard, G. (ed.) CRYPTO 1989. LNCS, vol. 435, pp. 212–216. Springer, Heidelberg (1990)
6. Damgård, I.: Efficient Concurrent Zero-Knowledge in the Auxiliary String Model. In: Preneel, B. (ed.) EUROCRYPT 2000. LNCS, vol. 1807, pp. 418–430. Springer, Heidelberg (2000)
7. Galbraith, S.D., Mao, W., Paterson, K.G.: RSA-Based Undeniable Signatures for General Moduli. In: Preneel, B. (ed.) CT-RSA 2002. LNCS, vol. 2271, pp. 200–217. Springer, Heidelberg (2002)
8. Gentry, C., Molnar, D., Ramzan, Z.: Efficient Designated Confirmer Signatures Without Random Oracles or General Zero-Knowledge Proofs. In: Roy, B. (ed.) ASIACRYPT 2005. LNCS, vol. 3788, pp. 662–681. Springer, Heidelberg (2005)
9. Huang, X., Mu, Y., Susilo, W., Wu, W.: Provably Secure Pairing-Based Convertible Undeniable Signature with Short Signature Length. In: Takagi, T., Okamoto, T., Okamoto, E., Okamoto, T. (eds.) Pairing 2007. LNCS, vol. 4575, pp. 367–391. Springer, Heidelberg (2007)
10. Jakobsson, M., Sako, K., Impagliazzo, R.: Designated Verifier Proofs and Their Applications. In: Maurer, U.M. (ed.) EUROCRYPT 1996. LNCS, vol. 1070, pp. 143–154. Springer, Heidelberg (1996)
11. Kudla, C., Paterson, K.G.: Non-interactive Designated Verifier Proofs and Undeniable Signatures. In: Smart, N.P. (ed.) Cryptography and Coding 2005. LNCS, vol. 3796, pp. 136–154. Springer, Heidelberg (2005)
12. Kurosawa, K., Heng, S.-H.: 3-Move Undeniable Signature Scheme. In: Cramer, R. (ed.) EUROCRYPT 2005. LNCS, vol. 3494, pp. 181–197. Springer, Heidelberg (2005)
13. Liskov, M., Micali, S.: Online-Untransferable Signatures. In: Cramer, R. (ed.) PKC 2008. LNCS, vol. 4939, pp. 248–267. Springer, Heidelberg (2008)
14. Michels, M., Stadler, M.: Generic Constructions for Secure and Efficient Confirmer Signature Schemes. In: Nyberg, K. (ed.) EUROCRYPT 1998. LNCS, vol. 1403, pp. 406–421. Springer, Heidelberg (1998)
15. Monnerat, J., Vaudenay, S.: Short 2-Move Undeniable Signatures. In: Nguyên, P.Q. (ed.) VIETCRYPT 2006. LNCS, vol. 4341, pp. 19–36. Springer, Heidelberg (2006)
16. Pedersen, T.P.: Non-interactive and information-theoretic secure verifiable secret sharing. In: Feigenbaum, J. (ed.) CRYPTO 1991. LNCS, vol. 576, pp. 129–140. Springer, Heidelberg (1992)
17. Steinfeld, R., Bull, L., Wang, H., Pieprzyk, J.: Universal Designated-Verifier Signatures. In: Laih, C.-S. (ed.) ASIACRYPT 2003. LNCS, vol. 2894, pp. 523–542. Springer, Heidelberg (2003)
18. Teranishi, I., Oyama, T., Ogata, W.: General Conversion for Obtaining Strongly Existentially Unforgeable Signatures. In: Barua, R., Lange, T. (eds.) INDOCRYPT 2006. LNCS, vol. 4329, pp. 191–205. Springer, Heidelberg (2006)
19. Waters, B.: Efficient Identity-Based Encryption Without Random Oracles. In: Cramer, R. (ed.) EUROCRYPT 2005. LNCS, vol. 3494, pp. 114–127. Springer, Heidelberg (2005)

Round-Efficient Sub-linear Zero-Knowledge Arguments for Linear Algebra

Jae Hong Seo

Department of Mathematical Sciences and ISaC-RIM,
Seoul National University, Seoul, 151-747, Korea
jhsbhs0@snu.ac.kr

Abstract. The round complexity of interactive zero-knowledge arguments is an important measure along with communication and computational complexities. In the case of zero-knowledge arguments for linear algebraic relations over finite fields, Groth proposed (at CRYPTO 2009) an elegant methodology that achieves sub-linear communication overheads and low computational complexity. He obtained zero-knowledge arguments of sub-linear size for linear algebra using reductions from linear algebraic relations to equations of the form $z = \mathbf{x} *' \mathbf{y}$, where $\mathbf{x}$, $\mathbf{y} \in \mathbb{F}_p^n$ are committed vectors, $z \in \mathbb{F}_p$ is a committed element, and $*' : \mathbb{F}_p^n \times \mathbb{F}_p^n \to \mathbb{F}_p$ is a bilinear map. These reductions impose additional rounds on zero-knowledge arguments of sub-linear size. We focus on minimizing such additional rounds, and we reduce the rounds of sub-linear zero-knowledge arguments for linear algebraic relations as compared with Groth's zero-knowledge arguments for the same relations. To reduce round complexity, we propose a general transformation from a t-round zero-knowledge argument, satisfying mild conditions, to a $(t-2)$-round zero-knowledge argument; this transformation is of independent interest.

Keywords: Round-efficient zero-knowledge arguments, sub-linear zero-knowledge arguments, linear algebra.

1 Introduction

The round complexity of interactive zero-knowledge arguments is an important measure along with communication and computation complexities. In computer networks, interactions between entities consume the most time; hence, it is important to reduce the round complexity of protocols, and many researches have attempted devise round-efficient protocols [1,4,6,8,13,3,17,2,5,14,18,7].

To optimize round complexity, interactive zero-knowledge arguments can be usually transformed into non-interactive zero-knowledge arguments by using the Fiat-Shamir heuristic, i.e., a cryptographic hash function is used by the prover to compute the verifier's challenges. To prove the soundness property in such non-interactive zero-knowledge arguments, we should assume the so-called random oracle model such that the cryptographic hash functions are viewed as random

D. Catalano et al. (Eds.): PKC 2011, LNCS 6571, pp. 387–402, 2011.

oracles. Recently, non-interactive zero-knowledge arguments have been proposed without the random oracle model [10,11]; however, they require communication overheads of quasi-linear size or non-standard assumptions in bilinear groups.

Groth proposed interactive zero-knowledge arguments of sub-linear size for linear algebra [9]. They attain sub-linear communication size and require fewer assumptions than the non-interactive zero-knowledge arguments mentioned above, i.e., they require only the discrete logarithm assumption and common reference strings. Therefore, we start from [9] to obtain round-efficient zero-knowledge arguments of sub-linear size under minimal assumptions.

The techniques used in [9] yield zero-knowledge arguments of sub-linear size; however, they require several additional rounds. We believe that such additional rounds are not necessary to obtain zero-knowledge arguments of sub-linear size for statements involving linear algebra. In addition, we attempt to obtain round-efficient zero-knowledge arguments of sub-linear size for linear algebraic relations.

Our Contributions. The proposed method for obtaining round-efficient zero-knowledge arguments of sub-linear size for linear algebraic relations involves two steps.

First, we reduce arguments for linear algebra to two types of equations that use two different types of bilinear maps; one bilinear map is defined from $\mathbb{F}_p^n \times \mathrm{Mat}_{n\times n}(\mathbb{F}_p)$ to $\mathbb{F}_p^n$, and the other, from $\mathbb{F}_p^n \times \mathbb{F}_p^n$ to $\mathbb{F}_p^n$. In [9], all arguments are reduced to one type of bilinear equations. On the other hand, we do not try to reduce one type of bilinear equations to the other; however, we construct specific short-round zero-knowledge arguments for each type of bilinear equations. Thus, we can obtain shorter rounds than those of [9]. As a result we reduce the three-to-one rounds of zero-knowledge arguments, as compared with those of [9].

Second, we propose a general transformation from a t-round zero-knowledge argument A to a $(t-2)$-round zero-knowledge argument A' if A satisfies some mild conditions that are satisfied by our zero-knowledge arguments and virtually the zero-knowledge arguments in [9]. We show that the rounds of all zero-knowledge arguments in [9] can be reduced by two using the proposed transformation for the same relation. Although the proposed transformation reduces the round complexity of zero-knowledge arguments, it increases the communication and computational complexities. However, for each reduction, the proposed transformation requires, at most 3 times more communication overheads than the original zero-knowledge arguments in [9].

Outline. In the next section, we introduce the notations, useful tools, and basic definitions used in this paper. In Section 3 we construct zero-knowledge arguments for two types of bilinear equations. In Section 4, we present a general transformation from t-round zero-knowledge arguments to $(t-2)$-round arguments with conditions when zero-knowledge arguments are eligible for transformation. In Section 5, we apply the general transformation to the zero-knowledge arguments obtained in Section 3. Finally, in Section 6, we compare the results with the zero-knowledge arguments of [9].

2 Preliminaries

Notation. We use $[a, b]$ to denote a set of integers, at least a and at most b. For a set S and an element $a \in S$, $a \stackrel{\$}{\leftarrow} S$ implies that a is randomly chosen from S. For an algorithm A, $A(x) \rightarrow s$ implies that A outputs s when its input is x.

Generalized Pedersen Commitment. We use the generalization of the Pedersen commitment scheme [16]. To commit to a vector in $\mathbb{F}_p^n$, we use a generalized Pedersen commitment $\mathsf{Com}(\cdot;\cdot) : \mathbb{F}_p^n \times \mathbb{F}_p \rightarrow G$, where p is a prime, $\mathbb{F}_p$ is a finite field of characteristic p, and G is a group of order p. The generalized Pedersen commitment scheme consists of two algorithms, the key generation algorithm K and the commitment algorithm C. K takes the security parameter λ as input, and it outputs $(G, g_1, \cdots, g_n, h)$, where G is a cyclic group of order p and $g_1, \cdots, g_n, h$ are randomly chosen generators of G. C takes a vector $\mathbf{x} = (\zeta^{(1)}, \cdots, \zeta^{(n)}) \in \mathbb{F}_p^n$ and a randomizer $r \in \mathbb{F}_p$ as input, and it outputs $\mathsf{Com}(\mathbf{x}; r) := h^r \prod_{j=1}^{n} g_j^{\zeta^{(j)}}$. The generalized Pedersen commitment scheme is a perfectly hiding and computationally binding commitment scheme under the discrete logarithm assumption in G.

The usefulness of the generalized Pedersen commitment scheme is attributed to its homomorphic property: For $\forall \mathbf{x}, \mathbf{y} \in \mathbb{F}_p^n$ and $r, s \in \mathbb{F}_p$,

$$\mathsf{Com}(\mathbf{x} + \mathbf{y}; r + s) = \mathsf{Com}(\mathbf{x}; r) \cdot \mathsf{Com}(\mathbf{y}; s).$$

Schwartz-Zippel Lemma. We use the Schwartz-Zippel lemma to prove the soundness property of arguments. The schwartz-Zippel lemma enables us to carry out a useful equality test for two multi-variate polynomials. Given two multi-variate d-degree polynomials, $f_1(x_1, \cdots, x_k)$ and $f_2(x_1, \cdots, x_k)$, we can test whether $f_1(e_1, \cdots, e_k) \stackrel{?}{=} f_2(e_1, \cdots, e_k)$ for randomly chosen $e_1, \cdots, e_k$ from $\mathbb{F}_p$. If $f_1 = f_2$, then the equality will always hold for all $e_1, \cdots, e_k$; however, if $f_1 \neq f_2$, then the equality will hold with probability at most d/p.

Lemma 1. *Let $f(x_1, \cdots, x_k)$ be a non-zero multivariate polynomial of degree d over $\mathbb{F}_p$. Then,*

$$Pr[f(e_1, \cdots, e_k) = 0] \leq \frac{d}{p},$$

where the probability goes over $e_1, \cdots, e_k$ randomly chosen from $\mathbb{F}_p$.

Special Honest Verifier Zero-Knowledge Arguments. In this paper, we are interested in Special Honest Verifier Zero-Knowlege (SHVZK) arguments of knowledge in the common reference string model. SHVZK arguments feature completeness, the SHVZK property, and witness-extended emulation. In particular, all the SHVZK arguments proposed in this paper have perfect completeness and perfect SHVZK. We refer to [12] for the formal definition of SHVZK arguments.

3 SHVZK Arguments for Equations with Vectors and Matrices

In this section we consider 6 types of equations over committed matrices $X_i, Y_i, Z \in \text{Mat}_{n\times n}(\mathbb{F}_p)$, committed vectors $\mathbf{x}_i, \mathbf{y}_i, \mathbf{z} \in \mathbb{F}_p^n$, and committed elements $z \in \mathbb{F}_p$, with public $a_i \in \mathbb{F}_p$.

$$Z = \sum_{i=1}^m a_i X_i Y_i, \qquad Z = \sum_{i=1}^m a_i X_i \circ Y_i, \qquad \mathbf{z}^\top = \sum_{i=1}^m a_i X_i \mathbf{y}_i^\top,$$
$$\mathbf{z} = \sum_{i=1}^m a_i \mathbf{x}_i Y_i, \qquad \mathbf{z} = \sum_{i=1}^m a_i \mathbf{x}_i \circ \mathbf{y}_i, \qquad z = \sum_{i=1}^m a_i \mathbf{x}_i \mathbf{y}_i^\top,$$

where $\circ$ is a entry-wise product, the so-called Hadamard product.

We propose SHVZK arguments for committed vectors and matrices of elements from $\mathbb{F}_p$ satisfying the equations stated above, which are typically used in linear algebra. In particular, we focus on the three lower equations because we can consider the three upper equations as a set of n equations of the corresponding types below. More precisely, there exists an 1-round reduction from several equations of the upper form to the corresponding lower equations without a public coefficient a_i [9]. Going a further, we can consider the last two lower equations as a bilinear equation of the form

$$\mathbf{z} = \sum_{i=1}^m a_i \mathbf{x}_i * \mathbf{y}_i,$$

where $* : \mathbb{F}_p^n \times \mathbb{F}_p^n \to \mathbb{F}_p^n$ is a bilinear map. One example of $*$ is the Hadamard product of two vectors in $\mathbb{F}_p^n$, and another example of $*$ is the standard inner product of two vectors in $\mathbb{F}_p^n$, obtained by filling zeros in $\mathbb{F}_p^{n-1}$ of the range $\mathbb{F}_p^n$. That is, $\mathbf{x} * \mathbf{y} := (\langle \mathbf{x}, \mathbf{y}\rangle, 0, \cdots, 0)$, where $\langle , \rangle$ is the standard inner product. Similarly, we can consider a product of a vector and a matrix as a bilinear map defined from $\mathbb{F}_p^n \times \text{Mat}_{n\times n}(\mathbb{F}_p)$ to $\mathbb{F}_p^n$. For simplification, we assume that all public coins a_i are 1 because both the prover and the verifier can compute the commitment to $a_i\mathbf{x}_i$ from the committed $\mathbf{x}$ and public a_i.

First, we consider an equation $\mathbf{z} = \sum_{i=1}^m \mathbf{x}_i Y_i$, and next, we will consider a bilinear equation $\mathbf{z} = \sum_{i=1}^m \mathbf{x}_i * \mathbf{y}_i$. Let us consider an equation $\mathbf{z} = \sum_{i=1}^m \mathbf{x}_i Y_i$. We provide a 3-round SHVZK argument for a minimal case $\mathbf{z} = \mathbf{x}Y$; then, we provide a 2-round reduction from a general case to the minimal case.

3.1 SHVZK Arguments for $\mathbf{z} = \sum_{i=1}^m \mathbf{x}_i Y_i$

The Minimal Case. As the first step, we consider the statement $\mathbf{z} = \mathbf{x}Y$. Let $\mathbf{y}^{(j)}$ be the j-th row vector of Y, and $\mathbf{x} = (\zeta^{(1)}, \cdots, \zeta^{(n)})$.
Common Input (CI): $C_\mathbf{x} = \mathsf{Com}(\mathbf{x}; r), C_{\mathbf{y}^{(i)}} = \mathsf{Com}(\mathbf{y}^{(i)}; s^{(i)})$ for $i \in [1, n]$, $C_\mathbf{z} = \mathsf{Com}(\mathbf{z}; t)$ where $\mathbf{x}, \mathbf{y}^{(1)}, \cdots, \mathbf{y}^{(n)} \leftarrow \mathbb{F}_p^n$ and $s^{(1)}, \cdots, s^{(n)}, t \leftarrow \mathbb{F}_p$.
Prover Input (PI): $\mathbf{x}, \mathbf{y}^{(1)}, \cdots, \mathbf{y}^{(n)}, \mathbf{z}, r, s^{(1)}, \cdots, s^{(n)}$ and t where $\mathbf{z} \leftarrow \mathbb{F}_p^n$ and $t \leftarrow \mathbb{F}_p$.

Goal: Prove the knowledge of $\mathbf{x}$, Y and $\mathbf{z}$ such that $\mathbf{z} = \mathbf{x}Y$.

Prover		Verifier
Choose $\hat{\mathbf{x}} = (\hat{\zeta}^{(1)}, \cdots, \hat{\zeta}^{(n)}), \mathbf{y}^{(0)} \xleftarrow{\$} \mathbb{F}_p^n,\ \hat{r}, s^{(0)}, \hat{t} \xleftarrow{\$} \mathbb{F}_p$.		
Compute $C_{\hat{\mathbf{x}}} := \mathsf{Com}(\hat{\mathbf{x}}; \hat{r})$, $C_{\mathbf{y}^{(0)}} := \mathsf{Com}(\mathbf{y}^{(0)}; s^{(0)})$, $C_{\hat{\mathbf{z}}} := \mathsf{Com}(0; \hat{t}) \prod_{j=1}^{n} (C_{\mathbf{y}^{(j)}})^{\hat{\zeta}^{(j)}}$.		
Send	$\xrightarrow{C_{\hat{\mathbf{x}}}, C_{\mathbf{y}^{(0)}}, C_{\hat{\mathbf{z}}}}$	
		Choose $e \xleftarrow{\$} \mathbb{F}_p$.
	$\xleftarrow{e}$	Send
Compute $\tilde{\mathbf{x}} := \hat{\mathbf{x}} + e\mathbf{x}$, $\tilde{r} := \hat{r} + er$, $\tilde{\mathbf{y}} := \sum_{j=0}^{n} e^j \mathbf{y}^{(j)}$, $\tilde{s} := \sum_{j=0}^{n} e^j s^{(j)}$, $\tilde{t} := \hat{t} + e(t - \sum_{j=1}^{n} \zeta^{(j)} s^{(j)})$.		
Send	$\xrightarrow{\tilde{\mathbf{x}}, \tilde{r}, \tilde{\mathbf{y}}, \tilde{s}, \tilde{t}}$	

The verifier accepts the argument if

$$
\begin{aligned}
&(1)\ \mathsf{Com}(\tilde{\mathbf{x}}; \tilde{r}) = C_{\hat{\mathbf{x}}} C_{\mathbf{x}}^{e},\\
&(2)\ \mathsf{Com}(\tilde{\mathbf{y}}; \tilde{s}) = \prod_{j=0}^{n} (C_{\mathbf{y}^{(j)}})^{e^j},\\
&(3)\ \mathsf{Com}(0; \tilde{t}) \prod_{j=1}^{n} (C_{\mathbf{y}^{(j)}})^{\tilde{\zeta}^{(j)}} = C_{\hat{\mathbf{z}}} C_{\mathbf{z}}^{e},
\end{aligned}
$$

where $\tilde{\mathbf{x}} = (\tilde{\zeta}^{(1)}, \cdots, \tilde{\zeta}^{(n)})$.

Theorem 1. *the above argument has perfect completeness, perfect SHVZK and witness-extended emulation.*

The proof of Theorem 1 is deferred to the full version of this paper.

2-Round Reduction to Minimal Case. We consider the statement $\mathbf{z} = \sum_{i=1}^{m} \mathbf{x}_i Y_i$. We follow Groth's 2-round reduction methodology from the general case $z = \sum_{i=1}^{m} \mathbf{x}_i \mathbf{y}_i$ to the minimal case $z = \mathbf{x}\mathbf{y}$. Let us briefly explain key idea before describing the argument in detail. First, we consider the product of following elements.

$$\mathbf{x}' = \sum_{k=1}^{m} e^{k-1} \mathbf{x}_k \text{ and } \mathbf{y}'^{(j)} = \sum_{i=1}^{m} e^{m-i} \mathbf{y}_i^{(j)} \text{ for } j \in [1, n],$$

where e is a random element in $\mathbb{F}_p$ and $\mathbf{y}_i^{(j)}$ be the j-th row vector of Y_i. Let $\mathbf{x}_i = (\zeta_i^{(1)}, \cdots, \zeta_i^{(n)})$. Then, the product is as follows:

$$\mathbf{x}'Y' = \sum_{j=1}^{n} (\mathbf{y}'^{(j)} \sum_{k=1}^{m} e^{k-1} \zeta_k^{(j)}) = \sum_{\ell \in [0, 2m-2]} e^{\ell} (\sum_{j=1}^{n} \sum_{\substack{i,j: \\ \ell = m-i+k-1}} \mathbf{y}_i^{(j)} \zeta_k^{(j)})$$

In the above equation, the part corresponding to $\ell = m-1$ is exactly equal to $\sum_{i=1}^{m} \mathbf{x}_i Y_i$. The prover sends commitments to $\mathbf{z}_\ell$ which is suppose to be $\sum_{j=1}^{n} \sum_{\substack{i,j: \\ \ell=m-i+k-1}} \mathbf{y}_i^{(j)} \zeta_k^{(j)}$ (set $\mathbf{z}_{m-1}$ by $\mathbf{z}$) before receiving the challenge e. Next, both the prover and the verifier compute commitments to $\mathbf{x}'$, Y', and $\sum_{\ell \in [0,2m-2]} e^\ell \mathbf{z}_\ell$ by using Com's additive homomorphic property; then, they run the minimal case with them as input. If the verifier accepts the transcript of the argument, then it means that the following equality holds with overwhelming probability:

$$\sum_{\ell \in [0,2m-2]} e^\ell \Big(\sum_{j=1}^{n} \sum_{\substack{i,j: \\ \ell=m-i+k-1}} \mathbf{y}_i^{(j)} \zeta_k^{(j)}\Big) = \sum_{\ell \in [0,2m-2]} e^\ell \mathbf{z}_\ell.$$

Since all commitments are chosen by the prover before he see the random challenge e, by Schwartz-Zippel lemma all coefficients of e^ℓ in the left side of above equality are equal to corresponding coefficients of the right side of above equality without probability at most $\frac{2m-2}{p}$. Therefore, $\mathbf{z} = \sum_{i \in [1,m]} \mathbf{x}_i Y_i$ with overwhelming probability.

Now, we provide the complete description of 2-round reduction from $\mathbf{z} = \sum_{i=1}^{m} \mathbf{x}_i Y_i$ to the minimal case $\mathbf{z} = \mathbf{x}Y$.

Common Input: $C_{\mathbf{x}_i} = \mathsf{Com}(\mathbf{x}_i; r_i), C_{\mathbf{y}_i^{(1)}} = \mathsf{Com}(\mathbf{y}_i^{(1)}; s_i^{(1)})$ for $i \in [1,m], j \in [1,n]$, $C_{\mathbf{z}} = \mathsf{Com}(\mathbf{z}; t)$.

Prover Input: $\mathbf{x}_i, r_i, \mathbf{y}_i^{(j)}, s_i^{(j)}$ for $i \in [1,m], j \in [1,n]$, and $\mathbf{z}, t$.

Goal: Prove the knowledge of $\mathbf{x}_i$, Y_i and $\mathbf{z}$ such that $\mathbf{z} = \sum_{i=1}^{m} \mathbf{x}_i Y_i$.

	Prover		Verifier
Choose	For $\ell \in [0, m-2] \cup [m, 2m-2]$, $t_\ell \xleftarrow{\$} \mathbb{F}_p$.		
Set	$t_{m-1} := t - \sum_{i=1}^{m} \sum_{j=1}^{n} \zeta_i^{(j)} s_i^{(j)}$.		
Compute	$C_\ell := \mathsf{Com}(0; t_\ell) \prod_{\ell=m+k-i-1} \prod_{j=1}^{n} (C_{\mathbf{y}_i^{(j)}})^{\zeta_k^{(j)}}$.		
Then,	$C_{m-1} = C_{\mathbf{z}}$.		
Send		$\xrightarrow{C_0, \cdots, C_{2m-2}}$	
			Choose $e \xleftarrow{\$} \mathbb{F}_p$.
		$\xleftarrow{e}$	Send
Define	$C_{\mathbf{x}'} := \prod_{k=1}^{m} (C_{\mathbf{x}_k})^{e^{k-1}}$,		Define $C_{\mathbf{x}'}$,
	$C_{\mathbf{y}'^{(j)}} := \prod_{i=1}^{m} (C_{\mathbf{y}_i^{(j)}})^{e^{m-i}}$,		$C_{\mathbf{y}'^{(j)}}$,
	$C_{\mathbf{z}'} := \prod_{\ell=0}^{2m-2} (C_\ell)^{e^\ell}$.		$C_{\mathbf{z}'}$
Compute	an opening of $C_{\mathbf{x}'}$,		
	$\mathbf{x}' = \sum_{k=1}^{m} e^{k-1} \mathbf{x}_k$, $r' = \sum_{k=1}^{m} e^{k-1} r_k$,		
	an opening of $C_{\mathbf{y}'^{(j)}}$,		
	$\mathbf{y}'^{(j)} = \sum_{i=1}^{m} e^{m-i} \mathbf{y}_i^{(j)}$, $s_i'^{(j)} = \sum_{i=1}^{m} e^{m-i} s_i^{(j)}$,		
	and a randomizer of $C_{\mathbf{z}'}$,		
	$t' = \sum_{\ell=0}^{2m-2} e^\ell (t_\ell + \sum_{\ell=m+k-i-1} \sum_{j=1}^{n} \zeta_k^{(j)} s_i^{(j)})$.		
Run	minimal case with $C_{\mathbf{x}'}, C_{\mathbf{y}'^{(j)}}$ and $C_{\mathbf{z}'}$		

Theorem 2. *The above argument has perfect completeness, perfect SHVZK and witness-extended emulation.*

Sketch of proof. Let $\mathbf{x}'$ be an opening vector of $C_{\mathbf{x}'}$, $\mathbf{y}'^{(j)}$ be an opening vector of $C_{\mathbf{y}'^{(j)}}$, and Y' be a matrix consisting of $\mathbf{y}'^{(j)}$ as its j-th row. Then, the product of a vector $\mathbf{x}'$ and a matrix Y' is equal to

$$\mathbf{x}' \cdot Y' = \sum_{j=1}^{n}((\sum_{k=1}^{m} e^{k-1}\zeta_k^{(j)}) \cdot (\sum_{i=1}^{m} e^{m-i}\mathbf{y}_i^{(j)})) = \sum_{j=1}^{n}\sum_{\ell=0}^{2m-2} e^{\ell}(\sum_{\ell=m+k-i-1} \zeta_k^{(j)}\mathbf{y}_i^{(j)}).$$

The right side of equality is equal to an opening vector of $C_{\mathbf{z}'}$; hence, the above argument has perfect completeness by perfect completeness of the SHVZK argument for the minimal case.

Now, we show that the above argument has perfect SHVZK. We describe a simulator $\mathcal{S}$ with inputs $CI = (C_{\mathbf{x}_i}, C_{\mathbf{y}_i^{(j)}}, C_{\mathbf{z}})$ and challenges, e_1 and e_2, that outputs the simulated argument with the same probability distribution to the real argument. First, $\mathcal{S}$ chooses $C_1, \cdots, C_{2m-2}$ as random commitment to $\mathbf{0}$ except $C_{m-1} = C_{\mathbf{z}}$, and it computes $C_{\mathbf{x}'}$, $C_{\mathbf{y}'^{(j)}}$ and $C_{\mathbf{z}'}$ by using e_1 according to their definition. Second, $\mathcal{S}$ feeds $C_{\mathbf{x}'}$, $C_{\mathbf{y}'^{(j)}}$, $C_{\mathbf{z}'}$, and e_2 to the simulator for the SHVZK argument for the minimal case. The simulated argument by $\mathcal{S}$ is identical to the real argument due to perfect SHVZK property of the minimal case and perfectly hiding property of Com,.

As the last step, we show that the argument has a witness-extended emulation. We can respectively obtain opening vectors $\mathbf{x}'$, $\mathbf{y}'^{(1)}, \cdots, \mathbf{y}'^{(n)}$ and $\mathbf{z}'$ of $C_{\mathbf{x}'}, C_{\mathbf{y}'^{(1)}}, \cdots, C_{\mathbf{y}'^{(n)}}$ and $C_{\mathbf{z}'}$ on random challenge e by using a witness-extended emulation for the SHVZK argument for the minimal case, so that these opening vectors satisfy $\mathbf{z}' = \mathbf{x}' \cdot Y'$. By definition of $C_{\mathbf{x}'}, C_{\mathbf{y}'^{(1)}}, \cdots, C_{\mathbf{y}'^{(n)}}$ and $C_{\mathbf{z}'}$, we have following equalities.

$$C_{\mathbf{x}'} = \prod_{k=1}^{m}(C_{\mathbf{x}_k})^{e^{k-1}}, \quad C_{\mathbf{y}'^{(j)}} = \prod_{i=1}^{m}(C_{\mathbf{y}_i^{(j)}})^{e^{m-i}}, \quad C_{\mathbf{z}'} = \prod_{\ell=1}^{2m-2}(C_{\ell})^{e^{\ell}}.$$

By binding property of Com,

$$\mathbf{x}' = \sum_{k=1}^{m} e^{k-1}\mathbf{x}_k, \quad \mathbf{y}'^{(j)} = \sum_{i=1}^{m} e^{m-i}\mathbf{y}_i^{(j)}, \quad \mathbf{z}' = \sum_{\ell=0}^{2m-2} e^{\ell}\mathbf{z}_{\ell},$$

where $\mathbf{x}_k$, $\mathbf{y}_i^{(j)}$, and $\mathbf{z}_\ell$ are openings vectors of $C_{\mathbf{x}_k}$, $C_{\mathbf{y}_i^{(j)}}$, and C_ℓ, respectively. We consider a vector $(1, e, \cdots, e^{2m-2})$ for random challenge e. When we have $2m-1$ such vectors for random challenge e, they are linearly independent with overwhelming probability, so that we can extract all $\mathbf{z}_0, \cdots, \mathbf{z}_{2m-2}$ from $2m-1$ equations on random e. Similarly, we can extract $\mathbf{x}_k$, $\mathbf{y}_i^{(j)}$ for all $k, i \in [1, m]$, $j \in [1, n]$ since m equations of $(1, e, \cdots, e^{m-1})$ for random e are linearly independent with overwhelming probability.

Now, we show that the extracted values $\mathbf{x}_k$, $\mathbf{y}_i^{(j)}$ and $\mathbf{z}$ satisfy $\mathbf{z} = \sum_{i=1}^{m} \mathbf{x}_i Y_i$. We already know that the extracted values satisfy $\mathbf{x}' = \sum_{k=1}^{m} e^{k-1}\mathbf{x}_k$, $\mathbf{y}'^{(j)} =$

$\sum_{i=1}^{m} e^{m-i}\mathbf{y}_i^{(j)}$, and $\mathbf{z}' = \sum_{\ell=0}^{2m-2} e^{\ell}\mathbf{z}_{\ell}$. The Schwartz-Zippel Lemma tells us the e^{m-1}'s coefficient of $\mathbf{x}'Y'$ should be equal to that of $\mathbf{z}'$ with overwhelming probability when we consider both $\mathbf{x}'Y'$ and $\mathbf{z}'$ as polynomials of e. The coefficient of e^{m-1} in $\mathbf{x}'Y'$ is $\sum_{j=1}^{n}\sum_{i=1}^{m}\zeta_i^{(j)}\mathbf{y}_i^{(j)}$ and the coefficient of e^{m-1} in $\mathbf{z}'$ is $\mathbf{z}_{m-1} = \mathbf{z}$. Therefore, we conclude that

$$\mathbf{z} = \sum_{j=1}^{n}\sum_{i=1}^{m}\zeta_i^{(j)}\mathbf{y}_i^{(j)} = \sum_{i=1}^{m}\mathbf{x}_i Y_i \qquad \square$$

3.2 SHVZK Arguments for $\mathbf{z} = \sum_{i=1}^{m} \mathbf{x}_i * \mathbf{y}_i$

We consider a SHVZK argument for $\mathbf{z} = \sum_{i=1}^{m}\mathbf{x}_i * \mathbf{y}_i$, where $* : \mathbb{F}_p^n \times \mathbb{F}_p^n \to \mathbb{F}_p^n$ is a bilinear map. Groth [9] proposed a SHVZK argument for $z = \sum_{i=1}^{m}\mathbf{x}_i *' \mathbf{y}_i$, where $*' : \mathbb{F}_p^n \times \mathbb{F}_p^n \to \mathbb{F}_p$ is a bilinear map. By simple extension from Groth's SHVZK argument for $z = \sum_{i=1}^{m}\mathbf{x}_i *' \mathbf{y}_i$, we can obtain 5-round SHVZK arguments for $\mathbf{z} = \sum_{i=1}^{m}\mathbf{x}_i * \mathbf{y}_i$. In particular, if we restrict $\mathbf{z} = \sum_{i=1}^{m}\mathbf{x}_i * \mathbf{y}_i$ to the case that $z = \sum_{i=1}^{m}\mathbf{x}_i\mathbf{y}_i^{\top}$, then it is totally same to that of [9]. Our key observation of this bilinear equation is that Hadamard product is already a bilinear map. In [9], the case that $\mathbf{z} = \sum_{i\in[1,m]}\mathbf{x}_i \circ \mathbf{y}_i$ is reduced to of the form $z = \sum_{i\in[1,m]}\mathbf{x}_i *' \mathbf{y}_i$, where $*'$ is a bilinear map, so that it requires total 6-round; however, we need just 5-round for zero-knowledge arguments.[1] We leave details of arguments in the full version of this paper.

4 General Transformation for Reducing Rounds of SHVZK Arguments

In this section, we present a general transformation from a t-round SHVZK argument $A = (K, \mathsf{P}, \mathsf{V})$, satisfying some condition, to a $(t-2)$-round SHVZK argument A', where K is a common reference string generator, and P and V are polynomial time interactive algorithms called the prover and the verifier, respectively. In Definition 1 we formally define the condition that is required to reduce an argument A to A'.

First, let us define the terms used in Definition 1. We write $< \mathsf{P}(x), \mathsf{V}(y) > \to tr$ for the public transcript tr produced by P and V with inputs x and y. The transcript tr can be written as $tr = (p_0; e_1; p_1; \cdots; e_m; p_m)$, where p_i is a value sent by the prover and e_i is a value sent by the verifier. We assumed that P is an interactive algorithm; hence, we can set p_i as an output of some function $P_i(PI, CI, \rho, e_1, \cdots, e_{i-1})$, where PI is the input prover's input PI, CI is the

[1] In [9], Groth constructed a SHVZK argument for the standard inner product and the Hadamard product by building a SHVZK argument for a bilinear equation $z = \sum_{i=1}\mathbf{x}_i * \mathbf{y}_i$ where $* : \mathbb{F}_p^n \times \mathbb{F}_p^n \to \mathbb{F}_p$ is a bilinear map. That is, $*$ can be considered as the standard inner product, or it can be considered as $\mathbf{x} * \mathbf{y} = \mathbf{x}(\mathbf{y} \circ \mathbf{t})^{\top}$, where $\mathbf{t}$ is a public vector chosen by the verifier. This approach requires one additional round for transmitting $\mathbf{t}$ to prove a SHVZK argument for the Hadamard product.

common input, ρ is the randomness of the prover, $e_1, \cdots, e_{i-1}$ are the challenges sent by V, and p_0 is an output of the function $P_0(PI, CI, \rho)$. Further, we assume that all e_i are independently and randomly chosen because we only consider SHVZK arguments. At the end of the interactions between P and V, the verifier accepts or rejects the statement. We write this procedure of the verifier as an algorithm with inputs CI and tr, $\mathsf{Ver}(CI, tr) \rightarrow \{0,1\}$, where 0 and 1 correspond to V rejecting and accepting the statements, respectively.

Definition 1. *Let $A = (K, \mathsf{P}, \mathsf{V})$ be an argument for a relation R with completeness and soundness. For A, we use following notation:*

$$\begin{cases} < \mathsf{P}, \mathsf{V} > & \rightarrow tr = (p_0; e_1; p_1; \cdots; e_m; p_m) \\ \mathsf{Ver}(CI, tr) & \rightarrow \{0,1\} \end{cases}$$

If there exist functions F_i and P'_i such that $F_i(p'_i, e_i) = P_i(PI, CI, \rho, e_1, \cdots, e_i)$ for some $i \in [1, m]$, where $p'_i = P'_i(PI, CI, \rho, e_1, \cdots, e_{i-1})$, then we define the reduced argument of A, $A' = (K', \mathsf{P}', \mathsf{V}')$ of A at i with F_i and P'_i for the relation R as follows:

$$\begin{cases} K' & = K \\ < \mathsf{P}', \mathsf{V}' > & \rightarrow tr' = (p_0; \cdots; e_{i-1}; p_{i-1}, p'_i; e_i, e_{i+1}; p_{i+1}; \cdots; p_m) \\ \mathsf{Ver}'(CI, tr') = & \mathsf{Ver}(CI, (p_0; \cdots; p_{i-1}; e_i; F_i(p'_i; e_i); e_{i+1}; \cdots; p_m)) \end{cases}$$

The following lemma shows that a reduced argument also has the completeness and soundness if the original argument has. Although the reduced arguments have completeness and soundness, we need additional requirements for the reduced arguments to have the SHVZK property. We will consider the extra requirements for the SHVZK property after we state Lemma 2.

Lemma 2. *Let A be an argument for a relation R with completeness and soundness, and let A′ be a reduced argument of A at i with F_i and P'_i for the relation R. Then, A′ has completeness and soundness. Moreover, if A has a witness-extended emulation, then A′ also has the same.*

Proof. Since $\mathsf{Ver}'(CI, tr') = \mathsf{Ver}(CI, tr)$ and A has completeness, A' also has completeness.

Let us consider the soundness of the argument A'. Assume that there exists a cheating prover $\widetilde{\mathsf{P}}'$ of argument A', with success probability ϵ. We construct a simulator $\mathcal{S}$ using $\widetilde{\mathsf{P}}'$ to cheat the verifier V of the argument A, with success probability ϵ. Now, we explain the role of $\mathcal{S}$. $\mathcal{S}$ interacts with $\widetilde{\mathsf{P}}'$ on behalf of the verifier of A', and it simultaneously interacts with a verifier V on behalf of the prover of A. $\mathcal{S}$ transfers $\widetilde{\mathsf{P}}'$'s message p_k to V and V's challenge e_{k+1} to $\widetilde{\mathsf{P}}'$ in order for $k \in [0, i-2]$. Then, $\mathcal{S}$ receives p_{i-1} and p'_i from $\widetilde{\mathsf{P}}'$, sends p_{i-1} to V, receives e_i from V, computes $p_i = F(p'_i, e_i)$, and sends it to V. V sends a challenge e_{i+1} to $\mathcal{S}$; then, $\mathcal{S}$ sends e_i and e_{i+1} to $\widetilde{\mathsf{P}}'$. Next, $\mathcal{S}$ transfers all the messages between $\widetilde{\mathsf{P}}'$ and V. The operations of $\mathcal{S}$' is shown below.

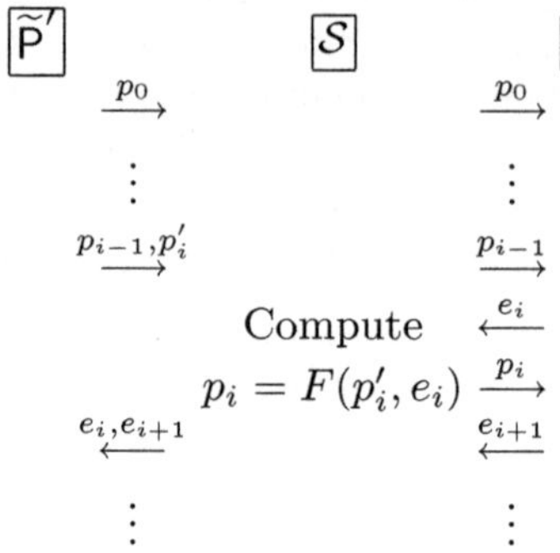

In $\widetilde{\mathsf{P}}'$'s view, simulated challenges are identical to real arguments. Therefore, the transcript tr' between $\widetilde{\mathsf{P}}'$ and $\mathcal{S}$ is accepted with probability ϵ, i.e., $\mathsf{Ver}'(CI, tr') = 1$ with probability ϵ. As aforementioned $\mathsf{Ver}'(CI, tr') = \mathsf{Ver}(CI, tr)$; hence V accepts with probability ϵ.

Now, we show that A' has a witness-extended emulation whenever A has a witness-extended emulation. We showed that we can construct a simulator $\mathcal{S}$ that performs the role of the prover of A by using the prover P' of A'. Therefore, the witness-extended emulation of A can extract a witness of A from $\mathcal{S}$. A and A' are arguments for the same relation R; hence, the lemma is complete. □

Now, we consider the SHVZK property of the reduced argument A' of A at i with F_i and P_i'. To show that A' has the SHVZK property, we should construct a simulator that can generate the prover's output $(p_0, \cdots, p_i', \cdots, p_m)$ according to the randomness ρ of the prover from given the common input CI and all challenges $e_1, \cdots, e_m$. The following lemma shows the requirements for the SHVZK property of A'. We introduce the simple notations used to concisely state Lemma 3. For fixed PI ,CI, and $e_1, \cdots, e_m$, we define two probabilities $X_{(p_0,\cdots,p_i,\cdots,p_m)}$ and $Y_{(p_0,\cdots,p_i',\cdots,p_m)}$. $X_{(p_0,\cdots,p_i,\cdots,p_m)}$ denotes the probability that the prover of the original argument A outputs $(p_0, \cdots, p_i, \cdots, p_m)$ for fixed $PI, CI, e_1, \cdots, e_m$. Similarly, $Y_{(p_0,\cdots,p_i,\cdots,p_m)}$ denotes the probability that the prover of the reduced argument A' outputs $(p_0, \cdots, p_i', \cdots, p_m)$ for fixed $PI, CI, e_1, \cdots, e_m$. The probability of $X_{(p_0,\cdots,p_i,\cdots,p_m)}$ and $Y_{(p_0,\cdots,p_i',\cdots,p_m)}$ goes over the prover's randomness ρ.

Lemma 3. *The reduced argument A' of A at i with F_i and P_i' has the SHVZK property if the original argument A has the SHVZK property, and there exists an algorithm Alg which runs as follows:*

Alg with input $CI, e_1, \cdots, e_n$, and $(p_0, \cdots, p_i, \cdots, p_m)$ outputs p_i' satisfying $p_i = F_i(p_i', e_i)$, with probability

$$\frac{Y_{(p_0,\cdots,p_i',\cdots,p_m)}}{X_{(p_0,\cdots,p_i,\cdots,p_m)}},$$

where Alg uses independent randomness.

Proof. We construct a simulator $\mathsf{Sim}^{A'}$ that generates all the output of the prover of argument A', $(p_0, \cdots, p_i', \cdots, p_m)$. Thereafter, we show that for each $(p_0, \cdots, p_i', \cdots, p_m)$, the probability that $\mathsf{Sim}^{A'}$ outputs $(p_0, \cdots, p_i', \cdots, p_m)$, $\Pr[\mathsf{Sim}^{A'} \to (p_0, \cdots, p_i', \cdots, p_m)]$, is equal to $Y_{(p_0, \cdots, p_i', \cdots, p_m)}$. Then, the proof is complete.

The original argument A has the SHVZK property; hence, there exists a simulator Sim^A that can generate all the output of the prover of A, $(p_0, \cdots, p_i, \cdots, p_m)$, from the given common input CI and challenges $e_1, \cdots, e_m$ with the probability $X_{(p_0, \cdots, p_i, \cdots, p_m)}$. Now, we construct $\mathsf{Sim}^{A'}$ by using Sim^A and Alg. First, $\mathsf{Sim}^{A'}$ runs Sim^A with inputs CI and $e_1, \cdots, e_m$, and $(p_0, \cdots, p_i, \cdots, p_m)$ is obtained. Then, it runs Alg with inputs CI, $e_1, \cdots, e_m$, and $(p_0, \cdots, p_i, \cdots, p_m)$, and it receives Alg's output p_i'. Lastly, $\mathsf{Sim}^{A'}$ outputs $(p_0, \cdots, p_i', \cdots, p_m)$.

We consider $\Pr[\mathsf{Sim}^{A'} \to (p_0, \cdots, p_i', \cdots, p_m)]$ for each $(p_0, \cdots, p_i', \cdots, p_m)$;

$$\begin{aligned}
&\Pr[\mathsf{Sim}^{A'} \to (p_0, \cdots, p_i', \cdots, p_m)] \\
&= \Pr[\mathsf{Sim}^{A} \to (p_0, \cdots, p_i, \cdots, p_m) \text{ such that } p_i = F_i(p_i', e_i), \text{ and } Alg \to p_i'] \\
&= \Pr[\mathsf{Sim}^{A} \to (p_0, \cdots, p_i, \cdots, p_m) \text{ such that } p_i = F_i(p_i', e_i)] \\
&\cdot \Pr[Alg(CI, e_1, \cdots, e_m, (p_0, \cdots, p_m)) \to p_i'] \\
&= X_{(p_0, \cdots, p_i, \cdots, p_m)} \cdot \frac{Y_{(p_0, \cdots, p_i', \cdots, p_m)}}{X_{(p_0, \cdots, p_i, \cdots, p_m)}} \\
&= Y_{(p_0, \cdots, p_i', \cdots, p_m)}.
\end{aligned}$$

The probabilities go over all the randomness used by Sim^A and Alg. The second equality holds because the two algorithms Sim^A and Alg use independent randomness. □

5 Applying Transformation to SHVZK Arguments for Linear Algebra

5.1 Application I: SHVZK Arguments for $\mathbf{z} = \sum_{i=1}^m \mathbf{x}_i Y_i$

In this section, we apply the general transformation to SHVZK arguments for $\mathbf{z} = \sum_{i=1}^m \mathbf{x}_i Y_i$ presented in Section 3.1. Before applying the general transformation, we modify SHVZK arguments for $\mathbf{z} = \sum_{i=1}^m \mathbf{x}_i Y_i$. This modification does not affect the argument itself; however, it adds redundant computations on the prover side, thereby enabling us to easily apply the general transformation. Let us briefly review 5-round SHVZK arguments for $\mathbf{z} = \sum_{i=1}^m \mathbf{x}_i Y_i$. P and V perform the 2-move reduction to the minimal case; then, P starts the minimal case SHVZK by sending $C_{\hat{\mathbf{x}}}, C_{\mathbf{y}'^{(0)}}, C_{\hat{\mathbf{z}}}$, where

$$C_{\hat{\mathbf{x}}} = \mathsf{Com}(\hat{\mathbf{x}}; \hat{r}), C_{\mathbf{y}'^{(0)}} = \mathsf{Com}(\mathbf{y}'^{(0)}; s'^{(0)}) \text{ for } \hat{\mathbf{x}}, \mathbf{y}'^{(0)} \xleftarrow{\$} \mathbb{F}_p^n, \hat{r}, s'^{(0)} \xleftarrow{\$} \mathbb{F}_p,$$

$$C_{\hat{\mathbf{z}}} = \mathsf{Com}(0; \hat{t}) \prod_{j=1}^{n} (C_{\mathbf{y}'^{(j)}})^{\hat{\zeta}^{(j)}} \text{ for } \hat{t} \xleftarrow{\$} \mathbb{F}_p, \text{ where } \hat{\mathbf{x}} = (\hat{\zeta}^{(1)}, \cdots, \hat{\zeta}^{(n)}).$$

This process is shown below.

P		V
	$\xrightarrow{C_1,\cdots,C_{2m-2}}$	
	$\xleftarrow{e_1}$	
	Define $C_{\mathbf{x}'}, C_{\mathbf{y}'^{(j)}}, C_{\mathbf{z}'}$ for $j \in [1,n]$	
	$\xrightarrow{C_{\hat{\mathbf{x}}},C_{\mathbf{y}'^{(0)}},C_{\hat{\mathbf{z}}}}$	
	$\xleftarrow{e_2}$	
	$\xrightarrow{\tilde{\mathbf{x}},\tilde{r},\tilde{\mathbf{y}},\tilde{s},\tilde{t}}$	

We modify this process by computing $C_{\hat{\mathbf{x}}}, C_{\mathbf{y}^{(0)}}$, and $C_{\hat{\mathbf{z}}}$ as follows:

Step (1). Compute $C_{\hat{\mathbf{x}}_k}$, $C_{\mathbf{y}'^{(0)}}$, and C'_ℓ as

$$\begin{aligned} C_{\hat{\mathbf{x}}_k} &= \mathsf{Com}(\hat{\mathbf{x}}_k, \hat{r}_k) \text{ for } k \in [1,m], \text{ where } \hat{\mathbf{x}}_k \xleftarrow{\$} \mathbb{F}_p^n, \hat{r}_k \xleftarrow{\$} \mathbb{F}_p, \\ C_{\mathbf{y}'^{(0)}} &= \mathsf{Com}(\mathbf{y}'^{(0)}; s'^{(0)}) \text{ where } \mathbf{y}'^{(0)} \xleftarrow{\$} \mathbb{F}_p^n, s'^{(0)} \xleftarrow{\$} \mathbb{F}_p, \\ C'_\ell &= \mathsf{Com}(0; \hat{t}_\ell)\Big(\prod_{i,k:\ell=m+k-i-1} \prod_{j=1}^{n} C_{\mathbf{y}_i^{(j)}}^{\hat{\zeta}_k^{(j)}}\Big), \text{ where } \hat{\mathbf{x}}_k = (\hat{\zeta}_k^{(1)}, \cdots, \hat{\zeta}_k^{(n)}). \end{aligned}$$

Step (2). Compute $C_{\hat{\mathbf{x}}}$ and $C_{\hat{\mathbf{z}}}$ using $C_{\hat{\mathbf{x}}_k}$, C'_ℓ, and the first challenge e_1 as follows:

$$C_{\hat{\mathbf{x}}} = \prod_{k=1}^{m} (C_{\hat{\mathbf{x}}_k})^{e_1^{k-1}}, \quad C_{\hat{\mathbf{z}}} = \prod_{\ell=0}^{2m-2} C_\ell'^{e_1^\ell}.$$

Then, $C_{\hat{\mathbf{x}}} = \prod_{k=1}^m (C_{\hat{\mathbf{x}}_k})^{e_1^{k-1}} = \mathsf{Com}(\sum_{k=1}^m e_1^{k-1}\hat{\mathbf{x}}_k; \sum_{k=1}^m e_1^{k-1}\hat{r}_k)$. $\sum_{k=1}^m e_1^{k-1}\hat{\mathbf{x}}_k$ and $\sum_{k=1}^m e_1^{k-1}\hat{r}_k$ are uniformly distributed; hence, $C_{\hat{\mathbf{x}}}$ has the same distribution as that before modification. The $C_{\hat{\mathbf{z}}}$ case is slightly complicated.

$$\begin{aligned} C_{\hat{\mathbf{z}}} &= \prod_{\ell=0}^{2m-2} C_\ell'^{e_1^\ell} \\ &= \prod_{\ell=0}^{2m-2} \mathsf{Com}(0; \hat{t}_\ell)^{e_1^\ell}\Big(\prod_{i,k:\ell=m+k-i-1} \prod_{j=1}^{n} C_{\mathbf{y}_i^{(j)}}^{\hat{\zeta}_k^{(j)}}\Big)^{e_1^\ell} \\ &= \mathsf{Com}(0; \sum_{\ell=0}^{2m-2} e_1^\ell \hat{t}_\ell) \prod_{\ell=0}^{2m-2} \prod_{i,k:\ell=m+k-i-1} \prod_{j=1}^{n} (C_{\mathbf{y}_i^{(j)}}^{\hat{\zeta}_k^{(j)}})^{e_1^\ell} \\ &= \mathsf{Com}(0; \sum_{\ell=0}^{2m-2} e_1^\ell \hat{t}_\ell) \prod_{i=1}^{m} \prod_{k=1}^{m} \prod_{j=1}^{n} C_{\mathbf{y}_i^{(j)}}^{\hat{\zeta}_k^{(j)} e_1^{m+k-i-1}} \\ &= \mathsf{Com}(0; \sum_{\ell=0}^{2m-2} e_1^\ell \hat{t}_\ell) \prod_{j=1}^{n} \Big(\prod_{i=1}^{m} C_{\mathbf{y}_i^{(j)}}^{e_1^{m-i}}\Big)^{\sum_{k=1}^m e_1^{k-1}\hat{\zeta}_k^{(j)}} \end{aligned}$$

$$= \mathsf{Com}(0; \sum_{\ell=0}^{2m-2} e_1^\ell \hat{t}_\ell) \prod_{j=1}^{n} (C_{\mathbf{y}'^{(j)}})^{\hat{\zeta}^{(j)}},$$

where $\hat{x} = (\hat{\zeta}^{(1)}, \cdots, \hat{\zeta}^{(n)})$. $\sum_{\ell=0}^{2m-2} e_1^\ell \hat{t}_\ell$ is uniformly distributed; hence, $C_{\hat{\mathbf{z}}}$'s distribution in this modification is identical to the original distribution.

Now, we are ready to apply the general transformation to this modified SHVZK argument. In the modified argument, we consider the third move as an output of the function $P_1(PI, CI, \rho, e_1)$, and we can separate $P_1(PI, CI, \rho, e_1)$ into two steps. All the computations associated with the challenge e_1 are contained in *Step (2)* and not in *Step (1)*; thus, the function $P_1(PI, CI, \rho, e_1)$ can be rewritten as $F_1(P_1'(PI, CI, \rho), e_1)$, where P_1' is a function that denotes *Step(1)* and F_1 is a function that denotes *Step(2)*. Therefore, we can apply the general transformation in Definition 1 so that the reduced argument at 1 with F_1 and P_1' for the relation $\mathbf{z} = \sum_{i=1}^m \mathbf{x}_i Y_i$ has perfect completeness and witness-extended emulation by Lemma 2.

Next, to show that the reduced argument has perfect SHVZK, we construct an algorithm *Alg* satisfying the condition in Lemma 3. First, we consider the probabilities $X_{(p_0,p_1,p_2)}$ and $Y_{(p_0,p_1',p_2)}$ that are defined in Lemma 3. Since we use the perfect hiding commitment scheme, p_0 and p_1 are independent. Similarly, p_0 and p_1' are independent. Therefore, we obtain

$$\begin{aligned}\frac{Y_{(p_0,p_1',p_2)}}{X_{(p_0,p_1,p_2)}} &= \frac{\Pr[P_0 = p_0] \cdot \Pr[P_1' = p_1'] \cdot \Pr[P_2 = p_2 | p_0, p_1']}{\Pr[P_0 = p_0] \cdot \Pr[P_1 = p_1] \cdot \Pr[P_2 = p_2 | p_0, p_1]} \\ &= \frac{\Pr[P_1' = p_1']}{\Pr[P_1 = p_1]} \\ &= \Pr[P_1' = p_1' | P_1 = p_1],\end{aligned}$$

where P_i and P_1' are functions with input PI, CI, the randomness ρ, and appropriate verifier's challenges, and the probabilities go over ρ. The second equality holds since $\Pr[P_2 = p_2 | p_0, p_1'] = \Pr[P_2 = p_2 | p_0, p_1]$, and the last equality holds since the event $P_1'(PI, CI, \rho) = p_1'$ implies $P_1(PI, CI, \rho, e_1) = F_1(p_1', e_1) = p_1$. Now, we construct an algorithm *Alg* that outputs $p_1' = (C_{\hat{\mathbf{x}}_k}, C_{\mathbf{y}'^{(0)}}, C_\ell')$ with above probability. *Alg* takes $C_{\hat{\mathbf{x}}}$, $C_{\hat{\mathbf{z}}}$ and e_1 as inputs, and it outputs uniform $C_{\hat{\mathbf{x}}_k}, C_{\mathbf{y}'^{(0)}}$ and C_ℓ' satisfying the equalities of *Step(2)*. More precisely, *Alg* randomly chooses $C_{\mathbf{y}'^{(0)}}$, $C_{\hat{\mathbf{x}}_k}$ and C_ℓ' for $k \in [2, m]$ and $\ell \in [1, 2m-2]$, and it computes $C_{\hat{\mathbf{x}}_1} = C_{\hat{\mathbf{x}}} \prod_{k=1}^m (C_{\hat{\mathbf{x}}_k})^{-e_1^{k-1}}$ and $C_0' = C_{\hat{\mathbf{z}}} \prod_{\ell=1}^{2m-2} (C_\ell')^{-e_1^\ell}$. Then, $C_{\hat{\mathbf{x}}_k}$ and C_ℓ' for $k \in [1, m]$, and $\ell \in [0, 2m-2]$ are uniformly distributed with the restrictions $C_{\hat{\mathbf{x}}} = \prod_{k=1}^m (C_{\hat{\mathbf{x}}_k})^{e_1^{k-1}}$ and $C_{\hat{\mathbf{z}}} = \prod_{\ell=0}^{2m-2} C_\ell'^{e_1^\ell}$. We can easily check that *Alg* satisfies the condition in Lemma 3, and the original argument has the SHVZK property so that, by Lemma 3, the reduced argument has perfect SHVZK.

Therefore, we obtain the following theorem. The complete description of the 3-round SHVZK for $\mathbf{z} = \sum_{i=1}^m \mathbf{x}_i Y_i$ is provided in the full version of this paper.

Theorem 3. *The reduced 3-round argument obtained by applying the general transformation to the 5-round SHVZK argument of the knowledge of committed values $\mathbf{x}_i$, Y_i, and $\mathbf{z}$ such that $\mathbf{z} = \sum_{i=1}^{m} \mathbf{x}_i Y_i$ has perfect completeness, perfect SHVZK, and witness-extended emulation.*

Complexity. The proposed argument requires 3 moves. The prover sends $5m-2$ commitments and $2n+3$ field elements to the verifier. The prover's computation is dominated by C_ℓ and C'_ℓ for $\ell \in [0, 2m-2]$, $C_{\mathbf{x}_k}$ for $k \in [1, m]$, and $\mathbf{y}'^{(j)}$. If we naively compute them, they require $2m^2n + mn$ group exponentiations and mn^2 field multiplications. However, if we sue the multi-exponentiation technique [15] for group exponentiation, they require less than $\frac{4m^2n\kappa}{\log m^2 n} + \frac{2mn\kappa}{\log n}$ multiplications in G and mn^2 field multiplications, where κ is the size of p. The verifier computes $\frac{12m\kappa}{\log m}$ group multiplications to define $C_{\mathbf{x}'}, C_{\mathbf{z}'}, C_{\hat{\mathbf{x}}}$, and $C_{\hat{\mathbf{z}}}$, $\frac{2mn\kappa}{\log m}$ group multiplications to define $C_{\mathbf{y}'^{(j)}}$ for $j \in [1, m]$, and $\frac{8n\kappa}{\log n}$ group multiplications during the verification procedure. We can use the batch verification technique for reduction from $\frac{8n\kappa}{\log n}$ to $\frac{6n\kappa}{\log n}$ group multiplications.

5.2 Application II: SHVZK Arguments for $\mathbf{z} = \sum_{i=1}^{m} \mathbf{x}_i * \mathbf{y}_i$

We can apply the general transformation to the 5-round SHVZK argument for $\mathbf{z} = \sum_{i=1}^{m} \mathbf{x}_i * \mathbf{y}_i$ to reduce the 3-round SHVZK argument for the same relation. The basic strategy is similar to that in the case of SHVZK arguments for $\mathbf{z} = \sum_{i=1}^{m} \mathbf{x}_i * Y_i$. The details are provided in the full version of this paper.

Theorem 4. *The reduced 3-round argument obtained by applying the general transformation to the 5-round SHVZK argument of the knowledge of committed values $\mathbf{x}_i$, Y_i and $\mathbf{z}$ such that $\mathbf{z} = \sum_{i=1}^{m} \mathbf{x}_i * \mathbf{y}_i$ has perfect completeness, perfect SHVZK, and witness-extended emulation.*

6 Comparison

We compare our results with the SHVZK arguments in [9]. First, we briefly explain the results of [9]. In [9], several SHVZK arguments for linear algebraic equations as well as the 6 types of equations considered in this paper are proposed. For example, an equation where the product of two committed matrices is equal to the identity matrix, an equation where one committed matrix is (known or hidden) a permutation of another committed matrix, the satisfiability of an arithmetic circuit, etc. All such SHVZK arguments for linear algebraic equations need at least one SHVZK argument for one of three types of equations mainly considered in this paper, i.e.,

$$\mathbf{z} = \sum_{i=1}^{m} \mathbf{x}_i Y_i, \qquad \mathbf{z} = \sum_{i=1}^{m} \mathbf{x}_i \circ \mathbf{y}_i, \qquad z = \sum_{i=1}^{m} \mathbf{x}_i \mathbf{y}_i^\top.$$

In particular, an SHVZK argument for $z = \sum_{i=1}^{m} \mathbf{x}_i \mathbf{y}_i^\top$ is used in all the SHVZK arguments for the linear algebraic equations in [9] since all the SHVZK arguments are reduced to the case where $z = \sum_{i=1}^{m} \mathbf{x}_i \mathbf{y}_i^\top$. In cases where $\mathbf{z} = \sum_{i=1}^{m} \mathbf{x}_i Y_i$ and $z = \sum_{i=1}^{m} \mathbf{x}_i \circ \mathbf{y}_i$, 3-round reduction and 1-round reduction to the case where $z = \sum_{i=1}^{m} \mathbf{x}_i \mathbf{y}_i^\top$ are used, respectively.

For the three types of equations stated above, we obtain 3-round SHVZK arguments that have smaller rounds than those in [9]; however, their communication overheads are similar to those in [9]. In the case where $z = \sum_{i=1}^{m} \mathbf{x}_i \mathbf{y}_i^\top$ we built a 3-round SHVZK argument with four times the communication and computational overheads of those in [9]. Therefore, we can reduce 2 rounds of each SHVZK argument for the linear algebraic equations in [9] with four times communication and computational overheads of those in [9]. Furthermore, in cases where $\mathbf{z} = \sum_{i=1}^{m} \mathbf{x}_i Y_i$ and $\mathbf{z} = \sum_{i=1}^{m} \mathbf{x}_i \circ \mathbf{y}_i$, we reduced one and three more rounds than $z = \sum_{i=1}^{m} \mathbf{x}_i \mathbf{y}_i^\top$, respectively, by trading small additional computational overheads. Exact comparisons are provided in Table 1.

SHVZK Argument	Rounds		Communication	
	Ours	[9]	Ours	[9]
$\mathbf{z} = \sum_{i=1}^{m} \mathbf{x}_i Y_i$	3	8	$5m\kappa' + 2n\kappa$	$7m\kappa' + 2n\kappa$
$\mathbf{z} = \sum_{i=1}^{m} \mathbf{x}_i \circ \mathbf{y}_i$	3	6	$8m\kappa' + 2n\kappa$	$2m\kappa' + 2n\kappa$
$z = \sum_{i=1}^{m} \mathbf{x}_i \mathbf{y}_i^\top$	3	5	$8m\kappa' + 2n\kappa$	$2m\kappa' + 2n\kappa$

Table 1. Comparisons

Prover Computation		Verifier Computation	
Ours	[9]	Ours	[9]
$mn^2 + \frac{2mn\kappa\tau}{\log n} + \frac{4m^2 n\kappa\tau}{\log mn}$	$9m^2 n + 12m\kappa\tau + \frac{4mn\kappa\tau}{\log n}$	$\frac{12m\kappa}{\log m} + \frac{2mn\kappa}{\log m} + \frac{6n\kappa}{\log n}$	$2m\kappa\tau + \frac{24m\kappa\tau}{\log m} + \frac{2mn\kappa\tau}{\log n}$
$4m^2 n + \frac{16mn\kappa\tau}{\log n}$	$2m^2 n + 4m\kappa\tau + \frac{4n\kappa\tau}{\log n}$	$\frac{20m\kappa}{\log m} + \frac{2n\kappa}{\log n}$	$\frac{8m\kappa\tau}{\log m} + \frac{2n\kappa\tau}{\log n}$
$4m^2 n + 12m\kappa\tau + \frac{4mn\kappa\tau}{\log n}$	$m^2 n + 4m\kappa\tau + \frac{4n\kappa\tau}{\log n}$	$\frac{20m\kappa\tau}{\log m} + \frac{2n\kappa\tau}{\log n}$	$\frac{8m\kappa\tau}{\log m} + \frac{2n\kappa\tau}{\log n}$

$\mathbf{x}_i$, $\mathbf{y}_i$, $\mathbf{z} \in \mathbb{F}_p^n$, $z \in \mathbb{F}_p$, $Y_i \in \mathrm{Mat}_{n\times n}(\mathbb{F}_p)$,
κ': size of group element in G, κ: size of field element in $\mathbb{F}_p$,
τ: cost of multiplication in G measured in multiplication in $\mathbb{F}_p$.

Acknowledgments. We would like to thank Jung Hee Cheon and Jens Groth for helpful discussions and suggestions. This research was supported by Basic Science Research Program through the National Research Foundation of Korea(NRF) funded by the Ministry of Education, Science and Technology (2010-0001655).

References

1. Bar-Ilan, J., Beaver, D.: Non-cryptographic fault-tolerant computing in a constant number of rounds. In: ACM PODC, pp. 201–209 (1989)
2. Beaver, D.: Minimal-latency secure function evaluation. In: Preneel, B. (ed.) EUROCRYPT 2000. LNCS, vol. 1807, pp. 335–350. Springer, Heidelberg (2000)

3. Beaver, D., Feigenbaum, J., Kilian, J., Rogaway, P.: Locally random reductions: Improvements and applications. Journal of Cryptology 10, 17–36 (1997)
4. Beaver, D., Micali, S., Rogaway, P.: The round complexity of secure protocols. In: STOC, pp. 503–513. ACM, New York (1990)
5. Cachin, C., Camenisch, J., Kilian, J., Müller, J.: One-round secure computation and secure autonomous mobile agents. In: Welzl, E., Montanari, U., Rolim, J.D.P. (eds.) ICALP 2000. LNCS, vol. 1853, p. 512. Springer, Heidelberg (2000)
6. Feige, U., Kilian, J., Naor, M.: A minimal model for secure computation. In: STOC, pp. 554–563 (1994)
7. Gennaro, R., Ishai, Y., Kushilevitz, E., Rabin, T.: The round complexity of verifiable secret sharing and secure multicast. In: STOC, pp. 580–589 (2001)
8. Goldreich, O., Kahan, A.: How to construct contant-round zero-knowledge proof systems for np. Journal of Cryptology 9, 167–190 (1996)
9. Groth, J.: Linear algebra with sub-linear zero-knowledge arguments. In: Halevi, S. (ed.) CRYPTO 2009. LNCS, vol. 5677, pp. 192–208. Springer, Heidelberg (2009)
10. Groth, J.: Short non-interactive zero-knowledge proofs. In: Abe, M. (ed.) ASIACRYPT 2010. LNCS, vol. 6477, pp. 341–358. Springer, Heidelberg (2010)
11. Groth, J.: Short pairing-based non-interactive zero-knowledge arguments. In: Abe, M. (ed.) ASIACRYPT 2010. LNCS, vol. 6477, pp. 321–340. Springer, Heidelberg (2010)
12. Groth, J.: Honest Verifier Zero-Knowledge Arguments Applied. PhD thesis, Department of Computer Science, University of Aarhus (June 2004)
13. Ishai, Y., Kushilevitz, E.: Private simultaneous messages protocols with applications. In: ISTCS, pp. 174–184 (1997)
14. Ishai, Y., Kushilevitz, E.: Randomizing polynomials: A new representation with applications to round-efficient secure computation. In: FOCS, pp. 294–304 (2000)
15. Lim, C.H.: Efficient multi-exponentiation and application to batch verification of digital signatures (2000), `http://dasan.sejong.ac.kr/~chlim/pub/multi_exp.ps`
16. Pedersen, T.: Non-interactive and information-theoretic secure verifiable secret sharing. In: Feigenbaum, J. (ed.) CRYPTO 1991. LNCS, vol. 576, pp. 129–140. Springer, Heidelberg (1992)
17. Sander, T., Young, A., Yung, M.: Non-interactive cryptocomputing for nc1. In: FOCS, pp. 554–567 (1999)
18. Tzeng, W.-G., Tzeng, Z.-J.: Round-efficient conference key agreement protocols with provable security. In: Okamoto, T. (ed.) ASIACRYPT 2000. LNCS, vol. 1976, pp. 614–627. Springer, Heidelberg (2000)

Signatures on Randomizable Ciphertexts

Olivier Blazy[1], Georg Fuchsbauer[2,⋆],
David Pointcheval[1], and Damien Vergnaud[1]

[1] École normale supérieure - CNRS - INRIA, Paris, France
[2] Dept. Computer Science, University of Bristol, UK

Abstract. Randomizable encryption allows anyone to transform a ciphertext into a fresh ciphertext of the same message. Analogously, a randomizable signature can be transformed into a new signature on the same message. We combine randomizable encryption and signatures to a new primitive as follows: given a signature on a ciphertext, anyone, knowing neither the signing key nor the encrypted message, can randomize the ciphertext and *adapt* the signature to the fresh encryption, thus maintaining public verifiability. Moreover, given the decryption key and a signature on a ciphertext, one can compute ("extract") a signature on the encrypted plaintext. As adapting a signature to a randomized encryption contradicts the standard notion of unforgeability, we introduce a weaker notion stating that no adversary can, after querying signatures on ciphertexts of its choice, output a signature on an encryption of a *new* message. This is reasonable since, due to extractability, a signature on an encrypted message can be interpreted as an encrypted signature on the message.

Using Groth-Sahai proofs and Waters signatures, we give several instantiations of our primitive and prove them secure under classical assumptions in the standard model and the CRS setting. As an application, we show how to construct an efficient non-interactive receipt-free universally verifiable e-voting scheme. In such a scheme a voter cannot prove what his vote was, which precludes vote selling. Besides, our primitive also yields an efficient round-optimal blind signature scheme based on standard assumptions, and namely for the classical Waters signature.

1 Introduction

Homomorphic cryptographic primitives have already found numerous applications. A nice side effect of homomorphic encryption is that ciphertexts can be *randomized*: given a ciphertext, anyone can—without knowing the encrypted message—produce a fresh ciphertext of the same message. E-voting schemes make use of homomorphic encryption: users encrypt their votes under such a scheme (and add proofs and signatures), so combining the ciphertexts leads to an encryption of the election result. All signed encryptions are then made public and *verifiable*, enabling the users to check that their vote was counted, and anybody to verify the correctness of the final tally. Now, if instead of directly using

⋆ Work done while at École normale supérieure, Paris, France.

D. Catalano et al. (Eds.): PKC 2011, LNCS 6571, pp. 403–422, 2011.

a user's ciphertext, the voting center first randomizes it and proves that it did so correctly, in a non-transferable way, then users are prevented from proving the content of their vote by opening it. This deters from *vote selling*, since someone buying a vote has no means to check whether the user voted as told.

However, such a (non-transferable) proof of correct randomization is costly, and the randomization breaks most of the proofs of validity of the individual ciphertexts and signatures, and thus universal verifiability. More efficient techniques are thus desirable, and this is precisely the motivation for our new primitive: it allows to adapt signatures and proofs on the content of a ciphertext when the ciphertext is randomized, that is, when the content itself is not modified. This makes proofs of correct randomization obsolete, since after receiving an encrypted vote with a validity proof and a signature from a user, the voting center can randomize the ciphertext as well as the proof and signature accordingly, which preserves universal verifiability. However, because of the randomization of the encryption of the vote, the voter is not able to open nor link this encrypted ballot to anything: no receipt can be built.

In contrast to e-voting, there are situations where encryption and signing are not performed by the same person; consider a user that encrypts a message and asks for a signature on the ciphertext. Assume now that the user can compute from this an actual signature on the message (rather than on an encryption thereof). The signature on the ciphertext could then be seen as an encrypted signature on the message, which can be decrypted by the user. This resembles a blind signature, as the signer made a signature on an unknown message; but not quite, since he may later recognize the signature (knowing the random coins he used) and thus break blindness. A possible remedy are *randomizable signatures*, which allow to transform a given signature into a new one on the same message. Such signatures, a classical example being Waters signatures [Wat05], do not satisfy *strong* unforgeability, which requires that it be impossible even to create a new signature on a signed message. As we show, this apparent weakness is actually a feature, as it can be exploited to achieve *unlinkability*: the blindness property is achieved by randomizing a signature after reception.

Fischlin [Fis06] gives a generic construction of *round-optimal* blind signatures which has been efficiently instantiated recently [Fuc09, AFG$^+$10]. To prevent the signer from linking a blind signature to the signing session, they define a blind signature as a (non-interactive) *proof of knowledge* of a signature. This makes blind signatures significantly longer than signatures of the underlying scheme, which can be avoided using randomizable signatures. In Fischlin's scheme a blind signature is a proof of knowledge of a signature on a ciphertext together with a proof that the ciphertext decrypts to the message. In the scheme in [Fuc09], the user obtains an actual signature on the message, of which he proves knowledge. We go one step further: again, the user can extract a signature on the message; but instead of making a proof of knowledge, it suffices to simply randomize it to make it unlinkable. A blind signature has therefore the same format as the underlying signatures and, in addition to being round-optimal, is thus short.

Getting back to the receipt-free voting schemes, unlinkability of ciphertexts after randomization is guaranteed by the semantic security of the encryption scheme. If at the same time of randomizing the ciphertext, the voting center adapts the proofs (validity of the ciphertext and signature by the voter), the ciphertexts remain unlinkable, under the conditions that they are valid ciphertext and signed by a given voter. As a consequence, two valid encrypted votes signed by the same voter are unlinkable: there is no way to know nor prove (even for the voter) if they contain the same vote or any specific vote, which prevents the voter from selling his vote.

Our contribution. We first introduce the notion of *signatures on randomizable ciphertexts*: given a signature on a ciphertext, anyone, knowing neither the signing key nor the encrypted message, can randomize the ciphertext and *adapt* the signature to the fresh encryption. A pair of a ciphertext and a signature on it can thus be randomized simultaneously and consistently.

Since adapting a signature on one ciphertext to a signature on another ciphertext contradicts the standard notion of unforgeability for signatures, we define a weaker notion, which still implies the security of our applications: unforgeability of signatures on randomizable ciphertexts means that the only thing an adversary can do is produce signatures on encryptions of messages of which he already knows a signature on an encryption; but he cannot make a signature on an encryption of a *new* message. Formally, no adversary can, after querying signatures on ciphertexts of its choice, output a signature on a ciphertext whose decryption is different from the decryptions of all queried ciphertexts.

We then extend our primitive to *extractable* signatures on randomizable ciphertexts: given the decryption key, from a signature on a ciphertext one can *extract* a signature on the encrypted plaintext. This enables the user in a blind-signature scheme to recover a signature on the message after the signer has signed an encryption of it.

Instantiations. We give several instantiations of extractable signatures on randomizable ciphertexts, all of which are based on weak assumptions. Our constructions use the following building blocks, from which they inherit their security: Witness-indistinguishable Groth-Sahai proofs for languages over pairing-friendly groups [GS08] and Waters signatures derived from the scheme in [Wat05] and used in [BW06]. Since verification of Waters signatures is a statement of the language for Groth-Sahai proofs, these two building blocks combine smoothly. The first instantiation of our new primitive is in symmetric pairing-friendly elliptic curves and additionally uses linear encryption [BBS04]. Both unforgeability and semantic security of this construction rely solely on the decision linear assumption (DLin). Due to space limitations, an instantiation with improved efficiency, in *asymmetric* bilinear groups, using ElGamal encryption and the SXDH variant of Groth-Sahai proofs is available in the full version [BFPV11]. This setting requires to transfer Waters' signature scheme to asymmetric groups. Whereas standard Waters signatures are secure under the computational Diffie-Hellman assumption (CDH), we prove our variant secure under a slightly stronger assumption, we term CDH^+, where some additional elements in the second group are given to the adversary. The following table details the size of a ciphertext-signature pair, where the parameter k denotes the bit length of a message:

Symmetric Pairing	$\mathbb{G}$
Waters + Linear	$9k + 33$

Asymmetric Pairing	$\mathbb{G}_1$	$\mathbb{G}_2$
Waters + ElGamal	$6k + 15$	$6k + 7$

Applications. Using our new primitive, we immediately obtain a reasonably efficient round-optimal blind-signature scheme based on standard assumptions. Moreover, exploiting the fact that our encryption is homomorphic, we construct a non-interactive receipt-free universally verifiable e-voting scheme as follows: the user encrypts his vote, proves its validity, and sends the encryption, a signature on it, and the proof to the voting center. The latter can now randomize the ciphertext, *adapt* both the proof and the user's signature, and publish them. After the results are announced, the user can verify his signature, which convinces him that the randomized ciphertext still contains his original vote due to our notion of unforgeability; however he cannot prove to anyone what his vote was.

Related work. The issue of signing messages that are only available as an encryption was already addressed by Fuchsbauer in [Fuc10]. He introduced *commuting signatures and verifiable encryption* where, given a ciphertext, a signer can produce a verifiably encrypted signature on the plaintext. These encrypted signatures can be randomized and used to construct the first delegatable anonymous credentials [BCC+09] with a non-interactive delegation protocol.

We avoid (randomizable) verifiable encryption of signatures by using signatures that are themselves randomizable. In our instantiation of round-optimal blind signatures, the blind signature is an *actual* signature rather than a verifiable encryption of it. Moreover, our construction is based on standard assumptions, whereas [Fuc10] relies on a "q-type" assumption. The efficiency of the two approaches is comparable when signing short messages, as required by our application to e-voting—since votes typically consist of only a few bits. We note however that the size of our ciphertexts is linear in the bit length of the message.

Organization. In the next section, we present the primitive and the security model. We then give two instantiations in symmetric bilinear groups based on the decision linear assumption. We first fix ideas using standard Waters signatures and then define a variant which yields a significant efficiency improvement of our instantiation, proven secure under the same assumptions. Due to space limitations, our instantiation in asymmetric groups based on ElGamal encryption and an asymmetric variant of Waters signatures is deferred to the full version [BFPV11]. In the last section, we illustrate applications of our primitive.

2 Definitions

This section presents the global framework and the security model for our new concept of signatures on ciphertexts (or commitments). We thus first briefly recall the basics of signatures and encryption. We then combine both into a single scheme.

2.1 Notations for Signature and Encryption

Definition 1 (Encryption Scheme). $\mathcal{E} = (\mathsf{Setup}, \mathsf{EKeyGen}, \mathsf{Encrypt}, \mathsf{Decrypt})$:

- $\mathsf{Setup}(1^k)$, *where* k *is the security parameter, generates the global parameters* param *of the scheme;*
- $\mathsf{EKeyGen}(\mathsf{param})$ *generates a pair of keys, the public (encryption) key* pk *and the associated private (decryption) key* dk*;*
- $\mathsf{Encrypt}(\mathsf{pk}, m; r)$ *produces a ciphertext* c *on the input message* $m \in \mathcal{M}$ *and the public key* pk*, using the random coins* $r \in \mathcal{R}_e$*;*
- $\mathsf{Decrypt}(\mathsf{dk}, c)$ *decrypts the ciphertext* c *under the private key* dk*; it outputs the plaintext, or* $\perp$ *if the ciphertext is invalid.*

Definition 2 (Signature Scheme). $\mathcal{S} = (\mathsf{Setup}, \mathsf{SKeyGen}, \mathsf{Sign}, \mathsf{Verif})$:

- $\mathsf{Setup}(1^k)$, *where* k *is the security parameter, generates the global parameters* param *of the scheme;*
- $\mathsf{SKeyGen}(\mathsf{param})$ *generates a pair of keys, the public (verification) key* vk *and the private (signing) key* sk*;*
- $\mathsf{Sign}(\mathsf{sk}, m; s)$ *produces a signature* σ *on the input message* m*, under the signing key* sk*, and using the random coins* $s \in \mathcal{R}_s$*;*
- $\mathsf{Verif}(\mathsf{vk}, m, \sigma)$ *checks whether* σ *is a valid signature on* m*, w.r.t. the public key* vk*; it outputs 1 if the signature is valid, and 0 otherwise.*

In Waters' signature scheme, the signing algorithm first transforms the message to $F = \mathcal{F}(M)$, where $\mathcal{F}$ is a hash function. Given F, the value of M is not required for signing and verification, but for the security guarantee. We could thus replace M by the pair (F, Π_M), where Π_M is a proof of knowledge of a preimage of F under the function $\mathcal{F}$ (which we assume implicitly). We define $\mathsf{Sign}(\mathsf{sk}, (F, \Pi_M); s)$ and $\mathsf{Verif}(\mathsf{vk}, (F, \Pi_M), \sigma)$ that extend the above definitions.

2.2 Signatures on Ciphertexts

We now define a scheme of signatures on ciphertexts. Note that this definition can be adapted for commitments, when one uses a perfectly binding commitment scheme, which uniquely defines the committed input.

Definition 3 (Signatures on Ciphertexts). $\mathcal{SC} = (\mathsf{Setup}, \mathsf{SKeyGen}, \mathsf{EKeyGen}, \mathsf{Encrypt}, \mathsf{Sign}, \mathsf{Decrypt}, \mathsf{Verif})$ *is defined as follows:*

- $\mathsf{Setup}(1^k)$, *where* k *is the security parameter, generates the global parameters* param_e *and* param_s *for the associated encryption and signature schemes;*
- $\mathsf{EKeyGen}(\mathsf{param}_e)$ *generates a pair of keys, the encryption key* pk *and the associated decryption key* dk*;*
- $\mathsf{SKeyGen}(\mathsf{param}_s)$ *generates a pair of keys, the verification key* vk *and the signing key* sk*;*
- $\mathsf{Encrypt}(\mathsf{pk}, \mathsf{vk}, m, r)$ *produces a ciphertext* c *on input the message* $m \in \mathcal{M}$ *and the encryption key* pk*, using the random coins* $r \in \mathcal{R}_e$*. This ciphertext is intended to be later signed under the signing key associated to the verification key* vk *(the field for* vk *can be empty if the signing algorithm is universal and does not require a ciphertext specific to the signer);*

$\mathsf{Exp}^{\mathsf{uf}}_{\mathcal{SC},\mathcal{A}}(k)$
$(\mathsf{param}_e, \mathsf{param}_s) \leftarrow \mathsf{Setup}(1^k)$; $\mathsf{SM} := \emptyset$
$\{(\mathsf{pk}_i, \mathsf{dk}_i)\} \leftarrow \mathsf{EKeyGen}(\mathsf{param}_e)$; $(\mathsf{vk}, \mathsf{sk}) \leftarrow \mathsf{SKeyGen}(\mathsf{param}_s)$
$(\mathsf{pk}_j, c, \sigma) \leftarrow \mathcal{A}^{\mathsf{Sign}(\mathsf{sk},\cdot,\cdot)}(\mathsf{param}_s, \mathsf{param}_e, \mathsf{vk}, \{(\mathsf{pk}_i, \mathsf{dk}_i)\})$;
$m \leftarrow \mathsf{Decrypt}(\mathsf{dk}_j, \mathsf{vk}, c)$
IF $m = \perp$ OR $m \in \mathsf{SM}$ OR $\mathsf{Verif}(\mathsf{vk}, \mathsf{pk}_j, c, \sigma) = 0$ RETURN 0
RETURN 1

Fig. 1. Unforgeability of signatures on ciphertexts

- $\mathsf{Sign}(\mathsf{sk}, \mathsf{pk}, c; s)$, *on input a ciphertext c and a signing key* sk, *using the random coins* $s \in \mathcal{R}_s$, *produces a signature* σ, *or* $\perp$ *if the ciphertext c is not valid (w.r.t.* pk, *and possibly* vk *associated to* sk*);*
- $\mathsf{Decrypt}(\mathsf{dk}, \mathsf{vk}, c)$ *decrypts the ciphertext c under the private key* dk. *It outputs the plaintext, or* $\perp$ *if c is invalid (w.r.t.* pk, *and possibly* vk*);*
- $\mathsf{Verif}(\mathsf{vk}, \mathsf{pk}, c, \sigma)$ *checks whether* σ *is a valid signature on c, w.r.t. the public key* vk. *It outputs 1 if the signature is valid, and 0 otherwise (possibly because of an invalid ciphertext c, with respect to* pk, *and possibly* vk*).*

Classical security notions could still be applied to this signature scheme, but we want ciphertexts and signatures to be efficiently malleable, as long as the plaintext is not affected. This will be useful for probabilistic schemes, and even more so for the randomizable scheme we will present below. In the classical definition of *existential unforgeability* (EUF) [GMR88], a new signature on an already signed message is not considered a valid forgery—as opposed to strong unforgeability (SUF). When signing ciphertexts, EUF would consider a signature on a randomized ciphertext as a valid forgery. But if the ciphertext is equivalent to an already signed ciphertext (i.e. it encrypts the same plaintext), this may not be critical in some applications; in particular if we decrypt later anyway and a decrypted message-signature pair is unforgeable. We thus define the most appropriate unforgeability (UF) notion for signatures on ciphertexts:

$\mathcal{SC}$ is *unforgeable* if, for any polynomial-time adversary $\mathcal{A}$, the advantage $\mathsf{Succ}^{\mathsf{uf}}_{\mathcal{SC},\mathcal{A}}(k) := \Pr[\mathsf{Exp}^{\mathsf{uf}}_{\mathcal{SC},\mathcal{A}}(k) = 1]$ is negligible, with $\mathsf{Exp}^{\mathsf{uf}}_{\mathcal{SC},\mathcal{A}}$ defined in Figure 1. There, $\mathsf{Sign}(\mathsf{sk}, \cdot, \cdot)$ is an oracle that takes as input a previously generated encryption key pk_i and a ciphertext c, and generates a signature σ on it (if the ciphertext is valid). It also updates the set SM of signed plaintexts with $m = \mathsf{Decrypt}(\mathsf{dk}_i, \mathsf{vk}, c)$, if the latter exists.

Unforgeability in the above sense thus states that no adversary is able to generate a new valid ciphertext-signature pair for a ciphertext that encrypts a new message, i.e. different to those encrypted in ciphertexts that were queried to the signing oracle.

2.3 Signatures on Randomizable Ciphertexts

Our primitive is based on an encryption scheme and a signature scheme. Since we want randomizability, we start by enhancing these schemes with randomization algorithms satisfying certain properties.

Definition 4 (Randomizable Encryption Scheme). *Let* (Setup, EKeyGen, Encrypt, Decrypt) *be an encryption scheme with the following additional algorithm:*

– Random(pk, $c; r'$) *produces a new ciphertext c', equivalent to the input ciphertext c, under the public key* pk, *using the additional random coins $r' \in \mathcal{R}_e$.*

An encryption scheme is called randomizable *if for any* param ← Setup(1^k), (pk, dk) ← EKeyGen(param), *message $m \in \mathcal{M}$, coins $r \in \mathcal{R}_e$, and ciphertext $c =$* Encrypt(pk, $m; r$), *the following distributions are statistically indistinguishable: $\mathcal{D}_0 = \{r' \xleftarrow{\$} \mathcal{R}_e : \mathsf{Encrypt}(\mathsf{pk}, m; r')\}$ and $\mathcal{D}_1 = \{r' \xleftarrow{\$} \mathcal{R}_e : \mathsf{Random}(\mathsf{pk}, c; r')\}$.*

Definition 5 (Randomizable Signature Scheme). *Let* (Setup, SKeyGen, Sign, Verif) *be a signature scheme, with the following additional algorithm:*

– Random(vk, $(F, \Pi_M), \sigma; s'$) *produces a new signature σ' valid under* vk *from σ on a message M given as $F = \mathcal{F}(M)$ and a proof Π_M of knowledge of M, using the additional random coins $s' \in \mathcal{R}_s$.*

A signature scheme is called randomizable *if for any* param ← Setup(1^k), (vk, sk) ← SKeyGen(param), *message $M \in \mathcal{M}$, proof of knowledge Π_M of the preimage M of $F = \mathcal{F}(M)$, random $s \in \mathcal{R}_s$, signature $\sigma =$* Sign(sk, $(F, \Pi_M); s$), *the following distributions are statistically indistinguishable: $\mathcal{D}_0 = \{s' \xleftarrow{\$} \mathcal{R}_s : \mathsf{Sign}(\mathsf{sk}, (F, \Pi_M); s')\}$ and $\mathcal{D}_1 = \{s' \xleftarrow{\$} \mathcal{R}_s : \mathsf{Random}(\mathsf{vk}, (F, \Pi_M), \sigma; s')\}$.*

The usual unforgeability notions apply (except strong unforgeability, since the signature is malleable, by definition). We now extend the randomization to signatures on randomizable ciphertexts:

Definition 6 (Randomizable Signature on Randomizable Ciphertexts). *Let* (Setup, SKeyGen, EKeyGen, Encrypt, Sign, Decrypt, Verif) *be a scheme of signatures on ciphertexts, with the following additional algorithm:*

– Random(vk, pk, $c, \sigma; r', s'$) *outputs a ciphertext c' that encrypts the same message as c under the public key* pk, *and a signature σ' on c'. Further inputs are a signature σ on c under* vk, *and random coins $r' \in \mathcal{R}_e$ and $s' \in \mathcal{R}_s$.*

A signature on ciphertexts is called randomizable *if for any global parameters* (param_e, param_s) ← Setup(1^k), *keys* (pk, dk) ← EKeyGen(param_e) *and* (vk, sk) ← SKeyGen(param_s), *$m \in \mathcal{M}$, and random coins $r \in \mathcal{R}_e$ and $s \in \mathcal{R}_s$, for $c =$* Encrypt(pk, vk, $m; r$) *and $\sigma =$* Sign(sk, pk, $c; s$) *the following distributions $\mathcal{D}_0$ are statistically indistinguishable:*

$$\mathcal{D}_0 = \{r' \xleftarrow{\$} \mathcal{R}_e; s' \xleftarrow{\$} \mathcal{R}_s : (c' = \mathsf{Encrypt}(\mathsf{pk}, \mathsf{vk}, m; r'), \sigma' = \mathsf{Sign}(\mathsf{sk}, \mathsf{pk}, c'; s'))\}$$

$$\mathcal{D}_1 = \{r' \xleftarrow{\$} \mathcal{R}_e; s' \xleftarrow{\$} \mathcal{R}_s : (c', \sigma') = \mathsf{Random}(\mathsf{vk}, \mathsf{pk}, c, \sigma; r', s')\}$$

We will denote by 1_e and 1_s the neutral elements in $\mathcal{R}_e$ and $\mathcal{R}_s$ that keep the ciphertexts and/or signatures unchanged after randomization. If $\mathcal{R}_e$ and $\mathcal{R}_s$ are groups (which will be the case for all our schemes, with addition being the group operation) and if we show that it is possible to additively update the randomness then this proves that the schemes are randomizable. The same unforgeability notion as above applies. If an additional extraction algorithm exists for the signature, we get extractable signatures on ciphertexts (defined below). Then, our above unforgeability notion for signatures on ciphertexts follows from the standard unforgeability notion on signatures.

2.4 Extractable Signatures on Randomizable Ciphertexts

For a scheme of signatures on randomizable ciphertexts (SRC) $\mathcal{SC}$, we define the following additional algorithm:

- $\mathsf{SigExt}_{\mathcal{SC}}(\mathsf{dk}, \mathsf{vk}, \sigma)$, which is given a decryption key, a verification key and a signature, outputs a signature σ'.

Let us assume that there is a signature scheme $\mathcal{S}$ where $\mathsf{Setup}_{\mathcal{S}}$ is the projection of $\mathsf{Setup}_{\mathcal{SC}}$ on the signature component and $\mathsf{SKeyGen}_{\mathcal{S}} = \mathsf{SKeyGen}$. The scheme $\mathcal{SC}$ is *extractable* if the following holds: for any $(\mathsf{param}_e, \mathsf{param}_s) \leftarrow \mathsf{Setup}_{\mathcal{SC}}(1^k)$, $(\mathsf{pk}, \mathsf{dk}) \leftarrow \mathsf{EKeyGen}(\mathsf{param}_e)$, $(\mathsf{vk}, \mathsf{sk}) \leftarrow \mathsf{SKeyGen}(\mathsf{param}_s) = \mathsf{SKeyGen}_{\mathcal{S}}(\mathsf{param}_s)$, $m \in \mathcal{M}$, random coins $r \in \mathcal{R}_e, s \in \mathcal{R}_s$, for $c = \mathsf{Encrypt}_{\mathcal{SC}}(\mathsf{pk}, \mathsf{vk}, m; r)$ and $\sigma = \mathsf{Sign}_{\mathcal{SC}}(\mathsf{sk}, \mathsf{pk}, c; s)$, the output $\sigma' = \mathsf{SigExt}_{\mathcal{SC}}(\mathsf{dk}, \mathsf{vk}, \sigma)$ is a valid signature on m under vk, that is, $\mathsf{Verif}_{\mathcal{S}}(\mathsf{vk}, m, \sigma')$ is true.

An extractable SRC scheme $\mathcal{SC}$ allows the following: a user can encrypt a message m and obtain a signature σ on the ciphertext c. From (c, σ) the owner of the decryption key can now not only recover the encrypted message m, but also a signature σ' on the *message* m, using the functionality $\mathsf{SigExt}_{\mathcal{SC}}$. The signature σ on the ciphertext c could thus be seen as an *encryption of a signature* on the message m: for extractable signatures on ciphertexts, encryption and signing can thus be seen as commutative (see Figure 2).

2.5 Strong Extractability

We can immediately apply the notion of extractable signatures on randomizable ciphertexts to build a one-round classical blind signature scheme, but we can even consider more complex scenarios, such as *three-player blind signature schemes* (see Section 5) with applications to e-cash systems.

As already sketched above, we may have an additional property: as for encryption, knowing the random coins used for encryption may suffice to decrypt. After encrypting a message m as c, one knows the random coins r used for the encryption. In all our instantiations we have that σ is the encryption of σ' with the same coins r used to encrypt the message. The user who encrypted m is thus able to extract σ', and not only the owner of the decryption key. A system $(\mathcal{SC}, \mathcal{S})$ with such a property will be called a *Strong Extractable (Randomizable) Signature on Ciphertexts* (augmented by the dotted lines in Figure 2).

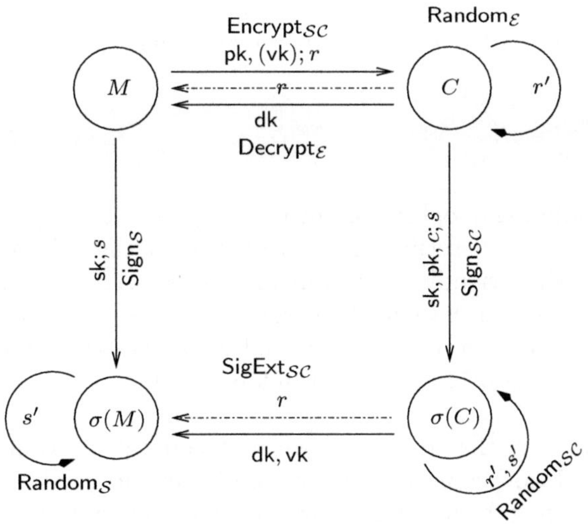

A message M can be encrypted using random coins r (Encrypt$_{\mathcal{SC}}$). The signer can sign this ciphertext (Sign$_{\mathcal{SC}}$) and anyone can randomize the pair (Random$_{\mathcal{SC}}$). A signature on the plaintext can be obtained using either dk (for SigExt$_{\mathcal{SC}}$) or the coins r (if $\sigma(C)$ has not been randomized); the result is the same as a signature of M by the signer (Sign$_{\mathcal{S}}$).

Fig. 2. (Strong) extractable signatures on randomizable ciphertexts

3 A First Instantiation

Our first construction combines linear encryption [BBS04] and Waters signatures [Wat05] as follows: given an encryption of the "Waters hash" $\mathcal{F}(M)$ of a message M (and some additional values), the signer can make an encryption of a signature on M. Decrypting the latter leads thus to a classical Waters signature on M, which will provide extractability.

Before presenting the final scheme in Section 4, we first fix ideas by combining linear encryption and standard Waters signatures. We then modify the Waters signature to significantly improve efficiency of the scheme. The constructions we give here make all use of a symmetric pairing, whereas we give an instantiation for (more efficient) asymmetric pairings in the full version [BFPV11].

3.1 Assumptions

Our constructions rely on classical assumptions: CDH for the unforgeability of signatures and DLin for the semantic security of the encryption scheme, as well as soundness of the proofs:

Definition 7 (Computational Diffie-Hellman assumption (CDH)). *Let $\mathbb{G}$ be a cyclic group of prime order p. The* CDH *assumption in $\mathbb{G}$ states that for a generator g of $\mathbb{G}$ and random $a, b \in \mathbb{Z}_p$, given (g, g^a, g^b) it is hard to compute g^{ab}.*

Definition 8 (Decision Linear assumption (DLin)). *Let $\mathbb{G}$ be a cyclic group of prime order p. The* DLin *assumption states that given $(g, g^x, g^y, g^{xa}, g^{yb}, g^c)$ for random scalars $a, b, x, y, c \in \mathbb{Z}_p$, it is hard to decide whether $c = a + b$.*

When $(g, u = g^x, v = g^y)$ is fixed, a tuple (u^a, v^b, g^{a+b}) is called a linear tuple w.r.t. (u, v, g), whereas a tuple (u^a, v^b, g^c) for a random and independent c is called a random tuple.

3.2 Basic Primitives

We briefly sketch the basic building blocks: commitments, linear encryption and the Waters signature. They are described in more detail in the full version [BFPV11]. They need a pairing-friendly environment $(p, \mathbb{G}, \mathbb{G}_T, e, g)$, where $e\colon \mathbb{G}\times\mathbb{G} \to \mathbb{G}_T$ is an admissible bilinear map, for two groups $\mathbb{G}$ and $\mathbb{G}_T$, of prime order p, generated by g and $g_t = e(g, g)$ respectively. From the basic descriptions, it follows immediately that the three primitives are randomizable.

Groth-Sahai Commitments. In the following, we will commit to values (group elements or scalars) and do proofs that they satisfy certain relations. We will use Groth-Sahai commitments that are secure under the DLin assumption: The commitment key is of the form $(\mathbf{u}_1 = (u_{1,1}, 1, g), \mathbf{u}_2 = (1, u_{2,2}, g), \mathbf{u}_3 = (u_{3,1}, u_{3,2}, u_{3,3})) \in (\mathbb{G}^3)^3$ and is set up by choosing $u_{1,1}, u_{2,2} \xleftarrow{\$} \mathbb{G}$ and $\lambda, \mu \xleftarrow{\$} \mathbb{Z}_p^*$ and setting $\mathbf{u}_3 = \mathbf{u}_1^\lambda \odot \mathbf{u}_2^\mu = (u_{3,1} = u_{1,1}^\lambda, u_{3,2} = u_{2,2}^\mu, u_{3,3} = g^{\lambda+\mu})$, which makes $\mathbf{u}_3$ a linear tuple w.r.t. $(u_{1,1}, u_{2,2}, g)$.

- To commit a group element $X \in \mathbb{G}$, choose random $s_1, s_2, s_3 \xleftarrow{\$} \mathbb{Z}_p$ and set

$$\mathcal{C}(X) := (1, 1, X) \odot \mathbf{u}_1^{s_1} \odot \mathbf{u}_2^{s_2} \odot \mathbf{u}_3^{s_3} = (u_{1,1}^{s_1} \cdot u_{3,1}^{s_3}, u_{2,2}^{s_2} \cdot u_{3,2}^{s_3}, X \cdot g^{s_1+s_2} \cdot u_{3,3}^{s_3}).$$

- To commit a scalar $x \in \mathbb{Z}_p$, choose random coins $\gamma_1, \gamma_2 \in \mathbb{Z}_p$ and set

$$\mathcal{C}'(x) := (u_{3,1}^x, u_{3,2}^x, (u_{3,3}g)^x) \odot \mathbf{u}_1^{\gamma_1} \odot \mathbf{u}_3^{\gamma_2} = (u_{3,1}^{x+\gamma_2} \cdot u_{1,1}^{\gamma_1}, u_{3,2}^{x+\gamma_2}, u_{3,3}^{x+\gamma_2} \cdot g^{x+\gamma_1}).$$

When $\mathbf{u}_3$ is a linear tuple these commitments are perfectly binding and the proofs will be perfectly sound. The committed values can even be extracted if the randomness of the commitment key is known (a scalar commitment allows extraction of x for small x only). However, if $\mathbf{u}_3$ is a random tuple (which is indistinguishable under DLin), the commitments become perfectly hiding and the proofs perfectly witness-indistinguishable.

Waters Signatures. The public parameters are a generator $h \xleftarrow{\$} \mathbb{G}$ and a vector $\boldsymbol{u} = (u_0, \dots, u_k) \xleftarrow{\$} \mathbb{G}^{k+1}$, which defines the *Waters hash* of a message $M = (M_1, \dots, M_k) \in \{0, 1\}^k$ as $\mathcal{F}(M) = u_0 \prod_{i=1}^k u_i^{M_i}$. A public key is of the form $\mathsf{vk} = Y = g^y$, with corresponding secret key $\mathsf{sk} = Z = h^y$, for a random $y \xleftarrow{\$} \mathbb{Z}_p$.

The signature on M is $\sigma = \big(\sigma_1 = Z \cdot \mathcal{F}(M)^s, \sigma_2 = g^{-s}\big)$, for some random $s \xleftarrow{\$} \mathbb{Z}_p$. It can be verified by checking $e(g, \sigma_1) \cdot e(\mathcal{F}(M), \sigma_2) = e(Y, h)$. We note that signing and verifying can be performed without knowing the message M itself; it suffices to know $F = \mathcal{F}(M)$. However, existential unforgeability [Wat05] (against chosen-message attacks under the CDH assumption) is for the pair (M, σ). As a consequence, if we work directly with $\mathcal{F}(M)$, we will need to add a proof of knowledge Π_M of M to guarantee unforgeability. Since our goal is to construct randomizable signatures and encryption, we will use Groth-Sahai proofs for a commitment C_M of M (bit-by-bit to make it extractable: $C_M = (\mathcal{C}'(M_1), \ldots, \mathcal{C}'(M_k)))$ and a proof that F is actually the evaluation of $\mathcal{F}$ on the committed M. Such a proof can be found in (the full version of) [FP09].

Linear Encryption. The secret key dk is a pair of random scalars (x_1, x_2) and the public key is $\mathsf{pk} = (X_1 = g^{x_1}, X_2 = g^{x_2})$. One encrypts a message $M \in \mathbb{G}$ as $c = \big(c_1 = X_1^{r_1}, c_2 = X_2^{r_2}, c_3 = g^{r_1+r_2} \cdot M\big)$, for random scalars $r_1, r_2 \stackrel{\$}{\leftarrow} \mathbb{Z}_p$. To decrypt, one computes $M = c_3/(c_1^{1/x_1} c_2^{1/x_2})$. As shown by Boneh, Boyen and Shacham [BBS04], this scheme is semantically secure against chosen-plaintext attacks (IND-CPA) under the DLin assumption.

3.3 Waters Signature on Linear Ciphertexts

Using Waters signatures, we will sign a linear encryption of $F = \mathcal{F}(M)$. We note that from a "ciphertext" using the decryption key, one can only extract $\mathcal{F}(M)$ (from which M can be obtained for small message spaces). As mentioned before, signatures remain unforgeable on F if in addition a proof Π_M of knowledge of M such that $F = \mathcal{F}(M)$ is given. The keys are independent Waters signature keys ($\mathsf{vk} = Y = g^y$ and $\mathsf{sk} = Z = h^y$), and linear encryption keys ($\mathsf{dk} = (x_1, x_2)$, $\mathsf{pk} = (X_1 = g^{x_1}, X_2 = g^{x_2})$). A first idea would be to define a signature on an encrypted message $c = \big(c_1 = X_1^{r_1}, c_2 = X_2^{r_2}, c_3 = g^{r_1+r_2} \cdot \mathcal{F}(M)\big)$ as $\sigma = (c_1^s, c_2^s, Z \cdot c_3^s)$. However, there are two problems:

- While the randomization of the signing coins s into $s + s'$ is easy from c, the randomization of the encryption coins r into $r + r'$ requires the knowledge of the values X_1^s, X_2^s and g^s (see Section 4.2 for how to randomize). We therefore include them in the signature.
- For the reduction of our notion of unforgeability to the security of Waters' scheme, we need to simulate the oracle returning signatures on ciphertexts having a Waters signature oracle. We can first extract M from the proof of knowledge Π_M and submit M to our oracle. From a reply $(Z \cdot \mathcal{F}(M)^s, g^{-s})$, we then have to generate $\sigma = (c_1^s, c_2^s, Z \cdot c_3^s; X_1^s, X_2^s, g^s)$ for an unknown s. We could do so if we knew the randomness (r_1, r_2) for c_1, c_2 and c_3; hence we add another proof to the extended ciphertext: Π_r proves knowledge of r_1 and r_2, used to encrypt $\mathcal{F}(M)$, which consists of bit-by-bit commitments $C_1 = (\mathcal{C}'(r_{1,1}), \ldots, \mathcal{C}'(r_{1,\ell}))$ and $C_2 = (\mathcal{C}'(r_{2,1}), \ldots, \mathcal{C}'(r_{2,\ell}))$, where ℓ is the bit-length of the order p, and proofs that each sub-commitment is indeed a bit commitment.

The global proof on the message and the randomness, which we denote by $\Pi = (\Pi_M, \Pi_r)$, can be done with randomizable commitments and proofs, using the Groth-Sahai methodology [GS08, FP09], and consists of $9k + 18\ell + 6$ group elements (where k and ℓ are the respective bit lengths of messages and of the order of $\mathbb{G}$). Such an extended ciphertext (c, Π) can then be signed, after a test of validity of the proof Π. Decryption and verification follow straight from the corresponding algorithms for Waters signatures and linear encryption. More interestingly, the above signature on randomizable ciphertexts is *extractable*: on a valid signature, if one knows the decryption key $\mathsf{dk} = (x_1, x_2)$, one can compute $\Sigma = (\Sigma_1 = \sigma_3/(\sigma_1^{1/x_1}\sigma_2^{1/x_2}), \Sigma_2 = \sigma_6^{-1})$, which is a valid signature on M:

$$\Sigma_1 = \sigma_3/(\sigma_1^{1/x_1}\sigma_2^{1/x_2}) = Z \cdot g^{s(r_1+r_2)} \cdot \mathcal{F}(M)^s/(g^{sr_1}g^{sr_2}) = Z \cdot \mathcal{F}(M)^s$$
$$\Sigma_2 = \sigma_6^{-1} = g^{-s}$$

Note that without knowing the decryption key, the same can be obtained from the coins (r_1, r_2) used for encryption: $\Sigma = (\Sigma_1 = \sigma_3/\sigma_6^{r_1+r_2}, \Sigma_2 = \sigma_6^{-1})$.

From the randomization formula of the basic schemes, we easily get the randomization property of the above Waters signature on linear ciphertexts. One shows unforgeability in the UF sense under the CDH assumption in $\mathbb{G}$: extractability provides a forgery on a new message, but only known as $F = \mathcal{F}(M)$. Since one also has to provide a valid proof Π_M that contains commitments to the bits of message M, the knowledge of the trapdoor (λ, μ) for the commitments allows to recover M too, which leads to an existential attack of the basic Waters signature scheme. The complete description and security analysis can be found in the full version [BFPV11].

4 An Efficient Instantiation

The construction in the previous section is a concrete and feasible signature on randomizable ciphertexts, which is furthermore extractable, and even in a strong way. We have thus achieved our goal, and all the applications we had in mind can benefit from it. The main drawback, from an efficiency point of view, are the bit-by-bit commitments C_M, C_1 and C_2 of M, r_1 and r_2, respectively. Whereas the message M to be signed could be short (and even a single bit for voting schemes), r_1 and r_2 are necessarily large (the bit length of the order of the group). For a k-bit long message M, Π_M (composed of C_M and a proof) consists of $9k + 2$ group elements. The random coins r_1 and r_2 being ℓ-bit long, Π_r (which includes C_1, C_2 and the proof) requires $18\ell + 4$ group elements. We now revisit the Waters signature scheme, which will allow us to remove the costly bit-by-bit commitments C_1 and C_2.

The main idea for the construction is to build a scheme which is unforgeable against a stronger kind of chosen-message attack under the same assumption: the adversary can submit "extended messages" $(M, R_1 := g^{r_1}, R_2 := g^{r_2}, T := \mathsf{vk}^{r_1+r_2})$ and the oracle replies with the tuple $(\mathsf{sk}\cdot(\mathcal{F}(M)R_1R_2)^s, g^{-s}, R_1^{-s}, R_2^{-s})$. We name this attack *chosen-extended-message attack* and note that this security notion implies the classical one, since querying $(M, 1_{\mathbb{G}}, 1_{\mathbb{G}}, \mathsf{vk})$ yields a signature on M. Intuitively, the extra parameters (R_1, R_2) will allow simulation of the signature on the ciphertext without having to know the random coins r_1 and r_2 explicitly.

4.1 Revisited Waters Signature

Our variant is defined by the four algorithms.

- $\mathsf{Setup}(1^k)$: The scheme is defined over a bilinear group $(p, \mathbb{G}, \mathbb{G}_T, e, g)$, where $e\colon \mathbb{G} \times \mathbb{G} \to \mathbb{G}_T$ is an admissible bilinear map, $\mathbb{G}$ and $\mathbb{G}_T$ are groups of prime order p, generated by g and $e(g, g)$ respectively.

  We will sign messages $M = (M_1, \ldots, M_k) \in \{0,1\}^k$. The parameters are a randomly chosen generator $h \xleftarrow{\$} \mathbb{G}$ and a vector $\boldsymbol{u} = (u_0, \ldots, u_k) \xleftarrow{\$} \mathbb{G}^{k+1}$, which defines the *Waters Hash* as $\mathcal{F}(M) = u_0 \prod_{i=1}^k u_i^{M_i}$. We set $\mathsf{param} := (p, \mathbb{G}, \mathbb{G}_T, e, g, h, \boldsymbol{u})$.
- $\mathsf{SKeyGen}(\mathsf{param})$: Choose a random scalar $y \xleftarrow{\$} \mathbb{Z}_p$, which defines the public key $\mathsf{vk} = Y = g^y$, and the secret key as $\mathsf{sk} = Z = h^y$.
- $\mathsf{Sign}(\mathsf{sk} = Z, M, R_1, R_2, T; s)$: First check the consistency of (R_1, R_2, T): if $e(R_1R_2, Y) = e(g, T)$ then this guarantees that there exists (r_1, r_2) such that $R_1 = g^{r_1}, R_2 = g^{r_2}, T = Y^{r_1+r_2}$. Choose a random $s \xleftarrow{\$} \mathbb{Z}_p$ and define the signature as $\sigma = \big(\sigma_1 = Z \cdot (\mathcal{F}(M)R_1R_2)^s, \sigma_2 = g^{-s}, \sigma_3 = R_1^{-s}, \sigma_4 = R_2^{-s}\big)$. Again, we may replace the input message M by the pair $(\mathcal{F}(M), \Pi_M)$.
- $\mathsf{Verif}(\mathsf{vk} = Y, M, R_1, R_2, T, \sigma)$: Check whether $e(g, \sigma_1) \cdot e(\mathcal{F}(M)R_1R_2, \sigma_2) = e(Y, h)$, $e(g, \sigma_3) = e(\sigma_2, R_1)$ and $e(g, \sigma_4) = e(\sigma_2, R_2)$, as well as the consistency of (R_1, R_2, T), via $e(R_1R_2, Y) = e(g, T)$.

To randomize a signature, we define $\mathsf{Random}(\mathsf{vk}, (F, \Pi_M), R_1, R_2, T, \sigma = (\sigma_1, \sigma_2, \sigma_3, \sigma_4); s')$ to output $\sigma' = (\sigma_1 \cdot (FR_1R_2)^{s'}, \sigma_2 \cdot g^{-s'}, \sigma_3 \cdot R_1^{-s'}, \sigma_4 \cdot R_2^{-s'})$, for a random $s' \xleftarrow{\$} \mathbb{Z}_p$. This simply changes the initial randomness s to $s + s' \bmod p$. Hence, if s' is uniform then the internal randomness of σ' is uniform in $\mathbb{Z}_p$.

Theorem 9. *Our variant of the Waters signature scheme is randomizable, and existentially unforgeable under chosen-extended-message attacks if the* CDH *assumption holds.*

The proof of unforgeability is similar to that for the original Waters scheme and can be found in the full version [BFPV11].

4.2 Signatures on Encrypted Messages

In our new scheme, we will sign a linear encryption of $F = \mathcal{F}(M)$ using our Revisited Waters signatures:

- $\mathsf{Setup}(1^k)$: The scheme is based on a bilinear group $(p, \mathbb{G}, \mathbb{G}_T, e, g)$, which constitutes the parameters param_e for encryption. For the signing part, we require moreover a vector $\boldsymbol{u} = (u_0, \ldots, u_k) \xleftarrow{\$} \mathbb{G}^{k+1}$, and a generator $h \xleftarrow{\$} \mathbb{G}$ and define $\mathsf{param}_s := (p, \mathbb{G}, \mathbb{G}_T, e, g, h, \boldsymbol{u})$.
- $\mathsf{EKeyGen}(\mathsf{param}_e)$: Choose two random scalars $x_1, x_2 \xleftarrow{\$} \mathbb{Z}_p$, which define the secret key $\mathsf{dk} = (x_1, x_2)$, and the public key as $\mathsf{pk} = (X_1 = g^{x_1}, X_2 = g^{x_2})$.
- $\mathsf{SKeyGen}(\mathsf{param}_s)$: Choose a random scalar $y \xleftarrow{\$} \mathbb{Z}_p$, which defines the public key as $\mathsf{vk} = Y = g^y$, and the secret key as $\mathsf{sk} = Z = h^y$.

– $\mathsf{Encrypt}(\mathsf{pk} = (X_1, X_2), \mathsf{vk} = Y, M; (r_1, r_2))$: For a message $M \in \{0,1\}^k$ and random scalars $r_1, r_2 \in \mathbb{Z}_p$, define the ciphertext as $c = (c_1 = X_1^{r_1}, c_2 = X_2^{r_2}, c_3 = g^{r_1+r_2} \cdot \mathcal{F}(M))$. To guarantee our notion of unforgeability of signatures on ciphertexts, we add proofs of knowledge of M and an image of r_1 and r_2:
  - Proof Π_r contains the commitments $C_r = (C_1 = \mathcal{C}'(r_1), C_2 = \mathcal{C}'(r_2), C_3 = \mathcal{C}(Y^{r_1+r_2}))$, from which the simulator can extract R_1, R_2 and T in the reduction (see below). Π_r moreover contains proofs of consistency: C_3 is a commitment to Y raised to the values r_1 and r_2 committed in C_1 and C_2, which in turn are the coins for the encryption: $c_1 = X_1^{\langle r_1 \rangle}$, $c_2 = X_2^{\langle r_2 \rangle}$, and $\langle Y^{r_1+r_2} \rangle = Y^{\langle r_1 \rangle + \langle r_2 \rangle}$. These equations are multi-scalar multiplication equations, from which only the last one is non-linear. We require 9 group elements for the commitments and 13 for the proofs, thus 22 group elements instead of $18\ell + 4$ in the previous construction.
  - Proof Π_M proves knowledge of M s.t. $\mathcal{F}(M)$ is encrypted in c. It consists of a bit-by-bit commitment $C_M = (\mathcal{C}'(M_1), \ldots, \mathcal{C}'(M_k))$ and proofs that each committed value is a bit ($6k$ group elements); moreover, a proof that c_3 is well-formed: $c_3 = (u_0 \prod_{i \in \{1,\ldots,k\}} u_i^{\langle M_i \rangle}) \cdot g^{\langle r_1 \rangle + \langle r_2 \rangle}$, which is a linear multi-scalar multiplication equation (2 additional group elements). Π_M is therefore composed of $9k + 2$ group elements.

  The global proof (containing the commitments) Π consists therefore of $9k + 24$ group elements (instead of $9k + 18\ell + 6$ when using the original Waters scheme), where k and ℓ are the bit lengths of the message M and elements of $\mathbb{G}$, respectively.

– $\mathsf{Sign}(\mathsf{sk} = Z, \mathsf{pk} = (X_1, X_2), (c = (c_1, c_2, c_3), \Pi); s)$: To sign a ciphertext $c = (c_1, c_2, c_3)$, first check if Π is valid, and if so, output

$$\sigma = (c_1^s,\ c_2^s,\ Z \cdot c_3^s;\ X_1^s,\ X_2^s, g^s) .$$

– $\mathsf{Decrypt}(\mathsf{dk} = (x_1, x_2), \mathsf{vk} = Y, (c = (c_1, c_2, c_3), \Pi))$: On a valid ciphertext (verifiable via Π), knowing the decryption key $\mathsf{dk} = (x_1, x_2)$, one can obtain $F = \mathcal{F}(M)$ since $F = c_3/(c_1^{1/x_1} c_2^{1/x_2})$.

– $\mathsf{Verif}(\mathsf{vk} = Y, \mathsf{pk} = (X_1, X_2), (c = (c_1, c_2, c_3), \Pi), \sigma = (\sigma_1, \sigma_2, \sigma_3; \sigma_4, \sigma_5, \sigma_6))$: In order to verify the signature, one verifies Π and checks whether the following pairing equations hold: $e(\sigma_3, g) = e(h, Y) \cdot e(c_3, \sigma_6)$ and

$$\begin{aligned} e(\sigma_1, X_1) &= e(c_1, \sigma_4) \qquad & e(\sigma_2, X_2) &= e(c_2, \sigma_5) \\ e(\sigma_1, g) &= e(c_1, \sigma_6) \qquad & e(\sigma_2, g) &= e(c_2, \sigma_6) \end{aligned}$$

– $\mathsf{Random}(\mathsf{vk} = Y, \mathsf{pk} = (X_1, X_2), (c = (c_1, c_2, c_3), \Pi), \sigma; r_1', r_2', s')$: In order to randomize the signature and the ciphertext, the algorithm outputs:

$$\begin{aligned} c' &= (c_1 \cdot X_1^{r_1'}, c_2 \cdot X_2^{r_2'}, c_3 \cdot g^{r_1'+r_2'}) \\ \sigma' &= \big(\sigma_1 \cdot c_1^{s'} \cdot \sigma_4^{r_1'} \cdot X_1^{r_1' s'}, \sigma_2 \cdot c_2^{s'} \cdot \sigma_5^{r_2'} \cdot X_2^{r_2' s'}, \sigma_3 \cdot c_3^{s'} \cdot \sigma_6^{r_1'+r_2'} \cdot g^{(r_1'+r_2')s'} \\ &\qquad \sigma_4 \cdot X_1^{s'}, \sigma_5 \cdot X_2^{s'}, \sigma_6 \cdot g^{s'}\big) \end{aligned}$$

together with a randomization Π' of Π.

– $\mathsf{SigExt}(\mathsf{dk} = (x_1, x_2), \mathsf{vk}, (c = (c_1, c_2, c_3), \Pi), \sigma)$: Return the following: $\Sigma = (\Sigma_1 = \sigma_3/(\sigma_1^{1/x_1}\sigma_2^{1/x_2}), \Sigma_2 = \sigma_6^{-1})$, which is a valid signature on M:

$$\Sigma_1 = \sigma_3/(\sigma_1^{1/x_1}\sigma_2^{1/x_2}) = Z \cdot g^{s(r_1+r_2)} \cdot \mathcal{F}(M)^s/g^{sr_1}g^{sr_2} = Z \cdot \mathcal{F}(M)^s,$$
$$\Sigma_2 = \sigma_6^{-1} = g^{-s}.$$

The same can be obtained from the coins (r_1, r_2) used for encryption.

Theorem 10. *The above scheme is randomizable and unforgeable (in the UF sense) under the* CDH *assumption in* $\mathbb{G}$.

Proof. Correctness of Random follows from inspection of the construction of c' and σ' and the fact that Groth-Sahai proofs are randomizable.

Since we have proved that our variant of Waters signatures is secure under a stronger kind of attack, we can use it for an appropriate simulation of the signing oracle. The full proof can be found in the full version [BFPV11]. As motivated when introducing the additional elements in the signature query, from a valid signing query, our simulator can extract M, but also $R_1 = g^{r_1}$, $R_2 = g^{r_2}$ and $T = Y^{r_1+r_2}$ from the commitments (but not the scalars r_1 and r_2). It adds M to the set SM and then queries $\mathsf{Sign}_{\mathcal{S}}(\mathsf{sk}, M, R_1, R_2, T)$ to the extended-message signing oracle to obtain $\sigma' = (\sigma_1' = \mathsf{sk} \cdot (\mathcal{F}(M)R_1R_2)^s, \sigma_2' = g^{-s}, \sigma_3' = R_1^{-s}, \sigma_4' = R_2^{-s})$. It then returns the following to the adversary:

$$\sigma = \begin{pmatrix} \sigma_1 = \sigma_3'^{-x_1} = X_1^{sr_1}, \sigma_2 = \sigma_4'^{-x_2} = X_2^{sr_2}, \sigma_3 = \sigma_1' = \mathsf{sk}\,\mathcal{F}(M)^s g^{s(r_1+r_2)} \\ \sigma_4 = \sigma_3'^{-x_1} = X_1^s, \quad \sigma_5 = \sigma_4'^{-x_2} = X_2^s, \quad \sigma_6 = \sigma_2'^{-1} = g^s \end{pmatrix}.$$

Finally, if the adversary wins by outputting a ciphertext and a signature, we can extract a signature on the plaintext and the plaintext from the proof of knowledge. □

5 Applications

We have introduced *extractable signatures on randomizable ciphertexts* (ESRC), a new primitive that has many applications to anonymity. A first straightforward application is to blind signatures, which yields a similar (yet more efficient) result to [MSF10, GK08]; however, this does not exploit all the power of our new tool. A more interesting application is to receipt-free voting schemes. We discuss this in the following and then show how to construct variants of blind signatures from our primitive.

5.1 Non-interactive Receipt-Free E-voting

In voting schemes, anonymity is a crucial property: nobody should be able to learn the content of my vote. This can be achieved with encryption schemes. However, this does not address the problem of *vote sellers*: a voter may sell his vote and then reveal/prove the content of his encrypted vote to the buyer. He could do so by simply revealing the randomness used when encrypting the vote, which allows to verify that a claimed message was encrypted.

A classical approach to prevent vote selling uses heavy interactive techniques based on randomizable encryption schemes and designated-verifier zero-knowledge proofs: the voter encrypts his vote v as c and additionally signs it to bar any modification by the voting center. But before doing so, the voting center randomizes c into c' (which cannot be opened by the voter anymore since he no longer knows the random coins) and then proves that c and c' contain the same plaintext. This proof must be non-transferable, otherwise the voter could open c (by revealing the random coins) and transfer the proof to the buyer, which together yields a proof of opening for c'. The used proof is thus a designated-verifier zero-knowledge proof. Finally, after receiving c' and being convinced by the proof, the voter signs c'.

Signatures on ciphertexts that can be randomized allow to avoid interactions altogether: all a voter does is encrypt his vote v as c and make a signature σ on c. The voting center can now consistently randomize both c and σ as c' and σ', so that the randomness used in c' is unknown to the signer, who is however guaranteed that the vote was not modified by the voting center because of the unforgeability notion for ESRC: nobody can generate a signature on a ciphertext that contains a different plaintext. We have thus constructed a non-interactive *receipt-free* voting scheme.

Since our ESRC candidates use not only randomizable but homomorphic encryption schemes (the encryption of the vote is actually the bit-commitments of the M_i's, which are either linear encryptions or ElGamal encryptions of g^{M_i}), classical techniques for voting schemes with homomorphic encryption and threshold decryption can be used [BFP+01]: there is no risk for the signature on the ciphertext to be converted into a signature on the plaintext if the board of authorities uses the decryption capability on the encrypted tally only.

If the vote consists of one box to be checked, the size of the ballot is only 42 group elements in the instantiation with linear encryption, and even smaller for the instantiation using ElGamal detailed in the full version [BFPV11]: 21 $\mathbb{G}_1$ elements and 13 $\mathbb{G}_2$ elements. Furthermore, if the vote consists of several (say k) boxes to be checked or not, with various constraints, the ballot size grows only slowly in k, since while the votes are committed bit by bit, the proofs can be global. Hence, the size basically corresponds to the signature on a ciphertext of a k-bit message. The extended ciphertexts already contains proofs that plaintexts are bits only, and all the proofs are randomizable.

5.2 Blind Signatures and Variants

Since the beginning of e-cash, blind signatures have been their most important tool. They provide an interactive protocol between a bank and a user, letting a user have a message signed by the bank without revealing it. Moreover, the message-signature pair obtained by the user is uncorrelated to the view of the protocol execution by the bank, which enables the user to withdraw anonymous coins. Several signature schemes have been turned into blind signature schemes.

The best-known is the first scheme by Chaum [Cha83], which is derived from RSA signatures [RSA78], and has been proven secure [BNPS01] under the one-more RSA assumption in the random-oracle model [BR93]. As defined in [PS96, PS00], for e-cash, the security requirement is the resistance to one-more forgeries: after interacting q times with the signer, an adversary should not be able to output more than q valid signed messages.

With ESRC, one can build a computationally blind signature scheme: the user encrypts the message m into the ciphertext c under his own key, and asks for a signature on c. He gets back a signature on the ciphertext c from which he can then extract σ, a valid signature on m. This signature is not yet blind, since the signer knows the coins used to compute it, and can thus link σ to the transcript. However, due to the randomizability of the signature, the user can randomize σ into σ' that is a secure blind signature:

- the blindness property relies on the semantic security of the encryption scheme (here DLin) and the randomization, which is information-theoretic;
- the one-more unforgeability relies on unforgeability of the signature scheme (here CDH), since the user cannot generate a signature for a message that has not been asked, encrypted, to the signer. Of course, we do not obtain strong one-more unforgeability (where several signatures on the same message would be counted several times), which is impossible with randomizable signatures.

This construction is similar to [GK08] but with better efficiency and much less bandwidth consumption since the latter relies on inefficient NIZK techniques [DFN06].

One-Round Fair Blind Signatures. With a *strong* extractable randomizable signature on ciphertexts, we get more than just standard blind signatures: we have *fair* blind signatures [SPC95]. Using a *strong* ESRC scheme, the user does not need to encrypt m under his own key, since the random coins suffice to extract the signature. He can thus encrypt the message m under a tracing authority's key. Using the decryption key, the authority can extract the message from c (or at least check if c encrypts a purported message) w.r.t. the signed message and thus revoke anonymity in case of abuse.

One-Round Three-Party Blind Signatures. Our primitive also allows to design a *three-party* blind signature scheme, which we define as follows: a party A makes a signer C sign a message m for B so that neither A nor C can later link the final message-signature pair (for A among all the signatures for the message m, and for C among all the valid message-signature pairs). To realize this primitive, the party A encrypts the message m under the key of B, and sends it to the signer C, who signs the ciphertext and applies the randomization algorithm to the ciphertext-signature pair (this is useful only in case A and B are distinct, as then A does not know the randomness for encryption and therefore cannot extract a signature). C sends the encrypted signature to B (possibly via A, who cannot decrypt anyway) and B also applies the randomization algorithm (so that C does not know the random coins used for signing) and then extracts the signature. With such a 2-flow scheme, B can obtain a signature, unknown to A, on a message chosen by A, unknown and even indistinguishable from any message-signature pair to C. Applied to group signatures, such a primitive allows a group manager A to add a new member B without learning his certificate provided by the authority C: A can define the rights in the message, but only B receives the certificate generated by C.

Additional Properties. Using our instantiation of ESRC, we can define an additional trapdoor: the extraction key for the commitments. It is not intended to be known by anybody (except the simulator in the security analysis), since the commitment key is in the CRS, but one could consider a scenario where it is given to a trusted authority that gets revocation capabilities.

Our construction is similar to previous efficient round-optimal blind signatures [Fuc09, AFG$^+$10] in that it uses Groth-Sahai proofs. However, we rely on standard assumptions only, and our resulting blind signature is a standard Waters signature, which is much shorter (2 group elements!) than the proof of knowledge of a signature used in all previous constructions.

Acknowledgments

This work was supported by the French ANR-07-TCOM-013-04 PACE Project, by the European Commission through the ICT Program under Contract ICT-2007-216676 ECRYPT II and by EADS.

References

[AFG$^+$10] Abe, M., Fuchsbauer, G., Groth, J., Haralambiev, K., Ohkubo, M.: Structure-preserving signatures and commitments to group elements. In: Rabin, T. (ed.) CRYPTO 2010. LNCS, vol. 6223, pp. 209–236. Springer, Heidelberg (2010)

[BBS04] Boneh, D., Boyen, X., Shacham, H.: Short group signatures. In: Franklin, M. (ed.) CRYPTO 2004. LNCS, vol. 3152, pp. 41–55. Springer, Heidelberg (2004)

[BCC$^+$09] Belenkiy, M., Camenisch, J., Chase, M., Kohlweiss, M., Lysyanskaya, A., Shacham, H.: Randomizable proofs and delegatable anonymous credentials. In: Halevi, S. (ed.) CRYPTO 2009. LNCS, vol. 5677, pp. 108–125. Springer, Heidelberg (2009)

[BFP$^+$01] Baudron, O., Fouque, P.-A., Pointcheval, D., Stern, J., Poupard, G.: Practical multi-candidate election system. In: 20th ACM Symposium Annual on Principles of Distributed Computing, pp. 274–283. ACM Press, New York (2001)

[BFPV11] Blazy, O., Fuchsbauer, G., Pointcheval, D., Vergnaud, D.: Signatures on randomizable ciphertexts. In: Gennaro, R. (ed.) Proceedings of PKC 2011. LNCS, vol. 6571. Springer, Heidelberg (2010), Full version available from the web page of the authors

[BNPS01] Bellare, M., Namprempre, C., Pointcheval, D., Semanko, M.: The Power of RSA Inversion Oracles and the Security of Chaum's RSA-Based Blind Signature Scheme. In: Syverson, P.F. (ed.) FC 2001. LNCS, vol. 2339, pp. 309–338. Springer, Heidelberg (2002)

[BR93] Bellare, M., Rogaway, P.: Random oracles are practical: A paradigm for designing efficient protocols. In: Ashby, V. (ed.) ACM CCS 1993: 1st Conference on Computer and Communications Security, pp. 62–73. ACM Press, New York (1993)

[BW06] Boyen, X., Waters, B.: Compact group signatures without random oracles. In: Vaudenay, S. (ed.) EUROCRYPT 2006. LNCS, vol. 4004, pp. 427–444. Springer, Heidelberg (2006)

[Cha83] Chaum, D.: Blind signatures for untraceable payments. In: Chaum, D., Rivest, R.L., Sherman, A.T. (eds.) Advances in Cryptology – CRYPTO 1982, pp. 199–203. Plenum Press, New York (1983)

[DFN06] Damgård, I., Fazio, N., Nicolosi, A.: Non-interactive Zero-Knowledge from Homomorphic Encryption. In: Halevi, S., Rabin, T. (eds.) TCC 2006. LNCS, vol. 3876, pp. 41–59. Springer, Heidelberg (2006)

[Fis06] Fischlin, M.: Round-optimal composable blind signatures in the common reference string model. In: Dwork, C. (ed.) CRYPTO 2006. LNCS, vol. 4117, pp. 60–77. Springer, Heidelberg (2006)

[FP09] Fuchsbauer, G., Pointcheval, D.: Proofs on Encrypted Values in Bilinear Groups and an Application to Anonymity of Signatures. In: Shacham, H., Waters, B. (eds.) Pairing 2009. LNCS, vol. 5671, pp. 132–149. Springer, Heidelberg (2009)

[Fuc09] Fuchsbauer, G.: Automorphic signatures in bilinear groups and an application to round-optimal blind signatures. Cryptology ePrint Archive, Report 2009/320 (2009), `http://eprint.iacr.org/`

[Fuc10] Fuchsbauer, G.: Commuting signatures and verifiable encryption and an application to non-interactively delegatable credentials. Cryptology ePrint Archive, Report 2010/233 (2010), `http://eprint.iacr.org/`

[GK08] Gjøsteen, K., Kråkmo, L.: Round-Optimal Blind Signatures from Waters Signatures. In: Baek, J., Bao, F., Chen, K., Lai, X. (eds.) ProvSec 2008. LNCS, vol. 5324, pp. 112–126. Springer, Heidelberg (2008)

[GMR88] Goldwasser, S., Micali, S., Rivest, R.L.: A digital signature scheme secure against adaptive chosen-message attacks. SIAM Journal on Computing 17(2), 281–308 (1988)

[GS08] Groth, J., Sahai, A.: Efficient non-interactive proof systems for bilinear groups. In: Smart, N.P. (ed.) EUROCRYPT 2008. LNCS, vol. 4965, pp. 415–432. Springer, Heidelberg (2008)

[MSF10] Meiklejohn, S., Shacham, H., Freeman, D.M.: Limitations on Transformations from Composite-Order to Prime-Order Groups: The Case of Round-Optimal Blind Signatures. In: Abe, M. (ed.) ASIACRYPT 2010. LNCS, vol. 6477, pp. 519–538. Springer, Heidelberg (2010)

[PS96] Pointcheval, D., Stern, J.: Provably secure blind signature schemes. In: Kim, K., Matsumoto, T. (eds.) ASIACRYPT 1996. LNCS, vol. 1163, pp. 252–265. Springer, Heidelberg (1996)

[PS00] Pointcheval, D., Stern, J.: Security arguments for digital signatures and blind signatures. Journal of Cryptology 13(3), 361–396 (2000)

[RSA78] Rivest, R.L., Shamir, A., Adleman, L.M.: A method for obtaining digital signature and public-key cryptosystems. Communications of the Association for Computing Machinery 21(2), 120–126 (1978)

[SPC95] Stadler, M.A., Piveteau, J.-M., Camenisch, J.: Fair Blind Signatures. In: Guillou, L.C., Quisquater, J.-J. (eds.) EUROCRYPT 1995. LNCS, vol. 921, pp. 209–219. Springer, Heidelberg (1995)

[Wat05] Waters, B.: Efficient Identity-Based Encryption Without Random Oracles. In: Cramer, R. (ed.) EUROCRYPT 2005. LNCS, vol. 3494, pp. 114–127. Springer, Heidelberg (2005)

Revocation for Delegatable Anonymous Credentials

Tolga Acar and Lan Nguyen

Microsoft Research
One Microsoft Way, Redmond, WA 98052, USA
tolga@microsoft.com, languyen@microsoft.com
http://research.microsoft.com/en-us/people/{tolga,languyen}

Abstract. This paper introduces and formalizes *homomorphic proofs* that allow 'adding' proofs and proof statements to get a new proof of the 'sum' statement. Additionally, we introduce a construction of homomorphic proofs, and show an *accumulator scheme with delegatable non-membership proofs* (ADNMP) as one of its applications with provable security. Finally, the proposed accumulator method extends the BC-CKLS scheme [1] to create a new provably secure *revocable delegatable anonymous credential* (RDAC) system. Intuitively, the new accumulator's delegatable non-membership (NM) proofs enable user A, without revealing her identity, to delegate to user B the ability to prove that A's identity is not included in a blacklist that can later be updated. The delegation is redelegatable, unlinkable, and verifiable.

1 Introduction

Proof systems play important roles in many cryptographic systems, such as signature, authentication, encryption, anonymous credential and mix-net. In a proof system between a prover and a verifier, an honest prover with a *witness* can convince a verifier about the truth of a *statement* but an adversary cannot convince a verifier of a false statement. Groth and Sahai [2] proposed a novel class of non-interactive proof systems (GS) with a number of desirable properties which are not available in previous ones. They are efficient and general. They do not require the random oracle assumption [3]. They can be randomized, i.e. one can generate a new proof from an existing proof of the same statement without knowing the witness. In this paper, we unveil another valuable feature of GS proofs: homomorphism.

Proof systems are used to construct *accumulators* [4,5,6,7,8]. An accumulator allows aggregation of a large set of elements into one constant-size *accumulator value*. The 'membership' proof system proves that an element is accumulated. An accumulator is *universal* if it has 'non-membership' proof system to prove that a given element is not accumulated in the accumulator value [9,10]. An accumulator is *dynamic* if the costs of adding and deleting elements and updating the accumulator value and proof systems' witnesses do not depend on the number of elements aggregated. Applications of accumulators include space-efficient time

D. Catalano et al. (Eds.): PKC 2011, LNCS 6571, pp. 423–440, 2011.

stamping, ad-hoc anonymous authentication, ring signatures, ID-Based systems, and membership *revocation* for identity escrow, group signatures and *anonymous credentials* [6].

In *anonymous credential* systems, a user can prove some credentials without revealing any other private information such as her identity. There have been several proposals [11,12,1]; applications such as in direct anonymous attestation (DAA) [13] and anonymous electronic identity (eID) token [14,15]; and implementations such as U-prove [15], Idemix [14] and java cards [16]. An anonymous credential system is *delegatable* [1] if its credential can be delegated from one user to another user so that a user can anonymously prove a credential which is delegated some levels away from the original issuer. Delegation is important for efficient credential management in organizations, as a person typically delegates certain authorities to colleagues to execute tasks on her behalf. *Revocation* is indispensable in credential systems in practice, as dispute, compromise, abuse, mistake, identity change, hacking and insecurity can make any credential become invalid before its expiration. The anonymity and delegation properties make revocation more challenging: the user must prove anonymously that her whole credential chain is not revoked. The primary revocation methods are based on accumulators [17,10], offering a constant cost for an unrevoked proof. However, the current schemes do not work for delegated anonymous credentials.

Contributions. We present three contributions in this paper, incrementally building on each other: *(i)* formal definition of homomorphic proofs and a construction based on GS proofs, *(ii)* dynamic universal accumulators with delegatable non-membership proof (ADNMP), and *(iii)* a revocable delegatable anonymous credential system (RDAC).

We first introduce and formally define the new notion of *homomorphic proofs*, which means there is an operation that 'adds' proofs, their statements and witnesses to produce a new proof of the 'sum' statement and the 'sum' witness. We present and prove a construction for homomorphic proofs from GS proofs [2]. The general nature of GS proofs partly explains the reason behind its numerous applications: group signatures, ring signatures, mix-nets, anonymous credentials, and oblivious transfers. Our homomorphic construction uses the most general form of GS proofs to maximize the range of possible applications.

Homomorphic proofs can be applied to homomorphic signatures [18], homomorphic authentication [19], that found applications in provable cloud storage [19], network coding [20,21], digital photography [22] and undeniable signatures [23]. Another possible application area is homomorphic encryption and commitment schemes that are used in mix-nets [24], voting [25], anonymous credentials [1] and other multi-party computation systems. Gentry's recent results on fully homomorphic encryption [26] allow computing any generic function of encrypted data without decryption and can be applied to cloud computing and searchable encryption.

Section 3.3 compares this work to the DHLW homomorphic NIZK (Non Interactive Zero Knowledge) recently proposed in [27]. While the DHLW scheme takes the traditional homomorphism approach, we employ Abelian groups and introduce a more general definition where proof systems satisfying the DHLW definition are a subset of the new proof systems. We note that DHLW's homomorphic NIZK definition and construction do not cover the new homomorphic proofs to build ADNMP and RDAC. From an application point of view, DHLW homomorphic NIZK targets leakage-resilient cryptography, and the new homomorphic proofs target accumulators and revocation.

Secondly, we introduce and build an *accumulator with delegatable non-membership proof* (ADNMP) scheme based on homomorphic proofs. We define security requirements for ADNMP, and give security proofs for the ADNMP scheme. The constructions in the SXDH (Symmetric External Diffie Hellman) or SDLIN (Symmetric Decisional Linear) instantiations of GS proofs allow the use of the most efficient curves for pairings in the new accumulator scheme [28].

To our knowledge, this is the first accumulator with a *delegatable* non-membership proof. Previously, there were only two accumulators with non-membership proofs, i.e. universal accumulators LLX [9] and ATSM [10]; both are not delegatable. Delegability allows us to construct delegatable revocation for delegatable anonymous credentials. Our accumulator uses GS proofs without random oracles where LLX and ATSM rely on the random oracle assumption for non-interactive proofs. LLX is based on the Strong RSA assumption and defined in composite-order groups, and ATSM is based on the Strong DH assumption and defined in prime-order bilinear pairing groups. Our scheme is also built in prime-order bilinear pairing groups that require storage much smaller than RSA composite-order groups. The new non-membership prover requires no pairing compared to ATSM's four pairings.

The main challenge in blacklisting delegatable anonymous credentials that can further be delegated is to create accumulators satisfying the following requirements. First, user A, without revealing private information, can delegate the ability to prove that her identity is not accumulated in any blacklist to user B so that such proofs generated by A and B are indistinguishable and the blacklist may change anytime. Second, the delegation must be unlinkable, i.e. it must be hard to tell if two such delegations come from the same delegator A. Third, user B is able to redelegate the ability to prove that A's credential is not blacklisted to user C, such that the information C obtains from the redelegation is indistinguishable from the information one obtains from A's delegation. Finally, any delegation information must be verifiable for correctness. The new ADNMP scheme satisfies these requirements.

By employing the ADNMP approach, our final contribution is to create the first *delegatable anonymous credential system with delegatable revocation* capability; an RDAC system. Traditionally, blacklisting of anonymous credentials relies on accumulators [8]. The identities of revoked credentials are accumulated in a blacklist, and verification of the accumulator's NM proof determines the credential's revocation status. A natural rule in a revoked delegatable credential, that our scheme also follows, is to consider all delegated descendants of the credential revoked. Applying that rule to delegatable anonymous credentials, a user must anonymously prove that all ancestor credentials are not revoked, even when the blacklist changes.

Homomorphic proofs bring delegability of proofs to another level. A proof's *statement* often consists of *commitments* of variables (witnesses) and *conditions.* Randomizable and malleable proofs introduced in [1] allows generation of a new proof and randomization of the statement's commitments without knowing the witness, but the statement's conditions always stay the same. Homomorphic proofs allow generating a new proof for a new statement containing new conditions, without any witness. A user can delegate her proving capability to another user by revealing some homomorphic proofs. A linear combination of these proofs and their statements allows the delegatee to generate new proofs for other statements with different conditions (e.g., an updated blacklist in ADNMP). In short, the BCCKLS paper [1] deals with delegating proofs of the same statements' conditions, whereas this paper deals with delegating proofs of changing statements' conditions.

2 Background

Tech Report [29] provides more details of existing cryptographic primitives: Bilinear Map Modules, $\mathcal{R}$-module, Bilinear pairings, SXDH, Composable zero-knowledge (ZK), Randomizing proofs and commitments, Partial extractability, Accumulator, and Delegatable anonymous credentials.

NOTATION. PPT stands for Probabilistic Polynomial Time; CRS for Common Reference String; Pr for Probability; NM for non-membership; ADNMP for Accumulator with Delegatable NM Proofs; RDAC for Revocable Delegatable Anonymous Credential; and $\leftarrow$ for random output. For a group $\mathbb{G}$ with identity $\mathcal{O}$, let $\mathbb{G}^* := \mathbb{G}\backslash\{\mathcal{O}\}$. $Mat_{m\times n}(\mathcal{R})$ is the set of matrices with size $m \times n$ in $\mathcal{R}$. For a matrix Γ, $\Gamma[i,j]$ is the value in i^{th} row and j^{th} column. A vector $\boldsymbol{z}$ of l elements can be viewed as a matrix of l rows and 1 column. For a vector $\boldsymbol{z}$, $z[i]$ is the ith element. For a function $\nu : \mathbb{Z} \rightarrow \mathbb{R}$, ν is *negligible* if $|\nu(k)| < k^{-\alpha}$, $\forall\alpha > 0$, $\forall k > k_0$, $\exists k_0 \in \mathbb{Z}^+$, $k \in \mathbb{Z}$.

PROOF SYSTEM. Let $\mathbf{R}$ be an efficiently computable relation of (*Para*, *Sta*, *Wit*) with setup parameters *Para*, a statement *Sta*, and a witness *Wit*. A non-interactive proof system for $\mathbf{R}$ consists of 3 PPT algorithms: a Setup, a prover Prove, and a verifier Verify. A non-interactive proof system (Setup, Prove, Verify) must be complete and sound. Completeness means that for every PPT adversary $\mathcal{A}$, $|\Pr[Para \leftarrow \mathsf{Setup}(1^k);\ (Sta,\ Wit) \leftarrow \mathcal{A}(Para);\ Proof \leftarrow \mathsf{Prove}(Para, Sta,\ Wit) : \mathsf{Verify}(Para, Sta, Proof) = 1 \text{ if } (Para, Sta,\ Wit) \in \mathbf{R}] - 1|$ is negligible. Soundness means that for every PPT adversary $\mathcal{A}$, $|\Pr[Para \leftarrow \mathsf{Setup}(1^k);\ (Sta, Proof) \leftarrow \mathcal{A}(Para) : \mathsf{Verify}(Para, Sta, Proof) = 0 \text{ if } (Para, Sta,\ Wit) \notin \mathbf{R}, \forall\, Wit] - 1|$ is negligible.

GS PROOFS. Tech Report [29] provides a comprehensive summary of GS proofs and its instantiation in SXDH. Briefly, the GS *setup* algorithm generates Gk and CRS σ. Gk contains L tuples, each of which has the form (A_1, A_2, A_T, f) where A_1, A_2, A_T are $\mathcal{R}$-modules with map $f : A_1 \times A_2 \rightarrow A_T$. L is also the number of equations in a statement to be proved. CRS σ contains L corresponding tuples of $\mathcal{R}$-modules and maps $(B_1, B_2, B_T, \iota_1, \iota_2, \iota_T)$, where ι_j :

$A_j \to B_j$. A GS *statement* is a set of L corresponding tuples ($\boldsymbol{a} \in A_1^n, \boldsymbol{b} \in A_2^m, \Gamma \in Mat_{m\times n}(\mathcal{R}), t \in A_T$) satisfying $\boldsymbol{a} \cdot \boldsymbol{y} + \boldsymbol{x} \cdot \boldsymbol{b} + \boldsymbol{x} \cdot \Gamma \boldsymbol{y} = t$; where ($\boldsymbol{x} \in A_1^m, \boldsymbol{y} \in A_2^n$) is the corresponding witness (there are L witness tuples), and denote $\boldsymbol{a} \cdot \boldsymbol{y} = \sum_{j=1}^{n} f(a[j], y[j])$. The *proof* of the statement includes L corresponding tuples, each of which consists of commitments $\boldsymbol{c} \in B_1^m$ of $\boldsymbol{x}$ and $\boldsymbol{d} \in B_2^n$ of $\boldsymbol{y}$ with values $\boldsymbol{\pi}$ and $\boldsymbol{\psi}$. In the SXDH instantiation of GS proofs, *Para* includes bilinear pairing setup $Gk = (p, \mathbb{G}_1, \mathbb{G}_2, \mathbb{G}_T, e, P_1, P_2)$ and CRS $\sigma = (B_1, B_2, B_T, F, \iota_1, p_1, \iota_2, p_2, \iota_1', p_1', \iota_2', p_2', \iota_T, p_T, \boldsymbol{u}, \boldsymbol{v})$ where $B_1 = \mathbb{G}_1^2$, $B_2 = \mathbb{G}_2^2$ and $B_T := \mathbb{G}_T^4$. The maps are $\iota_j : A_j \to B_j$, $p_j : B_j \to A_j$, $\iota_j' : \mathbb{Z}_p \to B_j$ and $p_j' : B_j \to \mathbb{Z}_p$. Vectors $\boldsymbol{u}$ of $u_1, u_2 \in B_1$ and $\boldsymbol{v}$ of $v_1, v_2 \in B_2$ are commitment keys for $\mathbb{G}_1$ and $\mathbb{G}_2$.

3 Homomorphic Proofs

3.1 Formalization

Recall that an Abelian group must satisfy 5 requirements: Closure, Associativity, Commutativity, Identity Element and Inverse Element.

Definition 1. *Let* (Setup, Prove, Verify) *be a proof system for a relation* $\mathbf{R}$ *and* $Para \leftarrow \mathsf{Setup}(1^k)$. *Consider a subset* Π *of all* $(Sta, Wit, Proof)$ *such that* $(Para, Sta, Wit) \in \mathbf{R}$ *and* $\mathsf{Verify}(Para, Sta, Proof) = 1$, *and an operation* $+_\Pi : \Pi \times \Pi \to \Pi$. Π *is a set of* ***homomorphic*** *proofs if* $(\Pi, +_\Pi)$ *satisfies the 3 requirements: Closure, Associativity and Commutativity.*

Consider an $I_\Pi := (Sta_0, Wit_0, Proof_0) \in \Pi$. Π *is a set of* ***strongly homomorphic*** *proofs if* $(\Pi, +_\Pi, I_\Pi)$ *forms an Abelian group where* I_Π *is the identity element.*

Note that if Π is strongly homomorphic, then Π is also homomorphic. If $+_\Pi$ $((Sta_1, Wit_1, Proof_1), (Sta_2, Wit_2, Proof_2)) \mapsto (Sta, Wit, Proof)$, we have the following notations:

$(Sta, Wit, Proof) \leftarrow (Sta_1, Wit_1, Proof_1) +_\Pi (Sta_2, Wit_2, Proof_2)$, $Sta \leftarrow Sta_1 +_\Pi Sta_2$, $Wit \leftarrow Wit_1 +_\Pi Wit_2$, and $Proof \leftarrow Proof_1 +_\Pi Proof_2$.

We also use the multiplicative notation $n(Sta, Wit, Proof)$ for the self addition for n times of $(Sta, Wit, Proof)$. Similarly, we also use $\sum_i n_i\ (Sta_i, Wit_i, Proof_i)$ to represent linear combination of statements, witnesses and proofs. These homomorphic properties are particularly useful for randomizable proofs: one can randomize a proof computed from the homomorphic operation to get another proof, which is indistinguishable from a proof generated by Prove.

3.2 GS Homomorphic Proofs

Consider a GS proof system (Setup,Prove,Verify) of L equations. Each map $\iota_i : A_i \to B_i$ satisfies $\iota_i(x_1 + x_2) = \iota_i(x_1) + \iota_i(x_2)$, $\forall x_1, x_2 \in A_1$ and $i \in \{1, 2\}$.

We first define the **identity** $I_{GS} = (Sta_0, Wit_0, Proof_0)$. Sta_0 consists of L GS equations $(\boldsymbol{a}_0, \boldsymbol{b}_0, \Gamma_0, t_0)$, Wit_0 consists of L corresponding GS variables $(\boldsymbol{x}_0, \boldsymbol{y}_0)$, $Proof_0$ consists of L corresponding GS proofs $(\boldsymbol{c}_0, \boldsymbol{d}_0, \boldsymbol{\pi}_0, \boldsymbol{\psi}_0)$, and there are L tuples of corresponding maps (ι_1, ι_2). They satisfy:

- Let m be the dimension of $\boldsymbol{b}_0$, $\boldsymbol{x}_0$ and $\boldsymbol{c}_0$. $\exists M \subseteq \{1, ..., m\}$ such that $\forall i \in M$, $b_0[i] = 0$; $\forall j \in \bar{M}$, $x_0[j] = 0$ and $c_0[j] = \iota_1(0)$, where $\bar{M} := \{1, ..., m\} \backslash M$.
- Let n be the dimension of $\boldsymbol{a}_0$, $\boldsymbol{y}_0$ and $\boldsymbol{d}_0$. $\exists N \subseteq \{1, ..., n\}$ such that $\forall i \in N$, $a_0[i] = 0$; $\forall j \in \bar{N}$, $y_0[j] = 0$ and $d_0[j] = \iota_2(0)$, where $\bar{N} := \{1, ..., n\} \backslash N$.
- For both $(\forall i \in \bar{M}, \forall j \in \bar{N})$ and $(\forall i \in M, \forall j \in N)$: $\Gamma_0[i, j] = 0$.
- $t_0 = 0$, $\boldsymbol{\pi}_0 = 0$, and $\boldsymbol{\psi}_0 = 0$.

We next define a **set** Π_{GS} of tuples (*Sta*, *Wit*, *Proof*) from the identity I_{GS}. *Sta* consists of L GS equations $(\boldsymbol{a}, \boldsymbol{b}, \Gamma, t)$ (corresponding to Sta_0's $(\boldsymbol{a}_0, \boldsymbol{b}_0, \Gamma_0, t_0)$ with m, n, M, N); *Wit* consists of L corresponding GS variables $(\boldsymbol{x}, \boldsymbol{y})$; *Proof* consists of L corresponding GS proofs $(\boldsymbol{c}, \boldsymbol{d}, \boldsymbol{\pi}, \boldsymbol{\psi})$; satisfying:

- $\forall i \in M$, $x[i] = x_0[i]$ and $c[i] = c_0[i]$. $\forall j \in \bar{M}$, $b[j] = b_0[j]$.
- $\forall i \in N$, $y[i] = y_0[i]$ and $d[i] = d_0[i]$. $\forall j \in \bar{N}$, $a[j] = a_0[j]$.
- If $(i \in \bar{M}) \vee (j \in \bar{N})$, then $\Gamma[i, j] = \Gamma_0[i, j]$. That means $\forall i \in \bar{M}, \forall j \in \bar{N}$: $\Gamma[i, j] = 0$.

We finally define **operation** $+_{GS} : \Pi_{GS} \times \Pi_{GS} \to \Pi_{GS}$. For $i \in \{1, 2\}$ and $(Sta_i, Wit_i, Proof_i) \in \Pi_{GS}$, Sta_i consists of L GS equations $(\boldsymbol{a}_i, \boldsymbol{b}_i, \Gamma_i, t_i)$ corresponding to Sta_0's $(\boldsymbol{a}_0, \boldsymbol{b}_0, \Gamma_0, t_0)$, Wit_i consists of L corresponding GS variables $(\boldsymbol{x}_i, \boldsymbol{y}_i)$, and $Proof_i$ consists of L corresponding GS proofs $(\boldsymbol{c}_i, \boldsymbol{d}_i, \boldsymbol{\pi}_i, \boldsymbol{\psi}_i)$. We compute (*Sta*, *Wit*, *Proof*) $\leftarrow$ $(Sta_1, Wit_1, Proof_1) +_{GS} (Sta_2, Wit_2, Proof_2)$ of corresponding $(\boldsymbol{a}, \boldsymbol{b}, \Gamma, t)$, $(\boldsymbol{x}, \boldsymbol{y})$ and $(\boldsymbol{c}, \boldsymbol{d}, \boldsymbol{\pi}, \boldsymbol{\psi})$ as follows.

- $\forall i \in M$: $x[i] := x_1[i]$; $c[i] := c_1[i]$; $b[i] := b_1[i] + b_2[i]$. $\forall j \in \bar{M}$: $b[j] := b_1[j]$; $x[j] := x_1[j] + x_2[j]$; $c[j] := c_1[j] + c_2[j]$.
- $\forall i \in N$: $y[i] := y_1[i]$; $d[i] := d_1[i]$; $a[i] := a_1[i] + a_2[i]$. $\forall j \in \bar{N}$: $a[j] := a_1[j]$; $y[j] := y_1[j] + y_2[j]$; $d[j] := d_1[j] + d_2[j]$.
- If $(i \in \bar{M}) \vee (j \in \bar{N})$, then $\Gamma[i, j] := \Gamma_1[i, j]$. Otherwise, $\Gamma[i, j] := \Gamma_1[i, j] + \Gamma_2[i, j]$.
- $t = t_1 + t_2$, $\boldsymbol{\pi} = \boldsymbol{\pi}_1 + \boldsymbol{\pi}_2$, and $\boldsymbol{\psi} = \boldsymbol{\psi}_1 + \boldsymbol{\psi}_2$.

Theorem 1. *In the definitions above, Π_{GS} is a set of strongly homomorphic proofs with operation $+_{GS}$ and the identity element I_{GS}.*

Proof of theorem 1 can be found in Tech Report [29]. The proof validates the closure, associativity, commutativity, identity element, and inverse element properties of abelian groups.

3.3 Comparison with the DHLW Homomorphic NIZK

We compare our homomorphic proof approach with the independently proposed DHLW homomorphic NIZK [27]. Intuitively, DHLW defines that a NIZK proof system is homomorphic if for any $(Para, Sta_1, Wit_1), (Para, Sta_2, Wit_2) \in$ **R**: $\mathsf{Prove}(Para, Sta_1, Wit_1)_{Rand_1} + \mathsf{Prove}(Para, Sta_2, Wit_2)_{Rand_2} = \mathsf{Prove}(Para, Sta_1 + Sta_2, Wit_1 + Wit_2)_{Rand_1 + Rand_2}$, where $\mathsf{Prove}(\ldots)_{Rand}$ is the output of Prove() with randomness *Rand*. The new definition in this paper requires homomorphism for a subset of proofs generated by Prove, and differs from DHLW's homomorphism requirement for all such proofs, covering more proof systems.

The DHLW's homomorphic NIZK construction a special case of our construction above. It is for statements of 'one-sided' GS equations $\{\boldsymbol{x}_k \cdot \boldsymbol{b}_k = t_k\}_{k=1}^{L}$ whereas our construction generalizes to statements of 'full' GS equations $\{\boldsymbol{a}_k \cdot \boldsymbol{y}_k + \boldsymbol{x}_k \cdot \boldsymbol{b}_k + \boldsymbol{x}_k \cdot \Gamma \boldsymbol{y}_k = t_k\}_{k=1}^{L}$. As shown later, the ADNMP and RDAC are based on a GS homomorphic proof system of 'full' equations $\{(y_1 + y_2)X_{j1} + y_{j3}A_1 = T_{j1} \wedge X_{j3} - y_{j3}A_2 = 0 \wedge y_{j3}X_{j2} = T_{j2}\}_{j=1}^{m}$.

4 Accumulator with Delegatable NM Proofs - ADNMP

We refer to a universal accumulator as (Setup, ProveNM, VerifyNM, CompNMWit, Accu), that consists of only algorithms for setup; generating, verifying and computing witnesses for **non-membership** proofs; and accumulating, respectively. This paper does not deal with membership proofs. Tech Report [29] provides more details on accumulators.

The delegating ability to prove statements allows another user to prove the statements on one's behalf without revealing the witness, even if the statements' conditions change over time. For privacy reasons, adversaries should not be able to distinguish different delegations from different users. The delegatee can verify a delegation and unlinkably redelegate the proving ability further to other users. Thus, delegating an accumulator's NM proofs should meet 4 conditions formalized in Definition 2. *Delegability* means that an element Ele's owner can delegate her ability to prove that Ele is not accumulated without trivially revealing Ele. Even if the set of accumulated elements change overtime, the delegatee does not need to contact the delegator again to generate the proof. The owner gives the delegatee a key De generated from Ele. The proof generated from De by CompNMProof is indistinguishable from a proof generated by ProveNM. *Unlinkability* means that a delegatee should not be able to distinguish whether or not two delegating keys originate from the same element. It implies that it is computationally hard to find an element from its delegating keys. *Redelegability* means that the delegatee can redelegate De as De' to other users, and still maintains indistinguishability of De and De'. *Verifiability* means that one is able to validate that a delegating key De is correctly built.

Definition 2. *A universal accumulator (*Setup, ProveNM, VerifyNM, CompNMWit, Accu*) is a secure ADNMP (Accumulator with Delegatable NM Proofs) if there exist PPT algorithms*

- Dele*: takes public parameters $Para$ and an element Ele and returns its delegating key De;*
- Rede*: takes $Para$ and a delegating key De and returns another delegating key De';*
- Vali*: takes $Para$ and a delegating key De and returns* 1 *if De is valid or* 0 *otherwise;*
- CompNMProof*: takes $Para$, De, an accumulator set $AcSet$ and its accumulator value $AcVal$ and returns an NM proof that the element Ele corresponding to De is not accumulated in $AcSet$;*

satisfying:

- *Delegability: For every PPT algorithm* $(\mathcal{A}_1, \mathcal{A}_2)$, $|Pr[(Para, Aux) \leftarrow \mathsf{Setup}(1^k); (Ele, AcSet, state) \leftarrow \mathcal{A}_1(Para); AcVal \leftarrow \mathsf{Accu}(Para, AcSet); Wit \leftarrow \mathsf{CompNMWit}(Para, Ele, AcSet, AcVal); Proof_0 \leftarrow \mathsf{ProveNM}(Para, AcVal, Wit); De \leftarrow \mathsf{Dele}(Para, Ele); Proof_1 \leftarrow \mathsf{CompNMProof}(Para, De, AcSet, AcVal); b \leftarrow \{0,1\}; b' \leftarrow \mathcal{A}_2(state, AcVal, Wit, De, Proof_b): (Ele \notin AcSet) \wedge b = b'] - 1/2|$ *is negligible.*
- *Unlinkability: For every PPT algorithm* $\mathcal{A}$, $|Pr[(Para, Aux) \leftarrow \mathsf{Setup}(1^k); (Ele_0, Ele_1) \leftarrow Dom_{Para}; De \leftarrow \mathsf{Dele}(Para, Ele_0); b \leftarrow \{0,1\}; De_b \leftarrow \mathsf{Dele}(Para, Ele_b); b' \leftarrow \mathcal{A}(Para, De, De_b): b = b'] - 1/2|$ *is negligible.*
- *Redelegability: For every PPT algorithms* $(\mathcal{A}_1, \mathcal{A}_2)$, $|Pr[(Para, Aux) \leftarrow \mathsf{Setup}(1^k); (Ele, state) \leftarrow \mathcal{A}_1(Para); De \leftarrow \mathsf{Dele}(Para, Ele); De_0 \leftarrow \mathsf{Dele}(Para, Ele); De_1 \leftarrow \mathsf{Rede}(Para, De); b \leftarrow \{0,1\}; b' \leftarrow \mathcal{A}_2(state, De, De_b): b = b'] - 1/2|$ *is negligible.*
- *Verifiability: For every PPT algorithm* $\mathcal{A}$, $|Pr[(Para, Aux) \leftarrow \mathsf{Setup}(1^k); Ele \leftarrow \mathcal{A}(Para); De \leftarrow \mathsf{Dele}(Para, Ele): \mathsf{Vali}(Para, De) = 1 \text{ if } Ele \in Dom_{Para}] - 1|$ *and* $|Pr[(Para, Aux) \leftarrow \mathsf{Setup}(1^k); De' \leftarrow \mathcal{A}(Para): \mathsf{Vali}(Para, De') = 0 \text{ if } De' \notin \{De|De \leftarrow \mathsf{Dele}(Para, Ele'); Ele' \in Dom_{Para}\}] - 1|$ *are negligible.*

Unlinkability combined with Redelegability generalizes the Unlinkability definition allowing an adversary $\mathcal{A}$ access an oracle $\mathcal{O}(Para, De)$ that returns another delegating key De' of the same element corresponding to De. That means $\mathcal{A}$ can get several delegating keys of Ele_0 and of Ele_b using $\mathcal{O}$. Rede can be used for such an oracle.

For any ADNMP, given an element Ele and a delegating key De, one can tell if De is generated by Ele as follows. First, she does not accumulate Ele and uses De to prove that De's element is not accumulated. Then she accumulates Ele and tries to prove again that De's element is not accumulated. If she cannot prove that anymore, she can conclude that Ele is De's element. Due to this restriction, in ADNMP's applications, Ele should be a secret that only its owner knows. This is related to the discussion in Tech Report [29] about the general conflict between delegability and anonymity.

5 An ADNMP Scheme

We propose a dynamic universal ADNMP. Its Setup, Accu and UpdateVal are generalized from [7,10].

◇ Setup: We need GS instantiations where GS proofs of this accumulator are composable ZK. We can use either the SXDH or SDLIN (Symmetric DLIN) [28] instantiations. We use SXDH as an example. Generate parameters $(p, \mathbb{G}_1, \mathbb{G}_2, \mathbb{G}_T, e, P_1, P_2)$ and CRS σ with perfectly binding keys for the SXDH instantiation of GS proofs (Sections 2), and auxiliary information $Aux = \delta \leftarrow \mathbb{Z}_p^*$. For the proof, generate $A \leftarrow \mathbb{G}_1$ and $\tau := \iota_2'(\delta)$. For efficient accumulating without Aux, a tuple $\varsigma = (P_1, \delta P_1, \ldots, \delta^{q+1} P_1)$ is needed, where

$q \in \mathbb{Z}_p^*$. The domain for elements to be accumulated is $\mathbb{D} = \mathbb{Z}_p^* \backslash \{-\delta\}$. We have $Para = (p, \mathbb{G}_1, \mathbb{G}_2, \mathbb{G}_T, e, P_1, P_2, A, \sigma, \varsigma, \tau)$.

◇ Accu: On input $AcSet = \{a_1, ..., a_Q\} \subset \mathbb{D}$, compute $m = \lceil Q/q \rceil$. If $Aux = \delta$ is available, the output $AcVal$ is a set of m component accumulator values $\{V_j\}_{j=1}^m$ computed as $V_j = \prod_{i=(j-1)q+1; i<Q}^{jq} (\delta + a_i)\delta P_1$. If Aux is not available, $AcVal$ is efficiently computable from ς and $AcSet$.

◇ UpdateVal: In case $a' \in \mathbb{D}$ is being accumulated; from 1 to m, find the first V_j that hasn't accumulated q elements, and update $V_j' = (\delta + a')V_j$; if such V_j isn't found, add $V_{m+1} = (\delta + a')\delta P_1$. In case a' is removed from $AcVal$, find V_j which contains a' and update $V_j' = 1/(\delta + a')V_j$.

In previous accumulators [7,10], the accumulator value is a single value $V = \prod_{a_i \in AcSet}(\delta + a_i)\delta P_1$ and they require that q of ς is the upper bound on the number of elements to be accumulated, i.e. $m = 1$. The above generalization, where the accumulator value is a set of V instead, relaxes this requirement and allows the ADNMP scheme to work even when q is less than the number of accumulated elements. It also allows smaller q at setup.

5.1 NM Proof

We need to prove that an element $y_2 \in \mathbb{D}$ is not in any component accumulator value V_j of $AcVal$ $\{V_j\}_{j=1}^m$. Suppose V_j accumulates $\{a_1, ..., a_k\}$ where $k \leq q$, denote $Poly(\delta) := \prod_{i=1}^k (\delta + a_i)\delta$, then $V_j = Poly(\delta)P_1$. Let y_{j3} be the remainder of polynomial division $Poly(\delta)$ mod $(\delta + y_2)$ in Z_p, and X_{j1} be scalar product of the quotient and P_1. Similar to [10], the idea for constructing NM proofs is that y_2 is not a member of $\{a_1, ..., a_k\}$ if and only if $y_{j3} \neq 0$. We have the following equation between δ, y_2, y_{j3} and X_{j1}: $(\delta + y_2)X_{j1} + y_{j3}P_1 = V_j$. Proving this equation by itself does not guarantee that y_{j3} is the remainder of the polynomial division above. It also needs to prove the knowledge of $(y_{j3}P_2, y_{j3}A)$ and the following Extended Strong DH (ESDH) assumption. It is a variation of the Hidden Strong DH (HSDH) assumption [30], though it is not clear which assumption is stronger. It is in the extended uber-assumption family [31] and can be proved in generic groups, similar to HSDH.

DEFINITION. ***q-ESDH:*** *Let $(p, \mathbb{G}_1, \mathbb{G}_2, \mathbb{G}_T, e, P_1, P_2)$ be bilinear parameters, $A \leftarrow \mathbb{G}_1^*$ and $\delta \leftarrow \mathbb{Z}_p^*$. Given P_1, δP_1, ..., $\delta^{q+1}P_1, A, P_2, \delta P_2$, it is computationally hard to output $(\frac{y_3}{\delta + y_2}P_1, y_2, y_3 P_2, y_3 A)$ where $y_3 \neq 0$.*

We will show later that if one can prove the knowledge of $(y_{j3}P_2, y_{j3}A)$ satisfying $(\delta + y_2)X_{j1} + y_{j3}P_1 = V_j$ and y_2 is accumulated in V_j but $y_{j3} \neq 0$, then she can break the assumption. To prove the knowledge of $(y_{j3}P_2, y_{j3}A)$, we need equation $X_{j3} - y_{j3}A = 0$. To verify $y_{j3} \neq 0$, we need equation $T_j = y_{j3}X_{j2}$ and the verifier checks $T_j \neq 0$. We now present the NM proof and its security in theorem 2. Proof of theorem 2 can be found in Tech Report [29].

◇ CompNMWit takes y_2, and for each component accumulator value V_j of $AcVal$ $\{V_j\}_{j=1}^m$, computes remainder y_{j3} of $Poly(\delta)$ mod $(\delta + y_2)$ in Z_p which is efficiently computable from $\{a_1, ..., a_k\}$ and y_2. It then computes $X_{j1} = (Poly(\delta) - y_{j3})/(\delta + y_2)P_1$, which is efficiently computable from $\{a_1, ..., a_k\}$, y_2 and ς. The witness includes y_2 and $\{(X_{j1}, X_{j3} = y_{j3}A, y_{j3})\}_{j=1}^m$. UpdateNMWit is for one V_j at a time and similar to [10] with the extra task of updating $X_{j3} = y_{j3}A$.

- ◇ ProveNM generates $X_{j2} \leftarrow \mathbb{G}_1^*$ and outputs $T_j = y_{j3}X_{j2}$ for each V_j and a GS proof for the following equations of variables $y_1 = \delta, y_2, \{(X_{j1}, X_{j3}, X_{j2}, y_{j3})\}_{j=1}^m$. $\bigwedge_{j=1}^m((y_1 + y_2)X_{j1} + y_{j3}P_1 = V_j \wedge X_{j3} - y_{j3}A = 0 \wedge y_{j3}X_{j2} = T_j)$. Note that the prover does not need to know y_1. From τ, it is efficient to generate a commitment of δ and the proof.
- ◇ VerifyNM verifies the proof generated by ProveNM and checks that $T_j \neq 0$, $\forall j \in \{1, \ldots, m\}$. It accepts if both of them pass or rejects otherwise.

Theorem 2. *The proof system proves that an element is not accumulated. Its soundness depends on the ESDH assumption. Its composable ZK depends on the assumption underlying the GS instantiation (SXDH or SDLIN).*

The proof in [29] follows the GS SXDH instantiation and shows that the NM proof system for this accumulator is *composable* ZK. The *completeness* comes from GS, and *soundness* relies on the ESDH assumption.

5.2 NM Proofs Are Strongly Homomorphic

We can see that for the same constant A, the same variables δ, y_2 and X_{j2} with the same commitments, the set of NM proofs has the form of strongly homomorphic GS proofs constructed in Section 3. For constructing delegatable NM proofs, we just need them to be homomorphic. More specifically, 'adding' 2 proofs of 2 sets of equations (with the same commitments for δ, y_2 and X_{j2}) $\bigwedge_{j=1}^m((\delta + y_2)X_{j1}^{(1)} + y_{j3}^{(1)}P_1 = V_j^{(1)} \wedge X_{j3}^{(1)} - y_{j3}^{(1)}A = 0 \wedge y_{j3}^{(1)}X_{j2} = T_j^{(1)})$ and $\bigwedge_{j=1}^m((\delta + y_2)X_{j1}^{(2)} + y_{j3}^{(2)}P_1 = V_j^{(2)} \wedge X_{j3}^{(2)} - y_{j3}^{(2)}A = 0 \wedge y_{j3}^{(2)}X_{j2} = T_j^{(2)})$ form a proof of equations
$\bigwedge_{j=1}^m((\delta + y_2)X_{j1} + y_{j3}P_1 = V_j \wedge X_{j3} - y_{j3}A = 0 \wedge y_{j3}X_{j2} = T_j)$
where $X_{j1} = X_{j1}^{(1)} + X_{j1}^{(2)}$, $X_{j3} = X_{j3}^{(1)} + X_{j3}^{(2)}$, $y_{j3} = y_{j3}^{(1)} + y_{j3}^{(2)}$, $V_j = V_j^{(1)} + V_j^{(2)}$ and $T_j = T_j^{(1)} + T_j^{(2)}$.

5.3 Delegating NM Proof

We first explain the idea behind the accumulator's delegatable NM proof construction. We write the component accumulator value $V = \prod_{i=1}^k(\delta + a_i)\delta P_1$ as $V = \sum_{i=0}^k b_i\delta^{k+1-i}P_1$ where $b_0 = 1$ and $b_i = \sum_{1 \le j_1 < j_2 < \ldots < j_i \le k} \prod_{l=1}^i a_{j_l}$. Thus, V can be written as a linear combination of $\delta P_1, \ldots, \delta^{k+1}P_1$ in ς.

Next, we construct homomorphic proofs for $(\delta + y_2)X_1^{(i)} + y_3^{(i)}P_1 = \delta^i P_1 \wedge X_3^{(i)} - y_3^{(i)}A = 0 \wedge y_3^{(i)}X_2 = T^{(i)}$ where $i \in \{1, ..., k+1\}$. Using the same linear combination of $\delta P_1, \ldots, \delta^{k+1}P_1$ for V, we linearly combine these proofs to get a proof for $(\delta + y_2)X_1 + y_3P_1 = V \wedge X_3 - y_3A = 0 \wedge y_3X_2 = T$, where

$X_1 = \sum_{i=0}^{k} b_i X_1^{(k+1-i)}$, $X_3 = \sum_{i=0}^{k} b_i X_3^{(k+1-i)}$, $y_3 = \sum_{i=0}^{k} b_i y_3^{(k+1-i)}$ and $T = \sum_{i=0}^{k} b_i T^{(k+1-i)}$. This is the same as the NM proof for each of the component accumulator value provided above.

We now provide the algorithms for delegating NM proofs and its security theorem. We also add UpdateProof to be used in place of CompNMProof when possible for efficiency.

◇ Dele($Para, Ele$). For each $i \in \{1, ..., q+1\}$, compute remainder $y_3^{(i)}$ of δ^i mod $(\delta + y_2)$ in Z_p, and $X_1^{(i)} = (\delta^i - y_3^{(i)})/(\delta + y_2)P_1$, which are efficiently computable from y_2 and ς. In fact, we have $y_3^{(i)} = (-1)^i y_2^i$ and $X_1^{(i+1)} = \sum_{j=0}^{i}(-1)^j y_2^j \delta^{i-j} P_1 = \delta^i P_1 - y_2 X_1^{(i)}$ (so the cost of computing all $X_1^{(i)}$, $i \in \{1, ..., q+1\}$ is about q scalar products). Generate $X_2 \leftarrow \mathbb{G}_1^*$, the delegation key De includes $\{T^{(i)} = y_3^{(i)} X_2\}_{i=1}^{q+1}$ and a GS proof of equations $\bigwedge_{i=1}^{q+1}((\delta + y_2)X_1^{(i)} + y_3^{(i)} P_1 = \delta^i P_1 \wedge X_3^{(i)} - y_3^{(i)} A = 0 \wedge y_3^{(i)} X_2 = T^{(i)})$.

◇ Rede($Para, De$). For each $i \in \{1, ..., q+1\}$, extract proof $Proof_i$ of $y_3^{(i)} X_2 = T^{(i)}$ in De. In each $Proof_i$, for the same $y_3^{(i)}$ and its commitment, $Proof_i$ is of homomorphic form. So generate $r \leftarrow Z_p^*$ and compute $Proof_i' = rProof_i$ which is a proof of $y_3^{(i)} X_2' = T'^{(i)}$, where $X_2' = rX_2$ and $T'^{(i)} = rT^{(i)}$. Note that commitments of $y_3^{(i)}$ stay the same. For every $i \in \{1, ..., q+1\}$, replace $T^{(i)}$ by $T'^{(i)}$ and $Proof_i$ by $Proof_i'$ in De to get a new GS proof, which is then randomized to get the output De'.

◇ Vali($Para, De$). A simple option is to verify the GS proof De. An alternative way is to use batch verification: Divide De into proofs $NMProof_i$ of $(\delta + y_2)X_1^{(i)} + y_3^{(i)} P_1 = \delta^i P_1 \wedge X_3^{(i)} - y_3^{(i)} A = 0 \wedge y_3^{(i)} X_2 = T^{(i)}$ for $i \in \{1, ..., q+1\}$. Generate $q+1$ random numbers to linearly combine $NMProof_i$s and their statements and verify the combined proof and statement.

◇ CompNMProof($Para, De, AcSet, AcVal$). Divide De into proofs $NMProof_i$ as in Vali. For each component accumulator value V of $\{a_1, ..., a_k\}$, compute b_i for $i \in \{0, ..., k\}$ as above. $NMProof_i$s belong to a set of homomorphic proofs, so compute $NMProof = \sum_{i=0}^{k} b_i NMProof_{k+1-i}$, which is a proof of $(\delta + y_2)X_1 + y_3 P_1 = V \wedge X_3 - y_3 A = 0 \wedge y_3 X_2 = T$ where X_1, X_3, y_3, T and V are as explained above.

Extract proof $SubProof$ of $y_3 X_2 = T$ in $NMProof$. For the same y_3 and its commitment, $SubProof$ is of homomorphic form. So generate $r \leftarrow Z_p^*$ and compute $SubProof' = rSubProof$ which is a proof of $y_3 X_2' = T'$, where $X_2' = rX_2$ and $T' = rT$. Note that y_3's commitment stays the same. Replace T by T' and $SubProof$ by $SubProof'$ in $NMProof$ to get a new proof $NMProof'$.

Concatenate those $NMProof'$ of all V in $AcVal$ and output a randomization of the concatenation.

◇ UpdateProof($Para$,De,$AcSet$,$AcVal$,$Proof$,$Opens$). $Proof$ is the proof to be updated and $Opens$ contains openings for randomizing commitments of $y_1 = \delta$ and y_2 from De to $Proof$. Suppose there is a change in accumulated elements of a component value V, we just compute $NMProof'$ for the updated V as in CompNMProof. Randomize $NMProof'$ so that its commitments of y_1 and y_2 are the same as those in $Proof$ and put it in $Proof$ in place of its old part. Output a randomization of the result.

To prove that this construction provides an ADNMP, we need the following Decisional Strong Diffie Hellman (DSDH) assumption, which is not in the uber-assumption family [31], but can be proved in generic groups similarly to the PowerDDH assumption [32]. Proof of theorem 3 is in Tech Report [29].

DEFINITION. ***q-DSDH:*** *Let* $(p, \mathbb{G}_1, \mathbb{G}_2, \mathbb{G}_T, e, P_1, P_2)$ *be bilinear parameters,* $B_0, B_1 \leftarrow \mathbb{G}_1^*$, $x_0, x_1 \leftarrow \mathbb{Z}_p^*$ *and* $b \leftarrow \{0,1\}$. *Given* $B_0, x_0B_0, \ldots, x_0^qB_0$, B_1, x_bB_1, $\ldots$, $x_b^qB_1$, *no PPT algorithm can output* $b' = b$ *with a probability non-negligibly better than a random guess.*

Theorem 3. *The accumulator is a secure ADNMP, based on ESDH, DSDH and the assumption underlying the GS instantiation (SXDH or SDLIN).*

6 Revocable Delegatable Anonymous Credentials - RDAC

6.1 Model

This is a model of RDAC systems, extended from BCCKLS [1] which is briefly described in Tech Report [29]. Participants include users and a Blacklist Authority (BA) owning a blacklist BL. For each credential proof, a user picks a new nym indistinguishable from her other nyms. We need another type of nym for revocation, called *r-nym*, to distinguish between two types of nyms. When an r-nym is revoked, its owner cannot prove credentials anymore. The PPT algorithms are:

- Setup(1^k) outputs public parameters $Para_{DC}$, BA's secret key Sk_{BA}, and an initially empty blacklist BL. Denote BL_e an empty blacklist.
- KeyGen$(Para_{DC})$ outputs a secret key Sk and a secret r-nym Rn for a user.
- NymGen$(Para_{DC}, Sk, Rn)$ outputs a new nym Nym with an auxiliary key $Aux(Nym)$. A user O can become a root credential issuer by publishing a nym Nym_O and a proof that her r-nym Rn_O is not revoked that O has to update when BL changes.
- Issue$(Para_{DC}, Nym_O, Sk_I, Rn_I, Nym_I, Aux(Nym_I), Cred, DeInf, Nym_U, BL, L) \leftrightarrow$ Obtain$(Para_{DC}, Nym_O, Sk_U, Rn_U, Nym_U, Aux(Nym_U), Nym_I, BL, L)$ lets user I issue a level $L+1$ credential to user U. Sk_I, Rn_I, Nym_I and $Cred$ are the secret key, r-nym, nym and level L credential rooted at Nym_O of issuer I. Sk_U, Rn_U and Nym_U are the secret key, r-nym and nym of user U. I gets no output and U gets a credential $Cred_U$.

  Delegation information $DeInf$ is optional. When it is included, U also gets delegation information $DeInf_U$ to later prove that r-nyms of all delegators in her chain are not revoked. If $L = 0$ then $Cred$ is omitted and $DeInf = 1$ is optionally included.

- Revoke($Para_{DC}, Sk_{BA}, Rn, BL$) updates BL so that a revoked user Rn can no longer prove, delegate or receive credentials. Denote $Rn \in BL$ or $Rn \notin BL$ that Rn is blacklisted or not, respectively.
- CredProve($Para_{DC}, Nym_O, Cred, DeInf, Sk, Rn, Nym, Aux(Nym), BL, L$) takes a level L credential $Cred$, Sk, Rn and optionally $DeInf$ to output $CredProof$, which proves that: (i) a credential level L is issued to Nym's owner. (ii) Nym's Rn is not revoked. (iii)(optional, when $DeInf$ is included) all r-nyms on the credential's chain are not revoked.
- CredVerify($Para_{DC}, Nym_O, CredProof, Nym, BL, L$) verifies if $CredProof$ is a valid proof of the above statements.

The differences with the model for delegatable anonymous credentials without revocation [1] are the introductions of BA with Sk_{BA} and BL; r-nyms; delegation information $DeInf$; Revoke; and the two $CredProof$'s conditions (ii) and (iii). Note that $DeInf$'s inclusion in the algorithms is optional and allows a user the choice to either just prove that she is not blacklisted or fully prove and delegate that all users on her credential chain are not blacklisted. We can use one of traditional methods for BA to obtain r-nyms to revoke (Tech Report [29]).

Tech Report [29] formally defines RDAC security. Briefly, there are 3 requirements extended from the security definition of delegatable anonymous credentials [1]: Correctness, Anonymity and Unforgeability. Tech Report [29] discusses the trade offs between delegability and anonymity.

7 An RDAC Scheme

7.1 Overview

We first describe intuitions of the BCCKLS delegatable anonymous credential scheme in [1], and then show how ADNMP extends it to provide revocation.

BCCKLS uses an F-Unforgeable certification secure authentication scheme $\mathcal{AU}$ of PPT algorithms AtSetup, AuthKg, Authen, VerifyAuth. AtSetup(1^k) returns public parameters $Para_{At}$, AuthKg($Para_{At}$) generates a key Sk, Authen($Para_{At}$, Sk, $\boldsymbol{m}$) produces an authenticator $Auth$ authenticating a vector of messages $\boldsymbol{m}$, and VerifyAuth($Para_{At}$, Sk, $\boldsymbol{m}$, $Auth$) accepts if and only if $Auth$ validly authenticates $\boldsymbol{m}$ under Sk. The scheme's *security* requirements include *F-Unforgeability* [12] for a bijective function F, which means ($F(\boldsymbol{m})$, $Auth$) is unforgeable without obtaining an authenticator on $\boldsymbol{m}$; and *certification security*, which means no PPT adversary, even after obtaining an authenticator by the challenge secret key, can forge another authenticator. An adversary can also have access to two oracles. $\mathcal{O}_{Authen}(Para_{At}, Sk, \boldsymbol{m})$ returns Authen($Para_{At}$, Sk, $\boldsymbol{m}$) and $\mathcal{O}_{Certify}(Para_{At}, Sk^*, (Sk, m_2, \ldots, m_n))$ returns Authen($Para_{At}, Sk^*, (Sk, m_2, \ldots, m_n)$). BCCKLS also uses a secure two party computation protocol (AuthPro) to obtain a NIZKPK of an authenticator on $\boldsymbol{m}$ without revealing anything about $\boldsymbol{m}$.

In BCCKLS, a user U can generate a secret key $Sk \leftarrow$ AuthKg($Para_{At}$), and many nyms $Nym =$ Com($Sk, Open$) by choosing different values $Open$. Suppose U has a level L+1 credential from O, let ($Sk_0 = Sk_O$, Sk_1, ... , Sk_L, $Sk_{L+1} = Sk$) be the keys such that Sk_i's owner delegated the credential to Sk_{i+1}, and let H :

$\{0,1\}^* \rightarrow Z_p$ be a collision resistant hash function. $r_i = H(Nym_O, atributes, i)$ is computed for a set of attributes for that level's credential. U generates a proof of her delegated credential as
$CredProof \leftarrow \mathsf{NIZKPK}[Sk_O \text{ in } Nym_O, Sk \text{ in } Nym]$
$\{(F(Sk_O), F(Sk_1), ..., F(Sk_L), F(Sk), auth_1, ..., auth_{L+1}) :$
$\mathsf{VerifyAuth}(Sk_O, (Sk_1, r_1), auth_1) \wedge$
$\mathsf{VerifyAuth}(Sk_1, (Sk_2, r_2), auth_2) \wedge ... \wedge$
$\mathsf{VerifyAuth}(Sk_{L-1}, (Sk_L, r_L), auth_L) \wedge$
$\mathsf{VerifyAuth}(Sk_L, (Sk, r_{L+1}), auth_{L+1})\}$.

Now we show how ADNMP extends BCCKLS to provide revocation. Using ADNMP, BA's blacklist BL includes an accumulated set of revoked Rns and its accumulator value. Beside a secret key Sk, user U has a secret r-nym Rn in the accumulator's domain, and generates nyms $Nym = (\mathsf{Com}(Sk, Open_{Sk}), \mathsf{Com}(Rn, Open_{Rn}))$. ADNMP allows delegation and redelegation of a proof that an Rn is not accumulated in a blacklist $Rn \notin BL$. U generates a proof of her delegated credential and validity of the credential's chain as follows.

$$
\begin{aligned}
CredProof \leftarrow \mathsf{NIZKPK}[&Sk_O \text{ in } Nym_O[1], Sk \text{ in } Nym[1], Rn \text{ in } Nym[2]] \\
&\{(F(Sk_O), F(Sk_1), F(Rn_1), ..., F(Sk_L), F(Rn_L), F(Sk), F(Rn), \\
&\quad auth_1, ..., auth_L, auth_{L+1}) : \\
&\mathsf{VerifyAuth}(Sk_O, (Sk_1, Rn_1, r_1), auth_1) \wedge (Rn_1 \notin BL) \wedge \\
&\mathsf{VerifyAuth}(Sk_1, (Sk_2, Rn_2, r_2), auth_2) \wedge (Rn_2 \notin BL) \wedge ... \wedge \\
&\mathsf{VerifyAuth}(Sk_{L-1}, (Sk_L, Rn_L, r_L), auth_L) \wedge (Rn_L \notin BL) \wedge \\
&\mathsf{VerifyAuth}(Sk_L, (Sk, Rn, r_{L+1}), auth_{L+1}) \wedge (Rn \notin BL)\}.
\end{aligned}
$$

Delegability allows a user, on behalf of the user's delegators without any witness, to prove that the user's ancestor delegators are not included in a changing blacklist. The proofs a user and a delegator generates are indistinguishable from each other. Redelegability allows a user to redelegate those proofs on the delegators to the user's delegatees. Unlinkability prevents colluding users to link delegations of the same delegator. Verifiability allows a user to validate the correctness of a delegation token.

7.2 Description

The RDAC scheme has several building blocks. (i) An ADNMP with a malleable NM proof system (NMPS) of AcSetup, ProveNM, VerifyNM, CompNMWit, Accu, Dele, Rede, Vali, CompNMProof, with commitment ComNM. (ii) Those from BCCKLS, including $\mathcal{AU}$; AuthPro; H; and a malleable NIPK credential proof system (CredPS) of PKSetup, PKProve, PKVerify, RandProof, with commitment Com. (iii) A malleable proof system (EQPS), with PKSetup and AcSetup in setup, to prove that two commitments Com and ComNM commit to the same value.

Assume that a delegating key De contains a commitment of element Ele. CompNMProof and Rede randomize the commitment in De and generate Ele's commitment. Elements of the accumulator domain and the authenticator's key space can be committed by Com. The following algorithm inputs are the same as in the model and omitted.

- Setup: Use PKSetup(1^k), AtSetup(1^k) and AcSetup(1^k) to generate $Para_{PK}$, $Para_{At}$, and ($Para_{Ac}$, Aux_{Ac}). The blacklist includes an accumulated set of revoked r-nyms and its accumulator value. Output an initial blacklist BL with an empty accumulator set and its initial accumulator value, $Para_{DC} = (Para_{PK}, Para_{At}, Para_{Ac}, H)$, and $Sk_{BA} = Aux_{Ac}$.
- KeyGen: Run AuthKg($Para_{At}$) to output a secret key Sk. Output a random r-nym Rn from the accumulator's domain.
- NymGen: Generate random $Open_{Sk}$ and $Open_{Rn}$, and output nym $Nym = (\mathsf{Com}(Sk, Open_{Sk}), \mathsf{Com}(Rn, Open_{Rn}))$ and $Aux(Nym) = (Open_{Sk}, Open_{Rn})$.
- The credential originator O publishes a Nym_O and a proof $NMProof_O$ that Rn_O is not revoked. O updates the proof when BL changes.
- Issue $\leftrightarrow$ Obtain: If $L = 0$ and $Nym_O \neq Nym_I$, aborts. Issue aborts if $Nym_I \neq (\mathsf{Com}(Sk_I, Open_{Sk_I}), \mathsf{Com}(Rn_I, Open_{Rn_I}))$ or PKVerify($Para_{PK}$, $(Nym_0, (\mathsf{Com}(Sk_I, 0), \mathsf{Com}(Rn_I, 0))), Cred$) rejects, or $Rn_I \in BL$, or Nym_U is invalid. Obtain aborts if $Nym_U \neq (\mathsf{Com}(Sk_U, Open_{Sk_U}), \mathsf{Com}(Rn_U, Open_{Rn_U}))$ or $Rn_U \in BL$. Otherwise, each of Issue and Obtain generates a proof and verifies each other's proofs that $Rn_I \notin BL$ and $Rn_U \notin BL$ using (ProveNM, VerifyNM) with EQPS (to prove that Com(Rn_I) in Nym_I and ComNM(Rn_I) generated by ProveNM commit to the same value Rn_I, and similarly for Rn_U). They then both compute $r_{L+1} = H(Nym_O, attributes, L + 1)$ for a set of attributes for that level's credential. They run AuthPro for the user U to receive: $Proof_U \leftarrow \mathsf{NIZKPK}[Sk_I \text{ in } Nym_I[1], Sk_U \text{ in } \mathsf{Com}(Sk_U, 0), Rn_U \text{ in } \mathsf{Com}(Rn_U, 0)]$ $\{(F(Sk_I), F(Sk_U), F(Rn_U), auth) : \mathsf{VerifyAuth}(Sk_I, (Sk_U, Rn_U, r_{L+1}), auth)\}$. U's output is $Cred_U = Proof_U$ when L = 0. Otherwise, suppose the users on the issuer I's chain from the root are 0 (same as O), 1, 2,..., L (same as I). I randomizes $Cred$ to get a proof $CredProof_I$ (containing the same Nym_I) that for every Nym_j on I's chain ($j \in \{1, ..., L\}$), Sk_j and Rn_j are authenticated by Sk_{j-1} (with r_j). U verifies that PKVerify($Para_{PK}$, $(Nym_0, Nym_I), CredProof_I$) accepts, then concatenates $Proof_U$ and $CredProof_I$ and projects Nym_I from statement to proof to get $Cred_U$.

  The optional $DeInf$ includes a list of delegating keys De_js generated by the accumulator's Dele to prove that each Rn_j is not accumulated in the blacklist, and a list of $EQProof_j$ for proving that two commitments of Rn_j in $Cred$ and De_j commit to the same value Rn_j, for $j \in \{1, ..., L-1\}$. Verifying $DeInf$ involves checking Vali($Para_{Ac}$, De_j) and $EQProof_j$, for $j \in \{1, ..., L-1\}$. When $DeInf$ is in the input, Issue would aborts without interacting with Obtain if verifying $DeInf$ fails. Otherwise, it uses CompNMProof to generate a proof $NMChainProof$ that each Rn_j's on I's chain of delegators is not accumulated in the blacklist. U aborts if its verification on $NMChainProof$ fails. Otherwise, I Redes these delegating keys, randomizes $EQProof_j$ to match commitments in the new delegating keys and $Cred_U$, and adds a new delegating key De_I to prove that Rn_I is not revoked and a proof $EQProof_I$ that two commitments of Rn_I in $Nym_I[2]$ and De_I commit to the same value. The result $DeInf_U$ is sent to and verified by U.

- Revoke: Add Rn to the accumulated set and update the accumulator value.
- CredProve: Abort if $Nym \neq (\mathsf{Com}(Sk, Open_{Sk}), \mathsf{Com}(Rn, Open_{Rn}))$, or $\mathsf{PKVerify}(Para_{PK}, (Nym_0, (\mathsf{Com}(Sk, 0), \mathsf{Com}(Rn, 0))), Cred)$ rejects, or verifying $DeInf$ fails. Otherwise, use ProveNM to generate a proof $NMProof$ that Rn is not blacklisted. Generate $EQProof'_L$ that Rn's commitments in $NMProof$ and in $Nym[2]$ both commit to the same value. Randomize $Cred$ to get a proof which contains Nym. Concatenate this proof with $NMProof$ and $EQProof'_L$ to get $CredProof'$. If the optional $DeInf$ is omitted, just output $CredProof'$.

  Otherwise, use CompNMProof to generate a proof $NMChainProof$ that each Rn_j's on the user's chain of delegators is not accumulated in the blacklist. For $j \in \{1, ..., L-1\}$, update and randomize $EQProof_j$ of $DeInf$ to get $EQProof'_j$ which proves Rn_j's commitments in $NMChainProof$ and $CredProof'$ both commit to the same value. Concatenate $NMChainProof$, $CredProof'$ and $EQProof'_j$ for $j \in \{1, ..., L-1\}$ to output $CredProof$ as described in (1).
- CredVerify runs PKVerify on the randomization of $Cred$, VerifyNM on $NMProof$ and

  $NMChainProof$, and verifies $EQProof'_j$ for $j \in \{1, ..., L\}$ to output accept or reject.

Theorem 4. *If the authentication scheme is F-unforgeable and certification-secure; the ADNMP is secure; CredPS, NMPS and EQPS are randomizable and composable ZK; CredPS is also partially extractable; and H is collision resistant, then this construction is a secure revocable delegatable anonymous credential system.*

Proof of theorem 4 is given in Tech Report [29]. Instantiation of the building blocks are given in Tech Report [29]. Briefly, a secure ADNMP is presented in Section 5; the BCCKLS building blocks can be instantiated as in [1]; and an EQPS can be constructed from [12,1].

References

1. Belenkiy, M., Camenisch, J., Chase, M., Kohlweiss, M., Lysyanskaya, A., Shacham, H.: Randomizable proofs and delegatable anonymous credentials. In: Halevi, S. (ed.) CRYPTO 2009. LNCS, vol. 5677, pp. 108–125. Springer, Heidelberg (2009)
2. Groth, J., Sahai, A.: Efficient non-interactive proof systems for bilinear groups. In: Smart, N.P. (ed.) EUROCRYPT 2008. LNCS, vol. 4965, pp. 415–432. Springer, Heidelberg (2008)

3. Bellare, M., Rogaway, P.: Random oracles are practical: A paradigm for designing efficient protocols. In: Ashby, V. (ed.) ACM CCS 1993: 1st Conference on Computer and Communications Security, Fairfax, Virginia, USA, November 3-5, pp. 62–73. ACM Press, New York (1993)
4. Benaloh, J.C., de Mare, M.: One-way accumulators: A decentralized alternative to digital sinatures (extended abstract). In: Helleseth, T. (ed.) EUROCRYPT 1993. LNCS, vol. 765, pp. 274–285. Springer, Heidelberg (1994)
5. Barić, N., Pfitzmann, B.: Collision-free accumulators and fail-stop signature schemes without trees. In: Fumy, W. (ed.) EUROCRYPT 1997. LNCS, vol. 1233, pp. 480–494. Springer, Heidelberg (1997)
6. Camenisch, J., Lysyanskaya, A.: Dynamic accumulators and application to efficient revocation of anonymous credentials. In: Yung, M. (ed.) CRYPTO 2002. LNCS, vol. 2442, pp. 61–76. Springer, Heidelberg (2002)
7. Nguyen, L.: Accumulators from bilinear pairings and applications. In: Menezes, A. (ed.) CT-RSA 2005. LNCS, vol. 3376, pp. 275–292. Springer, Heidelberg (2005)
8. Camenisch, J., Kohlweiss, M., Soriente, C.: An accumulator based on bilinear maps and efficient revocation for anonymous credentials. In: Jarecki, S., Tsudik, G. (eds.) PKC 2009. LNCS, vol. 5443, pp. 481–500. Springer, Heidelberg (2009)
9. Li, J., Li, N., Xue, R.: Universal accumulators with efficient nonmembership proofs. In: Katz, J., Yung, M. (eds.) ACNS 2007. LNCS, vol. 4521, pp. 253–269. Springer, Heidelberg (2007)
10. Au, M.H., Tsang, P.P., Susilo, W., Mu, Y.: Dynamic universal accumulators for DDH groups and their application to attribute-based anonymous credential systems. In: Fischlin, M. (ed.) CT-RSA 2009. LNCS, vol. 5473, pp. 295–308. Springer, Heidelberg (2009)
11. Camenisch, J., Lysyanskaya, A.: Signature schemes and anonymous credentials from bilinear maps. In: Franklin, M. (ed.) CRYPTO 2004. LNCS, vol. 3152, pp. 56–72. Springer, Heidelberg (2004)
12. Belenkiy, M., Chase, M., Kohlweiss, M., Lysyanskaya, A.: P-signatures and noninteractive anonymous credentials. In: Canetti, R. (ed.) TCC 2008. LNCS, vol. 4948, pp. 356–374. Springer, Heidelberg (2008)
13. Brickell, E.F., Camenisch, J., Chen, L.: Direct anonymous attestation. In: Atluri, V., Pfitzmann, B., McDaniel, P. (eds.) ACM CCS 2004, Washington D.C., USA, October 25-29, pp. 132–145. ACM Press, New York (2004)
14. Camenisch, J., Van Herreweghen, E.: Design and implementation of the idemix anonymous credential system. In: Atluri, V. (ed.) ACM CCS 2002, Washington D.C., USA, November 18-22, pp. 21–30. ACM Press, New York (2002)
15. Microsoft: U-prove community technology preview (2010), `https://connect.microsoft.com/`
16. Bichsel, P., Camenisch, J., Groß, T., Shoup, V.: Anonymous credentials on a standard java card. In: Al-Shaer, E., Jha, S., Keromytis, A.D. (eds.) ACM CCS 2009, Chicago, Illinois, USA, November 9-13, pp. 600–610. ACM Press, New York (2009)
17. Tsang, P.P., Au, M.H., Kapadia, A., Smith, S.W.: PEREA: towards practical TTP-free revocation in anonymous authentication. In: Ning, P., Syverson, P.F., Jha, S. (eds.) ACM CCS 2008, Alexandria, Virginia, USA, October 27-31, pp. 333–344. ACM Press, New York (2008)
18. Johnson, R., Molnar, D., Song, D., Wagner, D.: Homomorphic signature schemes. In: Preneel, B. (ed.) CT-RSA 2002. LNCS, vol. 2271, pp. 244–262. Springer, Heidelberg (2002)

19. Ateniese, G., Kamara, S., Katz, J.: Proofs of storage from homomorphic identification protocols. In: Matsui, M. (ed.) ASIACRYPT 2009. LNCS, vol. 5912, pp. 319–333. Springer, Heidelberg (2009)
20. Charles, D., Jain, K., Lauter, K.: Signatures for network coding. International Journal on Information and Coding Theory (2006)
21. Yun, A., Cheon, J., Kim, Y.: On homomorphic signatures for network coding. Transactions on Computer (2009)
22. Johnson, R., Walsh, L., Lamb, M.: Homomorphic signatures for digital photographs. Suny Stony Brook (2008)
23. Monnerat, J., Vaudenay, S.: Generic homomorphic undeniable signatures. In: Lee, P.J. (ed.) ASIACRYPT 2004. LNCS, vol. 3329, pp. 354–371. Springer, Heidelberg (2004)
24. Groth, J.: A verifiable secret shuffle of homomorphic encryptions. In: Desmedt, Y. (ed.) PKC 2003. LNCS, vol. 2567, pp. 145–160. Springer, Heidelberg (2002)
25. Fouque, P.-A., Poupard, G., Stern, J.: Sharing decryption in the context of voting or lotteries. In: Frankel, Y. (ed.) FC 2000. LNCS, vol. 1962, pp. 90–104. Springer, Heidelberg (2001)
26. Gentry, C.: Fully homomorphic encryption using ideal lattices. In: Mitzenmacher, M. (ed.) 41st ACM STOC Annual ACM Symposium on Theory of Computing, Bethesda, Maryland, USA, May 17-20, pp. 169–178. ACM Press, New York (2009)
27. Dodis, Y., Haralambiev, K., Lopez-Alt, A., Wichs, D.: Cryptography against continuous memory attacks (2010)
28. Ghadafi, E., Smart, N.P., Warinschi, B.: Groth sahai proofs revisited. In: Nguyen, P.Q., Pointcheval, D. (eds.) PKC 2010. LNCS, vol. 6056, pp. 177–192. Springer, Heidelberg (2010)
29. Acar, T., Nguyen, L.: Revocation for delegatable anonymous credentials. Technical Report MSR-TR-2010-170, Microsoft Research, One Microsoft Way, Redmond, WA 98052 (December 2010)
30. Boyen, X., Waters, B.: Full-domain subgroup hiding and constant-size group signatures. In: Okamoto, T., Wang, X. (eds.) PKC 2007. LNCS, vol. 4450, pp. 1–15. Springer, Heidelberg (2007)
31. Boyen, X.: The uber-assumption family (invited talk). In: Galbraith, S.D., Paterson, K.G. (eds.) Pairing 2008. LNCS, vol. 5209, pp. 39–56. Springer, Heidelberg (2008)
32. Camenisch, J., Neven, G., Shelat, A.: Simulatable adaptive oblivious transfer. In: Naor, M. (ed.) EUROCRYPT 2007. LNCS, vol. 4515, pp. 573–590. Springer, Heidelberg (2007)

Cryptanalysis of Multivariate and Odd-Characteristic HFE Variants

Luk Bettale*, Jean-Charles Faugère, and Ludovic Perret

INRIA, Paris-Rocquencourt Center, SALSA Project
UPMC Univ Paris 06, UMR 7606, LIP6, F-75005, Paris, France
CNRS, UMR 7606, LIP6, F-75005, Paris, France
luk.bettale@lip6.fr, jean-charles.faugere@inria.fr,
ludovic.perret@lip6.fr

Abstract. We investigate the security of a generalization of HFE (multivariate and odd-characteristic variants). First, we propose an improved version of the basic Kipnis-Shamir key recovery attack against HFE. Second, we generalize the Kipnis-Shamir attack to Multi-HFE. The attack reduces to solve a MinRank problem directly on the public key. This leads to an improvement of a factor corresponding to the square of the degree of the extension field. We used recent results on MinRank to show that our attack is polynomial in the degree of the extension field. It appears that multi-HFE is less secure than original HFE for equal-sized keys. Finally, adaptations of our attack overcome several variants (i.e. minus modifier and embedding). As a proof of concept, we have practically broken the most conservative parameters given by Chen, Chen, Ding, Werner and Yang in 9 days for 256 bits security. All in all, our results give a more precise picture on the (in)security of several variants of HFE proposed these last years.

Keywords: Hidden Field Equations, MinRank, Gröbner bases.

1 Introduction

Multivariate Public-Key Cryptography (MPKC) is the set of public-key schemes using multivariate polynomials. The concept of MPKC is very appealing since its security is related to the hardness of a post-quantum problem, namely solving a quadratic system of algebraic equations [23]. In addition, the encryption/decryption procedures are very efficient and can be done in constrained environments [6,10]. Among these cryptosystems, the Hidden Field Equations cryptosystem (HFE) is probably the most studied one. It has been proposed by Patarin [29] after his cryptanalysis [28] of the historical multivariate scheme C* [27]. In [26] Kipnis and Shamir proposed a key recovery attack on HFE, which reduces to the so-called MinRank [12] problem. Although the attack is not practical for the proposed parameters, it was conjectured to be sub-exponential. Later, Faugère and Joux [17,19] proposed an efficient message recovery attack based

* Luk Bettale is partially supported by DGA/MRIS (french secretary of defense).

D. Catalano et al. (Eds.): PKC 2011, LNCS 6571, pp. 441–458, 2011.

on Gröbner bases. This attack, which is "*quasi-polynomial*" [24], raises serious doubt about the security of HFE. To thwart both attacks on HFE, it has been proposed to use a multivariate system as the secret key [5] or odd-characteristic fields [14] or even both in a recent paper [11]. This new family of schemes is called multi-HFE in the rest of the paper.

Our contributions. We propose here a key recovery attack on HFE, multi-HFE and some of its variants. Our attack is an adaptation and improvement of the Kipnis-Shamir attack [26]. Precisely, we reduce the attack to the problem of finding a linear combination of the public quadratic forms of low rank. This problem is known as the MinRank (MR) problem (MR is usually defined for matrices, but the problem can be defined equivalently on quadratic forms). The coefficients in the linear relation that we are looking for are strongly related to one of the affine transforms used to hide the (multi-)HFE structure. We show that the MinRank can be expressed in the small field, which allows to considerably speed-up solving by approximately a factor corresponding to the square of the degree of the extension field. Thanks to recent results on MinRank [20,21] and bilinear systems [22], we conjecture that the attack is polynomial in the degree of the extension. Using this complexity analysis, we can prove that, for the same size of keys (a precise definition of this notion is given in Sect. 3.6), multi-HFE is always less secure than HFE. In addition, the large number of equivalent keys allows to attack the minus variant (this amounts to remove some equations in the public key) using the induced degrees of freedom of the MinRank. Finally, we present an attack on the embedding variant of (multi-)HFE. This variant consists in instantiating some variables of the public system. However, a low rank linear combination of the quadratic forms can still be found. In this case, solving the corresponding MinRank on truncated quadratic forms allows us to recover only a rectangular sub matrix of the linear transform; to overcome this difficulty we need to extend this matrix in a special way (details can be found in Sect. 5) to make it invertible. As a proof of concept, we practically broke several parameters proposed in [11], supposed to have up to 256 bits security (experiments are given in Sect. 6). We also mention that the second part of the attack of Kipnis and Shamir as presented in [26] does not apply in characteristic 2. It is possible to overcome this problem but due to space limitation, this will be presented in an extended version of this paper. Consequently, we assume in the rest of the paper that q (the size of the small field) is odd.

2 Multivariate HFE

Throughout this paper, we use the following conventions: an underlined letter denotes a vector, e.g. $\underline{v} = (v_1, \ldots, v_n)$. A capital bold font letter denotes a matrix, e.g. $\mathbf{M} = [m_{i,j}]$. A calligraphic capital letter denotes a general mapping, e.g. $\mathcal{F}$.

For Multi-HFE, the parameters considered are $(q, N, d, D) \in \mathbb{N}^4$. Here, q (odd) denotes the size of the ground field $\mathbb{F}_q$, d is the degree of the extension field $\mathbb{F}_{q^d}$, N is the number of variables and equations of the secret polynomials in the ring $\mathbb{F}_{q^d}[X_1, \ldots, X_N]$, and D their degree. Throughout the paper, we use capital letters for elements relative to the big field $\mathbb{F}_{q^d}$ (e.g. $V_i \in \mathbb{F}_{q^d}$, $F_i \in \mathbb{F}_{q^d}[X_1, \ldots, X_N]$), and small letters for elements relative to $\mathbb{F}_q$ (e.g. $v_i \in \mathbb{F}_q$, $f_i \in \mathbb{F}_q[x_1, \ldots, x_n]$). The secret internal transformation is $\mathcal{F}^* : (V_1, \ldots, V_N) \in (\mathbb{F}_{q^d})^N \mapsto (F_1(V_1, \ldots, V_N), \ldots, F_N(V_1, \ldots, V_N)) \in (\mathbb{F}_{q^d})^N$ with $\deg(F_i) \leqslant D$. The degree D is chosen such that $\mathcal{F}^*$ is easy to invert. In addition, the polynomials $F_1, \ldots, F_N$ are constructed in a specific way:

$$F_k = \sum_{1 \leqslant i \leqslant j \leqslant N} \sum_{\substack{0 \leqslant u,v < d \\ q^u + q^v \leqslant D}} A_{k,i,u,j,v} X_i^{q^u} X_j^{q^v} + \sum_{1 \leqslant i \leqslant N} \sum_{\substack{0 \leqslant u < d \\ q^u \leqslant D}} B_{k,i,u} X_i^{q^u} + C_k.$$

From now on, we say that such systems have (multi-)HFE-shape. For convenience, we denote $n = N\,d$. Let φ_N be the natural morphism $(\mathbb{F}_{q^d})^N \mapsto (\mathbb{F}_q)^n$ and $\mathcal{F}$ be the small field representation of the secret polynomials $\mathcal{F} = \varphi_N \circ \mathcal{F}^* \circ \varphi_N^{-1}$ with $\mathcal{F} : (v_1, \ldots, v_n) \in (\mathbb{F}_q)^n \mapsto (f_1(v_1, \ldots, v_n), \ldots, f_n(v_1, \ldots, v_n)) \in (\mathbb{F}_q)^n$. Due to the HFE-shape, each polynomial f_i has total degree 2. For the secret key, the mapping $\mathcal{F}^*$ is supplemented by two affine maps $\mathcal{S}, \mathcal{T} \in \mathrm{Aff}(n, \mathbb{F}_q)$ represented by matrices $\mathbf{S}$ and $\mathbf{T}$ which hide the internal structure. The public key $\mathcal{G} = \mathcal{T} \circ \mathcal{F} \circ \mathcal{S} : (v_1, \ldots, v_n) \in (\mathbb{F}_q)^n \mapsto (g_1(v_1, \ldots, v_n), \ldots, g_n(v_1, \ldots, v_n)) \in (\mathbb{F}_q)^n$ is then composed of polynomials $g_1, \ldots, g_n \in \mathbb{F}_q[x_1, \ldots, x_n]$ of total degree 2.

To encrypt, we evaluate $g_1, \ldots, g_n$ in the message $\underline{m} = (m_1, \ldots, m_n) \in (\mathbb{F}_q)^n$. With the knowledge of the private key, the decryption of a ciphertext $\underline{c} = (c_1, \ldots, c_n) \in (\mathbb{F}_q)^n$ is done by computing $\mathcal{S}^{-1} \circ \varphi_N \circ \mathcal{F}^{*-1} \circ \varphi_N^{-1} \circ \mathcal{T}^{-1}(\underline{c})$. As each part can be inverted efficiently, the decryption is done efficiently.

The original HFE scheme [29] is mostly used over $\mathbb{F}_2$ with a single univariate polynomial as a secret map. It is then an instantiation of multi-HFE with $q = 2$ and $N = 1$. The construction PHFE (for projected HFE) of [14] is an odd characteristic univariate HFE that uses the embedding modifier (see Sect. 5). The scheme IFS (for Intermediate Field System) from [5] is a multi-HFE in characteristic 2 and THFE from [11] is a multi-HFE in odd characteristic (possibly with embedding modifier). To make the decryption efficient, all instances of multi-HFE with $N > 1$ use quadratic polynomials as internal secret transformations. Parameters examples from the literature are given in the tables below.

	q	N	d	D	security
HFE [29]	2	1	128	513	128
PHFE [14]	7	1	67	56	201

	q	N	d	D	security
IFS [5]	2	8	16	2	128
THFE [11]	31	3	10	2	150

We now review two attacks on the original HFE: the direct algebraic attack (message recovery) of [19] and the key recovery attack of [26].

2.1 Direct Algebraic Attack

Let $(c_1, \ldots, c_n) \in (\mathbb{F}_q)^n$ be a ciphertext, a message-recovery reduces to solve a system of quadratic equations, i.e. $\{g_1 - c_1 = 0, \ldots, g_n - c_n = 0\}$. A classical method to solve algebraic systems is to compute a Gröbner basis [8,1,13]. The historical method for computing Gröbner bases has been proposed by Buchberger in his PhD thesis [8]. The algorithms F_4 [15] and F_5 [16] by Faugère permit to improve the basic Buchberger's algorithm. A good measure of the complexity for Gröbner bases is the so-called "*degree of regularity*" of a system. This is the maximum degree of the polynomials appearing during the computation (see [2,3]).

It appeared [17,19] that inverting the public key of the original HFE is much easier than expected (i.e. in comparison to a random system of the same size). For original HFE, the degree of regularity has been experimentally shown to be roughly $\log_q(D)$ (see [19]). This makes the attack sub-exponential in the number of variables. Further analysis of the Gröbner basis approach [24] confirmed this result. Note that the field equations (i.e. $x_1^q - x_1 = \ldots = x_n^q - x_n = 0$) are mandatory to achieve this complexity. Their role is to force the solutions to be only in the base field $\mathbb{F}_q$. To prevent a direct algebraic attack, it has been proposed [14] to use a field with a bigger characteristic. Field equations only intervene in degree at least q. Typically, a HFE system with $q > n$ seems very hard to solve with a direct approach (for n sufficiently big). Note that the hybrid approach described in [4] has been designed to solve such systems. However, for $n = 28$ and $q = 31$ the complexity of the hybrid approach is 2^{82}. It is better than a direct solving (2^{115}) but the attack remains exponential. For multi-HFE, the situation is almost similar. On characteristic 2, multi-HFE can still be attacked similarly. This confirms that the algebraic attack is somehow "optimal" over $\mathbb{F}_2$. However, the direct algebraic attack does not affect instantiations of multi-HFE with bigger odd characteristic as adding the field equations would not be useful.

2.2 Original Kipnis-Shamir (KS) Attack

We now describe the key recovery attack proposed in [26] for the original HFE scheme ($N = 1, n = d$). The starting idea is to remark that polynomials of the public key – as well as the transformations $\mathcal{S}, \mathcal{T}$ – can be viewed as mappings $\mathcal{G}^*, \mathcal{S}^*, \mathcal{T}^* : \mathbb{F}_{q^n} \mapsto \mathbb{F}_{q^n}$ and represented by the univariate polynomials $G, S, T \in \mathbb{F}_{q^n}[X]$. The public key relation then becomes $G = \mathcal{G}^*(X) = \mathcal{T}^*(\mathcal{F}^*(\mathcal{S}^*(X)))$. Kipnis and Shamir [26] proposed interpolation to recover a univariate representation of the public key. We present a more efficient and simpler way in Sect. 3 to perform this step.

Kipnis and Shamir [26] also showed that the univariate polynomials can be written as a "*non-standard quadratic form*". For instance, we have:

$$G = \sum_{i=0}^{n-1} \sum_{j=0}^{n-1} g_{i,j} X^{q^i+q^j} = \underline{X}\mathbf{G}\underline{X}^t, \text{ where } \underline{X} = (X, X^q, \ldots, X^{q^{n-1}})$$

and $\mathbf{G} = [g_{i,j}]$ is a symmetric matrix. Similarly, we define $\mathbf{F} = [f_{i,j}]$ the symmetric matrix representation of the secret univariate polynomial.

The Kipnis-Shamir attack is based on the remark that $\text{Rank}\left(\mathbf{F}\right) \leqslant \log_q(D)$. Indeed, the degree of the secret polynomial is smaller than D and the only non-zero entries in $\mathbf{F}$ are $f_{i,j}$, if $i, j \leqslant \log_q(D)$. In addition, if we write $\mathcal{T}^{*-1}(X) = \sum_{k=0}^{n-1} t_k X^{q^i}$ and $\mathcal{S}^*(X) = \sum_{k=0}^{n-1} s_k X^{q^i}$ the equation $\mathcal{G}^*(X) = \mathcal{T}^*(\mathcal{F}^*(\mathcal{S}^*(X)))$ implies this so-called "*Fundamental Equation*" (see [26] for the proof).

$$\sum_{k=0}^{n-1} t_k \mathbf{G}^{*k} = \mathbf{G}' = \widetilde{\mathbf{W}} \mathbf{F} \widetilde{\mathbf{W}}^t \tag{1}$$

where $\widetilde{\mathbf{W}} = [\widetilde{w}_{i,j}]$ is a specified invertible matrix (with $\widetilde{w}_{i,j} = s_{j-i}^{q^i}$) and $\mathbf{G}^{*k}$ the matrix such that its (i,j)-th entry is $g_{i-k,j-k}^{q^k}$. As the rank of $\mathbf{F}$ is bounded, so is the rank of $\mathbf{G}'$. Recovering the t_k's reduces to solve a MinRank problem:
MinRank (MR) in a finite field $\mathbb{K}$
Input: $n, r, k \in \mathbb{N}$ and matrices $\mathbf{M_1}, \ldots, \mathbf{M_k} \in \mathbb{K}^{n \times n}$.
Question: is there a k-tuple $(\lambda_1, \ldots, \lambda_k) \in \mathbb{K}^k$ such that Rank $\left(\sum_{i=1}^{k} \lambda_i \mathbf{M_i}\right)$ $\leqslant r$.
The MinRank problem is NP-complete [9]. From an algorithmic point of view, Kipnis and Shamir proposed to model the problem as a system of overdetermined quadratic equations and then to solve it with the so-called relinearization method [26]. This Kipnis-Shamir modeling – which turns to be a set of bilinear equations [21] – as well as the so-called Minors modeling have been further studied and improved in [20,21]. In both modelings, solving MinRank reduces to compute the solutions of a system of structured algebraic equations.

Once the t_k's of equation (1) are known, the s_k's are recovered by solving a linear system. From (1), we see that $\ker(\mathbf{G}') = \ker(\widetilde{\mathbf{W}}\mathbf{F})$ and thus $\ker(\mathbf{G}')\,\widetilde{\mathbf{W}} = \ker(\mathbf{F})$. Due to the special shape of $\mathbf{F}$, the first $\ell = \log_q(D)$ columns of its left kernel are 0. This gives rise to a linear system of equations of $\ell\,(n-\ell)$ equations in n^2 variables. Since $w_{i+1,j+1} = w_{i,j}^q$, Kipnis and Shamir proposed to reinterpret the equations over $\mathbb{F}_q$. This gives $n\,\ell\,(n-\ell)$ equations in n^2 variables over $\mathbb{F}_q$. Solving this overdetermined system completes the key recovery.

3 Improvement and Generalization of KS Attack

3.1 Improving the Univariate Case

To generalize the KS attack, it is convenient to interpret it as vector/matrix operations. In this paper, we denote by $\mathrm{Frob_k}$ the function raising all the components of a vector or a matrix to the power q^k in any field $\mathbb{K}$ of characteristic q. For example $\mathrm{Frob_k}(\underline{v}) = (v_1^{q^k}, \ldots, v_m^{q^k})$, for a vector $\underline{v} = (v_1, \ldots, v_m) \in \mathbb{K}^m$ and $\mathrm{Frob_k}(\mathbf{A}) = [a_{i,j}^{q^k}]$, for a matrix $\mathbf{A} = [a_{i,j}]$.

Proposition 1. *Let $(\theta_1, \ldots, \theta_n) \in (\mathbb{F}_{q^n})^n$ be a vector basis of $\mathbb{F}_{q^n}$ over $\mathbb{F}_q$ and $\mathbf{M}_n$ be the $n \times n$ matrix whose columns are the Frobenius powers of the basis:*

$$\mathbf{M}_n = \begin{pmatrix} \theta_1 & \theta_1^q & \ldots & \theta_1^{q^{n-1}} \\ \theta_2 & \theta_2^q & & \vdots \\ \vdots & & \ddots & \vdots \\ \theta_n & \theta_n^q & \ldots & \theta_n^{q^{n-1}} \end{pmatrix}.$$

We can express the morphism $\varphi_1 : \mathbb{F}_{q^n} \mapsto (\mathbb{F}_q)^n$ *as*

$$V \mapsto (V, V^q, \ldots, V^{q^{n-1}})\,\mathbf{M}_n^{-1}$$

and its inverse $\varphi_1^{-1} : (\mathbb{F}_q)^n \mapsto \mathbb{F}_{q^n}$ *as*

$$(v_1, \ldots, v_n) \mapsto V_1, \text{ with } (V_1, \ldots, V_n) = (v_1, \ldots, v_n)\,\mathbf{M}_n.$$

Furthermore, we have that $V^q_{(i \bmod n)+1} = V_{(i+1 \bmod n)+1}$.

Proof. The i-th entry of $(v_1, \ldots, v_n)\,\mathbf{M}_n$ is $(\sum_{j=1}^n v_j\theta_j)^{q^i}$, the q^i-th power of the representation of $(v_1, \ldots, v_n)$ in $\mathbb{F}_{q^n}$ with respect to the basis $(\theta_1, \ldots, \theta_n)$. □

The matrix $\mathbf{M}_n$ allows to go back and forth from the big ($\mathbb{F}_{q^n}$) to the small field ($\mathbb{F}_q$). It can be used to have the univariate representation of the public key in a simpler way than in [26]; we replace interpolation by matrix multiplication. For the sake of simplicity, from now on, we consider only linear transformations and homogeneous polynomials. What follows can easily be adapted to the affine case (as pointed in [26]).

Let $\mathbf{F}^{*k}$ be the matrix such that its (i, j)-th entry is $f^{q^k}_{i-k,j-k}$. The matrix $\mathbf{F}^{*k}$ is the "matrix representation" of the q^k-th power of the univariate polynomial F. Indeed, since $F = \sum_{i=0}^{n-1}\sum_{j=0}^{n-1} f_{i,j}X^{q^i+q^j}$, we have

$$\sum_{i=0}^{n-1}\sum_{j=0}^{n-1} f^{q^k}_{i-k,j-k}X^{q^i+q^j} = \sum_{i=0}^{n-1}\sum_{j=0}^{n-1} f^{q^k}_{i,j}X^{q^{i+k}+q^{j+k}} = F^{q^k}.$$

Then, $F^{q^k} = \underline{X}\mathbf{F}^{*k}\underline{X}^t$.

Consider now the symmetric matrices $(\mathbf{G_1}, \ldots, \mathbf{G_n})$ such that $g_i = \underline{x}\mathbf{G_i}\underline{x}^t$ for all i, $1 \leqslant i \leqslant n$, where $\underline{x} = (x_1, \ldots, x_n)$. Using the definition of φ_1 with the matrix $\mathbf{M}_n$, the equation $\mathcal{G} = \mathcal{T} \circ \mathcal{F} \circ \mathcal{S}$ becomes

$$(\mathbf{G_1}, \ldots, \mathbf{G_n}) = (\mathbf{S}\mathbf{M}_n\mathbf{F}^{*0}\mathbf{M}_n^t\mathbf{S}^t, \ldots, \mathbf{S}\mathbf{M}_n\mathbf{F}^{*n-1}\mathbf{M}_n^t\mathbf{S}^t)\mathbf{M}_n^{-1}\mathbf{T}.$$

As $\mathbf{T}$ and $\mathbf{M}_n$ are invertible, we have

$$(\mathbf{G_1}, \ldots, \mathbf{G_n})\,\mathbf{T}^{-1}\mathbf{M}_n = (\mathbf{S}\mathbf{M}_n\mathbf{F}^{*0}\mathbf{M}_n^t\mathbf{S}^t, \ldots, \mathbf{S}\mathbf{M}_n\mathbf{F}^{*n-1}\mathbf{M}_n^t\mathbf{S}^t). \quad (2)$$

In other words, we have a direct relation between the polynomials of the public key written as quadratic forms and the secret polynomial F or more precisely its matrices $\mathbf{F}^{*i}$. From now on, we denote by $\mathbf{U}$ the matrix $\mathbf{T}^{-1}\mathbf{M}_n$ and $\mathbf{W}$ the matrix $\mathbf{S}\mathbf{M}_n$ and rewrite (2) as

$$(\mathbf{G_1}, \ldots, \mathbf{G_n})\,\mathbf{U} = (\mathbf{W}\mathbf{F}^{*0}\mathbf{W}^t, \ldots, \mathbf{W}\mathbf{F}^{*n-1}\mathbf{W}^t). \quad (3)$$

By construction, $u_{i,j+1} = u^q_{i,j}$ and $w_{i,j+1} = w^q_{i,j}$. Thus, we only need to know one column of $\mathbf{U}$ to recover the whole matrix. By considering $(u_{0,0}, \ldots, u_{n-1,0})^t$, the first column of $\mathbf{U}$, we have

$$\sum_{k=0}^{n-1} u_{k,0}\mathbf{G_{k+1}} = \mathbf{W}\mathbf{F}^{*0}\mathbf{W}^t = \mathbf{W}\mathbf{F}\mathbf{W}^t. \quad (4)$$

The equation is similar to (1), but we have not used the univariate representation of $\mathcal{G}$. Here again, as the rank of $\mathbf{F}$ is $\log_q(D)$, so is the rank of $\mathbf{WFW}^t$. Contrarily to the initial attack, $\mathbf{G_i}$ are the public matrices and not matrices with coefficients in the big field. This leads to the following theorem.

Theorem 1. *For HFE, recovering* $\mathbf{U}$ *reduce to solve a MinRank with* $k = n$ *and* $r = \log_q(D)$ *on the public matrices* $\mathbf{G_1}, \ldots, \mathbf{G_n}$ *whose entries are in* $\mathbb{F}_q$.

Computing a Gröbner basis of a system over a smaller field ($\mathbb{F}_q$ instead of $\mathbb{F}_{q^n}$ is faster as the cost of arithmetic operations is decreased. The expected gain is a factor M(n) (the cost of the multiplication of two univariate polynomials of degree n) over the KS attack. In the table below, we compare the original KS Minrank attack and the new MinRank attack on HFE ($N = 1$) with parameters $q = 31$, $D = 31^2 + 31 = 992$. The implementation used is the same as in Sect. 6.

n	8	9	10	11	12	13	14	15	16
KS attack (in s.)	15.3	20.4	76.9	391	680	1969	2439	3197	13407
new attack (in s.)	0.75	1.25	2.05	4.45	8.80	16.9	30.2	68.5	103
ratio	20.4	16.3	37.5	87.9	77.3	117	80.8	46.7	130

3.2 Attacking Multi-HFE

The Kipnis-Shamir attack uses the univariate representation of the public key. In multi-HFE the degree of the univariate representation of the secret key is not bounded. This was in fact the initial motivation for the design of IFS [5]. As a consequence, there is no linear combination of the $\mathbf{G}^{*k}$ leading to a small rank, making the MinRank attack impossible. The hidden field structure exists but it can only be unveiled by working in the right field. To have the correct analogy with the univariate case, we introduce a new change of basis between the small field vector space $(\mathbb{F}_q)^n$ and the big field vector space $(\mathbb{F}_{q^d})^N$.

Proposition 2. *Let* $(\theta_1, \ldots, \theta_d) \in (\mathbb{F}_{q^d})^d$ *be a vector basis of* $\mathbb{F}_{q^d}$ *over* $\mathbb{F}_q$. *Let* $\mathbf{M}_{N,d}$ *be the* $(n \times n)$*-matrix such that* $\mathbf{M}_{N,d} = \mathrm{Diag}(\underbrace{\mathbf{M}_d, \ldots, \mathbf{M}_d}_{N})$. *We can express the morphism* $\varphi_N : (\mathbb{F}_{q^d})^N \mapsto (\mathbb{F}_q)^n$ *as*

$$(V_1, \ldots, V_N) \mapsto (V_1, V_1^q, \ldots, V_1^{q^{d-1}}, \ldots\ldots, V_N, V_N^q, \ldots, V_N^{q^{d-1}})\mathbf{M}_{N,d}^{-1}$$

and its inverse $\varphi_N^{-1} : (\mathbb{F}_q)^n \mapsto (\mathbb{F}_{q^d})^N$ *as*

$(v_1, \ldots, v_n) \mapsto (V_1, V_{d+1}, \ldots, V_{d(N-1)+1})$ *with* $(V_1, \ldots, V_n) = (v_1, \ldots, v_n)\mathbf{M}_{N,d}$.
Furthermore, we have that $V_{i\,d+(j \bmod d)+1}^q = V_{i\,d+(j+1 \bmod d)+1}$.

Proof. The $(d\,(i-1)+j)$-th entry of $(v_1, \ldots, v_n)\mathbf{M}_{N,d}$ is $\left(\sum_{\ell=1}^d v_{d\,(i-1)+\ell}\theta_\ell\right)^{q^j}$. Each N block of d values represents the vector $(V_i, V_i^q, \ldots, V_i^{q^{d-1}})$, for all i, $1 \leqslant i \leqslant N$. Thus $(v_1, \ldots, v_n)\mathbf{M}_{N,d}$ is $(V_1, V_1^q, \ldots, V_1^{q^{d-1}}, \ldots, V_N, V_N^q, \ldots, V_N^{q^{d-1}})$ with respect to the basis $(\theta_1, \ldots, \theta_d)$. □

Note that $\mathbf{M}_{1,d} = \mathbf{M}_d$ which generalizes Proposition 1. When $N > 1$, the q^k-th power of a polynomial $F_i \in \mathbb{F}_{q^d}[X_1, \ldots, X_N]$ is represented by the matrix $\mathbf{F_i}^{*d,k} = [f^{q^k}_{d\lfloor i/d \rfloor + (i-1 \mod d), d\lfloor j/d \rfloor + (j-1 \mod d)}]$ (this definition matches the case $N = 1$). Equation (3) can be generalized for multi-HFE. Let $\mathbf{F_i}^{(j)} = \mathbf{W}\mathbf{F_i}^{*d,j}\mathbf{W}^t$, with i, $1 \leqslant i \leqslant N$, and j, $0 \leqslant j < d$. We have the relation:

$$(\mathbf{G_1}, \ldots, \mathbf{G_n})\,\mathbf{U} = (\mathbf{F_1}^{(0)}, \ldots, \mathbf{F_1}^{(d-1)}, \ldots\ldots, \mathbf{F_N}^{(0)}, \ldots, \mathbf{F_N}^{(d-1)}).$$

Similarly to (4), as $\mathbf{F_i}^{*d,0} = \mathbf{F_i}$, when we consider the $(i\,d)$-th columns of $\mathbf{U}$ for $0 \leqslant i < N$ we have

$$\sum_{k=0}^{n-1} u_{k,0}\mathbf{G_{k+1}} = \mathbf{W}\mathbf{F_1}\mathbf{W}^t, \ldots, \sum_{k=0}^{n-1} u_{k,Nd}\mathbf{G_{k+1}} = \mathbf{W}\mathbf{F_N}\mathbf{W}^t. \tag{5}$$

As in the univariate case, the problem of finding correct values for $\mathbf{U}$ turns to be a simultaneous MinRank problem.

Theorem 2. *For multi-HFE, recovering* $\mathbf{U}$ *reduce to simultaneously solve* N *MinRank problems with* $k = n$ *and* $r = N\log_q(D)$ *on the public matrices* $\mathbf{G_1}, \ldots, \mathbf{G_n}$ *whose entries are in* $\mathbb{F}_q$.

Proof. Each polynomial F_i has degree bounded by D, thus each variable X_i has at most degree D. By construction of the matrix $\mathbf{M}$ of Proposition 2, the only non-zero entries of the matrix $\mathbf{F_i} = \mathbf{F_i}^{*d,0}$ are the ones in the upper-left $\log_q(D)$ square of each N diagonal $(d \times d)$ block. The rank of $\mathbf{F_i}$ is then $N\log_q(D)$. By construction, the rank of $\mathbf{F_i}^{*d,j}$ is left unchanged. □

Before discussing of the complexity of the MinRank attack for Multi-HFE, we introduce equivalent keys.

3.3 About Equivalent Keys and Induced Degrees of Freedom

Two keys are equivalent if they have the same public key. The subject has already been treated for original HFE [31,30]. It has been shown to have (at least) $(n\,q^{2n}(q^n-1)^2)$ equivalent keys. A larger number of equivalent keys in multi-HFE induces a degree of freedom when solving the MinRank that can be used to attack the minus variant. Due to space limitations, proofs of Propositions 3, 4, and 5 will be given in an extended version of this paper.

Definition 1. *Let* $(\mathcal{F}^*, \mathcal{S}, \mathcal{T})$ *be a multi-HFE private key with parameters* (q, N, d, D). *We say that* $(\mathcal{F}^{*\prime}, \mathcal{S}', \mathcal{T}')$ *is an equivalent key iff* $\mathcal{F}^{*\prime}$ *has a HFE-shape, and* $\mathcal{T}' \circ \varphi_N \circ \mathcal{F}^{*\prime} \circ \varphi_N^{-1} \circ \mathcal{S}' = \mathcal{G} = \mathcal{T} \circ \varphi_N \circ \mathcal{F}^* \circ \varphi_N^{-1} \circ \mathcal{S}$ *(same public key).*

Wolf and Preneel [31] introduced the notion of sustaining transformations which is a couple of affine transformations $(\mathcal{A}^*, \mathcal{B}^*)$ such that $\mathcal{B}^* \circ \mathcal{F}^* \circ \mathcal{A}^*$ preserves the "shape" of $\mathcal{F}^*$. For HFE, the "big sustainer" (multiplication in the big field), the "additive sustainer" and the "Frobenius sustainer" keep the HFE-shape unchanged. In multi-HFE, not only multiplication keeps the HFE-shape. We also have any affine transformation on the N variables. Thus, the two first sustainers can be generalized as follows.

Lemma 1. *Let* $(q, N, d, D) \in \mathbb{N}^4$ *and* $\mathcal{F}^* : (\mathbb{F}_{q^d})^N \mapsto (\mathbb{F}_{q^d})^N$ *a mapping with HFE-shape. Let* $\mathcal{A}^*, \mathcal{B}^*$ *be invertible affine transformations over* $(\mathbb{F}_{q^d})^N$*. Then* $\mathcal{B}^* \circ \mathcal{F}^* \circ \mathcal{A}^*$ *has the HFE-shape.*

Proof. The only exponents occurring in a single variable X_i is a power of q. The transformation $\mathcal{A}^*$ mixes the variables $X_1, \ldots, X_N$ by affine combinations. Thus by linearity of the Frobenius, we know that no other exponents can appear and the system keeps its HFE-shape. Trivially, as $\mathcal{B}^*$ only performs affine combinations of the polynomials $F_1, \ldots, F_N$ the shape is also unchanged. □

With lemma 1, we can produce HFE internal maps while keeping the same property. To build equivalent keys, we look at these affine transformations in the small field $\mathbb{F}_q$.

Proposition 3. *Let* $(\mathcal{F}^*, \mathcal{S}, \mathcal{T})$ *be a multi-HFE private key with parameters* (q, N, d, D)*. For any invertible affine transformations* $\mathcal{A}^*, \mathcal{B}^*$ *over* $(\mathbb{F}_{q^d})^N$*, let* $\mathcal{A} = \varphi_N \circ \mathcal{A}^* \circ \varphi_N^{-1}$ *and* $\mathcal{B} = \varphi_N \circ \mathcal{B}^* \circ \varphi_N^{-1}$*, then* $\left(\mathcal{B}^* \circ \mathcal{F}^* \circ \mathcal{A}^*, \mathcal{A}^{-1} \circ \mathcal{S}, \mathcal{T} \circ \mathcal{B}^{-1}\right)$ *is an equivalent key.*

The following proposition gives the structure of one of these transformations in the linear case. It has to be slightly adapted in the affine case.

Proposition 4. *Let* $\mathbf{A}^* = [a_{i,j}]$ *be the matrix representing a linear transformation* $\mathcal{A}^*$ *over* $(\mathbb{F}_{q^d})^N$*.* $\mathcal{A}^*$ *can be represented in the field* $\mathbb{F}_q$ *as* $\mathbf{A} = \mathbf{M}_{N,d}\widetilde{\mathbf{A}^*}\mathbf{M}_{N,d}^{-1}$ *where* $\mathbf{M}_{N,d}$ *is the matrix of Proposition 2 and* $\widetilde{\mathbf{A}^*}$ *is a matrix of* $N \times N$ *blocks of Frobenius powers of elements of* $\mathbf{A}^*$*, i.e.*

$$\widetilde{\mathbf{A}^*} = \begin{pmatrix} \begin{vmatrix} a_{0,0} & & & \\ & a_{0,0}^q & & \\ & & \ddots & \\ & & & a_{0,0}^{q^{d-1}} \end{vmatrix} & \cdots & \begin{vmatrix} a_{0,N-1} & & & \\ & a_{0,N-1}^q & & \\ & & \ddots & \\ & & & a_{0,N-1}^{q^{d-1}} \end{vmatrix} \\ \vdots & & \vdots \\ \begin{vmatrix} a_{N-1,0} & & & \\ & a_{N-1,0}^q & & \\ & & \ddots & \\ & & & a_{N-1,0}^{q^{d-1}} \end{vmatrix} & \cdots & \begin{vmatrix} a_{N-1,N-1} & & & \\ & a_{N-1,N-1}^q & & \\ & & \ddots & \\ & & & a_{N-1,N-1}^{q^{d-1}} \end{vmatrix} \end{pmatrix}$$

In addition, for any k, $0 \leqslant k < d$, the components polynomials of $(\mathrm{Frob_k} \circ \mathcal{F}^* \circ \mathrm{Frob_{d\text{-}k}})(X_1, \ldots, X_N) = \left(\mathcal{F}^*(X_1^{q^{d-k}}, \ldots, X_N^{q^{d-k}})\right)^{q^k}$ have the same monomials as $\mathcal{F}^*(X_1, \ldots, X_N)$ but their coefficients are raised to the power of q^k. That is, if $\mathcal{F}^*(X_1, \ldots, X_N)$ has HFE-shape, so is $(\mathrm{Frob_k} \circ \mathcal{F}^* \circ \mathrm{Frob_{d\text{-}k}})(X_1, \ldots, X_N)$.

Proposition 5. *Let* $(\mathcal{F}^*, \mathcal{S}, \mathcal{T})$ *be a multi-HFE private key with parameters* $(q, N, d, D) \in \mathbb{N}^4$*. For all* k*,* $0 \leqslant k < d$*,*

$$\left(\mathrm{Frob_k} \circ \mathcal{F}^* \circ \mathrm{Frob_{d\text{-}k}}, \varphi_N \circ \mathrm{Frob_k} \circ \varphi_N^{-1} \circ \mathcal{S}, \mathcal{T} \circ \varphi_N \circ \mathrm{Frob_{d\text{-}k}} \circ \varphi_N^{-1}\right)$$

is an equivalent key.

According to Proposition 5, we can derive $(d-1)$ other equivalent keys from any valid private key. This refers to the so-called Frobenius sustainer of [31]. We can count the number of equivalent keys.

Theorem 3. *For any multi-HFE private key* $(\mathcal{F}^*, \mathcal{S}, \mathcal{T})$ *with given parameters* $(q, N, d, D) \in \mathbb{N}^4$, *there are at least* $\left(q^{dN} \prod_{i=0}^{N-1}(q^{dN} - q^{d\,i})\right)^2$ *equivalent keys coming from affine transformations in* $\mathrm{Aff}(N, \mathbb{F}_{q^d})$.

Proof. There are exactly $\prod_{i=0}^{N-1}\left((q^d)^N - (q^d)^i\right)$ invertible $(N \times N)$-matrices over $\mathbb{F}_{q^d}$. We have to multiply this by $(q^d)^N$ to include the affine transformations. From Proposition 3, one can choose 2 invertible affine transformations over the big field to build an equivalent key, thus the previous value is squared. □

This number may actually be bigger (at most d times) using the Frobenius sustainer. An interesting particularity of the MinRank arising in HFE/multi-HFE is that the kernel of the matrices in (5) is independent on which equivalent key is used up to Frobenius transforms.

Theorem 4. *Let* $(\mathcal{F}^*, \mathcal{S}, \mathcal{T})$ *and* $(\mathcal{F}^{*\prime}, \mathcal{S}', \mathcal{T}')$ *be equivalent multi-HFE private keys and* $(\mathbf{G_1} \ldots \mathbf{G_n})$ *be the matrices of their associated public key. Let* $\mathbf{S}$, $\mathbf{T}$, $\mathbf{S}'$, *and* $\mathbf{T}'$ *be the matrix representation of respectively* $\mathcal{S}$, $\mathcal{T}$, $\mathcal{S}'$, *and* $\mathcal{T}'$. *Let* $\mathbf{U} = \mathbf{T}^{-1}\mathbf{M}_{N,d} = [u_{i,j}]$, $\mathbf{K} = \ker(\sum_{i=0}^{n} u_{i,0}\mathbf{G_i})$, $\mathbf{U}' = \mathbf{T}'^{-1}\mathbf{M}_{N,d} = [u'_{i,j}]$ *and* $\mathbf{K}' = \ker(\sum_{i=0}^{n} u'_{i,0}\mathbf{G_i})$, *then* $\exists k,\ 0 \leqslant k < d, \mathbf{K}' = \mathrm{Frob_k}\,(\mathbf{K})$.

Proof. By construction of equivalent keys, $u'_{i,j}$ are linear combinations of the $u_{i,\ell}^{q^k}$ for a given k. Linear combinations of $u_{i,j}$ do not change the kernel. By linearity, $u'_{i,j} = \sum_\ell \alpha_\ell u_{i,\ell}^{q^k} = \left(\sum_\ell \alpha_\ell u_{i,\ell}\right)^{q^k}$. Consequently, $\mathbf{K}' = \mathrm{Frob_k}(\mathbf{K})$. □

We discuss the complexity of our attack in the next section.

3.4 Complexity Analysis of the Attack

In this section, we study the particularities of the MinRank problems coming from (5). Here again we consider only linear maps and homogeneous polynomials for the sake of simplicity.

Let an instance of HFE with parameters $(q, N, d, D) \in \mathbb{N}^4$, and $\ell = \lceil \log D \rceil$. We have to solve the MinRank problem on the $(n \times n)$-matrices $\mathbf{G_1} \ldots \mathbf{G_n}$ whose entries lie in $\mathbb{F}_q$ with target rank $N\ell$. Using the Kipnis-Shamir modeling described in [26,20,21], it is equivalent to solve the algebraic system of the $(n\,(n - N\ell))$ bilinear equations in $(N\ell\,(n - N\ell) + n)$ variables given by the entries of the matrix

$$\begin{pmatrix} 1 & & & x_{1,1} & \ldots & x_{1,N\ell} \\ & \ddots & & \vdots & & \vdots \\ & & 1 & x_{n-N\ell,1} & \ldots & x_{n-N\ell,N\ell} \end{pmatrix} \cdot \left(\sum_{i=1}^{n} \lambda_i \mathbf{G_i}\right). \tag{6}$$

Note that we are looking for solutions in the field $\mathbb{F}_{q^d}$ rather than in $\mathbb{F}_q$.

From now on, and similarly to [20], these equations are called the KS (Kipnis-Shamir) equations. We denote by $\mathcal{I}_{\mathrm{KS}}$ the ideal generated by the KS equations and $\mathcal{V}_{\mathrm{KS}} \subset \mathbb{F}_{q^d}$ its associated variety.

Theorem 5. *The MinRank problem associated to HFE (resp. multi-HFE) can be solved by fixing one (resp. N) coefficients to random values. That is, the dimension of* $\mathcal{I}_{\mathrm{KS}} \cap \mathbb{F}_q[\lambda_1, \ldots, \lambda_n]$ *is at least one (resp. N).*

Proof. We know that any column of $\mathbf{U} = \mathbf{T}^{-1}\mathbf{M}_{N,d}$ is a solution of MinRank for $(\lambda_1, \ldots, \lambda_n)$. From Proposition 3, for any invertible matrix $\mathbf{A}^*$, the columns of the matrix $\mathbf{U}\widetilde{\mathbf{A}^*}$ give a solution $(\lambda_1, \ldots, \lambda_n)$ for the MinRank. As each column of $\widetilde{\mathbf{A}^*}$ has N non-zero entries, this allows to choose N coefficients λ_i arbitrarily. □

This means that for valid values $x_{i,j}$, there are $(q^d)^N$ possible vectors $(\lambda_1, \ldots, \lambda_n)$ such that the kernel of $\left(\sum_{i=1}^{n} \lambda_i \mathbf{G_i}\right)$ is the one induced by the $x_{i,j}$'s. Therefore, the values of N components (say $\lambda_1, \ldots, \lambda_N$) can be randomly chosen. The new system still has $(n\,(n-N\ell))$ equations but only $(N\ell\,(n-N\ell)+n-N)$ variables. As described in Sect. 3.1, the coefficients are in the small field $\mathbb{F}_q$. To keep this property, we fix variables with values over the small field. Experimentally, fixing one variable to 1 (or any value from $\mathbb{F}_q$) and the $(N-1)$ others to 0 gives the best results. After N variables $(\lambda_1, \ldots, \lambda_N)$ have been fixed, $\mathcal{V}_{\mathrm{KS}}$ has at least d elements. This property already noticed in [25] for HFE is a direct consequence of theorem 4. Once $\mathbf{K} = \ker\left(\sum_{k=1}^{n} \lambda_k \mathbf{G_k}\right)$ is recovered, finding a valid transformation $\mathbf{U}'$ is done by solving a linear system as entries of (6) become linear. Some experimental results of our attack are presented in Sect. 6.

It is interesting to remark that the degree of regularity experimentally observed seems to be constant when d grows. This behavior can be explained theoretically using the bound on the degree of regularity of MinRank given in [21].

Proposition 6 (Faugère, Safey El Din, Spaenlehauer [21]). *Let* (n, r, k) *be the parameters of a MinRank instance. Let* $\mathbf{A} = [a_{i,j}]$ *be the* $(r \times r)$*-matrix defined by* $a_{i,j}(t) = \sum_{\ell=0}^{n-\max(i,j)} \binom{n-i}{\ell}\binom{n-j}{\ell} t^{\ell}$. *The degree of regularity of the system associated to MinRank instance is bounded from above by* $1 + \deg(\mathrm{HS}(t))$ *where* $\mathrm{HS}(t)$ *is the polynomial obtained from the first positive terms of the series* $(1-t)^{(n-r)^2-k} \frac{\det \mathbf{A}(t)}{t^{\binom{r}{2}}}$.

Back to our specific MinRank problem, we have instantiate this theoretical bound with multi-HFE parameters for values of $N \leqslant 20$ and $\ell \leqslant 10$. When d, is sufficiently bigger than ℓ, we always obtain $(N\ell + 1)$ (verified for $N\,d$ up to 500). Since the parameter d is not involved we state the following conjecture.

Conjecture 1. The degree of regularity of the MinRank problem associated to a multi-HFE instance does not depend on d. When d grows to infinity, it is bounded from above by $(N\ell + 1)$.

The degree of regularity depends only in the number N of secret variables and the degree D of the secret polynomials. This is consistent with the observations on simple HFE where d_{reg} was observed to be $\log(D)$. We have the necessary material to evaluate the difficulty of MinRank involved in HFE/multi-HFE.

Proposition 7. *Assuming Conjecture 1, for N and ℓ fixed, the complexity of solving the multi-HFE MinRank problem is $\mathcal{O}\left(d^{(N\ell+1)\,\omega}\right)$ $(2 \leqslant \omega < 3$ being the linear algebra constant$)$ and thus polynomial in d.*

Proof. According to Conjecture 1, the degree of regularity is $(N\ell + 1)$ and thus independent of the degree of the extension d. When d grows to infinity, the complexity of the Gröbner basis computation [2,3] is $\mathcal{O}\left(\binom{Nd+N\ell+1}{N\ell+1}^{\omega}\right) \sim \mathcal{O}\left((Nd)^{(N\ell+1)\,\omega}\right) \sim \mathcal{O}\left(d^{(N\ell+1)\,\omega}\right)$. □

Once the matrix $\mathbf{U}$ has been found with the MinRank attack, we need to recover the matrix $\mathbf{W}$.

3.5 Recovering the Transformation on the Variables

Kipnis and Shamir [26] originally proposed a method for this step by solving an overdetermined system of $(n\,\ell\,(n-\ell))$ linear equations in n^2 variables over $\mathbb{F}_q$. Applied to multi-HFE, it would give $(n\,\ell\,(n-N\ell))$ equations in n^2 variables over $\mathbb{F}_q$. We propose here an alternative method which reduces the number of variables and equations by a factor d while it is done over the big field.

Lemma 2. *Let $(\mathbf{G_1},\ldots,\mathbf{G_n})$ be a multi-HFE public key and $\ell = \lceil \log_q(D) \rceil$. Suppose that* $\mathrm{Rank}(\sum_{k=1}^{n} \lambda_k \mathbf{G_k}) = N\ell$ *and let* $\mathbf{K} = \ker\left(\sum_{k=1}^{n} \lambda_k \mathbf{G_k}\right)$. *Once $\mathbf{K}$ is known, then we can recover a matrix $\mathbf{W}' = \mathbf{S}'\mathbf{M}_{N,d}$ such that $\mathbf{S}'$ is a valid matrix for the private key by solving a linear system of $(N\ell\,(n-N\ell))$ equations in $(N\,(n-N))$ variables.*

Proof. To find the coefficients $w_{i,j}$, it is enough to remark that from (5) one has $\mathbf{KW}' = \ker(\mathbf{F_i})$. We know by construction of the private key that $\ker(\mathbf{F_i})$ has $N\ell$ columns set to zero. By construction of $\mathbf{W}'$, N columns are needed to build the whole matrix. We build the corresponding linear system of $\left(N\,(n-N\ell)\right)$ equations in Nn variables. Proposition 3 tells us that one can randomly fix N variables on each of the N columns which gives $(N\,(n-N))$ variables left. If $\ell > 1$, the system is underdetermined. To find the matrix, we have to add the $\left((\ell-1)\,N\,(n-N\ell)\right)$ equations coming from $\mathrm{Frob_j}(\mathbf{K})\mathbf{W}' = \ker({\mathbf{F_i}}^{*d,j})$. For j, $(d-\ell+1) \leqslant j < d$, it can be verified that $\ker({\mathbf{F_i}}^{*d,j})$ has also $N\ell$ columns set to zero. The system has $\left(N\ell\,(n-N\ell)\right)$ linear equations. □

Recovering the polynomial system. Once the matrices $\mathbf{T}' = \mathbf{M}_{N,d}\mathbf{U}'^{-1}$ and $\mathbf{S}' = \mathbf{W}'\mathbf{M}_{N,d}^{-1}$ are recovered, we only need to reconstruct a private transformation. It is done simply by computing $\mathcal{F}^{*\prime} = \varphi_N^{-1} \circ \mathcal{T}'^{-1} \circ \mathcal{G} \circ \mathcal{S}'^{-1} \circ \varphi_N$. By construction of its components, the transformation $\mathcal{F}$ respects the HFE-shape.

3.6 Weaknesses of Multi-HFE Relative to the Original HFE

In order to compare instances of HFE/multi-HFE, we introduce the notion of "similarity" between instances. Two similar instances share the same size of public key and private key.

Definition 2. *Two (multi-)HFE instances of resp. parameters* (q_1, N_1, d_1, D_1) *and* (q_2, N_2, d_2, D_2) *are* similar *iff* $q_1 = q_2$ *and* $N_1 d_1 = N_2 d_2$ *and* $N_1 \log_{q_1}(D_1) = N_2 \log_{q_2}(D_2)$ *holds.*

The KS equations of two similar instances have the same number of variables and equations as the target rank is the same $N \log_q(D)$. According to the complexity of the MinRank given in Proposition 7, the bigger is d, the harder it is to mount our attack. In particular, the case $N = 1$ (original HFE) is the more resistant. This behavior has been verified experimentally. For similar keys, choosing $N = 1$ seems to be the optimal value for security. With respect to our attack, multi-HFE is then less secure than HFE.

As a side remark, speed of decryption has to be taken into account when designing a scheme. Choosing $N = 1$ and a big degree D of the inner univariate polynomial can sometimes dramatically slow down the decryption process for similar keys. Multi-HFE construction could still be competitive if a modification can prevent attacks. To this end, the minus modifier and the embedding modifier have been proposed. We study these variants in the next sections.

4 Multivariate HFE$^-$

In this section, we study a classical variant of multivariate schemes, the so-called "minus" modifier. It consists in removing some polynomials from the public key. This construction is only suitable for signature as the decryption (signature generation) is not unique.

Description. Let $(\mathcal{F}^*, \mathcal{S}, \mathcal{T})$ be a multi-HFE private key as defined in Sect. 2 with parameters $(q, N, d, D) \in \mathbb{N}^4$. We define the parameter $s \in \mathbb{N}$ and the projection $\pi : (\mathbb{F}_q)^n \mapsto (\mathbb{F}_q)^{n-s}$. The public key is the mapping $\mathcal{G} = \pi \circ \mathcal{T} \circ \varphi_N^{-1} \circ \mathcal{F}^* \circ \varphi_N \circ \mathcal{S}$ viewed as $(n-s)$ polynomials in n variables. To sign, s random values from $\mathbb{F}_q$ are appended to a message $\underline{m} = (m_1, \ldots, m_{n-s})$ before applying the basic decryption process. Verifying a signature consists in its evaluation in $\mathcal{G}$.

Attack. The goal is to find a valid private key with only $(n-s)$ public polynomials. Usually the minus modification is enough to prevent classical attacks as some information is missing. In particular it is the case for basic HFE ($N = 1$). In Sect. 3.4, we have shown that the problem has N degrees of freedom. Indeed, only $(n - N + 1)$ matrices are needed to recover the kernel. This means that if $s < N$, the kernel matrix $\mathbf{K}$ can still be found with no additional cost. Still, the recovering step has to be adapted. We know that there exists a vector $(\lambda_1, \ldots, \lambda_n)$ and symmetric $(n \times n)$-matrices $(\mathbf{\Gamma_1}, \ldots, \mathbf{\Gamma_s})$ such that

$$\ker\left(\sum_{i=1}^{n-s} \lambda_i \mathbf{G_i} + \sum_{i=1}^{s} \lambda_{n-s+i} \mathbf{\Gamma_i}\right) = \mathbf{K}.$$

The $\mathbf{\Gamma_i}$'s matrices are unknown and correspond to the removed polynomials. If we fix N values λ_i, we still have solutions to our system. For instance, let $(\lambda_{n-N+1}, \ldots, \lambda_n) = (\ell_1, \ldots, \ell_N)$. We write

$$\mathbf{K} \cdot \left(\sum_{i=1}^{n-N} \lambda_i \mathbf{G_i} + \sum_{i=1}^{N-s} \ell_i \mathbf{G_{n-N+i}} + \sum_{i=1}^{s} \ell_{N-s+i} \mathbf{\Gamma_i} \right) = \mathbf{0}. \tag{7}$$

The resulting system has $n\,(n - N\ell)$ *linear* equations in $\left((n-N) + s\frac{n\,(n+1)}{2}\right)$ variables. The system is greatly underdetermined and hence have many solutions. To find the correct entries, we use the following remark:

Proposition 8. *For any* j, $0 \leqslant j < d$, *we have* $\mathrm{Frob_j}(\mathbf{K}) \cdot \left(\sum_{i=1}^{n} \lambda_i^{q^j} \mathbf{G_i}\right) = \mathbf{0}$.

Proof. By definition, $\mathrm{Frob_j}\left(\mathbf{K} \cdot (\sum_{i=1}^{n} \lambda_i \mathbf{G_i})\right) = \mathbf{0}$. By linearity of the Frobenius, this is equal to $\mathrm{Frob_j}(\mathbf{K}) \cdot \left(\sum_{i=1}^{n} \lambda_i^{q^j} \mathrm{Frob_j}(\mathbf{G_i})\right)$. As $\mathbf{G_i}$ has its entries in $\mathbb{F}_q$, we also have that $\mathrm{Frob_j}(\mathbf{G_i}) = \mathbf{G_i}$. □

Solving together equations (7) and their Frobenius images forces the entries of $\mathbf{\Gamma_i}$ to be in $\mathbb{F}_q$. To avoid carrying equations of degree q^j (coming from $\lambda_i^{q^j}$), we add $(d-1)(n-N)$ new variables $(\lambda_1^{(1)}, \ldots, \lambda_{n-N}^{(1)}, \ldots, \lambda_1^{(d-1)}, \ldots, \lambda_{n-N}^{(d-1)})$. The new system then becomes:

$$\mathrm{Frob_j}(\mathbf{K}) \cdot \left(\sum_{i=1}^{n-N} \lambda_i^{(j)} \mathbf{G_i} + \sum_{i=1}^{N-s} \ell_i^{q^j} \mathbf{G_{n-N+i}} + \sum_{i=1}^{s} \ell_{N-s+i}^{q^j} \mathbf{\Gamma_i} \right) = \mathbf{0},$$

for all j, $0 \leqslant j < d$. The resulting system is overdetermined and has a solution if $(\ell_1, \ldots, \ell_N) \neq (0, \ldots, 0)$. We have to solve N times this linear system with different values for $(\ell_1, \ldots, \ell_N)$ to get a valid matrix $\mathbf{U}$. With this technique, the private key of a multi-HFE$^-$ can be recovered almost as efficiently as the standard construction if the number of withdrawn equations is less than $(N-1)$. Experimental results are presented in Sect. 6.

5 Multivariate HFE with Embedding

In [14], it has been proposed to use a variant of HFE with embedding. This so-called PHFE construction consists in removing few variables of the public key and is claimed to resist Kipnis-Shamir's attack. The authors of [11] use the same modification on multi-HFE and claim that it prevents a possible "big-field" based attack. Still, for both PHFE and its multivariate version a key recovery attack is possible.

Description. Let $(\mathcal{F}^*, \mathcal{S}, \mathcal{T})$ be a multi-HFE private key as defined in Sect. 2 with parameters $(q, N, d, D) \in \mathbb{N}^4$. We define a new parameter $r \in \mathbb{N}$ and the embedding $\rho : (\mathbb{F}_q)^{n-r} \mapsto (\mathbb{F}_q)^n$ which is part of the private key. Then the public key is the mapping $\mathcal{G} = \mathcal{T} \circ \varphi_N^{-1} \circ \mathcal{F}^* \circ \varphi_N \circ \mathcal{S} \circ \rho$. To encrypt a plaintext, we still evaluate $\mathcal{G}$. To decrypt, as in the standard scheme, one inverts each component separately. To simplify, we can assume w.lo.g. that the embedding is always $\rho_0 : (x_1, \ldots, x_{n-r}) \in (\mathbb{F}_q)^{n-r} \mapsto (x_1, \ldots, x_{n-r}, 0, \ldots, 0) \in (\mathbb{F}_q)^n$. Indeed, from any embedding ρ and any invertible transformation $\mathcal{S}$, one can find an invertible transformation $\mathcal{S}'$ such that $\mathcal{S} \circ \rho = \mathcal{S}' \circ \rho_0$; this gives equivalent keys.

Attack. The matrix representation $\mathbf{G_i}$ of the public key polynomials have $(n-r)$ rows and columns. However, the rank of $\sum_{i=0}^{n} u_{i,0}\mathbf{G_{i+1}}$ remains bounded by $N\log_q(D)$ (i.e. removing rows or columns does not increase the rank).

Let $\mathbf{K} = \ker\left(\sum_{i=0}^{n} u_{i,0}\mathbf{G_{i+1}}\right)$. As usual a matrix $\mathbf{U}'$ can still be recovered by solving a MinRank as soon as $\mathbf{K}$ is known. The problem appears when trying to recover the matrix $\mathbf{W}' = \mathbf{S}'\mathbf{M}_{N,d}$ where $\mathbf{S}'$ is an equivalent matrix for the private key. By following the method described in Sect. 3.5, we get a system having $N\ell\,(n-r-N\ell)$ equations with only $N\,(n-r-N)$ variables. Let the following matrix be a solution of this linear system

$$\mathbf{W}' = \begin{pmatrix} w_{0,0} & w_{0,0}^{q} & \cdots & w_{0,0}^{q^{d-1}} & \cdots\cdots & w_{0,N-1} & w_{0,N-1}^{q} & \cdots & w_{0,N-1}^{q^{d-1}} \\ \vdots & \vdots & \cdots & \vdots & \cdots\cdots & \vdots & \vdots & \cdots & \vdots \\ w_{n-r,0} & w_{n-r,0}^{q} & \cdots & w_{n-r,0}^{q^{d-1}} & \cdots\cdots & w_{n-r,N-1} & w_{n-r,N-1}^{q} & \cdots & w_{n-r,N-1}^{q^{d-1}} \end{pmatrix}.$$

The matrix $\mathbf{W}'$ has $(n-r)$ rows and thus is not invertible. However, such $\mathbf{W}'$ needs to be inverted in order to compute a full private key.

The first idea is to build a new invertible matrix $\mathbf{W_r}$ by appending to $\mathbf{W}'$ a $(r \times n)$-matrix $\mathbf{V} = [v_{i,j}]$ such that $v_{i,j}^{q} = v_{i,j+1}$. The secret transformation is reconstructed by computing $\mathbf{G_i}' = \mathbf{W_r}^{-1}\mathbf{G_i}\mathbf{W_r}^{-t}$. As the matrix $\mathbf{W_r}^{-1}$ has non-zero coefficients in its r last rows, so is $\mathbf{G}_i'$. Since the MinRank was done over $(n-r \times n-r)$-matrices, when we finally compute $\sum_{i=0}^{n} u_{i,0}\mathbf{G_{i+1}}'$, monomials in the last variables $(x_{n-r+1},\ldots,x_n)$ are mixed with the other monomials, which eventually leads to polynomials that are not in HFE-shape (and then hard to invert). To circumvent this issue, we no longer append a random matrix to $\mathbf{W}'$, we construct an invertible matrix $\mathbf{W_z}$ by appending vertically to $\mathbf{W}'$ the matrix

$$\mathbf{Z} = \begin{pmatrix} 0 & \cdots\cdots\cdots\cdots & 0 & 1 & \\ \vdots & & \vdots & & \ddots & \\ 0 & \cdots\cdots\cdots\cdots & 0 & & & 1 \end{pmatrix}.$$

We ensure the property that $\mathbf{W_z}$ is invertible. The variables $(x_{n-r+1},\ldots,x_n)$ do not appear when we build $\mathbf{G_i}' = \mathbf{W_z}^{-1}\mathbf{G_i}\mathbf{W_z}^{-t}$, and the rank property is preserved. The only difference is that the relation $w_{i,j}^{q} = w_{i,j+1}$ only holds for all i, $0 \leqslant i < n-r$. The consequence is that $\mathbf{S}' = \mathbf{W_z}\mathbf{M}_{N,d}^{-1}$ has coefficients in the big field $\mathbb{F}_{q^d}$. Still, $\mathbf{S}'$ can be inverted and a mapping $\mathcal{F}^*$ with HFE-shape can be recovered. Experimental results are given in Sect. 6.

6 Experimental Results

We present some experimental results for our attacks implemented in MAGMA [7] (V2.16-10). All the timings have been obtained on a 2.93 GHz Intel® Xeon® CPU. The MinRank's have been solved using the Kipnis-Shamir modeling.

The degree of regularity experimentally observed is noted d_{reg}. The theoretical degree of regularity is denoted by $d_{\mathrm{reg}}^{\mathrm{theo}}$. We applied our attack to the real-scale parameters proposed in [10] (multi-HFE with embedding). They are not secure since they are practically broken (9 days for the most conservative, i.e. 256 bits claimed security). One may get even better results using the minors modeling of MinRank and the F_5 implementation available in the FGb software [18]. The following results are obtained on the same computer.

q	N	d	D	security	$d_{\text{reg}}^{\text{theo}}$	time MAGMA	mem MAGMA	time FGb	d_{reg}
31	2	15	2	150 bits	3	2 min 27 s	434 MB	21.1 s	3
31	3	10	2	150 bits	4	1 h 38 min	1500 MB	24 min 56 s	3
31	3	15	2	192 bits	4	2 days 1 h	12 GB		3
31	3	18	2	256 bits	4	9 days 16 h	33 GB		3

We also compare the different steps of our attack to the minus and the embedding variants for multi-HFE with parameters $q = 31, N = 3, d = 8, D = 2$ ($\approx$ 120 bits security). The minus modifier does not change the time of the MinRank attack but recovering $\mathbf{W}$ will be slower. In practice, multi-HFE with the embedding takes more time to break but the degree of regularity is the same.

	MR time	MR d_{reg}	Finding $\mathbf{U}$	Finding $\mathbf{W}$
No variant (ref. time)	23.3 s	3	0.01 s	7.29 s
Minus ($s = 1$)	23.2 s	3	0.01 s	16.71 s
Minus ($s = 2$)	23.4 s	3	0.01 s	35.24 s
Minus ($s = 3$)		Not possible		
Embedding ($r = 1$)	788 s	3	0.01 s	6.14 s
Embedding ($r = 2$)	2811 s	3	0.01 s	5.25 s
Embedding ($r = 3$)	401 s	3	0.01 s	4.44 s

7 Conclusion

Multi-HFE over an odd-characteristic field seems to fix the weaknesses of HFE. The embedding modifier was also proposed to better hide the big field structure in the public key. These properties turn out to be weaknesses. Not only does our attack allow to do a complete key recovery in polynomial time, it is also more efficient on multi-HFE than on original HFE. On multi-HFE, key recovery on real-size parameters becomes practical. We broke parameter sets from [11] up to claimed 256 bits security. It is therefore insecure to use multi-HFE. Increasing the number N of secret variables/equations or their degree D may lead to a set of parameters out of reach of our attack but then, the rightful decryption would be very slow or infeasible. With respect to our attacks, among the studied constructions, only the minus variants of HFE/multi-HFE are secure if the removed equations is bigger than $(N - 1)$. Note that vinegar variants of HFE are not concerned.

Acknowledgments. We would like to thank C. Wolf for his helpful comments which helped to improve the paper. The work described in this paper has been supported in part by the European Commission through the ICT programme under contract ICT-2007-216676 ECRYPT II. The authors were also supported in part by the french ANR under the Computer Algebra and Cryptography (CAC) project ANR-09- JCJCJ-0064-01.

References

1. Adams, W.W., Loustaunau, P.: An Introduction to Gröbner Bases. Graduate Studies in Mahematics, vol. 3. AMS, Providence (1994)
2. Bardet, M., Faugère, J.-C., Salvy, B.: On the complexity of Gröbner basis computation of semi-regular overdetermined algebraic equations. In: Proc. International Conference on Polynomial System Solving (ICPSS), pp. 71–75 (2004)
3. Bardet, M., Faugère, J.-C., Salvy, B., Yang, B.-Y.: Asymptotic behaviour of the degree of regularity of semi-regular polynomial systems. In: Proc. of MEGA 2005, Eighth International Symposium on Effective Methods in Algebraic Geometry (2005)
4. Bettale, L., Faugère, J.-C., Perret, L.: Hybrid approach for solving multivariate systems over finite fields. Journal of Mathematical Cryptology, 177–197 (2009)
5. Billet, O., Patarin, J., Seurin, Y.: Analysis of Intermediate Field Systems. In: SCC 2008 (2008)
6. Bogdanov, A., Eisenbarth, T., Rupp, A., Wolf, C.: Time-area optimized public-key engines: MQ-cryptosystems as replacement for elliptic curves? In: Oswald, E., Rohatgi, P. (eds.) CHES 2008. LNCS, vol. 5154, pp. 45–61. Springer, Heidelberg (2008)
7. Bosma, W., Cannon, J.J., Playoust, C.: The Magma algebra system I: The user language. Journal of Symbolic Computation 24(3-4), 235–265 (1997)
8. Buchberger, B.: Ein Algorithmus zum Auffinden der Basiselemente des Restklassenringes nach einem nulldimensionalen Polynomideal. Ph.D. thesis, University of Innsbruck (1965)
9. Buss, W., Frandsen, G., Shallit, J.: The computational complexity of some problems of linear algebra. Journal of Computer and System Sciences (1999)
10. Chen, A.I.-T., Chen, M.-S., Chen, T.-R., Cheng, C.-M., Ding, J., Kuo, E.L.-H., Lee, F.Y.-S., Yang, B.-Y.: SSE implementation of multivariate PKCs on modern x86 CPUs. In: Clavier, C., Gaj, K. (eds.) CHES 2009. LNCS, vol. 5747, pp. 33–48. Springer, Heidelberg (2009)
11. Chen, C.H.O., Chen, M.S., Ding, J., Werner, F., Yang, B.Y.: Odd-char multivariate Hidden Field Equations. Cryptology ePrint Archive (2008), `http://eprint.iacr.org/2008/543`
12. Courtois, N.T.: Efficient zero-knowledge authentication based on a linear algebra problem MinRank. In: Boyd, C. (ed.) ASIACRYPT 2001. LNCS, vol. 2248, pp. 402–421. Springer, Heidelberg (2001)
13. Cox, D.A., Little, J.B., O'Shea, D.: Ideals, Varieties and Algorithms. Springer, Heidelberg (2005)
14. Ding, J., Schmidt, D., Werner, F.: Algebraic attack on HFE revisited. In: Wu, T.-C., Lei, C.-L., Rijmen, V., Lee, D.-T. (eds.) ISC 2008. LNCS, vol. 5222, pp. 215–227. Springer, Heidelberg (2008)

15. Faugère, J.-C.: A new efficient algorithm for computing Gröbner bases (F4). Journal of Pure and Applied Algebra 139, 61–88 (1999)
16. Faugère, J.-C.: A new efficient algorithm for computing Gröbner bases without reduction to zero (F5). In: Proceedings of the 2002 International Symposium on Symbolic and Algebraic Computation ISSAC, pp. 75–83. ACM Press, New York (2002)
17. Faugère, J.-C.: Algebraic cryptanalysis of HFE using Gröbner bases. Reasearch report RR-4738, INRIA (2003), `http://hal.inria.fr/inria-00071849/PDF/RR-4738.pdf`
18. Faugère, J.-C.: FGb: A Library for Computing Gröbner Bases. In: Fukuda, K., Hoeven, J., Joswig, M., Takayama, N. (eds.) ICMS 2010. LNCS, vol. 6327, pp. 84–87. Springer, Heidelberg (2010), `http://www-salsa.lip6.fr/~jcf/Papers/ICMS.pdf`
19. Faugère, J.-C., Joux, A.: Algebraic cryptanalysis of Hidden Field Equation (HFE) cryptosystems using Gröbner bases. In: Boneh, D. (ed.) CRYPTO 2003. LNCS, vol. 2729, pp. 44–60. Springer, Heidelberg (2003)
20. Faugère, J.-C., Levy-dit-Vehel, F., Perret, L.: Cryptanalysis of MinRank. In: Wagner, D. (ed.) CRYPTO 2008. LNCS, vol. 5157, pp. 280–296. Springer, Heidelberg (2008)
21. Faugère, J.-C., Safey El Din, M., Spaenlehauer, P.J.: Computing loci of rank defects of linear matrices using Gröbner bases and applications to cryptology. In: Proceedings of the International Symposium on Symbolic and Algebraic Computation 2010 – ISSAC 2010 (2010)
22. Faugère, J.C., Safey El Din, M., Spaenlehauer, P.J.: Gröbner Bases of Bihomogeneous Ideals Generated by Polynomials of Bidegree (1,1): Algorithms and Complexity. Journal of Symbolic Computation, 1–39 (2010)
23. Garey, M.R., Johnson, D.S.: Computers and Intractability: A Guide to the Theory of NP-Completeness. W.H. Freeman, New York (1979)
24. Granboulan, L., Joux, A., Stern, J.: Inverting HFE is quasipolynomial. In: Dwork, C. (ed.) CRYPTO 2006. LNCS, vol. 4117, pp. 345–356. Springer, Heidelberg (2006)
25. Jiang, X., Ding, J., Hu, L.: Kipnis-Shamir attack on HFE revisited. In: Pei, D., Yung, M., Lin, D., Wu, C. (eds.) Inscrypt 2007. LNCS, vol. 4990, pp. 399–411. Springer, Heidelberg (2008)
26. Kipnis, A., Shamir, A.: Cryptanalysis of the HFE Public Key Cryptosystem by Relinearization. In: Wiener, M. (ed.) CRYPTO 1999. LNCS, vol. 1666, pp. 19–30. Springer, Heidelberg (1999)
27. Matsumoto, T., Imai, H.: Public quadratic polynomial-tuples for efficient signature-verification and message-encryption. In: Günther, C.G. (ed.) EUROCRYPT 1988. LNCS, vol. 330, pp. 419–453. Springer, Heidelberg (1988)
28. Patarin, J.: Cryptoanalysis of the Matsumoto and Imai public key scheme of Eurocrypt 1988. In: Coppersmith, D. (ed.) CRYPTO 1995. LNCS, vol. 963, pp. 248–261. Springer, Heidelberg (1995)
29. Patarin, J.: Hidden Fields Equations (HFE) and Isomorphisms of Polynomials (IP): two new families of asymmetric algorithms. In: Maurer, U.M. (ed.) EUROCRYPT 1996. LNCS, vol. 1070, pp. 33–48. Springer, Heidelberg (1996)
30. Wolf, C., Preneel, B.: Equivalent keys in HFE, C*, and variations. In: Dawson, E., Vaudenay, S. (eds.) Mycrypt 2005. LNCS, vol. 3715, pp. 33–49. Springer, Heidelberg (2005)
31. Wolf, C., Preneel, B.: Large superfluous keys in multivariate quadratic asymmetric systems. In: Vaudenay, S. (ed.) PKC 2005. LNCS, vol. 3386, pp. 275–287. Springer, Heidelberg (2005)

Cryptanalysis of Cryptosystems Based on Non-commutative Skew Polynomials

Vivien Dubois[1] and Jean-Gabriel Kammerer[1,2]

[1] DGA/MI, Rennes, France
vivien.dubois@m4x.org
[2] IRMAR, Université de Rennes 1, France
jean-gabriel.kammerer@m4x.org

Abstract. Ten years ago, Ko *et al.* described a Diffie-Hellman like protocol based on decomposition with respect to a non-commutative semi-group law. Instantiation with braid groups had first been considered, however intense research on braid groups revealed vulnerabilities of the group structure and of the braid based DH problem itself.

Recently, Boucher *et al.* proposed a similar scheme based on a particular non-commutative multiplication of polynomials over a finite field. These so called skew polynomials have a well-studied theory and have many applications in mathematics and coding theory, however they are quite unknown in a cryptographic application.

In this paper, we show that the Diffie-Hellman problem based on skew polynomials is susceptible to a very efficient attack. This attack is in fact general in nature, it uses the availability of a one-sided notion of gcd and exact division. Given such tools, one can shift the Diffie-Hellman problem to a linear algebra type problem.

Keywords: Diffie-Hellman key exchange, skew polynomials.

1 Introduction

Since the proposal of the original Diffie-Hellman key exchange [5], many variations around the same principle have been proposed. The core structure of a Diffie-Hellman-like key exchange is as follows. Let $\mathcal{K}$ and $\mathcal{D}$ denote fixed domains and let F be a function from $\mathcal{D} \times \mathcal{K}$ to $\mathcal{D}$ such that

1. for any z in $\mathcal{D}$, $F(z, \cdot)$ is a one-way function,
2. the set of functions $F_a = F(\cdot, a)$, a in $\mathcal{K}$ is commutative for the composition of maps, that is, for any a, b in $\mathcal{K}$, $F_a \circ F_b = F_b \circ F_a$.

In a Diffie-Hellman like protocol the function F and a particular element z of $\mathcal{D}$ are public information. In a first pass, each party chooses a random element on his own in the set $\mathcal{K}$ (a for Alice's and b for Bob's), encrypts z with it and sends the result to the other party. Then each party encrypts again the received data with his element of $\mathcal{K}$. In the end, they both hold

$$F_a(F_b(z)) = F_b(F_a(z)).$$

D. Catalano et al. (Eds.): PKC 2011, LNCS 6571, pp. 459–472, 2011.

The security of the protocol assumes that given $F_a(z)$ and $F_b(z)$ it is infeasible to compute this common data (computational DH assumption). Of course, the assumption cannot be true if $F(\cdot, z)$ is not a one-way function since one can (for instance) recover a from $F_a(z)$ and stand in the same position as Alice.

The original proposal by Diffie and Hellman was to set F to the exponentiation of z by a in a multiplicative group [5]. When z is a fixed group element, its powers describe a cyclic (hence commutative) subgroup. The one-wayness of $F(z, \cdot)$ means the infeasability of identifying the power (discrete logarithm) of z corresponding with an arbitrary element of $\langle z \rangle$. This proposal seems to find satisfactory instantations and is by far the most widely used. On the other hand, there has been several attempts at building functions F for a Diffie-Hellman protocol from non-commutative algebraic structures. In general, these schemes rely on a particular factorization problem rather than on discrete logarithms. Furthermore, they all appear as variations of the following construction [6].

Generic Diffie-Hellman protocol based on a non-commutative semigroup. Let $(\mathcal{D}, \diamond)$ be a non-commutative semigroup (that is, it needs not have either a neutral element nor all elements to have an inverse). Elements of $\mathcal{D}$ may decompose in a large number of ways in general. Therefore it is assumed that given elements z, z' of $\mathcal{D}$ such that z' admits a factorization of the shape $u \diamond z \diamond v$, it is intractable to find such left and right factors u, v. Hence, one defines the set $\mathcal{K}$ to be ordered pairs $[u, v]$ and one defines $F : (z, [u, v]) \mapsto u \diamond z \diamond v$. When the aforementioned decomposition problem is intractable, F has Property 1. For Property 2 to be fulfilled too, we need functions $F(\cdot, [u, v])$ to commute. This is ensured by choosing elements u, v in a *commutative* sub-semigroup $\mathcal{S}$. Hence, $\mathcal{K} = \mathcal{S} \times \mathcal{S}$. Note that this specialization modifies the one-wayness property of F. It becomes: given an element z of $\mathcal{D}$ and an element z' in $\mathcal{S} \diamond z \diamond \mathcal{S}$ where $\mathcal{S}$ is a commutative sub-semigroup, find u and v in $\mathcal{S}$ such that $z' = u \diamond z \diamond v$.

At this point, it is not clear whether picking left and right factors in a commutative sub-semigroup weakens the decomposition problem. Either way, the cryptosystem can hardly save such a property. Encountered variations of the above description are choosing u and v in distinct subsets $\mathcal{L}, \mathcal{R}$ that are either both commutative or commute the one with the other (in this case, one party of the Diffie-Hellman protocol uses $\mathcal{L} \times \mathcal{R}$ while the other one uses $\mathcal{R} \times \mathcal{L}$). A more general setting may not require the commutative semigroup $\mathcal{S}$ to be a subset of $\mathcal{D}$: it simply needs to act in a different way on the left and on the right of $\mathcal{D}$. This is even more analogous to a general discrete log scheme where $\mathcal{D}$ may be an arbitrary cyclic group while $\mathcal{S}$ is a set of modular integers.

Instantiations. A well-known example of such a scheme based on a full group structure is the one on braid groups [6]. In this setting, pairs of the shape $[u, u^{-1}]$ are considered and the associated decomposition problem is called conjugacy problem. There had already been partial (still exponential) attacks on the general conjugacy problem in braid groups (see for instance [7] for a survey). It turned out a specific polynomial time algorithm exists to attack the Diffie-Hellman assumption in braid groups [3]. The attack uses the property that braid group elements can be represented by invertible matrices over some (complicated) ring.

For any element z of the braid group, we denote by Z the associated matrix. Then, the conjugacy problem arising in the DH-like protocol rewrites to finding U in the representation of the particular commutative set $\mathcal{S}$ such that $Z' = UZU^{-1}$ where Z, Z' are public matrices such that at least one such solution U exists. Candidate solutions can be found by solving the linear algebraic problem: find a matrix U such that $Z'U - UZ = 0$ and $S_iU - US_i = 0$ for any generator s_i of $\mathcal{S}$. The above system in general has many solutions that are not representatives of elements of the braid groups. When it is not possible to sieve them, it does not solve the conjugacy problem. However, the authors further observed that any invertible solution of the above linear system (a random solution is invertible with high probability) has the property of commuting with the elements of $\mathcal{S}$, and as a consequence is equally useful to uncover the shared output of the Diffie-Hellman protocol. Albeit polynomial, this attack did not yield a practical break as is. Yet, instantiation of the Diffie-Hellman protocol with braid groups does not seem to be still investigated.

Recently, Boucher *et al.* proposed a Diffie-Hellman scheme (and a companion ElGamal scheme) based on so called skew polynomials [1]. Skew polynomials are polynomials over a finite field with a particular non-commutative multiplication which uses the Frobenius field automorphism. They were introduced by Ore in 1933 [8] and have found many applications in applied mathematics and coding theory. The proposed Diffie-Hellman scheme follows the previous description with multiplication of skew polynomials as the non-commutative law. Hence, it is an instantiation of the non-commutative Diffie-Hellman protocol which is not based on a full group law.

Our contribution. In this paper, we show that the scheme of Boucher *et al.* is susceptible to a very efficient attack. The attack in fact only remotely uses the structure of skew polynomials, it only uses the availability of a notion of left (or right) gcd and of a related exact division procedure in the underlying domain. In any such setting, one can shift solving the relevant decomposition problem to a linear algebra type problem.

Similarly to the attack on the braid DH scheme, not all solutions of the linear type problem are solutions of the initial decomposition problem. *Real solutions* satisfy one additional condition (such as invertibility in the case of the braid DH scheme). Particular heuristics must be used, then, to find real solutions among the solutions of the linear problem. Based on an assumption which is satisfied in practice for skew polynomials, one can very easily get a real solution by means of gcds, and the attack is completely polynomial time.

We first describe the precise setting and working of the attack without reference to skew polynomials and then describe application to this particular case. Then we consider a possible variation of the scheme based on modular skew polynomials. While our attack seems unfit to this case, we show that density of invertible elements makes it completely weak. Finally we point out that examples based on matrix multiplication can be reduced to modular skew polynomials. Hence, although the attack looks general in nature, we could not produce another case of application of this attack, and we must leave it as an open problem.

2 General Strategy of the Attack

2.1 The Invertible Case

As a preliminary, let us consider the case when the commutative subsemigroup $\mathcal{S}$ is a group, that is, all its elements are invertible. In this case, a pair $[u, v]$ in $\mathcal{S} \times \mathcal{S}$ is a solution to the Diffie-Hellman decomposition problem for given z in $\mathcal{D}$ and z' in $\mathcal{S} \diamond z \diamond \mathcal{S}$ if and only if $z' = u \diamond z \diamond v$, that is $u' \diamond z' = z \diamond v$ where $u' = u^{-1}$. (Let us denote public variables in bold letters.) As a consequence, when $\mathcal{S}$ is a group, the quadratic looking equation $\boldsymbol{z}' = u \diamond \boldsymbol{z} \diamond v$ with unknown (u, v) can be directly turned into a linear looking equation $u' \diamond \boldsymbol{z}' = \boldsymbol{z} \diamond v$ with unknown $(u' = u^{-1}, v)$. Of course it is the case in braid groups, where one additionally has $u' = v$. Then, representing braid groups elements by matrices (see [3]), the linear looking equation is turned into a linear relation on matrices (over a ring), which can be solved as shown in [3]. Note that the attack may not solve the DH conjugacy problem. The same approach can also be used to find a linear invertible change of variables mapping two sets of quadratic multivariate equations over a field (which is a particular instance of the problem of 'isomorphism of polynomials' [9]). Quadratic multivariate polynomials can be represented by upper triangular matrices (with the same number of non-zero coefficients), and whenever a linear invertible change of coordinates U maps a quadratic polynomial $\boldsymbol{p}$ to a polynomial $\boldsymbol{p}'$, it translates into the matrix identity $\boldsymbol{P}' = U^t \boldsymbol{P} U$ (where superscript t denotes transposition). Since U is invertible, one can attack this problem by solving the linear equation $V\boldsymbol{P}' - \boldsymbol{P}U = 0$ with unknown $(V = (U^t)^{-1}, U)$. It can easily be seen that roughly 3 independent pairs $(\boldsymbol{p}, \boldsymbol{p}')$ with the same U only are heuristically needed to directly solve this problem (*i.e.* find a one-dimensional space of solutions). Higher degree cases can be attacked as well from the identity between degree 2 homogeneous parts. A less immediate attack to this problem was also developed in [10]. A similar attack was independently developed in [2].

2.2 The Setting of Our Attack

In this paper, we consider cases where the elements of $\mathcal{S}$ are not invertible, but breaking the Diffie-Hellman problem can also be shifted to solving a commutator-type (linear looking) equation. The basic structure of $\mathcal{D}$ that we need is that of *a domain with a computable notion of left (or right) gcd and computable left (right) exact division* (and $\mathcal{S}$ is a multiplicative commutative sub-semigroup). Here, domain means that it has no divisors of zero: $u \diamond v = 0 \Rightarrow u$ or $v = 0$. Note that it needs not be a form of Euclidean ring.

Before we describe the attack, let us recall the problem hierarchy on which the protocol relies. The Diffie-Hellman problem is: given $\boldsymbol{z}_A = u_A \diamond \boldsymbol{z} \diamond v_A$ and $\boldsymbol{z}_B = u_B \diamond \boldsymbol{z} \diamond v_B$, compute

$$z_{AB} = u_A \diamond \boldsymbol{z}_B \diamond v_A = u_B \diamond \boldsymbol{z}_A \diamond v_B.$$

A sufficient way of breaking the DH problem is the one of solving the decompositional problem arising in the Diffie-Hellman protocol: given $\boldsymbol{z}$ in $\mathcal{D}$ and $\boldsymbol{z}'$

in $\mathcal{S} \diamond \boldsymbol{z} \diamond \mathcal{S}$, compute u, v in $\mathcal{S}$ such that $\boldsymbol{z}' = u \diamond \boldsymbol{z} \diamond v$. However, it was noted in [3,12] that breaking a relaxed variant of this problem is enough to break the DH problem: given $\boldsymbol{z}$ in $\mathcal{D}$ and $\boldsymbol{z}'$ in $\mathcal{S} \diamond \boldsymbol{z} \diamond \mathcal{S}$, compute u, v *commuting with the elements of* $\mathcal{S}$ such that $\boldsymbol{z}' = u \diamond \boldsymbol{z} \diamond v$. Indeed, if an attacker can recover u'_A, v'_A commuting with the elements of $\mathcal{S}$ and such that $\boldsymbol{z}_A = u'_A \diamond \boldsymbol{z} \diamond v'_A$, she can compute

$$u'_A \diamond \boldsymbol{z}_B \diamond v'_A = u_B \diamond u'_A \diamond \boldsymbol{z} \diamond v'_A \diamond v_B = u_B \diamond \boldsymbol{z}_A \diamond v_B = z_{AB}.$$

Hence, with obvious notation, the problem hierarchy is

$$\text{DH} \quad \leq \quad \text{RelaxedDecomposition-DH} \quad \leq \quad \text{Decomposition-DH}.$$

2.3 The Attack

Now we describe the attack. Assume that $\mathcal{D}$ is a domain such that the left (or right) greatest common divisor of two elements always exists and can be computed efficiently and left (or right) exact division can performed efficiently. The attack originates from the following observation. Elements in the set $\mathcal{S} \diamond \boldsymbol{z} \diamond \mathcal{S}$ are changed in a particular way when multiplying on the left (or on the right) by an element of $\mathcal{S}$. This property is used to create the Diffie-Hellman protocol. Here we use it to attack the scheme. Let indeed $\boldsymbol{\lambda}$ be an arbitrary element of $\mathcal{S}$. Then,

$$\boldsymbol{\lambda} \diamond (u \diamond \boldsymbol{z} \diamond v) = u \diamond \boldsymbol{\lambda} \diamond \boldsymbol{z} \diamond v.$$

Hence, we obtain another element which too has u as a left divisor. As a consequence, taking the left gcd of $\boldsymbol{z}' = u \diamond \boldsymbol{z} \diamond v$ and $\boldsymbol{\lambda} \diamond \boldsymbol{z}'$, one obtains a (non-zero) multiple of u. The same can be done for any element $\boldsymbol{\lambda}$ of $\mathcal{S}$. Let $\boldsymbol{\lambda}_1, \ldots, \boldsymbol{\lambda}_s$ be generators of $\mathcal{S}$. For $\mathcal{S}$ to be transmittable data, these generators must be in very practical number (which we do not consider). Let $\boldsymbol{g}$ be the left gcd of $\{\boldsymbol{z}', \boldsymbol{\lambda}_1 \diamond \boldsymbol{z}', \ldots, \boldsymbol{\lambda}_s \diamond \boldsymbol{z}'\}$, obtained iteratively from pairwise left gcds. Hopefully, relying on the non-commutativity of $\boldsymbol{z}$ with the elements of $\mathcal{S}$, $\boldsymbol{g}$ might be u itself. It can happen that we actually have a way to distinguish between u and its non-trivial multiples. In this case and if $\boldsymbol{g}$ actually equals u, then we use the left exact division algorithm and get v, and the decomposition problem is already broken. Otherwise, let anyway a be such that $\boldsymbol{g} = u \diamond a$. By using the exact division algorithm, we obtain $\boldsymbol{m}$ and $\boldsymbol{m}_i$, $i = 1, \ldots, s$ such that

$$\begin{cases} \boldsymbol{z}' = \boldsymbol{g} \diamond \boldsymbol{m} \\ \boldsymbol{\lambda}_i \diamond \boldsymbol{z}' = \boldsymbol{g} \diamond \boldsymbol{m}_i. \end{cases} \tag{1}$$

Since there are no divisors of zero in $\mathcal{D}$, this system of equations rewrites

$$\begin{cases} \boldsymbol{z} \diamond v - a \diamond \boldsymbol{m} = 0 \\ \boldsymbol{\lambda}_i \diamond \boldsymbol{z} \diamond v - a \diamond \boldsymbol{m}_i = 0. \end{cases} \tag{2}$$

Hence we obtain a set of linear looking equations in the unknown (v, a). Since relevant solutions v commute with $\mathcal{S}$, one has the additional linear looking equations $\boldsymbol{\lambda}_i \diamond v - v \diamond \boldsymbol{\lambda}_i$ for any $i = 1, \ldots, s$.

Not all solutions to these linear conditions, however, are solutions of the initial decomposition problem. This is because when shifting from the initial system (1) to the linear system (2), one loses the information that $\boldsymbol{z} \diamond v$ is a right divisor of $\boldsymbol{z}'$. For instance, any linear combination of such solutions satisfies the linear conditions while not necessarily satisfying the divisibility condition.

Let divisor solutions refer to the linear solutions (v, a) such that $\boldsymbol{z} \diamond v$ is a right divisor of $\boldsymbol{z}'$. We first show that any divisor solution is enough to break the relaxed decomposition problem. Assume indeed that the linear-looking system can be solved, and let (v', a') be an arbitrary divisor solution. By using the exact division algorithm, we get u' such that $\boldsymbol{z}' = u' \diamond \boldsymbol{z} \diamond v'$. From $\boldsymbol{z} \diamond v' = a' \diamond \boldsymbol{m}$, we find $\boldsymbol{z}' = u' \diamond a' \diamond \boldsymbol{m}$, and therefore we also have $\boldsymbol{g} = u' \diamond a'$. Furthermore, u' commutes with all generators of $\mathcal{S}$: from $\boldsymbol{\lambda}_i \diamond \boldsymbol{z} \diamond v' = a' \diamond \boldsymbol{m}_i$, we get

$$u' \diamond \boldsymbol{\lambda}_i \diamond \boldsymbol{z} \diamond v' = \boldsymbol{g} \diamond \boldsymbol{m}_i = \boldsymbol{\lambda}_i \diamond \boldsymbol{z}' = \boldsymbol{\lambda}_i \diamond u' \diamond \boldsymbol{z} \diamond v',$$

and therefore $u' \diamond \boldsymbol{\lambda}_i = \boldsymbol{\lambda}_i \diamond u'$. As a consequence, (u', v') is a pair of elements that both commute with $\mathcal{S}$ and satisfy $\boldsymbol{z}' = u' \diamond \boldsymbol{z} \diamond v'$. Hence the relaxed decomposition problem is broken.

The only unproven step in the attack is the one of finding a divisor solution among the solution linear space. This part may only be heuristically approached. We first need to understand the structure of the solution linear space itself. Observe the following property.

Property 1. Let $\tilde{\mathcal{C}}(\boldsymbol{z})$ denote the set of pairs (c, c') such that $\boldsymbol{z} \diamond c = c' \diamond \boldsymbol{z}$. For any solution (v, a) of the linear looking equation $\boldsymbol{z} \diamond v - a \diamond \boldsymbol{m} = 0$ and any (c, c') in $\tilde{\mathcal{C}}(\boldsymbol{z})$, the pair $(c \diamond v, c' \diamond a)$ is also a solution. As a consequence, the solutions of the equation $\boldsymbol{z} \diamond v - a \diamond \boldsymbol{m} = 0$ are closed under left (coordinate-wise) multiplication by $\tilde{\mathcal{C}}(\boldsymbol{z})$. Of course, they are also closed under addition.

One easily sees that the property generalizes to solutions (v, a) of the complete linear system and left multiplication by $I = \tilde{\mathcal{C}}(\boldsymbol{z}) \cap_i \tilde{\mathcal{C}}(\boldsymbol{\lambda}_i \diamond \boldsymbol{z}) \cap (\cdot, \bar{\mathcal{S}})$ where $\bar{\mathcal{S}}$ is the elements that commute with $\mathcal{S}$. One easily checks that $I = \tilde{\mathcal{C}}(\boldsymbol{z}) \cap (\bar{\mathcal{S}}, \bar{\mathcal{S}})$. Also note that I is a ring for coordinate-wise addition and multiplication.

Additivity and left multiplication by I are degeneracies that are independent of the existential solution. Save these degeneracies, we expect the system of equations to be characteristic of the existential solution. Hence, we expect:

Claim. the solutions of the linear system are all spanned by a single generator through addition and left multiplication by I. Let (v_g, a_g) denote this generator.

Then, since I is a ring, the linear solutions write simply $I \diamond (v_g, a_g)$. For the existential solution $(\hat{v}, \hat{a})$ in particular, there exists $(\hat{c}, \hat{c}')$ in I such that $(\hat{v}, \hat{a}) = (\hat{c}, \hat{c}') \diamond (v_g, a_g)$. This means that (v_g, a_g) is $(\hat{v}, \hat{a})$ purged out of its factors in I. This shows that (v_g, a_g), just like $(\hat{v}, \hat{a})$, is a divisor solution. The other ones are spanned by factors (c, c') related to left factors c of v or right factors c' of u. Finally note that (v_g, a_g) is a common right factor of all linear solutions and since it includes itself, it is in fact the right gcd of the linear solutions.

3 Application to Skew Polynomials Cryptosystems

Diffie-Hellman and ElGamal-like schemes based on skew polynomials were recently presented at PQCrypto 2010 [1]. The Diffie-Hellman protocol follows the general construction described in the introduction and developed by the earlier group-based proposals. We first recall the particularities of skew polynomials and review the setting up of the cryptosystem. Then, we describe unrolling the attack in this particular case. Since the ElGamal scheme relies on the DH problem, we only consider the DH protocol.

3.1 Skew Polynomials

Skew polynomials are polynomials over a finite field with a particular non-commutative inner product. Let $\mathbb{F}_q$ denote the finite field with q elements, and p be the characteristic of the field. Automorphisms of $\mathbb{F}_q$ are the so-called Frobenius maps which are powering to a power of p. Let θ be such an automorphism. We denote by $\star$ the inner product of skew polynomials. It is defined inductively for all $a \in \mathbb{F}_q$ by $X \star a = \theta(a)X$. The ring of skew polynomials is sometimes denoted $\mathbb{F}_q[X, \theta]$.

The ring of skew polynomials is a left and right Euclidean domain, that is, there are both a left and a right Euclidean division algorithm. Using the Euclidean algorithms one can thus compute left and right greatest common divisors, and also perform exact division.

As priorly addressed, due to the non-commutativity of the inner product, skew polynomials admit many factorizations instead of a single one. The cardinality of the number of possible factorizations is expected to be exponential in the degree of the polynomial.

3.2 Generation of the Scheme

For the sake of completeness, we recall part of the specification given in [1]. However the attack is not tied to any particular key generation.

A brute-force approach is suggested to construct the commutative subset $\mathcal{S}$. One iteratively constructs a set of generators $G_0, \ldots, G_s$ of small degree δ. At each step, a polynomial of degree δ is picked at random and tested for commuting with the current set of generators. If it does, it increments the set of generators, otherwise repeat. The set $\mathcal{S}$ is the commutative algebra spanned by these generators.

Let d be the security parameter of the protocol. A public polynomial $\boldsymbol{Z}$ of degree d is generated. At the execution of the Diffie-Hellman protocol, each participant selects two elements U and V in $\mathcal{S}$ with degree d through combination of the generators of $\mathcal{S}$. More precisely, any picked element is a product of sums of products of the generators.

All tasks performed during the protocol can clearly be made practical, however the cost of generating a set of generators for $\mathcal{S}$ is quoted as a long computation without further detail. The proposed instantiation is with skew polynomials over

$\mathbb{F}_4$, generators of $\mathcal{S}$ have degree $\delta = 8$ or 9, and the protocol uses polynomials of degree $d = 600$. For these parameters, they give two examples of $\mathcal{S}$ through generator sets with ≥ 90 polynomials.

3.3 Commutativity among Skew Polynomials

Before we go on with the attack, it is useful to investigate commutation properties of skew polynomials.

There are particular skew polynomials that commute with any other. This subset is called the center and we denote it by $\mathcal{C}$. A characterization of these elements can be found in [1]. Let m be the order of θ (say the degree of $\mathbb{F}_q$ over $\mathbb{F}_p$ to simplify). Then, the center polynomials are the polynomials over $\mathbb{F}_p$ and in the only powers of X^m.

$$\mathcal{C} = \mathbb{F}_p[X^m].$$

Also, for any polynomial P, let $\mathcal{C}_P$ denote the set of polynomials that commute with P. Of course, P commutes with itself and the elements of the center. As a consequence, $\mathcal{C}_P$ contains the algebra generated by P and $\mathcal{C}$, that is, the set of sums and multiples of elements of $\mathcal{C}$ and P. This algebra is a vector space over $\mathbb{F}_p$. On the other hand, if P has all its coefficients in $\mathbb{F}_p$, it commutes with any polynomial with coefficients in $\mathbb{F}_p$, not only itself and $\mathcal{C}$. Hence,

$$\begin{cases} \text{if } P \in \mathbb{F}_p[X], & \mathcal{C}_P \supseteq \mathbb{F}_p[X] \\ \text{otherwise,} & \mathcal{C}_P \supseteq \langle \mathcal{C}, P \rangle = \mathcal{C}[P]. \end{cases}$$

On the other hand, for any degree bound r, one can easily compute the elements of $\mathcal{C}_P$ with degree $\leq r$. Indeed, the equation $P \star Q = Q \star P$ in the degree $\leq r$ indeterminate Q is a linear system over $\mathbb{F}_p$, and one can find its solutions through linear algebra. We ran experiments for many random choices of P (not with all coefficients in $\mathbb{F}_p$) of degree $\simeq \delta = 8$ and we found that at least up to degree $r = 30$ the elements of $\mathcal{C}_P$ in fact all were in $\mathcal{C}[P]$. It suggests that when one picks generators for $\mathcal{S}$ during the key generation, one actually obtains polynomials in $\mathcal{C}[P_0]$, where P_0 is any smallest degree polynomial in $\mathcal{S}$ modulo $\mathcal{C}$ (modulo is well defined for central polynomials). Hence it is pointless to generate this set by brute force. This is confirmed for the proposals of $\mathcal{S}$ in [1]: we respectively found $P_0 = X^5 + X^3 + \alpha$ and $P_0 = X^3 + X + \alpha$.

3.4 Unrolling the Attack

Let $\boldsymbol{Z}' = U \star \boldsymbol{Z} \star V$ be the data transmitted by one of the participants of the protocol. Our first step is to take the gcd of $\boldsymbol{Z}'$ and $\boldsymbol{A}_i \star \boldsymbol{Z}'$ over all generators $\boldsymbol{A}_i$ of $\mathcal{S}$. Due to our previous comment on these generators having a common generator P_0, we actually take the gcd of $\boldsymbol{Z}'$ and $P_0 \star \boldsymbol{Z}'$. We find a polynomial $\boldsymbol{G}$ which is a right multiple of U: there exists a polynomial A such that $\boldsymbol{G} = U \star A$. Also we compute $\boldsymbol{M}, \boldsymbol{M}_0$ such that

$$\begin{cases} \boldsymbol{Z}' = \boldsymbol{G} \star \boldsymbol{M} \\ P_0 \star \boldsymbol{Z}' = \boldsymbol{G} \star \boldsymbol{M}_0. \end{cases}$$

Since the ring is a domain, one deduces from these equations,

$$\begin{cases} \boldsymbol{Z} \star V = A \star \boldsymbol{M} \\ P_0 \star \boldsymbol{Z} \star V = A \star \boldsymbol{M}_0. \end{cases}$$

In addition, one has $P_0 \star V = V \star P_0$ since V commutes with $\mathcal{S}$. These three equations are not linear over $\mathbb{F}_q$, as it would be with the usual product of polynomials, however they are linear alright over $\mathbb{F}_p$. Expanding these equations over $\mathbb{F}_p$, and degree bounding the search space according to the expected degree of the existential (V, A), we can solve the system through linear algebra.

The output of the previous phase is a degree bounded restriction of the entire solution subspace of the linear system. Recall from Section 2.3 that the entire solution space is closed under left multiplication by $I = \tilde{\mathcal{C}}(\boldsymbol{Z}) \cap (\bar{\mathcal{S}}, \bar{\mathcal{S}})$. If the entire solution space indeed is monogeneous under linearity and left multiplication by I, then this generator is the lowest degree solution (V, A). Then, it can also be found as the right gcd of a linear basis of the bounded degree solution subspace.

We checked the previous expectations in practice with the recommended parameters and beyond. In any tested case, the bounded degree solution admitted only one lowest degree solution (up to $\mathbb{F}_p$ multiples) and any other solution was a multiple of it by a central factor. It incidentally shows that $I = \mathcal{C} \star (1, 1)$ up to the fixed degree bound. Let (V_g, A_g) denote the found generator. We verified that it is a divisor solution and also that it is the right gcd of a linear basis of the bounded degree solutions. We also checked that (V_g, A_g) is the greatest central factor of the original solution $(\hat{V}, \hat{A})$. This greatest central factor can be extracted by taking the left gcd of $(\hat{V}, \hat{A})$ and arbitrary left multiples of it.

The attack has theoretical complexity in $(md)^3$. It takes about a minute with the recommended parameters and with a straightforward implementation in `C++` using the NTL library [11].

4 The Case of Modular Skew Polynomials

We consider a possible variation of the Diffie-Hellman protocol based on modular skew polynomials.

4.1 Constructing Modular Skew Polynomial Rings

Let $\mathcal{R}$ denote a skew polynomial ring $\mathbb{F}_q[X, \theta]$ and let N be an arbitrary element of $\mathcal{R}$. Let $\star N$ denote the set of left multiples of N. Congruence modulo $\star N$ is an equivalence relation over $\mathcal{R}$ and the associated partition elements are called left classes modulo N. Obviously right classes can be defined all the same from $N\star$. Reduction modulo $\star N$ (resp. $N\star$) can be performed by using the left (resp. the right) Euclidean division algorithm.

One awkward property of left classes is that they cannot be multiplied the ones with the others unless N commutes with their elements in a certain sense. Indeed, let $U + \Lambda \star N$ and $V + \Lambda' \star N$ be arbitrary reprensentatives of two classes. Their product

$$(U + \Lambda \star N) \star (V + \Lambda' \star N) = U \star V + \Lambda \star N \star V + (U \star \Lambda' + \Lambda \star N \star \Lambda') \star N$$

does not equal $U \star V$ modulo $\star N$ unless $\Lambda \star N \star V$ is right divisible by N. Since Λ may be chosen arbitrarily, it means $N \star V$ is right divisible by N, or again that there exists W such that $N \star V = W \star N$. When such a W exists, we say that N quasi-commutes with V.

When we want the set of classes to itself be a (non-commutative) ring, we need N to quasi-commute with all elements of $\mathcal{R}$. Let $\mathcal{N}$ denote the set of such N's; we call it the quasi-center of $\mathcal{R}$. Observe that the elements with which N quasi-commutes is closed under the ring operations. Therefore, N quasi-commutes with all elements of $\mathcal{R}$ if and only it quasi-commutes with X and all the constants. This yields an easy characterization of such polynomials N. If a constant a quasi-commutes with N, because of the degree, its dual element is also a constant b. For any a,

$$N \star a = \sum_i n_i \theta^i(a) X^i = b \star N = \sum_i n_i b X^i$$

implies that there is k such that all non-zero terms are at $i \equiv k \bmod m$ (where m is the order of θ, assumed equal to the degree of $\mathbb{F}_q$ over $\mathbb{F}_p$). Next, we do the same with X: there are constants a, c such that

$$N \star X = \sum_i n_i X^{i+1} = (aX + c) \star N = \sum_i (a\theta(n_i) + cn_{i+1}) X^{i+1} + cn_0 \,.$$

Let j be the smallest i such that $n_i \neq 0$. Then, $c.n_j = 0$ implies $c = 0$, and for any i such that $n_i \neq 0$, $a = n_i/\theta(n_i)$. Let $\bar{a}$ satisfy $\bar{a}/\theta(\bar{a}) = a$. Finally the quasi-center $\mathcal{N}$ is the union of the sets $\mathcal{N}_{k,a} = \bar{a}X^k\mathcal{C}$, $k \in \{0, \ldots, m-1\}$, $a \in \mathbb{F}_q$, where $\mathcal{C} = \mathbb{F}_p[X^m]$ is the center of $\mathcal{R}$. More concisely, this is $\cup_{k=0}^{m-1} \mathbb{F}_q X^k \mathcal{C}$.

For any polynomial N of the quasi-center, left and right multiples of N are just the same sets and classes modulo these sets are simply said classes modulo N. These classes form a ring, which we denote by $\mathcal{R}_N$.

4.2 The Modified Scheme

Modular skew polynomial rings might be considered at the basis of a non-commutative Diffie-Hellman protocol following exactly the same construction as proposed in [1]. Let d denote the degree of N. Any class admits a unique representative with degree $< d$. Multiplication of classes is realized through multiplication of canonical representatives and subsequent reduction mod N. We denote this operation by $\circ$.

A commutative set $\mathcal{S}$ may again be constructed by selecting commuting classes with canonical representatives of small degree δ. Since δ is a small constant while d is the security parameter, one can assume $2\delta < d$. In this case, the picked canonical representatives in fact commute without modulo. Then, following 3.3, these representatives are spanned over the center by a single polynomial P_0. Elements of $\mathcal{S}$ are arbitrary combinations of P_0 over $\mathcal{C}$ reduced mod N.

4.3 First Remarks

It is immediate that $\mathcal{R}_N$ is not a domain: for any pairwise factorization $U \star V$ of N, we get $U \star V = 0 \bmod N$ while neither U or V is divisible by N.

Now, by definition a class $\bar{U}$ is a left factor of a class $\bar{P}$ if $\bar{P}$ lies in the image space of the map

$$\begin{aligned} \mu(\bar{U}) : \mathcal{R}_N &\longrightarrow \mathcal{R}_N \\ \bar{V} &\longmapsto \bar{U} \circ \bar{V}. \end{aligned}$$

Recall the product $\bar{U} \circ \bar{V}$ equals $U \star V \bmod N$ where U and V are arbitrary elements of $\bar{U}$ and $\bar{V}$. Observe that the set $U \star + N\star$ is independent of the particular U in $\bar{U}$. As a consequence, the left gcd G of U and N is independent of U in $\bar{U}$. Now, from $U\star + N\star = G\star$, we get that right multiples of U and G are the same mod N. As a consequence, the image of $\mu(\bar{U})$ is the set $\bar{G}\circ$. Finally,

Property 2. the set of left factors of a class $\bar{P}$ is the set of classes $\bar{U}$ whose left gcd G with N is a left factor of the canonical representative $\hat{P}$ of $\bar{P}$.

In particular, classes that are left coprime with N are left factors of any class (they are the units of the ring). As one can see, the relationship between divisors and multiples is very loose in $\mathcal{R}_N$. Then, the fact that a class $\bar{\boldsymbol{Z}}'$ is computed as $\bar{U} \circ \bar{\boldsymbol{Z}} \circ \bar{V}$ hardly carries information on the particular $\bar{U}$ and $\bar{V}$. Therefore, there is not much information to be obtained in using gcds on $\bar{\boldsymbol{Z}}'$ and $\bar{P}_0 \circ \bar{\boldsymbol{Z}}'$. Instead, since $\bar{P}$ is loosely related to the initial $(\bar{U}, \bar{V})$, one may take advantage of the many equivalent pairs $(\bar{U}, \bar{V})$.

4.4 Attacking the Modular Decomposition

Given a class $\bar{\boldsymbol{Z}}'$, we search for a decomposition $\bar{U} \circ \bar{\boldsymbol{Z}} \circ \bar{V}$ where $\bar{U}$ and $\bar{V}$ are in $\bar{\mathcal{S}}$ (that is, commute with P_0 mod N). At least one exists by construction of $\bar{\boldsymbol{Z}}'$ and we expect many others.

We target pairs $(\bar{U}, \bar{V})$ in $\bar{\mathcal{S}} \times \bar{\mathcal{S}}$ such that (e.g.) $\bar{U}$ is left coprime with N. Before going on, observe that being left or right coprime with N is the same when N is in the quasi-center. From $\bar{U}$ being left coprime with N, we get $\bar{W}$ such that $\bar{U} \circ \bar{W} = \bar{1}$. Then right multiplication by $\bar{U}$ is clearly injective (right multiply by $\bar{W}$), and therefore bijective, and $\bar{U}$ is right coprime with N. Using associativity, one also gets that the left and right inverses are the same.

Now, for any $(\bar{U}, \bar{V})$ in $\bar{\mathcal{S}} \times \bar{\mathcal{S}}$ where $\bar{U}$ is coprime with N,

$$\bar{\boldsymbol{Z}}' = \bar{U} \circ \bar{\boldsymbol{Z}} \circ \bar{V} \qquad \Longleftrightarrow \qquad \bar{W} \circ \bar{\boldsymbol{Z}}' = \bar{\boldsymbol{Z}} \circ \bar{V}.$$

Also, $\bar{U}$ commutes with $\bar{P}_0$ iff $\bar{W}$ commutes with $\bar{P}_0$. Therefore, the decompositional pairs $(\bar{U}, \bar{V})$ in $\bar{\mathcal{S}} \times \bar{\mathcal{S}}$ with $\bar{U}$ invertible are in bijection with the solutions $(\bar{W}, \bar{V})$ in $\bar{\mathcal{S}} \times \bar{\mathcal{S}}$ of the linear equation $\bar{W} \circ \bar{\boldsymbol{Z}}' = \bar{\boldsymbol{Z}} \circ \bar{V}$ with $\bar{W}$ invertible.

For the attack to be successful, we simply need to extract a pair $(\bar{W}, \bar{V})$ with an invertible $\bar{W}$ from the solutions of the linear system. For this, we rely on the density of such pairs. First observe that restricting elements of pairs in $\bar{\mathcal{S}}$ can only negligibly impact the density of invertible elements. Indeed, the modular condition is exact for all classes whose canonical representative has degree under $\deg(N) - \deg(P_0)$. Since $\deg(P_0)$ asymptotically remains a small constant, the fraction of classes for which the condition involves N is negligible. Besides, we see no reason why the density of $\bar{W}$ among solution pairs $(\bar{W}, \bar{V})$ (not restricted to $\bar{\mathcal{S}}$) of the equation $\bar{W} \circ \bar{\boldsymbol{Z}}' = \bar{\boldsymbol{Z}} \circ \bar{V}$ should differ from the global density.

Claim. The density of invertible $\bar{W}$ among solutions $(\bar{W}, \bar{V})$ in $\bar{\mathcal{S}} \times \bar{\mathcal{S}}$ of $\bar{W} \circ \bar{\boldsymbol{Z}}' = \bar{\boldsymbol{Z}} \circ \bar{V}$ is the same as the density of invertible classes in $\mathcal{R}_N$.

Although N can have many distinct irreducible left factors, these are a subcollection of all possible irreducible left factors of degree $\deg(N)$ polynomials. As a consequence, we expect the fraction of classes left coprime with N to be asymptotically a constant close to 1. For the sake of intuition, for any $u < d$, the fraction of right multiples of degree u monic polynomial L among degree $< d$ polynomials is q^{-u}. Then, the probability that two degree $< d$ polynomials be both right multiples of L is q^{-2u}. Then, the probability that they not share a common left multiple of degree u is $1 - q^u q^{-2u} = 1 - q^{-u}$. We estimate the probability of they be coprime by the probability not to share degree 1 left factors: $1 - 1/q$. This is our expectation of the density of invertible elements.

We checked the above properties in practice. We checked (through sampling) the density of invertible classes both among the left coordinates $\bar{W}$ of the solution space and among all classes. We found densities of the same order in both cases: equal in large characteristic and indeed close to $1 - 1/q$, but slightly different in small characteristic. Hence, we could in any case extract a decompositional solution almost at once.

Interestingly, the attack can be slightly generalized. We may more generally target decompositional solutions $(\bar{U}, \bar{V})$ such that $\bar{U}$ has a right gcd with N not necessarily 1 but a target right factor G of N which commutes with P_0 (for instance a central polynomial). Then, for any such $\bar{U}$, there exist $\bar{W}$ such that $\bar{W} \circ \bar{U} = \bar{G}$. Since both $\bar{U}$ and $\bar{G}$ commute with P_0, the same holds for $\bar{W}$. Then, we simply compute solutions $(\bar{W}, \bar{V})$ in $\bar{\mathcal{S}} \times \bar{\mathcal{S}}$ of $\bar{W} \circ \bar{\boldsymbol{Z}}' = \bar{G} \circ \bar{\boldsymbol{Z}} \circ \bar{V}$, and extract $\bar{W}$ such that the left gcd of $\bar{W}$ and N is G.

5 Beyond the Case of Skew Polynomials?

Another usual example of a non-commutative algebra are square matrices over a finite field. Then, the question arises as to whether this algebra can be used to build a non-commutative Diffie-Hellman protocol. We answer this question negatively by simply describing a well-known connection between square matrices over a finite field and modular skew polynomials (see for instance [4]): square matrices are modular skew polynomials with particular moduli.

Let $\mathbb{F}_p$ be an arbitrary finite field and let $\mathbb{F}_q = \mathbb{F}_{p^n}$ be the degree m extension field of $\mathbb{F}_p$. It is well-known that $\mathbb{F}_q$ is an m-dimensional vector space over $\mathbb{F}_p$. Hence, fixing arbitrary basis elements of $\mathbb{F}_q$ over $\mathbb{F}_p$, one can encode any element of $\mathbb{F}_q$ into an n-dimensional vector. This correspondence also induces a one-to-one correspondence between $\mathbb{F}_p$-linear maps on $\mathbb{F}_q$ and $\mathbb{F}_p$-linear maps on $(\mathbb{F}_p)^m$. The latter simply are represented by $m \times m$ matrices over $\mathbb{F}_p$. Hence, composition of $\mathbb{F}_p$-linear maps on $\mathbb{F}_q$ is the same as matrix multiplication. We can now regard the ring of $m \times m$ matrices over $\mathbb{F}_p$ as the ring of $\mathbb{F}_p$-linear maps on $\mathbb{F}_q$ for $+$ and the composition of maps, which we denote by $L_p(\mathbb{F}_q)$.

We now describe a generator basis for $L_p(\mathbb{F}_q)$. It is well-known that powerings to the power of p are $\mathbb{F}_p$-linear bijections on $\mathbb{F}_q$ called the Frobenius maps. We let

θ be the first Frobenius map. Multiplication by an element of $\mathbb{F}_q$ is also $\mathbb{F}_p$-linear. Therefore, linear combinations over $\mathbb{F}_q$ of Frobenius maps (powers of θ) also are $\mathbb{F}_p$-linear. It can be seen that this representation is injective and finally, by a simple cardinality argument, one-to-one. As a consequence, $L_p(\mathbb{F}_q)$ is the set of linear combinations over $\mathbb{F}_q$ of the Frobenius maps, $\sum_{i=0}^{m-1} a_i \circ \theta^i$. These maps can be identified with polynomials $\sum_{i=0}^{m-1} a_i X^i$. Mapping the ring structure of $L_p(\mathbb{F}_q)$ to these polynomials, one obtains the usual + law but multiplication following the identities $X \circ a = \theta(a)X$ and $X^n = 1$. This is exactly the identities defining skew polynomials $\mathbb{F}_q[X, \theta]$ modulo the center polynomial $X^m - 1$.

$$\{m \times m \text{ matrices over } \mathbb{F}_p\} = L_p(\mathbb{F}_q) = \mathbb{F}_q[X, \theta]/(X^m - 1).$$

Acknowledgements

We wish to thank Pierre Loidreau for helpful discussion on the subject of skew polynomials at the early stage of this work. We also wish to thank the anonymous referees of PKC 2011 for their detailed and helpful comments.

References

1. Boucher, D., Gaborit, P., Geiselmann, W., Ruatta, O., Ulmer, F.: Key exchange and encryption schemes based on non-commutative skew polynomials. In: Sendrier, N. (ed.) PQCrypto 2010. LNCS, vol. 6061, pp. 126–141. Springer, Heidelberg (2010)
2. Bouillaguet, C., Faugère, J.-C., Fouque, P.-A., Perret, L.: Differential-algebraic algorithms for the isomorphism of polynomials problem. Cryptology ePrint Archive, Report 2009/583 (2009), `http://eprint.iacr.org/`
3. Cheon, J.H., Jun, B.: A polynomial time algorithm for the braid Diffie-Hellman conjugacy problem. In: Boneh, D. (ed.) CRYPTO 2003. LNCS, vol. 2729, pp. 212–225. Springer, Heidelberg (2003)
4. Coulter, R.S., Havas, G., Henderson, M.: Giesbrecht's algorithm, the HFE cryptosystem, and Ore's ps-polynomials. In: Shirayanagi, K., Yokoyama, K. (eds.) Computer Mathematics: Proceedings of the Fifth Asian Symposium (ASCM 2001). Lecture Notes Series on Computing, vol. 9, pp. 36–45. World Scientific, Singapore (2001)
5. Diffie, W., Hellman, M.E.: New Directions in Cryptography. IEEE Transactions on Information Theory IT–22(6), 644–654 (1976)
6. Ko, K.H., Lee, S.-J., Cheon, J.H., Han, J.W., Kang, J.-s., Park, C.-s.: New public-key cryptosystem using braid groups. In: Bellare, M. (ed.) CRYPTO 2000. LNCS, vol. 1880, pp. 166–183. Springer, Heidelberg (2000)
7. Mahlburg, K.: An overview of braid group cryptography (2004)
8. Ore, O.: Theory of non-commutative polynomials. Annals of Mathematics 34, 480–508 (1933)
9. Patarin, J.: Hidden fields equations (HFE) and isomorphisms of polynomials (IP): Two new families of asymmetric algorithms. In: Maurer, U.M. (ed.) EUROCRYPT 1996. LNCS, vol. 1070, pp. 33–48. Springer, Heidelberg (1996)

10. Perret, L.: A fast cryptanalysis of the isomorphism of polynomials with one secret problem. In: Cramer, R. (ed.) EUROCRYPT 2005. LNCS, vol. 3494, pp. 354–370. Springer, Heidelberg (2005)
11. Shoup, V.: NTL: A library for doing number theory, http://www.shoup.net/ntl
12. Shpilrain, V., Ushakov, A.: The conjugacy search problem in public key cryptography: unnecessary and insufficient. Cryptology ePrint Archive, Report 2004/321 (2004), http://eprint.iacr.org/

Practical Cryptanalysis of the Identification Scheme Based on the Isomorphism of Polynomial with One Secret Problem

Charles Bouillaguet[1], Jean-Charles Faugère[2,3], Pierre-Alain Fouque[1], and Ludovic Perret[3,2]

[1] Ecole Normale Supérieure, Paris, France
{charles.bouillaguet,pierre-alain.fouque}@ens.fr
[2] INRIA, Paris-Rocquencourt Center, SALSA Project
UPMC Univ. Paris 06, UMR 7606, LIP6, F-75005, Paris, France
[3] CNRS, UMR 7606, LIP6, F-75005, Paris, France
jean-charles.faugere@inria.fr, ludovic.perret@lip6.fr

Abstract. This paper presents a practical cryptanalysis of the Identification Scheme proposed by Patarin at Crypto 1996. This scheme relies on the hardness of the Isomorphism of Polynomial with One Secret (IP1S), and enjoys shorter key than many other schemes based on the hardness of a combinatorial problem (as opposed to number-theoretic problems). Patarin proposed concrete parameters that have not been broken faster than exhaustive search so far. On the theoretical side, IP1S has been shown to be harder than Graph Isomorphism, which makes it an interesting target. We present two new deterministic algorithms to attack the IP1S problem, and we rigorously analyze their complexity and success probability. We show that they can solve a (big) constant fraction of all the instances of degree two in polynomial time. We verified that our algorithms are very efficient in practice. All the parameters with degree two proposed by Patarin are now broken in a few seconds. The parameters with degree three can be broken in less than a CPU-month. The identification scheme is thus quite badly broken.

1 Introduction

Multivariate cryptography is concerned with the use of multivariate polynomials over finite fields to design cryptographic schemes. The use of polynomial systems in cryptography dates back to the mid eighties with the design of C* [33], and many others proposals appeared afterwards [37,38,39,28,47]. The security of multivariate schemes is in general related to the difficulty of solving random or structured systems of multivariate polynomial equations. This problem has been proved to be NP-complete [22], and it is conjectured [2] that systems of random polynomials are hard to solve in practice. As usual when a trapdoor must be embedded in a hard problem, easy instances are transformed into random-looking instances using secret transformations. In multivariate cryptography, it is common to map an easily-invertible collection of polynomials a into an apparently random one b. It is then assumed that, being supposedly indistinguishable from random, b should be hard to solve. The structure-hiding transformation is

D. Catalano et al. (Eds.): PKC 2011, LNCS 6571, pp. 473–493, 2011.

very often the composition with linear (or affine) invertible mappings S and T, namely $\mathbf{b} = T \circ \mathbf{a} \circ S$. The matrices S and T are generally part of the secret-key.

The Isomorphism of Polynomials (IP) is the problem of recovering the secret transformations S and T given $\mathbf{a}$ and $\mathbf{b}$. It is a fundamental problem of multivariate cryptography, since its hardness implies the difficulty of the key-recovery for various multivariate cryptosystems. Notorious examples include C* [33], the traitor tracing scheme proposed by Billet and Gilbert [8], the SFLASH signature scheme [38], the ℓ-IC signature scheme [12], the square-vinegar signature scheme [1] and the Square encryption scheme [11][1]. All these schemes have been broken, because the structure of the central map was not hidden well enough. The corresponding IP problem was then not random, but structured. However, when no apparent structure exists in both $\mathbf{a}$ and $\mathbf{b}$, then the IP problem is fairly difficult. This motivated Patarin to introduce it as an intractable assumption by itself in [35]. So far only exponential algorithms [40,17] are known to attack the general IP problem.

An important special case of IP is the *IP problem with one secret* (IP1S for short), where T is the identity matrix. Patarin suggested in 1996 [36] to construct a zero-knowledge identification scheme relying on the hardness of IP1S, inspired by the Zero-Knowledge proof system for Graph Isomorphism of [25]. The proposed parameters lead to relatively small key sizes (for instance to secret and public keys of 256 bits each and no additional information), as the complexity of the problem was believed to be exponential. The proposed parameters have not been broken so far, and no technique better than exhaustive search is known to attack the scheme. The IP1S problem is also interesting from a complexity-theoretic point of view. It has been proved in [40] that IP1S is *Graph Isomorphism-hard* (GI-hard for short). This leads Patarin *et al.* to claim that IP1S is unlikely to be solvable in polynomial time, because no polynomial algorithm is known for GI in spite of more than forty years of research. On the other hand, GI is not known to be NP-complete. Generating hard instances GI is pretty non-trivial, and there are powerful heuristics as well as expected linear time algorithms for random graphs [19]. This compromises the use of GI as an identification mechanism, and was part of the motivation for introducing IP1S as an alternative. Moreover, when used in this context, instances of the IP problem are random, which presumably avoids all the attacks on the cryptographic schemes mentioned above.

Previous and Related Work. The identification scheme based on IP1S is not based on number-theoretic assumptions, unlike for instance the well-known Fiat-Shamir protocol [18]. Many other identification schemes are not based on number theoretic assumptions [42,43,44,45,31]. However, the IP1S-based identification scheme enjoys shorter keys than most others.

To our knowledge, the first algorithm dedicated to IP1S can be found in Geiselmann *et al.* [23]. The authors of [23] remarked that each row of a matrix solution of IP1S verifies an algebraic system of equations. They then used an exhaustive search to find the solutions of such system. Soon after, this technique has been improved by Levy-dit-Vehel and Perret [13] who replaced this exhaustive search by a Gröbner basis

[1] In the description of some of these schemes, the easily-invertible central map contains parameters that are part of the secret-key. However, in this case there exists an equivalent secret key where these parameters have a fixed value. This is notoriously the case of C*.

computation. This still yields exponential algorithms, and the improvement induced by this replacement is is as significant as the gain obtained when comparing Gröbner basis and exhaustive search for solving random algebraic systems. It is negligible over small field (*i.e.*, typically, $\mathbb{F}_2$), but significant for instances of IP1S over large fields. However, the complexity of those algorithms remains exponential by nature.

Finally, Perret [41] shows that the affine and linear variants of IP1S are equivalent, *i.e.*, one can without loss of generality restrict our attention to the case where S is linear (as opposed to affine). In addition, a new approach for solving IP1S using the Jacobian matrix was proposed. The algorithm is polynomial when the number u of polynomials in $\mathbf{a}$ and $\mathbf{b}$ is equal to the number of variables n. However, when $u < n$, the complexity of this approach is not well understood. Moreover, when the number of polynomials is very small, for instance $u = 2$, this algorithm is totally inefficient.

The main application of IP1S is the identification scheme proposed in [40]. The public key being composed of two sets of u polynomials, it is interesting to keep the number of polynomials as small as possible (1 or 2). For such parameters, the authentication mechanism based on IP1S looks appealing in terms of key size. Additionally, it does not require hash functions or commitments.

All in all, the existing literature on the IP and IP1S problem can be split in two categories: *heuristic* algorithms with (more or less vaguely) "known" complexity and unknown success probability [40], and *rigorous* algorithms that always succeeds but with unknown complexity [17,41,13,23]. This situation makes it very difficult, if not plainly impossible to compare these algorithms based on their theoretical features. The class of instances that can be solved by a given algorithm of the first type is in general not known. Conversely, the class of instances over which an algorithm of the second type terminates quickly are often not known as well. This lead the authors of IP/IP1S algorithms to measure the efficiency of their techniques in practice, or even not to measure it at all. Several sets of concrete parameters for IP and IP1S were proposed by Patarin in [36], and can be used to measure the progress accomplished since their introduction. The techniques presented in this paper allow to break all these challenges in practice.

Techniques. The algorithms presented here are deterministic, and rely on the two weapons that have dealt a severe blow to multivariate cryptography: linear algebra and Gröbner bases. Our ideas borrow to the recent differential cryptanalysis of multivariate schemes. While the algorithms are not very complicated, analyzing their running time is fairly non-trivial, and requires the invocation of not-so-well-known results about linear algebra (such as the dimension of the commutant of a matrix, or the properties of the product of two skew-symmetric matrices), as well as known results about random matrices, most notably the distribution of the rank and the probability of being cyclic. The two most delicate steps of the analysis involve lower-bounding the dimension of the kernel of a homogeneous system of matrix equations, and upper-bounding the degree of polynomials manipulated by a Gröbner-basis algorithm.

Our Results. We present two new "rigorous" and deterministic algorithms. On the practical side, these algorithms are efficient: random quadratic IP1S instances and random cubic inhomogeneous IP1S instances can be broken in time $\mathcal{O}\left(n^6\right)$ for any size of the parameters. In particular, all the quadratic IP1S challenges proposed by Patarin are now broken in a few seconds. The biggest homogeneous cubic IP1S challenge can be broken in less than 1 CPU-month. The IP1S identification scheme is thus broken beyond repair in the quadratic case. In the case of cubic IP1S, our attack runs in time $\mathcal{O}\left(n^6 \cdot q^n\right)$, and the security parameter have to be seriously reconsidered, which makes the scheme much less attractive, since the key size is cubic in n.

A rigorous analysis of our algorithms is both necessary and tricky. When generating linear equations, special care has to be taken to count how many of them are independent. The recent history of algebraic cryptanalysis taught us that failure to do so may have drastic consequences. Additionally, the complexity of Gröbner bases computation, even though a bit more well-understood now in the generic case, is still often a delicate matter for structured systems.

A unique and distinctive feature of our algorithms compared to the previous state of affairs, and one of our main theoretical contribution, is that we characterize the class of instances that can be solved by our techniques in polynomial time. We show, for instance, that a (big) constant fraction of all quadratic IP1S instances can be solved in polynomial time.

This break however has little consequences in the multivariate cryptology ecosystem, except that it brings the IP1S-based identification scheme down. The security of UOV [27] in particular is not related to the hardness of IP1S, because in UOV the vector of polynomial composed with a linear change of variable (the "a" part) is kept secret.

Organisation of the paper. In section 2, we recall some useful facts about the IP1S problem. Then, in section 3, we introduce the identification scheme based on the hardness of IP1S and compare it to other non-number theoretic based ID schemes. We then introduce our algorithms to break IP1S in the quadratic case in section 4, and in the cubic case in section 5.

2 The IP1S Problem

We recall the definition of the IP1S problem. Given two families of u polynomials $\mathbf{a}$ and $\mathbf{b}$ in $\mathbb{F}_q[x_1, \ldots, x_n]$ the task is to find an invertible matrix $S \in \mathrm{GL}_n(\mathbb{F}_q)$ and a vector $c \in (\mathbb{F}_q)^n$ such that:

$$\mathbf{b}(\mathbf{x}) = \mathbf{a}(S \cdot \mathbf{x} + c). \tag{1}$$

We will denote by $f^{(k)}$ the homogeneous component of degree k of f, and by extension $\mathbf{a}^{(k)}$ denotes the vector of polynomials obtained by taking the homogeneous components of degree k of all the coordinates of $\mathbf{a}$. We define the derivative of $\mathbf{a}$ in c to be the function $\frac{\partial \mathbf{a}}{\partial c} : \mathbf{x} \mapsto \mathbf{a}(\mathbf{x}+c) - \mathbf{a}(\mathbf{x})$. The following lemma, which is very similar to [17, lemma 4] is very useful.

Lemma 1. *i) For all $k \geq 1$, we have:*

$$\mathbf{b}^{(k)} = \left(\mathbf{a} + \frac{\partial \mathbf{a}}{\partial c}\right)^{(k)} \circ S.$$

ii) If d is the degree of $\mathbf{a}$ *and* $\mathbf{b}$*, then* $\mathbf{b}^{(d)} = \mathbf{a}^{(d)} \circ S$.
iii) S transforms the set of common zeroes of $\mathbf{a}^{(d)}$ *into the set of common zeroes of* $\mathbf{b}^{(d)}$.

Proof. It follows from the definition of the derivative that:

$$\mathbf{b} = \left(\mathbf{a} + \frac{\partial \mathbf{a}}{\partial c}\right) \circ S$$

This equality also holds if only the degree-k homogeneous component is considered. The point is that since S is linear (and thus not "degree-changing"), if P is a multivariate polynomial we have:

$$(P \circ S)^{(k)} = P^{(k)} \circ S$$

This establishes the first statement of the lemma. The second statement follows from the fact that if $\mathbf{a}$ is of degree d, then the function $\frac{\partial \mathbf{a}}{\partial c}$ is of degree $d - 1$. Thus the homogeneous component of degree d of $\frac{\partial \mathbf{a}}{\partial c}$ is identically zero. The third statements is a direct consequences of the second one. □

A useful consequence of lemma 1 is that without loss of generality we may assume c to be the null vector[2]. A consequence of point ii) is that from any instance of the problem we can deduce a *linear homogeneous* instance by considering only the homogeneous component of highest degree. If this instance can be solved, and S can be retrieved, then recovering c is not difficult, using a slight generalization of the idea shown in [24]. If S is known, then $\frac{\partial \mathbf{a}}{\partial c}$ can be explicitly computed, and c can usually be deduced therefrom. More specifically, focusing on the homogeneous component of degree one yields a system of $u \cdot n$ linear equations in n variables that admits c as a solution. In most cases, it will in fact admit *only* c as a solution, which enables recovering c.

It was pointed out in [40] that if there is only one quadratic polynomial, then the problem is easily solved in polynomial time. This follows from the fact that quadratic forms admit a canonical representation (see for instance [30]). The change of coordinate can then be easily computed. We will therefore focus on the case of $u \geq 2$ when the polynomials are quadratic.

For various reasons, the IP1S problem becomes *easier* when u is close to n, and *harder* when u is small. For instance, the algorithm given in [41] deals with the case $u = n$ in polynomial time, but cannot tackle the case where $u = 2$ and n is big, which prevented it from breaking the parameters proposed by Patarin. Additionally, small values of u leads to smaller public keys. Therefore, we will restrict our attention to the case where $u = 2$ when the polynomials are quadratic, and where $u = 1$ when they are cubic. These are the most cryptographically relevant cases, and the most challenging. We will also consider the case where $\mathbb{F}_q$ is a field of characteristic two. It can be shown that this makes the problem a bit harder, but again this is the most cryptographically relevant case. The quadratic and cubic IP1S problems are very different and lead to specific approaches, therefore we will discuss them separately.

---

[2] This was already observed in [41].

3 Patarin's IP1S-Based Identification Scheme

Zero-Knowledge proofs were introduced in 1985 by Goldwasser, Micali and Rackoff in [26]. Soon afterwards, Fiat and Shamir [18] used the hardness of quadratic residuosity to build an efficient identification scheme. Many other identification schemes appeard afterwards, all relying on the hardness of number-theoretic assumptions. Some cryptographers took a different line of research, and tried to design identification scheme from different computational assumptions, not relying on number theory, but instead on the NP-hardness of some specific combinatorial problems.

One of the very-first combinatorial identification scheme was proposed by Shamir [43], and relied on the hardness of the *Permuted Kernel Problem* (PKP). Later on, Stern proposed in [44] a scheme based on the intractability of *Syndrome Decoding* (SD), and in [45] a scheme based on the intractability of *Constrained Linear Equations* (CLE). Finally, Pointcheval [42] proposed a scheme related to the hardness of the *Perceptron Problem*, originating from the area of learning theory. All these problems are NP-complete (as opposed to IP1S). The designers proposed practical parameters, aiming for a security level of 2^{64} or more, which are summarized in table 1. In all these schemes, it is required that all users share a public common set of information, a "common setting", usually describing the instance of the hard problem. For instance, in number-theoretic problems, the description of the curve, or of the group over which a discrete logarithm problem is considered is a common public information. While this information is not a "key" *stricto sensu*, it must nevertheless be stored by the prover and by the verifier, leading to higher memory requirements. However, in some case it can be chosen randomly, or generated online from a small seed using a PRNG.

Table 1. Key sizes in bits corresponding to practical parameters proposed in [42,43,44,45,36] in order to obtain a security level of roughly 2^{64}

Scheme	Common Setting	Public Key	Secret Key
PKP	2048	256	374
	7992	512	808
SD	131 072	256	512
	524 288	512	1024
CLE	3600	80	80
	3600	96	96
Perceptron	10807	144	117
IP1S	0	256	272

On the contrary, the IP1S-based identification scheme proposed by Patarin in [35,36] does not need the prover and the verifier to share additional information (except maybe the description of the finite field, which is very small). It works very similarly to the original identification scheme based on a zero-knowledge proof system for Graph-Isomorphism (GI) by Goldreich, Micali and Wigderson [25]. One of the reasons for replacing GI by IP1S is the existence of efficient heuristic algorithms for GI, capable of solving efficiently random instances. The generation of hard instances of GI is a delicate matter [19]. Replacing the GI problem by IP1S yields shorted key, and random

Table 2. Concrete parameters for IP1S. Patarin proposed challenges A,B,C and D in [36]. We introduce challenge E.

Challenge	n	q	Degree	Polynomial(s)	Public Key	Private Key
A	16	2	2	2	272 bits	256 bits
B	16	2	3	1	816 bits	256 bits
C	6	16	2	2	168 bits	144 bits
D	6	16	3	1	224 bits	144 bits
E	32	2	2	2	1056 bits	1024 bits

instances of IP1S were *a priori* secure. Patarin proposed concrete parameters, which are shown in table. 2. The PKP and SD schemes lead to bigger keys than IP1S, while the Perceptron scheme leads to comparable key-sizes, and CLE yields smaller keys than IP1S, if we neglect the additional memory requirement imposed by the common string shared between all the participants.

Additionnaly, the IP1S-based identification scheme does not makes use of either hash functions or commitment schemes. This is in strong contrast with all the other proposals.

The IP1S challenges described in table 2 cannot be attacked using the existing techniques [17,23,41]. As such, the best attack remains exhaustively searching for the secret key. As a final note, let us mention that Lyubashevsky recently proposed in [31] to build an identification scheme using the hardness of lattice problems, but did not propose concrete parameters.

4 Cryptanalysis of Quadratic IP1S

The main observation underlying our quadratic IP1S algorithm is that by *differentiating* equation (1), it is possible to collect linear equations between the coefficients of S and those of S^{-1}.

We denote by $\mathrm{D}f : (\mathbb{F}_q)^n \times (\mathbb{F}_q)^n \rightarrow {\mathbb{F}_q}^u$ the *differential* of a function $f : \mathbb{F}_q^n \rightarrow \mathbb{F}_q^u$. $\mathrm{D}f$ is defined by:

$$\mathrm{D}f(\mathbf{x}, \mathbf{y}) = f(\mathbf{x} + \mathbf{y}) - f(\mathbf{x}) - f(\mathbf{y}) + f(0)$$

It is easy to see that $\mathrm{D}f(\mathbf{x}, \mathbf{y}) = \mathrm{D}f(\mathbf{y}, \mathbf{x})$. If f is a polynomial of total degree d, then $\mathrm{D}f$ is a polynomial of total degree d, but of degree $d-1$ in $\mathbf{x}$ and $\mathbf{y}$. Thus, when f is quadratic, then $\mathrm{D}f$ is a symmetric *bilinear* mapping.

Going back to the quadratic IP1S problem, for all vectors $\mathbf{x}, \mathbf{y} \in (\mathbb{F}_q)^n$, we have:

$$\forall \mathbf{x}, \mathbf{y} \in (\mathbb{F}_q)^n, \quad \mathrm{D}\mathbf{b}(\mathbf{x}, \mathbf{y}) = \mathrm{D}\mathbf{a}\left(S \cdot \mathbf{x}, S \cdot \mathbf{y}\right).$$

Using the change of variable $\mathbf{y}' = S \cdot \mathbf{y}$, this equation becomes:

$$\forall \mathbf{x}, \mathbf{y}' \in (\mathbb{F}_q)^n, \quad \mathrm{D}\mathbf{b}(\mathbf{x}, S^{-1} \cdot \mathbf{y}') = \mathrm{D}\mathbf{a}(S \cdot \mathbf{x}, \mathbf{y}'). \tag{2}$$

Since $\mathbf{a}$ and $\mathbf{b}$ are of total degree 2, then $\mathrm{D}\mathbf{a}$ and $\mathrm{D}\mathbf{b}$ are *bilinear* (symmetric) mappings. In this case, since equation (2) is valid for all $\mathbf{x}$ and $\mathbf{y}$, then in particular it is valid on a basis of $(\mathbb{F}_q)^n \times (\mathbb{F}_q)^n$, and substituting fixed basis vectors for $\mathbf{x}$ and $\mathbf{y}$ yields *linear equations* between the coefficients of S and those of S^{-1}.

This idea for obtaining linear equations can also be described relatively simply using the usual theory of quadratic forms. If $\mathbb{F}_q$ is a field of even (resp. odd) characteristic, then the set of homogeneous quadratic polynomials in n variables over $\mathbb{F}_q$ is in one-to-one correspondance with the set of symmetric matrices with zero diagonal (resp. of symmetric matrices). Let $\mathcal{P}(\mathbf{a}_k)$ denote the matrix of the symmetric bilinear form associated with $\mathbf{a}_k$ (it is related to the *polar form* of $\mathbf{a}_k$ in odd characteristic). Recall that the coefficient of index (i,j) of $\mathcal{P}(\mathbf{a}_k)$ is $\mathrm{D}\mathbf{a}_k(e_i, e_j)$, where $(e_i)_{1\leq i\leq n}$ is a basis of $(\mathbb{F}_q)^n$. We then have:

$$\mathcal{S}: \quad \begin{cases} S^{-1} \cdot \mathcal{P}(\mathbf{b}_1) = \mathcal{P}(\mathbf{a}_1) \cdot {}^tS \\ \vdots \\ S^{-1} \cdot \mathcal{P}(\mathbf{b}_u) = \mathcal{P}(\mathbf{a}_u) \cdot {}^tS \end{cases} \tag{3}$$

Each one of these u matrix equations yields n^2 linear homogeneous equations between the $2n^2$ coefficients of S and those of S^{-1}. These last $u \cdot n^2$ homogeneous linear equations cannot be linearly independent as they admit a non-trivial solution $\left(S^{-1}, S\right)$. The kernel of $\mathcal{S}$ is thus non-trivial, and our hope would be that it describes *only* one solution. When u is strictly greater than two, we then have much more linear equations than unknowns, and we empirically find only one solution (when the polynomials are randomly chosen). When $u = 2$, which is again the most relevant case, the situation is unfortunately not as nice; Theorem 1 below shows that the kernel of $\mathcal{S}$ is of dimension higher that $2n$ in characteristic two (at least n in odd characteristic). This means that solving the linear equations cannot by itself reveal the solution of the IP1S problem, because $\mathcal{S}$ admits at least q^n solutions, out of which only very few are actual solutions of the IP1S instance[3]. However, the linear equations collected this way can be used to simplify the resolution of the IP1S problem.

When looking at one coordinate of (1), we have an equality between two multivariate polynomials that holds for any value of the variables. Therefore the coefficients of the two polynomials can be identified (this is essentially the algorithm presented in [17]). This yields a system $\mathcal{S}_{\text{quad}}$ of $u \cdot n^2/2$ quadratic equations in n^2 unknown over $\mathbb{F}_q$. With $u = 2$, this precisely gives n^2 equations in n^2 unknown, which cannot be solved by any existing techniques faster than exhaustive search.

However, we now know that $\left(S, S^{-1}\right)$ lives in the kernel of $\mathcal{S}$, and therefore S can be written as the sum of $k = \dim\ker\mathcal{S}$ matrices that can be easily computed using standard linear algebra. Identifying coefficients in (1) then yields a system $\mathcal{S}_{\text{quad}}$ of $u \cdot n^2/2$ quadratic equations in k unknown. Our hope is that k is small enough for the system to be very overdetermined, so that computing a Gröbner basis of $\mathcal{S}_{\text{quad}}$ is polynomial in theory, and feasible in practice.

The analysis of the attack then proceeds in two steps:

1. Estimate the rank of $\mathcal{S}$ (*i.e.*, the value of k).
2. Estimate the complexity of the Gröbner basis computation.

[3] We note that this contradicts the hope expressed in section 9 of [40].

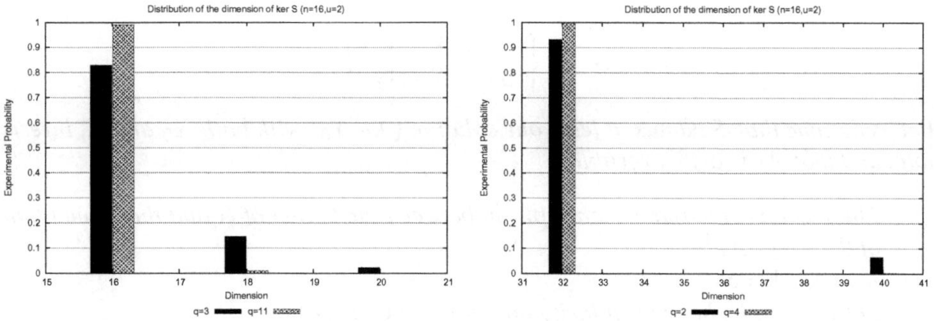

Fig. 1. Experimental distribution of dim ker $\mathcal{S}$

For the sake of simplicity, we will analyze the attack algorithm under some assumptions on the input system. For instance, we will assume that that n is even, and that one of the two quadratic forms we are dealing with is non-degenerate. We will then argue that a random instance satisfies this assumption with high probability, but we are well aware that some structured instance may not. This is in fact quite logical, because a worst-case polynomial algorithm for IP1S would imply a worst-case polynomial for Graph-Isomorphism (a fact that would be quite surprising). The situation of the IP1S problem is in this respect quite similar to that of GI: heuristics are capable of dealing efficiently with the random case, while some very special instances make them fail (interestingly, hard instances for GI are transformed into hard instances for IP1S through the reduction). Lastly, we mention that our algorithm does not necessarily fail on an instance that does not meet our assumptions. However, we no longer have a guarantee on its running time. Random instances fail to meet the assumption with a small probability, but we empirically observed that the algorithm solves them in reasonable time as well.

4.1 Counting Linearly Independent Equations

Obtaining guarantees on the number of linearly independent equations in $\mathcal{S}$ is the most important and the most delicate part of the attack. Since dim ker $\mathcal{S}$ is a function of the instance, it makes sense to consider the random variable giving dim ker $\mathcal{S}$ assuming the instance was randomly chosen. Fig. 1 above shows its (experimentally observed) distribution for various sizes of the base field. We immediately see that in odd characteristic, dim ker $\mathcal{S}$ is often n, while in characteristic two it is often $2n$. In the sequel we provide mathematical arguments to back this observation up. We will focus on the (harder) case of fields of characteristic two, since this is the more cryptographically relevant case.

Our results are expressed in terms of the *similarity invariants* $P_1, \ldots, P_s$ of a matrix M. Their product is the characteristic polynomial of M, P_s is the minimal polynomial of M, and P_i divides P_{i+1}. The main technical result needed to understand the rank of $\mathcal{S}$ is the following theorem.

Theorem 1. *Let* A_1, A_2, B_1, B_2 *be four given matrices of size* $n \times n$ *with coefficients in* $\mathbb{F}_q$*. Let us consider the set of all pairs* (X, Y) *of* $n \times n$ *matrices satisfying the following linear equations:*

$$\mathcal{S}: \begin{cases} B_1 = X \cdot A_1 \cdot Y \\ B_2 = X \cdot A_2 \cdot Y \end{cases}$$

Let us assume that $\mathcal{S}$ admits at least one solution (X_0, Y_0) with both X_0 and Y_0 invertible, and that A_1 is also invertible.

i) There is a vector-space isomorphism between the kernel of $\mathcal{S}$ and the commutant of $\mathcal{C} = A_2 \cdot A_1^{-1}$.
ii) $n \leq \dim \ker \mathcal{S}$.
iii) Let $P_1, \ldots, P_s$ be the similarity invariants of $\mathcal{C}$. Then:

$$\dim \ker \mathcal{S} = \sum_{j=1}^{s} (2s - 2j + 1) \cdot \deg P_j$$

Proof. Because a solution of $\mathcal{S}$ exists, then B_1 is invertible. Thanks to this, we can write:

$$\mathcal{S}: \begin{cases} Y = A_1^{-1} \cdot X^{-1} \cdot B_1 \\ B_2 \cdot B_1^{-1} \cdot X = X \cdot A_2 \cdot A_1^{-1} \end{cases}$$

Using the particular solution X_0 then gives:

$$\mathcal{S}: \begin{cases} Y = A_1^{-1} \cdot X^{-1} \cdot B_1 \\ \mathcal{C} \cdot \left(X_0^{-1} \cdot X\right) = \left(X_0^{-1} \cdot X\right) \cdot \mathcal{C} \end{cases}$$

From there, it is not difficult to see that the kernel of $\mathcal{S}$ is in one-to-one correspondance with the commutant of $\mathcal{C}$, the isomorphism being $(X, Y) \mapsto X_0^{-1} \cdot X$. The second point of the theorem follows from the well-known fact that n lower-bounds the dimension of the commutant of any endomorphism on a vector space of dimension n (see for instance [7, Fact 2.18.9]). The third point follows from a general result on the dimension of the commutant [20, chapter 6, exercise 32]. □

Theorem 1 directly applies to our study of the rank of $\mathcal{S}$ with $A_i = \mathcal{P}(\mathbf{a}_i)$ and $B_i = \mathcal{P}(\mathbf{b}_i)$. However, it holds only if $\mathcal{P}(\mathbf{a}_1)$ or $\mathcal{P}(\mathbf{a}_2)$ is invertible (we may swap them if we wish, or even take a linear combination). Note that since $\mathcal{P}(\mathbf{a}_1)$ is a random skew-symmetric matrix, it cannot be invertible if n is odd, and the analysis is more complicated in that case. This is why we focus on the case where n is even, and where one of the two quadratic forms is non-degenerate. The following lemma gives us the probability that $\mathcal{P}(\mathbf{a}_i)$ (or $\mathcal{P}(\mathbf{b}_i)$) is invertible.

Lemma 2 ([32], theorem 3). *Let $N_0(n, r)$ denote the number of symmetric matrices of size $n \times n$ over $\mathbb{F}_q$ with zeros on the diagonal and of rank r.*

$$N_0(n, 2s) = \prod_{i=1}^{s} \frac{q^{2i-2}}{q^{2i} - 1} \cdot \prod_{i=1}^{2s-1} \left(q^{n-i} - 1\right)$$
$$N_0(n, 2s + 1) = 0$$

If n is even, the probability that $\mathcal{P}(\mathbf{a}_1)$ is invertible if $q = 2$ is about 0.419 (this probability increases exponentially with q). The probability that either $\mathcal{P}(\mathbf{a}_1)$ or $\mathcal{P}(\mathbf{a}_2)$ is invertible is then about 0.662 when $q = 2$.

Theorem 1 is then applicable in more than half of the cases when $q = 2$ (and we expect this proportion to grow very quickly with q). When it is applicable, what guarantee does it exactly offer? We would need to know something about the similarity invariants of $\mathcal{C}$. An easy case would be when the minimal and characteristic polynomials are the same (then there is only one invariant factor, and it is precisely the characteristic polynomial). Then Theorem 1 tells us that the dimension of $\ker \mathcal{S}$ is n. For random matrices, the probability of this event is given by the following lemma.

Lemma 3 ([21], theorem 1). *Let $c(n,q)$ be the proportion of cyclic $n \times n$ matrices (i.e., matrices for which the minimal polynomial is of degree n). We have:*

$$\frac{1}{q^2(q+1)} < 1 - c(n,q) < \frac{1}{(q^2-1)(q-1)}$$

And asymptotically, we have:

$$\lim_{n \to \infty} c(n,q) = \frac{q^5 - 1}{q^2(q-1)(q^2-1)} \cdot \prod_{i=1}^{\infty} \left(1 - \frac{1}{q^i}\right)$$

For random matrices over $\mathbb{F}_2$, and for n big enough, the proportion of cyclic matrices approaches 0.746. Unfortunately, $\mathcal{C}$ is hardly a random matrix. In odd characteristic it is the product of two symmetric matrices, while in characteristic two it is the product of two symmetric matrices with null diagonal (and these are in fact *skew-symmetric* matrices). The product of two skew-symmetric matrices is *very far* from being random, and it is in fact *never* cyclic, as the following result shows.

Theorem 2 ([6]). *Let M be a non-singular matrix of even dimension. Then the two following conditions are equivalent:*

i) M can be written as the product of two skew-symmetric matrices .
ii) M has an even number of similarity invariants $P_1, \dots P_{2\ell}$, and $P_{2i+1} = P_{2i+2}$.

Corollary 1. *In characteristic two, if n is even and $\mathcal{C}$ is invertible, then* $\ker \mathcal{S}$ *has dimension at least $2n$.*

Proof. By theorem 2, $\mathcal{C}$ has at least two invariants, both equal to the minimal polynomial of $\mathcal{C}$ (which thus happens to be of degree $n/2$). Then theorem 1, point *iii)* shows that $\ker \mathcal{S}$ has dimension $2n$. If $\mathcal{C}$ has more invariants, $\ker \mathcal{S}$ can only be of higher dimension. □

Corollary 1 shows that with a constant probability (when the two quadratic forms are non-degenerate) $\dim \ker \mathcal{S}$ is greater than $2n$, which sounds like bad news. When $\mathcal{C}$ is not invertible, theorem 2 no longer holds (there are counter-examples), but what does apparently still hold is the fact that the minimal polynomial of $\mathcal{C}$ has degree at most $n/2$, and this would be sufficient to show that in all cases $\dim \ker \mathcal{S} \geq 2n$, in accordance with Fig. 1.

What we would in fact need to know is the probability that $\ker \mathcal{S}$ is exactly of dimension $2n$. Theorem 1 still connects this dimension to the similarity invariants of $\mathcal{C}$, even though $\mathcal{C}$ is not a uniformly random matrix. It seems plausible that $\mathcal{C}$ is unlikely to have a very high number of similarity invariants, and that the most common situation is that it has only two invariants (twice the minimal polynomial). We could not compute explicitly this probability, and we could not find ways to obtain it in the available literature. We measured it experimentally and found 0.746 (after 10^5 trials) when $q = 2$. This is strikingly close to the result brought by lemma 3 in the random case. Under the conjecture that $\mathcal{C}$ has two invariant factors with this probability, then theorem 1 tells us that in about 75% of the cases, $\dim \ker \mathcal{S} = 2n$. The empirical probability seems to be even higher, as shown by Fig. 1.

4.2 Solving Very Overdefined Quadratic Systems

The solution of the IP1S instance (1) is systematically the solution of a system $\mathcal{S}_{\text{quad}}$ of n^2 quadratic equations. In the previous section, we argued that we can reduce this system to n^2 equations in $2n$ unknowns with high probability, and (much) more unknowns with negligible probability. The system is so overdefined that it can almost be resolved by linearization. Indeed, it has $N^2/4$ equations in N unknowns. In practice, computing a Gröbner basis of the ideal generated by $\mathcal{S}_{\text{quad}}$ terminates very quickly, and allows to recover the actual solutions of the problem.

This last fact can be theoretically justified. It is well-known that Gröbner basis algorithms [15,16] are more efficient on overdefined systems. The complexity of most algorithms strongly depend on a parameter of the ideal called the *degree of regularity*. Indeed, the cost of computing a Gröbner basis is polynomial in the degree of regularity D_{reg} of the system when the ideal has dimension zero, *i.e.*, when the number of solutions is finite. The computation of a Gröbner basis essentially amounts to solve a system of M sparse linear equations in M variables, where M is the number of monomials of degree D_{reg} in N variables. The complexity of this process is roughly $\mathcal{O}\left(N^{\omega \cdot D_{\text{reg}}}\right)$, with $2 < \omega \leq 3$ the linear algebra constant, and N the number of variables of ideal considered (in our case, $N = 2n$).

The behavior of the degree of regularity is well understood for "random" systems of equations [3,4,5] (*i.e.*, *regular* or *semi-regular* systems). It is conjectured that the proportion of semi-regular systems on N variables goes to 1 when N goes to $+\infty$. Therefore, we can assume that for large N a random system is almost surely semi-regular. This is to some extent a worst-case assumption, as it usually means that our system is not easier to solve than the others. The coefficients of the Hilbert series associated with the ideal generated by a semi-regular sequence of m equations in N variables coincide with those of the series expansion of the function $f(z) = \left(1 - z^2\right)^m / (1 - z)^N$, up to the degree of regularity. The degree of regularity is the smallest degree d such that the coefficient of degree d in the series expansion of $f(z)$ is not strictly positive. This property enables an explicit computation of the degree of regularity for given values of m and N.

Furthermore, the available literature readily provide asymptotic estimates of the degree of regularity for semi-generic ideals of $N + k$ or $\alpha \cdot N$ equations in N variables, but unfortunately not for the case of $\alpha \cdot N^2$ in N variables, which is the situation we

Table 3. Degree of regularity of random with the same parameters as those occuring in our attack

n	2	3	4	5	6	7	8	...	16	...	32
N	4	6	8	10	12	14	16	...	32	...	64
m	4	9	16	25	36	49	64	...	256	...	1024
D_{reg}	5	4	3	3	3	3	3	...	3	...	3

are facing here. We thus tabulated in table. 3 the degree of regularity for semi-regular systems of equations having the same number of equations and unknowns as those occurring in our attack. From this table, we conclude that for any reasonable value of the parameters, the degree of regularity will be 3, and thus computing a Gröbner basis of $\mathcal{S}_{\text{quad}}$ should have complexity at most $\mathcal{O}\left(n^9\right)$. In practice, the maximal degree reached by the F_4 algorithm on our equations is two, which is even better.

4.3 Implementation

We demonstrated that the algorithm described in this section terminates in time $\mathcal{O}\left(n^6\right)$ on a constant fraction of the instances. This reasoning is backed up by empirical evidence: we implemented the algorithm using the computer algebra system MAGMA [9]. Solving the equations of $\mathcal{S}_{\text{quad}}$ is achieved by first computing a Gröbner basis of these equations for the Graded-Reverse Lexicographic order using the F_4 algorithm [15], and then converting it to the Lexicographic order using the FGLM algorithm [14]. This implementation breaks the random instances of IP1S in very practical time. For instance, Challenges A and C are solved in a few seconds. Random instances with $n = 24, u = 2$ require about a minute. Challenge E takes about 10 minutes, but the dominating part in the execution of the algorithm is in fact the symbolic manipulation of polynomials required to write down the equations of $\mathcal{S}_{\text{quad}}$. Actually solving the resulting quadratic equations turns out to be easier than generating them. We never generated a random instance that we could not solve with our technique, for any choice of the parameters.

There are only public parameter sets, and no public challenges to break, so we unfortunately cannot provide the solution of an open challenge to prove that our algorithm works. However, the source code of our implementation is available on the webpage of the first author.

5 Cryptanalysis of Cubic IP1S

In this section, we focus on the case where $\mathbf{a}$ and $\mathbf{b}$ are composed of a single cubic polynomial. We assume that $\mathbf{a}$ and $\mathbf{b}$ are given explicitly, i.e.:

$$\mathbf{a} = \sum_{i=1}^{n}\sum_{j=i}^{n}\sum_{k=j}^{n} A_{i,j,k} \cdot x_i x_j x_k, \qquad \mathbf{b} = \sum_{i=1}^{n}\sum_{j=i}^{n}\sum_{k=j}^{n} B_{i,j,k} \cdot x_i x_j x_k.$$

As already explained, we can restrict our attention to the homogenous case. The techniques developed previously for the quadratic case cannot directly applied in this setting. Indeed, the differential is no longer a bilinear mapping, and then there is no obvious linear equations between the coefficients of a solution and those of its inverse. However, we can combine the use of the differential together with the Gröbner basis approach proposed in [17]. We denote by $S_0 = \{s^0_{i,j}\}_{1 \le i,j \le n}$ a particular solution of IP1S between $\mathbf{a}$ and $\mathbf{b}$, *i.e.*, it holds that $\mathbf{b} = \mathbf{a} \circ S_0$. For all vectors $\mathbf{x}, \mathbf{y} \in (\mathbb{F}_q)^n$, we have:

$$\mathrm{D}\mathbf{a}(S_0 \cdot \mathbf{x}, \mathbf{y}) = \mathrm{D}\mathbf{b}(\mathbf{x}, S_0^{-1} \cdot \mathbf{y}).$$

$\mathbf{a}$ and $\mathbf{b}$ being of total degree 3, the coefficients of S_0 and S_0^{-1} appear with degree two in the expression of $\mathrm{D}\mathbf{a}$ and $\mathrm{D}\mathbf{b}$ above. Let R be the ring $\mathbb{K}[s_{1,1}, \ldots, s_{n,n}, u_{1,1}, \ldots, u_{n,n}]$. We consider the algebra $\mathcal{A}^s$ of all $n \times n$ matrices over R. Let $S = \{s_{i,j}\}$ and $U = \{u_{i,j}\}$ in $\mathcal{A}^s$ be symbolic matrices. We denote by $\mathcal{I}_{\mathbf{a},\mathbf{b}}$ the ideal generated by all the coefficients in R of the equations:

$$\mathrm{D}\mathbf{a}(S \cdot \mathbf{x}, \mathbf{y}) - \mathrm{D}\mathbf{b}(\mathbf{x}, U \cdot \mathbf{y}) = 0, \qquad U \cdot S - 1_n = 0_n, \qquad S \cdot U - 1_n = 0_n.$$

It is easy to see that $U = S_0^{-1}$ and $S = S_0$ is a particular solution of this system, and also a solution of IP1S between $\mathbf{b}$ and $\mathbf{a}$. Our goal is to provide an upper bound on the maximum degree reached during a Gröbner basis computation of $\mathcal{I}_{\mathbf{a},\mathbf{b}}$.

We prove here that $D_{\mathrm{reg}} = 2$ for $\mathcal{I}_{\mathbf{a},\mathbf{b}}$ under the hypothesis that we know one row of a particular solution S_0, *i.e.*, we assume then that we know the following ideal $\mathcal{J} = \left\langle s_{1,j} - s^{(0)}_{1,j} \mid j = 1, \ldots, n \right\rangle$.

Theorem 3. *The degree of regularity of $\mathcal{I}_{\mathbf{a},\mathbf{b}} + \mathcal{J}$ is 2. Therefore, computing a Gröbner basis of this ideal takes time $\mathcal{O}\left(n^6\right)$.*

Proof. We use the fact that the degree of regularity of an ideal is generically left invariant by any linear change of the variables or generators [29]. In particular, we consider the ideal $\mathcal{I}'_{\mathbf{a},\mathbf{b}}$ generated by all the coefficients in $\mathbb{K}[x_1, \ldots, x_n, y_1, \ldots, y_n]$ of the equations:

$$\mathrm{D}\mathbf{a}(S_0(S + I_n)\mathbf{x}, \mathbf{y}) - \mathrm{D}\mathbf{b}(\mathbf{x}, (U + I_n)S_0^{-1}\mathbf{y}) = 0, \qquad U \cdot S = 0_n, \qquad S \cdot U = 0_n.$$

It is clear that $\mathcal{I}'_{\mathbf{a},\mathbf{b}}$ is obtained from $\mathcal{I}_{\mathbf{a},\mathbf{b}}$ by replacing S (resp. U) by $S_0(I_n + S)$ (resp. $(U + I_n)S_0^{-1}$). Thus, the degree of regularity of $\mathcal{I}'_{\mathbf{a},\mathbf{b}}$ and $\mathcal{I}_{\mathbf{a},\mathbf{b}}$ are equal. Using the same transformation, the ideal $\mathcal{J}$ becomes

$$\mathcal{J}' = \langle s_{1,j} \mid j = 1, \ldots, n \rangle.$$

We now estimate the degree of regularity of the ideal $\mathcal{I}'_{\mathbf{a},\mathbf{b}} + \mathcal{J}'$. For a reason which will become clear in the sequel, it is more convenient to work with $\mathcal{I}'_{\mathbf{a},\mathbf{b}} + \mathcal{J}'$. In what follows, F will denote the generators of $\mathcal{I}'_{\mathbf{a},\mathbf{b}} + \mathcal{J}'$. We will show that many new linear equations appear when considering equations of degree 2. To formalize this, we introduce some definitions related to the F_4 algorithm [16]. In particular, we will denote by $I_{d,k}$ the linear space generated during the k-th step of F_4 when considering polynomials of degree d.

Definition 1. *We have the following recursive definition of $I_{d,k}$:*

$$
\begin{aligned}
I_{d,0}(F) &= \mathit{Vect}_{\mathbb{K}}\,(F) \\
I_{d,1}(F) &= \mathit{Vect}_{\mathbb{K}}\,(s_{i,j}f \mid 1 \leqslant i,j \leqslant n \textit{ and } f \in I_{d,0}(F)) \\
&\quad + \mathit{Vect}_{\mathbb{K}}\,(u_{i,j}f \mid 1 \leqslant i,j \leqslant n \textit{ and } f \in I_{d,0}(F)) \\
I_{d,k}(F) &= \mathit{Vect}_{\mathbb{K}}\,(s_{i,j}f \mid 1 \leqslant i,j \leqslant n \textit{ and } f \in I_{d,k-1}(F) \textit{ and } \deg(f) \le d-1) \\
&\quad + \mathit{Vect}_{\mathbb{K}}\,(u_{i,j}f \mid 1 \leqslant i,j \leqslant n \textit{ and } f \in I_{d,k-1}(F) \textit{ and } \deg(f) \le d-1)\,.
\end{aligned}
$$

Roughly speaking, the index k is the number of steps in the F_4/F_5 [16] algorithm to compute an element $f \in I_{d,k}(F)$. We show that $I_{2,1}(F)$ contains exactly n^2+2n linear equations. This means that we have already many linear equations generated during the first step of a Gröbner basis computation of F.

Lemma 4. *$I_{2,1}(F)$ contains the following linear equations:*

$$\{u_{1,j} \mid j = 1, \dots, n\}. \tag{4}$$

Proof. From the first row of the following zero matrix $S \cdot U$ we obtain the following equations:

$$
\begin{cases}
s_{1,1}\,u_{1,1} + s_{1,2}\,u_{2,1} + s_{1,3}\,u_{3,1} + \cdots + s_{1,n}\,u_{n,1} = 0, \\
s_{1,1}\,u_{1,2} + s_{1,2}\,u_{2,2} + s_{1,3}\,u_{3,2} + \cdots + s_{1,n}\,u_{n,2} = 0, \\
s_{1,1}\,u_{1,3} + s_{1,2}\,u_{2,3} + s_{1,3}\,u_{3,3} + \cdots + s_{1,n}\,u_{n,3} = 0, \\
\dots \\
s_{1,1}\,u_{1,n} + s_{1,2}\,u_{2,n} + s_{1,3}\,u_{3,n} + \cdots + s_{1,n}\,u_{n,n} = 0
\end{cases}
$$

Using the equations $s_{1,j} = 0$ from the ideal $\mathcal{J}'$, we obtain then $u_{1,1} = 0, u_{1,2} = 0, \dots, u_{1,n} = 0$. □

We can also predict the existence of other linear equations in $I_{2,1}(F)$.

Lemma 5. *For all $(i,j) \in \{1,\dots,n\}^2$ the coefficient of $y_1 y_i x_j$ in $\mathrm{Da}(S_0(S + I_n)\mathbf{x},\mathbf{y}) - \mathrm{Db}(\mathbf{x},(U + I_n)S_0^{-1}\mathbf{y})$ is a non zero[4] linear equation modulo the equations of the ideal $\mathcal{J}'$ and (4). Among these equations, there are n which depend only of the variables $\{s_{k,\ell} \mid 1 \le k,\ell \le n\}$.*

Proof. We consider the coefficient of the monomial $m = y_1 y_i x_j$ in the expression

$$\Delta = \Delta_a - \Delta_b = \mathrm{Da}(S_0(S + I_n)\mathbf{x},\mathbf{y}) - \mathrm{Db}(\mathbf{x},(U + I_n)S_0^{-1}\mathbf{y}).$$

Since the monomial m is linear in x_j it is clear that the corresponding coefficient in $\Delta_a = \mathrm{Da}(S_0(S+I_n)\mathbf{x},\mathbf{y})$ is also linear in the variables $s_{i,j}$; moreover this coefficient is non zero. We have now to consider the coefficient of m in Δ_b. Since $\mathrm{Db}(\mathbf{x},\mathbf{y})$ is the differential of an homogenous polynomial of degree 3 we can always write:

$$\mathrm{Db}(\mathbf{x},\mathbf{y}) = \sum_{i=1}^{n}\sum_{j=i}^{n} \ell_{i,j}(y_1,\dots,y_n)\,x_i x_j + \sum_{i=1}^{n} q_i(y_1,\dots,y_n) x_i \tag{5}$$

[4] More precisely, generically non zero.

where $\ell_{i,j}$ (resp. q_i) is a polynomial of degree 1 (resp. 2). Consequently, the coefficient of m in Db is also the coefficient of $y_1 y_i$ in $q_j((U+I_n)S_0^{-1}\mathbf{y})$. That is to say, in $q_j(\mathbf{y})$ we have now to replace $\mathbf{y}=(y_1,\ldots,y_n)$ by $((U+I_n)S_0^{-1}\mathbf{y})$. Thus, modulo the equations of the ideal $\mathcal{J}'$ and (4), we can write the product $((U+I_n)S_0^{-1}\mathbf{y})$ as

$$= \begin{pmatrix} y_1 \\ \vdots \\ \vdots \\ y_n \end{pmatrix} \begin{pmatrix} 1 & 0 & 0 & 0 \\ u_{2,1} & \cdots & \cdots & u_{2,n} \\ \vdots & \cdots & \cdots & \vdots \\ u_{n,1} & \cdots & \cdots & u_{n,n} \end{pmatrix} \begin{pmatrix} * & * & * & * \\ * & * & * & * \\ * & * & * & * \\ * & * & * & * \end{pmatrix}$$

$$= \begin{pmatrix} y_1 \\ \vdots \\ \vdots \\ y_n \end{pmatrix} \begin{pmatrix} * & * \quad * & * \\ (*u_{2,1}+\cdots+*u_{2,n}) & \cdots\cdots & (*u_{2,1}+\cdots+*u_{2,n}) \\ \vdots & \cdots\cdots & \vdots \\ (*u_{n,1}+\cdots+*u_{n,n}) & \cdots\cdots & (*u_{2,1}+\cdots+*u_{n,n}) \end{pmatrix}$$

$$= \begin{pmatrix} *y_1+(*u_{2,1}+\cdots+*u_{2,n})y_2+\cdots+(*u_{n,1}+\cdots+*u_{n,n})y_n \\ *y_1+(*u_{2,1}+\cdots+*u_{2,n})y_2+\cdots+(*u_{n,1}+\cdots+*u_{n,n})y_n \\ \vdots \\ *y_1+(*u_{2,1}+\cdots+*u_{2,n})y_2+\cdots+(*u_{n,1}+\cdots+*u_{n,n})y_n \end{pmatrix}$$

Hence the coefficient of $y_1 y_i$ in $q_j((U+I_n)S_0^{-1}\mathbf{y})$ is linear in the variables $u_{k,l}$ when $i \neq 1$ and the coefficient of y_1^2 is a constant. □

To summarize:

Lemma 6. *$I_{2,1}(F)$ contains exactly n^2+2n linear equations.*

Proof. In $I_{2,1}(F)$, we have n linear equations from lemma 5, n linear equations from the very definition of $\mathcal{J}'$, and n^2 linear equations from lemma 5 □

As explained before, we obtain n^2+2n linear equations for $I_{2,1}(F)$. However, we have $2n^2$ variables. So, we have to consider $I_{2,2}(F)$, *i.e.*, the equations generated at degree 2 during the second step. Thanks to lemma 6, we can reduce the original system to a quadratic system in $2n^2-(2n+n^2)=(n-1)^2$ variables. W.l.o.g we can assume that we keep only the variable $u_{i,j}$ where $2\le i,j\le n$. Let F' be the system obtained from F after substituting the $2n+n^2$ linear equations of lemma 6. All the monomials in $\mathbb{K}[x_1,\ldots,x_n,y_1,\ldots,y_n]$ of $\mathbf{Da}(S_0(S+I_n)\mathbf{x},\mathbf{y})-\mathbf{Db}(\mathbf{x},(U+I_n)S_0^{-1}\mathbf{y})$ have the following shape:

$$x_i y_j y_k \text{ or } y_i x_j x_k \text{ with } 1\le i,j,k\le n.$$

Hence the number of such monomials is $2n\frac{n(n+1)}{2}=n^2(n+1)\approx n^3$, which implies that the number of equations in F' is also n^3.

Thanks to this remark, we will now prove that we can linearize F'. Let $T(F')$ be the set of all monomials occurring in F'. We can assume that $T(G')=[t_1<t_2<\cdots<t_N]$. It is important to remark that $t_1=u_{2,2}$ up to $t_{(n-1)^2}=u_{n,n}$ are in fact variables. Now, let M be the matrix representation of G' w.r.t. $T(G')$. Since we know precisely the shape of the equations from the proof of lemma 5, it is possible to establish that:

1. most of the equations are very sparse, namely each equation contains about n^2 non-zero terms.
2. all the variables $t_1, \ldots, t_{(n-1)^2}$ occur in *all* the equations.

After a Gaussian elimination of the matrix M, we obtain the following shape:

$$\widetilde{M} = \begin{bmatrix} 1_{(n-1)^2} & 0 & 0 & 0 \\ 0 & \times & \cdots & \cdots \\ 0 & \times & \ddots & \vdots \\ 0 & \times & \cdots & \ddots \end{bmatrix}$$

Hence, we obtain after a second step of computation in degree 2 the equations $u_{2,2} = \cdots = u_{n,n} = 0$. This means that after 2 steps of computation at degree 2, we obtain $(n-1)^2 + 2n + n^2 = 2n^2$ linear equations in $2n^2$ unknowns. This explains why the maximum degree reached during the Gröbner basis computation of $\mathcal{I}'_{\mathbf{a},\mathbf{b}} + \mathcal{J}'$ is bounded by 2, and concludes the proof of theorem 3. □

5.1 Application to the Linear Inhomogeneous Case

If $c = 0$ in equation (1), and if $\mathbf{a}$ has a non-trivial homogeneous component of degree one, then looking at the homogeneous component of degree one yields the image of S on one point. We are then in a situation where theorem 3 is applicable, and S can be determined though a Gröbner basis computation which terminates in time $\mathcal{O}\left(n^6\right)$.

5.2 Implementation and Application to the Other Cases

All the other cases reduce to the linear homogeneous case, as mentioned in section 2. In this setting, the problem is that we do not have enough knowledge on S to make the Gröbner basis computation efficient. A simple idea would be to guess a column of S then compute the Gröbner basis. This approach has complexity $\mathcal{O}\left(n^6 \cdot q^n\right)$ as explained before. It is possible to reduce this complexity by a factor of q, by discarding guesses for the column of S that yields different values of $\mathbf{a}$ and $\mathbf{b}$ on the corresponding points.

The biggest proposed cubic IP1S challenge (Challenge C in fig. 2) has $u = 1, n = 16$ and $q = 2$. Given one relation on S, the computation of the Gröbner basis takes 90 seconds on a 2.8Ghz Xeon computer using the publicly available implementation of F_4 in MAGMA. Since this has to be repeated 2^{15} times, the whole process takes about one CPU-month (and can be parallelized at will). For challenge D, the Gröbner basis is computed in 0.1 second, and the whole process takes about 2 hours.

5.3 An Interesting Failure

We conclude this section with a simple idea that could have lead to an improvement, by efficiently giving a relation on S, but which fails in an interesting manner. Let us assume that $\mathbf{a}$ and $\mathbf{b}$ are homogeneous, and that $c = 0$ (in that setting, if $c \neq 0$, then c can be retrieved following the observation of [24]). Let us denote by $Z_{\mathbf{a}}$ (resp. $Z_{\mathbf{b}}$) the set of zeroes of $\mathbf{a}$ (resp. $\mathbf{b}$). Because of lemma 1, and since S is linear, we have:

$$S\left(\sum_{\mathbf{x}\in Z_{\mathbf{a}}} \mathbf{x}\right) = \sum_{\mathbf{y}\in Z_{\mathbf{b}}} \mathbf{y}$$

This yields a relation on S, which is enough to use theorem 3. $\mathbf{a}$ and $\mathbf{b}$ may be assumed to have about q^{n-1} zeroes. Finding them requires time $\mathcal{O}(q^n)$. The complexity of the attack could thus be improved to $\mathcal{O}\left(n^6 + q^n\right)$. Surprisingly, this trick fails systematically, and this happen to be consequence of the Chevalley-Warning theorem [10,46].

Lemma 7. *The sum of the zeroes of a cubic form on 5 variables or more over* $\mathbb{F}_q$ *is always zero.*

Proof. Let us consider the elements of $Z_{\mathbf{a}}$ having α as their first coordinate, and let us denote by n_α their number. These are in fact the common zeroes of $(\mathbf{a}, x_1 - \alpha)$. By the Chevalley-Warning theorem [10,46], if $\mathbf{a}$ has at least 5 variables, then the characteristic of the field divides n_α. Therefore, their sum has zero on the first coordinate. Applying this result for all values of α shows that the sum of zeroes of $\mathbf{a}$ has a null first coordinate. We then just consider all coordinates successively. □

6 Conclusion

In this paper, we present algorithms for the IP problem with one secret for two random quadratic equations and one cubic equation. As already explained, there are the most cryptographically relevant instances. Moreover, we explain the complexity, success probability and give sufficient conditions so that the algorithms work. We combine the use of the differential and the computation of Gröbner bases of very overdefined systems. All the proposed IP1S challenges can be broken in practice by the technique we describe, as the following table shows.

Challenge	Attack time on one core
A	3 seconds
B	1 month
C	0 seconds
D	1 hours
E	3 minutes

In view of these results, we conclude that Patarin's IP1S-Based identification scheme is no longer competitive with respect to others combinatorial-based identification schemes [42,43,44,45].

Acknowledgement. The first author thanks Mehdi Tibouchi for pointing out the existence of the Chevalley-Warning theorem, and for interesting discussions in general. The work described in this paper has been supported in part by the European Commission through the ICT programme under contract ICT-2007-216676 ECRYPT II. The two last authors were also supported in part by the french ANR under the Computer Algebra and Cryptography (CAC) project ANR-09- JCJCJ-0064-01.

References

1. Baena, J., Clough, C., Ding, J.: Square-vinegar signature scheme. In: Buchmann, J., Ding, J. (eds.) PQCrypto 2008. LNCS, vol. 5299, pp. 17–30. Springer, Heidelberg (2008)
2. Bardet, M., Faugère, J.C., Salvy, B., Yang, B.Y.: Asymptotic Behaviour of the Degree of Regularity of Semi-Regular Polynomial Systems. In: MEGA 2005, Eighth International Symposium on Effective Methods in Algebraic Geometry, Porto Conte, Alghero, Sardinia (Italy), May 27-June 1 (2005)
3. Bardet, M.: Étude des systèmes algébriques surdéterminés. Applications aux codes correcteurs et à la cryptographie. PhD thesis, Université de Paris VI (2004)
4. Bardet, M., Faugère, J.-C., Salvy, B.: On the complexity of Gröbner basis computation of semi-regular overdetermined algebraic equations. In: Proc. International Conference on Polynomial System Solving (ICPSS), pp. 71–75 (2004)
5. Bardet, M., Faugère, J.-C., Salvy, B., Yang, B.-Y.: Asymptotic behaviour of the degree of regularity of semi-regular polynomial systems. In: Proc. of MEGA 2005, Eighth International Symposium on Effective Methods in Algebraic Geometry (2005)
6. Bennett, A.A.: Products of skew-symmetric matrices. American M. S. Bull. 25, 455–458 (1919)
7. Bernstein, D.S.: Matrix mathematics. Theory, facts, and formulas, 2nd expanded edn., vol. xxxix, p. 1139. Princeton University Press, Princeton (2009)
8. Billet, O., Gilbert, H.: A traceable block cipher. In: Laih, C.S. (ed.) ASIACRYPT 2003. LNCS, vol. 2894, pp. 331–346. Springer, Heidelberg (2003)
9. Bosma, W., Cannon, J.J., Playoust, C.: The Magma Algebra System I: The User Language. J. Symb. Comput. 24(3/4), 235–265 (1997)
10. Chevalley, C.: Démonstration d'une hypothèse de M. Artin. Abh. Math. Semin. Hamb. Univ. 11, 73–75 (1935)
11. Clough, C., Baena, J., Ding, J., Yang, B.-Y., Chen, M.-s.: Square, a new multivariate encryption scheme. In: Fischlin, M. (ed.) CT-RSA 2009. LNCS, vol. 5473, pp. 252–264. Springer, Heidelberg (2009)
12. Ding, J., Wolf, C., Yang, B.-Y.: ℓ-invertible cycles for multivariate quadratic public key cryptography. In: Okamoto, T., Wang, X. (eds.) PKC 2007. LNCS, vol. 4450, pp. 266–281. Springer, Heidelberg (2007)
13. dit Vehel, F.L., Perret, L.: Polynomial Equivalence Problems and Applications to Multivariate Cryptosystems. In: Johansson, T., Maitra, S. (eds.) INDOCRYPT 2003. LNCS, vol. 2904, pp. 235–251. Springer, Heidelberg (2003)
14. Faugère, J.-C., Gianni, P., Lazard, D., Mora, T.: Efficient Computation of Zero-Dimensional Gröbner Bases by Change of Ordering. Journal of Symbolic Computation 16(4), 329–344 (1993)
15. Faugère, J.-C.: A new efficient algorithm for computing Gröbner bases (F4). Journal of Pure and Applied Algebra 139(1-3), 61–88 (1999)
16. Faugère, J.-C.: A New Efficient Algorithm for Computing Gröbner Bases Without Reduction to Zero (F5). In: ISSAC 2002: Proceedings of the 2002 International Symposium on Symbolic and Algebraic Computation, pp. 75–83. ACM, New York (2002)
17. Faugère, J.-C., Perret, L.: Polynomial Equivalence Problems: Algorithmic and Theoretical Aspects. In: Vaudenay, S. (ed.) EUROCRYPT 2006. LNCS, vol. 4004, pp. 30–47. Springer, Heidelberg (2006)
18. Fiat, A., Shamir, A.: How to prove yourself: Practical solutions to identification and signature problems. In: Odlyzko, A.M. (ed.) CRYPTO 1986. LNCS, vol. 263, pp. 186–194. Springer, Heidelberg (1987)
19. Fortin, S.: The graph isomorphism problem. Technical report, University of Alberta (1996)

20. Fuhrmann, P.A.: A polynomial approach to linear algebra. Springer, New York (1996)
21. Fulman, J.: Random matrix theory over finite fields. Bull. Amer. Math. Soc. (N.S) 39, 51–85
22. Garey, M.R., Johnson, D.S.: Computers and Intractability, A Guide to the Theory of NP Completeness. Freeman, New York (1979)
23. Geiselmann, W., Meier, W., Steinwandt, R.: An Attack on the Isomorphisms of Polynomials Problem with One Secret. Int. J. Inf. Sec. 2(1), 59–64 (2003)
24. Geiselmann, W., Steinwandt, R., Beth, T.: Attacking the Affine Parts of SFLASH. In: Honary, B. (ed.) Cryptography and Coding 2001. LNCS, vol. 2260, pp. 355–359. Springer, Heidelberg (2001)
25. Goldreich, O., Micali, S., Wigderson, A.: Proofs that yield nothing but their validity and a methodology of cryptographic protocol design (extended abstract). In: FOCS, pp. 174–187. IEEE, Los Alamitos (1986)
26. Goldwasser, S., Micali, S., Rackoff, C.: The knowledge complexity of interactive proof-systems (extended abstract). In: STOC, pp. 291–304. ACM, New York (1985)
27. Kipnis, A., Patarin, J., Goubin, L.: Unbalanced Oil and Vinegar Signature Schemes. In: Stern, J. (ed.) EUROCRYPT 1999. LNCS, vol. 1592, pp. 206–222. Springer, Heidelberg (1999)
28. Koblitz, N.: Algebraic Aspects of Cryptography. Algorithms and Computation in Mathematics, vol. 3. Springer, Heidelberg (1998)
29. Lazard, D.: Gröbner-bases, gaussian elimination and resolution of systems of algebraic equations. In: van Hulzen, J.A. (ed.) ISSAC 1983 and EUROCAL 1983. LNCS, vol. 162, pp. 146–156. Springer, Heidelberg (1983)
30. Lidl, R., Niederreiter, H.: Finite fields. Cambridge University Press, New York (1997)
31. Lyubashevsky, V.: Lattice-based identification schemes secure under active attacks. In: Cramer, R. (ed.) PKC 2008. LNCS, vol. 4939, pp. 162–179. Springer, Heidelberg (2008)
32. MacWilliams, J.: Orthogonal matrices over finite fields. The American Mathematical Monthly 76(2), 152–164 (1969)
33. Matsumoto, T., Imai, H.: Public quadratic polynomial-tuples for efficient signature-verification and message-encryption. In: Günther, C.G. (ed.) EUROCRYPT 1988. LNCS, vol. 330, pp. 419–453. Springer, Heidelberg (1988)
34. Naccache, D. (ed.): CT-RSA 2001. LNCS, vol. 2020. Springer, Heidelberg (2001)
35. Patarin, J.: Hidden fields equations (hfe) and isomorphisms of polynomials (ip): Two new families of asymmetric algorithms. In: Maurer, U.M. (ed.) EUROCRYPT 1996. LNCS, vol. 1070, pp. 33–48. Springer, Heidelberg (1996)
36. Patarin, J.: Hidden Fields Equations (HFE) and Isomorphisms of Polynomials (IP): Two New Families of Asymmetric Algorithms. In: Maurer, U.M. (ed.) EUROCRYPT 1996. LNCS, vol. 1070, pp. 33–48. Springer, Heidelberg (1996), http://www.minrank.org/hfe.pdf
37. Patarin, J.: The Oil and Vinegar signature scheme. Presented at the Dagstuhl Workshop on Cryptography (1997)
38. Patarin, J., Courtois, N., Goubin, L.: Flash, a fast multivariate signature algorithm. In: [34], pp. 298–307
39. Patarin, J., Courtois, N., Goubin, L.: QUARTZ, 128-Bit Long Digital Signatures. In: [34], pp. 282–297
40. Patarin, J., Goubin, L., Courtois, N.: Improved Algorithms for Isomorphisms of Polynomials. In: Nyberg, K. (ed.) EUROCRYPT 1998. LNCS, vol. 1403, pp. 184–200. Springer, Heidelberg (1998)
41. Perret, L.: A Fast Cryptanalysis of the Isomorphism of Polynomials with One Secret Problem. In: Cramer, R. (ed.) EUROCRYPT 2005. LNCS, vol. 3494, pp. 354–370. Springer, Heidelberg (2005)

42. Pointcheval, D.: A new identification scheme based on the perceptrons problem. In: Guillou, L.C., Quisquater, J.-J. (eds.) EUROCRYPT 1995. LNCS, vol. 921, pp. 319–328. Springer, Heidelberg (1995)
43. Shamir, A.: An efficient identification scheme based on permuted kernels (extended abstract). In: Brassard, G. (ed.) CRYPTO 1989. LNCS, vol. 435, pp. 606–609. Springer, Heidelberg (1990)
44. Stern, J.: A new identification scheme based on syndrome decoding. In: Stinson, D.R. (ed.) CRYPTO 1993. LNCS, vol. 773, pp. 13–21. Springer, Heidelberg (1994)
45. Stern, J.: Designing identification schemes with keys of short size. In: Desmedt, Y. (ed.) CRYPTO 1994. LNCS, vol. 839, pp. 164–173. Springer, Heidelberg (1994)
46. Warning, E.: Bemerkung zur vorstehenden Arbeit von Herrn Chevalley.. Abh. Math. Semin. Hamb. Univ. 11, 76–83 (1935)
47. Wolf, C., Preneel, B.: Taxonomy of Public Key Schemes Based on the Problem of Multivariate Quadratic Equations. Cryptology ePrint Archive, Report 2005/077 (2005)

Author Index

GPSR Compliance

The European Union's (EU) General Product Safety Regulation (GPSR) is a set of rules that requires consumer products to be safe and our obligations to ensure this.

If you have any concerns about our products, you can contact us on ProductSafety@springernature.com

In case Publisher is established outside the EU, the EU authorized representative is:

Springer Nature Customer Service Center GmbH
Europaplatz 3
69115 Heidelberg, Germany

Batch number: 10704693

Printed by Printforce, the Netherlands